W0050128

Reint de Boer

Theory of Porous Media

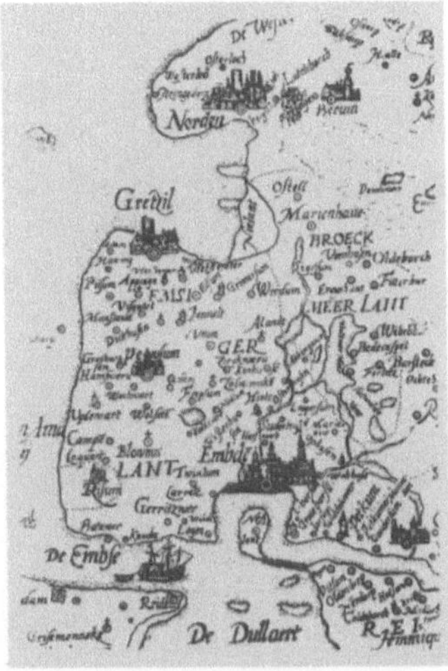

Map of the western part of Ostfriesland (Germany)

Made by David Fabricius in 1592 (Printed in 1613)

HOMAGE

to

David Fabricius

1564–1617

Johannes Fabricius

1587–1616 (?)

Great Astronomers and Discoverers of the Sun Spots

Springer
Berlin
Heidelberg
New York
Barcelona
Hong Kong
London
Milano
Paris
Singapore
Tokyo

Reint de Boer

Theory of Porous Media

Highlights in Historical Development and Current State

With 176 Figures

Springer

Professor Dr.-Ing. Reint de Boer
o. Professor für Mechanik
an der Universität Essen
Institut für Mechanik
Fachbereich 10 – Bauwesen
D-45117 Essen

URL: http://mechanik.bauwesen.uni-essen.de/pup.html
EMAIL: deboer@mechanik.bauwesen.uni-essen.de

ISBN-13: 978-3-642-64062-9 Springer Verlag Berlin Heidelberg New York

Library of Congress Cataloging-in-Publication Data

Boer, Reint de:
Theory of porous media: Highlights in Historical Development and Current State / Reint de Boer. – Berlin ; Heidelberg ; New York ; Barcelona ; Hong Kong ; London ; Milano ; Paris ; Singapore ; Tokyo : Springer, 2000

ISBN-13:978-3-642-64062-9 eISBN-13:978-3-642-59637-7
DOI:10.1007/978-3-642-59637-7

This work is subject to copyright. All rights are reserved, whether the whole or part of the material is concerned, specifically the rights of translation, reprinting, reuse of illustrations, recitation, broadcasting, reproduction on microfilm or in other ways, and storage in data banks. Duplication of this publication or parts thereof is permitted only under the provisions of the German Copyright Law of September 9, 1965, in its current version, and permission for use must always be obtained from Springer-Verlag. Violations are liable for prosecution act under German Copyright Law.

© Springer-Verlag Berlin Heidelberg 2000
Softcover reprint of the hardcover 1st edition 2000

The use of general descriptive names, registered names, trademarks, etc. in this publication does not imply, even in the absence of a specific statement, that such names are exempt from the relevant protective laws and regulations and therefore free for general use.

Typesetting with LaTeX: PTP – Berlin, Stefan Sossna
Cover: Medio GmbH, Berlin
Printed on acid-free paper SPIN: 10715209 62/3020 - 5 4 3 2 1 0 -

Dedicated to
Jan and Claas

Preface

Porous media theories play an important role in many branches of engineering, including material science, the petroleum industry, chemical engineering, and soil mechanics, as well as in biomechanics. When, in the early 1980s, investigations into these theories were begun at the University of Essen, it was soon recognized that some known theories were incomplete, unclear, and even partially incorrect. The original plan to write a book on the theory of porous media was quickly abandoned. The chief reason for this was the completely insufficient constitutive theory of granular and brittle materials (of which most porous solids consist) within the framework of the geometrically-linear and non-linear theory. Therefore, a program was embarked upon in order to develop a consistent theory for the complex field of liquid and/or gas saturated porous solids. This program initially included two strategies.

The first strategy involved the creation of clearly defined mechanical and thermodynamic terms for saturated porous bodies, and the description of the material-independent fundamental equations of these media (i.e., the kinematics of deformations, the balance equations, and the entropy inequality), avoiding any contradictions. It was revealed that, for this purpose, only elements of the mixture theory, restricted by the concept of volume fractions, were suitable, because they proceeded from the axioms of mechanics and thermodynamics and applied them to the individual constituents, under consideration of all coupling mechanisms as well as the bulk body. Existing classical models – for example, the models of *K. von Terzaghi* and *M.A. Biot* – seem to have been developed more intuitively. They are not derived from the basic equations of mechanics, contain partially unclear definitions, neglect important mechanical quantities in certain parts, and sometimes introduce the coupling terms between the constituents in an obscure way. These models do not admit further scientific developments; indeed, they hinder them.

The second strategy involved the development of consistent constitutive equations, in particular for the porous solid in the plastic range. In this range, immense difficulties arose in describing material behavior. "Natural" porous bodies, such as rock and soil, as well as artificially created ones, such as concrete and sinter metals, consist of granular and brittle materials. In comparison to ductile materials, these materials, known as frictional materials, show a distinct dependence of the material behavior on the hydrostatic pressure in the plastic range. The influence of hydrostatic pressure causes new and different effects, which are unknown in the plasticity theory of ductile materials and which make the creation of a consistent plasticity theory difficult. It is true that, in the early 1980s, extensive investigations were already being made into the material behavior of frictional materials, in particular in experimental work. However, these investigations were mainly restricted to the development of failure functions and flow rules within the framework of the geometrically-linear theory for one-component materials. Publications on the hardening of plastically-deformed granular and brittle materials referred almost entirely to isotropic hardening. The effect of kinematic hardening, in experiments confirmed and supported by thermodynamic investigations, was dealt with in only a few papers. Thus, it was recognized that, in the constitutive theory of frictional materials, a long-awaited demand existed, in particular a demand for the embedding of existing results into the general porous media theory.

Moreover, in the 1980s, attention was almost exclusively focused on incompressible porous media. It took a long time to also develop a theory for porous media with individual compressible constituents.

During the processing of both strategies, it became evident that a third strategy was necessary, namely the investigation of the historical development of the porous media theory. Such investigations are not only interesting from a purely historical point of view, they also clarify issues and complicated relations that exist in a theory as complex as that of porous media, and intensify the search for the recognition of specific and important mechanical and thermodynamic effects which can occur in saturated porous bodies. Finally, historical studies give us access to several published contributions which have been almost completely ignored and forgotten.

The consistent treatment of the material-independent fundamental equations of the theory of porous media, the development of constitutive equations for frictional materials in the elastic and plastic range, and the task of tracing the historical development of porous media theory all involved a large amount of effort during the 1980s through the first half of the 1990s on the part of the mechanics group of the University of Essen, Germany. The results of these intensive investigations are included in this book. Thus, for the first time, a unique treatment of fluid-saturated porous solids is presented containing the historical development of the corresponding theory, almost from its inception, as well as the current state of the theory of porous media.

Many persons have supported me to enable me to write this book. I would like to thank Dr. techn. A. Lechner and Dipl.-Ing. E. Jiresch from the Archive of the Technical University of Vienna for their cooperation. Mrs. Enerson and the staff from the Norwegian Geotechnical Institute (N.G.I.) supported my work in the Terzaghi Library in every respect.

Several colleagues have given me valuable hints concerning the theory of porous media and its historical development. In particular, I would like to thank Professor Dr.-Ing. W. Ehlers, University of Stuttgart, former co-worker, and Professor Dr. Z. Liu, University of Chongqing (PR China), former Humboldt scholar, for valuable stimulating discussions over many years in Essen. I also thank Privatdozent Dr.-Ing. J. Bluhm, University of Essen, for his constant willingness to discuss and clarify issues as well as for correcting and proofreading the section on the current state of the porous media theory.

The research work on the historical development and the current state of the theory of porous media could only be carried out with the financial support of the Deutsche Forschungsgemeinschaft (DFG), the Volkswagen-Stiftung, the Forschungspool of the University of Essen, and the Fördervereinigung für die Stadt Essen. This support is gratefully acknowledged.

Extensive discussions with my co-workers in the mechanics group in Essen during the last 15 years have assisted me in clarifying several issues in porous media theory, and for which I am eternally grateful.

Parts of the first chapters were translated by Mrs. S. Rozvany. Mrs. V. Lennartz, responsible for the word processing, has brought the manuscript to its present form and Mrs. V. Jorisch has taken care of the large number of historical photographs and of the drawings. I would like to express my deep gratitude to them.

I record here also my heartfelt thanks to Springer-Verlag for the careful publishing and the pleasant cooperation.

Essen, September, 1999 Reint de Boer

Contents

Chapter 1: Introduction

In general, many solids show an internal structure. On the one hand, this is due to the fact that several solids consist of different solid components, such as dense concrete without pores. On the other hand, solids can contain closed and open pores, such as ceramics and soils, as well as concrete. In the past, these materials were, in general, treated as one-component materials lacking an internal structure. Using such a model - one which consists of an ideal material - classical continuum mechanics has been very successful. All defined mechanical terms (for example, stress) are thereby understood as average quantities. Classical continuum mechanics can therefore not answer questions concerning the change of pores and the different motions belonging to the phases of liquid-saturated porous solids.

Materials with empty pore spaces, such as concrete, can be treated relatively easily because all the concrete ingredients have the same motion if the concrete body is deformed. However, the circumstances are much more complicated if the porous solid, containing open pore spaces (for example, soil or concrete) is filled with liquid. In this case, the solid and liquid constituents have different motions. Due to these different motions and the different material properties, there is interaction between the constituents, making the description of the mechanical (or the thermodynamic) behavior difficult. Above all, the problems become more complex because the internal pore structure shows, in general, a complicated geometry. At first sight, it seems to be possible to use two different strategies to solve the complex problem of saturated porous media. The first strategy may be seen as an exact solution to the respective problem. For this purpose, it would be necessary to separate the constituents with Euler's cut principle. Then, the axioms of mechanics and thermodynamics would be applied to the separated constituents in consideration of all boundary and initial conditions. Such a procedure may, in special cases, be successful. However, for the development of a general theory of saturated porous bodies, this procedure is completely unsuitable because of the complex internal geometric structures involved. Moreover, for many engineering problems, this procedure is not at all necessary, for these problems require a macroscopic description of the physical phenomena rather than a description on the microscopic scale. For example, in describing seepage, the planning engineer is not interested in the real velocity of the water in the single pores of the porous soil; for his design, only a knowledge of the seepage velocity is needed, i.e., the average velocity of the water flow in all pores. Thus, it depends on the problem as to whether investigations in the micro or macro range are necessary, whereby the macro range is understood as being the range in which all defined and measured quantities in the micro range are settled as statistical average values.

The above-considerations show that a second strategy is needed, which can be found in the macroscopic description of the problem. Thus, it is convenient to describe liquid-saturated porous media by means of a substitute model consisting of heterogeneously composed continua with internal interactions. This procedure involves the concept of volume fractions to distribute the mass of the solid skeleton and of the liquid over the total control space, which is shaped by the solid skeleton. The distribution takes place with the help of volume porosity numbers, which fix the ratios of the volumes of the constituents to the volume of the control space. This procedure assumes that the pores are statistically distributed over the control space. In this case, the equality of volume and surface porosity is given as a statistical necessity. The concept of volume fractions results in the appearance of "smeared" substitute continua with reduced densities for the solid and liquid phases which fill the control space simultaneously and which can be treated with the methods of continuum mechanics. It is obvious that the same properties appear which characterize the mixture theory, because in mixture theories, as well, it is assumed that the individual parts of the mixture cover the total control space. Well-founded conclusions concerning the theory of mixtures have been made in the last few decades. Thus, with the continuum theory of mixtures containing different constituents with their own degree of freedom, an ensured basis is available to treat the mechanical and thermodynamic behavior of liquid-saturated porous solids. However, it must be noted that in principle the assumptions of the theory of mixtures (for example, the complete mixing of all substances in the control space) are not valid for the real saturated solid skeleton, since the solid and the liquid phases are in reality immiscible. Thus, the second strategy described above does not exactly solve the problem, although it is an answer to problems of the substitute continuum model.

The mixture theory, restricted by the concept of volume fractions (porous media theory), yields the most consistently developed framework for the treatment of liquid-saturated porous solids. Other approaches to investigate such bodies are very often based on partially obscure assumptions. However, the porous media theory is extremely complicated, and as of yet not all of the problems have been solved.

For a better understanding of such complex theories as that of porous media theory, the investigation into their historical development is very helpful. On the one hand, it can reveal valuable contributions which have been almost completely forgotten or ignored. On the other hand, it sharpens the view on certain problems which are characteristic of saturated porous solids and which had previously been treated in the older literature. The investigation of the historical background of porous media theories found in this book does not, however, claim to be complete; rather, it will concentrate on some of the highlights which have occurred in the past, highlights which mark decisive steps in the development of the theory of porous media. These steps are sometimes connected with personal conflicts, as for example the controversy between the Viennese professors P. Fillunger and K. von Terzaghi in the 1930s.

There are three major periods in the historical development of the porous media theory. In the early era, during the 18th century, important principles of mechanics were developed, the theory of ideal fluids was founded, and the concept of volume fractions was first stated. In the classical era of the 19th century, linear elasticity was derived, fundamental laws of continuum mechanics and thermodynamics that cannot be separated from the development of the porous media theory were discovered, plasticity was founded, and the mixture theory was developed. The current century has seen the beginning of the modern era. In the period between 1910 and 1960, the first attempts to clarify the mechanical interaction of liquids, gases, and rigid porous solids were performed and for the first time deformable saturated porous solids were treated. In the 1970s and 1980s, theories of immiscible mixtures were developed which are still under study today. Following this scheme, the history of the porous media theory will be presented and the most important contributions will be subsequently traced.

Today, a consistent macroscopic porous media theory is available which seems to be physically and mathematically well-balanced. The main efforts to complete the mathematical model are, in particular, concerned with the development of constitutive equations for incompressible and compressible constituents in the elastic, viscoelastic, plastic, and viscoplastic ranges as well as with the calculation of initial- and boundary-value problems. Some recent findings in porous media theory will also be discussed in the section concerning the current state of porous media theory.

In this treatise many quotations will be cited in French and German, followed immediately by the translation in English.

There are three appendices at the end of the book which the reader may find of use in further exploring several concepts in porous media theory. In Appendix A, the entropy equality will be evaluated in detail; Appendix B contains an introduction to tensor and vector calculus; and Appendix C explores concepts important for an introduction into plasticity, specifically the graphic representation of the stress state and yield functions in the principal stress-space.

Chapter 2: The Early Era

Porous media theory is based on the fundamental axioms and principles of mechanics and thermodynamics. The development of the scientific treatment of mechanical problems started around 1600. It was Galileo Galilei who founded mechanics as a branch of science. He laid the corner stone for a scientific approach to the natural world – namely to gather experience of certain mechanical problems, to order them, to create a theory, and to prove this through selected experiments as he had done in the treatments of strength of material and projectile motion.

Christian Hygens went further by publishing, in 1673, valuable papers on free fall and center of oscillation. Moreover, he formulated thirteen statements on the centrifugal force involved rotation.

Important contributions also came from Jakob and Johann Bernoulli from Basel. Both were excellent exponents of the differential and integral calculus, just developing at that time. Jakob Bernoulli's solution to the elastic line problem was of great importance to engineering mechanics. Johann Bernoulli was a man of merit in mechanics, particularly in the field of hydraulics, see Section 2.2.

Isaac Newton, without a doubt, has exercised the greatest influence on physics. His most remarkable achievement stems from the year 1687, when he published his book *Philosophiae naturalis principia mathematica* in which he described the motion of mass points. It called for the outstanding intellect of Isaac Newton to collect all of the findings of his predecessors and contemporaries and to interpret them in the right way. This is true for the law of inertia which had been described by Galileo (though not, however, sufficiently developed), for the law of centrifugal force which Huygens had restricted solely to circular motions, and for the law of gravitation which had already been posited by Hooke. Newton's book was hard to read. The geometric-synthetic treatment of the motion of a mass point hampered the study of the book in the time to follow, so that it took a long time until it won the recognition it so richly deserved.

However, all scientists of merit, those already mentioned as well as those not, were predecessors of one man who pioneered the field of mechanics on a grand scale in the early era, namely Leonhard Euler. We shall deal with the particulars of his life's work in the next section.

Essentially, the early era spanned the entire eighteenth century. From the point of view of porous media theory, this period was characterized by the

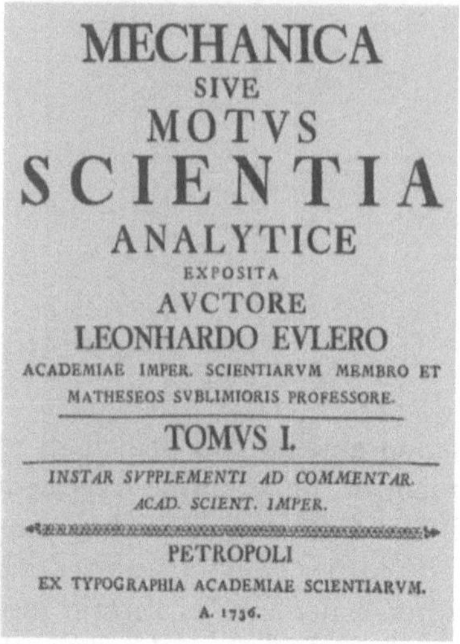

MECHANICA
SIVE
MOTVS
SCIENTIA
ANALYTICE
EXPOSITA
AVCTORE
LEONHARDO EVLERO
ACADEMIAE IMPER. SCIENTIARVM MEMBRO ET
MATHESEOS SVBLIMIORIS PROFESSORE.
TOMVS I.
INSTAR SVPPLEMENTI AD COMMENTAR.
ACAD. SCIENT. IMPER.
PETROPOLI
EX TYPOGRAPHIA ACADEMIAE SCIENTIARVM.
A. 1736.

Fig. 2.1. Front page of *Mechanica*

development of the axioms of mechanics as well as the description of the rigid body motion, the development of the ideal fluids theory, the description of a porous body, the treatment of the earth pressure, and the formulation of the concept of volume fractions. Since the development of the description of the axioms of mechanics is nearly fully documented in various other works (see Truesdell, 1968 and Szabó, 1987), only a few fundamental remarks will be made here on this topic. The same applies to the development of the theory of rigid bodies and of the ideal fluids theory. The earth pressure theory, however, as well as the volume fraction concept, will be examined in detail.

2.1 The Development of the Principles of Mechanics

The formulation and creation of the axioms of continuum mechanics, that is

- the balance of mass,
- the balance of momentum,
- the balance of moment of momentum,
- the cut principle,

are largely due to the efforts of one of the greatest mathematicians and mechanics specialists of all time, Leonhard Euler. Already by the age of 29, Euler (1736) had published his lucid two volume work *Mechanica.*

In this book, Euler gave an outline suggesting different topics in mechanics to be investigated in the future. It is amazing that Euler went on to solve nearly all of these problems in the course of his own scientific career.

Furthermore, Euler strongly criticized the lack of a principle for the treatment of the motion of bodies of finite size (the following quotations are taken from Truesdell, 1954):

"Those laws of motion which a body observes when left to itself in continuing rest or motion pertain properly to infinitely small bodies, which can be considered as points. Indeed, in bodies of finite magnitude, whose several parts are endowed with various motions, a given part will try to observe these laws, which however on account of the state of the body is not always possible. Therefore the body will follow that motion which is composed from the endeavors of the several parts, and from insufficiency of the principles this cannot yet be determined, but its treatment is to be left to the sequels. The diversity of bodies therefore will supply the primary division of our work. First indeed we shall consider infinitely small bodies... Then we shall attack bodies of finite magnitude which are rigid... Thirdly, we shall treat of flexible bodies. Fourthly, of those which admit extension and contraction. Fifthly, we shall subject to examination the motions of several separated bodies, some of which hinder [each other] from executing their motions as they attempt them. Sixthly, at last, the motion of fluids will have to be treated."

From Euler's remarks, one realizes that this clear-thinking man was exceptionally disturbed by the near-total lack of principles of mechanics. In fact, when Euler wrote his book, the definitions and axioms did not yet exist or were inadequately formulated. One must bear in mind that at this time only a few important definitions and terms (e.g., velocity, acceleration, and the term force) were known. These, in turn, had been clarified by Isaac Newton and other scientists after much debate in the seventeenth century. Only the balance of momentum for the point mass was known, which had also been formulated by Newton, although it had not yet been converted into real formulas. Euler's main concern in his book was applying an analytical treatment to the motion of point masses. He succeeded marvellously in doing so, thanks to his excellent knowledge of differential and integral calculus, which he had learned from Johann Bernoulli. In his book, he also addressed the cut principle. However, his reflections have little in common with the formulations used today, which state that all balance equations are valid not only for the total body but also for every part which is imaginarily separated from the bulk body, where the freed interactions on the cut surfaces are to be attached as external interactions. In the course of his long research career, Euler never gave a clear statement of the cut principle. However, in his considerable works he employed the cut principle with mastery.

In 1752, Euler made a decisive breakthrough in the development of the principles of mechanics for continua with his work *Découverte d'un nouveau principe de méchanique* (Euler, 1752a).

Here, for the first time, he explicitly formulated the balance equation of momentum, introduced by a statement from Newton, for a differential small body, imaginary cut out of a finite body:

$$2M\,ddx = Pdt^2\;;^{1}\quad 2M\,ddy = Qdt^2\;;\quad 2M\,ddz = Rdt^2\;. \tag{2.1.1}$$

[1] The factor on the left-hand side was produced due to the fact that Euler established the acceleration due to gravity with $g = \frac{1}{2}$ via the weight, M. Moreover, x, y, and z are coordinates, t the time, and P, Q, and R the force resultants.

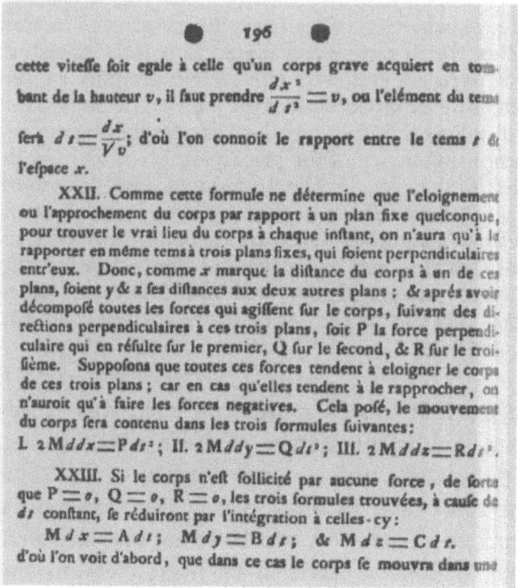

196

cette viteſſe ſoit egale à celle qu'un corps grave acquiert en tombant de la hauteur v, il faut prendre $\frac{dx^2}{dt^2} = v$, ou l'elément du tems ſera $dt = \frac{dx}{\sqrt{v}}$; d'où l'on connoit le rapport entre le tems t & l'eſpace x.

XXII. Comme cette formule ne détermine que l'eloignement ou l'approchement du corps par rapport à un plan fixe quelconque, pour trouver le vrai lieu du corps à chaque inſtant, on n'aura qu'à le rapporter en même tems à trois plans fixes, qui ſoient perpendiculaires entr'eux. Donc, comme x marque la diſtance du corps à un de ces plans, ſoient y & z ſes diſtances aux deux autres plans ; & aprés avoir décompoſé toutes les forces qui agiſſent ſur le corps, ſuivant des directions perpendiculaires à ces trois plans, ſoit P la force perpendiculaire qui en réſulte ſur le premier, Q ſur le ſecond, & R ſur le troiſième. Suppoſons que toutes ces forces tendent à eloigner le corps de ces trois plans ; car en cas qu'elles tendent à le rapprocher, on n'auroit qu'à faire les forces negatives. Cela poſé, le mouvement du corps ſera contenu dans les trois formules ſuivantes:

I. $2M\,ddx = P\,dt^2$; II. $2M\,ddy = Q\,dt^2$; III. $2M\,ddz = R\,dt^2$.

XXIII. Si le corps n'eſt ſollicité par aucune force, de ſorte que $P = o$, $Q = o$, $R = o$, les trois formules trouvées, à cauſe de dt conſtant, ſe réduiront par l'intégration à celles-cy:

$M\,dx = A\,dt$; $M\,dy = B\,dt$; & $M\,dz = C\,dt$.

d'où l'on voit d'abord, que dans ce cas le corps ſe mouvra dans une

Fig. 2.2. Euler's new principle of mechanics

He wrote: "And this alone is the one formula which contains all principles of mechanics." Euler (1752a) also clarified the general motion of a rigid body. He recognized that this is made up of translatoric and rotatoric motions.

This breakthrough in formulating the principles of mechanics for continua sharpened his view on the clarification of the principles of motion. In his lucid book *Theoria motus corporum solidorum seu rigidorum*, Euler (1765) (see also Wolfers, 1853) distinguished clearly between *internal* and *external principles* of motion. The internal principles were understood to include all that which was contained in the body itself and in which the cause lay for its repose or its motion. All external causes which could contribute something to its motion or repose were hereby excluded. Moreover, Euler remarked:

"That property of the body, which contains the cause for persisting in the same state is called the inertia or the force of inertia.... If one asks, therefore, why a body at absolute repose tends to stay at repose or a steady and straight motion tends to move steady and straight one cannot state another cause than its inertia. One may not look for another cause of this phenomenon elsewhere out of the body."

These are the internal principles of motion, which are based on the general property, called inertia. With the help of these principles, we can determine the motion of mass points if no external agency is acting. Furthermore, Euler (1765) pointed out that the reflections on the motion of mass points could be transferred to bodies of finite size.

Then, Euler explained what was to be understood by the external principles of motion, namely all agencies which acted outside of the body and which influenced its motion. Moreover, he specified his statement:

"...the agency which can change the absolute state of the body is called a force. One has to consider the force as an external cause because the body would remain in its state due to the internal causes."

He then raised the question: "Where do the forces come from?" He recognized that forces which changed the state of the bodies sprang from their impenetrability. The amount of these forces was not determined by the impenetrability, which was not measurable, but by the change of the state.

In what followed, Euler developed the balance of momentum for small bodies. He pointed out that, in order to investigate the effect of the forces on the change of the state of the bodies, it would be easier to proceed from infinitely small bodies and to investigate only the present effectiveness of the forces. He concluded:

"...if we according to this, set up the principles for infinite small bodies and an infinite small time interval, it will not be difficult to go over by integration to the motion of the bodies which changes in finite time."

Finally, Euler (1765) formulated the balance of momentum for small bodies; he stated that this balance equation was the only principle of the whole mechanics:

"If a small body with the mass A is driven by a force p and if the body covers, after decomposition of the motion into the direction of this force, during a small time interval dt the small displacement ds with the velocity $\frac{ds}{dt} = v$, then we have [2] $\frac{d\,ds}{dt^2} = \frac{\lambda p}{A}$ or $dv = \frac{\lambda p dt}{A}$. Thus, the increase of the velocity in the direction of the driving force, is directly proportional to the driving force in the time interval and indirectly to the mass of the small body."

In the final paragraph of his treatise on the external causes of the motion, Euler (1765) remarked:

"In the earlier developments of mechanics I had indeed laid down the principles of this science already in such a way that their truth could not be doubted; however, in this place it appears to be fitting to derive these from a more exact consideration of the nature of the body and to reduce to one single principle from which one can derive everything, concerning the motion, in a much easier way. Although I have there, in addition, already extensively investigated the motion of infinitely small bodies or only points, it will be after all convenient to briefly discuss as to how this comes out from a single principle, and I will treat this in such a way that the way will be more paved for the investigation of the motion of finite bodies."

Euler (1765) showed, indeed for several examples, how the motion of rigid bodies could be described by a sole principle, namely the balance of momentum.

After he had completed his fundamental work *Theoria motus corporum solidorum seu rigidorum* (Euler, 1765), he wrote the *Lettres à une princesse d'Allemagne sur divers sujets de physique & de philosophie* (Euler, 1768), which appeared in German only a year later (Euler, 1769a) and which, in the period

[2] λ is a scalar which regulates the dimensions.

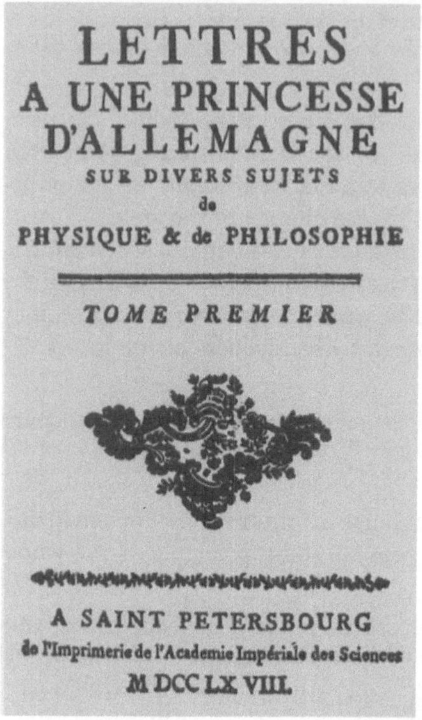

LETTRES
A UNE PRINCESSE
D'ALLEMAGNE
SUR DIVERS SUJETS
de
PHYSIQUE & de PHILOSOPHIE

TOME PREMIER

A SAINT PETERSBOURG
de l'Imprimerie de l'Academie Impériale des Sciences
M DCC LX VIII.

Fig. 2.3. Front page of lettres a une princesse

that followed, was translated into the most influential European languages of the time.

In these letters, Euler explained the fundamentals of mechanics, among other things, to Princess Friederike Charlotte Leopoldine Luise (1745–1808), the oldest daughter of Markgrave Heinrich von Brandenburg-Schwedt and future abbess of Herford. He developed the basic concepts of the principles of mechanics with impressive clarity. In his seventy-first and seventy-second letters, he described the repose and the motions of bodies, before arriving at two fundamental statements in his seventy-third letter of November 4, 1760 (Citation is taken from the English translation).

It is remarkable that Euler considered these statements to be foundations of mechanics. Nothing is mentioned about the rotation of bodies nor about the balance of moment of momentum, although he had described the motion of rigid bodies so ingeniously just two years earlier.

In 1776, he turned once again to the description of the motion of rigid bodies with the two lucid works *Formulae generales pro translatione quacumque corporum rigidorum* and *Nova methodus motum corporum rigidorum determinandi* (Euler, 1776a, b). The fundamentally new aspects in his treatises, the highlight of his work in mechanics, were the description of finite rotations of rigid bodies by a linear mapping, the explicit formulation of the balances of

Fig. 2.4. Princess of Brandenburg-Schwedt (Abbess of Herford)

momentum and moment of momentum, and the statement of the six equations which mathematically describe the general spatial motion of a rigid body. We will discuss these papers in Section 2.2.

We conclude from the proceeding investigations:

- Euler did not formulate the balance of mass. Obviously, it was for him a given that for small material points and for rigid bodies these balance equations had to be fulfilled. With this in mind, he developed the single principle.
- Eulers's single principle (balance of momentum) was a result of Newton's second law for mass points applying the balance of mass. However, Euler formulated the second law for the first time in an analytical way, avoiding the

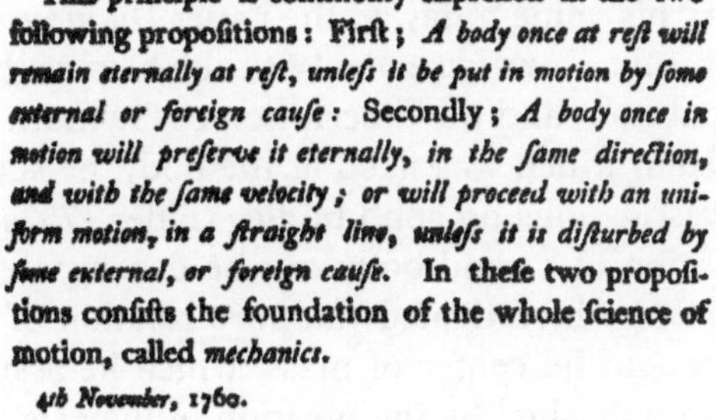

This principle is commonly expreſſed in the two following propoſitions: Firſt; ***A body once at reſt will remain eternally at reſt, unleſs it be put in motion by ſome external or foreign cauſe:*** **Secondly;** ***A body once in motion will preſerve it eternally, in the ſame direction, and with the ſame velocity; or will proceed with an uniform motion, in a ſtraight line, unleſs it is diſturbed by ſome external, or foreign cauſe.*** **In theſe two propoſitions conſiſts the foundation of the whole ſcience of motion, called** ***mechanics.***

4th November, 1760.

Fig. 2.5. Euler's fundamental statements

geometrical-synthetical procedure. Moreover, he demanded that this single principle be valid for mass elements imaginarily cut out from a finite body.
- There was no hint in Eulers work up to 1765 that he had considered the balance of moment of momentum as an axiom.
- Although Euler did not explicitly mention the cut principle, he applied this principle with great skill.

We can state that, up to 1765, Euler had done a considerable amount of work concerning the principles of mechanics.

2.2 The Dynamics of Rigid Bodies

One of the greatest achievements in mechanics was the creation of the theory of rigid bodies under the action of external forces and moments. The rigid body – which is an idealized substitutive body because everything is deformed under load – plays an important role in many branches of engineering, in particular in describing the motion of gyroscopes and its applications, namely gyrocontrolled processes. The main difficulty in describing the finite motion of rigid bodies is that its rotation cannot be described by vectors. It is true that each rotational motion can be given by a directed quantity in Euclidian space. However, this quantity does not obey the commutative law of vector algebra, so therefore the vector calculus cannot be applied. The rotational motion must rather be described by a proper orthogonal tensor. Such a tensor does not only allow the description of the finite rotational motion, but also the description of the finite rotational changes of a coordinate-system.

With the orthogonal tensor used to describe the finite motion of a rigid body, it is relatively easy to describe the balance of moment of momentum for a rigid body. This balance equation creates the connection between the time change of the moment of momentum and the resulting moment of the external forces and moments with respect to a spatial point at repose.

The achievement of creating a theory for the motion of rigid bodies was another highlight in the career of the ingenious Leonhard Euler.

The first great achievement was the derivation of the famous gyrostat-equations by Euler (1758), which carry his name today, in the paper *Du mouvement de rotation des corps solides entor d'un axe variable*. With the help of the special properties of the three principal axes, and the description of the motion of a rigid body with a coordinate system which was fixed at the body, he succeeded in improving his previous investigations on solid bodies (Euler 1752a). Proceeding from the fact that the motion of a rigid body can be decomposed into translatoric and rotatoric motions, he investigated the pure rotation of a body around an axis which went through the center of mass, which he assumed to be at repose. The motion was established by the position of the axis of rotation and the angular velocity. In order to determine the motion caused by forces, Euler (1758) chose the reverse way: he related forces to known motions.

In order to describe the motion, Euler introduced an orthogonal coordinate system with the origin in the center of mass which rotates at the moment with an angular velocity α around an axis. This axis was laid down by three angles of direction, from which he gained (by differentiation with respect to the time) the change of the rotation axis during an infinitesimal time unit. Using the cut principle, Euler (1758) considered the velocity of a mass element which he had imaginarily cut out of the rigid body. From the velocities, he gained the accelerations. From further procedures, it is evident why Euler (1758) described the motion first. In order to obtain the forces which cause the motion, he applied the balance of momentum. Then, he decomposed the force as well as the acceleration in three orthogonal directions. An integration over the whole body would yield the forces which caused the motion of the whole body. In order to simplify the integration, Euler (1758) assumed that at time t his orthogonal coordinate system coincided with the principal axis in the center of mass. This approach with the body fixed coordinate system, which corresponded with the principal axes, had two important advantages. On the one hand, the moments of inertia did not depend on time while, on the other hand, the centrifugal moment disappeared. Both facts simplified the formula considerably. After lengthy calculations, Euler (1758) arrived at his famous system of three differential equations to describe the rotation of a rigid body. The components of the angular velocities and of the external moments were referred to the body fixed coordinate system. The solution of the system of differential equations was difficult. Obviously, Euler (1758) was aware of this difficulty. So, he calculated, after the derivation of his differential system, a relatively easy example with vanishing components of the external moments. However, the calculations were also complex here, but he was convinced of his method and he wrote at the end of his general derivation:

"I would like to stress that the method, which I have applied, has brought me to the aim. This method deserves, therefore, great attention and is an example that one should not think that certain problems are insoluble although one meets many difficulties. The fact is that the calculations are complex."

Although Euler was convinced of his method, he was not happy with the kinematics. After his lucid work on the rigid body motion, (Euler, 1765), he again attacked the problem in two remarkable and fundamental papers (Euler 1776a, b): *Formulae generales pro translatione quacumque corporum rigidorum* and *Nova methodus motum corporum rigidorum determinandi.*

In the first paper, seldom cited in the literature, Euler (1776a) treated exclusively the kinematics of finite rotations of rigid bodies. As has already been mentioned, he was unhappy with the kinematics used in his treatise from 1758, in which he had investigated the causes for the motion and the kinematics simultaneously. Euler (1776a) separated the causes for the motions, namely the forces, from the kinematics. He was obviously of the opinion that he could develop the equations of motion for the rigid body in a more simple way. Indeed, the kinematic relations derived in his treatise deserve the greatest attention because he succeeded in developing the description of kinematic relations for the finite rotation of a rigid body by means of orthogonal transformation equations.

At first, he adopted the special property of a rigid body, namely that the distance between two arbitrary material points remained constant during motion. From the consideration of special displacement positions of straight connection lines between arbitrarily chosen material points, he concluded that the mapping had to be a linear one and had to obey certain relations. With these, the nine original coefficients required in Euler's transformation relations were reduced to only three.

One can show (see Blanc, 1968 and Höckel, 1997) that Euler's transformation relations contain the orthogonal tensor, which is today used for the description of rigid body rotation. Euler (1776a) revealed the main properties of his transformation relations, namely that there were three equations which were equal to unit and three which were equal to zero. These relations are nothing other than the well-known property of the orthogonal tensor by which the tensor product between the orthogonal tensor and its transpose is equal to the identity tensor (see de Boer, 1982). Moreover, his investigations concerning the position of the rotation axis by finite rotations led him to a basic relation for his transformation equations which contained the statement for an orthogonal tensor, namely that the determinant of an orthogonal tensor was equal to unit, see Höckel (1997). However, Euler did not succeed in evaluating this basic relation.

In his treatise *Nova methodus motum corporum rigidorum determinandi*, Euler (1776b) established the balances of momentum and moment of momentum by using new kinematic relations. In this regard, his work was essentially based on findings in his aforementioned paper (Euler, 1776a) and was marked by the search for a rotation axis and rotation angle. He supposed that the property of a rotation axis, namely to keep its direction, would help to simplify the equations of motion. As the translatoric motion of the center of mass was easy to describe, Euler refrained from discussing it and restricted his investigations to the rotation of the body around an axis crossing the center of mass. In order to describe this motion, he at first used spherical trigonometry. Starting from a sphere, in whose center of mass the origin of an orthogonal coordinate system is located, Euler (1776b) investigated the displacement of a point at the surface of the sphere. He described the new placement with angles which could be seen in the spherical angles at the surface of the sphere. The corners of these triangles appeared at those places where the coordinate axes, the rotation axis, and the radius of the considered point in the reference and actual placement pushed through the surface of the sphere. Euler (1776a) succeeded in gaining, through the consideration of special cases, nine equations which described the rotation of the body. The nine equations corresponded to the coefficients of an orthogonal tensor. However, he was able to reduce his system to four relations; three determined the location of the rotation axis with respect to the coordinate axes and one quantity represented the rotation angle. The three angles for describing the location of the axis of rotation had to, however, fulfill the constraint that the sum of the squares of their cosines be equal to unity. Therefore, only three quantities for describing the rotation of a rigid body remained.

After the investigation of the kinematics of rigid bodies, Euler (1776b) treated the motion of rigid bodies, which were driven by arbitrary forces. His goal was to determine the unknown quantities introduced by the kinematic considerations, by the motion at the beginning and by the driving forces. He considered for this purpose the acceleration of an individual element whose coordinates were given in the kinematics which he had developed. Euler (1776b) calculated the accelerations in the direction of the space-fixed coordinates. If one also decomposed all forces which drove the body in the directions of the space-fixed coordinates, then Euler's following statement had to be fulfilled:

"After the principles of motion it is necessary that these forces must be equal to the sum of all accelerating forces which come from the connection of all elements *dm* of the body."

Furthermore, starting from the balance equation of momentum, Euler stated:

"Moreover, all moments of the acceleration forces referring to the three fixed axes, in the sum must, however, be equal to the moments, which are derived from all driving forces referring to the same axis"

With these principles, Euler (1776b) derived six equations for the description of the translatoric and rotatoric motions of a rigid body from the balance equations of momentum and moment of momentum. The coordinates x, y, and z resulted, in this manner, from a linear mapping via the orthogonal tensor of the coordinates in the reference placement.

In this place it should be discussed whether Euler did in fact consider the balance of moment of momentum as an independent axiom. This has been stated by Truesdell (1964) and Szabó (1987). However, in Euler's work nothing is mentioned regarding the balance of moment of momentum as an independent axiom of mechanics. From the preceeding arguments in Section 2.1 and in this section, it is evident that, for Euler, the only axiom to describe the motion was the balance of momentum, not only for the whole body but also for each mass element which was imaginarily cut out of the body. Euler did not bother about the internal forces between the body elements. Obviously, it was clear to him that the norms of the cut forces or cut moments were the same, according to Newton's third law, though opposite in direction at the corresponding surfaces of the mass elements.

He considered the fact that the cut forces neutralized each other if the single mass elements were combined without, however, specifically mentioning this. Moreover, in no place in his works on the principles of mechanics or on the rigid body motions is there an allusion to single moments. Only in this case, when single moments are applied to a body, can the balance of moment of momentum be considered as an independent axiom. Euler's (1776b) treatise was exclusively concerned with an improvement of his description of the motion of rigid bodies

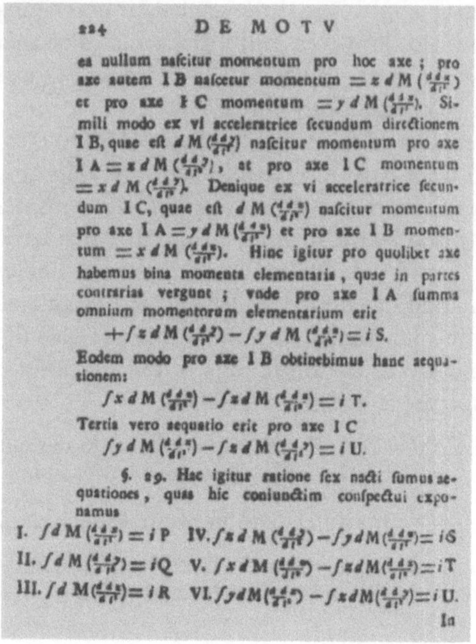

224 DE MOTV

ea nullum naſcitur momentum pro hoc axe ; pro axe autem IB naſcetur momentum $= z\,dM\left(\frac{ddx}{dt^2}\right)$ et pro axe IC momentum $= y\,dM\left(\frac{ddx}{dt^2}\right)$. Simili modo ex vi acceleratrice ſecundum directionem IB, quae eſt $dM\left(\frac{ddy}{dt^2}\right)$ naſcitur momentum pro axe IA $= z\,dM\left(\frac{ddy}{dt^2}\right)$, et pro axe IC momentum $= x\,dM\left(\frac{ddy}{dt^2}\right)$. Denique ex vi acceleratrice ſecundum IC, quae eſt $dM\left(\frac{ddz}{dt^2}\right)$ naſcitur momentum pro axe IA $= y\,dM\left(\frac{ddz}{dt^2}\right)$ et pro axe IB momentum $= x\,dM\left(\frac{ddz}{dt^2}\right)$. Hinc igitur pro quolibet axe habemus bina momenta elementaria, quae in partes contrarias vergunt ; vnde pro axe IA ſumma omnium momentorum elementarium erit

$$+\int z\,dM\left(\frac{ddy}{dt^2}\right) - \int y\,dM\left(\frac{ddz}{dt^2}\right) = iS.$$

Eodem modo pro axe IB obtinebimus hanc aequationem:

$$\int x\,dM\left(\frac{ddz}{dt^2}\right) - \int z\,dM\left(\frac{ddx}{dt^2}\right) = iT.$$

Tertia vero aequatio erit pro axe IC

$$\int y\,dM\left(\frac{ddx}{dt^2}\right) - \int x\,dM\left(\frac{ddy}{dt^2}\right) = iU.$$

§. 29. Hac igitur ratione ſex nacti ſumus aequationes, quas hic coniunctim conſpectui exponamus

I. $\int dM\left(\frac{ddx}{dt^2}\right) = iP$ IV. $\int z\,dM\left(\frac{ddy}{dt^2}\right) - \int y\,dM\left(\frac{ddz}{dt^2}\right) = iS$

II. $\int dM\left(\frac{ddy}{dt^2}\right) = iQ$ V. $\int x\,dM\left(\frac{ddz}{dt^2}\right) - \int z\,dM\left(\frac{ddx}{dt^2}\right) = iT$

III. $\int dM\left(\frac{ddz}{dt^2}\right) = iR$ VI. $\int y\,dM\left(\frac{ddx}{dt^2}\right) - \int x\,dM\left(\frac{ddy}{dt^2}\right) = iU$.

In

Fig. 2.6. Description of the motion of rigid bodies

(as the title of his treatise indicates). One can guess that, if Euler had invented a new principle (axiom), he would have explicitly mentioned this expansion of the basic principles as he had done in 1752.

Therefore, the claim by Truesdell (1964) that Euler (1776b) had formulated the balance equation of moment of momentum as an independent axiom must be rejected. This is also true for repeated statements of Szabó (1987). It seems that Szabó (1987) did not read Euler's original papers. Otherwise, one cannot understand why he cited Euler's work, introducing the center of gravity in Euler's (1776b) description of the basic equations of the rigid body motion instead of introducing the center of mass as Euler did.

2.3 The Theory of Ideal Fluids

By the early beginning of the development of porous media theory (the eighteenth century), the theory of ideal fluids had already been completed. Ideal fluids as independent constituents of water-filled earth bodies play only a subordinate role in porous media theory. Therefore, this theory will only be dealt with briefly here. The reader is referred to Truesdell (1968) and Szabó (1987) for more detailed accounts. The following description of the historical development of the theory of fluids is based primarily on these two treatises.

The forerunners of the ideal fluids theory will only be mentioned here in passing (see Szabó, 1987). Archimedes (287-212 BC), who discovered the buoyancy law, is one example. Similarly, the reader should recall Simon Stevin (1548–1620), who found Archimedes' laws in his own way and calculated the pressure at the bottom of a container. A further example is Blaise Pascal, who clearly stated the propagation of internal pressure. Referring to the theory of ideal fluids, we hereby understand it to be the theory which can only be derived from the principles of mechanics found in this chapter.

The first scientist to work in this sense was Isaac Newton (1643–1724). In the first book of his *Principia* (Newton, 1687), for example, he formulated his equations of motion and the general principle of mass attraction. Truesdell (1956a) elaborates:

"Newtons *Principia* ist ein Meisterwerk, das heute nicht mehr gelesen wird. Bereits im ersten Buch sind nahezu all die Dinge enthalten, derentwegen das gesamte Werk berühmt wurde. Jedoch zeigte Newton in diesem Buch wenig Originalität, vielmehr eine andere Eigenschaft, die ebenso groß ist: Die Fähigkeit, die früheren Ergebnisse in streng mathematischer Weise zu ordnen und aus einem Minimum von Voraussetzungen herzuleiten. Das zweite Buch, welches die Flüssigkeiten behandelt, ist hingegen fast vollkommen eigenständig und beinahe ganz falsch. Das deduktive Verfahren, welches das erste Buch in so hervorragender Weise kennzeichnet, wird hier beiseite gelassen, und bei jedem neuen Gedankengang wird eine neue Hypothese aufgestellt. Hier offenbart Newton sein höchst schöpferisches Genie. Wohl sind seine Lösungen nicht immer richtig; dennoch ist er der erste, der diese Grundprobleme ausgewählt und anzupacken gewagt hat."

"Newton's *Principia* is a master piece which is no longer read today. The first book contains nearly all those things which made the entire work famous. The fact that Newton showed less originality in this book is more than compensated for by his superior ability to order the former results in a strict mathematical manner, and to derive from them a minimum of presuppositions. The second book, which treats fluids, is, on the other hand, nearly completely autonomous and almost entirely wrong. The deductive procedure, which characterizes the first book in such an excellent manner, is here abandoned and from each new train of thought a new hypothesis is made. Here Newton reveals his highest creative genius. No doubt his solutions are not always correct; however, he was the first person who selected the basic problems and dared to handle them."

Isaac Newton's failure is not surprising. Apart from his formulation of the balance of momentum for the mass point, the principles of continuum mechanics were as yet unknown. In his second book, Newton treated generally the resistance of motion which solids undergo in fluids and gases. We will return to Newton's *Principia* in connection with the development of the theory of viscous fluids in the nineteenth century.

Daniel Bernoulli was the first scientist to explain, with the help of the principles of mechanics, the mechanical behavior of incompressible and compressible fluids, which flow without friction. His pioneering book *Hydrodynamica* (Bernoulli, 1734), which appeared in Strassburg in 1738, contained much which was new, including the term, hydrodynamics.

Daniel Bernoulli proceeded from two fundamental principles, the first being the balance of living forces and the second being the equation of continuity. Daniel Bernoulli was also the first to consider velocity in his description of fluid

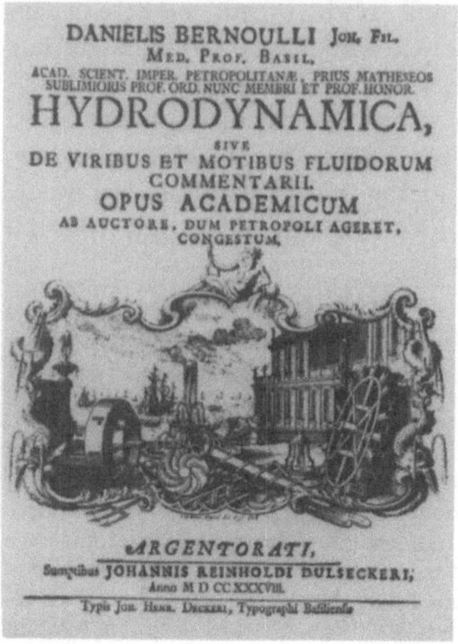

Fig. 2.7. Front page of Daniel Bernoulli's *Hydrodynamica*

motion. However, nowhere in his *Hydrodynamica* is the principle of balance of energy of ideal incompressible fluids to be found as the Bernoulli equation. Neither does the extended form, which is valid for instationary flow, appear in Daniel Bernoulli's book (see Szabó, 1987).

Daniel Bernoulli's *Hydrodynamica* attracted great attention. In the "Vorrede" of his treatise *Neue Grundsätze der Artillerie* Euler (1745) wrote:

"Denn die Bewegung der flüßigen Körper ist eine von den schweresten und verwirrtesten Materien, welche in der Mathematic und Physic immer vorkommen können, und mit einer gemeinen Erkentniß der Mathematic ist darinne nicht das geringste auszurichten. Die berühmten Herren Bernoulli sind die ersten gewesen, welche diese so dunkle Materie auf eine gründliche Art abgehandelt haben. Der Hr. Professor Daniel Bernoulli in Basel hat darüber zuerst sein unvergleichliches Werk unter dem Titel der *Hydrodynamic* herausgegeben, worinne er durch die subtilsten Rechnungen so wohl die Kräfte, als die Bewegungen der flüßigen Cörper, so gründlich bestimmet, daß allenhalben die schönsten Übereinstimmung mit der Erfahrung hervorleuchtet."

"The motion of liquids is one of the most difficult and most confusing subjects that can occur in mathematics and physics, where a common knowledge of mathematics is of no use. The famous Bernoulli gentlemen were the first to treat this dark subject thoroughly. In Basel, Professor Daniel Bernoulli first published his incomparable work under the title of the *Hydrodynamic*, where via subtle calculation he very carefully determined the forces as well as the motions of liquid bodies, the results of which are in perfect correspondence with the experience."

After the publication of Daniel Bernoulli's *Hydrodynamica* (see Szabó, 1987), Euler asked Johann Bernoulli (Daniel's father) on December 20, 1738 to send him his,

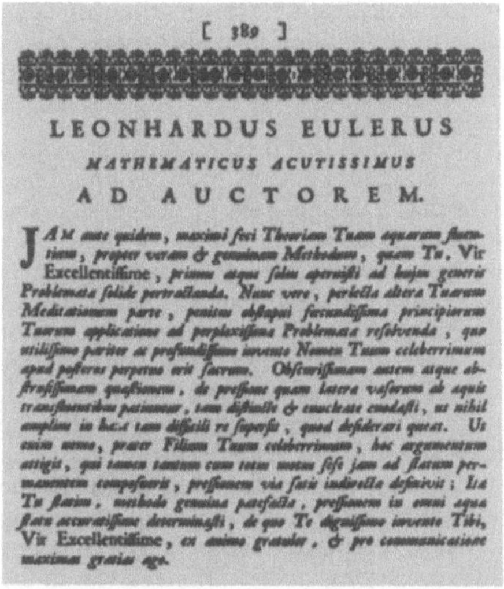

[389]

LEONHARDUS EULERUS

MATHEMATICUS ACUTISSIMUS

AD AUCTOREM.

J*AM ante quidem, maximi feci Theoriam Tuam aquarum fluentium, propter veram & genuinam Methodum, quam Tu*, Vir Excellentissime, *primus atque solus aperuisti ad hujus generis Problemata solide pertractanda. Nunc vero, perlecta altera Tuarum Meditationum parte, penitus obstupui foecundissima principiorum Tuorum applicatione ad perplexissima Problemata resolvenda, quo utilissimo pariter ac profundissimo invento Nomen Tuum celeberrimum apud posteros perpetuo erit sacrum. Obscurissimam autem atque abstrusissimam quaestionem, de pressione quam latera vasorum ab aquis transfluentibus patiuntur, tam distincte & enucleate enodasti, ut nihil amplius in hac tam difficili re supersit, quod desiderari queat. Ut enim nemo, praeter Filium Tuum celeberrimum, hoc argumentum attigit, qui tamen tantum cum totus motus sese jam ad statum permanentem composuerit, pressionem via satis indirecta definivit; Ita Tu statim, methodo genuina patefacta, pressionem in omni aquae statu accuratissime determinasti, de quo Te dignissimo invento Tibi,* Vir Excellentissime, *ex animo gratulor, & pro communicatione maximas gratias ago.*

Fig. 2.8. Euler's introduction to Johann Bernoulli's *Hydraulica*

"new and incomparable theory of the motion of liquids... that I am aware of the insufficiencies for a long time with which these were treated until now."

(Apparently Johann Bernoulli had indicated the premise of the new theory in his correspondence). Euler had obviously noticed that Daniel Bernoulli's axiomatic theory in *Hydrodynamica* was not fundamental, but rather that the principle of the balance of living forces could be derived from another fundamental principle. On March 7, 1739, Johann Bernoulli fulfilled Euler's wish. Euler further urged Johann Bernoulli to send him the remaining parts of his theory on the motion of fluids. Johann Bernoulli took more than one year to answer. On August 31, 1740, he sent Euler an additional part of his theory. Euler replied enthusiastically on October 18, 1740. Among other things, he wrote:

"After having read the second part of your examinations, I was astonished to see that your principles are so suitable for solving such complicated problems."

He continued:

"You have solved the very difficult and concealed problem of the pressure which is put on the receptacle walls by flowing water so clearly and convincingly, that it leaves nothing more to be desired. Nobody else dared to tackle this problem, other than your very famous son. He, however, only ascertained the pressure in an indirect way, when the whole motion had slowed to a stationary state. You yourself, on the other hand, discovered the direct method of precisely ascertaining the water pressure in its various states. I congratulate you on this discovery with all my heart and thank you for informing me."

Johann Bernoulli was so impressed by Euler's letter that he included it in his volume (Bernoulli, 1732) on the theory of the motion of fluids, *Hydraulica*, which appeared in 1742. Johann Bernoulli wrote in the preface of his book:

"The hydrodynamical work, recently published by my son, began with this material under happier auspices. My son relied upon an indirect foundation, i.e., the living forces, which was not accepted by all philosophers although I had proven it.

As yet no one has discovered a direct method for investigating a priori the principles of dynamics alone as well as regarding the nature of the motion of water coming out of openings in the container, or water flowing through the pipe of an unequal transverse cross-section.

I have also been wondering from whence the problem comes that the dynamical principles are not used for fluids as they are for solids. As I thought about the problem even more I discovered its real origin. It consists of the fact that a certain part of the compressive forces which are used to form an eddy (this is my term, others do not use it), have been neglected as if they did not exist. Additionally, they are considered as being infinitely small, only for the reason that the eddy is made up of a small part of the fluids. An eddy is formed when a fluid from a larger cross-sectional area flows into a smaller one or vice versa. In the first case the eddy is formed before the transition, in the second case afterwards. What an eddy is and how it is formed seldom arises from the examinations; it is evident that it is formed without any real loss of living forces. Thus it is clear why the principle of living forces can be used successfully and without error in hydraulics, even when the eddy is not taken into account. I will write up this study in two parts. In the first, I will consider the manifestation of flowing water and the occurrence of its flowing out from cylindrical and prismatic receptacles. These receptacles can be single or many put together, like systems made up of various pipes of varying cross-sectional areas. In the second part, I will examine everything quite generally; how the receptacles are formed, whether they be regular or irregular, or perforated and those to which canals and pipes are attached."

In the first part of *Hydraulica*, Johann Bernoulli extended this theme:

"Nobody should think that this motion force (which propels an infinitely small part of the fluid through the eddy) can be neglected. It has a very precise quantity. In spite of the small size of the matter, the acceleration force[3] must be large enough to reach the speed in the cross-section *GF* (see Fig. 2.9) during the short space of time for which it takes the fluid to pass through the small gap *HG*.

It is because this force has been neglected that nobody up until today has been able to successfully derive the laws of flowing fluids in non-uniform canals. Those who have tried, however, followed my example and went back to the principle of living forces. If I had not gone ahead, it is unlikely that they would have considered using it for this or for the other problems which occur in solids and fluids. I was the first to teach the use of the conservation of living forces. However, I myself was not yet satisfied with this indirect method. I did not give up searching for the direct method which had to be supported by the undisputed dynamical principles. Finally, after having racked my brains for quite some time, I realized in 1729 that the salient point of the problem lies in the observation of the eddy; nobody had as yet paid any attention to it before this. I shall now publish my discoveries, having already explained them privately to some friends, so that they will be discussed. With this end in view, after my description of the eddy I will pursue further, as clearly as possible, what I have already begun."

In the first part of *Hydraulica*, Johann Bernoulli examines the motion of water in containers and pipes joined to each other. In the second part, he studies:

"...the direct and general method of all sorts of hydraulic problems when water flows through canals of any form and position."

Szabó (1987) is of the opinion that the second part of *Hydraulica* contains the "Bernoulli equation" of instationary flow. The second part also contains

[3] According to Szabó (1987), this is acceleration as it occurs in today's terminology.

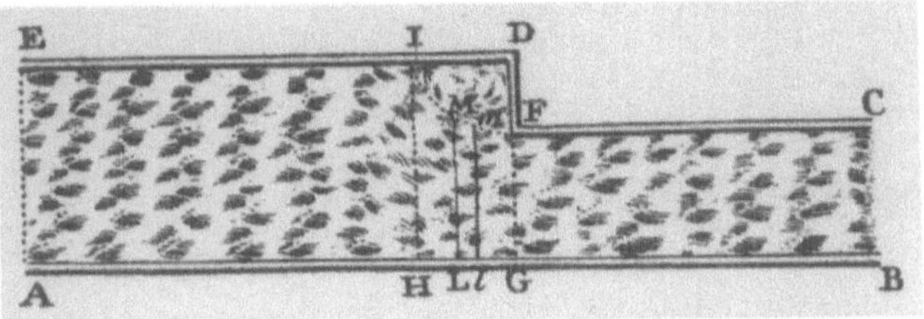

Fig. 2.9. Container consisting of two cylindrical tubes

Johann Bernoulli's calculation of hydraulic pressure on container walls. As Szabó says, Johann Bernoulli had great difficulties with this problem. However, unlike Daniel Bernoulli, he succeeded in formulating explicit formulas for hydraulic pressure on container walls.

The reactions to Johann Bernoulli's *Hydraulica* were varied. Euler, as we know, was full of praise. In a letter, Daniel Bernoulli spoke of his deep disappointment, believing that his father had taken most of the material in *Hydraulica* from his own book *Hydromechanica*, and D'Alembert's objections were of a factual nature. The mathematician and physicist A.G. Kästner[4] (1719–1800) (see Szabó, 1987) praised Johann Bernoulli, who "wanted another foundation" for his son's principle of the conservation of living forces "and therefore thought about his *Hydraulica*". W.G. Karsten (1732–1787) wrote in his book *Hydraulik* (see Szabó, 1987):

"Johann Bernoulli ist der erste, der die allgemeinen Grundsätze der Mechanik auf die Bewegung flüssiger Körper angewandt und aus völlig überzeugenden Gründen bewiesen hat, nach welchen Gesetzen ihre Bewegung von gegebenen Kräften beschleunigt wird."

"Johann Bernoulli was the first to apply the general principles of mechanics to the motion of liquid bodies and for totally convincing reasons he has proved due to which laws their motions are accelerated by given forces."

Without doubt, credit must be given to Szabó (1987) for his research on the history of fluid mechanics, which put an end to the subjective and irresponsible prejudice against Johann Bernoulli which occurred in the literature.

From Daniel and Johann Bernoulli's treatises, one-dimensional flow processes of incompressible, ideal fluids could be calculated. Although Johann Bernoulli did not describe his hypotheses and principles very well (his concept of the "eddy" was not very successful, as his notion of inner pressure was not clearly formulated), the ingenious Euler recognized the tremendous progress of Johann Bernoulli's publication. Truesdell (1954) writes:

"In JOHN BERNOULLI's hydraulics, not in its progenitor, the hydraulico-statics of DANIEL BERNOULLI, lie the roots of EULER's hydrodynamics."

After his fundamental work (Euler, 1752a) in which he explicitly formulated the balance of momentum for a differential small body, Euler began his treatises

[4] We will come back to Kästner in Section 2.6.

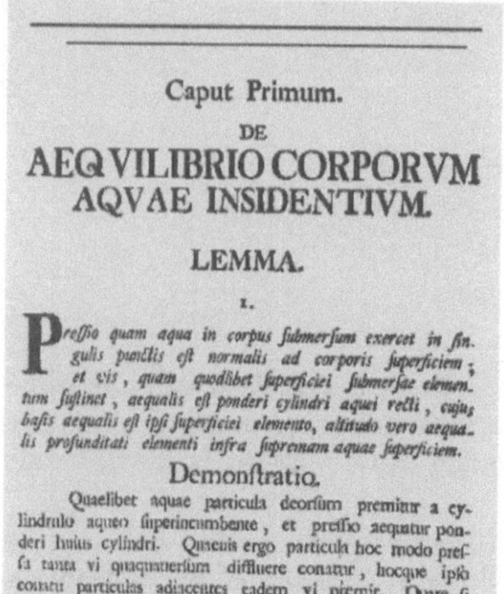

Caput Primum.

DE

AEQVILIBRIO CORPORVM AQVAE INSIDENTIVM.

LEMMA.

I.

Preſſio quam aqua in corpus ſubmerſum exercet in ſingulis punctis eſt normalis ad corporis ſuperficiem; et vis, quam quodlibet ſuperficiei ſubmerſae elementum ſuſtinet, aequalis eſt ponderi cylindri aquei recti, cujus baſis aequalis eſt ipſi ſuperficiei elemento, altitudo vero aequalis profunditati elementi infra ſupremam aquae ſuperficiem.

Demonſtratio.

Quaelibet aquae particula deorſum premitur a cylindrulo aqueo ſuperincumbente, et preſſio aequatur ponderi huius cylindri. Quaevis ergo particula hoc modo preſſa tanta vi quaquauerſum diffluere conatur, hocque ipſo conatu particulas adiacentes eadem vi premit. Quare ſi corpus fuerit aquae ſubmerſum, id in ſingulis ſuae ſuperficiei punctis a particulis aquae tanta vi premetur, quanta ipſae particulae premuntur, idque normaliter in ſuperficiem. Vnde veritas lemmatis conſtat, quae autem plenius in hydroſtatica euincitur. Q. E. D.

A Coroll.

Fig. 2.10. Begin of "Caput Primum" of Euler's *Scienta Navalis*

on the foundation of hydromechanics. In his first work, Euler (1752b) criticized existing theories. Moreover, he remarked that the principles of hydrostatics had been established, but the principles of hydraulics had not yet been. He praised Daniel Bernoulli for being the first to shed light on the motion of fluids with the help of the law of conservation of living forces. Similarly, he praised the father, Johann Bernoulli, who, in laying the foundation of the first principles of mechanics, had handled the same material with the same success. Further conclusive works on the foundation of hydromechanics are: *Principes généraux de l'état d'équilibre des fluides* (Euler, 1757a), *Principes généraux du mouvement des fluides* (Euler, 1757b), *Sectio prima de statu aequilibri fluidorum* (Euler, 1769b), and *Sectio secundo de principis motus fluidorum* (Euler, 1770a) as well as *Sectio tertia de motu fluidorum lineari potissimum aquae* (Euler 1771). These works all appear in the reports of the Berlin and Saint Petersburg Academies. Moreover, they also appear in *Opera omnia*, Ser. Secunda XII and XIII, and together with other works on hydrodynamical problems. In these works, Euler completed, among other things, the theory of ideal fluids. Another two-volume work, published earlier, *Scienta navalis* (Euler 1749), contains the definition of pressure (see Fig. 2.10 and Truesdell, 1954).

"The pressure which the water exerts upon a submerged body in its several points is normal to the surface of the body; and the force which an arbitrary element of the submerged surface sustains was equal to the weight of a right aqueous cylinder whose base is equal to the element of surface itself, and whose altitude is equal to the depth of the element below the upper surface of the water."

This lemma makes no previous reference to Stevin's theorem, which more precisely and more clearly stated that the pressure acted vertically to the surface and was equal in all directions. This proposition was at first only applied to water, but Euler later proved its validity for all fluids and also for gases under the influence of external forces.

After Euler (1749) had clarified the pressure concept and explicitly formulated the equation of motion (Euler, 1752b), he was ready to develop the equations of motion for fluids. After having dealt with hydrostatics in *Scienta navalis* (Euler, 1749), Euler completed the description of the motion of fluids (Euler, 1757a, b, 1770a, 1771) (in the aforementioned Berlin and Saint Petersburg reports); he wrote (see Truesdell, 1954):

"Having established in my preceding memoir the principles of the equilibrium of fluids in full generality..., I propose here to treat the motion of fluids on the same footing. It is easy to see that this matter is much more difficult, and that it includes researches which are incomparably deeper. Nevertheless I hope to succeed in so far that if there remain any difficulties, they shall not be on the side of mechanics, but solely on the side of analysis: for this science has not yet been carried to the degree of perfection which would be necessary in order to develop analytic formulae including the principles of the motion of fluids."

After introducing (in modern terminology) the material time derivative, Euler formulated the fundamental equations of hydrodynamics, namely the continuity equation and the equations of motion:[5]

$$\begin{aligned}
&\left(\frac{dq}{dt}\right) + \left(\frac{d\cdot qu}{dx}\right) + \left(\frac{d\cdot qv}{dy}\right) + \left(\frac{d\cdot qw}{dz}\right) = 0\,,\\
&P - \frac{1}{q}\left(\frac{dp}{dx}\right) = \left(\frac{du}{dt}\right) + u\left(\frac{du}{dx}\right) + v\left(\frac{du}{dy}\right) + w\left(\frac{du}{dz}\right),\\
&Q - \frac{1}{q}\left(\frac{dp}{dy}\right) = \left(\frac{dv}{dt}\right) + u\left(\frac{dv}{dx}\right) + v\left(\frac{dv}{dy}\right) + w\left(\frac{dv}{dz}\right),\\
&R - \frac{1}{q}\left(\frac{dp}{dz}\right) = \left(\frac{dw}{dt}\right) + u\left(\frac{dw}{dx}\right) + v\left(\frac{dw}{dy}\right) + w\left(\frac{dw}{dz}\right).
\end{aligned} \tag{2.3.1}$$

In these equations, q is the density, p is the pressure, P, Q, and R are the external accelerations relating to the density in the direction of the coordinates x, y, and z, and u, v, and w are the velocity components in the x, y, and z directions.

With the above continuity equation and the equations of motion as well as two articles at the beginning of the 1770s (Euler, 1770a, 1771), Euler completed the theory of ideal fluids.

2.4 Euler's Description of a Porous Body

The first descriptions of a porous body worth mentioning are those of the ingenious Leonhard Euler (1762). In the posthumously published *Anleitung zur*

[5] Euler did not yet use a special mathematical symbol for the partial derivative.

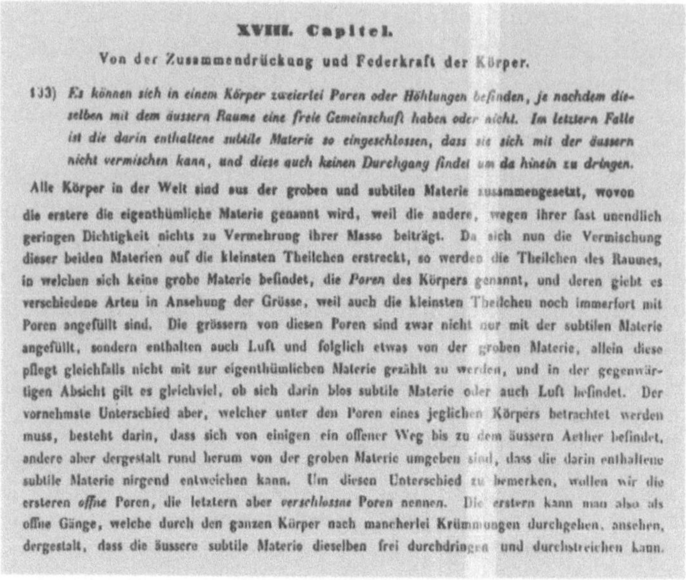

XVIII. Capitel.

Von der Zusammendrückung und Federkraft der Körper.

133) *Es können sich in einem Körper zweierlei Poren oder Höhlungen befinden, je nachdem dieselben mit dem äussern Raume eine freie Gemeinschaft haben oder nicht. Im letztern Falle ist die darin enthaltene subtile Materie so eingeschlossen, dass sie sich mit der äussern nicht vermischen kann, und diese auch keinen Durchgang findet um da hinein zu dringen.*

Alle Körper in der Welt sind aus der groben und subtilen Materie zusammengesetzt, wovon die erstere die eigenthümliche Materie genannt wird, weil die andere, wegen ihrer fast unendlich geringen Dichtigkeit nichts zu Vermehrung ihrer Masse beiträgt. Da sich nun die Vermischung dieser beiden Materien auf die kleinsten Theilchen erstreckt, so werden die Theilchen des Raumes, in welchen sich keine grobe Materie befindet, die *Poren* des Körpers genannt, und deren giebt es verschiedene Arten in Ansehung der Grösse, weil auch die kleinsten Theilchen noch immerfort mit Poren angefüllt sind. Die grössern von diesen Poren sind zwar nicht nur mit der subtilen Materie angefüllt, sondern enthalten auch Luft und folglich etwas von der groben Materie, allein diese pflegt gleichfalls nicht mit zur eigenthümlichen Materie gezählt zu werden, und in der gegenwärtigen Absicht gilt es gleichviel, ob sich darin blos subtile Materie oder auch Luft befindet. Der vornehmste Unterschied aber, welcher unter den Poren eines jeglichen Körpers betrachtet werden muss, besteht darin, dass sich von einigen ein offener Weg bis zu dem äussern Aether befindet, andere aber dergestalt rund herum von der groben Materie umgeben sind, dass die darin enthaltene subtile Materie nirgend entweichen kann. Um diesen Unterschied zu bemerken, wollen wir die ersteren *offne* Poren, die letztern aber *verschlossne* Poren nennen. Die erstern kann man also als offne Gänge, welche durch den ganzen Körper nach mancherlei Krümmungen durchgehen, ansehen, dergestalt, dass die äussere subtile Materie dieselben frei durchdringen und durchstreichen kann.

Fig. 2.11. Euler's description of a porous body

Naturlehre, he pointed out, in Chapter XVII of *Von der Zusammendrückung und Federkraft der Körper*, the nature of elasticity. He attributed the elasticity of a solid to a certain subtle matter in closed pores. This explanation was very unsatisfactory. However, Euler's remarks on porous bodies are interesting:

"All bodies in the world are composed of rough and subtile matter, where of the first is called the characteristic matter whereas the other due to its nearly infinitely small density contributes nothing to the increase of their mass. Since the mixture of both matters extends to the smallest parts, those parts of the space, in which no rough matter is contained, are called the pores of the bodies, and there are different kinds concerning the size, because also the smallest parts are still filled up with pores...The most distinct difference, however, which must be considered for the pores of any body, is that some form an open path to the others, whereas other ones are surrounded by the rough matter in such a way that the subtle matter therein contained can not escape. In order to denote this difference we will call the first *open pores* and the last *closed pores*."

Euler (1768) also dealt with the porous body in the aforementioned *Lettres á une princesse d'Allemagne*. In his sixty-ninth letter of October 21, 1760, after he had disregarded the statement of the Cartesians "that the nature of bodies consists of expansion", Euler arrived at the essential definition of what a body consists of:

"We easily discover, however, a general character, inseparable from all matter, and, consequently, pertaining to all bodies; it is impenetrability, the impossibility of being penetrated by other bodies, or the impossibility that two bodies should occupy the same place at once. In truth, *impenetrability* is what a vacuum wants in order to be a body."

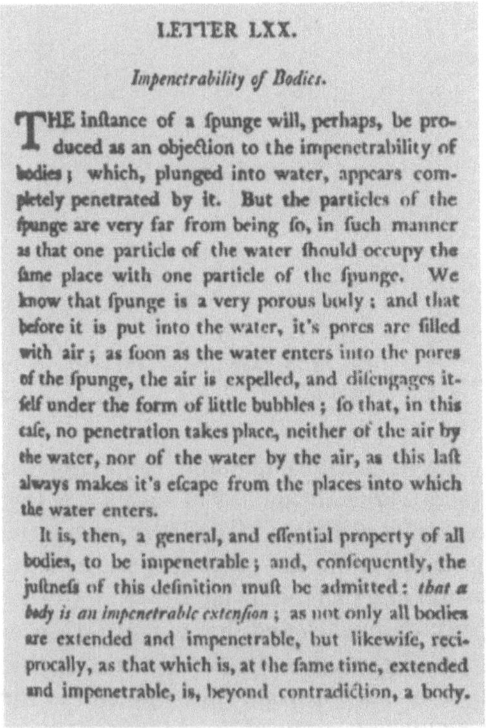

LETTER LXX.

Impenetrability of Bodies.

THE inſtance of a ſpunge will, perhaps, be produced as an objection to the impenetrability of bodies; which, plunged into water, appears completely penetrated by it. But the particles of the ſpunge are very far from being ſo, in ſuch manner as that one particle of the water ſhould occupy the ſame place with one particle of the ſpunge. We know that ſpunge is a very porous body; and that before it is put into the water, it's pores are filled with air; as ſoon as the water enters into the pores of the ſpunge, the air is expelled, and diſengages itſelf under the form of little bubbles; ſo that, in this caſe, no penetration takes place, neither of the air by the water, nor of the water by the air, as this laſt always makes it's eſcape from the places into which the water enters.

It is, then, a general, and eſſential property of all bodies, to be impenetrable; and, conſequently, the juſtneſs of this definition muſt be admitted: *that a body is an impenetrable extenſion*; as not only all bodies are extended and impenetrable, but likewiſe, reciprocally, as that which is, at the ſame time, extended and impenetrable, is, beyond contradiction, a body.

Fig. 2.12. Euler's description of a water-saturated porous body (sponge)

In his seventieth letter, of October 25, 1760, Euler gave an example of this definition, the water-saturated porous solid (sponge) which is of immediate relevance to our topic:
It is impressive how clearly Euler described the impenetrability of the constituents of a multi-component continuum. It seems that Euler did not go on to treat the porous solid in further contributions.

2.5 Coulomb's Earth Pressure Theory

The systematic study of soil bodies as one-component continua began in 1773, when Coulomb (1773) developed a method for calculating the earth pressure on retaining walls. He was inspired by his work as a military engineer on Martinique, where he dealt with the building of fortifications.

Coulomb's method for the calculation of failure and fracture was based on existing theoretical reflections and observations. The first to describe such observations of slope failure seems to have been Marshall Vauban (see Poncelet, 1840), who was also working on fortifications. At the time, however, it was not possible to calculate the state of failure in advance, because the principles of mechanics for solid bodies had not yet been fully developed.

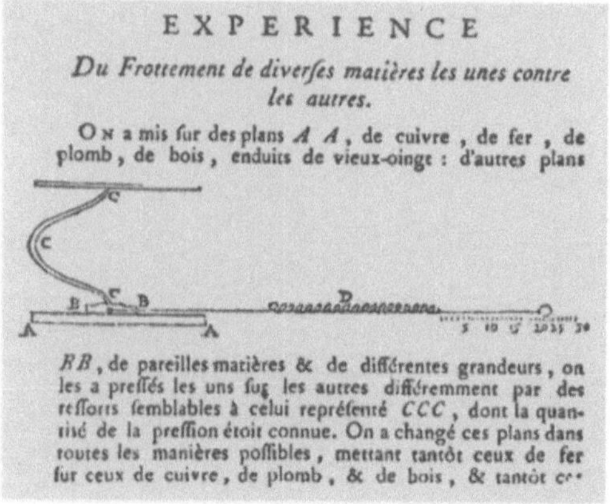

EXPERIENCE

Du Frottement de diverſes matières les unes contre les autres.

On a mis ſur des plans *A A*, de cuivre, de fer, de plomb, de bois, enduits de vieux-oinge : d'autres plans

BB, de pareilles matières & de différentes grandeurs, on les a preſſés les uns ſur les autres différemment par des reſſorts ſemblables à celui repréſenté *CCC*, dont la quantité de la preſſion étoit connue. On a changé ces plans dans toutes les manières poſſibles, mettant tantôt ceux de fer ſur ceux de cuivre, de plomb, & de bois, & tantôt ce-

Fig. 2.13. Amonton's experiment of friction

The biggest impetus to develop the mechanics of solid bodies came from the cut principle. At first, the stress resultants were only applied in such a way that they acted perpendicularly to the cross-section. Parent (1713) first introduced the notion of inner force acting *tangentially* upon the cross-section. In 1726, Couplet (1726) considered the equilibrium of a ball model (statics were already at this time surprisingly well-developed), and concluded that tangential forces had to be operating at the cross-section of a sand body. The knowledge that the inner force acted obliquely to the cross-section contributed greatly to the development of the rod theory in the eighteenth century.

The friction phenomenon, which occurs when bodies glide away from each other, had also been known for some time. Amontons (1699) had carried out experiments in this area as early as 1699; he recognized that the frictional force was proportional to the pressure:

"Primo, que la résistance causée par le frottement n'augmente ne diminue qu'à proportion des pressions plus ou moins grandes suivant que les parties qui frottent ont plus on moins d'étandue, ..."

"First the resistance, which is caused by friction, is not larger or smaller but proportional to the pressure, which is according to the more or less large surface of the parts of friction more or less large, ..."

Amontons' experiments, however, had one big disadvantage. They had been carried out on steel, copper, lead, and timber plates which had been treated with wagon grease. With a spring-balance, he measured the force which needed to be raised in order to actuate plates under loads (see Fig. 2.13).

In this way, Amontons always established the same proportionality factor between the spring rate and the surcharge. It was the famous Leibniz (see Gollub, 1989) who pointed out that this result could not possibly be correct, because the proportionality rate in bodies consisting of various materials gliding onto each other had to be different.

After extensive explanations about the use of the cut principle and the stress resultant, as well as about the friction phenomenon, the way was clear for

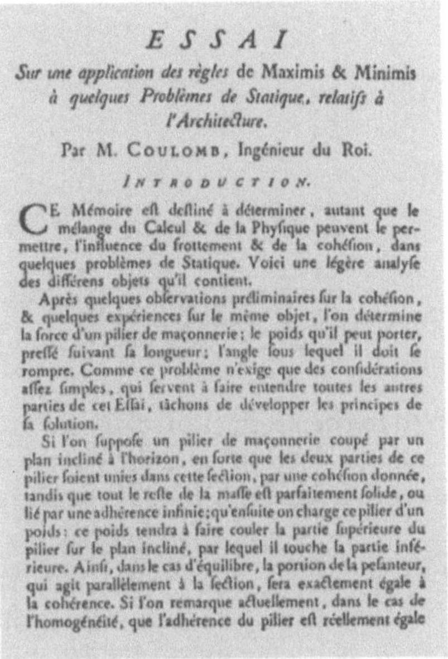

ESSAI

Sur une application des règles de Maximis & Minimis *à quelques Problèmes de Statique, relatifs à l'Architecture.*

Par M. Coulomb, Ingénieur du Roi.

Introduction.

Ce Mémoire eſt deſtiné à déterminer, autant que le mélange du Calcul & de la Phyſique peuvent le permettre, l'influence du frottement & de la cohéſion, dans quelques problèmes de Statique. Voici une légère analyſe des différens objets qu'il contient.

Après quelques obſervations préliminaires ſur la cohéſion, & quelques expériences ſur le même objet, l'on détermine la force d'un pilier de maçonnerie; le poids qu'il peut porter, preſſé ſuivant ſa longueur; l'angle ſous lequel il doit ſe rompre. Comme ce problème n'exige que des conſidérations aſſez ſimples, qui ſervent à faire entendre toutes les autres parties de cet Eſſai, tâchons de développer les principes de ſa ſolution.

Si l'on ſuppoſe un pilier de maçonnerie coupé par un plan incliné à l'horizon, en ſorte que les deux parties de ce pilier ſoient unies dans cette ſection, par une cohéſion donnée, tandis que tout le reſte de la maſſe eſt parfaitement ſolide, ou lié par une adhérence infinie; qu'enſuite on charge ce pilier d'un poids: ce poids tendra à faire couler la partie ſupérieure du pilier ſur le plan incliné, par lequel il touche la partie inférieure. Ainſi, dans le cas d'équilibre, la portion de la peſanteur, qui agit parallèlement à la ſection, ſera exactement égale à la cohérence. Si l'on remarque actuellement, dans le cas de l'homogénéité, que l'adhérence du pilier eſt réellement égale

Fig. 2.14. Coulomb's famous work

Coulomb (1773) to formulate his theory. In his excellent and lucid work *Essai sur une application des règles de maximis et minimis à quelques problèmes de statique, relatifs à l'architecture*, written in 1773 and published in 1776 in the *Mémoires de mathematiques et de physiques, présentés a l'academie royale des sciences par divers savants et lus dans ses assembles*, Coulomb presented his ideas. At first, he developed his earth pressure theory only for his own private use at work. He later gave it to the French Academy for publication.

In his treatise (we are here also citing the German translation from 1779), he introduced the topics which he intended to treat the determination of the influence of friction and its association to certain problems in statics. These problems included the study of the failure condition for a brick column and the establishing of earth pressure. In the introduction, he worked out in detail that the resistance was composed of cohesion and friction together, and that the establishment of earth pressure represented a minimum/maximum problem.

Coulomb then made a few equilibrium considerations and proceeded to discuss friction, whereby he referred to Amontons' work. He subsequently portrayed the attempts to determine cohesion, and found that the tensile strength approached the shear strength. In the evaluation of his experiments, Coulomb introduced a totally new idea. He related the tensile and shear strengths to the area where the respective force was exerted. In this way, he was the first scientist to work with stress, although he did not use this term. In a mere three pages, he developed the determination of the support load of a cantilever beam with a concentrated load and thus made a decisive contribution to the completion of the beam theory.

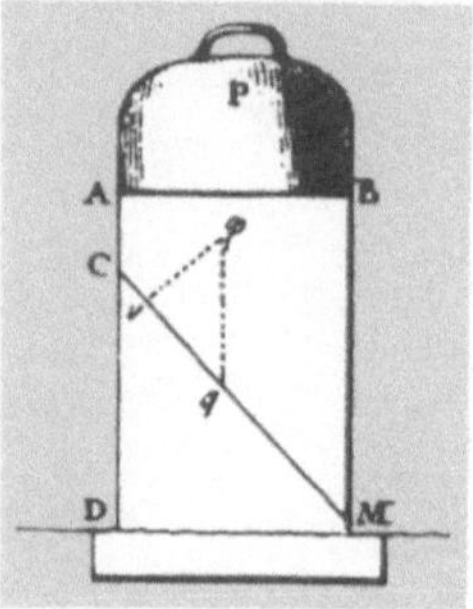

Fig. 2.15. Coulomb's determination of the collapse load

The following sections were of importance to the determination of the failure condition. Here, Coulomb determined the collapse load of an axially loaded brick column (see Fig. 2.15). There are three remarkable steps in his procedure (see Gollub, 1989):

- The consideration of the gliding fracture,
- the formulation of the equilibrium conditions for those partial bodies which could probably slip off the failure surfaces in consideration of the friction and cohesion resistance, and
- the maximum/minimum calculation to determine the possible failure surface and the limit external load.

To begin with, Coulomb carried out his experiment on a purely cohesive material and came to the correct conclusion that the "angle of the smallest resistance or fracture = 45°". After this, he also considered friction. He called the coefficient of friction (proportionality factor) $1/n$ [6]. From equilibrium considerations and the setting of a limit on the tangential inner forces (in the fracture assumed under the angle x) through the sum of the cohesion δ and the frictional force for the axial load P, he found:

$$P = \frac{\delta a^2}{\cos x \left(\sin x - \dfrac{\cos x}{n}\right)} . \tag{2.5.1}$$

From the minimum demand for P, he obtained the location of the fracture surface:

$$\tan x = \frac{1}{\sqrt{1 + \dfrac{1}{n^2}} - \dfrac{1}{n}} . \tag{2.5.2}$$

He explained his solution through an example and compared his result to the test results of the Dutchman, Musschenbroëk. He remarked:

[6] The fact that he calls the proportionality factor $1/n$ and not n (R. Woltman does the same, see the following section), probably leads to the practice of various contemporary researchers always indicating the proportionality factor by a fraction.

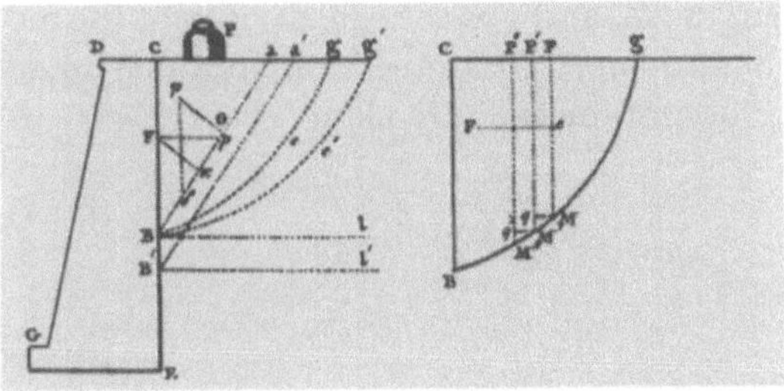

Fig. 2.16. Coulomb's determination of the earth-pressure against a retaining wall

"Au reste, je suis obligé d'avertir que la manière dont M. Musschenbroëk détermine la force d'un pilier de maçonnerie, n'a aucun rapport avec celle que je viens d'employer. Un pilier, pressé par une force dirigée suivant sa longueur, ne se rompt, dit ce Physicien célèbre, que parce qu'il commence à se courber; autrement il supporteroit toute sorte de poids. En partant de ce principe, il détermine la force des piliers quarrés, en raison inverse du quarré de leur longueur, & triplée de leurs cotés;..."

"Otherwise I have to point out that Musschenbroëk's method of calculating the strength of a brick column has nothing in common with mine. According to him, a column pressed in the direction of its length breaks only when it bends, otherwise it can bear any type of load. He proceeds from this principle and finds the strength of square columns in inverse proportion to the square of the length and the cube of their sides."

It would appear that Coulomb misunderstood this method. Musschenbroëk was actually dealing with stability problems.

The determination of earth pressure takes up the most space in Coulomb's treatise. He had already formulated his procedure in the introduction:

"Si l'on remarque ensuite que les terres étant supposées homogènes, peuvent se séparer dans le cas de rupture, non-seulement suivant une ligne droite, mais suivant une ligne courbe quelconque; il s'ensuit que pour avoir la pression d'une surface de terre contre un plan vertical, il faut trouver parmi toutes les surfaces décrites dans un plan indéfini vertical, celle qui, sollicitée par sa pesanteur, & retenue par son frottement & sa cohésion, exigeroit, pour son équilibre, d'être soutenue par une force horizontale, qui fut une *maximum*; car, pour lors il est évident que toute autre figure demandant une moindre force horizontale, dans le cas d'équilibre, la masse adhérente ne pourroit se diviser."

"If one remarks, at first, that the earth in a homogeneous soil can not only tear away in a straight but also in any curved line, it follows that, in order to determine the pressure against a vertical lateral surface, one must find within the described equilibrium surfaces an undefined vertical plane, which driven by its gravity and retained by its friction and cohesion, can be kept in equilibrium by a horizontal force which is a *maximum*: For, it is obvious that the connected mass cannot separate, since any other figure demands a smaller horizontal force."

In calculating the earth pressure, he proceeded from Fig. 2.16. Analogously to the procedure used for an axially loaded brick wall, he calculated the active earth pressure A taking a fracture plane:

$$A = \frac{\frac{gax}{2}\left(a - \frac{x}{n}\right) - \delta\left(a^2 + x^2\right)}{x + \frac{a}{n}}, \tag{2.5.3}$$

where a is the height BC, x is the length CA, and g is the specific weight (he did not consider the load P in Fig. 2.16 until later). From the minimum/maximum requirement, he designated the position of the fracture plane as:

$$x = -\frac{a}{n} + a\sqrt{1 + \frac{1}{n^2}} , \qquad (2.5.4)$$

and the accompanying earth pressure

$$A = ma^2 - \delta la , \qquad (2.5.5)$$

"where m and l are constant coefficients, consisting of n." He further remarked

"L'on peut conclure de la formule précédente, que l'adhérence n'influe point sur la valeur de x, ou que les dimensions du triangle qui produit la plus grande pression, dépendent absolument du frottement.

Si le frottement est nul, quelle que soit l'adhérence, le triangle de la plus grande pression sera isoscèle, ou celui dont l'angle sera de 45 degrés."

"From this formula it is apparent that the cohesion has no influence on the x-value , and that the dimensions of the triangle of the greatest pressure are only dependent on the friction.

If the friction is zero, then the triangle of the greatest pressure (no matter what the cohesion is) is isosceles, or the slope of angle is 45°."

After this, Coulomb considered the load P in Fig. 2.16, as well as the wall friction, in his formulas.

After a few additional comments about factors which could influence the determination of the size of the retaining wall, Coulomb treated fracture shapes with curved gliding surfaces.

He also determined the passive earth pressure (the horizontal force A') which was necessary in order to raise a soil wedge. He did realize that, in his calculation of the active earth pressure, he had only found an approximate value for the actual occuring earth pressure, exactly like the modern limit design procedure, something which differentiates him from many of his successors. He wrote:

"Il est donc démontré que lorsque la cohésion & le frottement contribuent à l'état de repos du triangle, que les limites de la force que l'on peut appliquer en F, perpendiculairement à CB, sans mettre le triagle en mouvement, seront comprises entres A & A'."

"Thus it is proven that in the case where cohesion and friction contribute to a state of repose, the limits of the force in F vertically operating on CB, which does not set the triangle into motion, fall between A and A'."

Although this collapse theorem is not valid according to today's knowledge, it can be seen that Coulomb was well aware of the approximate value of the earth pressure A.

Coulomb's formulations, although they are only used for special boundary-value problems, were pioneering in their description of the fracture behavior of brittle and granular media. This will become more apparent in the following sections.

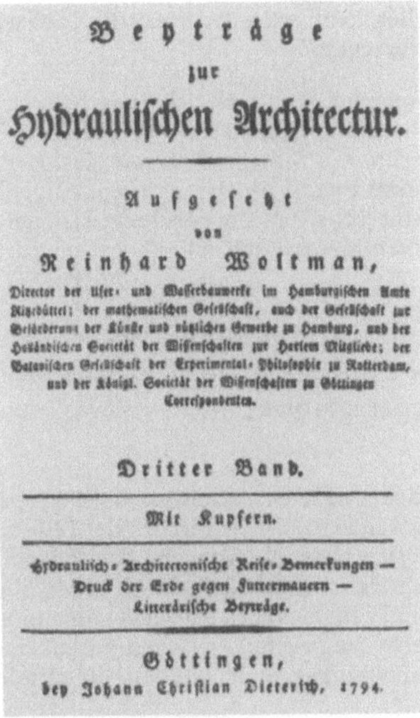

Beyträge
zur
Hydraulischen Architectur.

Aufgesetzt
von
Reinhard Woltman,
Director der Ufer- und Wasserbauwerke im Hamburgischen Amte Ritzebüttel; der mathematischen Gesellschaft, auch der Gesellschaft zur Beförderung der Künste und nützlichen Gewerbe zu Hamburg, und der Holländischen Societät der Wissenschaften zu Harlem Mitgliede; der Batavischen Gesellschaft der Experimental-Philosophie zu Rotterdam, und der Königl. Societät der Wissenschaften zu Göttingen Correspondenten.

Dritter Band.

Mit Kupfern.

Hydraulisch-Architectonische Reise-Bemerkungen — Druck der Erde gegen Futtermauern — Litterärische Beyträge.

Göttingen,
bey Johann Christian Dieterich, 1794.

Fig. 2.17. Front page of Woltman's work

2.6 Woltman's Contribution to the Porous Media Theory: The Introduction of the Angle of Internal Friction and the Volume Fraction Concept

The outstanding contributions to porous media theory made by one excellent engineer and scientist do not seem to have been noticed in the past. Reinhard Woltman was a harbor construction director (1757–1837) from Hamburg. In the third volume of his four-volume work *Beyträge zur Hydraulischen Architectur* from the years 1791 to 1799, he expanded his ideas on soil mechanics and porous bodies (Woltman, 1794/99).

Woltman first separated soil into four types: sand, lime, clay, and compost-earth. He then pointed out that the friction of all earth branches is common and that this differentiates soil from fluids. Only by total saturation with water can the friction effect be lost. Referring to this state, Woltman called the mixtures quicksand, drifting sand, mud, bog, or morass. Furthermore, he accepted the incompressibility of soils, whereby he pointed out that this is most true for sand. Woltman regarded cohesion as a result of "mixed-in moisture":

"Trocken und fein, stäubt der Thon; feucht hängt er zusammen. Umgekehrt hat der Thon noch die besondere hygrometrische Eigenschaft, daß er durch hinzukommende Feuchtigkeit an Volumen zu und durch Austrocknung abnimmt, daß er nach vorhergehender gänzlicher Durchnässung bey der nachfolgenden **Austrocknung** zur ziemlich festen Masse wird. Doch ist zu merken, daß wenn die Austrocknung schnell erfolgt, alsdann die zusammenziehende Kraft größer werden kann als die Cohäsion, folglich diese durch allerley Risse und Borsten pflegt unterbrochen zu werden. Zuweilen kann auch der **Frost** die Cohäsion der Erdmassen vermehren, zuweilen sie ganz unterbrechen."

"Dry and fine, the clay is dusty; if it is moist it stays together. Contrarily, clay has the special hygrometric property that its volume increases with additional moisture and decreases by drying, and that, after becoming totally wet, it becomes a rather solid mass during the subsequent **drying**. However, it must be born in mind that, if the drying occurs quickly, then the contracting force can become larger than the cohesion. Consequently, clay tends to be broken up by many cracks. Sometimes **frost** can increase the cohesion of the earth masses, and can sometimes completely stop it."

Woltman then arrived at the problem of the earth pressure. The incentive to solve the problem was a competition at the Imperial Academy of Sciences in St. Petersburg, as Woltman wrote in the first paragraph of his work (Woltman, 1794):

"Die Preisfrage kenne ich nur aus den Göttingischen gelehrten Anzeigen vom 30. April 1791, wo es heißt: 'auf 1792 verlangt die Kaiserliche Academie für den Druck der Erde auf Futtermauern eine vollkommenere Theorie als die bisher bekannte. Vornehmlich sollen die Gründe aus der Physik mehr dargestellt werden und was auf die Zähigkeit der Erde, ihre unterschiedene Feuchtigkeit auch Zusammenhang und Festigkeit des Bauzeuges ankommt. Dazu, so viel sich thun läßt, Versuche und practische Bemerkungen, schon vorhandene oder noch anzustellende, auf welche Hypothesen müssen gegründet werden, die mit der Natur besser übereinstimmen als die bisherigen'."

"I know the prize question only from the Göttingischen gelehrten Anzeigen of April 30, 1791, where it is stated that: 'in 1792 the Imperial Academy is demanding a more perfect theory than that known until now for the earth pressure on retaining walls. Especially the foundations of physics should be described in more detail; also matters relating to the viscosity of earth and its various moistures, and the cohesion and strength of the construction materials. In this respect, whenever possible, tests should be carried out, practical remarks (already existing or those yet to be made) should be formulated from which hypotheses that are in agreement with nature must be made'."

Woltman developed an earth pressure theory before the deadline but did not send it to St. Petersburg because he found the postage too expensive. In the preface to the third volume of his book he explained:

"Der Termin zur Concurrenz des Preises war vlt. Dec. 1792, und mein Manuscript war medio Nov. in Hamburg, konnte aber nicht anders als mit der reitenden Post gegen die bestimmte Zeit nach Petersburg kommen; und weil ich auf diesem Wege das Loth mit 38 Schil., folglich das Manuscript welches 22 Loth wog, mit 52 Mark 4 Schil. bezahlen sollte: so bewog mich dieser Umstand es zurück zu nehmen und hier abdrucken zu lassen; als wogegen wie ich hoffe, um so weniger etwas zu erinnern seyn wird, da diese Materie ihrer Natur nach vielleicht mit mehrerm Rechte zur Hydraulischen Architectur, als zur Fortification mag gerechnet werden."

"The date for the competition was December 1792 and my manuscript was in Hamburg by the middle of November, but could in no other way arrive on time in Petersburg than with the riding post. I was supposed to pay 38 schillinge per loth, and the manuscript weighed 22 loth which entailed a sum of 54 marks 4 schillings. Therefore this fact led me to take back the manuscript and let it be printed here, to which, I think, there will be all the less objections, since this matter due to its nature may be related with more right to hydraulic architecture than to fortification."

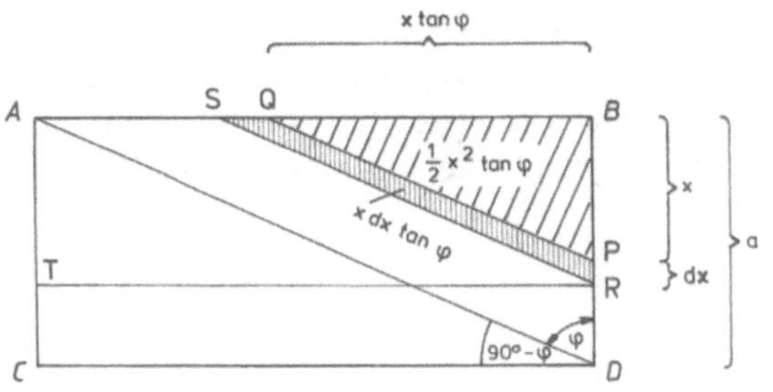

Fig. 2.18. Kästner's calculation of the earth pressure

After these introductory remarks, Woltman listed the conditions of his earth pressure theory.

"Wir haben also den Druck der Erde in einem vierfach verschiedenen Zustande der drückenden Masse zu betrachten:

1. wenn sie **trocken** und **locker** ist, wie Flugsand, Stauberde. - Hierher können auch die Getreidearten oder Samenkörner gerechnet werden;
2. wenn sie **feucht** und **locker** ist, wie Grubenerde, Bauerde;
3. wenn sie **naß** (mit Wasser ganz gesättigt) ist, wie Triebsand, Schlamm, Morast;
4. wenn sie **fest** ist, wie z.B. fester Thon."

"The earth pressure is to be considered by observing four different states of the pressing mass:

1. when it is **dry** and **loose**, e.g., quicksand, dust. - Cereals or seed grains can also be counted in this group;
2. when it is **damp** and **loose**, e.g., pit soil and mining soil;
3. when it is **wet** (totally saturated with water), e.g., drifting sand, mud, bog;
4. when it is **firm**, such as firm clay, for example."

Subsequently, Woltman discussed the calculation of earth pressure according to B.F. de Bélidor (1697–1761), who became famous through his four-volume work *Architectura Hydraulica* (de Bélidor, 1737–1753). He wrote:

"Bisher ist man in der Berechnung dieses Druckes wohl meistens **Bélidor** gefolgt, der nach Erfahrung und Muthmaßung annimmt, die Erde strebe unter einem Winkel von im Durchschnitt 45 Gr. abzusinken; und die Kraft, welche sie hierbey **seitwärts** anwendet, betrage etwa die Hälfte des Gewichtes von dem absinkenden Triangel oder Prisma."

"Up to now, when calculating the earth pressure, one has mostly followed Bélidor. Bélidor assumed by experience and conjecture that earth tends to sink at an angle of 45°, on average. The force which the earth thereby laterally uses, amounts to about half of the weight of the sinking triangle or prism."

Before he explained his own principles on the earth pressure theory, Woltman summarized the calculation of Privy Councillor Kästner, whom he highly admired. Kästner had made these calculations a couple of years earlier and had communicated them to Woltman. He proceeded from Fig. 2.18.

The task is to calculate the pressure of the soil wedge *ADB*, which slides down the gliding surface *AD* on the vertical line *DB*. Therefore, at the point

P, a differential trapezium $PQSR$ with the vertical height $PR = dx$ is examined. Kästner considered this trapezium as a weight whose size is expressed by the size of the trapezium. Kästner then carried out a force analysis on the differential trapezium. He thereupon determined the weight dG to the trapezium:

$$dG = x\,dx\,\tan\varphi\,. \tag{2.6.1}$$

The tangential components $dG_{||}$ are:

$$dG_{||} = x\,dx\,\tan\varphi\,\sin SRT\,, \tag{2.6.2}$$

or

$$dG_{||} = x\,dx\,\sin\varphi\,, \tag{2.6.3}$$

which is also known by the term respective gravity. From this respective gravity, Kästner established the components dG_H which act horizontally on PR:

$$dG_H = dG_{||}\,\sin\varphi\,, \tag{2.6.4}$$

$$dG_H = x\,dx\,\sin\varphi\,\sin\varphi\,. \tag{2.6.5}$$

We will return to this calculation later, since the relation (2.6.5) represents the essential point of Kästner's consideration.

Subsequently, the moment relative to the point of the reference D is to be considered and established. In particular, Kästner compared his considerations with those of Count Kinsky (*Beyträge zur Ingen. Wissensch. 1 Stück § 3*) and also considered the special case of fluids. Finally, he expressed his wish for the realization of experiments:

"Wären also nicht über die Sache Versuche zu wünschen? Dazu gehörte nun ein Kasten von dem eine Seitenwand beweglich ist; ihn mit Erde gefüllt, von der man weiß, wie viel sie abschießt; untersucht wie viel Kraft nötig ist, die Seitenwand zu halten, und das mit Höhe, Raum, Gewicht der Erde vergleichen."

"Would not experiments in this matter be desirable? To this end a box could be used, where one side wall is movable. It could be filled with soil, where one knows how much could be sealed in. Then one would examine how much force is necessary to support the side wall and compare it to the height, volume and weight of the earth."

Woltman praised Kästner's plans, although he mildly criticized his calculation of earth pressure (Eqn. 2.6.5) because he was not fully convinced of this method. He developed his own concept (we will come back to this later) against Kästner's ideas. Before coming finally to his own derivation of earth pressure, Woltman discussed a short excerpt on the matter in Kästner's Italian book. More precisely, he discussed the test results published by Delanges in 1779.

He began his thoughts on the earth pressure theory in the following manner:

"**Lehrsatz:** Die Friktion der Erdteilchen kann ihren lothrechten Druck nicht vermindern."

"**Theorem:** The friction of soil particles cannot decrease its vertical pressure."

Subsequently, he proved this theorem. With regard to the friction law, he referred to one of Euler's works (Euler, 1765). He then formulated the following "natural law":

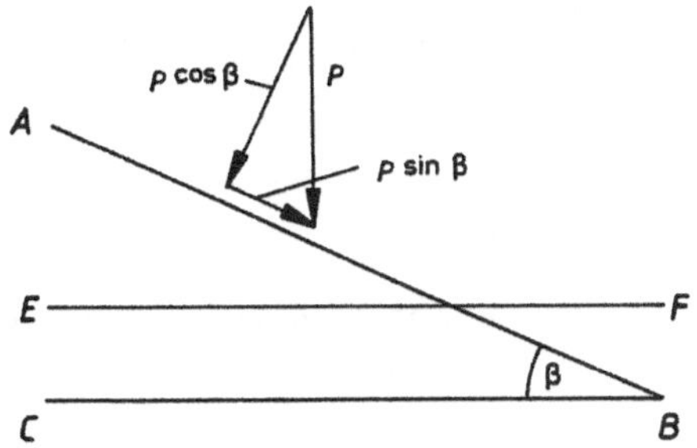

Fig. 2.19. Woltmann's analysis of forces

"Naturgesetz: Die Friktion fester Körper an einander ist ihrem Druck proportional, hängt nicht von ihrer Gestalt, auch nicht von der Größe ihrer Oberfläche ab, wenn nicht etwa der Druck selbst davon abhängt."

"Natural law: The friction of solid bodies on each other is proportional to their pressure, it does not depend on their shape or on the size of their surface, if the pressure itself does not depend on it."

He then continued to formulate further theorems.

"Lehrsatz: Die Friktion der Erdteilchen mindert ihren schiefen Druck."

"Theorem: The friction of the soil particles reduces its oblique pressure."

He proved this theorem through the use of force analysis (see Fig. 2.19).

A soil particle with the weight p lies on the sloping plane AB. The downwards-directed accelerating tangential force is $p \sin\beta$, as long as no friction is active. If the friction is activated, then, according to the above "law of nature", the frictional force $\frac{1}{n} p \cos\beta$ counteracts the accelerating force. Consequently, the accelerating force F remains,

$$F = p \sin\beta - \frac{1}{n} p \cos\beta \,. \tag{2.6.6}$$

However, if the pressure in direction AB decreases, any other oblique pressure towards EF, or another direction, will result from the force towards AB.

In the following clause, Woltman introduced an extremely important concept for the calculation of failure conditions in soil mechanics, i.e., the angle of internal friction[7]:

"Zusatz: Wenn das Theilchen bloß vermöge der Friktion auf der geneigten Ebene ruht, und man diese allmählich erhöhet; β größer nimmt, so lange bis das Theilchen beginnt herab zu fallen: so ist für das letzte Gleichgewicht

$$p \sin\beta = \frac{1}{n} p \cos\beta \,, \tag{2.6.7}$$

[7] According to M. Rühlmann (1885), it was Parent who substituted the friction-coefficient by the tangent of the angle of inclination of a wedge for the first time.

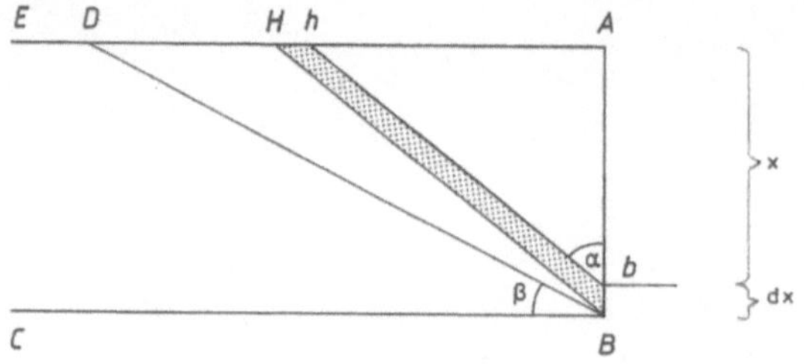

Fig. 2.20. Earth pressure against a vertical wall

also

$$\frac{1}{n} = \tan\beta\,, \tag{2.6.8}$$

und so werde ich im folgenden die Friktion ausdrücken. Nähmlich wenn ein Körper dessen Gewicht $= p$ und Friktion $= f$ auf der geneigten Ebene ruht, so verhält sich $p : f = 1 : \tan\beta$."

"**Clause:** If the particle rests on the inclined plane only by virtue of the friction and this is gradually increased, i.e., β made bigger, for so long until the particle begins to fall downwards, then it is equilibrium for the last one:

$$p \sin\beta = \frac{1}{n}\, p \cos\beta\,, \tag{2.6.7}$$

therefore

$$\frac{1}{n} = \tan\beta\,, \tag{2.6.8}$$

which is the way I will subsequently express the friction. If a body, whose weight $= p$ and friction $= f$, rests on the inclined plane, then $p : f = 1 : \tan\beta$."

With the angle of internal friction, Woltman gave his formulas a more elegant touch in comparison to those of Coulomb later. After Woltman had proven that a banked-up pile of earth had to be conical, he presented a method for experimentally determining the internal friction. In one particular table, he explicitly stated the angle of internal friction for various types of soils. In his treatise, Woltman finally formulated the problem of the calculation of earth pressure:

"**Problem:** An einer lotrechten Wand *AB* (...) liege Erde, deren Oberfläche die Horizontallinie *AE* ist, die Friction der Erde ist bekannt, man soll ihren Seitendruck senkrecht auf die Wand bestimmen."

"**Problem:** On a vertical wall *AB* (...) lies soil, whose surface is the horizontal line *AE*, and the friction of the soil is fixed. The side pressure on the wall is to be calculated."

In calculating the earth pressure, Woltman proceeded from a layer model. He then established the weight of the trapezium *HhbB* through the specific weight p, as (the symbols can be seen in Fig. 2.20 and Fig. 2.21)

$$P_\Delta = p\, x\, dx\, \tan\alpha\,. \tag{2.6.9}$$

The tangential and normal components P_T and P_N yield (see Fig. 2.20)

$$\begin{aligned} P_T &= p\,x\,dx \tan\alpha \cos\alpha \\ &= p\,x\,dx \sin\alpha\,, \end{aligned} \tag{2.6.10}$$

$$P_N = p\,x\,dx \tan\alpha \sin\alpha\,. \tag{2.6.11}$$

According to the "natural law" given above, Woltman obtained for the frictional force $R_{(P_N)}$ with the reaction of P_N

$$R_{(P_N)} = p\,x\,dx \tan\alpha \sin\alpha \tan\beta\,. \tag{2.6.12}$$

He established equilibrium by introducing a horizontal force v. He decomposed it into tangential and normal components (see Fig. 2.21). With the action of the pressure force $v \cos\alpha$, Woltman calculated the friction force,

$$R_{(V)} = v \cos\alpha \tan\beta\,. \tag{2.6.13}$$

Thereafter he established v from the equilibrium condition that the sum of all tangential forces must be equal to zero:

$$p\,x\,dx \sin\alpha - p\,x\,dx \tan\alpha \sin\alpha \tan\beta - v \cos\alpha \tan\beta = v \sin\alpha\,, \tag{2.6.14}$$

or

$$p\,x\,dx - p\,x\,dx \tan\alpha \tan\beta = v\,(1 + \cos\alpha \tan\beta)\,, \tag{2.6.15}$$

thus

$$p\,x\,dx \left(\frac{1 - \tan\alpha \tan\beta}{1 + \cot\alpha \tan\beta}\right) = v\,. \tag{2.6.16}$$

After integration, Woltman obtained the earth pressure E_w as:

$$E_w = \frac{1}{2}\,p\,x^2 \left(\frac{1 - \tan\alpha \tan\beta}{1 + \cot\alpha \tan\beta}\right), \tag{2.6.17}$$

which agrees with Coulomb's calculation.

Woltman then calculated the earth pressure according to Privy Councillor Kästner. In consideration of the relation (2.6.5), he found

$$(p\,x\,dx \sin\alpha - p\,x\,dx \tan\alpha \sin\alpha \tan\beta) \sin\alpha = v\,, \tag{2.6.18}$$

or

$$p\,x\,dx \sin^2\alpha\,(1 - \tan\alpha \tan\beta) = v\,. \tag{2.6.19}$$

After integration, he obtained the entire pressure E_k on a wall:

$$E_k = \frac{1}{2}\,p\,x^2 \sin^2\alpha\,(1 - \tan\alpha \tan\beta)\,. \tag{2.6.20}$$

The essential difference between Woltman's and Kästner's considerations is the fact that Kästner did not proceed from the equilibrium of the system's soil particles and retaining wall, and that he did not correctly apply the cut principle. Woltman carefully criticized Kästner's method, although he greatly admired him. In particular, he showed that Kästner's earth pressure formula for frictionless material, see (2.6.20), produced an incorrect result.

Woltman subsequently calculated the maximal earth pressure on the basis of a maximum/minimum requirement. Without giving a reason, he left out

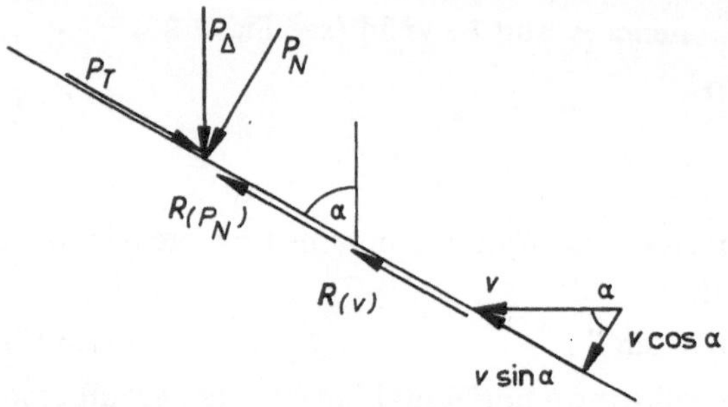

Fig. 2.21. On the derivation of the earth pressure by Woltman

the specific weight in the first part of his calculation. In my view, this must obviously have been a printing error. In his earth pressure calculations, Woltman proceeded from the following consideration:

"Wenn also die Höhe einer lothrechten Wand $= a$, ihre Breite $= b$, so ist der Druck den sie von der Erde leidet gleich dem Gewicht einer Masse

$$\frac{1}{2} a\, a\, b \left(\frac{1 - \tan\alpha \, \tan\beta}{1 + \cot\alpha \, \tan\beta} \right) \tag{2.6.21}$$

von eben der Erde. In diesem Ausdruck ist α unbekannt. Wäre es möglich durch Versuche den Druck wirklich für einen gewissen Fall zu bestimmen, so könnte daraus α gefunden werden; aber die Folge wird ergeben, daß das unmöglich sey. Wir müssen also durch Vernunftschlüsse α zu bestimmen suchen."

"If the height of a vertical wall $= a$, its width $= b$, then the pressure it conducts from the soil is equal to the weight of a mass $\frac{1}{2}\, a\, a\, b\, (\frac{1 - \tan\alpha \, \tan\beta}{1 + \cot\alpha \, \tan\beta})$ from the plane of the soil. In this expression α is unknown. If it were possible through experiments to actually determine the pressure for a certain case, then α could be found; but the result will show that this is impossible. Thus we must try to determine α through syllogisms."

He then established:

"Demnach muß in unserer Formel $\alpha = HBA$ nothwendig so genommen werden, daß der Druck in parallelen festen Schichten ein Maximum wird, weil er nur in diesem Zustande dem Druck der lockern Erde gleich werden kann."

"Consequently, in our formula, $\alpha = HBA$ must be taken in such a way that the pressure in parallel solid layers is a maximum, as only in this state can it equal the pressure of loose soil."

"Also (Therefore):

$$\frac{1}{2} a\, a\, b \left(\frac{1 - \tan\alpha \, \tan\beta}{1 + \cot\alpha \, \tan\beta} \right) = \text{max.} \tag{2.6.22}$$

$$d.\frac{1}{2} a\, a\, b \left(\frac{1 - \tan\alpha \, \tan\beta}{1 + \cot\alpha \, \tan\beta} \right) =$$

$$- \frac{(1 + \cot\alpha \, \tan\beta)\frac{1}{2} a\, a \, \tan\beta \, d \, \tan\alpha}{(1 + \cot\alpha \, \tan\beta)^2}$$

$$- \frac{\frac{1}{2} a\, a \, \tan\beta \,(1 - \tan\alpha \, \tan\beta) d \, \cot\alpha}{(1 + \cot\alpha \, \tan\beta)^2} = 0 \tag{2.6.23}$$

$$d \, \tan\alpha = d\,\alpha \sec\alpha^2 , \quad d \, \cot\alpha = - \, d\,\alpha \, \operatorname{cosec}\alpha^2 \; {}^{8} \tag{2.6.24}$$

Diese Werthe substituirt, erhält man nach vollendeter Reduction alles dessen was zu beiden Seiten sich aufhebt. (Substitution of these values after reduction on both sides yields)

$$\frac{1+\cot\alpha\,\tan\beta}{\cos\alpha^2}=\frac{1-\tan\alpha\,\tan\beta}{\sin\alpha^2}\,; \tag{2.6.25}$$

also (therefore)

$$\tan\alpha^2+2\tan\alpha\,\tan\beta=1\,; \tag{2.6.26}$$

$$\tan\alpha^2+2\tan\alpha\,\tan\beta+\tan\beta^2=1+\tan\beta^2\,; \tag{2.6.27}$$

folglich (therefore)

$$\tan\alpha+\tan\beta=\sqrt{(1+\tan\beta^2)}=\sec\beta\,. \tag{2.6.28}$$

Also (therefore)

$$\tan\alpha=\sec\beta-\tan\beta\,; \tag{2.6.29}$$

Und so wird (and therefore)

$$\cot\alpha=\sec\beta+\tan\beta \tag{2.6.30}$$

seyn; denn es ist (since it is)

$$\cot\alpha=\frac{1}{\tan\alpha}=\frac{\sec\beta^2-\tan\beta^2}{\sec\beta-\tan\beta}=\sec\beta+\tan\beta\,. \tag{2.6.31}$$

Setzen wir also in unserer Formel für $\tan\alpha$ nun $\sec\beta-\tan\beta$; und $\sec\beta+\tan\beta$ statt $\cot\alpha$, so wird sie den Druck am größten geben und so aussehen: (Thus if we replace $\tan\alpha$ by $\sec\beta-\tan\beta$, and $\cot\alpha$ by $\sec\beta+\tan\beta$ in our formula, then the greatest pressure results)

$$\frac{1}{2}pa^2b\left(\frac{1-\tan\beta(\sec\beta-\tan\beta)}{1+\tan\beta(\sec\beta+\tan\beta)}\right)^{9} \tag{2.6.32}$$

$$=\frac{1}{2}paab\left(\frac{\sec\beta^2-\tan\beta^2-\sec\beta\tan\beta+\tan\beta^2}{\sec\beta^2-\tan\beta^2+\tan\beta\sec\beta+\tan\beta^2}\right)$$

$$=\frac{1}{2}paab\left(\frac{\sec\beta\,(\sec\beta-\tan\beta)}{\sec\beta\,(\sec\beta+\tan\beta)}\right) \tag{2.6.33}$$

$$=\frac{1}{2}paab\left(\frac{\dfrac{1}{\cos\beta}-\dfrac{\sin\beta}{\cos\beta}}{\dfrac{1}{\cos\beta}+\dfrac{\sin\beta}{\cos\beta}}\right)=\frac{1}{2}paab\left(\frac{1-\sin\beta}{1+\sin\beta}\right)\text{."} \tag{2.6.34}$$

After having derived this fundamental formula for the maximum earth pressure, Woltman checked his results by means of exceptions. He finally arrived at the observation:

"Ich halte demnach dafür, daß die gefundene Formel abseiten der Theorie keinem Zweifel unterworfen sey."

"I consequently believe that the formula cannot be doubted as far as the theory goes."

Woltman then communicated his "experience with earth pressure." He described in detail the testing procedure (set up by himself) and then imparted

---

[8] $\sec\alpha^2=(\sec\alpha)^2=\left(\frac{1}{\cos\alpha}\right)^2$ $\quad$ $\operatorname{cosec}\alpha^2=(\operatorname{cosec}\alpha)^2=\left(\frac{1}{\sin\alpha}\right)^2$

[9] At this point in the calculation, the specific weight p of the soil appears once more.

Wir wollen annehmen, daß die leeren Zwischenräume der lockern Erde, welche das Wasser anfüllt, sich zu dem ganzen Volumen derselben Erde wie r : 1 verhalte, wo denn r ein eigentlicher Bruch seyn wird. Das Gewicht der lockern Erde P; des Wassers p, das Volumen der Erde v; des beygemischten Wasser u, so ist das Gewicht der Mischung vP + up; und dessen Volumen v + u − rv; also die specifische Schwere des Schlamms $= \frac{vP + up}{v + u - rv}$. Wenn der Schlamm nicht mehr Wasser enthält als die Zwischenräume der lockern Erde fassen können, so ist u = rv, und die Schwere des Schlamms $= \frac{vP + up}{v} = \frac{v(P + rp)}{v}$ = P + rp und am größten. Wird aber mehr Wasser zugemischt als die Zwischenräume enthalten können, u > rv, so wird der Schlamm leichter wenn P > p; bleibt unverändert wenn P = p; wird schwerer wenn P < p. Welches letztere nur bey Moor oder Torferde statt haben kann; ich werde bald eine Tafel beyfügen, die hierüber noch mehr Licht geben wird.

Fig. 2.22. Woltman's volume fraction concept

his test results for various granular substances. The care with which Woltman carried out his experiments is impressive. In particular, he listed all of the influences which might have falsified the results. He thereupon compared the test results with those of his theory. He subsequently showed that the earth pressure resultants, which, according to his theory, lay at one third of the height of the retaining wall, also approached this height in the test results. With respect to the size of the earth pressure, Woltman ascertained that the test values were much smaller than those calculated in his theory. He pointed out that his test results had been falsified by the friction on the walls of his box and therefore had to be corrected. He arrived at the following assessment:

"Wie dem auch sey, so traue ich noch zur Zeit der Theorie mehr, als dergleichen Versuchen."

"Whatever the case may be, at this time I believe more in the theory than in the tests."

After strongly criticizing the results of the Delanges, Woltman turned his attention to partially and totally water-saturated substances. He pointed out, in particular, that the moisture content augmented the cohesion but did not, however, attribute any greater importance to the cohesion. He attentively observed totally water-saturated loose soil, or mud, and noted that his theory was the most appropriate, since mud behaves like a liquid mass. In this connection, Woltman spoke of a mixture and, surprisingly, introduced the concept of the volume fraction. He was probably the first scientist to use this concept.

"Mud results from all loose kinds of earth if they are completely saturated with water or are in surplus with that mixed up. Its weight depends on the more or less admixture of the water.

We will assume that the empty spaces between the loose earth, which are filled with water, are to the bulk volume of that earth as r : 1, where r will be a proper ratio. The weight of the loose earth P; the water p, the volume of the earth v; of the added water u, then the weight of the mixture is $vP+up$; and its volume $v+u-rv$; therefore the specific gravity of the mud $= \frac{vP+up}{v+u-rv}$. If the mud does not contain more water than the spaces between the loose earth can hold, then $u = rv$, and the gravity of the mud $= \frac{vP+up}{v} = \frac{v(P+rp)}{v} = P+rp$ is largest. If, however, more water is added than the spaces between can contain $u > rv$, the mud becomes lighter if $P > p$; remains unchanged if $P = p$, becomes heavier if $P < p$. The latter can only occur for moor or peaty mold; I will soon add a table which can give more light hereupon"

Obviously, Woltman did not continue his interesting studies on porous bodies. There is no hint in any of his publications that he followed up on his original work on water-saturated porous solids with further studies.

After his remarks on partially and totally water-saturated loose soil, Woltman stated some design principles for dams and walls. He warned against bringing cohesion into the formulas, because this could be disconnected by becoming soft through frost and rain. In addition, Woltman gave some ideas for the construction of brick retaining walls, and addressed the influence of additional live loads behind the retaining walls and of the wall friction (soil-retaining wall) on the calculation. Woltman concluded his excellent treatise by pointing out that, for the achievement of a perfect theory of retaining walls, further investigations would be necessary. He further announced a second part to his treatise. This second part was published in the fourth volume of his work *Beyträge zur hydraulischen Architektur*. In his preface, he referred to a "certain reviewer" who had strongly criticized his earth pressure theory. Woltman defended himself, at times in a very polemical way, against the suppositions of the reviewer. Whether or not the criticism of this reviewer or that of the physics and mathematics professor from Göttingen, Privy Councillor Kästner, whom he greatly admired, made him unsure of himself, can no longer be established. In any case, on the basis of the aforementioned considerations by Kästner, Woltman developed a second earth pressure theory. In this second theory, the test results were in closer agreement than they were with the theory published previously in the third volume of his work. He still preferred his own theory, for theoretical reasons. However, he was no longer as convinced as he had been in the third volume, and he left it to the reader to choose freely between the two earth pressure theories. Woltman's attitude in this matter is inexplicable, especially as his friend, the mathematician C.L. Brünings from Alphen near Leiden in Holland, ecstatically praised the original earth pressure theory in a letter. This letter, written by Brünings in German, reads as follows:

"Die absolute Größe des Drucks der Erde sey αc^m also c^m das Gesetz des Drucks, nach welchem er von der Höhe $= c$ abhängt. So hängt von der absoluten Größe allerdings die Abmessung der Futtermauer ab; ist jene unrichtig bestimmt, so wird diese fehlerhaft seyn. Aber der Fehler, der aus der unrichtigen Bestimmung von α entspringt, ist beständig; dagegen variiert der Fehler, der aus der unrichtigen Bestimmung von c^m entspringt, unablässig mit der Mauer. Daher war mir das Gesetz des Drucks der Grund zur Classification der verschiedenen Theorien. Ich habe auf eine allgemeine Art erwiesen, daß der Druck der Erde dem Quadrat ihrer Höhe proportional sey: wenn man voraussetzt, daß der Durchschnitt der drückende Erde ein Dreieck ist, und daß alle Erdtheile nach geraden und gleichlaufenden Richtungen abschießen würden. – Dem zur Folge habe ich die verschiedenen Theorien von **Bélidor, Ypey, von Clasen, Kinsky, Couplet** und **Goudriaan** vorgetragen, welche alle den Druck der Erde dem Quadrat ihrer Höhe proportinal setzen, und außerdem (für einen Böschungswinkel der Erde $\beta = 45$ Gr.) einen gleichen Werth für das **Moment des Drucks** geben; so verschieden auch die Gründe sind, auf welchen diese Theorien beruhen. – Alle diese Schriftsteller haben den Widerstand der Mauer gerade so, wie **Belidor**, berechnet, daher wird überhaupt jede Abmessung der Futtermauer der Höhe proportional und für $\beta = 45$ Gr. erhält in der That jede Abmessung den nämlichen absoluten Werth nach allen der erwähnten Theorien. – Auch **Lorgna** findet zwar den Druck dem Quadrat der Höhe, das Moment des Drucks dem Cubus der Höhe proportional; aber dieß Moment ist ihm $\frac{c^3}{6}$ (wie bey flüssigen Massen), anstatt daß es nach den bereits erwähnten Theorien $\frac{c^3}{12}$ war; und dieß ist ein wesentlicher Unterschied. Überdieß **Lorgna** bringt in die Berechnung des Widerstandes der Futtermauern ihren Zusammenhang mit ihrer Grundfläche; daher, daß die Abmessungen der Mauer nach ihm der Quadratwurzel aus der Höhe proportional seyn müßten. – **Stahljwerds** Theorie verdient den Namen einer Theorie nicht; sie ist übrigens von den bisher erwähnten Theorien sehr unterschieden, weil **Stahljwerd** eine progressive und keine rotatoire Bewegung der Mauer voraussetzt. – Die besagten Theorien beruhen auf der unrichtigen Voraussetzung, daß alle die drückenden Erdteile in den ihrer natürlichen Böschung gleichlaufenden Richtungen abschließen würden, und befassen die Reibung u.s.w. entweder gar nicht, oder nur *in vago*. – Diesen Mängeln hat **Coulomb** (Mémoires presentés a. 1773) in einer ingeniösen Abhandlung über diesen Gegenstand abgeholfen: erst seit kurzem bin ich mit seiner Theorie bekannt geworden. Er ist ungefähr den nämlichen Weg mit Ihnen gegangen, hat außerdem den Zusammenhang der abschießenden Erde in seine Rechnung aufgenommen. Wenn ich in seinen Formeln, denen ich eine geschmeidigere Gestalt gegeben habe, den Zusammenhang $= 0$ setze, so erhalte ich mit Ihnen (S. 175 Ihrer Abhandlung) $\frac{1}{2}\,p\,a\,a\,b\,[\tan(45° - \frac{1}{2}\,\beta°)]^2$. Freylich enthält Ihre reichhaltige und schöne Abhandlung für die Anwendung noch so vieles, daß jeder unbefangene Leser ihr den Vorzug vor der des **Coulomb** geben muß. – Ich habe es gewagt, Ihre oder des **Coulomb** Theorie auf ein ideologisches Principium zu gründen: wozu mich die Abhandlung **Eulers des Einzigen** (methodus inveniendi lineas curvas etc.) veranlaßt hat: 'Die Erfahrung lehrt, daß eine gewisse Menge Erde abschießen würde, wenn die Futtermauer sie nicht zurück hielte; Ferner würden zuverlässig alle die abschießenden Erdtheile (wie verwickelt auch ihre Bewegung seyn möge) mit der größesten Geschwindigkeit abschießen, welche ihre Reibung und ihr Zusammenhang gestattet, weil der Gesetzgeber der Natur immer seine Absichten auf dem kürzesten Wege erreicht: daher wird auch die horizontale Kraft, welche sie zurückhält, ein **Maximum** seyn müssen. Nun ist es sehr wohl möglich, anstatt jener unbekannten Erdmasse ein trianguläres Erd-Prisma zu fingiren, dessen Theile alle in, mit seiner Grundfläche gleichlaufenden Richtungen gleiten würden, mit gleichem Erfolge: das heißt, so daß aus ihrem gesammten Drucke **eine gleich große horizontale Kraft** entspringt. Folglich müssen auch die Abmessungen dieses fingirten Prisma's dergestalt beschaffen seyn, daß die mehr besagte horizontale Kraft ein **Maximum** wird.' Hat man nun auf diese Weise $\tan 2\alpha = - \cot\beta$ gefunden, so wird durch α keineswegs weder die wirklich drückende Erdmasse, noch die Richtung bestimmt nach der sie in der That abschießen wird, sondern nur das Resultat einer Erscheinung, deren besondere Modificationen wir weder a priori noch a posteriori jemals erforschen können. – Dieser teleologische Grundsatz ist mir ein überzeugender Beweis, daß Ihre Theorie richtig ist: da vordem, ich will es Ihnen gerne gestehen Ihr Verfahren nicht so einleuchtend für mich gewesen ist, zumalen nachdem ich unter den unzähligen grundlosen Einwürfen Ihres Recensenten in der *Allgem. Lit. Zeit.* einige erhebliche gegen Ihrer Vorstellungsart gelesen hatte. – Weil mir die Vervollkommung nützlicher Kenntnisse Pflicht ist, so bitte ich Sie, mich Ihres Beyfalls zu vergewissern oder eines bessern zu belehren. – Uebrigens wünsche ich recht sehr, daß ich die Fortsetzung Ihrer Abhandlung bald erhalten möge: und ich melde Ihnen mit Vergnügen, daß erst nach Verlauf eines Jahres das **belidorsche** Werk gedruckt wird. – Der statische Satz, daß der wagerechte Druck eines auf einer schiefen Ebene liegenden Körpers dem Producte aus seinem Gewichte in die Tangente des Neigungswinkels gleich ist – ist so einleuchtend, so übereinstimmend mit allen den besonderen statischen Theorien der festen Körper von **Archimedes** bis zum **de la Grange**, daß ich vermuthe, der verehrungswürdige **Kästner** müsse auf die Abweichung etwa durch einige Nebengriffe über die Beschaffenheit der Erdmasse geleitet seyn, welche gleichsam ein Mittelding zwischen festen und flüssigen Massen ist. Bey Voraussetzung fester prismatischer Körper sehe ich nicht, daß der mindeste Zweifel hierüber statt haben könne.

Alphen bey Leiden,
den 19. Decemb. 1798 C.L. Brünings."

"The absolute magnitude of the earth pressure is $\alpha\, c^m$, i.e., c^m is the pressure which is dependent on the height = c. The dimension of the retaining wall is, however, dependent on the absolute magnitude, and if this is incorrectly determined, then it will be defective. However, the error which occurs when α is incorrectly determined is constant, whereas the error which occurs when c^m is incorrectly determined varies incessantly with the wall. Therefore, the law of pressure was my reason for classifying various theories. I have proved in a general way that the earth pressure is proportional to the square of its height, if one assumes that the average of the pressing earth is a triangle and that all soil particles are sealed in straight and parallel directions. As a result of this, I have expanded on the various theories of **Bélidor, Ypey, von Clasen, Kinsky, Couplet** and **Goudriaan**, who all put the earth pressure proportional to its height and moreover (for an angle of repose of the soil $\beta = 45°$) give the same value for the **moment of pressure**, however different the reasons may be upon which these theories are based. They all have the resistance of the wall in the same way as **Belidor**. Consequently, any measurement of the retaining wall will be proportional to the height, and for $\beta = 45°$, each measurement in fact obtains the absolute theories. Also **Lorgna** finds the pressure as the square of the height, the moment of pressure as proportional to the cube of height, but this moment is for him $c^3/6$ (as with fluid masses), in rather than $c^3/12$ as in the already-mentioned theories, and this is a great difference. Moreover, in the calculation of the resistance of retaining walls, **Lorgna** includes their connection with their area; thus the measurement of the wall must, according to him, be proportional to the square root of the height. **Stahljwerd's** theory does not deserve to be called a theory; it is, moreover, very different from the aforementioned theories because **Stahljwerd** assumes a progressive, and not a rotational, movement of the wall. The said theories are based on the incorrect assumption that all the pressing earth particles in them surround natural gradients of parallel directions and do not deal with friction or only *in vago*. **Coulomb** (*Mémoires presentés* a. 1773) has helped remedy these deficiencies in an ingenious treatise – I have only recently become acquainted with his theory. He has gone in approximately the same direction as you have and has additionally considered in his calculation the cohesion of the soil which slides away. If we make cohesion = 0 in his formula, which I have given a more flexible form, then I obtain, as you do (p. 175 of your treatise), $\frac{1}{2}\, p\, a\, a\, b\,[\tan(45° - \frac{1}{2}\,\beta°)]^2$. Certainly, your comprehensive and fine treatise contains so much more that any unbiased reader would prefer it to **Coulomb's**. I have dared to base your or Coulomb's theory on an ideological principle; I was inspired to do this by the **one and only Euler** (methodus inveniendi lineas curvas etc.): 'Practice has proven that a certain amount of earth would slide away if the retaining wall did not hold it back. Moreover, all the remaining soil particles (however complicated their movements might be) would slide away with the greatest possible speed allowed by their friction and cohesion because nature's legislator always attains his goals in the shortest way. Thus the horizontal force which holds them back must also be a **maximum**.' It is quite possible to establish, instead of an unknown earth mass, a triangular soil-prisma, whose parts would all slide in the same direction, so that **an equally large horizontal force** arises. Consequently, the measurements of this established prisma must be formed in such a way that the said horizontal force is a maximum. If one has found $2\alpha = -\cot\beta$ in this way, then neither the real pressing earth mass nor the direction in which it would actually slide will be determined by α. In fact, either *a priori* or *posteriori*, only the result of an occurance will be found whose particular modifications we will not be able to discover. This teleological axiom is for me convincing proof that your theory is correct, because I must admit that before this your method of procedure was not so clear to me, especially after I had read some of the critiques of your method of procedure in the countless unfounded comments made by your reviewer in the *Allgem. Lit. Zeit.* Since the perfection of useful knowledge is a duty for me, I ask you to confirm your approval or to teach me otherwise. Moreover, I would be very glad to receive the continuation of your treatise soon, and it is with pleasure that I inform you that **Belidor's** work will be printed in only a year's time. The static principle that the horizontal pressure of a body lying on an inclined plane is equal to the product of its weight on the tangents of the angle of gradient, is so obvious and in such agreement with all the particular statical theories of solids from **Archimedes** to **de la Grange**, that I surmise that the honorable **Kästner** must have been led to digress by some errors about the creation of the earth mass, which is something in-between solid and fluid masses. With the assumption of solid prismatic bodies, I do not see how there could be the slightest doubt about this.

Alphen bey Leiden,
den 19. Decemb. 1798 C.L. Brünings."

With this letter, Woltman concluded his *Theorie vom Druck der Erde*. In the course of his treatise *Über die beste Construction der Futtermauern*, Woltman (1799) dealt with the strength problems and construction principles of the retaining wall itself. It would appear that Woltman was unaware of Coulomb's treatise. In fact, this treatise seems to have been unknown to everybody in this field for quite some time, see Kötter (1893). The competition task proposed in 1791 by the Imperial Academy of Sciences in St. Petersburg on the question of earth pressure can be explained in no other way. The letter of 1798 from Brünings supports this conjecture. Woltman's achievements are even more admirable when one takes into consideration that he was working in the Hamburg office Ritzebüttel, but still found the time, supported by some preliminary work by Privy Councillor Kästner, to develop a complete earth pressure theory. This great engineer had such extensive mathematical knowledge (e.g., the use of the then relatively new differential calculus) and talent that he was able to work in this difficult area. It is a pity that he obviously let himself be influenced by Kästner's considerations and developed a second, and according to him, equal, theory. Maybe Kästner intervened; in the preface to the third volume of his work, Woltman asked for a verification of his theory:

"Ich gestehe, daß diese Theorie für meine beschränkten Kenntnisse schwer ist; und daß ich gern Urteile einsichtsvoller Männer über dieselbe vernehmen möchte, bevor ich mich der Mühe unterziehe, sie zu vollenden. Insbesondere wünschte ich, daß mein verehrungswürdiger Lehrer, Hr. Hofrath Kästner, durch die schon ehemals ihr gewidmete Aufmerksamkeit, sich bewogen fände, sie auch in der jetzigen Gestalt noch einer Prüfung zu unterwerfen."

"I admit that this theory is difficult for my limited knowledge and that I would like to know the opinion of discerning men on this subject before I take the pains to complete it. Especially, I wish that my honourable teacher, Herr Privy Council Kästner, would find himself disposed to look after my theory in its present form, due to all the attention which has been paid to it already."

Similarly, it was unskillful of Woltman to publish his earth pressure theory in his four-volume work *Beyträge zur hydraulischen Architektur*, because the largest part of the book attracts a different circle of readers than those who would be interested in earth pressure. In this way, it was Coulomb's theory, being very lucid in its mathematical representation, which was accepted. In soil mechanics, however, the angle of friction introduced by Woltman has proven itself invaluable.

2.7 Concluding Remarks

Coulomb's and Woltman's treatises on earth pressure outshined those of later scientists as far as the basics of the earth pressure theory were concerned, until 1856, when Rankine's fundamental work (Rankine 1856, 1857) appeared (we will return to him later). In the works to follow, simplifications of Coulomb's theory were made and various boundary-value problems treated.

In 1802, de Prony (1802) considered the same case as Coulomb had, i.e., the earth pressure exerted upon a vertical wall on horizontal ground. He, however (as Woltman had done before him), introduced the angle of internal friction or its complementary angle, instead of the friction coefficient of sand on sand, and, instead of the cohesion coefficient, he used the depth to which the ground can be vertically dug to. He thereby obtained simpler and higher yielding formulas. De Prony also extended this theory to the case of a non-vertical wall. Here, however, he made an erroneous assumption: that the earth pressure was horizontal, as in the case of a non-inclined wall.

Further contributions to the earth pressure theory came from J.-H. Mayniel (1808). His treatise did not, however, contain anything really new. Similarly, Francais (1820) calculated the earth pressure of an inclined retaining wall. He also used the friction angle of sand on sand.

The history of earth pressure was not, from this point, a closed matter. It has continued into the classical and modern eras, and we will from time to time return to it.

With these remarks, we will conclude our survey of the Early Era: The development of the principles of mechanics, the dynamics of rigid bodies, and the theory of ideal fluids were completed by Euler's ingenious calculations. Similarly, the first systematical studies of soil bodies as one-component continua were carried out by Coulomb and Woltman. Moreover, Woltman introduced the volume fraction concept, an essential part of the theory of porous media. The studies of Coulomb and Woltman were, however, limited to special boundary-value problems and could not be transferred to general three-dimensional continua because a decisive step in the formulation of the stress concept was missing. This step was taken by the famous mathematician A. Cauchy (1823). With Cauchy's discovery, it was possible to complete the formulation of material-independent fundamental equations and to take in hand the development of constitutive relations, as well as stress and strain relations. Thereafter, a new phase in mechanics and porous media theory began.

2.8 Biographical Notes

Johann Bernoulli (1667–1748)

Johann Bernoulli, born in Basel on July 27, 1667, came from one of the most famous learned families of Europe. The Bernoulli family, whose various members were highly talented in mathematics and physics, came from the Netherlands. In the second half of the sixteenth century, a certain Jakob Bernoulli (the name Jakob often reappears later in the family history) was banished from Antwerpen by the Duke of Alba because of his religious beliefs, and moved to Frankfurt am Main. A grandson of this Jakob Bernoulli, also named Jakob, lived in Basel, where he died in 1634 at the mere age of thirty-six. His son Nikolaus (1623–1708) was a town-councillor in Basel. He was the father of Jakob (1654–1705) (the famous mathematician, physicist, and theologian, best known in mechanics through his examination of transverse beams), Nikolaus, Johann, and Hieronymus. It was his father's wish that Johann become a merchant, so he had to spend one year in a merchant house in Neuchatel. After this, it was no longer possible to keep him away from science studies. His older brother Jakob gave him mathematics lessons. On Jakob's advice, Johann also studied medicine. He received his licentiate in this last subject in September, 1690, after the publication of his first work on fermentation. Immediately afterwards, Johann Bernoulli travelled to Geneva, to Lyons and finally to Paris. In all of the places that he stayed, he studied and taught. His most famous students from this time were the Marquis de l'Hospital and Varignon. He later accused l'Hospital of plagiarism.

The Marquis de l'Hospital had published his book *Analyse des infiniment petits* in Paris in 1696. In a confidential letter to Leibniz of February 8, 1698, Johann Bernoulli claimed that l'Hospital simply published the material he (Johann Bernoulli) had drawn up for lessons.

In December, 1692, Johann Bernoulli, already a famous mathematician, returned to Basel and went back to his studies in medicine. He graduated in the spring of 1694.

Due to Leibniz's negotiation, he was to be appointed as a mathematician at the Academy of Wolfenbüttel. This plan was put to an end with Johann Bernoulli's marriage to Dorothea Falkner in Basel at about this time. Later he was recommended by Huygens for the Chair for Mathematics and Physics at Groningen. He took up this position in September, 1695, and occupied the chair until 1705, when, due to his father-in-law's wish and his own poor health, he resigned. Already in October, 1703, his father-in-law had procured for him the Chair for Greek in Basel. In those days, it was not an uncommon practice to appoint a famous scientist to a university by giving him a professorship which was not in his field, until the appropriate chair became free.

After the death of his famous brother, Jakob, on August 16, 1705, the entire university council visited Johann Bernoulli to ask him to take up the vacant chair; the salary increase was considerable. On November 17, 1705, he took up the position, which he was to hold for the forty-two years until his death. In his inaugural lecture, he spoke on analysis and higher geometry.

He turned down offers of employment from Leyden, Padua, Groningen, and Berlin which he received during his time in Basel. His scientific reputation was so great that, in 1699, the Paris Academy named him one of only eight foreign members. Additionally, he became a member of the Berlin Academy in 1701, the London Academy in 1712, the Bologne Academy in 1724, and the St. Petersburg Academy in 1725. He won the renowned Paris Academy prize three times.

Between 1716 and 1741, Johann Bernoulli was vice-chancellor of Basel University and, in 1725, he was assigned the task of directing the improvement of the lower schools, a task to which he eagerly applied himself. One of his students was Leonhard Euler. The Euler and Bernoulli families were closely acquainted. Leonhard Euler's father, Paul Euler, had attended mathematics classes given by Johann's older brother, Jakob. Leonhard Euler was lucky enough to be permitted to attend a private class for special students given by Johann Bernoulli. Johann Bernoulli quickly recognized Euler's talent, and he admired him noticably throughout the course of many years (see Section 2.1). Johann's achievements can be found in many areas: medicine, chemistry, physics, and especially, in pure and applied mathematics. Contemporary writers nicknamed him the "Archimedes of his time". His greatest achievement in mechanics is undoubtedly his *Hydraulica*.

According to certain biographers, Johann Bernoulli was a quarrelsome and egotistical man. For example, in one instance, he was in disagreement over a scientific matter, namely a priority question about the sail curve. In March, 1692, Jakob Bernoulli put the task of locating the sail curve as a competition question for the mathematicians of his time. He had corresponded with Johann about the curves prior to this. In April 1692, Johann, then living in Paris, published a note on the nature of sail curves and catenaries. At first, Jakob made no comment regarding this publication. Johann left Basel on September 1, 1695, to take up the professorship for mathematics in Groningen. Around this time, something must have happened between the brothers. In the December, 1695, issue of the

Fig. 2.23. Johann Bernoulli (1667–1748)

famous journal *Acta Eruditorum*, Jakob violently attacked his brother, mocking the article which had been published in Paris in 1692, as well as the article which had appeared in 1694. He used phrases like: "Johann brings in boiled eggs when the dinner is already over." Johann did not answer publicly. In June 1696, Johann put forward the problem of the brachistochrones in the *Acta Eruditorum*, becoming one of the founders of the calculus of variations. In florid language, Johann Bernoulli reported in January, 1697 (see Stäckel, 1894):

"Johann Bernoulli, Professor of Mathematics, greets the most sagacious mathematicians of the whole globe. Experience shows that noble minds cannot be more incited to further their knowledge than when they are given difficult as well as useful tasks, the solution of which will give them fame and eternity with future generations. In the same way I hope to earn the gratitude of the mathematical world, following the examples of men like Mersenne, Pascal, Fermat, Viviani and others [10], who have done the same before me, by giving a problem to the excellent analysts of our times. Enabling them through this test, to judge their methods, test their powers and if they do not find anything, to inform me and publicly receive their well-earned praise from me."

He then declared that he wanted to postpone the deadline for the competition until Easter, on the advice of Leibniz.

[10] Mersenne put forward the task of the oscillation centre. Pascal's famous competiton of 1658 was based on the cycloide. Fermat gave the English geometers various problems about the theory of numbers. The "Florentine" problem comes from Viviane. Amongst others, Leibniz, who posed the problem of the isochrome in 1687, and Jakob Bernoulli, who posed the problem of the catenary, in 1690, should be also mentioned.

In the May, 1697, issue of *Acta Eruditorum*, Jakob Bernoulli (see Stäckel, 1894) replied:

"Although I was indifferent to my brother's challenge, I could not resist solving the problem when the famous Leibniz amicably invited me to do so. After he informed me in a letter of September 13 that he had solved the problem and wished for others to try it, I set about it although I would otherwise have let it be. I was at once successful, already by October 6 I had the solution and started showing it to my friends."

Jakob's words reveal his contempt for his brother as well as a certain egotism varity that he had been asked by the famous Leibniz to take part in the competition. At the end of the above-quoted essay from May, 1697, Jakob insultingly explained the so-called isoperimetrical problem to his brother:

"It is unfair when someone is not compensated for work which he has done on behalf of someone else, at the expense of his own time and to the detriment of his own affairs. Thus, my brother, for whom I vouch, wants the assurance of a fee of fifty ducats, in addition to the well-earned praise, under the condition that he tries to solve the problem within three months of this publication and produces the solutions by means of quadratures, which is possible, before the end of the year."

The fact that Johann replied violently to this publication is not surprising and should not be interpreted as a character weakness of him. The reasons for the argument probably lay with Jakob Bernoulli, himself. Jakob was often sick and as such easily irritable. He had worked his way into pure mathematics through autodidactical studies, struggling all the way. One can easily imagine that he was a little envious of his talented brother, whom he had partially taught himself.

The other examples which support the theory asserting Johann Bernoulli's supposed character weakness do not appear to be very sound. Johann Bernoulli's conduct in the priority question between Leibniz and Newton has also been criticized. On June 7, 1713, Johann wrote to Leibniz, giving his opinion about the unfairness of the verdict laid down by the London Society of Science. He finished his letter with the words:

"Make the right use of this letter without exposing me to Newton and his countrymen. I do not want to become involved in these quarrels, let alone seem ungrateful to Newton who has overwhelmed me with demonstrations of his goodwill."

Leibniz ignored the request. Instead, he published Johann Bernoulli's letter, although without the signature. However, it was clear to the specialists who the author was. Johann made haste to write to Newton:

"I implore you, and as all that is holy to mankind is my witness, be convinced that everything which was published without a name was unjustly attributed to me."

Even if one adheres to the idea that Johann Bernoulli's letter to Newton (July 5, 1719) was to be taken as a response to Leibniz's compromising publication, one can still understand his reaction to Leibniz's act.

The dispute for which Johann Bernoulli has been most criticized, that is, the priority argument with his son Daniel, has only recently been brought into a new light by Szabó (1987). The accusation of plagiarism by various authors is also to be found in Truesdell's (1954) two statements:

"According to the judgement of the day, adopted by posterity, it was the father who plagiarized the son. ... His miserable attempt to steal his son's masterpiece."

In his review of Szabó's book, however, Truesdell (1980a) distances himself from these two statements:

"I am not sure that Szabó's arguments absolve the old father entirely, but certainly they make me wish I could soften those two statements."

Thus, there are valid doubts as to whether Johann Bernoulli should be described as an envious and quarrelsome egotist. In his field, he was held in high esteem, a fact made evident by his membership in many renowned academies and in his honourable appointments. The great L. Euler, to whom he had given private lessons when Euler was young and with whose parents he was on friendly terms, was a friend and highly revered him.

Daniel Bernoulli (1700–1782)

Daniel Bernoulli, born in Groningen on January 29, 1700, was Johann's second son. He was a mathematician, physicist, doctor, and botanist. At the age of five, he moved with his parents to Basel when his father took over the mathematics professorship from his deceased uncle, Jakob Bernoulli.

At the age of eleven, he started learning mathematics, first from the lessons given to him by his brother, Nikolaus, then later as a student of his father. He was supposed to become a merchant or practice medicine. After having received his Bachelor degree in 1716, he studied medicine, first in Basel, then in Heidelberg in 1718, and finally in Strassburg in 1719. With the work *On breathing*, he presented himself successfully for his doctor's examination in September, 1721. Immediately afterwards, he applied for professorship anatomy, botany, and logic at Basel University. His search for employment was, however, without success. In 1723 he went to Venice to obtain some practical experience under Doctor P.A. Michelotti. He still engaged in mathematics and his study *Exercitationes qualdom mathematicae* appeared in 1724 (the cost of the printing having been subsidized by some friends). The book caused quite a sensation, especially due to a polemical piece in which he defended his father and uncle against scientific attacks. Towards the end of 1724, he became perilously ill in Padua. During his recovery, the St. Petersburg Academy offered him an appointment. The proceeding was a little bit strange. The Academy of Sciences, which had just been founded according to a plan of Peter the Great's, appointed Daniel's brother, Nikolaus, as a member in 1725; however, the precise requirements could only have been fulfilled by Daniel. Thus, it was unclear which of the two brothers was actually meant to be appointed. The problem was solved when it was decided to appoint them both.

Fig. 2.24. Daniel Bernoulli (1700–1782)

Daniel Bernoulli was attracted by the prospect of living with his beloved brother and of concentrating solely on mathematics. Daniel Bernoulli was twenty-four years old at this time, and already a member of the newly-founded Bologne Institute. However, he turned down a chairmanship offer from an academy which was to be founded in Genoa, in favor of the St. Petersburg position.

Together with his brother, Daniel arrived in the Russian capital in October, 1725. Sadly, their time together was only to last for a short while. Nikolaus died a year later from an intestinal ulcer. Fortunately, at around the same time, Euler, a friend from their youth in Basel, accepted an appointment to the St. Petersburg Academy. Daniel had also influenced this appointment. Euler and Daniel Bernoulli remained friends throughout their lives.

Before Daniel went to St. Petersburg he won the Paris Academy prize for his work on the uniformity of movement of hour-glasses on ships. During his lifetime, he won the renowned prize a total of ten times.

Daniel had pledged himself to the St. Petersburg Academy for a total of five years. In 1730, he decided to return to Basel. The Academy, however, was very much interested in keeping the famous scientist, and through their splendid offers, they succeeded in doing so for another three years. He spent the last year in the company of his brother Johann II, who came to visit him and then accompanied him on the return trip to Basel. During his trip, he applied for a vacant professorship in anatomy and botany at the University of Basel. His application was successful and he took up the chair in 1733, after having graduated as a doctor of medicine. He stayed in Basel and refused offers of employment from Berlin and St. Petersburg.

His scientific fame grew and his achievements were universally acknowledged. In 1747, he was made a member of the Berlin Academy and subsequently, in 1750, of the London Society. In 1748, the Paris Academy named him a foreign member, as a successor to his deceased father. He was also one of the seven scientists given the golden coin of gratitude by the Empress Catherine at the time of her first peace with the Turks.

In addition to his existing chair, he took over the experimental physics professorship in 1750. This was an honor given to him by the university, which in this case gave him the chair without drawing lots for it. He taught in this position with great success almost until the end of his life. Taking his continuous salary as a university lecturer into consideration, he also wrote treatises from time to time. Their contents were mostly in the field of mechanics and he informed the Academy of them. He was Vice-Chancellor of the University of Basel twice. Daniel enjoyed good health and only in the last years of life, from 1776 to 1782, did he let his grandsons Daniel and Jakob represent him at his lectures. He died on March 17, 1782, in Basel; he was found by his servants in his bed, having died peacefully while asleep.

Daniel Bernoulli's scientific achievements are to be found in the most varying areas of mathematics and mechanics. The theory of probabilities owes its development to him. In mechanics, his main influence lies in the theory of oscillating chords and in the merits of his *Hydrodynamica*. This book, completed by 1733, was printed in Strassburg in 1738.

Daniel Bernoulli was obviously very well-balanced and amiable. Apart from the argument with his father, any other disputes with his colleagues are unknown.

Leonhard Euler (1707–1783)

Leonhard Euler, the son of the priest, Paul Euler, and Margareta Brucker, was born in Basel on April 15, 1707. His father was very interested in mathematics and had at one stage eagerly and successfully attended lectures given by the great Jakob Bernoulli (1654–1705). His mother came from an old learned Basel family. Leonhard Euler's family was close to the Basel mathematician, Hermann, as well as with Jakob Bernoulli's thirteen year younger brother, Johann Bernoulli (1667–1748). Euler was introduced to mathematics by his father. He hurried through the schools of Basel; the high school was at that time in a poor condition and had almost nothing to offer the young Euler by way of mathematics. At the mere age of thirteen, he enrolled on October 9, 1720 at the University of Basel, in the faculty of philosophy. There he acquired the knowledge which is today taught in high schools. Three years later, he enrolled in the theological faculty, because of his father's wish that he become a theologian. However, his interest in mathematics was already too great, especially as he had had the great fortune to attend a private class for special students taught by the loquacious Johann Bernoulli, who was, at that time the greatest mathematician in Europe. Euler engaged in lively discussions with Johann Bernoulli's three sons (Nikolaus II 1695–1726, Daniel I 1700–1782, and Johann II 1710–1790).

Johann Bernoulli thoroughly encouraged the young Euler. Euler's scholarly successes and progress were very impressive. At the age of sixteen, he compared the systems of Descartes and Newton in his first public lecture for his bachelor degree. He wrote his first mathematical essays at the ages of eighteen and nineteen. Johann Bernoulli praised the twenty-two year old sagacity and ingenuity. This admiration for Euler is obvious from the way he addressed Euler in his letters. In 1728: "the very learned and ingenious young man"; in 1729: "the highly famed and learned man"; and finally in 1745: "the incomparable Leonhard Euler, the prince of all mathematicians".

Affected by the influence of the Bernoullis, Euler's father finally gave up the idea that his son should continue his theological studies.

At the age of nineteen, Euler took part in an open competition of the Paris Academy of Sciences for the design of the best ship mast. He did not win the competition, but, being a typical inlander, he had at this time probably never seen a ship with masts. In the course of his life, Euler won the renowned prize twelve times.

In 1726, at the age of nineteen, on Johann Bernoulli's advice, Euler applied for the vacant physics professorship in Basel and, as proof of his capability, he wrote his dissertation on resonance. However, he did not get on to the short list. This failure was in fact very fortunate, as Euler then was not tied to his narrow native country, but could look for positions abroad. In 1727 he found this position in Saint Petersburg. Saint Petersburg, founded by Peter the Great, was at the time the intellectual centre of the Russian enlightenment. Stimulated by Leibniz, Peter the Great founded the Academy of Science. In her attempts to procure the best learned men for the Academy, Catherine I, Peter the Great's widow, appointed Daniel and Nikolaus Bernoulli as well-endowed professors in 1725. Euler missed his study companions immensely. When Daniel Bernoulli wrote that the chair for physiology and anatomy at the Academy was expected to become vacant and that Euler should prepare himself for it, Euler enrolled in the faculty of medicine at the University of Basel and began to study keenly.

On April 5, 1727, Euler, merely twenty years old, travelled by boat down the Rhein to Mainz and then through Frankfurt, Marburg (where he visited the great philosopher Christian Wolf), Kassel, Hamburg and Lübeck, to St. Petersburg. He arrived on May 17, 1727.

In Saint Petersburg, however, the situation had changed drastically. The companion of his youth, Nikolaus Bernoulli, had died a year previously. Thus, in St. Petersburg Euler had only Daniel Bernoulli remaining (with whom he had an excellent relationship his whole life). Moreover, on the day of his arrival the Empress had passed away and the countinuing existence of the Academy was put seriously in question. Euler's situation was difficult and so he was happy to have the prospect of taking up a position as lieutenant in the Russian navy, which was being built up at the time. In 1730, however, the physics professor, Jakob

Hermann, returned to Basel, and his chair became vacant. Euler was named as his successor. Only in 1733 did he get the mathematics professorship, when Daniel Bernoulli left this position to return to his home town. Thus, by the age of twenty six, Euler had attained a brilliant position and was able to start his own household. On December 27, 1733, he married Katarina Gsell from St. Gallen. Of their thirteen children, only three sons and two daughters survived past infancy.

At this time, Euler was already deeply engrossed in research, although the results were meager when compared to those of later years. His treatises caused great sensations and brought him a world-wide reputation. In these years, he started something quite new, something nobody before him had ever done on such a scale: he started writing extensive textbooks. In order to write the textbooks, Euler was obliged to fill in any gaps he found by using his own research. The textbooks on mechanics, (Part I 1734, Part II 1736) (see Wolfers, 1848) and his algebra book (Euler, 1770b) became the most well-known. His creative power appears all the more astonishing if one takes into account that, in 1738, he lost his right eye as the result of a dangerous illness.

In 1740, Empress Anna died and one of her favorites became regent. A year later, Elizabeth I, Peter the Great's daughter, took over the throne. The Academy's future was uncertain. Understandably, because of this Euler entered into correspondence with Frederick II in 1740. The eccentric Frederick II was strongly under the influence of French culture and planned to decorate his "enlightened land" with an academy. The old academy, which had been founded by Leibniz, had been totally spoiled under the reign of the soldier-king and was no longer of any practical importance. Nothing characterizes its position at the time better than the fact that its expenses were kept under the household title "for all the royal fools".

Frederick II endeavored to appoint Frenchmen to the Academy. He wrote to Voltaire, who recommended Maupertuis as director. D'Alembert, a highly renowned mathematician and mechanics specialist, recommended Euler to the King as the greatest mathematician. After tough negotiations, the appointment was settled. Euler left St. Petersburg with his family and, on July 25, 1741 he arrived in Berlin, where he stayed until 1766. The relationship between Euler and the King remained very cold during the time that Euler was in Berlin. Deep down, Frederick II despised Germans, and he looked upon Euler as being one.

During this time in Berlin, an incident occurred in which Euler involuntarily became involved, namely the dispute between the president of the Academy, Maupertuis, and the mathematician from Bern, S. Koenig. The ambitious and vain Maupertuis had developed the so-called "principle of least action" at the end of the 1740s. He believed that he had found a comprehensive principle of mechanics. In actual fact, according to Szabó (1987), it was only a dalliance with the principle of virtual work. Already at this time, that principle was disputed in the scientific world. Worse then this was the question of priority. A much better formulation was to be found in a letter dated October 16th,

Fig. 2.25. Leonhard Euler (1707–1783)

1707, from Leibniz to an unknown recipient. In the autumn of 1750, Koenig drew Maupertuis' attention to this fact. Maupertuis, who had been in favour of Koenig's appointment to the Berlin Academy, reacted indignantly to Koenig's suggestion that his objection be published in the Academy's report. Thereupon, Koenig published a summary of his opinion in *Nova acta eruditorium*, in the distinguished Leipzig monthly scientific magazine in 1751.

His contribution caused such a storm that the Academy felt obliged to intervene. The president demanded to see the original letter, which Koenig could not deliver, as he only possessed one single copy of it. Mainly, at Maupertuis' instigations, the Academy *unanimously* (half of the members were absent) declared this copy to be a forgery. As a result, Koenig sent his diploma back to the Academy and defended his action in various publications. Koenig found a collaborator in Voltaire. In a satirical poem, *Diatribe du docteur Akakia*, Voltaire makes fun of Maupertuis, Euler, and the Berlin Academy. Thereupon, Frederick II himself intervened, causing Voltaire to flee from Prussia and Maupertuis to tender his resignation.

Euler had been on the side of his president all throughout this time. His great mathematical intelligence should have made him recognize the weakness of Maupertuis' principle. He probably acted out of a false sense of loyalty to his superior.

During his time in Berlin, Euler took a great interest in the development of the Academy. He conducted the mathematics course for twenty years. After Maupertuis' death, Euler even took over the administration of the whole Academy, although he was never made president by the King. The King still hoped to win over d'Alembert. D'Alembert, however, preferred to stay in Paris and direct the Academy from there through his constant correspondence with the King. This was an insult to Euler, who was a far superior mathematician to d'Alembert. As his relations with the King were not good anyway, he again cultivated friendly relations (which had never been completely broken off) with St. Petersburg.

In Russia, Catherine II had ascended to the throne in 1763. She promoted art and science and made Euler generous offers in order to entice him to come back to the St. Petersburg Academy. In the negotiations, he managed to secure a salary of three thousand rubles and a furnished house as well as a widow's pension of one thousand rubles. His eldest son obtained a secretarial position with a salary of two thousand rubles at the Academy, and his youngest son also procured a secure position there. Frederick II made Euler hand in his resignation three times before he agreed to it. On June 9, 1766, Euler and his family left Berlin and arrived in St. Petersburg on July 17, 1766. Euler was very satisfied with his new position, especially as the Academy's budget had been increased. Unfortunately, his remaining eye was threatened by cataract. In 1771, he was operated on by the most famous cataract surgeon of the time, Baron von Wenzel. After the operation, he was able to see a little, but in the course of a few days his old condition returned.

Thus, scientific work at this point in Euler's life was not easy. He dictated his studies of mathematics and mechanics to his students, who in turn read new publications out loud to him. In 1773, Euler's wife died; he remarried in 1776.

On September 10, 1783, Euler suffered a stroke while working and died a few hours later. His family stayed in Russia, highly respected by everyone. All of his friends praised Euler's sense of justice and his altruistic character. He did not bother about questions of priority, but generously gave away new discoveries and knowledge. Moreover, he was very modest.

Euler's publications are overwhelming. He published almost nine hundred papers and books. Amazingly, thirty-four percent of his work was written between 1775 and 1783, after he became blind. Furthermore, three thousand of his letters are known to exist. The volume of his work stands up to any comparison with other great intellects.

Abraham Gotthalf Kästner (1719–1800)

Abraham Gotthalf Kästner was born on September 27, 1719, in Leipzig. His father was a Professor and Doctor of Law at Leipzig University. His father and uncle educated him through private lessons; he never attended a public school. Kästner was very perceptive, had a good memory, and was very resolute.

At the early age of ten, he was permitted to attend his father's law lectures and, at the age of twelve, he enrolled in the Faculty of Law, although he did not neglect his mathematical studies. In addition, he studied the old and new classical authors and attended mathematics, physics, history, and philosophy

Fig. 2.26. Abraham Gotthalf Kästner (1719–1800)

lectures. He had a particular preference for the subject of anatomy. In 1735, Kästner received his "Baccalaureus" degree from the Philosophy Faculty and two years later he graduated. He still continued his studies and attended lectures in anatomy, botany, and chemistry. In 1739, at the age of twenty, Kästner became a lecturer at the University of Leipzig. His lectures were very successful and, in 1746, he was made Associate Professor of Mathematics. He kept this position for ten years, after which he was appointed as Professor of Mathematics and Physics in Göttingen under very favourable conditions.

Shortly before moving to Göttingen, Kästner married Rosina Baumann, the sister of one of his friends. She died barely two years later. Käster's second wife was the widow of a French officer.

Kästner was a professor in Göttingen for forty-four years. His most active and brilliant period took place during the 1760s and 1770s. He earned merits for the reputation of the Georgia Augusta, Göttingen University. At this time, there were also many people from other faculties attending his lectures. The reputation of his lectures and his scientific publications brought him great fame. His mathematical textbooks surpassed the ones which had been used previously. Moreover, he was made a member of almost all of the prominent scholarly societies, including the Göttingen, Berlin, St. Petersburg, and Bologne Academies of Science.

He carried out extensive correspondence with the scholars of his time, including some science-loving princes, such as the Duke of Braunschweig and the

Duke von der Lippe. In 1765, he was appointed Royal Privy Councillor of Great Britain and Braunschweig.

In spite of his many diverse activities in mathematics and physics, he still found the time to write poetry and long prose essays, as well as to review literary works. In these fields he was also successful, a fact proven by Lessing's praise for him. Some of his epigrams were feared for their biting wit. However, in spite of his sharp tongue, he was regarded as totally honest and respectable.

Kästner was fortunate enough to remain healthy until an advanced age. He died on June 20, 1800.

Charles Augustin Coulomb (1736–1806)

Charles Augustin Coulomb was born in Angoûlème on June 19, 1736. He came from a family whose members had been in the public service in Montpellier for quite some time. He was very young when he moved to Paris. There he discovered his love for mathematics, and decided to devote himself to the science. However, certain circumstances arose which prevented him from carrying out this plan and, in 1762, he was obliged to enter the Royal French Engineering Corps. In order to be promoted quickly, Coulomb let himself be transferred to America. In the West-Indian colony of Martinique, he was in charge of the building of fortifications. He carried out his first experiments on the strength of brick walls. Unfortunately, his health suffered due to the climate of Martinique, and he decided to return to Europe. In 1776, he returned to Paris, having been promoted to lieutenant-colonel.

Coulomb had submitted his notes (see previous section) to the French Academy of Sciences as early as 1773; they were published in 1776. After his return, he worked as an engineer in, among others, La Rochelle and Cherbourg, and dealt almost exclusively with scientific problems. He informed the Academy of his results through various publications. The Academy found his work so profound that they made him a corresponding member.

In 1775, Coulomb gave a lecture at the Academy on human strength and how it could be advantageously put to use in machines. The lecture was first published twenty-five years later, supplemented by several additions arising from the practical experience he had gathered throughout these years. In 1777, Coulomb won the Academy prize, together with von Swindon, for his treatise on the construction of the compass. In 1781, he won the double prize given by the Academy for the best work on the resistance of bodies through friction. It appeared in 1785 under the title *Theorie des machines simples, en ayant égard au frottement de leurs parties et la voideur des cordages* in the tenth volume of the *Mémoires de mathematique et de physique, présentés à l'academie royale des sciences* (Coulomb, 1785).

With his unique experimental methods, he was far ahead of his predecessors, especially Amontons. He was aware of the inadequacy of the model test. Therefore, he made efforts to carry out large-scale experiments during his stay

Fig. 2.27. Charles Augustin Coulomb (1736–1806)

in Rochefort, efforts in which the arsenal inspector, La Touche-Treville, willingly supported him. His test results served as the basis of understanding the friction phenomenon for a long time.

In 1781, Coulomb, still a member of the army, returned to Paris. He became a full member of the Academy and director of river constructions. In the time which followed, he dedicated himself especially to the construction of the torsional balance. This made it possible for him to carry out experimental checks on the electro and magno-statical laws (which had been named after him) for the attraction or repulsion of two point charges with much more precision than ever before.

When Coulomb opposed the construction of navigable canals in Brittany, he was imprisoned for some time. As a result he requested a discharge from the army, although this was denied. When it was insisted that he change his opinion, Coulomb still refused. The province Diet finally acknowledged his sincerity.

At the outbreak of the Revolution, Coulomb gave up his posts as Inspector-General of wells and supervisor of the plans chamber. After the dissolution of the Academy, he became a member of the commission for the determination of mass and weight. In particular, he was to calculate the length of the second pendulum in various parts of France. After he was expelled from this commission by edict of the national convent, he retired with some friends to his country estate not far from Blois in 1792. There he carried out several experiments on plant physiology, in particular on the circulation of sap in trees. He did this much more accurately than his predecessors. These works, however, were only partially made known to the general public.

Under Napoleon, Coulomb returned to Paris, after having been made a member of the National Institute. Soon after, he was appointed as Inspector-General of the universities and public education system as a whole. He took a great interest in the reform of the education system. These activities entailed extensive travelling, which in turn affected his health. Coulomb died on August 23, 1806.

Coulomb was an engineer with great mathematical capabilities. He intuitivly recognized physical problems and derived his solutions by proceeding from recognized principles of physics. He wrote:

"I have tried, as far as possible, to arrange the principles which I have used in such a way that an expert who has been instructed to a certain degree can understand and use them."

His works are outstanding for their clarity of thought, and the experiments which he carried out were all the more remarkable for the high degree of precision reached. Without a doubt, he was one of the creators of mathematical physics.

Moreover, Coulomb was a very modest scientist. In the preface to his treatise, he wrote (see Timoshenko, 1953):

"This memoir, written some years ago, was at first meant for my own individual use in the work upon which I was engaged in my profession. If I dare to present it to this Academy it is only because the feeblest endeavors are kindly welcomed by it when they have a useful objective. Besides, the sciences are monuments consecrated to the public good. Each citizen ought to contribute to them according to his talents. While great men will be carried to the top of the edifice where they can mark out and construct the upper storeys, ordinary artisans who are scattered through the lower storeys or are hidden in the obscurity of the foundations should seek only to perfect that which cleverer hands have created."

An obituary written by one of his contemporaries says:

"He left his two sons no other inheritance than a respected name, the example of his virtues and the memory of his brilliant merits in the sciences."

Reinhard Woltman (1757–1837)

Reinhard Woltman, the son of a peasant, was born in December, 1757, in Axstedt, in what was then the Dukedom of Bremen. He was one of the most interesting personalities in eighteenth century engineering. He grew up in poor conditions and worked conscientiously to acquire enough knowledge to become a school teacher in the town of his birth. After a short time he applied for the position of foreman and clerk in the Hamburg district of Ritzebüttel (which today belongs to Cuxhaven).

On May 14, 1779, the dike delegation, consisting of members of the council, delegates of the admirality, and the treasury department as well as members of the shipping industry, voted him into the position. The change to the Hamburg council was a decisive step for his career, due to the important function of the position in Ritzebüttel.

Already, in 1725, the admirality and the treasury department had taken over the maintenance of sea signals and the harbor of refuge in Ritzebüttel, which were of great importance to the Hamburg shipping industry. The inhabitants of the Ritzebüttel district could not be expected to pay for the expensive shore fortification out of their own pockets. Thus, help had to come from the public sector. For this purpose, the dike delegation was called in to take care of the shore protection.

When Woltman began working for the dike delegation, whose president at the time was the Syndic, Jakob Schuback (1728-1784), his immediate superiors were V. Grumkow and H. Zitting. The latter was the first to recognize and appreciate Woltman's capabilities. His unceasing thirst for knowledge and his never tiring diligence enabled him to acquire the necessary knowledge for studies in mathematics and dike construction at the university. His superior actively encouraged him in all areas. Meanwhile, the members of the delegation, especially the president, had become aware of his gifts. Syndic Schuback himself, and, upon his recommendation, various senators and figures in public life, put means at Woltman's disposal to enable him to study. The sole condition that they required was that after completing his studies, Woltman give his services first and foremost to Hamburg.

From Easter, 1780, onwards, Woltman attended high school in Hamburg. After this, he enrolled at Kiel University in order to attend lectures given by Professor Johann Nicolaus Tentens, the only professor in Germany who was at that time treating dike construction scientifically. During this time, he also had the opportunity to gain practical experience.

During Christmas, 1782, the position of his superior, H. Zitting, became vacant and on April 25, 1783, the dike delegation expressed their desire to make Woltman his successor. Syndic Schuback had supported Woltman and had influenced the delegation to share his opinion and appoint Woltman. At the same time, Woltman asked for leave to continue his academic studies and to take an educational trip. The delegation approved of a leave of one and a half years with full pay; he was even granted a monthly subsidy for his trip.

Woltman stayed in Kiel for a while and then studied for a semester in Göttingen (here he most likely attended the lectures of the Privy Councillor and Professor of Mathematics and Physics A. G. Kästner). On Easter, 1784, Woltman set out from Göttingen. He decribed in detail the stops on his trip and interesting dike construction problems and structures in the third volume of *Beyträge zur Hydraulischen Architektur*, in which his literary talent was very useful. Via Kassel, Marburg, Gießen, and "Frankfurt am Mayn", he travelled to "Manheim". This city, with its two rivers, the Neckar and the Rhein, and its inhabitants, charmed him. In "Strasburg" (a city with 48,000 inhabitants at the time) he was particularly impressed by the cathedral, which he described in detail.

As a dike construction specialist, he criticized the poor condition of the bridges. He found that the inhabitants spoke better French than German, "they appear polite, courteous and not very luxurious." Via Nancy, Woltman arrived at his first long stop in Paris. He described this magnificent city in detail, paying

Fig. 2.28. Reinhard Woltman (1757–1837)

particular attention to the most important sights, such as the Louvre and the Opera. He was appalled by the dirty streets. Although he wrote that he had "not been in Paris long enough to judge the morality and customs of social life in this city," he was still very critical. For example:

"Alles schmiert, pudert und bemahlt sich, die Wilden thun auch deßgleichen, die Mode entschuldigt es und selbst die Natur, sagte mir Jemand, mit dem ich über diese Eitelkeit mich unterhielt, scheint solchen Schmuck zu lieben. ... Einen alten Mann ehret ein graues Haar, und ein Geschäftsman von mittlerem Alter mag dieß mit Puder nachahmen, aber zehnjährige Jünglinge mit weißen Köpfen haben mir immer lächerlich geschienen, und ich weiß nicht, ob die Mode solche gothische Verzierungen rechtfertigen kann."

"Everybody smears, powders and paints themselves, as savages do. I talked to somebody about this vanity and he told me that fashion excuses this and even nature seems to love such adornments. ... An old man appears distinguished with grey hair and thus a middle-aged business man may copy this with powder, but I have always found ten-year old youths with white heads ridiculous, and I don't know that fashion can justify such Ghotic decorations."

In Paris, Woltman also found the support of distinguished personalities, for example, the royal minister, Chev de Viviers. He was even granted a visit to Versailles. In florid language he described the palace with its halls and gardens, portraying court life as he saw it.

Woltman then travelled to Cherbourg via Normandy. He occupied himself fully with the hydraulic constructions on the French coast and with special hydraulic problems. He also made suggestions as to how various constructions should be built in order to withstand the waves from the sea.

From Calais, whose port was in poor shape, Woltman travelled to London via Dover. Being a hydraulics specialist, he was impressed by the bridges over the Thames: "...these three bridges appear to be so massively and durably built, as if they were intended for eternity."

He described the city in detail and praised London's inhabitants as being dependable and hard-working. From his travel description, one feels that the British character was much closer to his north German character than that of the French.

Woltman stayed in Holland for a longer period of time in order to study intensely. He returned via Friesland, Oldenburg, and Bremen and arrived in Ritzebüttel on November 20, 1784, to take over his office.

Until 1810, he developed his main capabilities in Ritzebüttel, in practical as well as scientific and literary respects. The work of Woltman's predecessor had been accompanied by many failures. Dike construction in Ritzebüttel had been carried out since 1733 with varying success. All constructions built up until 1740 were lost. The period from 1740 to 1756 saw the most solid constructions being erected, constructions which, apart from repairs, lasted for one hundred years and more. However, because these constructions were mostly built from stone, they were so costly, that for economic reasons they could not be continued. Therefore, fascine constructions took over. However, this construction method was found to be even more expensive. This was due to the fact that the fortifications, which were made up of bushes and stakes, broke constantly. Thus, fascine construction was abandoned in favor of stone construction, although the procedure used had not yet been perfected.

The failures of the previous thirty years and the constant change in the construction systems used for the building of embankments came to an end when Woltman took over the direction of the dike concerns. He laid a sure foundation for protecting the works of Ritzebüttel, which, although they had been fully extended and built out, had been only minimally maintained for a long period of time. Woltman was able to achieve these excellent results due to the introduction of, and strict adherence to, a few principles recognized as essential, which aimed at the greatest possible simplicity and strength of the constructions. He founded dike building on a few irrevocable principles. Woltman's great talent for abstraction was here made most evident; he had already proven this before (see his earth pressure theory).

Woltman's significance was not only due to the fact that he recognized what was technically correct and suggested that it could be realized, but also because he understood how to present his ideas clearly to the Deputation and to have them accepted in the political decision process without being hindered by consideration of the cost.

His reputation was also based on his great success in his office. It was further enhanced by the recognition he received from the scientific world. This was due, in particular, to his four-volume book *Beyträge zur hydraulischen Architektur*, which appeared between 1791 and 1799. The first volume, dealing with sea dike economy and shore fortifications (and also including literary contributions) quickly made him known throughout the scientific world.

In 1792, Woltman was made a member of the Dutch Society of Sciences of Harlem and the Butavical Society of Experimental Philosophy in Rotterdam. In addition, the Royal Society of Sciences in Prague offered him membership. The royal Society of Sciences in Göttingen made him a corresponding member in 1793. These close ties with the academic world led Woltman to seek for a change in his official title of Conductor, which appeared outwardly to be a subordinate one. The council complied with his request and agreed on the suggestion of the Deputation to the title of a Director of Shore and Hydraulic Constructions in the district of Ritzbüttel at the estuary of the Elbe.

Meanwhile, Woltman's achievements had brought about further consequences. In August, 1792, he received an offer for an important position as a dike-reeve with a respectable salary in the Dukedom of Oldenburg. Woltman could not just turn down this honourable offer. This lifetime position came with a very high salary, which he could never hope to receive in Ritzbüttel. On the other hand, he had received much support and many benefits from his adopted town. However, he felt that he had more or less paid this back through his long years of service and his accomplishments in dike constructions. These considerations made him decide to apply for his discharge. Understandably, the Deputation did not consent to this, having extended his office for a further ten years only a year earlier. However, they did promise to expand his sphere of activity, ameliorate his financial position, and make his position a lifetime one.

The position in Oldenburg continued to remain vacant, and the Oldenburg Government urgently asked the President of the Deputation to let Woltman take up the position of dike-reeve. In order to be able to refuse this request emphatically, a change in Woltman's position in Ritzbüttel was undeniable. Thus, on May 8, 1793, the Council endorsed the promises already made by the Deputation. With the Council's knowledge, Woltman committed himself to remaining in the Hamburg office, to inspecting all Hamburg dikes at least twice a year and to drawing up a report on their condition.

Woltman occupied himself primarily with raising family in later years. On October 1, 1797, he married Johanna Elisabeth Schuback, a daughter of his first patron, Syndic Jakob Schuback. All five of the children that resulted from this marriage were born healthy and lived a full life.

In 1810, Hamburg was incorporated into the French Empire. On the request of the French authorities, Woltman was obliged to move from Ritzbüttel to Hamburg and to take part in the planned canal works. However, the plans were never realized because the liberation wars (1813) set in and French rule was put to an end. The dike deputation was not re-established. Its tasks were taken over by the newly-founded shipping and port deputation. Woltman became the director of river and shore works and canals after turning down an offer from Prussia to become head director of all sea ports.

Woltman's tasks were further extended, so that finally he supervised the complete hydraulic constructions of the Elbe. He moved to Hamburg and let himself be represented in Ritzbüttel by a conductor. His new position meant that Woltman could not utilize the experience gathered in Ritzbüttel for the dike constructions and embankments of the Higher Elbe.

In spite of the demanding work in his office, Woltman still found peace and leisure in pursuing his literary work. In particular, he reported on the hydraulic works constructed, published a *Handbook of Navigation*, and occupied himself with the dredging of rivers. The Senate showed its great appreciation in this matter by giving him a respectable honorary gift. On October 27, 1836, Woltman was honorably pensioned off with a full salary. He died on April 20, 1837.

Woltman was characterized by his north German homeland, his ascent from humble conditions and, later, by Hamburg society. He could never deny his north German character and it gave him a certain stiffness in his manner, but also rendered him open and honest. His straight-forward character made him capable of carrying out solutions forcefully, once they had been recognized as correct. Similarly, he always managed to procure the necessary financial means for the dike constructions in Ritzbüttel from the Hamburg Senate. He was open and honest with his colleagues and superiors, which in turn made him well-liked and trusted. Supported by a good team, he set right the dike works in the Hamburg district. His superiors, as we have already seen, supported him in all matters, something for which he in turn was very thankful.

Woltman was very settled; he never left the Hamburg district, excepting for his educational trip. Moreover, he was very thrifty. How else can one explain the fact that he avoided the more expensive mounted post service which would have caused his manuscript on the earth pressure theory to arrive on time in St. Petersburg?

Basically, Woltman was very meticulous. This can also be seen in his careful travel descriptions and his treatment of engineering problems, especially in dike constructions. He based these works on a few principles which he had recognized as being correct, and from which he never deviated. This quality was, of course, also very useful in his scientific work. He was excellent at observing and abstracting, in shaping the observed material into an order and drawing conclusions from it, conclusions which he then accommodated in his theories.

His scientific work made him known far outside the Hamburg area. Yet, although he became very famous, he remained a very modest scientist.

His political views were conservative in every respect, and he believed very strongly in authority. In spite of his high position, he remained naive in his opinion of political conditions and social systems. Nothing characterized his views in these matters better than his comments on the French Revolution in the preface to his *Beyträge zur Hydraulischen Architektur* (Vol. 3) (Woltman, 1794/99).

"Man kann gegenwärtig kaum an Frankreich denken ohne sich seiner unglücklichen Staats-Revolution zu erinnern. Vielleicht wird mancher erwarten, daß in Bemerkungen auf einer

Reise durch Frankreich einer so wichtigen Begebenheit billig Erwähnung geschehe. Aber meine Reise geschehe Anno 1784 wo man mit den Gouvernement wohl zufrieden, und an Revolution noch nicht zu denken schien; überdem ist auch Staats-Politik nicht mein Fach. Sollte ich als Mensch, als Weltbürger, über diese Begebenheit meine einfältige Meinung äußern, so dürfte sie am füglichsten hier stehen. (Es ist wenigstens nicht ungewöhnlich, daß Schriftsteller in den Vorreden ihrer Bücher manches erwähnen, was mit dem Inhalt des Buchs selbst nicht in Verbindung steht). Hier mögen ein Paar unbedeutende Bemerkungen, welche hinreichend sind, meine Gesinnungen über diesen Punct zu äußern, mir hier zu Gute gehalten werden.

Ich verstehe unter Staats-Revolution das Verfahren, nach welchem ein Teil des Gouvernements, oder der Bürger, oder der Staatsbedienten, den anderen durch Zwang nöthiget, die bisher unter ihnen bestandene Verfassung und Ordnung aufzuheben, und eine neue anzunehmen.

Alle Staatsverfassungen können Mängel haben; sie zu verbessern, genügt der Gebrauch der Vernunft und des freyen Willens, welche die edelsten Kräfte und Vorzüge des Menschen sind.

Menschen, welche die Verschiedenheit ihrer Meinung nach dem Maße ihrer körperlichen Stärke entscheiden, Beyfall und Gehorsam sich erzwingen wollen, würdigen in meinen Augen sich fast zu den Thieren herab.

Revolutionen sind traurige Beweise von Zerrüttung der Vernunft, Tugend, Treue und Gerechtigkeitsliebe, durch welche allein doch nur das menschliche Geschlecht sich der möglichst vollkommenen Glückseligkeit auf dieser Erde nähern kann.

In eigenen Angelegenheiten unparteyisch zu urtheilen erfordert eine Stärke der Vernunft und Tugend, die wenige Menschen erreichen; daher wäre es zu wünschen, daß streitige Meinungen unparteyischen Schiedsrichtern unterworfen würden, und statt der Anarchie zwischen den Cabinetten der Fürsten, Gesetze und Gerechtigkeitspflege, wie zwischen einzelnen Bürgern, eingeführt würde. Man sehe die vortreffliche Denkschrift auf Leopold II, über die Größe und Ruhm von dem Hrn. Abbeé Graber, Prag 1792, Seite 30.

Revolutionen mögen zuweilen glücklich ausfallen. Zuweilen mag der Raub den Räuber glücklich machen; oder auch einer unterdrückten Partey mag zuweilen die Befreyung gelingen; aber das Glück, die Freyheit, welche durch Vernunft und Tugend im ordentlichen Wege erworben werden, sind dauerhaft, beruhigend, ehrenvoll und gemeinnützlich für das menschliche Geschlecht. Jenes wird alle Mahl theuer erkauft und nur durch immerwährendes Mißtrauen und Machiavellismen behalten.

Ein guter Weltbürger kann kein Freund von Revolutionen seyn; auch nicht von der Französischen. Dieß letztere verdient noch einer nähern Erklärung.

Die französische Revolution scheint nicht aus Notwendigkeit wegen Ungerechtigkeit und Unterdrückung, sondern abseiten des Gouvernements aus dem Wohlwollen des Königs und seinem löblichen Bestreben, die Finanz-Verfassung zu verbessern; abseiten der ersten National-Versammlung aber aus gewissen neuen Begriffen über die bürgerlichen Verhältnisse der Mitglieder eines Staats unter einander auch mit dem Gouvernement, entstanden zu seyn.

Diese neuen Begriffe rouliren auf zwey Principal-Angeln. **Freyheit** und **Gleichheit** nicht nur aller Franzosen, sondern aller Menschen.

Der Stifter dieser neuen Lehre ist unstreitig **Jean Jacques Rosseau** [11]; ihre gegenwärtigen Bekenner und Verteidiger nennen sich **Jacobiner-Philosophen.** Man könnte sie, um die Philosophie nicht zu compromittiren, Jeanjacquisten oder Neufranken nennen.

Wie weit die Freyheit und Gleichheit unter den Menschen nach dieser neuen Lehre gehen solle, habe ich nicht deutlich bestimmt gefunden; ich urtheile bloß nach den Fortschritten, welche die Neufranken bis jetzt damit gemacht haben.

---

[11] Discours sur l'inégalité parmi les hommes; the same in le Code de nations, ou du Contract social.

Es gibt gewisse heilsame und für das menschliche Geschlecht beglückende Institutions-Wohltaten, die der Mensch im Stande der Natur nicht hat, und nicht haben kann; die aber in der bürgerlichen Verfassung zugleich Mittel und Zweck sind, z.B. Sicherheit des Lebens, der Erwerbmittel und des erworbenen Eigenthums; vernünftige Verehrung eines höchsten Wesen; Menschenliebe: gerade diese *Sacrosancta beneficia* sind von den Neufranken in ihren Revolutions-Decreten und Handlungen dergestalt angegriffen und verletzt worden, daß man schließen muß, ihre Freyheit und Gleichheit könne mit dem, was dem Menschen am allerwichtigsten seyn muß, nicht bestehen.

Und so gibt es denn doch bis jetzt wohl keine bürgerliche Verfassung, wenigstens nicht in Europa, in welcher diese Institutions-Wohltaten, das heiligste und wichtigste, was der Mensch besitzen kann, dergestalt ihm entzogen würden, daß er Ursache hätte, sie mit **Rousseau**'s Freyheit und Gleichheit zu vertauschen.

Aber die Möglichkeit ist freylich wohl nicht zu leugnen, daß bürgerliche Verfassungen dergestalt entarten, verderbt, unnützlich und lästig werden können, daß der größte Theil ihrer Bürger es gerathener finden möchte, ihnen den Stand der Natur vorzuziehen. Die Sicherheit des Lebens kann durch vielfältige Kriege; die Sicherheit der Erwerbsmittel durch ein unzähliges Heer von Real- und Personal-Privilegien und unnützen Interdicten die Sicherheit des Eigenthums durch Vermehrung der Contributionen, die gleichfalls durch Kriege, große stehende Heere und Kriegsflotten entstehen, geschmälert und verloren werden. Die Religion kann durch Unwissenheit, Trägheit und schlechte Gesinnung ihrer Diener in Verachtung und Verspottung kommen; Falschheit, Untreue und Unglauben, kann überhand nehmen; und Menschenliebe unter allen solchen Gräueln erlöschen, wie ein Licht in giftigen Dünsten.

Gegen eine solch höchst unglückliche Catastrophe dürfte das menschliche Geschlecht doch auf ewig gesichert seyn. Die Fürsten, Regenten und Magisträte, dürften immer mehr die Nothwendigkeit erkennen, sich einander Treue, Glauben, Gerechtigkeit und friedfertige Gesinnungen zu beweisen, sich als Väter ihrer Staaten und die Bürger als erwachsene Kinder derselben, anzusehen und zu behandeln; und diese dürften an Frankreichs Bürgern ein unvergeßliches Beyspiel lernen, wie traurig ein solches abscheuliches Vergehen, das Zutrauen eines guten Fürsten zu mißbrauchen, und seinen Vater und Wohlthäter aufs Blutgerüste zu führen, in unglücklichen Folgen ist."

"One can currently hardly think of France without remembering its unhappy state revolution. Maybe some people would expect that in the comments on a trip through France such an important event would only rightly be mentioned. However, my trip took place in 1784 where one seemed to be satisfied with the government and did not yet seem to be thinking about the Revolution. Moreover, state politics are not my field. If I, as a person and world citizen, were to state my simple opinion of this great event, then it would be most suitable to do so here (it is at least not unusual for authors to mention things in the preface to books which do not have any connection with the contents of the book itself). Here, I hope that a few important comments, which are enough to utter my sentiments on this point, will be looked upon leniently.

With state revolution I understand the process where a part of the government, the citizens or the civil servants coerce others to abolish the constitution endured until then and to adopt a new one. All state constituents can have faults, to correct these, the use of common sense and free will, which are the most noble strengths and qualities of men, are sufficient.

People who make the differences of their opinion according to their physical strength, and who want to gain approval and obedience by force, degrade themselves almost to animals.

Revolutions are sad evidence of the disorder of reason, virtue, faithfulness and the love of justice, and only through these can man attain the most complete happiness on earth.

To objectively judge matters which concern oneself, one needs such good common sense and virtue which few people possess. Thus it would be desirable that contestable opinions be subjected to objective umpires and that, instead of anarchy between the cabinets of the princes, laws and cultivation of justice be introduced like between single citizens. See the excellent memorandum to Leopold II about greatness and fame by the Abbé Graber, Prague 1792, page 30.

Revolutions may occasionally turn out well. Sometimes robbery makes the robber happy, or a suppressed faction may once in a while be delivered, but the happiness and the freedom that are obtained through reason and virtue in the proper manner are permanent, soothing, honourable and of public benefit for mankind. The former is expensively bought and can only be kept through mistrust and Machiavellism.

A world citizen cannot be a friend of revolutions, also not of the French one. This latter statement deserves a closer explanation.

The French Revolution does not appear to have orginated because of injustice and suppression, apart from the government's attempts due to the benevolence of the King and his praiseworthly endeavour to ammeliorate the financial constitution, and apart from the first national committee, but from certain new concepts of the bourgeois relationships of the members of a state between each other and the government.

These new concepts are based on two principal points. **Freedom** and **equality** not only for all Frenchmen but for mankind.

The orginator of this teaching is without a doubt **Jean Jaques Rousseau**; its present followers and defenders call themselves **jacobine-philosophers**. In order not to compromise the philosophy, one could call them jeanjacquists or neo-franks.

I did not find it evident to what extent the freedom and equality of man goes within this new teaching, I only judge it from the progress which the neo-franks have made with it until now.

There are certain salutary institutional charities which make man happy but are not found in nature, indeed could not be found there, which in the bourgeois constitution, however, are both the end and the means, i.e. security of life, means of livelihood and the acquiring of property, the reasonable worship of the highest being, love of man. These *sacrosancta beneficia* have been so attacked and damaged by the neo-franks in their revolutionary decrees and actions, that one must conclude that their freedom and equality could not exist with that which is the most important to man.

As yet there does not exist, at least in Europe, a bourgeois constitution in which these institutional charities, the holiest and most important things man can possess, are taken away from him in such a way that he would have the need to exchange them for Rousseau's freedom and equality.

The possibility, however, can certainly not be denied that bourgeois institutions degenerate in such a way as to become useless and cumbersome, so that the majority of its citizens may find it advantageous to prefer the state of nature. The security of life can be impaired and lost through multifarious wars; the security of means of income by countless armies of real personal privileges and useless prohibitions; the security of property through increase of contribution, which also grow from war, big armies and naval forces. Religion can become a source for contempt and mockery through the ignorance, indolence and bad disposition of its servant. Falseness, unfaithfulness and disbelief can gain the upper hand and love of man can be extinguished by such ghastliness like light in poisonous fumes.

However, mankind will be saved for ever from such an unhappy catastrophe. Princes, regents and magistrates will always recognize the necessity to prove to each other their faithfulness, faith, justice and peace-loving attitude; to regard themselves as fathers of their states and the citizens as adult children of them and to act accordingly. This should teach France an unforgettable example as to the sad repercussions of such an abominable offence of misusing the confidence of a good prince and leading one's father and benefactor to the bloody scaffold."

Chapter 3: The Classical Era

The classical era of the development of porous media theory spanned almost one hundred years, from 1822 to 1913. The term *classical* era is justified in the sense that, during this time, fundamental ideas from the eighteenth century were made concrete and various theories completed. As was mentioned at the end of the previous chapter, this is true, on the one hand, for the concrete formulation of the stress concept and for the derivation of the material-independent fundamental equations; and on the other hand, for the completion of the theory of linear-elastic bodies, that of viscous fluids, the earth pressure theory, and the failure condition for granular material. This is also true for the formulation of a plasticity theory for ductile materials as well as the development of thermodynamics.

After the development of the stress concept is presented, the history of the linear elasticity theory will be examined, in which the extensive contribution of Todhunter and Pearson (1886/1893) will be used as a starting point.

Around the mid-19th century, fundamental effects concerning porous media were studied and described by Delesse, Fick, and Darcy, namely the equality of surface and volume fractions in porous media with statistically distributed pores, the diffusion phenomenon, and the interaction between the constituents. Darcy was the first scientist to study the interaction between two constituents, i.e., between the skeleton (porous soil body) and water. Although studies on the diffusion problem and water flow in porous solids were mostly intuitive, and purely experimental, they have been so important that the porous media theory cannot dispense these constitutive relations.

Following that, the development of the theory of viscous fluids will be discussed. The completion of the earth pressure theory and the failure condition for granular material will be also then described. The motion of liquids in tubes (including capillary motion) and porous solids will be outlined. Thereafter, the foundation of the mixture theory by Stefan – an essential part of the porous media theory – will be reviewed and discussed.

Finally, highlights in the early development of thermodynamics will conclude the description of the classical era.

3.1 Cauchy's Formulation of the Stress Concept

Out of all the fundamental concepts of mechanics, the stress concept took the longest to develop. Although the concept of stress does not appear difficult today, the way to this concept was anything but easy. The question of the *inner forces* in loaded continua had been posed since the creation of scientific mechanics.

3.1.1 Cauchy's Predecessors

Galileo (1638) had already gathered important knowledge, with regard to the inner forces, in his *Discorsi*.

In his examination of the rupture of tensile rods, Galileo assumed that failure occurs when a certain force in the tensile direction per area unit is attained. After this, it was adhered to that the "inner force" acts vertically to the cross-sectional area. In a remarkable work dating from 1713, Parent (see Timoshenko, 1953) mentioned in a single sentence that for reasons of equilibrium, tangential forces (shear forces) also had to play a part. Parent's observation remained unnoticed and without influence. A possible reason for this is surmised by Timoshenko (1953):

"The cause of this may perhaps lie in the fact that Parent's principal results were not published by the Academy and appeared in the volumes of his collected papers which were poorly edited and contain many misprints. Moreover, Parent was not a clear writer, and it is difficult to follow his derivations. In his writing he was very critical of the work of other investigators, and doubtless this made him unpopular with scientists of his time."

Couplet (1726) obtained the same result as Parent that, in the case of equilibrium on a cross-sectional area, tangential forces must act. His method was produced by considerations he had applied to a totally different problem; the earth pressure problem. In order to gain insight into the state of the inner forces in soils, Couplet developed, in several works, starting in 1726 (see Couplet, 1726) a mathematical model. In his model, Couplet assumed that sand consists of equally sized balls, which are regularly piled in layers. Each grain rests on three others with which it forms a regular tetrahedron (see Fig. 3.1). From here, the slope is derived theoretically. If a smooth wall is erected in front of the sand, then the grains of sand situated directly behind it are supported differently from those situated on the inside. From this, Couplet derived that the weight of the sand enclosed by the wall; the slope plane is to be separated into two components. One of these components, the earth pressure, is vertical to the wall; the direction of the other one is independent of the slope. Couplet's studies had no influence on the earth pressure theory. However, the knowledge that the pressure on the cut surface did not act vertically was very important for following investigations.

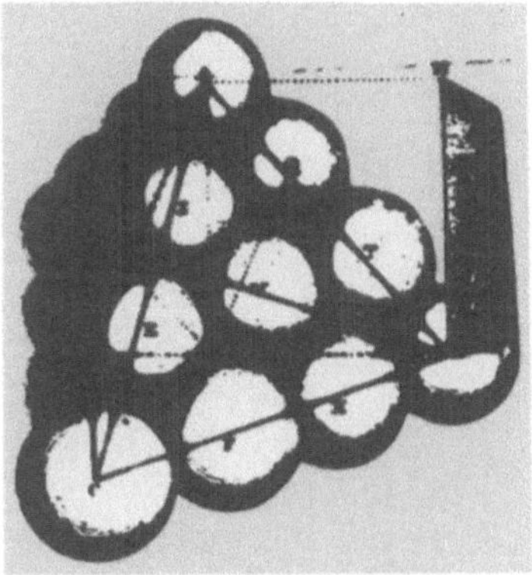

Fig. 3.1. Couplet's considerations on the equilibrium of a regular tetrahedron

The next important step in the creation of the stress concept was L. Euler's work in hydrodynamics. Inspired by Johann Bernoulli (see Section 2.3), Euler introduced the general concept of internal pressure and the general principle of linear momentum. Truesdell (1956a) supposes that D'Alembert's prize essays (which Euler, as director of the Berlin Academy, had to read) gave him the final impulse for the creation of his general hydrostatics and hydrodynamics. Euler explained the concept of inner pressure in several publications. In 1750, he spoke of an element of water as being "subject to the pressures of the particles of water surrounding it"; in his next treatment (Euler, 1752b), he spoke further of "the pressures with which the particles of water everywhere act upon each other" (see Truesdell, 1956a). The pressure is a field varying in space and time; the direction of the pressure is assumed to be normal to the surface on which it acts. Truesdell writes further:

"Although Euler never gives any fuller or clearer explanation in words except in the case of equilibrium, the mathematical treatment of pressure is entirely explicit and clear."

In works on a general theory of deformable lines written between 1771 and 1774, Euler found, through studying the dynamic behavior of these rods, that the introduction of shear forces, in addition to the normal forces, was generally necessary (see Truesdell, 1956b). In 1771, he succeeded in deriving the complete system of equations of motion for elastic lines with finite displacements.

Coulomb's ideas (1773) converged with Parent's and Couplet's considerations. In his studies of the flexure of beams, the failure state of a brick column, and the development of the earth pressure theory, he assumed that normal as well as shear forces acted on the cut surfaces. In order to assess the strength, he referred these forces consistently to the loaded area, and consequently was the first to introduce shear stress. From experiments on sandstone from the Bordeaux area and bricks from Provence, Coulomb established that the tensile strength was approximately equal to the shear strength.

Contrary to Parent and Couplet, Coulomb (1773) also considered the inner forces on different planes and established that, in frictionless materials, the

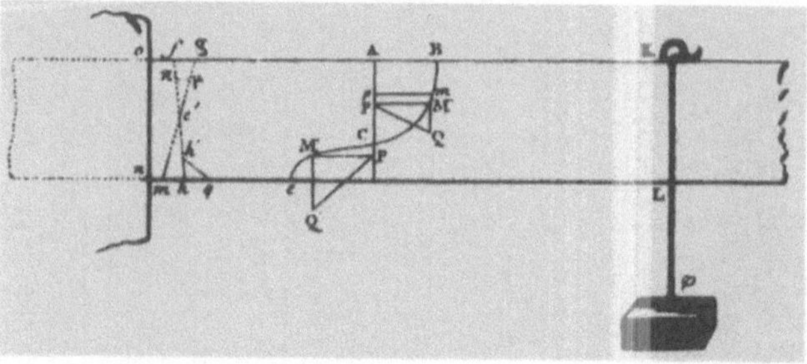

Fig. 3.2. Coulomb's investigations on the flexure of beams

failure occurred at 45°. In spite of the clarity of his discourses, he still did not go further than examining special boundary-value problems, such as transverse beams, columns, and the earth pressure on a retaining wall. He did not carry over his excellent idea regarding the introduction of the stress vector to general continuum. With the creation of the shear stress, Coulomb went further than Euler's concept of inner pressure, yet his contribution by no means reached the level of Euler's treatises on hydromechanics. It must be noted that Coulomb was an excellent engineer with great intuitive talent who left behind valuable ideas for his successors, however, his contribution should not be exaggerated in such a way that it is seen as having influenced the creation of the general stress concept.

3.1.2
The Final Step

The complete analysis of "inner forces" in loaded continua, as well as both the creation of the stress concept and the stress tensor, are due to the famous mathematician Cauchy (1823). In the course of his work on the elasticity theory, he quickly succeeded in carrying out the complete analysis of the inner force state in loaded solid continua, and also in deriving the stress tensor. In 1822, Cauchy presented the corresponding considerations from the later work of 1823 to the Academie Royal des Sciences.

"Moreover, in his calculations he assumed that all forces are vertical to the lines and areas on which they act. It seemed to me that both sorts of forces could be reduced to one single one, which should be called stress or pressure. This force seemed to be of the same nature as the hydraulic pressure of a stationary fluid on the surface of a solid body. However, this new pressure did not always remain vertical to the areas to which it was exposed and was not the same in all directions at one given point. In developing this idea, I soon arrived at the following conclusions.

If one envisages a solid element in a solid, elastic or non-elastic body which is restricted by certain areas and in some way loaded, then this element is under stress (traction or pressure) at every point of the surface. This stress is similar to that which occurs in fluids; the only difference being that the fluid pressure at one point is always vertical to the area (however orientated), while the stress at a given point of a solid body is generally oblique to

(10)

libre du plan élastique, avait considéré deux espèces de forces produites, les unes par la dilatation ou la contraction, les autres par la flexion de ce même plan. De plus, il avait supposé, dans ses calculs, les unes et les autres perpendiculaires aux lignes ou aux faces contre lesquelles elles s'exercent. Il me parut que ces deux espèces de forces pouvaient être réduites à une seule, qui devait constamment s'appeler tension ou pression, et qui était de la même nature que la pression hydrostatique exercée par un fluide en repos contre la surface d'un corps solide. Seulement la nouvelle pression ne demeurait pas toujours perpendiculaire aux faces qui lui étaient soumises, ni la même dans tous les sens en un point donné. En développant cette idée, j'arrivai bientôt aux conclusions suivantes.

Si dans un corps solide élastique ou non élastique on vient à rendre rigide et invariable un petit élément du volume terminé par des faces quelconques, ce petit élément éprouvera sur ses différentes faces, et en chaque point de chacune d'elles, une pression ou tension déterminée. Cette pression ou tension sera semblable à la pression qu'un fluide exerce contre un élément de l'enveloppe d'un corps solide, avec cette seule différence, que la pression exercée par un fluide en repos contre la surface d'un corps solide, est dirigée perpendiculairement à cette surface de dehors en dedans, et indépendante en chaque point de l'inclinaison de la surface par rapport aux plans coordonnés, tandis que la pression ou tension exercée en un point donné d'un corps solide contre un très-petit élément de surface passant par ce point, peut être dirigée perpendiculairement ou obliquement à cette surface, tantôt de dehors en dedans, s'il y a condensation, tantôt de dedans en dehors, s'il y a dilatation, et peut dépendre de l'inclinaison de la surface par rapport aux plans dont il s'agit. De plus, la pression ou tension exercée contre un plan quelconque se déduit très-facilement, tant en grandeur qu'en direction, des pressions ou tensions exercées contre trois plans rectangulaires donnés. J'en étais à ce point, lorsque M. Fresnel, venant à me parler des travaux auxquels il se livrait sur la lumière, et dont il n'avait encore présenté qu'une partie à l'Institut, m'apprit que, de son côté, il avait obtenu sur les lois, suivant lesquelles l'élasticité varie dans les diverses directions qui émanent d'un point unique, un théorème analogue au mien. Toutefois le théorème dont il s'agit était loin de me suffire pour l'objet que je me proposais, dès cette époque, de former les équations générales de l'équilibre et du mouvement intérieur d'un corps; et c'est uniquement dans ces derniers temps que je suis parvenu à établir de nouveaux principes propres à me conduire à ce résultat, et que je vais faire connaître.

Fig. 3.3. Cauchy's formulation of the stress concept

the surface element placed at this point and independent of the position of surface element. This stress can easily be derived from the stresses occurring in the three coordinate levels.

I spoke to M. Fresnel about this point, and he informed me that he had obtained an analogous theorem to mine own, with respect to the laws according to which the elasticity is variable in different directions. However, for the object which I had decided to tackle, this theorem was not enough to form general equations for equilibrium and motion inside a body. It is only recently that I have arrived at new principles which are suitable to lead me to this result.

From the theorem the very important statement results that the pressure or stress at each point equals unity, which is divided by the ray vectors of an ellipsoid. The three axes of this ellipsoid correspond to three pressures, or stresses, which we call principal stresses."

Cauchy linked his studies directly to the results in hydromechanics which had been completed by Euler. He did not mention Coulomb's studies or those of other authors (apart from Fresnel). In fact, Cauchy's analysis of the inner forces of solid continua is the logical continuation of Euler's studies of the inner pressure of ideal fluids. His considerations can be represented mathematically as follows. First, he introduced the stress vector $\mathbf{t}_n$ which acts obliquely to the cut surface. This stress vector is dependent on the material point considered, which is described by the local vector $\mathbf{x}$ and the position of the cut surface (determined by the unit area vector $\mathbf{n}$).

One could now develop a continuum theory with the stress vector $\mathbf{t}_n$, since the stress vector completely describes the state of the inner forces in a continuum. This would have the aggravating disadvantage that the quantity used in the calculations would be dependent on a special cut surface. Cauchy, however,

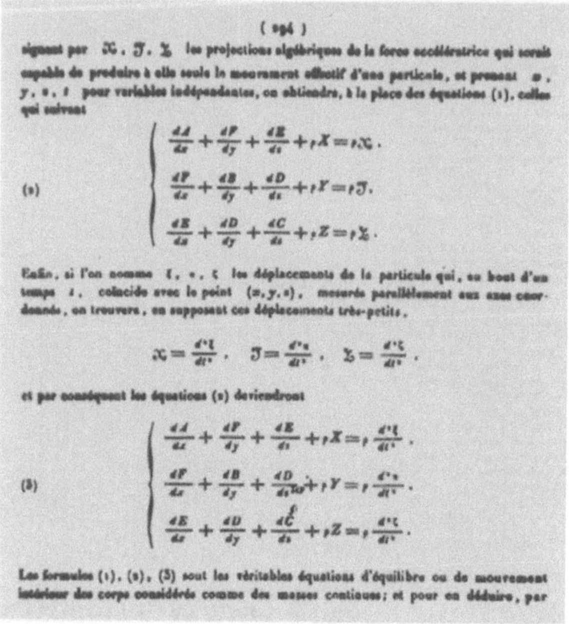

(294)

signant par $\mathfrak{X}, \mathfrak{Y}, \mathfrak{Z}$ les projections algébriques de la force accélératrice qui serait capable de produire à elle seule le mouvement effectif d'une particule, et prenant x, y, z, t pour variables indépendantes, on obtiendra, à la place des équations (1), celles qui suivent

$$(2)\quad \begin{cases} \dfrac{dA}{dx}+\dfrac{dF}{dy}+\dfrac{dE}{dz}+\rho X=\rho\mathfrak{X}, \\ \dfrac{dF}{dx}+\dfrac{dB}{dy}+\dfrac{dD}{dz}+\rho Y=\rho\mathfrak{Y}, \\ \dfrac{dE}{dx}+\dfrac{dD}{dy}+\dfrac{dC}{dz}+\rho Z=\rho\mathfrak{Z}. \end{cases}$$

Enfin, si l'on nomme ξ, η, ζ les déplacements de la particule qui, au bout d'un temps t, coïncide avec le point (x, y, z), mesurés parallèlement aux axes coordonnés, on trouvera, en supposant ces déplacements très-petits,

$$\mathfrak{X}=\frac{d^2\xi}{dt^2}, \quad \mathfrak{Y}=\frac{d^2\eta}{dt^2}, \quad \mathfrak{Z}=\frac{d^2\zeta}{dt^2}.$$

et par conséquent les équations (2) deviendront

$$(3)\quad \begin{cases} \dfrac{dA}{dx}+\dfrac{dF}{dy}+\dfrac{dE}{dz}+\rho X=\rho\dfrac{d^2\xi}{dt^2}, \\ \dfrac{dF}{dx}+\dfrac{dB}{dy}+\dfrac{dD}{dz}+\rho Y=\rho\dfrac{d^2\eta}{dt^2}, \\ \dfrac{dE}{dx}+\dfrac{dD}{dy}+\dfrac{dC}{dz}+\rho Z=\rho\dfrac{d^2\zeta}{dt^2}. \end{cases}$$

Les formules (1), (2), (3) sont les véritables équations d'équilibre ou de mouvement intérieur des corps considérés comme des masses continues; et pour en déduire, par

Fig. 3.4. Cauchy's development of the equations of motion

ingeniously succeeded in replacing the stress vector $\mathbf{t}_n$ by a linear mapping with the stress tensor $\mathbf{T}$ in the form:

$$\mathbf{t}_n = \mathbf{T}\mathbf{n}\,, \tag{3.1.1}$$

where the stress tensor $\mathbf{T}$ is the field function of the placement (described by the position vector $\mathbf{x}$) and thus a pure field function like the inner pressure introduced by Euler.

Truesdell (1956b) commented on Cauchy's introduction of the stress vector acting obliquely on the cut surface:

"Daß Cauchys Gedanke einfach ist, zeigt seine Originalität um so mehr. Ihn gefaßt zu haben, ist eine Leistung von Eulerischer Tiefe und Klarheit."

"That Cauchy's idea is simple shows his originality all the more. To have formed it, is an achievement of truly Eulerian profoundness and clarity."

This comment is undoubtely even more valid for the second part of the study, the creation of the stress tensor. With the stress tensor, Cauchy (1828a, b) was able to complete Euler's ideas in his following works. He evaluated Euler's principle of linear momentum and the principle of moment of momentum (Cauchy, 1828a, b) and obtained with the mass density ρ, the external acceleration $\mathbf{b}$, and the acceleration $\ddot{\mathbf{x}}$, the partial differential equation:

$$\operatorname{div}\mathbf{T} + \rho\mathbf{b} = \rho\ddot{\mathbf{x}}\,, \tag{3.1.2}$$

and the symmetry of the stress tensor,

$$\mathbf{T} = \mathbf{T}^T\,. \tag{3.1.3}$$

Fig. 3.5. Augustin-Louis Cauchy (1789–1857)

These two equations are now called Cauchy's first and second laws of motion. They are valid for smooth fields with the absence of moment stresses and thus cover a large range of continuum mechanics. Cauchy's studies represent a landmark in the history of mechanics; they stand as a victory for physical-mathematical research.

3.1.3 Biographical notes

Baron Augustin-Louis Cauchy (1789–1857)

The French mathematician Baron Augustin-Louis Cauchy was born on April 21, 1789, in Paris. His father was a successful lawyer in government service. During the French Revolution, his father had to leave Paris and take refuge in the small village of Arcueil (near Paris), where the famous scientists Laplace and Bertholet were also living at the time. After Napoleon had taken control of France, Laplace's house became the meeting place for many outstanding scientists from Paris. The young Cauchy got to know many of them, including J.L. Lagrange, who very quickly recognized his great mathematical talent.

After having received private lessons from his father for some years, at the age of thirteen Cauchy was admitted to the l'École Centrale du Panthéon, where his achievements were brilliant and extraordinary. At the age of fifteen, he won the "grand prix d'humanités", which had been endowed by the Emperor. The following year, 1805, he passed the entrance exam to the École Polytechnique after a preparatory period of ten months. During his years of study, he proved to have a great competence for mathematics. After having completed this curriculum, Cauchy decided in 1807 to enroll in the École des Ponts et Chaussées in order to study engineering problems. He completed his studies in 1810 as the best student of that year. He was thereupon employed as an engineer in Cherbourg, where he worked in the area of harbor construction.

He continued, however, to dedicate himself to mathematics and published some important contributions. For health reasons, he took a leave of absence in 1813 and, after his return to Paris, he declared that he wanted to give up working as an engineer on building sites and to dedicate himself totally to mathematics henceforth. He was not yet twenty-four years old, but his reputation as an excellent scholar was already widespread; he was soon to astonish Europe's greatest mathematicians with an abundance of brilliant discoveries. He became a member of the Academy of Sciences in 1816.

From 1816 to 1830, he simultaneously gave courses on mechanics at l'École Polytechnique, on higher algebra at the *faculté des sciences*, as well as courses on physics and mathematics at the *Collège de France*. In his lectures on analysis, he tried to bring a more rigorous form to this subject, in a way that had not been done previously.

The originality of his presentation did not attract only students, but also professors and scientists from outside. The publishing of his Cours d'analyse de l'École Polytechnique had an important influence on the further development of mathematics.

At this time, Navier presented his first work on the elasticity theory to the Academy of Sciences. Cauchy took great interest in this work and he began his own studies in this field. In a short period of time he produced, as we have already seen, the foundations of continuum mechanics. On September 30, 1822, he presented his discourse on the stress concept to the Academy.

In 1826, Cauchy edited the first volumes of the *Exercises mathématiques*, in which he had solved several difficult mathematical problems by new and original methods. These publications were interrupted in 1830, the year that the so-called "July-Revolution" broke out. Cauchy's religious beliefs hindered him from swearing the oath of allegiance to the new government. Rather, he preferred to leave the university. Moreover, he did not even want to stay in France while "his king" lived in exile. In spite of all requests by his parents and his wife to stay in his country, he moved to Fribourg and then to Turin in 1831, where the king of Sardinia created a professorship in "physique sublime" for him. He lived there for two years. During this time Cauchy wrote his papers *Résumés analytiques* and *Memoir sur la mechanique céleste et sur un nouveau calcul appelé calcul des limites*. In 1833, Charles X invited him to give the Duke of Bordeaux (Count of Chambord) private lessons. Cauchy accepted willingly and for the next five years he lived in Taplitz, Kirchberg, Goritz, and Prague. He became a member of the Academy of Science in Prague. There he edited his memorandum on the dispersion of light, which later became famous. In 1838, the lessons with the Duke of Bordeaux were finished and Charles X conferred on Cauchy the title of Baron. Cauchy then moved back to France. He was able to return to his old position at the Institute, since the members were no longer forced to pledge the oath of allegiance. He taught mathematics at various places and was busy in his leisure time with editing the collections of the Academy of Sciences and other publications. He also published a great number of memoranda, notes, and reports on various problems of pure and applied mathematics.

In 1838, his colleagues offered him a free position at the Collège de France which, however, he refused. In the same year, the members of the Bureau des longitudes elected him as the successor of Prony, known as one of the founders of the École Polytechnique and Professor of Mechanics. For four years, Cauchy carefully exercised the functions of his new position. He was appointed Professor of Mathematical Astronomy by the Republic in 1848. However, the oath of allegiance was re-introduced and Cauchy had to leave his chair again. Finally, he gave up his professorship in 1854. Three years later, in 1857, Cauchy died.

Cauchy had two brothers. The first, Alexandre-Laurant, was born in 1792 in Paris, and died there in 1857. At first, he studied mathematics; later, he studied law. He had a brilliant career, becoming president of the chamber of the court of appeal in 1847 and, in 1849, consultant at the Supreme Court.

The second brother, Eugène-Francois, was born in Paris in 1802, and died there in 1877. He was a lawyer and publicist and also achieved great eminence.

Augustin-Louis Cauchy was, without doubt, one of the most brilliant mathematicians and mechanicians ever. His contributions to mechanics are overwhelming. He not only rounded off Euler's mechanics, but also made valuable contributions to the elasticity theory, as will be seen in the next chapter. He, however, did not write voluminous articles. Rather, he published his findings in the form of memoirs and notes; 785 of these he published in the *Correspondance et le journal de l'école, Bullentin des sciences, Bullentin de la société philomatique, Journal de mathematique* and in different collections of the Academy of Sciences. The collection of the Academy contains more than 500 memoirs and notes from 1839 to 1857. The budget of the Academy was too small to have them printed.

During his life-time, Cauchy became a famous man. He was a member of foreign academies of sciences in Berlin, Saint Petersburg, Boston, Stockholm, Modena, Naples, and Palermo, and the royal societies of Edinburgh, Copenhagen, and Göttingen.

Cauchy was a very dogmatic man. He preferred to leave his country than disclose his political conviction. He also sympathized with the Jesuits and defended their claims. His uncompromising behavior brought him criticism, as well as praise. Apart from his papers on mathematics and mechanics, Cauchy wrote articles such as "Remarks on The Religious Order" in 1843, and "Reflections on The Freedom of Lessons" in 1844. Moreover, Cauchy was actively involved in the organization of the foundation of the holy Francois-Régis, and his functions at the Catholic institute, where he led the committee of sciences, made him unpopular. He was considered a fanatic and intolerant clergyman.

The attacks against Cauchy did not spare his scientific work; this was attacked even after his death. He was also criticized, with some justification, for following his brilliant intellect too much, and formulating his articles too quickly, instead of letting his ideas rest and thinking them over. One must also state that he sometimes jumped from one place to another in science and that he scattered his findings over a great number of articles.

However, Cauchy's character weaknesses fade in view of his great discoveries in mechanics and mathematics. Although he did not always formulate his findings exactly, he gave his successors sufficiently exact descriptions of his discoveries.

With Cauchy's creation of the stress concept, the way to the development of continuum theories of materials was clear. At the same time, the foundation equations for linear elasticity theory and viscous fluids were being worked upon. Much later, Coulomb's ideas on the description of the failure state of brittle and granular materials would be taken up again and, with the use of Cauchy's stress tensor, they would be transferred to three-dimensional continua. In the following period, boundary-value problems, of importance in engineering, were treated. Some of these will be discussed in the following chapters.

3.2 The Development of the Linear Elasticity Theory

The classical linear elasticity theory includes the investigation of motion and, as a special case, the examination of the equilibrium of linear-elastic bodies, where the relation between the stresses and deformations which occur due to external forces are given by the generalized Hooke's law. The mathematical formulation of this theory is a creation of the nineteenth century. The mathematical description of the linear-elastic behavior of an elastic body consists of the derivation of so-called fundamental equations, which are gained as follows: The first of Cauchy's equations of motion serves as a basis. The stress components therein are replaced by distortion components with the help of the generalized Hooke's law. Finally, geometrical relations, for example distortion-displacement relations, replace the distortion components by the derivation of displacements, which results in a system of three partial differential equations. Taking a system with the coordinates x, y, z as a basis, the system of partial differential equations for the displacement components u, v, w (in the x, y, z direction) of a material point inside an elastic body is as follows:

$$
\begin{aligned}
\frac{E}{2(1+\nu)}\left(\Delta u+\frac{1}{1-2\nu}\frac{\partial\Theta}{\partial x}\right)+\rho X &= \rho\frac{\partial^2 u}{\partial t^2}, \\
\frac{E}{2(1+\nu)}\left(\Delta v+\frac{1}{1-2\nu}\frac{\partial\Theta}{\partial y}\right)+\rho Y &= \rho\frac{\partial^2 v}{\partial t^2}, \qquad (3.2.1)\\
\frac{E}{2(1+\nu)}\left(\Delta w+\frac{1}{1-2\nu}\frac{\partial\Theta}{\partial z}\right)+\rho Z &= \rho\frac{\partial^2 w}{\partial t^2},
\end{aligned}
$$

with the Laplace Operator $\Delta = \frac{\partial^2}{\partial x^2} + \frac{\partial^2}{\partial y^2} + \frac{\partial^2}{\partial z^2}$, and the volume strain $\Theta = \frac{\partial u}{\partial x} + \frac{\partial v}{\partial y} + \frac{\partial w}{\partial z}$.

Furthermore, in (3.2.1), E is Young's modulus and ν Poisson's ratio. The quantities X, Y, and Z denote the external accelerations in the x, y, and z directions. The remaining quantities ρ and t denote density and time. For a complete description of the mechanical behavior of a linear-elastic body, the corresponding relations existing on the boundary of the body must be stated (i.e., boundary conditions). However, this will not be dealt with in the narrative of the historical developments.

In order to derive the fundamental equations (3.2.1), several elements are necessary (as indicated above): Namely, Cauchy's first equation of motion, Hooke's generalized law and geometrical relations (for example, relations between distortions and displacements). The first of Cauchy's equations of motion (see previous section) was established in the early 1820s. Hooke's generalized law and the relation of distortion and displacement were obtained in connection with the completion of the linear elasticity theory.

Although the way to the creation of the linear elasticity theory seems to have been paved by previous versions, its historical development actually proceeded in a very roundabout way. Evidently, there was not sufficient insight into fundamental physical facts and the accompanying constants. The explanation for all of this was apparently to be the molecular theory of the elastic material behavior corresponding to the celestial mechanics which had been completed perfectly by Laplace.

3.2.1 Theoretical Molecular Formulations

The first important contribution to the linear-elastic theory to go in the direction of the molecular theory originated when Navier presented his work on this topic to the French Academy of Sciences on May 14, 1821 (see Navier, 1827).

Navier presented the foundation equations of elasticity as equations of motion for the individual particles. His assumptions were as follows: In the deformed state of an elastic body, a force Π acts between two molecules. This force is cumulatively split up into a central force π and a force π', which are activated by applied forces. Navier did not consider the central forces any further because their summation disappears over all the molecules.

According to his new "principle", Navier required a "constitutive equation", for the forces π', i.e., that these forces be proportional to the relative distances between the molecules of the deformed body. These relative distances fall, however, approximately into the directions of the original distances due to the smallness of the displacements. He then expanded the displacements u, v, w in x, y, z direction of an orthogonal coordinate system (which mark the relative distances) in series and considered quantities of the third-order and higher as being disregardably small.

Moreover, by assuming that the molecular forces were the same in all directions, i.e. that the body was isotropic, Navier succeeded in determining the force π' after introducing polar coordinates by integration over the molecular sphere of operation. As a result of his assumption of isotropy, Navier's proportionality factors, which are dependent on the distance of the molecules, are reduced to only one factor. Thus, Navier succeeded in expressing the molecular forces which are activated by outside forces as the constant ε and as displacement equations of the second-order.

With the help of the principle of linear momentum, Navier was then able to state the three equations of motion in the displacements u, v, and w:

$$\begin{aligned} \varepsilon\left(\Delta u + 2\frac{\partial\Theta}{\partial x}\right) + \rho X &= \rho\frac{\partial^2 u}{\partial t^2}, \\ \varepsilon\left(\Delta v + 2\frac{\partial\Theta}{\partial y}\right) + \rho Y &= \rho\frac{\partial^2 v}{\partial t^2}, \\ \varepsilon\left(\Delta w + 2\frac{\partial\Theta}{\partial z}\right) + \rho Z &= \rho\frac{\partial^2 w}{\partial t^2}. \end{aligned} \tag{3.2.2}$$

With the principle of virtual work, Navier derived (rather laboriously) the boundary condition on the body surface. The boundary conditions contain the quantity ε as the only material-dependent constant.

With his fundamental equations (3.2.2), Navier was very close to the actual fundamental equations of the linear elasticity theory (3.2.1). The main shortcoming was the lack of a second material-dependent constant.

The fundamental equations for an anisotropic elastic medium can already be found in Navier's aforementioned studies, if one understands the fifteen coefficients mentioned therein as fifteen different constants.

The concept of the anisotropic elastic body was first developed by A. Fresnel (see Todhunter and Pearson, 1886/1893). In the course of his studies on the double refraction of light in crystals, Fresnel came to the conclusion that "elasticity" can act differently in different directions.

The explicit derivation of the fundamental equations for anisotropic elastic continua is due to Cauchy (1828b). He retained the model of the crater-shaped molecules but, with respect to the central force Π, he went beyond Navier's assumption and tried to relate the central force as acting in a natural condition. Cauchy restricted himself to small alterations of form and assumed that, with regard to each molecule, the body had a central symmetry. In this manner, he succeeded in setting up a system of fundamental equations of three partial differential equations with twenty-one constants. The twenty-one constants are reduced in special cases of symmetry. In the case of isotropy, they are reduced to only two. From Cauchy's studies, the fundamental equations for isotropic

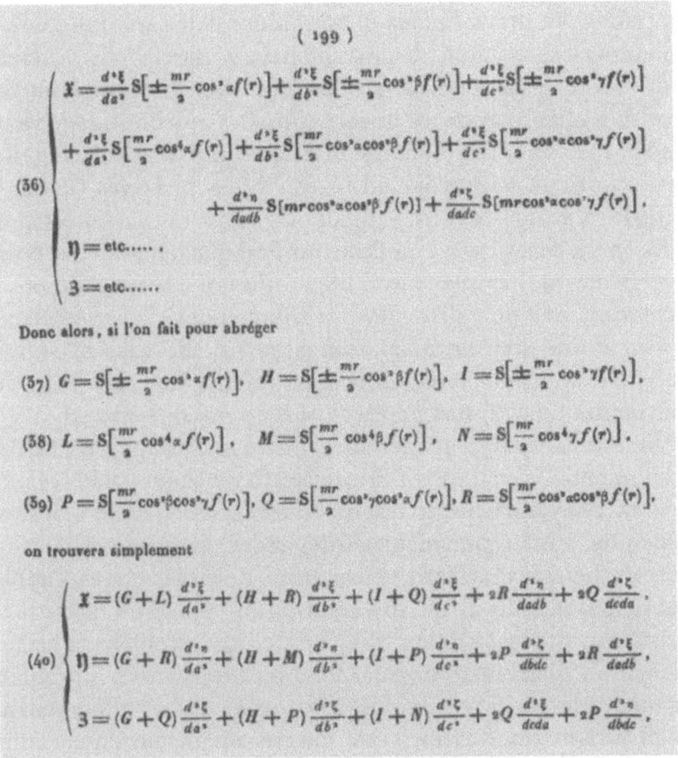

(199)

$$(36)\quad \begin{cases} \mathfrak{X} = \frac{d^2\xi}{da^2} S\left[\pm\frac{mr}{2}\cos^2\alpha f(r)\right] + \frac{d^2\xi}{db^2} S\left[\pm\frac{mr}{2}\cos^2\beta f(r)\right] + \frac{d^2\xi}{dc^2} S\left[\pm\frac{mr}{2}\cos^2\gamma f(r)\right] \\ \quad + \frac{d^2\xi}{da^2} S\left[\frac{mr}{2}\cos^4\alpha f(r)\right] + \frac{d^2\xi}{db^2} S\left[\frac{mr}{2}\cos^2\alpha\cos^2\beta f(r)\right] + \frac{d^2\xi}{dc^2} S\left[\frac{mr}{2}\cos^2\alpha\cos^2\gamma f(r)\right] \\ \quad + \frac{d^2\eta}{dadb} S[mr\cos^2\alpha\cos^2\beta f(r)] + \frac{d^2\zeta}{dadc} S[mr\cos^2\alpha\cos^2\gamma f(r)], \\ \mathfrak{Y} = \text{etc.}\ldots, \\ \mathfrak{Z} = \text{etc.}\ldots \end{cases}$$

Donc alors, si l'on fait pour abréger

$$(37)\quad G = S\left[\pm\frac{mr}{2}\cos^2\alpha f(r)\right],\quad H = S\left[\pm\frac{mr}{2}\cos^2\beta f(r)\right],\quad I = S\left[\pm\frac{mr}{2}\cos^2\gamma f(r)\right],$$

$$(38)\quad L = S\left[\frac{mr}{2}\cos^4\alpha f(r)\right],\quad M = S\left[\frac{mr}{2}\cos^4\beta f(r)\right],\quad N = S\left[\frac{mr}{2}\cos^4\gamma f(r)\right].$$

$$(39)\quad P = S\left[\frac{mr}{2}\cos^2\beta\cos^2\gamma f(r)\right],\ Q = S\left[\frac{mr}{2}\cos^2\gamma\cos^2\alpha f(r)\right],\ R = S\left[\frac{mr}{2}\cos^2\alpha\cos^2\beta f(r)\right].$$

on trouvera simplement

$$(40)\quad \begin{cases} \mathfrak{X} = (G+L)\frac{d^2\xi}{da^2} + (H+R)\frac{d^2\xi}{db^2} + (I+Q)\frac{d^2\xi}{dc^2} + 2R\frac{d^2\eta}{dadb} + 2Q\frac{d^2\zeta}{dcda}, \\ \mathfrak{Y} = (G+R)\frac{d^2\eta}{da^2} + (H+M)\frac{d^2\eta}{db^2} + (I+P)\frac{d^2\eta}{dc^2} + 2P\frac{d^2\zeta}{dbdc} + 2R\frac{d^2\xi}{dadb}, \\ \mathfrak{Z} = (G+Q)\frac{d^2\zeta}{da^2} + (H+P)\frac{d^2\zeta}{db^2} + (I+N)\frac{d^2\zeta}{dc^2} + 2Q\frac{d^2\xi}{dcda} + 2P\frac{d^2\eta}{dbdc}. \end{cases}$$

Fig. 3.6. Cauchy's derivation of the fundamental equations

continua are as follows:

$$\begin{aligned} (R+A)\Delta u + 2\frac{\partial\Theta}{\partial x} + \rho X &= \rho\frac{\partial^2 u}{\partial t^2}, \\ (R+A)\Delta v + 2\frac{\partial\Theta}{\partial y} + \rho Y &= \rho\frac{\partial^2 v}{\partial t^2}, \\ (R+A)\Delta w + 2\frac{\partial\Theta}{\partial z} + \rho Z &= \rho\frac{\partial^2 w}{\partial t^2}. \end{aligned} \tag{3.2.3}$$

Cauchy explicitly remarks that for $A = 0$, his equations agree with those of Navier, where $R = \varepsilon$ (see (3.2.2)).

Cauchy made a mistake with regard to the number of constants in his molecular theory studies. Six constants in anisotropic elastic behavior and one constant in isotropic elastic behavior do not depend on the elastic nature of the continua, but represent the initial stresses. These disappear if the natural state is chosen as the starting point, since the natural state is marked as being stress-free.

Further works on the molecular theory of elastic bodies came from S. D. Poisson. The most important results are to be found in two memoirs (Poisson, 1829,

1831) which he also published in his *Course of Mechanics.* Poisson's fundamental equations and boundary conditions do not differ very much from Navier's and Cauchy's derivations. The value of his studies, however, lies partly in the fact that, apart from his examination of special initial- and boundary-value problems, he succeeded in solving the fundamental equations and showing that interference in elastic continua is propagated in two types of waves (Poisson, 1830). One type of wave is propagated as a dilatation wave (i.e., connected to volume changes). This is the fastest wave, and the motion of each material point is normal to the wave front. In the other wave, the motion of a material point is tangential to the wave front, and only distortions without volume changes occur.

Poisson (1829) also found that an axial elongation in the case of simple tension of a prismatic bar was associated with cross-contraction. He fixed the size of the cross-contraction at 1/4, independent of the type of material.

Similarly, in the first section of their memoirs, Lamé and Clapeyron (1833) also derived the fundamental equations of linear elasticity theory with regard to the molecular theory. They did not expand on Navier's and Cauchy's derivations; their equations only contain one material-dependent constant. The value of these memoirs lies in the fact that, for the first time, a whole series of applications of the fundamental equations for the solution of problems relevant to practice were discussed. In the last section, they treated the problem of elastic semi-infinite bodies with a distributed normal load on the surface, a problem of some interest in theoretical soil mechanics. They used Navier's theorem to depict the surface load and obtained explicit relations for the displacement components. Finally, they carried out a similar analysis for an infinitely large body limited by infinitely long planes.

These memoirs were very important for their time. They contained all the results already gained in the linear elasticity theory, clearly and lucidly presented. Lamé's and Clapeyron's memoirs later served Lamé (1852) as a base for his famous book *Leçons sur la théorie mathematique de l'élasticité des corps solides*, which can be regarded as the first book on the elasticity theory. It is divided into twenty-four lectures. This set-up is extremely clever didactically, because each lecture treats a fully defined part of the elasticity theory. The contents of the book can be roughly divided into four branches. The first branch contains the general equations of the linear elasticity theory with respect to rectangular coordinates. The second deals with various applications of the previously introduced equations. In the third part, the general equations are made relevant for cylindrical and spherical coordinates. The final part consists of applications of the elasticity theory to the double refraction of light.

Although Lamé used the memoirs written by himself and Clapeyron as a base, in his own book he changed some of the equations, having by then arrived at the conclusion that two constants were necessary for the description of the isotropic material attributes. Of all the examples of the application treated in Lamés book, the wave propagation in an elastic medium is the most interesting for theoretical soil mechanics.

Fig. 3.7. Gabriel Lamé (1795–1870)

With this work, Lamé (1852) produced an excellent textbook. Todhunter and Pearson (1886/1893) commented:

"The mathematical investigations are clear and convincing, while the general reflections which are given so liberally at the beginning and the end of the Lectures are conspicuous for their elegance of language and the depth of thought. The work is eminently worthy of a writer whom Gauss is reported to have placed at the head of French mathematicians, and whom Jacobi described as *un des mathématiciens les plus pénétrants*."

Lamé continued to be interested in the elasticity theory after the publication of his book. After he had studied the deformations of spherical shells under the effect of random surface forces (Lamé, 1854), he published a further book titled *Leçons sur les coordonées curvilignes et leurs diverses applications* (see Todhunter and Pearson, 1886/1893). Here, he developed a general theory of curvilinear coordinates and their applications in mechanics and the elasticity theory. As an example of such an application, he treated the deformation of spherical shells. In the last chapter, Lamé discussed the principles of the elasticity theory. His main idea here seems to have been that one really does not know anything about molecules or molecular effects, and that one cannot therefore rely on a theory which is based on molecular theory considerations. In this way, he departed from Navier's derivations relating to this subject.

In his conclusions, Lamé shared Cauchy's opinion that the stress components of more than thirty-six components were connected to the first derivations of the displacement of linear molecules. With respect to the homogeneous isotropic body, he remarked that the two lemmas stated in his earlier book (Lamé, 1852), which were based on old ideas, had not been correctly introduced. These two lemmas involved simple traction and simple torsion. Lamé reformulated them and remarked that the introduction of formulas based on homogeneous, isotropic bodies "est complétement dégagé de toute hypothése, toute idée préconçue." Todhunter and Pearson (1886/1893) remarked critically:

"Lamé's statement here, together with that on p. 359, with regard to the easy establishment of the linearity of the stress-strain relation seem to me unsatisfactory. His lemmas do not *definitely* appeal to any physical axiom, and we have, precisely as in the case of Green, the apparent miracle of the theory of an important physical phenomenon springing created from the brain of the mathematician without any appeal to experience. The physical axiom or hypothesis of molecular force which Lamé uses in his *Lecons sur l'Elasticité* and which would undoubtedly have led him to rari-constancy if carried out is here dropped, and the only bridge over the void between the pure theory of quantity and the physical phenomenon is formed by these two lemmas, based upon considerations of symmetry, and a tacit assumption that the most sensible terms in stress are linear in strain. I have dwelt on this point, because we find even in mathematicians of the standing of Lamé not infrequently an omission to state clearly the physical principle upon which they based their calculations of a physical phenomenon."

Another leading advocate of the rari-constancy theory on the basis of the molecular theory was B. de Saint-Venant, who did so very much to convert the scientific results of the elasticity theory into practice. He was also the first university lecturer to explain the new developments of the elasticity theory to his students. His studies on flexure, and in particular on torsion, are well-known. In his treatment of the torsion of prismatic rods, he corrected older ideas and allowed for warping of the cross-section. His studies were so original that his name was given to the theory of torsion with allowance for warping. There is no area of the elasticity theory to which de Saint-Venant did not make a contribution, whether it be a discussion of the foundations of the elasticity theory or the treatment of oscillation on thrust problems.

In later years, de Saint-Venant turned to a totally new field: the plasticity theory. He recognized the fundamental importance of H. Tresca's experimental and theoretical studies and discussed them in several publications. He then went even further than Tresca by postulating a flow rule for two-dimensional flow. We will return to this problem later. Apart from the area of the kinematics of motion, de Saint-Venant's contributions to the general equations of the elasticity theory were not very great. In 1861, he was the first to derive the compatibility conditions for finite deformations (see Todhunter and Pearson, 1886/1893). Although St. Venant never published his numerous studies in book form, his considerations on elasticity are well-documented. More specifically, they can be found in the form of extensive annotations in Navier's book *Résumé des leçons...*(see Timoshenko, 1953) and of Clebsch's *Théorie de l'élasticité des corps solides* (see Timoshenko, 1953), which St. Venant edited completely and translated. In addition, he described the history of the development of the equations of the elasticity theory in Moigno's book *Leçons de mécanique analytique statique* (St. Venant, 1868).

Further advocates of the molecular theory, although this was partially during the early stages of their scientific activity, were J.M.C. Duhamel (1797–1872) and F. Neumann (1798–1895).

3.2.2 Continuum Mechanics Approach

The first continuum mechanics approach to the description of linear-elastic behavior can be credited to Cauchy. As was discussed in Section 3.1, on September 30, 1822, Cauchy read his treatise on the stress concept, and later, in 1824, stated what are today called Cauchy's first and second laws. The analysis of the stress concept necessitates the analysis of the deformation concept, which flows into the concept of the distortion tensor.

Cauchy (1828a, b) first developed this concept for the arbitrary displacements u, v, and w which, like the stress tensor, he related to the deformed system. Cauchy later went on to the assumption of small displacements and the displacement derivations. He explicitly stated that the six components of the distortion tensor in an orthogonal system were functions of the displacement derivatives: thus creating the geometrical relations (distortion and displacement relations). Cauchy also showed the existence of principal distortion and distortion ellipsoids in the distortion tensor.

In order to formulate the fundamental equations in the displacement components u, v, and w, a suitable relation between the stress and the distortion components (thus, also, the displacement components) was needed, as has already been mentioned in Section 3.1. Such a relation (also called a constitutive equation) was gained from a suitable interpretation, or generalization, of the law of linear dependence of force and displacement of a spring, which was first established by Hooke (1678). There are various interpretations of this law (*Ut tensio sic vis*), depending on how one interprets the words *tensio* and *vis*. If one interprets Hooke's law in such a way that only the linear dependence of a deformation and its cause are determined, then Jakob Bernoulli's (1691, 1694) law of the proportionality of the curvature of a bent beam and the acting bending moment can be seen as a suitable interpretation of Hooke's law.

Generally, Young (1807) is considered to have formulated Hooke's law precisely and to be the first to introduce the elasticity model. This, however, is hotly disputed by Truesdell (1969).

For isotropic linear-elastic continua, Cauchy (1828a, b) was the first to develop "Hooke's law". He concluded that the axes of principal stresses and principal strains coincide. He then placed the principal stresses proportional to the principal distortions, and obtained a material constant. He later modified this statement in so far as he assumed the principal stresses to be composed of two terms of a sum, the first of which was proportional to the principal distortion and the second to the volume strain (see Todhunter and Pearson, 1886/1893). In today's terminology, this means:

$$\mathbf{T} = k\mathbf{E} + K(\mathbf{E} \cdot \mathbf{I})\mathbf{I} \,, \tag{3.2.4}$$

where **T** and **E** are the stress and strain tensors and **I** is the identity tensor. The material-dependent constants desired by Cauchy are termed k and K. They are connected with the material-dependent values used today, the shear modulus G, Young's modulus E, and Poisson's ratio ν, as well as Lamé's constants λ and μ, as follows:

$$k = 2G = \frac{E}{1+\nu} = 2\mu\,; \quad K = \frac{E\nu}{(1+\nu)(1-2\nu)} = \frac{2G\nu}{1-2\nu} = \lambda\,. \tag{3.2.5}$$

Cauchy remarked that the constitutive relation (3.2.4) could be extended by another arbitrary constant in order to determine initial stress.

With the constitutive equation (3.2.4), in addition to the stress and distortion tensors and Cauchy's first equation of motion, the system of equations for deriving the fundamental equations for linear elastic continua was complete. The creation of the stress and distortion tensors, the development of Cauchy's first and second equations of motion, as well as the presentation of the constitutive relation (3.2.4), are of great historical importance in the development of mechanics. In the time to follow, it only remained to develop the linear elasticity theory with respect to general formulas and to solve concrete boundary-value problems.

The constitutive equations for anisotropic linear-elastic bodies were given by Poisson (1831) and Cauchy (1829). Both came to the conclusion that, for the description of anisotropic behavior, thirty-six material-dependent constants were necessary, as long as initial stress was excluded. Assuming the existence of the elastic potential functions (see Green, 1839), the number of material-dependent constants can be reduced to twenty-one.

3.2.3 Completion of the Theory

The dispute as to whether either the molecular theory or continuum mechanics was the correct way to proceed lasted almost the entire nineteenth century. The questions were: Firstly, was elastic isotropy marked by one or two constants, and secondly, was elastic anisotropic marked by fifteen or twenty-one constants? A rari-constancy theory and a multi-constancy theory were both discussed (see Todhunter and Pearson, 1886/1893). Most of the older scientists (Navier, Cauchy, Poisson, as well as Lamé and Clapeyron in their first works) were followers of the rari-constancy theory. It was only through the influence of the English school of thought, which brought the concept of the elastic potential function into the foreground (e.g., Green, Stokes, and Thomson), that it was later decided to adopt the multi-constancy theory, especially as test results also confirmed this theory (see Voigt, 1887, 1888a, b).

Why was so much time spent on molecular theory considerations, in particular, by the most outstanding mechanics specialists and mathematicians of the epoch? One of the reasons must have been the temptation of gaining the constitutive relation for isotropic and anisotropic elastic continua directly from pure mathematical studies and simple mechanical principles. It was only later realized that Hooke's *generalized law* is an assumption, and that the foundation of the linear relation had to be supported by experiments.

The completion of the linear elasticity theory took several courses. First, several extensive energy considerations were made. In this respect, studies on the setting up of the elastic potential were extensive. On the basis of a molecular-theoretical foundation, Navier (1827) and Green (1839, 1842) postulated the existence of an elastic potential function whose negative differential expressed the sum of the mechanical work which was carried out by the inner forces of small virtual displacements of a material point.

Green assumed that the elastic potential was a function of the six distortion components and finally presented the fundamental equations through the elastic potential function, rather than through the displacements. The proof of the existence of the elastic potential function for the two special changes of state of the isothermal and adiabatic deformations orginated with W. Thomson (1859), who proved its existence by thermodynamic studies.

Kirchhoff's (1859) proof of the uniqueness of the solutions of boundary-value problems was based on energy relations. In his demonstration, Kirchhoff limited himself to the case of a simple continuous body. Volterra (1905a, b), however, extended the study to multiple continuous elastic systems. Additionally, the question as to whether the solution of the fundamental equations of the linear elasticity theory even existed was posed in the last century.

The principle of the minimum of potential energy appears to go back to Castigliano (1875, 1879), who based his studies on Green's work. Castigliano's studies began to have some influence in engineering only after they were summarized by Lorenz (1913) and applied to numerous problems.

The principle of the minimum of potential energy can be formulated as follows: of all the displacements that take the given values of the surface displacements, it is the one that actually occurs which makes the potential energy a minimum. Here, potential energy should be taken to mean the inner deformation energy (elastic potential function), minus the mechanical work of the volume and surface forces, where only the surface forces find entry into the expression for the potential energies which are known on the surface of the body. The minimum principle has proven to be excellent in approximate solutions in continuum and structural mechanics. It is the basis for the Ritz (1909) method.

While the principle of the smallest potential energy states a minimum property of elastic displacements, in the so-called Castigliano's principle, a minimum property of stresses is formulated: from all stress systems which are at equilibrium with the given volume forces on the inside of an elastic body, and on the surface with the given surface forces, that system actually occurs for which the supplementary mechanical work will be a minimum.

Betti's (1872) reciprocity theorem has found useful applications in the elasticity theory. If an elastic body is subject successively to two systems of external forces, then the mechanical work of the forces of the first system (if as displacements those of the second system are chosen) is equal to the mechanical work of the forces of the second system (if the mechanical work is now formed with the displacements of the first system).

In the studies undertaken up until then (apart from Castigliano's principle), the fundamental equations and the energy considerations were formulated in the displacement components. If the displacement components were gained as solutions from the fundamental equations, then the stress components had to be determined from the constitutive equation, i.e., Hooke's law.

The idea then occurred to formulate the fundamental equations so that the solutions directly yielded the stress components. The first scientist to take this path seems to have been the Englishman, Airy (1863), who presented a work concerning a rectangular beam as a two-dimensional problem to the Royal Society. Airy showed that both equilibrium conditions were fulfilled by one function if suitable second derivatives referring to the two coordinates were set equal to the stress components. He developed the stress function in the form of a polynomial series and determined the coefficients of that polynomial in such a way as to satisfy the condition of the boundary. With the function determined in this way, the stress components were directly determined. Airy did not notice that the compatibility condition (we will come back to this later) was yet to be fulfilled. However, because Airy was the first to use the stress function, it rightly bears his name.

The final partial differential equation of the fourth-order of Airy's stress function, with regard to the compatibility condition, came from Mitchell (1900). In addition, Maxwell (see Love and Timpe, 1907) showed that the general solution of the fundamental equations of the linear elasticity theory could be determined by three stress functions.

The fundamental equations were formulated consistently in the stress components by E. Beltrami (1892). The points of departure for the formulation are the compatibility conditions which link the distortion components. The compatibility conditions, a total of six equations between which, as E. Beltrami (1892) remarked, three identical relations exist, were first derived by G. Kirchhoff (1859), for the linear theory, and by B. de Saint-Venant (see Todhunter and Pearson, 1886/1893), for the finite theory. The method for deriving Beltrami's equations is as follows: in the six compatibility conditions, the distortion components are expressed, with the help of the generalized Hooke's law, by the stress components. These six equations, however, only count as three, because, as has already been mentioned, they are linked by three differential relations. Thus, Beltrami's equations with the three equilibrium conditions are just enough to determine the six unknown stress components, whereby the boundary conditions must naturally be considered.

Energy considerations, which we have followed here only in the case of equilibrium, have also been employed for the state of motion. However, we will not go into this any further. The same applies for the inclusion of thermal effects, approaches for the expansion of the linear theory to physical and geometrically non-linear problems, and special cases such as structures. In the final chapter on the linear elasticity theory, we will deal with the historical development of the creation of some solutions important in some ranges of the theory of porous media.

3.2.4
Some Solutions of the Fundamental Equations

Two solutions of the fundamental equations which had already been obtained in the previous century are still being used today in soil mechanics. They are, first, the solution of the boundary-value problem for a semi-infinite body with a concentrated load in the statical case and, second, the solution of the wave problem in infinitely expanded media. The way in which the linear elasticity theory is used to solve these problems will not be examined here any further. In the case of statics, the linear-elastic area is very small in soils, and the relevant deformations are of a plastic nature. However, test results seem to indicate that the solutions of the linear elasticity theory provide at least a qualitatively good result for the stress distribution in soils.

The use of the linear elasticity theory appears the most justified in the dynamic case. It is well known that, by dynamic loading, the onset of plastic deformations is delayed, so that the linear-elastic deformations (naturally dependent on the rate of load application) can be much bigger than in the static case, something which can be seen clearly by observing earthquakes.

First, let us consider the static problem. Let a body be expanded in all directions, so that it can be considered as being infinitely large (in comparison to the area to which the distribution of load is applied), and its surface in the infinitely expanded plane $z = 0$. On this boundary, a normal load P (see Fig. 3.8) is active.

The most important solutions in soil mechanics are the vertical stresses in the arbitrary point Q:

$$\sigma_Z = \frac{3P}{2\pi} \frac{z^3}{R^5} , \tag{3.2.6}$$

and the vertical displacement

$$w = \frac{P(1+\mu)}{2\pi E} \left[\frac{2(1-\nu)}{R} + \frac{z^2}{R^3} \right] . \tag{3.2.7}$$

For a load per unit area, the stress and deformations result from an integration process.

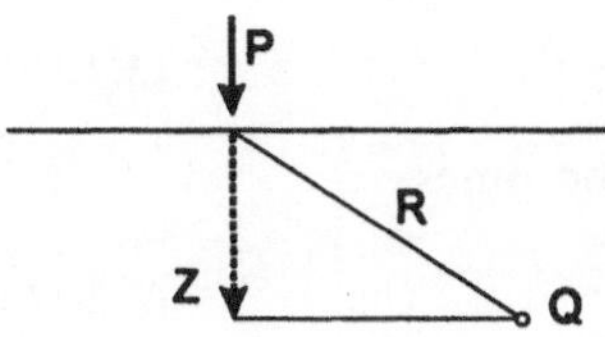

Fig. 3.8. Loaded half space

The first solution to the aforementioned problem, published by Lamé and Clapeyron (1833) (see Section 3.2.1), was obtained by means of expansion in series. A great improvement was made by J. Boussinesq (1885), with the introduction of the various types of direct and inverse potential functions, whereby he succeeded in representing the solutions by means of definite integrals. Cerruti (1888) also studied the case of arbitrary loading and obtained the solution through definite integrals, using the method of integration by Green's functions.

The study of the motion of unlimited isotropic elastic media is of certain importance for soil dynamics. Here, the fundamental equations (3.2.1) are valid and can be written in absolute notation as follows neglecting the external accelerations:

$$\rho\frac{\partial^2 \mathbf{u}}{\partial t^2} = \frac{G}{1-2\nu}\,\text{grad div}\,\mathbf{u} + G\,\Delta\,\mathbf{u}\,, \tag{3.2.8}$$

where $\mathbf{u}$ is the vector of elastic displacement, ρ the density, G the shear modulus, ν is the transverse contraction ratio (Poisson's ratio), and Δ denotes the Laplace operator. Due to

$$\Delta\mathbf{u} = \text{grad div}\,\mathbf{u} - \text{rot rot}\,\mathbf{u}\,, \tag{3.2.9}$$

with div and rot as operators (see de Boer, 1982), the above fundamental equation can be written as follows

$$\rho\frac{\partial^2 \mathbf{u}}{\partial t^2} = \frac{2(1-\nu)}{1-2\nu}G\,\text{grad div}\,\mathbf{u} - G\,\text{rot rot}\,\mathbf{u}\,. \tag{3.2.10}$$

If a scalar quantity Φ is determined from

$$\rho\frac{\partial^2 \Phi}{\partial t^2} = \frac{2(1-\nu)}{1-2\nu}G\,\Delta\,\Phi\,, \tag{3.2.11}$$

and a vector $\mathbf{q}$ is determined from

$$\rho\frac{\partial^2 \mathbf{q}}{\partial t^2} = G\,\Delta\,\mathbf{q}\,, \tag{3.2.12}$$

then, due to

$$\text{div grad}\,\Phi = \Delta\Phi\,, \quad \text{div rot}\,\mathbf{u} = 0\,, \quad \text{rot grad}\,\Phi = \mathbf{0}\,, \tag{3.2.13}$$

the vector,

$$\mathbf{u} = \text{grad}\,\Phi + \text{rot}\,\mathbf{q}\,, \tag{3.2.14}$$

becomes a solution of (3.2.10).

For the special case of irrotational motion with rot $\mathbf{u} = \mathbf{0}$, taking (3.2.10) and (3.2.9) into consideration, (3.2.8) becomes

$$\rho\frac{\partial^2 \mathbf{u}}{\partial t^2} = \frac{2(1-\nu)}{1-2\nu}G\,\Delta\,\mathbf{u}\,. \tag{3.2.15}$$

For non-dilatational motion with div $\mathbf{u} = 0$, (3.2.8) becomes

$$\rho\frac{\partial^2 \mathbf{u}}{\partial t^2} = G\,\Delta\,\mathbf{u}\,. \tag{3.2.16}$$

Thus, the dilatation propagates with the constant velocity a, whereas the rotation is propagated with the smaller velocity b, whereby

$$a^2 = \frac{2(1-\nu)}{\rho(1-2\nu)} G \quad \text{and} \quad b^2 = \frac{G}{\rho} \,. \tag{3.2.17}$$

The fundamental equation (where φ is an arbitrary scalar quantity and c a constant)

$$\frac{\partial^2 \varphi}{\partial t^2} = c^2 \, \Delta \, \varphi \,, \tag{3.2.18}$$

is due to Poisson (1808), who had already stated a solution to this partial differential equation. By applying Green's theorem to functions which satisfy the wave equation (3.2.18) within a finite area, Kirchhoff (1876) succeeded in obtaining a general solution which contained Poisson's solution. Further studies of the wave problem were carried out by Stokes (1849) and Lord Rayleigh (1877/78) (amongst others), who also examined the surface waves on the free surface of an elastic semi-infinite body.

3.2.5
Final Remarks

Within the scope of the description of the history of porous media theory, it is here only possible to discuss the historical development of some areas of the linear elasticity theory. The treatment of structures, shells, plates, and rods (uninteresting for, e.g., soil mechanics) has been completely omitted. On the other hand, the historical developments of the fundamental equations of the linear elasticity theory have been described in detail. We have seen that these had already been perfectly formulated mathematically by the middle of last century. This is due to the scientists mentioned in the previous chapter, who were mainly mathematicians or physicists well trained in mathematics.

However, for the conversion of this knowledge into practice, this was a disadvantage. Thus, a book excellent for its time, by Clebsch (1862), *Theorie der Elastizität fester Körper*, was almost disregarded by engineers. In order to bring the knowledge of the elasticity theory, and mechanics in general, closer to students and engineers working in the field, A. Föppl wrote a series of textbooks. It can be claimed that he was the first scientist to write a book on the elasticity theory for engineers (Föppl, 1898). The book was translated into both Russian and French, and it had a great influence in German-speaking districts, especially on engineering education. The claim that A. Föppl was one of the founders of so-called technical mechanics is fully justified. In the introduction to the first volume of his book *Vorlesungen über* Technische Mechanik (Föppl, 1898) he said:

"Bis jetzt habe ich immer nur von der Mechanik im allgemeinen gesprochen. Bei der Technischen Mechanik tritt als bestimmender Beweggrund für ihre Fassung zu der Absicht einer Erforschung der Wirklichkeit (in dem vorher erklärten Sinne) noch die andere

Absicht, ihre Lehren nutzbringend in der Technik zu verwerten. Auch dieser Zweck setzt freilich voraus, daß wir die Naturtatsachen zunächst richtig erkennen. Die praktischen Anforderungen der Technik haben jedoch vielfach bestimmend auf den weiteren Ausbau der Mechanik eingewirkt. Gar viele Vorstellungsreihen sind auf diesem Wege entstanden, von denen manche schon seit längerer Zeit dem Lehrinhalt der allgemeinen Mechanik einverleibt wurden, während andere auch jetzt noch ausschließlich in der technischen Mechanik zur Sprache kommen.

Der tiefere Grund für diese Absonderung der technischen Mechanik als eines besonderen Zweiges der Wissenschaft liegt darin, daß die allgemein gültigen Lehren der Mechanik keineswegs dazu ausreichen, alle Fragen, die sich im Gebiet der Mechanik überhaupt aufstellen lassen, streng und genau zu lösen. Solchen Fällen steht aber der Naturforscher anders gegenüber als der Techniker. Jener hat zwar auch den Wunsch, die noch bestehenden Zweifel aufzuhellen; er hat aber mit der Beantwortung irgendeiner einzelnen Frage keine Eile und stellt sie ohne Bedenken einstweilen zurück, wenn es ihm nicht gelingt, eine befriedigende Lösung dafür zu finden. Der Techniker dagegen steht unter dem Zwang der Notwendigkeit; er muß ohne Zögern handeln, wenn ihn irgendeine Erscheinung hemmend oder fördernd in den Weg tritt, und er muß sich daher unbedingt auf irgendeine Art, so gut es eben gehen will, eine theoretische Auffassung davon zurechtlegen. Den strengen Anforderungen, die man sonst an die Lehren der Mechanik stellt, können solche durch die Not geborenen Schöpfungen zunächst zwar nicht entsprechen; zuweilen gelingt es aber doch, sie bei weiterer Ausarbeitung allmählich so umzugestalten, daß sie mit Fug und Recht als gute Theorien bezeichnet werden können. Im anderen Falle müssen sie, um dem unabweisbaren praktischen Bedürfnis zu genügen, einstweilen unter der Bezeichnung als Näherungstheorien in der technischen Mechanik fortgeführt werden, aber mit der ausdrücklichen Warnung, daß ihre Aussagen nicht unbedingt zuverlässig sind, und mit dem Vorbehalt, sie, sobald es gelingt, durch genauer ausgearbeitete Theorien zu ersetzen, .."

"Until now I have always only generally spoken of mechanics. For the formulation of technical mechanics there appear, as the determining motivation, not only the intention to investigate the reality (in the aforementioned sense) but also the other intention being to utilize its theories in engineering. This purpose assumes also, of course, that we recognize the facts of nature correctly. The practical requirements of engineering have, however, in many cases decidedly influenced the further development of mechanics. Many ideas are created in this way, of which some have been incorporated into the curriculum of mechanics for a long time, whereas others are still exclusively applied in technical mechanics today.

The deeper reason for this separation of technical mechanics as an individual branch of science lies in the fact that the generally valid theories of mechanics are in no way sufficient to solve rigorously and exactly all questions which can be set up in mechanics. The scientist faces such cases, however, in another way than the engineer. The scientist does indeed also want to clarify the still remaining doubts; there is, however, no hurry for him to answer the single question and he can postpone it for a certain time without scruples if he does not succeed in finding a sufficient solution. The engineer, however, is forced by this requirement; he must act without hesitation whenever a phenomenon stands in the way, and he must arrange a theoretical conception of this problem in some sufficient way. Such creations which are made under duress cannot at first meet the rigorous standards which one usually demands of the theories of mechanics; sometimes one is, however, successful enough in gradually rearranging these creations in the further development that they can be denoted with full right as correct theories. In the other case they must be continued for the moment under the notation of approximation theories in technical mechanics in order to meet the necessary practical need, with however, the explicit warning that their statements are not reliable in any case, and with the reservation to replace the approximation theories by precisely formulated theories as soon as is possible, ..."

Due to the turbulent development of technology in the first half of this century, mechanics has developed, in some parts, one-sidedly in the direction of such technical mechanics. The requirement that such concrete problems were to be treated in mechanics, however, led to a strong dissipation of mechanics. It must be recognized that, in the scope of technical mechanics, important results could be won in the theoretical mechanics of structures, in elasticity theory, and in fluid mechanics as well. Reflections on the foundation of mechanics, and especially on the elasticity theory, have only taken place in the last decades. The results form the framework of modern continuum mechanics. It is a great merit of research in this area to have cleared up mistakes and inadequate presentations in the continuum theory which had been brought in by technical mechanics.

3.2.6 Biographical Notes

Because the information concerning some scientists mentioned in the preceding sections is quite brief, some encyclopedias have been used to complete the picture.

Louis-Marie-Henri Navier (1785–1836)

Louis-Marie-Henri Navier was born into the family of a well-to-do lawyer in Dijon. He lost his father when he was fourteen years old and was taken into the house of his uncle, the famous French engineer Ganthey. His uncle paid great attention to Navier's education and, in 1802, he passed the competitive entrance

Fig. 3.9. Louis-Marie-Henri Navier (1785–1836)

examination to enter the École Polytechnique. Later, in 1804, he attended the École des Ponts et Chaussées. He became famous due to a new edition of annotations to the works of his uncle, and to Belidor's *Architecture hydraulique*. In 1820, Navier presented a memoir on the bending of plates to the Académie des Sciences, and, in 1821, he published his famous paper presenting the fundamental equations of the mathematical theory of elasticity. In 1824, the Académie des Sciences chose him as member. After this he became professor of analysis and mechanics at the École Polytechnique.

Navier dealt with many interesting problems in mathematical physics, particularly, in the case of the beam theory. He is considered as the founder of the beam theory in the geometrically-linear case.

Navier died in Paris in 1836.

Adhémar-Jean-Claude Barré de Saint-Venant (1797–1886)

Barré de Saint-Venant was born in Villiers-en-Brie (Seine-et-Marne) on April 23, 1797. At the age of sixteen, he entered the École Polytechnique, after taking the competitive examinations.

The political events of 1814 had a great effect on Saint-Venant's career. As the first sergeant of a detachment, he refused to fight with the words: "My conscience forbids me to fight for an usurper ... ". His schoolmates resented this action very much and de Saint-Venant was proclaimed a deserter and forever barred from resuming his studies at the École Polytechnique. During the eight years following this incident, de Saint-Venant worked as an assistant in the powder industry. Then, in 1823, the government permitted him to enter the École des Ponts et Chaussées without an examination. He graduated from the school as the first in his class. For some years (1825–1830) he worked on the channel of Nivernais and, later, on the channel of Ardennes. In his spare time he did theoretical work and, in 1834, he presented two papers to the Académie des Sciences, one dealing with some theorems of theoretical mechanics, and the other with fluid dynamics. These papers made him known to the French scientific community and, in 1837–1838, during Professor Coriolis' illness, he was asked to give lectures on the strength of materials at the École des Ponts et Chaussées. Barré de Saint-Venant was the first scientist to bring the new developments in the elasticity theory to the attention of his students.

While lecturing at École des Ponts et Chaussées, de Saint-Venant also undertook practical work for the Paris municipal authorities. He also became interested in hydraulics and its applications in agriculture. For two years (1850–1852), he lectured on mechanics at the agronomical Institute in Versailles. However, he also continued his research in the field of theory of elasticity. In 1843, he presented a memoir on the bending of curved bars to the Académie, while his first memoir on torsion appeared in 1847. His final ideas on the treatment of torsional and bending problems were published in two famous memoirs in 1855 and 1856. Moreover, he studied the dynamical action of loads moving along a

Fig. 3.10. Barré de Saint-Venant (1797–1886)

beam and the impact action of a load falling down onto a bar and producing lateral or longitudinal vibration therein.

He published a series of other important papers, in particular, in the field of elasticity. These works made him famous, and, in 1868, he was elected a member of the Académie des Sciences de Paris as the successor to General Poncelet. He published his articles in various journals, especially, however, in the *Comptes rendus de l'Académie des Sciences de Paris* and in the *Journal de mathématiques de Liouville*. With Flamant, he translated Clebsch's fundamental book on elasticity *Theorie der Elastizität fester Körper* (Clebsch, 1862) in which he added several clarifying notes.

Barré de Saint-Venant died in Saint-Quen (Loir-et-Cher) on January 6, 1886.

The president of the French Académie, on announcing de Saint-Venant's death, used the following words: "Old age was kind to our great colleague. He died, advanced in years, without infirmities, occupied up to the last hour with problems which were dear to him and supported in the great passage by the hopes which had supported Pascal and Newton." (after Timoshenko, 1953).

3.3 Discovery of Fundamental Laws (Delesse, Fick, Darcy)

In the mid-19th century the mixture theory, as well as the theory of porous media, were unknown. After the development of the concept of volume fractions by R. Woltman (1794/99), obviously no substantial contribution to the theory of heterogeneously composed bodies was published for some time to follow. However, around the mid-19th century, some pieces of the mosaic, important for the mixture theory and the theory of porous bodies were discovered. The fundamental findings of Delesse, Fick, and Darcy are indebted to these discoveries.

3.3.1 The Delessian Law

R. Woltman's concept of volume fractions (1794/99) relates the volume elements of the single constituents of a water-saturated porous body to the volume element of the bulk body. With the birth of this concept, an important step towards the development of a theory of porous media had already been made – this still being the early days of the history of mechanics itself – as was later recognized. Using the concept of volume fractions, one is able to treat heterogeneously composed bodies with continuum mechanical methods. However, in order to evaluate the balance of momentum, it is necessary to also develop a corresponding concept for the surface elements of a saturated porous body. For the creation of such a concept, the mining engineer A. Delesse (1848) achieved a decisive contribution. Although he worked on a very different problem, he succeeded in proving that the surface fractions were, under certain circumstances, equal to the volume fractions. As a mining engineer, he was interested in determining the ratios of the single minerals in a piece of rock. He wrote that it was not sufficient to know only the single minerals. Rather, the fractions of the minerals in comparison to the bulk volume also had to be known. He complained that the determination of the volume fractions was nearly impossible, as he had learnt from experiments. Then he looked for a method to determine the ratios of the single minerals without destroying the piece of rock. He assumed that the volume of the rock was related to an orthogonal coordinate-system x, y, z and that p was the surface of a mineral-constituent in a section which was composed by a surface parallel to the plane $x - y$. In order to obtain the exact volume of the mineral in the rock, the subsequent values of p had to be known. Then, the integral $\int p\,dz$ would yield the value of the unknown volume.

Furthermore, A. Delesse (1848) remarked that p, as a function of z, could increase or decrease and also has a different minimum/maximum. However, if one denotes the smallest and the largest value of p by m and M, the integral $\int p\,dz$ will lie between mz and Mz, where z is the height of the considered volume. In addition, the minimum/maximum values m and M will vary less from each other the more uniformly the minerals are distributed in the rock. In this case, the ratios of the volumes of the mineral-constituents will coincide with the ratios of the surfaces of the corresponding sections or, at least, one can be sure that the ratios are located between the minimum and the maximum values of the surface.

Then, A. Delesse (1848) stated the concept of surface fractions, which is of course equal to the concept of volume fractions according to the explanation above. If p, p', p'', etc., are the surfaces of the mineral constituents in a

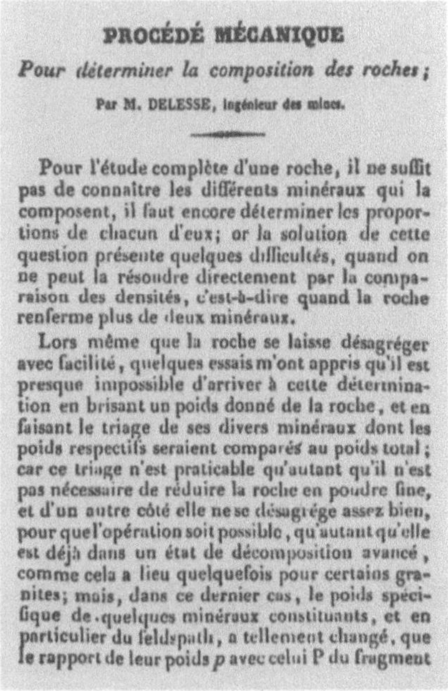

PROCÉDÉ MÉCANIQUE

Pour déterminer la composition des roches ;

Par M. DELESSE, ingénieur des mines.

Pour l'étude complète d'une roche, il ne suffit pas de connaître les différents minéraux qui la composent, il faut encore déterminer les proportions de chacun d'eux; or la solution de cette question présente quelques difficultés, quand on ne peut la résoudre directement par la comparaison des densités, c'est-à-dire quand la roche renferme plus de deux minéraux.

Lors même que la roche se laisse désagréger avec facilité, quelques essais m'ont appris qu'il est presque impossible d'arriver à cette détermination en brisant un poids donné de la roche, et en faisant le triage de ses divers minéraux dont les poids respectifs seraient comparés au poids total; car ce triage n'est praticable qu'autant qu'il n'est pas nécessaire de réduire la roche en poudre fine, et d'un autre côté elle ne se désagrége assez bien, pour que l'opération soit possible, qu'autant qu'elle est déjà dans un état de décomposition avancé, comme cela a lieu quelquefois pour certains granites; mais, dans ce dernier cas, le poids spécifique de quelques minéraux constituants, et en particulier du feldspath, a tellement changé, que le rapport de leur poids *p* avec celui P du fragment

Fig. 3.11. Deless' treatise on the surface porosity

homogeneous rock in the section P, then volume fractions of these minerals are:

$$\frac{p}{P}\,, \quad \frac{p'}{P}\,, \quad \frac{p''}{P}\,, \quad \dots \tag{3.3.1}$$

so that

$$\frac{p}{P} + \frac{p'}{P} + \frac{p''}{P} + \quad \dots \quad = \quad 1\,. \tag{3.3.2}$$

A. Delesse's reflections played an important role in the very beginning of the development of a porous media theory in the first half of this century. All competent researchers in this field, namely P. Fillunger (1913, 1935), K. von Terzaghi (1934a, b), and G. Heinrich and K. Desoyer (1955) referred to the fundamental statements of A. Delesse (1848), although these had to be considered as a statistical necessity (see K. von Terzaghi, 1934a, b).

3.3.2 Fick's Law

The first attempts to develop a phenomenological theory of mixture were made by Fick (1855), who studied the problem of diffusion. Fick, as a physician, was inspired by hydro-diffusion through membranes. He remarked that this problem was not only important for organic life, but also for a very interesting physical process, and he complained that only four contributions on this subject had been published up to that time. The reason for such few treatments probably

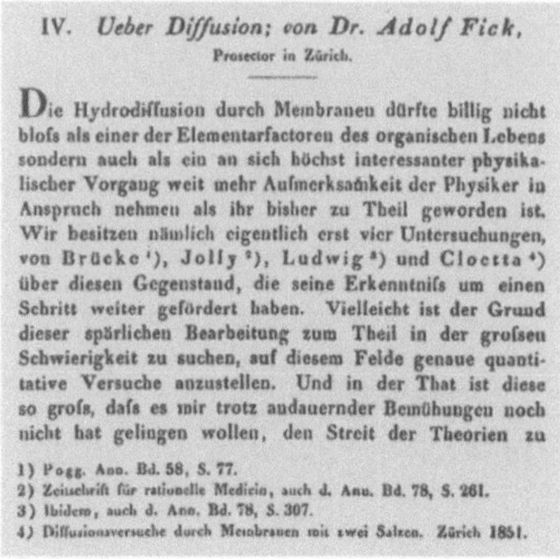

IV. *Ueber Diffusion; von Dr. Adolf Fick,*
Prosector in Zürich.

Die Hydrodiffusion durch Membranen dürfte billig nicht blofs als einer der Elementarfactoren des organischen Lebens sondern auch als ein an sich höchst interessanter physikalischer Vorgang weit mehr Aufmerksamkeit der Physiker in Anspruch nehmen als ihr bisher zu Theil geworden ist. Wir besitzen nämlich eigentlich erst vier Untersuchungen, von Brücke [1]), Jolly [2]), Ludwig [3]) und Cloetta [4]) über diesen Gegenstand, die seine Erkenntnifs um einen Schritt weiter gefördert haben. Vielleicht ist der Grund dieser spärlichen Bearbeitung zum Theil in der grofsen Schwierigkeit zu suchen, auf diesem Felde genaue quantitative Versuche anzustellen. Und in der That ist diese so grofs, dafs es mir trotz andauernder Bemühungen noch nicht hat gelingen wollen, den Streit der Theorien zu

1) Pogg. Ann. Bd. 58, S. 77.
2) Zeitschrift für rationelle Medicin, auch d. Ann. Bd. 78, S. 261.
3) Ibidem, auch d. Ann. Bd. 78, S. 307.
4) Diffusionsversuche durch Membranen mit zwei Salzen. Zürich 1851.

Fig. 3.12. Fick's treatise on diffusion

lay in the great difficulties that exist in order to perform quantitative tests in this field. Indeed these difficulties were so great that, despite substantial effort on his part, he did not succeed in clarifying the arguments of the theories. He would also have liked to have published his own findings from his new test results, and to have emphasized certain mechanical aspects, namely the relation between the actual diffusion through porous bodies and the simple expansion of a soluble body in a solvent. Fick pointed out that this last mentioned phenomenon had been treated by another author in a rather long contribution which contained many test observations of qualitative and quantitative experiments. However, the complete investigation was not based on the investigation of the elementary process. Therefore, Fick felt obliged to study the simple diffusion of mixtures again without the presence of a porous membrane, and to find the basic law which must determine the basic process from layer to layer.

Fick then discussed the motion of the molecules in detail; his work here is in some parts obscure. He stated that, in general, there is a stronger attraction between different molecules than between homogeneous molecules.

Finally, he arrived at his essential statements:

"Es wäre jetzt die erste Aufgabe, das Grundgesetz für diesen Bewegungsvorgang aus den allgemeinen Bewegungsgesetzen herzuleiten; und dieß wäre auch, glaube ich, wohl möglich ohne die Funktionen $f(r)$ und $\varphi(r)$ zu kennen ($f(r)$ and $\varphi(r)$ are forces and r the distance between atoms, the author). Meine dahin gerichteten Bestrebungen haben indessen keinen Erfolg gehabt. Dahingegen drängte sich mir beim ersten Überlegen jenes Grundgesetzes eine sehr naheliegende Vermuthung auf, die es mir experimentell außer allen Zweifel zu stellen gelungen ist. In der That wird man zugeben, daß von vorn herein nichts wahrscheinlicher sey als dieß: Die Verbreitung eines gelösten Körpers im Lösungsmittel geht, wofern sie ungestört unter dem ausschließlichen Einfluß der Molekularkräfte stattfindet, nach demselben Gesetz vor sich, welches Fourier für die Verbreitung der Wärme in einem Leiter aufgestellt hat, und welches Ohm bereits mit so glänzendem Erfolge auf die Verbreitung der Elektricität (wo es freilich bekanntlich nicht streng richtig ist) übertragen hat. Man darf nur in dem Fourier'schen Gesetz das Wort Wärmequantität mit dem Worte

Quantität des gelösten Körpers, und das Wort Temperatur mit Lösungsdichtigkeit vertauschen. Der Leitungsfähigkeit entspricht in unserem Falle eine von der Verwandschaft der beiden Körper abhängige Constante."

"Now, it would be the first task to describe the fundamental law for this motion from the general laws of motion; and this would also surely be possible, I believe, without knowing the functions $f(r)$ and $\varphi(r)$ ($f(r)$ and $\varphi(r)$ are forces and r the distance between atoms, the author). My concentrated efforts to solve this problem have had, however, no success. On the other hand, a very obvious supposition suggested itself during the first reflections on such a fundamental law, which I have proved experimentally. Indeed one must agree that apriori nothing is more probable than this: The propagation of a soluble body in a solution takes place, so far as it occurs under the exclusive influence of the molecular forces, according to the same law which Fourier has developed for the propagation of heat in a conductor and which Ohm has already transferred with such excellent success to the propagation of electricity (where it is of course not quite correct). One must only exchange in Fourier's law the term heat quantity for the term quantity of the soluble body, and the term temperature for the density of the solution. The conductivity corresponds in our case to a constant which depends on the similarity of both bodies."

According to the development of the Fourier equation of heat propagation, Fick (1855) arrived at the differential equation of the diffusion stream:

$$\frac{\partial y}{\partial t} = - k\left(\frac{\partial^2 y}{\partial x^2} + \frac{1}{Q}\frac{dQ}{dx}\frac{\partial y}{\partial x}\right), \tag{3.3.3}$$

where y is the concentration, t the time, k a constant which depends on the nature of the constituents, x a measure for the height in a container, and Q the cross-section of the container. Furthermore, Fick (1855) remarked that in the case of a constant cross-section, the above differential equation simplifies to:

$$\frac{\partial y}{\partial t} = - k\frac{\partial^2 y}{\partial x^2} \,. \tag{3.3.4}$$

This relation is today known as Fick's second diffusion law. There is no hint in Fick's paper as to a constitutive equation for the diffusive flux vector which is known as Fick's first law.

In the remaining part of his paper, Fick (1855) was concerned with the experimental proof of his fundamental law for the diffusion process.

3.3.3 Darcy's Law

Darcy (1856) observed, in tests with natural sand, the proportionality of the total volume of water running through the sand and the loss of pressure. Although his investigations were of a purely experimental nature, his results are essential for a continuum mechanical treatment of the motion of a liquid in a porous solid. In Darcy's (1856) contribution, the interaction of different constituents in a multiphase continuum – a binary model consisting of a rigid porous solid and a liquid in motion – was studied for the first time. Today, Darcy's law is theoretically well-founded by thermodynamic restrictions.

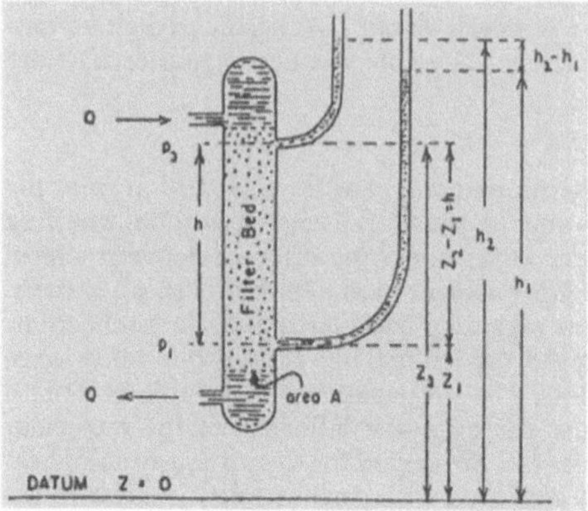

Fig. 3.13. Test to varify Darcy's Law

After Scheidegger (1963), Darcy gained his results from the depicted test. A liquid percolates through a homogeneous filter bed of the height h which is bounded by plane surfaces. A. Scheidegger (1963) explains:

"If open manometer tubes are attached at the upper and lower boundaries of the filter bed, the liquid rises to the heights h_2 and h_1 respectively above an arbitrary datum level. By varying the various quantities involved, one can deduce the following relationship:

$$Q = -\,KA(h_2 - h_1)/h\,, \tag{3.3.5}$$

where Q is the total volume of fluid percolating in unit time and K is a constant depending on the properties of the fluid and of the porous medium. This relationship (3.3.5) is known as *Darcy's law*. The minus sign in the expression for Q indicates that the flow is in the opposite direction of increasing h.

Darcy's law can be restated in terms of the pressure p and the density ρ of the liquid. At the upper boundary of the bed (elevation above the datum level denoted by z_2), the pressure is $p_2 = \rho g(h_2 - z_2)$, and at the lower boundary (elevation above datum level denoted by z_1), the pressure is $p_1 = \rho g(h_1 - z_1)$. Inserting this statement into (3.3.5), one obtains (as $z_2 - z_1 = h$)

$$Q = -\,KA\Big[(p_2 - p_1)/(\rho g h) + 1\Big]\,; \tag{3.3.6}$$

or, upon introduction of a new constant K', assuming ρ and g to be constants,

$$Q = -\,K'A(p_2 - p_1 + \rho g h)/h\,. \tag{3.3.7}$$

The equations (3.3.6) and (3.3.7) are equivalent statements of Darcy's law."

The validity of Darcy's law has been tested, and theoretically confirmed, by many authors. It has been discovered that for liquids at high velocities, and for gases at very low and at very high velocities, Darcy's law becomes invalid.

Although Delesse (1848), Fick (1855), and Darcy (1856) discoverd their fundamental laws more or less heuristically, rather then by a development from the fundamental relations of mechanics and thermodynamics, they created very important statements which were usefully applied in theories to follow, such as the mixture theory and the porous media theory.

3.3.4 Biographical Notes

Henri-Philibert Darcy (1803–1858)

Henri-Philibert Darcy was born in Dijon in 1803. He passed the École Polytechnique and entered in the Corps des Ponts et Chaussées as an engineering student in 1823, and became an engineer in 1828. After a short vacation in the Jura, he was called to Dijon as a civil engineer. He built two large bridges over the Saone. Moreover, he designed the drinking water system for the city of Dijon in 1834, and after that he began with its construction. As a chief engineer of the Cote-d'Or in 1840, Darcy was put in charge of tracing the railroad track from Paris to Lyon. He was worried about the problem of paving the roads in big cities. In order to study this problem, he was sent to London in 1850 to observe how the English engineers managed it. After his return, he published a report in which he compared the construction of the pavements in London and in Paris. In 1850, he asked for retirement on health grounds. He died in Paris, on January 3, 1858.

Achille Ernest Delesse (1817–1881)

Achille Ernest Delesse, born in Metz on February 3, 1817, studied mineralogy and geology. He was then offered a professorship at the Sorbonne. Later, he received a chair at the École des Mines. Finally, he became Chief of Staff of all French mines. At that time, he was a famous mining engineer, and the mineral Delessit was named in honor of him. On March 24, 1881, he died in Paris.

Adolf Fick (1829–1901)

Adolf Fick, born in Kassel, Germany, on September 3, 1829, first studied mathematics in Marburg. His friendship with the physiologist Carl Ludwig, at that time Privatdozent at the Institute of Anatomy, which was run by Adolf Fick's oldest brother Ludwig, induced him to change to medicine. After his 1851 promotion in Marburg, he followed Carl Ludwig to Zürich in 1852. There were already two famous scientists there, namely E. Du Bois-Reymond, whose lectures Fick had heard in Berlin in 1849, and J. Moleschott. Fick became one of the leading physiologists of the branch of medicine which was governed by physics and mathematics. This direction created the methadological base of nearly all progress to follow in physiology. E. Du Bois-Reymonds gave Fick the hint to investigate the diffusion problem. Fick became a professor in 1855 and, in 1862, full professor of physiology at Zürich; he was concerned with the physical-chemical fundamentals of different problems in medicine. Moreover, he proved the validity of the two laws of thermodynamics for muscle contraction, recognizing that, in the muscle, the conversion of chemical into mechanical energy

Fig. 3.14. Adolf Fick (1829–1901)

occurs directly, not indirectly via heat. In 1868, he was offered a professorship at the University of Würzburg, where he worked until 1899. Adolf Fick died in Blankenberghe (West-Flanders) on August 21, 1901.

3.4 The Development of the Theory of Viscous Fluids

The theories of viscous fluids do not have the same importance in porous media theory as linear elasticity theory. However, specialists in engineering may need to deal with this area, for example, in the pursuit of the filtration of oil.

3.4.1 Introduction: The Navier-Stokes Equations

The viscous fluids theory is marked by the Navier-Stokes equations. The basis for these equations is the constitutive equation for linear-viscous, compressible fluids:

$$\begin{aligned} \mathbf{T} &= -\,p\mathbf{I} + \gamma(\mathbf{D}\cdot\mathbf{I})\mathbf{I} + 2\mu\mathbf{D} \\ &= -\,p\mathbf{I} + (\gamma + \tfrac{2}{3}\mu)(\mathbf{D}\cdot\mathbf{I})\mathbf{I} + 2\mu\mathbf{D}^D\,, \end{aligned} \tag{3.4.1}$$

where the material constants γ and μ are the volume and shear viscosity, $\mathbf{T}$ is the stress tensor, $\mathbf{D}$ is the symmetrical part of the velocity gradients, $\mathbf{D}^D$ is the deviator of $\mathbf{D}$, p is the static pressure, and $\mathbf{I}$ is the identity tensor.

The constitutive relation for the stress tensor is used in Cauchy's first equation of motion (see Section 3.1):

$$\operatorname{div} \mathbf{T} + \rho \mathbf{b} = \rho \ddot{\mathbf{x}} \,. \tag{3.4.2}$$

After mathematical transformations, one arrives with the velocity vector **v** at the Navier-Stokes equation:

$$-\nabla p + (\gamma + \mu) \nabla \operatorname{div} \mathbf{v} + \mu \Delta \mathbf{v} + \rho \mathbf{b} = \rho \ddot{\mathbf{x}} \tag{3.4.3}$$

(where ∇ is the gradient and Δ is the Laplace operator), which must be expanded by the initial and boundary conditions in order to make the theory of viscous fluids complete.

3.4.2
The Historical Development of the Theory

The notion that the motion of fluids past other bodies is held back by friction arose from the following observations: The velocity of a stream leaving a receptacle turns out to be a little smaller than that predicted by Torricelli's theorem. The velocity of water flowing out of an inclined pipe or a canal is approximately the same at various cross-sections. In the first case, the loss of velocity was attributed to the resistance of air and, in the second case, it was assumed that the tangential components of the weight forces were at equilibrium with the frictional forces on the surface of the pipe or on the bed of the canal (see M. Rühlmann, 1880).

Isaac Newton (1687) was the first to recognize internal friction. He used his theory of the circulation of incompressible fluids to describe the concept of internal resistance between fluid layers which glide past each other and he assumed that this resistance was proportional to the relative velocity:

"Hypothesis. Resistentiam, quae oritur ex defectu lubricitatis partium fluidi, caeteris paribus, proportionalem esse velociatati, qua partes fluidi separantur ab invicem."

"Hypothesis. The resistance, which comes from the imperfect glide capability, is proportional to that velocity (under the same circumstances) with which these parts separate from each other."

The first attempts to introduce both types of friction into the equations of motion, for the material points inside as well as on the boundary, were made by Navier (1823). Navier presented his *Mémoir sur les lois du mouvement des fluides* on March 18, 1822 to the French Academy of Sciences. His studies on the theory of viscous fluids were based on molecular theoretical considerations. As we have already seen in the previous chapter, long into the nineteenth century many scientists were of the opinion that only the molecular theory was able to describe the mechanical phenomena of continua. As late as 1901, Love (1907/14) wrote:

"Die Theorie einer Flüssigkeit als eines mechanischen Systems von kontinuierlich den Raum erfüllenden Teilchen, die vermittelst eines geeigneten Drucks durch die sie trennenden Oberflächen hindurch auf einander wirken, scheint für die Erklärung der Phänomene, wie sie bei dem Gleichgewicht und den Bewegungen von Flüssigkeiten unter gewöhnlichen Bedingungen auftreten, im ganzen ausreichend zu sein. Doch gestattet dieselbe keine Anwendung auf solche Phänomene, wie Diffussion und osmotischen Druck. Eine vollständige Theorie der Flüssigkeitsbewegung würde notwendigerweise auf einer molekularen Grundlage aufgebaut werden müssen, und die beiden fundamentalen Begriffe der gewöhnlichen Theorie – Druck und Geschwindigkeit in einem Punkt – würden dann eine präzisere Definition erfordern. Selbst bei denjenigen Erscheinungen, auf die sich die gewöhnliche Theorie sehr wohl anwenden läßt, muss man notwendigerweise einen Kompromiss zwischen dem molekularen und dem mechanischen Standpunkte schließen, insbesondere was die Erscheinungen der Kapillarität und Zähigkeit angeht. Die erstere wird in der Kontinuitätstheorie durch die Annahme einer hinzutretenden Oberfächenspannung als einer Thatsache, die sich aus der Erfahrung ergiebt, unter den mechanischen Gesichtspunkt gebracht, die letztere durch die Annahme, dass die Energie nach einem bestimmten Gesetze dissipiert wird. Aber nur von Seiten der molekularen Theorie kann man eine Erklärung für den Ursprung der Oberfächenspannung und der mit der beobachteten Dissipation der Energie verbundenen Transformation derselben geben."

"The theory of a liquid considered as a mechanical system of continuous particle-filled space which by means of a suitable pressure act on each other through surfaces separating the systems, seems to be entirely sufficient to explain the phenomena which occur in equilibrium and motion of liquids under normal conditions. However, the theory allows no application to such phenomena like diffusion and osmotic pressure. A complete theory of liquid motions must necessarily be built on a molecular basis and both fundamental concepts of the common theory – pressure and velocity in a point – would then require a more precise definition. Even for such phenomena for which the common theory can be applied without doubts, one must necessarily make a compromise between the molecular and the mechanical standpoint, in particular, regarding the capillarity and the viscosity. The first one is in the continuum theory brought under the mechanical approach by the assumption of an additional surface stress as a fact, which is drawn out through experience, brought under the mechanical standpoint, the last one by the assumption that the energy is dissipated according to a determinate law. However, only from the side of the molecular theory can one give an explanation for the origin of the surface stress and the transformation connected with the dissipation of the energy observed."

Navier's theory, as has already been mentioned, was founded on the same ideas of intermolecular effects as he had used for the derivations of his fundamental equations (see Section 3.2.1), i.e., that forces act between the molecules of the fluid. After the introduction of assumptions with respect to these forces (the forces should act proportionally to the relative velocity of the molecules) and after lengthy calculations, Navier set up the fundamental equations for incompressible, viscous fluids, using the viscosity parameter ε:

$$-\nabla p + \varepsilon \triangle \mathbf{v} + \varrho \mathbf{b} = \varrho \ddot{\mathbf{x}} , \qquad (3.4.4)$$

which agrees with (3.4.3), if one takes into consideration that, in the case of incompressibility, the deviator of the tensor of the velocity gradient $\mathbf{D}^D$ agrees with the symmetrical part of the velocity gradient itself, and the divergence of $\mathbf{D}$ is equal to the Laplace operator of the velocity vector $\mathbf{v}$.

Thus, the Navier constant ε is equal to the shear viscosity constant 2μ in the continuum mechanics description. Navier determined the quantity ε from

$$
\frac{8.f(\rho)}{\rho^2}\left\{\begin{array}{l}
\left(\frac{du}{dx}\frac{\delta du}{dx}\alpha^4+\frac{du}{dy}\frac{\delta du}{dy}\alpha^2\beta^2+\frac{du}{dz}\frac{\delta du}{dz}\alpha^2\gamma^2\right)+\\
\left(\frac{dv}{dy}\frac{\delta du}{dx}\alpha^2\beta^2+\frac{dv}{dx}\frac{\delta du}{dy}\alpha^2\beta^2\right)+\\
\left(\frac{dw}{dz}\frac{\delta du}{dx}\alpha^2\gamma^2+\frac{dw}{dx}\frac{\delta du}{dz}\alpha^2\gamma^2\right)+\\
\left(\frac{du}{dx}\frac{\delta dv}{dy}\alpha^2\beta^2+\frac{du}{dy}\frac{\delta dv}{dx}\alpha^2\beta^2\right)+\\
\left(\frac{dv}{dx}\frac{\delta dv}{dx}\alpha^2\beta^2+\frac{dv}{dy}\frac{\delta dv}{dy}\beta^4+\frac{dv}{dz}\frac{\delta dv}{dz}\beta^2\gamma^2\right)+\\
\left(\frac{dw}{dy}\frac{\delta dv}{dz}\beta^2\gamma^2+\frac{dw}{dz}\frac{\delta dv}{dy}\beta^2\gamma^2\right)+\\
\left(\frac{du}{dx}\frac{\delta dw}{dz}\alpha^2\gamma^2+\frac{du}{dz}\frac{\delta dw}{dx}\alpha^2\gamma^2\right)+\\
\left(\frac{dv}{dy}\frac{\delta dw}{dz}\beta^2\gamma^2+\frac{dv}{dz}\frac{\delta dw}{dy}\beta^2\gamma^2\right)+\\
\left(\frac{dw}{dx}\frac{\delta dw}{dx}\alpha^2\gamma^2+\frac{dw}{dy}\frac{\delta dw}{dy}\beta^2\gamma^2+\frac{dw}{dz}\frac{\delta dw}{dz}\gamma^4\right)
\end{array}\right\}
$$

Fig. 3.15. Navier's contribution to the theory of viscous fluids

molecular forces. It eludes any distinct meaning. The theory becomes even more complicated with the introduction of further constants in the boundary conditions.

Nine years later, Poisson (1831) published his contribution to the theory of viscous fluids, which was also based on molecular considerations. With respect to the basic hypotheses, Poisson's theory differs greatly from Navier's (see Stokes, 1846). His main idea was as follows: He divided the time t into n equal intervals τ. In the first interval, he assumed that the fluid acts like an elastic solid body. If the causes of the displacements of the material points would cease to act, then the molecules would quickly take up a new arrangement which would lead to a hydrostatic stress state. While this rearranging was taking place, the stress would change in an unknown way, i.e., from a stress state belonging to a deformed elastic body to one for the fluid in the new state. The causes of the displacements still remain in the second interval τ, but as these small varying motions occur independently of one another, the new displacement in the second interval τ will be the same as if the molecules had not rearranged themselves. If, however, n approaches infinity, then we will have a case where the fluid is displaced like an elastic solid body and rearranges itself continuously so that a hydrostatic stress state occurs. Poisson finally obtained such equations of motion for homogenous, compressible, and elastic fluids, which correspond to (3.4.3), where the change of density is small, and which contain two material-dependent constants. His equations correspond to Navier's in the case of incompressibility.

With his studies, Poisson went further than Navier. With respect to the introduction of two material dependent constants, this is also true in comparison to the studies of de Saint-Venant and Stokes, which we will deal with next.

De Saint-Venant (1843) introduced new ideas to the development of the theory of viscous fluids. In the memoir which he presented to the French Academy on April 14, 1834, *Mémoire sur la dynamique de fluides*, he linked the stress

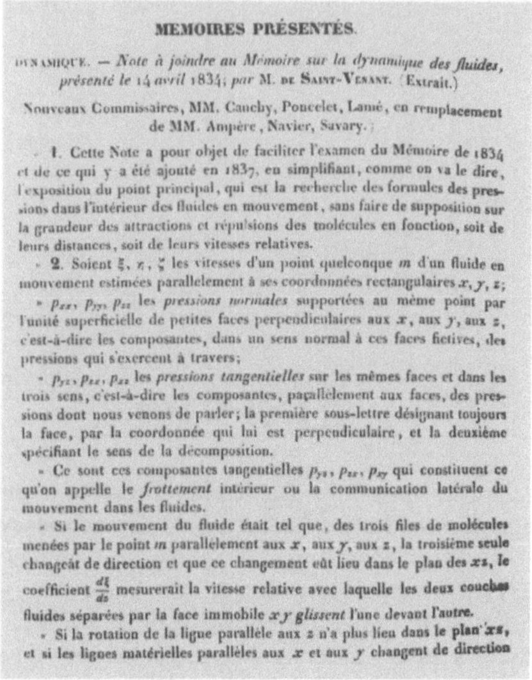

MEMOIRES PRÉSENTÉS.

DYNAMIQUE. — *Note à joindre au Mémoire sur la dynamique des fluides, présenté le 14 avril 1834; par* M. DE SAINT-VENANT. (Extrait.)

Nouveaux Commissaires, MM. Cauchy, Poncelet, Lamé, en remplacement de MM. Ampère, Navier, Savary.

« 1. Cette Note a pour objet de faciliter l'examen du Mémoire de 1834 et de ce qui y a été ajouté en 1837, en simplifiant, comme on va le dire, l'exposition du point principal, qui est la recherche des formules des pressions dans l'intérieur des fluides en mouvement, sans faire de supposition sur la grandeur des attractions et répulsions des molécules en fonction, soit de leurs distances, soit de leurs vitesses relatives.

« 2. Soient ξ, η, ζ les vitesses d'un point quelconque m d'un fluide en mouvement estimées parallèlement à ses coordonnées rectangulaires x, y, z;

» p_{xx}, p_{yy}, p_{zz} les *pressions normales* supportées au même point par l'unité superficielle de petites faces perpendiculaires aux x, aux y, aux z, c'est-à-dire les composantes, dans un sens normal à ces faces fictives, des pressions qui s'exercent à travers;

» p_{yz}, p_{zx}, p_{xz} les *pressions tangentielles* sur les mêmes faces et dans les trois sens, c'est-à-dire les composantes, parallèlement aux faces, des pressions dont nous venons de parler; la première sous-lettre désignant toujours la face, par la coordonnée qui lui est perpendiculaire, et la deuxième spécifiant le sens de la décomposition.

» Ce sont ces composantes tangentielles p_{yz}, p_{zx}, p_{xy} qui constituent ce qu'on appelle le *frottement* intérieur ou la communication latérale du mouvement dans les fluides.

» Si le mouvement du fluide était tel que, des trois files de molécules menées par le point m parallèlement aux x, aux y, aux z, la troisième seule changeât de direction et que ce changement eût lieu dans le plan des xz, le coefficient $\frac{d\xi}{dz}$ mesurerait la vitesse relative avec laquelle les deux couches fluides séparées par la face immobile xy *glissent* l'une devant l'autre.

» Si la rotation de la ligne parallèle aux z n'a plus lieu dans le plan xz, et si les lignes matérielles parallèles aux x et aux y changent de direction

Fig. 3.16. De Saint-Venant's treatise on the theory of viscous fluid

components with the velocity gradient, without considering the inter-molecular effects. He thereby proceeded from various principles (see Szabó, 1987):

1. The shear stress components p_{xy}, p_{xz}, and p_{yz} in an orthogonal coordinates system x, y, z are the determining quantities for the internal friction.
2. The derivatives of the velocity components ξ, η, ζ in an orthogonal coordinate system x, y, z with respect to these coordinates correspond to certain relative velocities between two neighbouring fluid elements.
3. As shear velocities the following quantities are introduced, at first, $\frac{d\xi}{dz} + \frac{d\zeta}{dx}$ and $\frac{d\eta}{dz} + \frac{d\zeta}{dy}$. Rotation velocity is not considered.
4. If the fluid is in a state of rest the shear stresses disappear and only normal stresses occur.
5. In moving fluids, the shear stresses occur in the direction of the shear strain. Correspondingly, the principle shear stresses occur in the direction of the principle shear strain. B. de Saint-Venant wrote on this: "this is the only hypothesis that I make".
6. The shear stresses are proportional to the shear velocities.

De Saint-Venant then pointed to Cauchy's theory, with respect to the creation of the stress concept, and made considerations upon dilatation velocity. Finally, he formulated his constitutive equations for viscous fluids:

$$\frac{p_{xx}-p_{yy}}{2\left(\frac{d\xi}{dx}-\frac{d\eta}{dy}\right)}=\frac{p_{zz}-p_{xx}}{2\left(\frac{d\zeta}{dz}-\frac{d\xi}{dx}\right)}=\frac{p_{yy}-p_{zz}}{2\left(\frac{d\eta}{dy}-\frac{d\zeta}{dz}\right)}=$$

$$\frac{p_{yz}}{\left(\frac{d\eta}{dz}+\frac{d\zeta}{dy}\right)}=\frac{p_{zx}}{\left(\frac{d\zeta}{dx}+\frac{d\xi}{dz}\right)}=\frac{p_{xy}}{\left(\frac{d\xi}{dy}+\frac{d\eta}{dx}\right)}=\varepsilon\,, \tag{3.4.5}$$

$$\frac{p_{xx}+p_{yy}+p_{zz}}{3}-\frac{2\varepsilon}{3}\left(\frac{d\xi}{dx}+\frac{d\eta}{dy}+\frac{d\zeta}{dz}\right)=\pi\,, \tag{3.4.6}$$

$$p_{xx}=\pi+2\varepsilon\frac{d\xi}{dx}\,,\quad p_{yy}=\pi+2\varepsilon\frac{d\eta}{dy}\,,\quad p_{zz}=\pi+2\varepsilon\frac{d\zeta}{dz}\,, \tag{3.4.7}$$

$$p_{yz}=\varepsilon\left(\frac{d\eta}{dz}+\frac{d\zeta}{dy}\right)\,,\quad p_{zx}=\varepsilon\left(\frac{d\zeta}{dx}+\frac{d\xi}{dz}\right)\,,\quad p_{xy}=\varepsilon\left(\frac{d\xi}{dy}+\frac{d\eta}{dx}\right)\,. \tag{3.4.8}$$

De Saint-Venant introduced only one material dependent constant into his constitutive equations and, in this respect, he remained behind Poisson's solutions, where the dilatation velocity is provided with an isolated constant. The novelty of de Saint-Venant's work is the lack of any molecular theory basis; he chose a pure continuum mechanics consideration of the complicated subject.

Stokes (1845) arrived at equations similar to those of Poisson and de Saint-Venant. Stokes developed the fundamental equations of the theory of viscous fluids, in a different way to de Saint-Venant, using continuum mechanics as a basis. In the treatise presented on April 14, 1845, he explained the essentials of viscous fluids and, on the basis of experimental results, he pointed out that in many cases friction occurred in the motion of fluids. He then briefly discussed Poisson's work:

"I afterwards found that Poisson had written a memoir on the same subject, and on referring to it I found that he had arrived at the same equations. The method which he employed was however so different from mine that I feel justified in laying the latter before this society."

He did not mention the work by de Saint-Venant in connection with this. Apparently it was unknown to him at this stage, having been published only two years earlier. One year later, in 1846, he gave a detailed report in on de Saint-Venant's treatise (Stokes, 1846).

In the first section of his treatise, where he developed the theory of viscous fluids, he put the following principle at the head of his exposition:

"That the difference between the pressure on a plane in a given direction passing through any point *P* of a given fluid in motion and the pressure which would exist in all directions about *P* if the fluid in its neighborhood were in a state of relative equilibrium depends

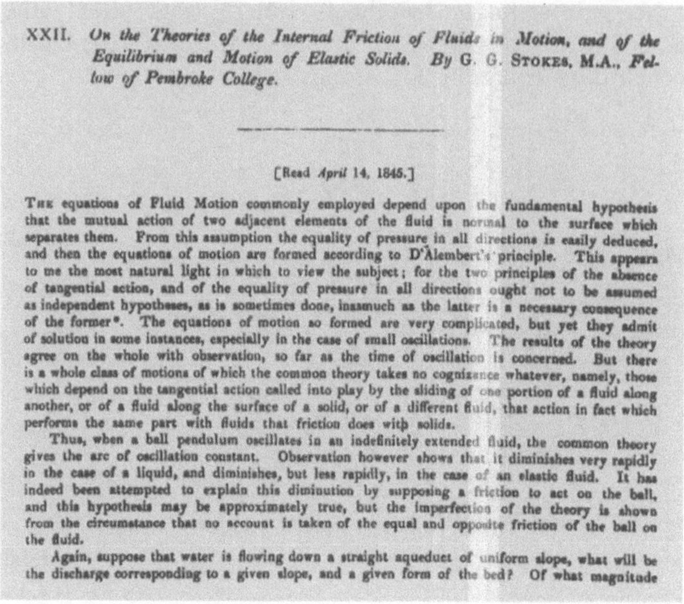

XXII. *On the Theories of the Internal Friction of Fluids in Motion, and of the Equilibrium and Motion of Elastic Solids.* *By* G. G. STOKES, M.A., *Fellow of Pembroke College.*

[Read *April* 14, 1845.]

THE equations of Fluid Motion commonly employed depend upon the fundamental hypothesis that the mutual action of two adjacent elements of the fluid is normal to the surface which separates them. From this assumption the equality of pressure in all directions is easily deduced, and then the equations of motion are formed according to D'Alembert's principle. This appears to me the most natural light in which to view the subject; for the two principles of the absence of tangential action, and of the equality of pressure in all directions ought not to be assumed as independent hypotheses, as is sometimes done, inasmuch as the latter is a necessary consequence of the former*. The equations of motion so formed are very complicated, but yet they admit of solution in some instances, especially in the case of small oscillations. The results of the theory agree on the whole with observation, so far as the time of oscillation is concerned. But there is a whole class of motions of which the common theory takes no cognizance whatever, namely, those which depend on the tangential action called into play by the sliding of one portion of a fluid along another, or of a fluid along the surface of a solid, or of a different fluid, that action in fact which performs the same part with fluids that friction does with solids.

Thus, when a ball pendulum oscillates in an indefinitely extended fluid, the common theory gives the arc of oscillation constant. Observation however shows that it diminishes very rapidly in the case of a liquid, and diminishes, but less rapidly, in the case of an elastic fluid. It has indeed been attempted to explain this diminution by supposing a friction to act on the ball, and this hypothesis may be approximately true, but the imperfection of the theory is shown from the circumstance that no account is taken of the equal and opposite friction of the ball on the fluid.

Again, suppose that water is flowing down a straight aqueduct of uniform slope, what will be the discharge corresponding to a given slope, and a given form of the bed? Of what magnitude

Fig. 3.17. Stoke's approach

only on the relative motion of the fluid immediately about P; and that the relative motion due to any motion of rotation may be eliminated without affecting the differences of the pressures above mentioned."

Stokes here clearly stated that the rotation must not influence the difference of the stress. Truesdell (1980a) remarks:

"I regard it as one of the early, creeping steps toward reduction of constitutive equations by applying the principle of material-indifference, ..."

Subsequently, Stokes completed, in a rather laborious way, the kinematics of the state of equilibrium. He then turned to the stress vector in the areas vertical to the axes of the orthogonal coordinate systems and divided this up into the static pressure p vertical to the considered area and an additional stress vector with the value of the components p', p'', p''', which he determined to be linear functions of the velocity derivatives from the three coordinates x, y, z. This occurs partially with the help of molecular theory considerations. Stokes then quickly arrived at a continuum mechanics point of view:

"Consequently we have only to consider the average effect of such starts and moreover we may without sensible error replace the impulsive forces ..., which succeed one another with great rapidity, by continuous forces. For planes perpendicular to the axes of extension these continuous forces will be the normal pressure p', p'', p'''."

He then, however, returned to some molecular theory considerations and arrived at the following connections between p', p'', p''' and e', e'', e''', which represent the derivations of the velocity components u, v, w from the coordinates x, y, z:

$$\begin{aligned} p' &= \frac{2}{3}\mu\,(e'' + e''' - 2e')\,, \\ p'' &= \frac{2}{3}\mu\,(e''' + e' - 2e'')\,, \\ p''' &= \frac{2}{3}\mu\,(e' + e'' - 2e''')\,, \end{aligned} \tag{3.4.9}$$

where μ is a material dependent constant and the expressions in parentheses are the deviators of e', e'', and e'''.

In the course of his treatise, Stokes made an interesting remark which made him the real founder of the theory of viscous fluids with the help of continuum mechanics:

"If we had started with assuming $\Phi(e', e'', e''')$ [12] to be a linear function of e', e'' and e''', avoiding all speculation as to the molecular constitution of a fluid we should have had at once $p' = C\,e' + C'\,(e'' + e''')$ since p' is symmetrical with respect to e'' and e'''; or, changing the constants,

$$p' = \frac{2}{3}\,\mu\,(e'' + e''' - 2e') + \kappa\,(e' + e'' + e''')\,. \tag{3.4.10}$$

The expressions for p'' and p''' would be obtained by interchanging the requisite quantities. Of course we may at once put $\kappa = 0$ if we assume that in the case of a uniform motion of dilatation the pressure at any instant depends only on the actual density and temperature at that instant, and not on the rate at which the former changes with the time. In most cases to which it would be interesting to apply the theory of the friction of fluids the density of the fluid is either constant, or may without sensible error be regarded as constant, or else changes slowly with the time. In the first two cases the results would be the same, and in the third case nearly the same, whether κ were equal to zero or not."

After Stokes somewhat laboriously explained the stress state with the stress transformations, which were already known through Cauchy's work, he stated the complete constitutive law for viscous fluids, using only one material-dependent constant, as de Saint-Venant had done in a previous treatise. With the developed constitutive relations, he formulated the equations of motion and the boundary conditions for some special boundary-value problems.

In the fourth section of his treatise, Stokes (1846) investigated Poisson's work more closely. He declared:

"Poisson himself has not made this reduction of his equations, nor any equivalent one, so that his equations, as he has left them, involve two arbitrary constants. The reduction of these two to one depends on the assumption that a uniform expansion of any particle does not require a rearrangement of the molecules, as it leaves the pressure still equal in all directions. If we do not make this assumption, but retain the two arbitrary constants, the equations will be the same as those which would be obtained by the method of this paper, supposing the quantity κ of Art. 3 not to be zero."

[12] In his studies, he had proceeded from the statement $p' = \phi(e', e'', e''')$, $p'' = \phi(e'', e''', e')$, $p''' = \phi(e''', e', e'')$.

Szabó (1987) criticized Stoke's work so strongly that he raised doubts as to whether the equations of motion for viscous fluids should also be named after him; for Szabó places de Saint-Venant's work above that of Stokes. Truesdell (1980a) has already rejected some of his critique. Here one should bear in mind, however, that it was Stokes who, on the basis of continuum mechanics, introduced two material constants to the constitutive equations for viscous fluids. This second constant is nowhere to be found in de Saint-Venant's work. Similarly, Stokes precisely described the limits of validity of constitutive equations with the two material values. Thus, Szabó's criticism appears to be totally incomprehensible.

We will not go into the solution of complicated differential equations here (Navier-Stokes equations); the reader is referred to Szabó (1987) for more information.

With the theory of ideal fluids (see Section 2.3.), and the theory of viscous fluids, the basis for the fluid motions which are to be treated in porous media are given. We will come back to this in Sections 3.6. We will complete this chapter (regarding the theory of viscous fluids) with these comments:

Several remarks concerning the development of the theory of gases are worth adding here. Already in 1662, Boyle (see M. Rühlmann, 1880) had published the constitutive law that the pressure of a gas is proportional to the density. Obviously independently, Mariotte developed the same law in 1679, and founded it by carefully performed experiments.

The second important law, concerning the expansion of gases for different temperatures (Gay-Lussac) was discovered by Charles (see M. Rühlmann, 1880), who was at that time professor of physics at the Conservatoire des art et métiers in Paris, and famous as the inventor of the hydrogen balloon. The reason that this law carries the name of Gay-Lussac (or Dalton) may lie in the fact that these physicists performed such extensively precisely and careful tests that their results were considered valid and completely sufficient for a long time.

3.4.3 Biographical Notes

George Gabriel Stokes (1819–1903)

Sir George Stokes was the son of a clergyman. His father, Gabriel Stokes, who was the Rector of Skreen, County Sligo, married Elizabeth Haughtone and had eight children with her, of whom George was the youngest.

Stokes was taught at home; he learned reading and arithmetic from the Parish Clerk, and Latin from his father who had been a scholar at Trinity College, Dublin. In 1835, he was sent to Bristol College for two years.

In 1837, he commenced residence at Cambridge, which he was to call his home, almost without interruption, for sixty-six years. At Pembroke College, his mathematical abilities soon attracted attention and, in 1841, he graduated

Fig. 3.18. George Gabriel Stokes (1819–1903)

as Senior Wrangler and first Smith's Prizeman. In the same year, he was elected Fellow of his College.

After his degree, Stokes lost little time in applying his mathematical abilities to original investigations. During the next three to four years, papers appeared dealing with hydrodynamics. A memoir of great importance on the *Friction of Fluids in Motion, etc.* followed a little later (1845). In applying his purely kinematical analysis to viscous fluids, Stokes laid down the following principle: "That the difference between the pressure on a plane passing through any point P of a fluid in motion and the pressure which would exist in all directions about P if the fluid in its neighborhood were in a state of relative equilibrium depends only on the relative motion of the fluid immediately about P; and that the relative motion due to any motion of rotation may be eliminated without affecting the differences of the pressures mentioned above."

In 1846, Stokes communicated to the British Association a report on recent researches in hydrodynamics. This was a model of what such a survey should consist of. In 1847 and 1849, he investigated anew the theory of oscillatory waves and wrote another great memoir on the Dynamical Theory of Diffusion.

In 1857, Stokes married Miss Robinson, daughter of Dr. Romney Robinson, an astronomer. Their first residence was in the Trumpington Road; afterwards, they took Lensfield Cottage, where they resided until her death in 1899. In 1902, he was chosen Master of Pembroke. Stokes died on the first of February, 1903.

A consideration of Stoke's work (even though it is limited only to what has been touched upon here) can only lead to the conclusion that in many subjects – and especially in Hydrodynamics and Optics – the advances which we owe to him are fundamental. Due to his basic findings, many scientific honors were showered upon him. He was a Foreign Associate of the French Institute and Knight of the Prussian Order Pour le mérite. He was awarded the Gauss Medal in 1877, the Arago on the occasion of the Jubilee Celebration in 1899, and the Helmholtz in 1901. In 1889, he was made a Baronet on the recommendation of Lord Salisbury. From 1887 to 1891, he represented the University of Cambridge in Parliament. He was Secretary of the Royal Society from 1854 to 1885 and President from 1885 to 1890; he received the Rumford Medal in 1852, and the Copley in 1893 (after Lord Rayleigh, 1903).

3.5 The Mohr-Coulomb Failure Condition and other Plasticity Theory Studies

After Cauchy had analyzed the state of forces on the inside of a continuum and created the stress concept, which enabled him to give the final form to Euler's axioms (balance of momentum and balance of moment of momentum), and after the basic relations of the linear elasticity theory had been developed, investigations into the strength of the solid body, and the question as to what load the solid body would fail under, began. Generally, Galileo (1638), is considered to have undertaken the first examinations on the strength of solids. Indeed, he propagated ingenious remarks on this subject in the first and second days in the *Discorsi*. He recognized that there had to be an absolute resistance against failure but was not yet able to formulate a failure condition. In Galileo (1638) it is also indicated that Aristotle had already also thought about strength. Leonardo da Vinci (see Zammattio *et al.*, 1981) also had some fundamental thoughts on this subject. One can assume that the problem of strength has never since disappeared from the consciousness of researchers, engineers, and builders.

The first researcher to publish a failure condition was Coulomb (see Section 2.5). Within the framework of his publications on the earth pressure on vertical surfaces, he introduced a condition which is today still named after him. However, in the time to follow, the application of Coulomb's condition remained restricted to the determination of the maximum load of certain boundary-value problems. It was Macquorn Rankine (1856/57) who succeeded in transferring Coulomb's ideas to the general three-dimensional stress and deformation state. However, Rankine limited himself to granular material without cohesion. Another interesting approach to the formulation of a failure condition for granular materials for a three-dimensional stress and deformation states was carried out by Holtzmann (1856), in an article which has been completely ignored and forgotten in the literature. Mohr (1900) was the first to extend Rankine's failure condition and also to take cohesion into consideration. The knowledge that the specification of a failure condition alone is not sufficient for the description of motion is due to Lévy (1871) and De Saint Venant (1871a) or (1871b). In the following section, we will examine the development of the so-called Mohr-Coulomb failure condition and further plasticity theory approaches.

3.5.1 W.J. Macquorn Rankine's Fundamental Failure Condition for Granular Material

The desired expansion of the study of the state of forces inside a granular medium in the case of failure was first indicated by the building contractor H. Scheffler[13] (1851) in his treatise *über den Druck im Inneren einer Erdmasse*:

"Das Gesetz, nach welchem sich der statische Druck im Inneren einer im Gleichgewicht befindlichen Erdmasse vertheilt, liegt noch völlig im Dunkeln. Es ist zwar in der vor kurzem erschienenen Schrift von **Ortmann** (*'Die Statik des Sandes'*) versucht worden, dieses Gesetz aufzuhellen, jedoch ohne Erfolg, da diese Theorie auf Hypothesen gegründet ist, welche aus der Natur der Erden nicht nachgewiesen und auch derselben gewiß nicht gemäß sind.

Bei Anwendungen in der Ausübung hat man sich damit begnügt, den Gesammtdruck zu suchen, welchen eine Erdmasse gegen einen endlichen Teil ihrer Umfangswände ausübt. Zu diesem Ende hat man von der hinter der Wand lagernden Erde denjenigen Theil ermittelt, welcher von allen das größte Bestreben äußert, auf seiner Basis herabzugleiten und die widerstehende Wand aus ihrer Lage zu drängen. Jedoch auch diese Aufgabe, obgleich sie nur einen speciellen Fall der Statik der Erden ausmacht, ist nicht allgemein und vollständig gelöst, indem die Untersuchungen von der ohne Beweis hingestellten Hypothese ausgehen, daß die Basis des Erdtheils vom größten Schube eine *Ebene* sei.

Um die Statik der Erden gründlicher und umfassender zu entwickeln und die dabei vorkommenden Erscheinungen eben so genügend zu erklären, wie in der Hydrostatik die Gleichgewichts-Erscheinungen vollkommener Flüssigkeiten, ist wesentlich noch die Kenntniß des Gesetzes nöthig, nach welchem sich der auf irgend einen Punct einer Erdmasse ausgeübte Druck durch die Masse *fortpflanzt.*"

"The law, which governs the static pressure in the inner part of an earth mass, lies still completely in darkness. Attempts have indeed been made by **Ortmann** to reveal this law in his paper *Die Statik des Sandes*, however, without success, because this theory is founded on hypotheses which from the nature of the earth are not proved and are certainly not in accordance with that.

By application in the practice one has made to do with the total pressure, which an earth mass applies to a finite part of its surrounding walls. Finally, one has determined, from the part of the earth stored behind the wall, which makes the most effort from all to glide on its basis and to push the retaining wall from its position. However, also this task, although it is only a special case of the static of the earth, is not generally and completely solved, because the investigations start from hypotheses, not proved, that the basis of the part of the earth with the maximum shear is a plane.

In order to develop the static of the earth more profoundly and comprehensively and to explain the phenomena appearing in this connection (as the equilibrium-phenomena of ideal fluids in hydrostatics) the knowledge of a law is more important, through which the pressure acting at any arbitrary point of the earth mass propagates through the mass."

The approaches following the quoted remarks are very disappointing; they do not live up to what the title and the introductory comments promise. Thus, H. Scheffler, like all other scientists before him, dealt with a special boundary-value problem and, in a sometimes obscure way, he examined the earth pressure.

[13] Hermann Scheffler, born on 10 October, 1820, dedicated himself to engineering, mathematics, and science, as well as philosophy. He wrote numerous papers, for example: *Die mechanischen Prinzipien der Ingenieurkunst, Theorie der Gewölbe, Futtermauern und eiserne Brücken, Körper und Geist, Theorie der Augenfehler und der Brille*, and *Die Grundlagen der Wissenschaft.*

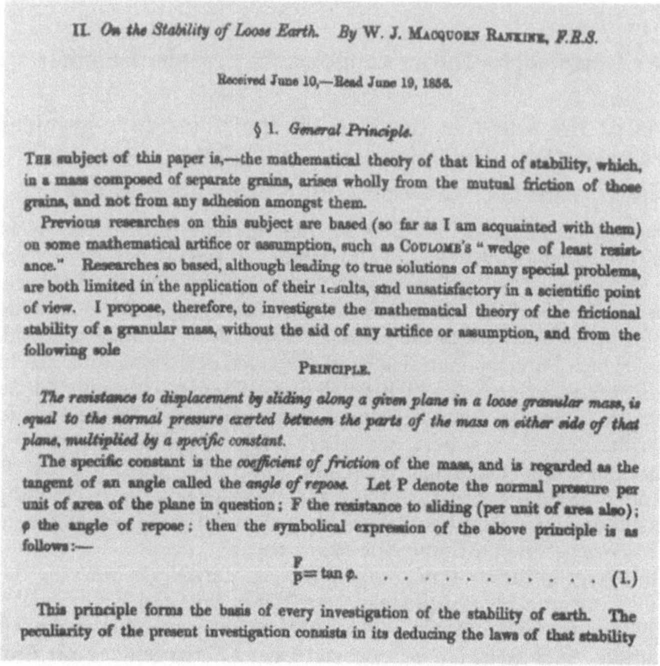

II. *On the Stability of Loose Earth.* By W. J. Macquorn Rankine, *F.R.S.*

Received June 10,—Read June 19, 1856.

§ 1. *General Principle.*

The subject of this paper is,—the mathematical theory of that kind of stability, which, in a mass composed of separate grains, arises wholly from the mutual friction of those grains, and not from any adhesion amongst them.

Previous researches on this subject are based (so far as I am acquainted with them) on some mathematical artifice or assumption, such as Coulomb's "wedge of least resistance." Researches so based, although leading to true solutions of many special problems, are both limited in the application of their results, and unsatisfactory in a scientific point of view. I propose, therefore, to investigate the mathematical theory of the frictional stability of a granular mass, without the aid of any artifice or assumption, and from the following sole

Principle.

The resistance to displacement by sliding along a given plane in a loose granular mass, is equal to the normal pressure exerted between the parts of the mass on either side of that plane, multiplied by a specific constant.

The specific constant is the *coefficient of friction* of the mass, and is regarded as the tangent of an angle called the *angle of repose.* Let P denote the normal pressure per unit of area of the plane in question; F the resistance to sliding (per unit of area also); φ the angle of repose; then the symbolical expression of the above principle is as follows:—

$$\frac{F}{P}=\tan\varphi. \quad \ldots \ldots \ldots \ldots \quad (1.)$$

This principle forms the basis of every investigation of the stability of earth. The peculiarity of the present investigation consists in its deducing the laws of that stability

Fig. 3.19. Macquorn Rankine's study on the stability of loose earth

Evidently, he had no knowledge of the great progress which had been made in continuum mechanics. Cauchy's brilliant successes (see Section 3.1) seem to have been totally unknown to him; Cauchy's stress concept, for example, does not appear anywhere in his work.

W.J. Macquorn Rankine's (1856/57) studies are, however, written much more clearly and are mathematically faultless. In a letter to Professor Stokes of February 19, 1856, he had already formulated his:

Principle of the Stability of Earth

"At each point in a mass of earth the directions of greatest and least compressive stress are at right angles to each other; and the condition of stability is, that at each point the ratio of the difference of those stresses to their sum shall not exceed the sine of the angle of natural slope of earth."

As early as June, 1856, four months after the written communication to Professor Stokes, Macquorn Rankine presented his detailed examinations *On the Stability of Loose Earth* to the Royal Society. In the section §1 *General Principle* he formulated the aim of his work:

"The subject of this paper is, – the mathematical theory of that kind of stability, which, in a mass composed of separate grains, arises wholly from the mutual friction of those grains, and not from any adhesion amongst them."

He therefore did not bring cohesion into his considerations. Furthermore, in regard to the state of research in his time, he remarked:

"Previous researches on this subject are based (so far as I am acquainted with them) on some mathematical artifice or assumption, such as COLOUMB's 'wedge of least resistance'. Researches so based, although leading to true solutions of many special problems, are both limited in the application of their results, and unsatisfactory in a scientific point of view. I propose, therefore to investigate the mathematical theory of the frictional stability of a granular mass, without the aid of any artifice or assumption, and from the following sole

PRINCIPLES

The resistance to displacement by sliding along a given plane in a loose granular mass, is equal to the normal pressure exerted between the parts of the mass on either side of that plane, multiplied by a specific constant."

Macquorn Rankine explained the specific constant as a friction coefficient and identified it with the tangent of the angle of repose, but he did not discuss who first chose this method of procedure. Obviously, this idea of replacing the friction coefficients by the tangent of the angle of repose was being used by mechanics specialists and engineers. He wrote further:

"Let P denote the normal pressure per unit of area of the plane in question; F the resistance to sliding (per unit of area also); φ the angle of repose; then the symbolical expression of the above principle is as follows:

(1)[14] $$\frac{F}{P} = \tan\varphi \quad \dots \qquad (3.5.1)$$

This principle forms the basis of every investigation of the stability of earth."

He further showed:

"§2 *Corollary as to Limit of Obliquity of Pressure*

..., let R be the total pressure, per unit of area, at any point of the given plane, making with the normal to the plane the angle of obliquity θ; let P be the normal and Q the tangential component of R; so that

$$P = R\cos\theta\,; \quad Q = R\sin\theta\,; \qquad \frac{Q}{P} = \tan\theta\,; \qquad (3.5.2)$$

then it is necessary to stability that

(2) $$Q \leq F = P\tan\varphi$$

and consequently that

$$\theta \leq \varphi\,." \qquad (3.5.3)$$

Macquorn Rankine then analyzed the stress state. He must have been familiar with the stress concept and the further foundations of continuum mechanics. Before he presented his contribution *On the Stability of Loose Earth* to the Royal

[14] The numbers on the left-hand side of the page come from the original text

Society in 1856, he had been working in the area of the elasticity theory for many years. According to Todhunter and Pearson (1886/1893), the creation of the word *strain* is, among other things, due to him. He introduced the principal stresses and principal axes and then explained in §3:

"Let P_x be the greatest and P_y the least of the three principal pressures at a given point O, and let On, making with Ox the angle $x\,On = \Psi$, be a line in the plane $x\,y$. Let R_n be the total pressure on unity of area of a plane normal to On, and let the direction of this pressure make with On the angle θ on the side of On towards x, so that the components of R_n are respectively, normal, $P_n = R_n \cos\theta$; tangential, $Q_n = R_n \sin\theta$.

Let the half-sum of the greatest and least principal pressure be denoted by

$$M = \frac{P_x + P_y}{2}\,, \tag{3.5.4}$$

and their half-difference by

$$D = \frac{P_x - P_y}{2}\,. \tag{3.5.5}$$

Then the magnitude and direction of the pressure exerted at the plane normal to On are given by the following equations: –

(5)
$$R_n = \sqrt{M^2 + D^2 + 2MD\cos 2\Psi}\,, \qquad \tan\theta = \frac{D\sin 2\Psi}{M + D\cos 2\Psi}\,, \tag{3.5.6}$$

or otherwise by the following : –

(6)
$$P_n = M + D\cos 2\Psi\,, \qquad Q_n = D\sin 2\Psi\,.\text{''} \tag{3.5.7}$$

In the last equations, the well-known transformation relations are used to determine the normal and shear stresses in any cut surface where the principal stresses are given.

Macquorn Rankine then states the maximum value Θ of the angle of obliquity without giving any proof. This refers to an maximum/minimum determination of θ from equation $(3.5.6)_2$:

"The *maximum* value Θ of the obliquity θ, and the corresponding position of the normal On, are given by the following equations:

(7)
$$\Psi = \frac{\pi}{4} + \frac{1}{2}\sin^{-1}\frac{D}{M}\,, \qquad \Theta = \sin^{-1}\frac{D}{M}\,, \tag{3.5.8}$$

to which correspond the following pressures, total, normal, and tangential:

(8)
$$R(\Psi) = \sqrt{M^2 - D^2}\,, \qquad P(\Psi) = M\left(1 - \frac{D^2}{M^2}\right), \qquad Q(\Psi) = D\sqrt{1 - \frac{D^2}{M^2}}\,.\text{''} \tag{3.5.9}$$

Macquorn Rankine graphically depicted the determined relations in the principal stress plane with the axes of the greatest and smallest principal stresses. He then stated the general stresses transformation equations, explained the equilibrium conditions, and considered some special boundary-value problems. After this, he arrived at the main point of his treatise, the formulation of the failure condition:

"φ being, as in §2, the angle of repose of a given kind of earth, and Θ, as in §3, the greatest obliquity of the pressure at any point in any plane traversing that mass, it appears from equation (2) that the condition of the stability of the mass is

$$\Theta \leq \varphi . \tag{3.5.10}$$

From this equation, the following propositions are deduced:

THEOREM I: *At each point in a mass of earth, the ratio of the difference of the greatest and least pressure to their sum cannot exceed the sine of the angle of repose.*

This theorem[15] follows from the second of the equations (7), its symbolical expression being

$$\frac{P_x - P_y}{P_x + P_y} = \frac{D}{M} \leq \sin \varphi . \tag{3.5.11}$$

THEOREM II: *The following is the expression of the condition of stability of a mass of earth in terms of the pressures at a point, referred to any pair of rectangular axes, O'_x, O'_y, in the plane of greatest and least pressures: –*

$$\frac{(P_{x'} - P_{y'})^2 + 4Q'^2}{(P_{x'} + P_{y'})^2} = \frac{D^2}{M^2} \leq \sin^2 \varphi ." \tag{3.5.12}$$

With the formulation of the above failure condition, Macquorn Rankine had succeeded in moving away from special problems and in giving a generally valid formulation of the failure condition. The consequent use of the stress concept, and the principal stresses introduced by Cauchy, enabled him to state the orientation of the planes of failure, see Eqn. (3.5.8). As possible planes of failure, he only considered those areas whose normal vectors lie in the plane of the greatest and smallest principal stresses P_x and P_y. He justified his assumptions with the following comment:

"The equations given above solve a particular case only of the general problem viz. the case in which the given elementary stresses act in the plane of greatest and least pressure. But in all actual problems respecting the stability of the earth, the plane of greatest and least stress is known; and it is therefore unnecessary to apply to that subject the general problem as to the finding of the axes of pressure in space of three dimensions; a problem which requires the solution of a cubic equation."

That Macquorn Rankine's assumptions, in regard to the formulation of the failure condition in stress components and to the orientation of possible planes of failure, are valid for a further area of the general three-dimensional stress state has recently been shown analytically by W. Gollub (1989).

Evidently unaware of Macquorn Rankine's excellent treatise, Lévy (1873) presented a contribution on the same theme to the French Academy of Sciences

[15] Already published in the *Proceedings of the Royal Society* for the 6th of March, 1856.

in 1867, with results corresponding to Macquorn Rankine's. Lévy's paper was directed to a committee which was to report on it, and to which de Saint-Venant (1870) also belonged. This report appeared in the *Comptes rendus*, T. XX, pp. 217-228, 1870. Later (Tome XV of the journal, pp. 271-280), de Saint-Venant acknowledged Macquorn Rankine's importance in a footnote (see Todhunter and Pearson, 1886/1893 and, for a detailed discussion, see Section 3.5.3).

A. Considere (1870) and E. Winkler (1872) also published studies similar to those undertaken by Rankine and Lévy. We will discuss Winkler's ideas here in more detail, as Considere's work did not contain anything essentially new. In the preface to his book, E. Winkler stated that he had already written down a new theory of earth pressure in a dissertation which he had handed in to the university in 1860. He was publishing it to ensure priority. In the first section, he examined the plane stress state and obtained already well-known results. In a following chapter, he formulated the *Cohäsionsgesetz in Verbindung mit dem Reibungsgesetz* and from minimum/maximum considerations he arrived at the statement that the cohesion did not have any influence on the position of the failure plane against the principal stresses. He further concluded that there existed two failure planes, which together formed an angle of $90° - \varphi$ (φ is the angle of repose), which was halved by the greatest principal stress. In the third chapter, he mentioned some examples of applications. In the second part of his book, he wrote a short history of the earth pressure theory which is in parts, however, imprecise and incorrect.

The description of some contents of Winkler's book concludes this section, and we will now turn to Mohr's work.

3.5.2 O. Mohr's Contributions to the Determination of the Elasticity and Failure Limits

Mohr's first contributions to the failure condition for granular materials originated in the years 1871 and 1872 (Mohr, 1871, 1872), with his work on the earth pressure theory. He tried to depict the pressure distribution with the help of graphic, geometrical methods. There was, however, nothing fundamentally new in his treatise, so it need not be discussed here. Moreover, his statements were controversial, as the partially polemical dispute between him and Winkler shows. Winkler (1872) criticized Mohr's work with sharp comments. For example:

"Mohr sagt, daß die Ingenieure zu dieser Annahme kein volles Vertrauen haben, ... Wir glauben indeß , dass die Mohrsche Annahme noch weniger Vertrauen verdient ..."

"Mohr says that the engineers do not have confidence in this assumption... We believe, however, that Mohr's assumption deserves even less confidence..."

Mohr (1900) answers no less polemically:

"Herr Winkler leitet seine Bemerkungen in einer etwas sonderbaren Weise ein, indem er nämlich mir eine Behauptung unterschiebt, die ich nicht gemacht habe und diese Behauptung als dann mit einer Angabe berichtigt, die aus meinem Aufsatz entnommen ist."

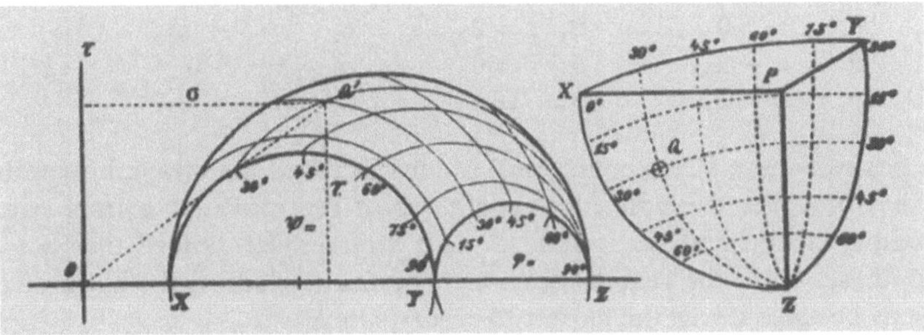

Fig. 3.20. Graphical depiction on the stress state by Mohr

"Mr. Winkler introduces his remarks in a somewhat strange way, namely by contributing a statement to me which I did not make and then correcting this with data which was taken from my article."

The entire contents of the essay are an attempt to deny Winkler's accusations. However, these polemical essays did not contribute to the explanation of earth pressure.

In this respect, Mohr (1900) was much more successful in the development of a condition for defining elasticity and failure limits. In the work which appeared in two parts in the *Zeitschrift des Vereines deutscher Ingenieure* in 1900, he laid down his fundamental thoughts on this topic. Primarily, he pointed out that he had already referred to the indurability of the older hypotheses with respect to the determination of the elasticity and failure limits in a paper (Mohr, 1882), though this had not met with much success in the scientific world. In fact, in this work, which contains the stress circuit later to be named after him (the graphical depiction of the transformation equations of tensor components in various rotated coordinate systems), he did give the first indications of the formulation of realistic elasticity and failure limits. In this manner, he consistently used the graphic method that he had developed for the depiction of the stress state (Fig. 3.20). However, his explanations do not attain the same clarity as his later treatise (Mohr, 1900).

In this later treatise, Mohr (1900) first explained the stress concept and presented his graphic depiction of the stress state. He then introduced four special principal stress states which should lie at the failure limit:

$$\begin{array}{lllll}
\text{I} & \sigma_x = 0, & \sigma_y = 0, & \sigma_z = +\kappa_1, & \\
\text{II} & \sigma_x = -\kappa_2, & \sigma_y = 0, & \sigma_z = 0, & \\
\text{III} & \sigma_x = -\kappa_3, & \sigma_y = 0, & \sigma_z = +\kappa_3, & \\
\text{IV} & \tau_{max} = +\kappa_4, & & &
\end{array} \quad (3.5.13)$$

where σ_x, σ_y, and σ_z represent the principal stresses, $\tau_{\max}$ represents the maximal shear stress, and κ_1 through κ_7 represent response parameters. The principal stress components at the elasticity limit are:

$$
\begin{array}{llll}
\text{V} & \sigma_x = 0, & \sigma_y = 0, & \sigma_z = +\kappa_5, \\
\text{VI} & \sigma_x = -\kappa_6, & \sigma_y = 0, & \sigma_z = 0, \\
\text{VII} & \sigma_x = -\kappa_7, & \sigma_y = 0, & \sigma_z = +\kappa_7.
\end{array}
\tag{3.5.14}
$$

On the basis of these stress conditions, Mohr discussed the strength hypotheses of that time, for example, the hypotheses of the maximal normal stress, the maximal strain, and the maximal shear stresses. He proved that not all the hypotheses could be realistic, although he acknowledged the shear stress hypothesis because it was the closest to reality, since:

"...sie die Spannungen der Gleit- und Bruchflächen als die maßgebenden Größen inbetracht zieht. Denn das Gleiten und Brechen wird doch wohl zunächst abhängig sein von den Spannungen derjenigen Flächen, in welchen diese Bewegungen wirklich stattfinden."

"...it considers the stresses of the gliding and failure surfaces as the governed quantities. For the gliding and breakage will at first be dependent on the stresses of those areas where these motions actually occur."

He then founded his new theory, namely that no experience could justify the assumption that the occurrence of motions would be dependent on the shear stress, and independent of the normal stress, of those areas. He proceeded to formulate, quite generally, his condition for the determination of the elasticity and the failure limits:

"Die Elastizitätsgrenze und die Bruchgrenze eines Materials werden bestimmt durch die Spannungen der Gleit- und Bruchflächen."

"The elasticity limit and the failure limit are determined by the stresses of the gliding and failure surfaces."

Moreover, Mohr pointed out that this condition would also be valid if only the shear stresses were governing. He further asked which changes in the shear stresses would lead to a transgression of the limits. It was most probable that such a change could only be achieved by an increase in the shear stress. Through this consideration, Mohr obtained the following extension of his hypothesis:

"Die Schubspannung der Gleitfläche erreicht an der Grenze einen von der Normalspannung und von der Materialbeschaffenheit abhängigen größten Wert."

"The shear stress of the gliding surfaces reaches at the limit a maximal value which depends on the normal stress and on the state of the material."

With the help of his graphic method of depicting the stress state (which he did so masterly), he evaluated his own hypothesis and thus arrived at a remarkable finding:

"Jede in einem Körperpunkte entstehende Gleitfläche oder Bruchfläche geht durch die y-Achse dieses Punktes, also durch die Richtung der mittleren Hauptspannung σ_y."

"Any gliding or failure surface occurring in a material point developed, cross the y-axis at this point and thus they have the direction of the intermediate principal stress σ_y."

With this, Mohr proved Macquorn Rankine's assumptions about the failure surface orientation, using graphical representations. The complete analytical proof has been carried out by W. Gollub (1989).

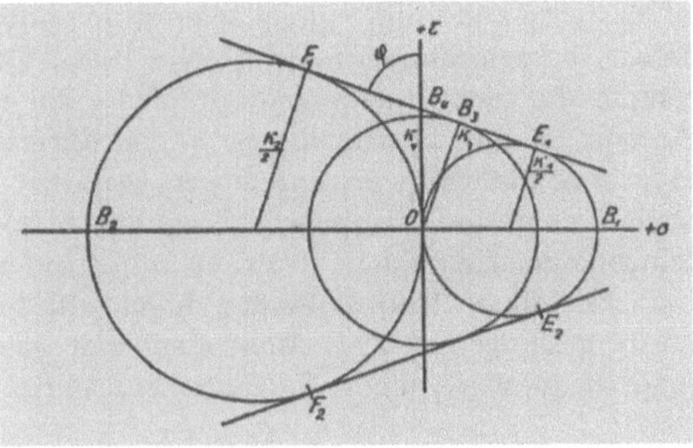

Fig. 3.21. Mohr's formulation of the failure condition

In the graphic (see Fig. 3.21) and verbal formulations of the condition for the elasticity and failure limits, Mohr also considered non-linear relations between shear and normal stress. In the mathematical formulation, however, he limited himself exclusively to one linear relation:

$$\frac{\sigma_z - \sigma_x}{2} - \frac{\kappa_1}{2} = \left(\frac{\kappa_1}{2} - \frac{\sigma_z + \sigma_x}{2}\right) \cos\varphi \,. \tag{3.5.15}$$

These specifications can be seen in Fig. 3.21. With σ_z and σ_x, the largest and smallest principal stresses are designated. The values κ_1 and φ are material constants whose significance is shown in Fig. 3.21.

If one replaces Mohr's material-dependent constants with the material values used today, i.e. with the angle of internal friction ρ and cohesion c, with help of the relations:

$$\varphi = 90^\circ - \rho, \quad \frac{\kappa_1}{2}(1 + \cos\varphi) = c\cos\rho \tag{3.5.16}$$

(see Fig. 3.21), then one obtains the known form of the so-called Mohr-Coulomb fracture condition:

$$\frac{\sigma_z - \sigma_x}{2} + \frac{\sigma_z + \sigma_x}{2}\sin\rho - c\cos\rho = 0 \,. \tag{3.5.17}$$

Mohr's failure hypothesis was a consistent further development of Coulomb's ideas. By transferring Coulomb's statement to the three-dimensional stress state, Mohr completed Coulomb's theory. In this way, he went one step further than Macquorn Rankine by including cohesion in his failure condition.

In the conclusion of the first part of his fundamental work, he remarked that similar considerations could be made regarding the stress states at the elasticity limit.

In the second part of his 1900 treatise, he verified his theory empirically. He wrote in on yield figures, fracture planes, elasticity, and failure limits. At the end of the second part, he pointed to a newer experimental study by Guest, who had carefully investigated stress states on the failure limit for steel, ingot steel, copper, and brass and consequently found that the intermediate principal stress did not influence the quantities of the boundary stresses.

Mohr took up the theme of elasticity and failure limits again in his book *Abhandlungen aus dem Gebiete der Technischen Mechanik* (Mohr, 1914). The explanations for the theory contain the statements already decided upon in 1900. What is new lies in the report on the experiments of T. von Kármán (1911), which in part speaks in favor of Mohr's theory and in part against it.

The importance of Mohr's failure condition was quickly recognized in soil mechanics and has been used since to determine limit states. Mohr's considerations are, however, far-reaching (see Gollub, 1989). Without a doubt, with his treatise on the failure condition for granular materials, Mohr made a decisive contribution to the development of the plasticity theory for frictional materials.

3.5.3 Extension of the Plasticity Theory

Experimental studies of the plastic behavior of metals under high hydrostatic pressure were carried out by H. Tresca (1864) between 1864 and 1872. The results led him to recognize that a metal will go over to the yield state when the maximum shear stress takes on a critical value. With this finding, it was explained for the first time that the onset of failure in a material is independent of the hydrostatical stress state. De Saint-Venant, who wrote a report on Tresca's work for the French Academy of Sciences, immediately recognized the importance of Tresca's discovery and thus, in the following year, he turned to plasticity theory problems.

At the same time that Macquorn Rankine, Mohr, Tresca, and others were trying to establish a failure condition (as has already been discussed in the preceding sections), de Saint-Venant and Lévy were producing new and decisively important ideas for the creation of a general plasticity theory in 1870. De Saint-Venant (1871a, b) published these ideas, in the form of interesting memoirs, in the *Journal de mathématique*.

In the first part of these memoirs, de Saint-Venant (1871a) recalled the new work of Cauchy, Poisson, and Navier on the foundations of continuum mechanics, the elasticity theory, and the theory of fluids. He gave an interesting overview of the historical development of the plasticity theory, in which he referred to Tresca's memoirs from the years 1864 to 1870. He reported on the attempts by Tresca and himself to gain solutions by purely kinematical considerations. He concluded that the problem was, however, more than kinematic; he stated that it was, in fact, mechanical and that the stresses had to be included in the considerations.

De Saint-Venant then arrived at the formulation of the foundation equations for the plasticity theory; in this manner, he regarded the plastic state as being incompressible. First, he stated the equations of motion of the ideal fluid:

$$(1) \qquad \frac{dp}{dx} = \rho\left(X - \frac{du}{dt} - u\frac{du}{dx} - v\frac{du}{dy} - w\frac{du}{dz}\right),$$

$$\frac{dp}{dy} = \rho\left(Y - \frac{dv}{dt} - u\frac{dv}{dx} - \qquad \ldots\right), \tag{3.5.18}$$

$$\frac{dp}{dz} = \rho\left(Z - \frac{dw}{dt} - \qquad \ldots\right),$$

and the incompressibility condition:

(2) $$\frac{du}{dx} + \frac{dv}{dy} + \frac{dw}{dz} = 0\,.$$

He then replaced the LHSs in (1) by the divergence of the stress tensor, by:

(3) $$\frac{dp_{xx}}{dx} + \frac{dp_{xy}}{dy} + \frac{dp_{zx}}{dz}, \frac{dp_{xy}}{dx} + \frac{dp_{yy}}{dy} + \frac{dp_{zy}}{dz}, \frac{dp_{xz}}{dx} + \frac{dp_{yz}}{dy} + \frac{dp_{zz}}{dz}\,. \tag{3.5.19}$$

An orthogonal coordinate system x, y, and z is used as a basis. X, Y, and Z are the applied accelerations, and u, v, and w are the velocities in the directions x, y, and z. The density is marked by ρ and the time by t; p is the hydrostatic pressure and $p_{..}$ are the stress components. The four equations in (1) and (2) represent relations between the three velocity components and the six stress components. Thus, five defining equations are missing.

Tresca showed that when a material reached the plastic state, the maximum shear stress took on a constant value K, which he measured for various materials. This yield condition is known in the literature as Tresca's yield condition. Tresca showed further that the maximum shear stress fell in the direction of the maximum gliding velocity.

De Saint-Venant only treated the case of two-dimensional plasticity, the case where the motion in all planes parallel to the x-z-plane is the same. He first stated the well-known transformation relations for the shear stress components $p_{x'z'}$ in a coordinate system x'-z' turned around the angle α:

$$\begin{aligned} p_{x'z'} &= -\,p_{xx}\sin\alpha\cos\alpha + p_{zz}\sin\alpha\cos\alpha + p_{zx}(\cos^2\alpha - \sin^2\alpha) \\ &= \frac{p_{zz} - p_{xx}}{2}\sin 2\alpha + p_{zx}\cos 2\alpha\,. \end{aligned} \tag{3.5.20}$$

This takes on the maximum value for

(5) $$\tan 2\alpha = \frac{p_{zz} - p_{xx}}{2\,p_{xz}}, \tag{3.5.21}$$

and is then of the intensity

$$\frac{1}{2}\sqrt{4\,p_{xz}^2 + (p_{zz} - p_{xx})^2}\,. \tag{3.5.22}$$

Then Tresca's failure condition becomes

(6) $$p_{zx}^2 + \left(\frac{p_{zz} - p_{xx}}{2}\right)^2 = K^2\,. \tag{3.5.23}$$

De Saint-Venant made analogous considerations for the shear velocity components and arrived at the transformation relation:

$$(7)\qquad \frac{dw'}{dx'} + \frac{du'}{dz'} = \left(\frac{dw}{dz} - \frac{du}{dx}\right)\sin 2\alpha + \left(\frac{dw}{dx} + \frac{du}{dz}\right)\cos 2\alpha\,, \qquad (3.5.24)$$

which reaches its maximum value at

$$(8)\qquad \tan 2\alpha = \frac{\dfrac{dw}{dz} - \dfrac{du}{dx}}{\dfrac{dw}{dx} + \dfrac{du}{dz}}\,. \qquad (3.5.25)$$

With Tresca's hypothesis that the direction of the maximum shear stresses and the maximum gliding velocities would coincide, de Saint-Venant arrived at the complete set of fundamental equations for the uniplanar plasticity theory, after replacing p_{xx}, p_{zz}, and p_{xz} by Lamés terms N_x, N_z, and T:

$$\frac{dN_x}{dx} + \frac{dT}{dz} = -\rho\left(X - \frac{du}{dt} - u\frac{du}{dx} - w\frac{du}{dz}\right),$$

$$\frac{dT}{dx} + \frac{dN_z}{dz} = -\rho\left(Z - \frac{dw}{dt} - u\frac{dw}{dx} - w\frac{dw}{dz}\right), \qquad (3.5.26)$$

$$\frac{du}{dx} + \frac{dw}{dz} = 0\,,$$

$$T^2 + \left(\frac{N_z - N_x}{2}\right)^2 = K^2\,, \qquad (3.5.27)$$

$$(9)\qquad \frac{N_z - N_x}{2T} = \frac{\dfrac{dw}{dz} - \dfrac{du}{dx}}{\dfrac{dw}{dx} + \dfrac{du}{dz}}\,. \qquad (3.5.28)$$

These five equations of the uniplanar plasticity theory (*Cinq équations d'hydro-stéréodynamique ou de plasticodynamique*) are the defining equations for the five unknown quantities u, w, N_x, N_z, and T.

De Saint-Venant remarked that even in this simple case of uniplanar plasticity, it would be difficult to solve the equations. He surmised, however, that it should be possible for cylindrical flow.

In a concluding section on page 316 (see Fig. 3.22), he remarked that the six stress components:

$$(10)\qquad p_{xx} = N_x\,,\ p_{yy} = N_y\,,\ p_{zz} = N_z\,,\ p_{yz} = T_x\,,\ p_{zx} = T_y\,,\ p_{xy} = T_z\,, \qquad (3.5.29)$$

should be assigned to corresponding expressions,

$$(13)\left\{\begin{array}{l}
\frac{dN_x}{dx}+\frac{dT_z}{dy}+\frac{dT_y}{dz}=-\rho\left(X_0-\frac{du}{dt}-u\frac{du}{dx}-v\frac{du}{dy}-w\frac{du}{dz}\right),\\
\frac{dT_z}{dx}+\frac{dN_y}{dy}+\frac{dT_x}{dz}=-\rho\left(Y_0-\frac{dv}{dt}-u\frac{dv}{dx}-v\frac{dv}{dy}-w\frac{dv}{dz}\right),\\
\frac{dT_y}{dx}+\frac{dT_x}{dy}+\frac{dN_z}{dz}=-\rho\left(Z_0-\frac{dw}{dt}-u\frac{dw}{dx}-v\frac{dw}{dy}-w\frac{dw}{dz}\right),\\
4\left(K^2+q\right)\left(4K^2+q\right)+27r^2=0,\\
\frac{du}{dx}+\frac{dv}{dy}+\frac{dw}{dz}=0,\\
\frac{T_x}{\frac{dv}{dz}+\frac{dw}{dy}}=\frac{T_y}{\frac{dw}{dx}+\frac{du}{dz}}=\frac{T_z}{\frac{du}{dy}+\frac{dv}{dx}}=\frac{N_y-N_z}{2\left(\frac{dv}{dy}-\frac{dw}{dz}\right)}=\frac{N_z-N_x}{2\left(\frac{dw}{dz}-\frac{du}{dx}\right)}.
\end{array}\right.$$

Fig. 3.22. Lévy's fundamental equations of the three-dimensional plasticity theory

(11)

$$2\varepsilon\frac{du}{dx},\ 2\varepsilon\frac{dv}{dy},\ 2\varepsilon\frac{dw}{dz},\ \varepsilon\left(\frac{dv}{dz}+\frac{dw}{dy}\right),\ \varepsilon\left(\frac{dw}{dx}+\frac{du}{dz}\right),\ \varepsilon\left(\frac{du}{dy}+\frac{dv}{dx}\right), \tag{3.5.30}$$

which completed the equations for plastic bodies. He pointed out that the same equations could be used to describe the motion of viscous fluids.

In the second treatise, as previously mentioned, Lévy (1871) first developed the fundamental equations of three-dimensional plasticity theory. From here, he applied the theory specifically to cylindrical plastic flow. The displacement components in an orthogonal coordinate system are, in turn, termed u, v, and w, and the applied accelerations, X_0, Y_0, and Z_0. For the normal and shear stress components, Lévy used the notation N_x, N_y, N_z and T_x, T_y, T_z. Moreover, he introduced the deviators:

$$\begin{aligned}
\Delta x &= N_x-\frac{1}{3}(N_x+N_y+N_z),\\
\Delta y &= N_y-\frac{1}{3}(N_x+N_y+N_z),\\
\Delta z &= N_z-\frac{1}{3}(N_x+N_y+N_z).
\end{aligned} \tag{3.5.31}$$

With the density ρ and the constant K in Tresca's yield condition, as well as the abbreviations:

(12)

$$\begin{aligned}
&\Delta y\,\Delta z+\Delta z\,\Delta x+\Delta x\,\Delta y-T_x^2-T_y^2-T_z^2=q,\\
&\Delta x\,T_x^{\,2}+\Delta y\,T_y^2+\Delta z\,T_z^2-\Delta x\,\Delta y\,\Delta z-2T_xT_yT_z=r,
\end{aligned} \tag{3.5.32}$$

and the hypothesis on the coaxiality of shear stresses and gliding velocities, Lévy established a total of nine equations for the determination of the six unknown stress components and the three unknown displacement components:

$$\frac{dN_x}{dx} + \frac{dT_z}{dy} + \frac{dT_y}{dz} = -\rho\left(X_0 - \frac{du}{dt} - u\frac{du}{dx} - v\frac{du}{dy} - w\frac{du}{dz}\right),$$
$$\frac{dT_z}{dx} + \frac{dN_y}{dy} + \frac{dT_z}{dz} = -\rho\left(Y_0 - \frac{dv}{dt} - u\frac{dv}{dx} - v\frac{dv}{dy} - w\frac{dv}{dz}\right), \quad (3.5.33)$$
$$\frac{dT_y}{dx} + \frac{dT_x}{dy} + \frac{dT_z}{dz} = -\rho\left(Z_0 - \frac{dw}{dt} - u\frac{dw}{dx} - v\frac{dw}{dy} - w\frac{dw}{dz}\right),$$

(13) $$4(K^2 + q)(4K^2 + q) + 27r^2 = 0, \quad (3.5.34)$$
$$\frac{du}{dx} + \frac{dv}{dy} + \frac{dw}{dz} = 0, \quad (3.5.35)$$

$$\frac{T_x}{\frac{dv}{dz} + \frac{dw}{dy}} = \frac{T_y}{\frac{dw}{dx} + \frac{du}{dz}} = \frac{T_z}{\frac{du}{dy} + \frac{dv}{dx}} = \frac{N_y - N_z}{2(\frac{dv}{dy} - \frac{dw}{dz})} = \frac{N_z - N_x}{2(\frac{dw}{dz} - \frac{du}{dx})}. \quad (3.5.36)$$

After stating the fundamental equations of the plasticity theory on the basis of Tresca's yield condition, Lévy remarked that de Saint-Venant had not considered the stress N_y in his development of the uniplanar plasticity theory. From the last of the above equations, taking the incompressibility theory into consideration, he obtained with $\frac{dv}{dy} = 0$:

$$N_y = \frac{N_z + N_x}{2}. \quad (3.5.37)$$

In concluding his excellent treatise, Lévy considered the special case of cylindrical plastic flow.

The abbreviations q and r in Tresca's yield condition, chosen by Lévy, represent the second and third invariants of the stress deviator $\mathbf{T}^D$:

$$q = \mathrm{II}_{\mathbf{T}^D} = -\frac{1}{2}\mathbf{T}^D \cdot \mathbf{T}^D,$$
$$r = \mathrm{III}_{\mathbf{T}^D} = \mathbf{T}^D\mathbf{T}^D \cdot \mathbf{T}^D. \quad (3.5.38)$$

Lévy was obviously the first scientist to formulate a yield condition, here Tresca's yield condition, in the invariants of the stress state. The best known failure conditions up to now, Mohr-Coulomb's and Tresca's conditions, have been "stress conditions". With these conditions, the possible position of the failure must be additionally stated in order to be able to apply the corresponding stresses to the failure condition.

The above system of nine equations is neither clearly nor completely formulated. First, Tresca's flow condition contains a clerical error – in the second factor, the square is missing. Moreover, it has not been taken into consideration that as a result of the assumed incompressibility, the hydrostatic pressure enters into the set of unknown quantities as an additional unknown. Finally, in the last equation of the above system, the following relation is missing:

$$\frac{N_x - N_y}{2\left(\frac{du}{dx} - \frac{dv}{dy}\right)} \,. \tag{3.5.39}$$

However, the fact remains that M. Lévy's treatise of 1871, presented to the French Academy on June 20, 1870, stands among the highlights of the creation of a general mathematical plasticity theory.

In a further treatise, de Saint-Venant (1871b) introduced the stress function Ψ for uniplanar deformations in the case of equilibrium in the following way:

$$N_x = \frac{d^2\,\Psi}{dz^2}\,,\; N_z = \frac{d^2\,\Psi}{dx^2}\,,\; T = -\frac{d^2\,\Psi}{dxdz}\,. \tag{3.5.40}$$

With these relations, both equilibrium conditions of the plane deformation state are fulfilled.

After entering these stress relations into Tresca's yield condition, he obtained a partial differential equation for the determination of the stress function Ψ:

$$4\left(\frac{d^2\,\Psi}{dx\,dz}\right)^2 + \left(\frac{d^2\,\Psi}{dx^2} - \frac{d^2\,\Psi}{dz^2}\right)^2 = 4\,K^2\,. \tag{3.5.41}$$

He suggested that this differential equation be solved approximately. De Saint-Venant proceeded to formulate the boundary conditions for various problems. In doing so, he differentiated three classes: (1) those in which the material on the surface remained in the elastic domain, (2) those in which the material on the surface lay in the plastic domain, and (3) those in which the material on the surface changed from the plastic to the elastic state. He called the boundary conditions *équations définies ou déterminées*, whereas he termed the above-mentioned fundamental equations *équations indéfinies*. As examples, de Saint-Venant treated the torsion of a cylinder in the elasto-plastic state and the bending of a beam with a rectangular cross-section.

With this, we will conclude the discussion of de Saint-Venant's and Lévy's work. De Saint-Venant treated the plasticity theory further in other works, and these contributions can be regarded as supplements to the three excellent works discussed above.

In 1885, Beltrami (1885) introduced a new idea into the discussion about the formulation of the yield condition. He assumed that the failure of a material occured when the entire specific strain energy of the linear-elastic body took on a limited value, i.e., he introduced an energy criterium. Thus, the onset of the yield state depended on the hydrostatic stress state, a rule which could not be accepted for many metallic materials.

In 1904, Huber (1904) improved on Beltrami's hypothesis by demanding that, with deformation states connected with a decrease in volume, only the deviatoric deformation work be considered. If, on the other hand, an increase

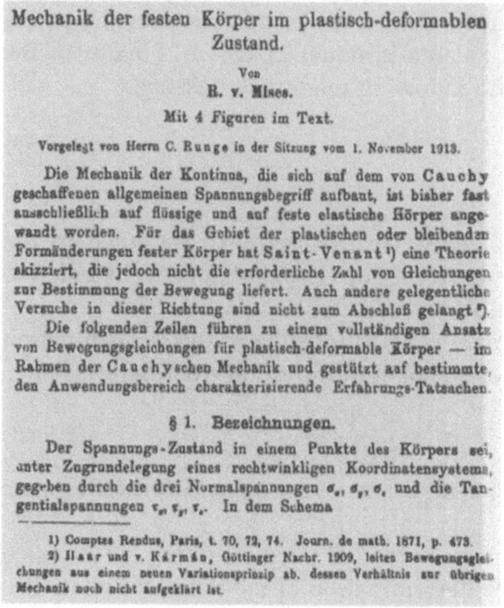

Mechanik der festen Körper im plastisch-deformablen Zustand.

Von

R. v. Mises.

Mit 4 Figuren im Text.

Vorgelegt von Herrn C. Runge in der Sitzung vom 1. November 1913.

Die Mechanik der Kontinua, die sich auf dem von Cauchy geschaffenen allgemeinen Spannungsbegriff aufbaut, ist bisher fast ausschließlich auf flüssige und auf feste elastische Körper angewandt worden. Für das Gebiet der plastischen oder bleibenden Formänderungen fester Körper hat Saint-Venant[1]) eine Theorie skizziert, die jedoch nicht die erforderliche Zahl von Gleichungen zur Bestimmung der Bewegung liefert. Auch andere gelegentliche Versuche in dieser Richtung sind nicht zum Abschluß gelangt[2]).

Die folgenden Zeilen führen zu einem vollständigen Ansatz von Bewegungsgleichungen für plastisch-deformable Körper — im Rahmen der Cauchyschen Mechanik und gestützt auf bestimmte, den Anwendungsbereich charakterisierende Erfahrungs-Tatsachen.

§ 1. Bezeichnungen.

Der Spannungs-Zustand in einem Punkte des Körpers sei, unter Zugrundelegung eines rechtwinkligen Koordinatensystems, gegeben durch die drei Normalspannungen $\sigma_x, \sigma_y, \sigma_z$ und die Tangentialspannungen τ_x, τ_y, τ_z. In dem Schema

1) Comptes Rendus, Paris, t. 70, 72, 74. Journ. de math. 1871, p. 473.

2) Haar und v. Kármán, Göttinger Nachr. 1909, leiten Bewegungsgleichungen aus einem neuen Variationsprinzip ab, dessen Verhältnis zur übrigen Mechanik noch nicht aufgeklärt ist.

Fig. 3.23. Von Mises' pioneering work (1913)

in volume was present, then Beltrami's hypothesis was to be used. Huber's hypothesis went unnoticed for a long time; this may be partially due to the fact that the treatise was written in Polish.

In Huber's work, the conjecture appeared, for the first time, that only the amount of the deviatoric deformation work of a linear-elastic body was responsible for the onset of the plastic state, even if only for a certain deformation state.

In a pioneering work of 1913, von Mises (1913) developed what is today known as the theory of rigid-ideal plastic material, on the basis of purely mathematical considerations. In the introductory comments, he pointed out that in "the area of plastic or permanent deformations of solids," de Saint-Venant had outlined a theory which did not, however, give the required amount of equations for the determining of motion. He was referring to a work of de Saint-Venant from 1871 (*Comptes Rendus*, Paris, t.70, 72, 74; *Journ. de Math.* 1871, p.443), and was obviously not familiar with de Saint-Venant's other basic works or, more particularly, those of Lévy.

After formulating the material-independent basic equations – introduction of the stress, distortion, and velocity gradient tensors with the accompanying kinematic relations as well as the introduction of the principal stresses and the invariants of the tensors – he turned to some empirical data:

"(a) Alle festen Körper verhalten sich bei hinreichend kleinen Spannungen wie elastische: es besteht eine ein-eindeutige Zuordnung zwischen Spannung und Deformation."

"(a) All solids behave, for sufficiently small stresses, like elastic: there is a unique relationship between stress and deformation"

Von Mises remarked that, with isotropic material behavior where no direction in the space was preferred, the stress and deformation tensor had the same principal directions and, the principal values were connected over two material constants:

"(b) Ist die Elastizitätsgrenze erreicht, so verhält sich der feste Körper wesentlich wie eine zähe, nahezu inkompressible Flüssigkeit."

"(b) If the elastic limit is reached, the solid behaves essentially like a viscous, nearly incompressible liquid."

This behavior of liquids is characterized by the fact that it is not the deformation state, as in the case of elastic bodies, but the deformation process which causes the stresses. Von Mises further pointed out that the volume changes occuring in the plastic range were of the magnitude of the elastic ones and could thus be neglected. He concluded that only the deviator of the stress state could be regarded as a linear function. He thereby obtained the following constitutive relations, where $\mathbf{T}^D$ is the deviator of the stress tensor $\mathbf{T}$, $\mathbf{D}^D$ the deviator of the symmetric part of the velocity gradients $\mathbf{D}$ ($\mathbf{D}^D$ is equal to $\mathbf{D}$ because of the assumed incompressibility), and k a magnitude which will be explained later:

$$\mathbf{T}^D = k\,\mathbf{D}^D\,. \tag{3.5.42}$$

This equation corresponds to the relation for viscous liquids (see Section 3.3). The difference lies in the significance of the magnitude k. This difference is explained by the following empirical data:

"(c) Verändert man unter Aufrechterhaltung aller Verhältnisse den absoluten Wert der Geschwindigkeiten, mit denen eine Bewegung vor sich geht, so ändert sich, bei plastisch deformablen Körpern, die Arbeit nicht, die zur Erziehlung einer bestimmten Formänderung verbraucht wird."

"(c) If one changes under maintenance of all proportions the absolute value of the velocities, with which a motion proceeds, for plastically deformable bodies the mechanical work, which is consumed for gaining a certain deformation, does not change."

Taking the previous constitutive relations into consideration, the deviatoric deformation work per unit is given by:

$$\mathbf{T}^D \cdot \mathbf{D}^D = k\,\mathbf{D}^D \cdot \mathbf{D}^D\,. \tag{3.5.43}$$

"Multipliziert man alle Geschwindigkeiten mit einem Faktor c, so ändert sich dieser Ausdruck proportional $k\,c^2$. Zugleich wird aber die Dauer des Deformations-Vorganges im Verhältniss $1 : c$ verkürzt, die ganze Arbeit wird also proportional kc. Daher muß der Proportionalitätsfaktor k umgekehrt proportional der Geschwindigkeit sein. Oder auch: Der Spannungstensor $\mathbf{T}^D$ bleibt derselbe, wenn alle Komponenten von $\mathbf{D}^D$ im gleichen Verhältnis verändert werden."

"If one multiplies all velocities with a factor c, this term changes proportionally kc^2. At the same time, the period of the deformation-proceedings is shortened with the proportion $1 : c$; the total work is thus proportional kc. Therefore, the proportional factor k must be inversely proportional to the velocity. Or also: the stress tensor $\mathbf{T}^D$ remains the same when all components of $\mathbf{D}^D$ are changed in the same proportion."

From the last sentence, von Mises concluded:

"(c') Bei plastischen Deformationen bleibt die Spannung stets an der Elastizitätsgrenze."

"(c') By plastic deformations the stress remains always at the elasticity limit."

Finally, he formulated the last empirical data:

"(d) In einem Koordinatensystem, das die Haupt-Tangentialspannungen zu Koordinaten hat, erscheint die Elastizitätsgrenze als eine geschlossene, den Nullpunkt einschließende, Kurve in der Ebene."

"(d) In a coordinate system which has the principal tangential stresses as coordinates, the elasticity limit appears as a closed curve in the plane including the zero-point."

Von Mises then showed his appreciation of Mohr's research and discussed the failure condition where the maximum shear stress was limited. In the graphic representation, on the so-called deviator level, this led to a hexagon. Through tests, only the corner points of the hexagon were determined. The straight-line connection came from the assumption that the intermediate middle principal stress did not have any influence on the failure. This assumption does not appear to be very plausible, to the extent that one would not attempt to replace the hexagon with a simpler figure such as the enveloped circle. In the spatial representation, this leads to:

$$\tau_1^2 + \tau_2^2 + \tau_3^2 = 2K^2 \,. \tag{3.5.44}$$

This is the famous von Mises yield condition which limits the second invariant of the stress deviators (here expressed by the principal values of the stress deviator τ_1, τ_2, and τ_3) by the constant K. Von Mises pointed out that his yield condition allowed a much simpler analytical treatment than Tresca's yield condition, without the difference being any greater than the margin left by the experiments that had been carried out up until then.

In the concluding chapter of his work, von Mises stated the complete system of equations for the description of the motion of a body in the plastic state. These were the three equations of motion, the six stresses, and velocity gradient relations, the incompressibility condition for the determining of the unknown hydrostatic pressure, and the yield condition, which he stated for the general stress state, and which today are written as:

$$\mathrm{II}_{\mathbf{T}^D} = K^2 \,, \quad \mathrm{II}_{\mathbf{T}^D} = \frac{1}{2}\mathbf{T}^D \cdot \mathbf{T}^D \,. \tag{3.5.45}$$

With this, the "complete system of equations of motion for plastic deformable bodies" became known. As a "boundary condition", the following was added: the specification of the velocity components u, v, and w for each surface point. It can be replaced by the specification of the surface stress on the entire surface area, or by a part of it.

Von Mises pointed out additionally that, in the case of a plane motion, his relations were to be reduced to those of de Saint-Venant. He crowned his excellent treatise with the depiction of his fundamental equations in vector and

Man kann die Gl. (I) bis (IV) sehr einfach mit Benützung von Vektor-Symbolik schreiben. Bezeichnet $\bar{v}$ den Geschwindigkeitsvektor, $\bar{x}$ den Vektor der spezifischen Kraft, so hat man:

$$(\mathrm{I}') \qquad \varrho \frac{d\bar{v}}{dt} = \bar{x} - \operatorname{grad} p + \nabla \bar{\bar{\sigma}}',$$

$$(\mathrm{II}') \qquad \bar{\bar{\sigma}}' = k \bar{\bar{\lambda}},$$

$$(\mathrm{III}') \qquad \operatorname{div} \bar{v} = 0,$$

$$(\mathrm{IV}') \qquad -(\bar{\bar{\sigma}}')_2 = \frac{4K^2}{3}.$$

Dabei bedeutet das Symbol ∇ in (I') die an der Dyade auszuführende Differentiation, die durch (I) bestimmt ist. Der Index 2 in (IV') soll darauf hinweisen, daß von der Dyade $\bar{\bar{\sigma}}'$ die zweite der in § 1, Gl. (8) angeschriebenen Orthogonal-Invarianten zu nehmen ist.

Aus (I') bis (IV') läßt sich auch leicht $\bar{\bar{\sigma}}'$ eliminieren und man erhält:

$$(\mathrm{a}) \qquad \varrho \frac{d\bar{v}}{dt} = \bar{x} - \operatorname{grad} p + \nabla (k \bar{\bar{\lambda}}),$$

$$(\mathrm{b}) \qquad \operatorname{div} \bar{v} = 0,$$

$$(\mathrm{c}) \qquad k^2 = -\frac{4K^2}{3(\bar{\bar{\lambda}})_2}.$$

Multipliziert man (I') skalar mit $\bar{v}$ und integriert nach dem Volumen, so findet man nach entsprechender Umformung, daß die Dissipationsfunktion durch (21) dargestellt wird, wodurch die Übereinstimmung des Ansatzes mit unserer Annahme c) in § 2 erwiesen ist.

Straßburg i. E., 4. Oktober 1913.

Fig. 3.24. Von Mises' fundamental equations for an ideal-plastic ductile solid

tensor calculus, and stated the whole theory of ideal plastic deformations in half a page (see Fig. 3.24).

He remarked on the first three equations of his system of fundamental equations (see Fig. 3.24):

"Der Ansatz (I ') bis (III ') stimmt völlig überein mit dem für zähe Flüssigkeiten. Allein dort ist die Größe *k* die gegebene Zähigkeitszahl, bei uns ist es eine Reaktionsgröße, die sich aus der Kenntnis der Bewegung selbst errechnen läßt. Hierzu dient die Aussage, daß die Spannung während der plastischen Deformation an der Elastizitätsgrenze bleibt."

"The relations (I ') to (III ') coincide completely with that for viscous liquids. Only, there, the quantity *k* is the given viscosity-coefficient, whereas for us it is a reaction quantity, which can be calculated from the knowledge of the motion itself. For this, the statement that the stress remains during the plastic deformation at the elasticity limit, is necessary."

Corresponding comments, as we have seen, are to be found in de Saint-Venant's work.

It appears that von Mises did not know the three previously discussed works of de Saint-Venant and Lévy, as his rather critical remark in the introduction to his derivation of de Saint-Venant's theory would otherwise be incomprehensible.

Usually, von Mises' treatise of 1913 is only mentioned in connection with the quotation of his yield condition. In addition to the development of the yield condition, the work contains the complete system of fundamental equations of the theory of ideal plastic bodies. Von Mises' expositions are of a rare clarity and lucidity, both from the point of view of mechanics as well as of mathematics.

In their correctness, clear representation, and completeness they go beyond de Saint-Venant's and Lévy's contributions.

3.5.4 Biographical Notes

Maurice Lévy (1838–1910)

Maurice Lévy, born, in Ribauville (Upper-Rhine) on February 28, 1838, attended the École Polytechnique and the École des Ponts et Chaussées. After graduating from the École Polytechnique (1858) and the École des Ponts et Chaussées (1861), he was appointed as an ordinary engineer in 1862. In the time to follow, he split his time between administration duties and research in mechanics. In 1872, after he had spent eight years in the Departments, he got a position as an engineer for bridges and roads. The city of Paris gave him the position of a chief engineer in 1880, and of a general inspector in 1894.

As a professor, he was a coach in mechanics at the École Polytechnique between 1862 and 1883. From 1875, he occupied the chair for applied mechanics at the École Centrale des Arts et Manufactures. The Académie des Sciences chose him as a member of the mechanics department. As a mathematician with great abilities, he contributed scholarly works to various branches of mechanics, in particular to kinematics, hydraulics, hydrodynamics, and elasticity. As already mentioned, Lévy wrote excellent articles on the strength of material (plasticity), which were far ahead of their time.

Maurice Lévy died in 1910.

William John Macquorn Rankine (1820–1872)

Macquorn Rankine was born in Edinburgh on July 5, 1820. He was educated at Ayr academy from 1828 to 1829, and at the high school of Glasgow in 1830. From 1836 to 1838, he was enrolled as a student at the university of Edinburgh. In 1838, he became a pupil of John Benjamin MacNeil, a surveyor of the north of Ireland under the railway commission. For four years, Rankine was employed in surveys and schemes for river improvements, water works, and harbor works. Between 1844 and 1845, and afterwards until 1848, he was employed on various railway projects.

Around 1848, he commenced a series of research into molecular physics which was to occupy him at intervals during the rest of his life. At the end of 1849, he sent his great paper *On a formula for calculating the expansion of liquids by heat* to the Royal Society of Edinburgh. He was made a fellow of the Royal Society of Edinburgh in 1849, and awarded the Keith medal in 1854. He was appointed to the chair of civil engineering and mechanics at Glasgow University in 1855. In 1857, he resigned from the associateship of the Institution of Civil Engineers and shortly afterwards was elected the first president of the

Fig. 3.25. William John Macquorn Rankine (1820–1872)

Institute of Engineers in Scotland upon its establishment. He died in Glasgow on Christmas Eve, 1872.

Besides writing in various newspapers, he contributed more than one hundred and fifty papers to scientific journals, many of them exhaustive essays on mathematical or physical questions and genuine contributions to the advancement of science.

Christian Otto Mohr (1835–1918)

Christian Otto Mohr was born on October 8, 1835, in Wesselburen/Holstein (the North Sea Coast). At the age of sixteen, he enrolled at the Polytechnikum in Hannover to study engineering. At the age of twenty, he joined the Hannover (and later the Oldenburg) Railway Company. In 1860, he published his first substantial scientific paper on a problem in structural mechanics. During this time, Christian Otto Mohr designed the first iron truss bridge, which was built close to Lüneburg.

He married Anna Buresch, daughter of the director of the Oldenburg railway company.

After ten years of practical occupation, he was offered the professorship of Technical Mechanics, Laying out of Lines, and Earth Work at the Polytechnikum of Stuttgart, which he accepted. In 1873, Christian Otto Mohr became the successor to Claus Koepcke as the chair of Railway Construction, Hydraulic Engineering, and Graphostatics at what was later to become Technische Hochschule of Dresden. From 1876 on, he taught, in addition to the above-mentioned fields, on the strength of materials and, from 1894 on, he lectured on general technical mechanics and strength of materials. He retired in 1900, after 33 years of teaching and research.

Fig. 3.26. Christian Otto Mohr (1835–1918)

In the year of his retirement, he published his treatise on the strength problem of material, which belongs, without any doubt, among his best scientific works.

Christian Otto Mohr died on October 2, 1918, in Dresden (after Steiding, 1985).

Richard von Mises (1883–1953)

Richard von Mises, also known by his full name, Richard Martin Edler von Mises, was born on April 19, 1883, in Lemberg, which at that time belonged to the Austro-Hungarian monarchy. He spent his childhood and youth in Vienna, where he studied mechanical engineering at the Technische Hochschule. After his studies, he became an assistant at the Deutsche Technische Hochschule in Brno. Under the instruction of Professor Hamel, who later became Professor of Mechanics in Berlin, he completed his dissertation. Then, he did his habilitation thesis in Brno and, in 1909, he was offered a professorship for applied mathematics in Strasbourg.

At the very beginning of the First World War, he joined the Austrian army as a volunteer. During the war, he worked as an officer on the airfield of Aspern where von Terzaghi, Fröhlich, and Fillunger were also on duty (see Chapter 4). In Aspern, he designed and built an aeroplane. In 1919, he acquired the position of full professor at the Technische Hochschule of Dresden. One year later, von Mises took over the direction of the Institute of Applied Mathematics at Berlin University. Here, in 1921, he founded the Journal for Applied Mathematics and Mechanics (ZAMM). After the Nazis came to power, he moved to Constantinople

Fig. 3.27. Richard von Mises (1883–1953)

in 1934 and, in 1938, to Harvard University. On July 14, 1953, von Mises died in Boston of cancer.

Richard von Mises did not do research solely in the field of plasticity. Rather, he published basic contributions on the stability theory and hydrodynamics, as well as on probability calculus and statistics. Moreover, he was an excellent authority on the works of the German poet Rainer Maria Rilke and also published contributions in the field of philosophy.

3.6 Motion of Liquids in Rigid Porous Solids

With the description of the constitutive behavior of ideal and viscous liquids, as well as, in some parts, the solid matrix, the investigations of the single constituents of the multiphase continua had achieved a well-founded basis. It followed, automatically, to deal with the interaction of liquid and porous solids. This happened in many papers of the 19th century, which were restricted to the treatment of the motion of liquids in rigid porous bodies. The pores of such solid matrices may consist of tube-like forms which have a distinct structure (as in fissured rocks) or they may be statistically distributed with no dependence on a special direction (as in sand or clay). Like the complicated pore structures, the theories are also not simple.

3.6.1 Motion of Liquids Through Narrow Tubes

First, the development of the basic equations of the motion of a liquid in tubes will be discussed. It was Stokes (1845) who first applied the equations of motion of a viscous liquid to the problem of the flow of an incompressible liquid in oblique pipes or canals with regular cross-sections:

"Although the discharge of a liquid through a long straight pipe or canal, under given circumstances, cannot be calculated without knowing the conditions to be satisfied at the surface of contact of the fluid and solid, it may be well to go a certain way towards the solution."

With z as the coordinate in the connection-line of the cross-sections and y, x perpendicular to z, and with α as the oblique-angle, Stokes derived from the basic equations of the motion of viscous liquid, which he had substantially formed, the differential equation to describe the motion of a liquid in oblique pipes and canals. With the assumption that the motion should be homogeneous, the velocity components u, v, w in x, y, z directions reach the following values:

$$u = 0 \ , \qquad v = 0 \ , \qquad w = f(x, y) \ . \tag{3.6.1}$$

Then, from the fundamental equations of the theory of viscous liquids,

$$\frac{dp}{dx} = 0 \ , \qquad \frac{dp}{dy} = g\,\rho\cos\alpha \ ,$$
$$\frac{dp}{dz} = g\,\rho\sin\alpha + \mu\left(\frac{d^2w}{dx^2} + \frac{d^2w}{dy^2}\right) , \tag{3.6.2}$$

are obtained, whereby p is the pressure, g the gravity acceleration, ρ the density, and μ a material-dependent constant.

Then Stokes considered the case that $\frac{dp}{dz} = 0$, an assumption which is only valid for an open canal. With this assumption, he solved the above differential equation in the case of a pipe with a circular cross-section. The complete solution for the case of a horizontal arranged pipe can, however, be obtained from Kirchhoff (1876).

In the above-considered motions, no change in the propagation of vortexes occurs. The liquid is ordered in layers of constant velocities; such motions are called *laminar* motions, and are as a rule restricted to small velocities. When, however, the velocity increases, irregular and turbulent motions are observed. This problem was experimentally investigated by Reynolds (1883). Since a laminar motion with any given average velocity in an arbitrary pipe is mathematically always possible, he concluded that the laminar motions in wide pipes, and at high velocities, had to be unstable. He found that a laminar motion at the average velocity c in a pipe with the radius a became unstable with increasing c. In general, the motion will be stable or unstable whether or not the value

$$R = \frac{ca}{\nu} , \tag{3.6.3}$$

is smaller or larger than a critical numerical value. (In the above equation, ν is the kinematic viscosity and an abbreviation for μ/ρ). Reynolds (1883) found that the critical value of R lay between 950 and 1000. Since the kinematic viscosity depends on the temperature, it is important for the onset of the turbulance at which temperature the liquid flows. After the onset of the turbulence, the friction-resistance is dependent solely on the density, and not on other properties; therefore, it is also not dependent on its temperature.

In the case of slow motions, one introduces, for the friction-resistance W between two layers,

$$W = \mu \frac{du}{dn} , \tag{3.6.4}$$

whereby n is measured normal to the layer, which consists of particles with equal velocity u. The *Poiseuille-Hagen* law can easily be derived from the above relation. This law states that in circular small pipes with the radius a and the cross-section A, the rate of flow is given by:

$$Q = \frac{\pi a^4}{8\mu} \gamma i , \tag{3.6.5}$$

where i is the pressure gradient and γ the specific weight. Furthermore, the average velocity c can be calculated from:

$$c = \frac{Ai\gamma}{8\pi\mu} = 0.03979 \frac{Ai\gamma}{\mu} . \tag{3.6.6}$$

It has been shown by Lamb (1895) that the average velocity of a liquid in a pipe with a triangular cross-section A yields:

$$c = \frac{1}{20\sqrt{3}} \frac{Ai\gamma}{\mu} = 0.02887 \frac{Ai\gamma}{\mu} . \tag{3.6.7}$$

The usefulness of the above formula has become more important since Boussinesq (see Lamb, 1895) showed that small deviations from the triangular cross-section do not essentially change the value of the velocity c.

If the diameter of a narrow tube reaches a critical small value, the well-known capillary phenomenon occurs. Capillarity describes the phenomenon when liquids in narrow tubes, cracks, and pores take on a motion caused by the surface tension of the liquid. Capillarity is based on the intermolecular forces of cohesion and adhesion which depend on the material of the constituents involved. If the forces of adhesion between the liquid and the tube wall are greater than the forces of cohesion between the molecules of the liquid, then a capillary motion occurs. This phenomenon was already described at the early stage of the development of mechanics. According to an encyclopedia (Ed.: Zedler, 1733), the phenomenon was discovered in France or in Italy. In particular, it was recognized that, in the gravity field, the capillary rise of the water in a narrow tube was inversely proportional to the diameter of the tube. Moreover, the cause of capillary motion, adhesion, had been known as early as 1733. It was also found that the capillary effect does not occur for nonmoistening liquids, like mercury. Although all these phenomena and their explanations were known, it was the sharp-witted Laplace (1806) who classified these phenomena. In particular, he introduced the hypothesis that the surface of the liquid shows a tension caused by the water lying beneath the surface.

3.6.2 Flow of a Liquid Through Porous Bodies with Statistically-Distributed Pores

As described in Section 3.3, it was Darcy (1856) who observed, in saturated natural sands, the proportionality of the rate of flow to the pressure gradient:

$$Q = -KA(h_2 - h_1)/h \tag{3.6.8}$$

(the terms in the above relation were explained in Section 3.3.)

If one divides the rate of flow Q through the cross-section of the space where the water runs, one obtains the filter velocity $\bar{c}$ which is, of course, only a fraction of the true average pore velocity c in the pores. Seelheim (1880) discovered that at 12°C the filter velocity in pure quartz-sand with an average diameter of the grains d was determined by:

$$\bar{c} = 0.38\, di\,, \tag{3.6.9}$$

where i is the loss of the pressure head.

Since, for permeability, the small grains which are settled between the larger grains are important, Hazen (1895) introduced the term for the effective size of grains, d_w. This term is characterized by the fact that all grains whose volumes are smaller than the content of a sphere $\pi/6\, d_w^3$ should have a weight of 1/10 of the total weight of the sand. The relation of Hazen (1895) is related to the highest possible permeability and to 10°C:

$$\bar{c} = 1.16\, d_w^2\, i\,. \tag{3.6.10}$$

If there are larger grains or if the motion is faster, then the average velocity does not increase, according to Kröber (1884), as much as the pressure gradient i. The reason for this fact is that a turbulent motion sets in.

New considerations in the discussion of the permeability were introduced by Forchheimer (1901). He extended Darcy's relation by non-linear terms:

$$i = \alpha\bar{c} + \beta\bar{c}^2 + \gamma\bar{c}^3\,, \tag{3.6.11}$$

where α, β, and γ are positive material-dependent constants. Furthermore, he proved that this relation described the test results better than the following relation:

$$i = \alpha\,\bar{c}^{\beta}\,, \tag{3.6.12}$$

with β as a positive constant.

In regard to further modifications of Darcy's law, which are partially based on experimental and partially on theoretical investigations, the reader is referred to Forchheimer's (1930) book, where a comprehensive overview of the historical development of the theory of groundwater flow can be found.

Darcy's linear law is widely applied. However, one must bear in mind that Darcy's law is only valid for stationary flow within a homogeneous velocity field. For general instationary processes, Darcy's law must be considered more or less as a simplifying assumption.

3.6.3 Application

As a simple example for the motion of liquids in rigid porous solids, the development of the so-called "well-formula" will be considered. The well-formula is a relation used to calculate the productivity of a perfect well in a stationary state. The model of such a perfect well is depicted schematically in the following figure.

The model consists of a hollow cylindrical ground-water carrier. It is a rigid porous solid with homogeneous porosity. The ground-water carrier is located at a horizontal, impermeable layer and is surrounded by reposing water at the height H. The assumption of reposing water beyond the distance R from the center of the well represents thereby a useful idealization of the real state, which is characterized by a slow propagation of the sinking beyond R. This, and the assumption that the depth of the water (s_u) in the well must be kept constant by pumping, has led, in the field of engineering sciences, to the assumption of a stationary state.

The first relation to calculate the productivity of a perfect well was developed by Dupuit (1863), immediately after Darcy's findings. Dupuit assumed that the pressure head $h(r, x_3)$, measured from the level of the impermeable layer, was independent of x_3, i. e. , $h(r, x_3) = h(r)$. This assumption corresponds with the presupposition of an anisotropic ground-water carrier with infinite permeability

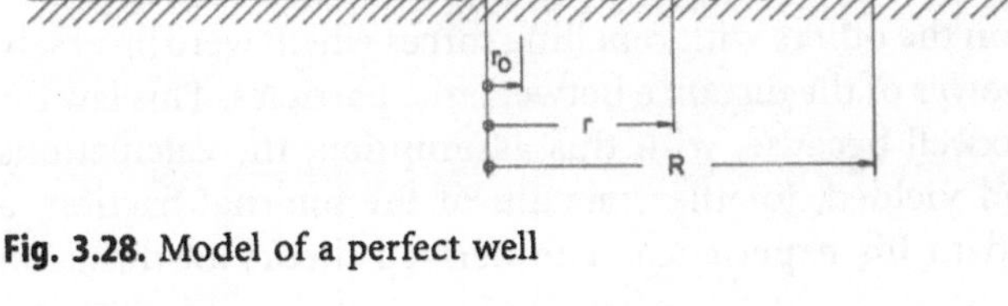

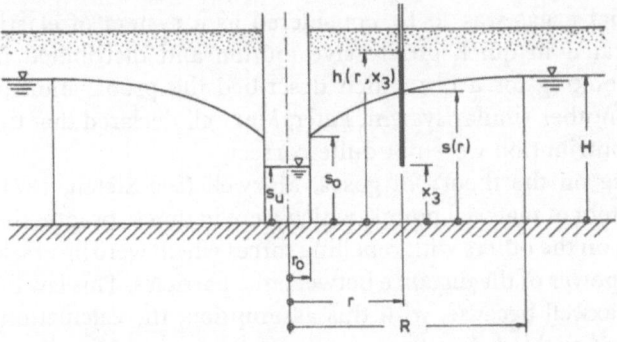

Fig. 3.28. Model of a perfect well

in the vertical direction. Therefore, Dupuit let run the unknown ordinate $s(r)$ of the rotational symmetric well, not into the height s_0, but into the height s_u of the well.

Dupuit's (1863) formula to determine the rate of the water flow Q is given by:

$$Q = \pi k \frac{H^2 - s_u^2}{\log R - \log r_0}, \tag{3.6.13}$$

where k is the permeability coefficient. Later, this relation was considered to be too imprecise, and was improved upon. In particular, Forchheimer (1930), a very famous scientist in hydraulics at that time, solved different boundary value problems with this relation.

3.7 Foundation of the Mixture Theory

3.7.1 Introduction

Already in the classical era, the development of a branch of mechanics of essential importance for the creation of a consistent porous media theory had begun, namely the development of the mixture theory. As we have seen in Section 3.3, it was Fick (1855) who discovered the differential equation for the diffusion problem of liquids with different concentrations. However, Fick (1855) did not proceed from established mechanical principles; rather, he derived his differential equation by an analogy-procedure from the Fourier equation for heat propagation.

A completely different approach to the diffusion problem, to the diffusion of gases, was founded by Maxwell (see Stefan, 1871). Stefan declared that Maxwell was the first scientist to develop the hydrodynamic equations for gas mixtures from the basic principles of mechanics. He developed the dynamic gas theory and treated the theory of diffusion in two papers. In the first, Maxwell proceeded from the assumption that a gas was to be considered as a system of elastic spheres which were located in quick progressive motion and distributed in all directions. He was looking for a law which described the propagation of such a system through another similar system. Later, Maxwell declared that the considerations in this contribution were not quite correct.

In his second treatise on the theory of gases, Maxwell (see Stefan, 1871) considered a gas as a system of material points, which were in quick, progressive motion, and which acted on the others with repelling forces which were inversely proportional to the fifth power of the distance between the particles. This law for forces was chosen by Maxwell because, with this assumption, the calculations were relatively simple and yielded, for the constant of the internal friction, a term which corresponded to his experience. The derived theory of diffusion

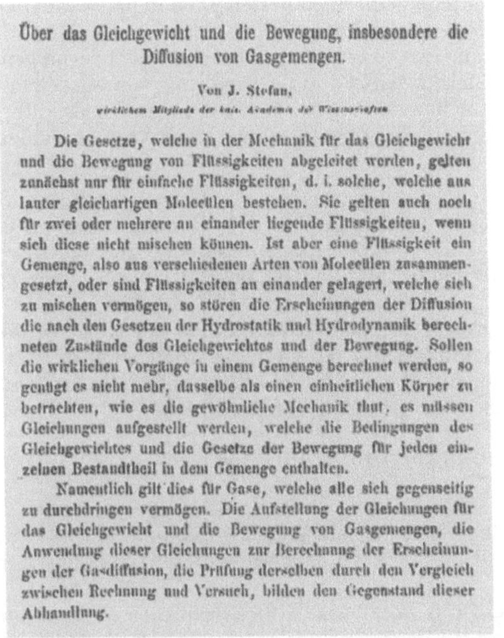

Über das Gleichgewicht und die Bewegung, insbesondere die Diffusion von Gasgemengen.

Von J. Stefan,

wirklichem Mitgliede der kais. Akademie der Wissenschaften

Die Gesetze, welche in der Mechanik für das Gleichgewicht und die Bewegung von Flüssigkeiten abgeleitet werden, gelten zunächst nur für einfache Flüssigkeiten, d. i. solche, welche aus lauter gleichartigen Molecülen bestehen. Sie gelten auch noch für zwei oder mehrere an einander liegende Flüssigkeiten, wenn sich diese nicht mischen können. Ist aber eine Flüssigkeit ein Gemenge, also aus verschiedenen Arten von Molecülen zusammengesetzt, oder sind Flüssigkeiten an einander gelagert, welche sich zu mischen vermögen, so stören die Erscheinungen der Diffusion die nach den Gesetzen der Hydrostatik und Hydrodynamik berechneten Zustände des Gleichgewichtes und der Bewegung. Sollen die wirklichen Vorgänge in einem Gemenge berechnet werden, so genügt es nicht mehr, dasselbe als einen einheitlichen Körper zu betrachten, wie es die gewöhnliche Mechanik thut, es müssen Gleichungen aufgestellt werden, welche die Bedingungen des Gleichgewichtes und die Gesetze der Bewegung für jeden einzelnen Bestandtheil in dem Gemenge enthalten.

Namentlich gilt dies für Gase, welche alle sich gegenseitig zu durchdringen vermögen. Die Aufstellung der Gleichungen für das Gleichgewicht und die Bewegung von Gasgemengen, die Anwendung dieser Gleichungen zur Berechnung der Erscheinungen der Gasdiffusion, die Prüfung derselben durch den Vergleich zwischen Rechnung und Versuch, bilden den Gegenstand dieser Abhandlung.

Fig. 3.29. Stefan's mixture theory

resulted in the velocity of the mixture of two gases being determined by a constant, the so-called diffusion-constant. This is proportional to the square of the absolute temperature. This law was confirmed in tests by Loschmidt (see Stefan, 1871). Moreover, Maxwell's formula for the diffusion constant also yielded the dependence of this constant on the densities of both penetrating gases, and on the repelling force between two gas particles. However, in this case Maxwell's formula cannot be brought into accordance with Loschmidt's tests.

Aside from this last point, Maxwell's theory meets all requirements. One can calculate all phenomena which are related to the mixture of two simple gases. Maxwell's treatise is very complicated. It may be that this complicated treatment discouraged many people and caused doubts concerning the conclusions it drew from past work.

3.7.2 Stefan's Development of the Mixture Theory

A decisive step towards a continuum mechanical theory of mixtures was accomplished by Stefan (1871). He clearly stated:

"Sollen die wirklichen Vorgänge in einem Gemenge berechnet werden, so genügt es nicht mehr, dasselbe als einen einheitlichen Körper zu betrachten, wie es die gewöhnliche Mechanik thut; es müssen Gleichungen aufgestellt werden, welche die Bedingungen des Gleichgewichts und die Gesetze der Bewegung für jeden einzelnen Bestandtheil in dem Gemenge enthalten.

Namentlich gilt dies für Gase, welche alle sich gegenseitig zu durchdringen vermögen. Die Aufstellung der Gleichungen für das Gleichgewicht und die Bewegung von Gasgemengen, die Anwendung dieser Gleichungen zur Berechnung der Erscheinungen der Gasdiffusion, die Prüfung derselben durch den Vergleich zwischen Rechnung und Versuch, bilden Gegenstand dieser Abhandlung."

"If the real processes in a mixture should be calculated, it is not sufficient anymore, to consider the mixture as a uniform body as the common mechanics does; equations must be set up which contain the condition of equilibrium and the laws of motion for every individual constituent in the mixture.

In particular, this is valid for gases, which can penetrate each other. The set up of the equations for the equilibrium and the motion of gas mixtures, the application of these equations to the calculation of the phenomena of gas diffusion, the proof of this by comparison between calculation and test, constitute the subject of this treatise."

Then he introduced the main assumption concerning the interaction forces between the constituents:

"In einem Gemenge erfährt jedes einzelne Theilchen eines Gases, wenn es sich bewegt, von jedem andern Gase einen Widerstand proportional der Dichte dieses Gases und der relativen Geschwindigkeit beider."

"In a mixture, each particle of a gas, if it is in motion, suffers from each of the other gases a resistance which is proportional to the density of this gas and the relative velocity of both."

After the introduction, Stefan (1871) formulated the equations of the equilibrium and motion. For a mixture of two gases he arrived at two equations of motions for the two gases 1 and 2 in the x-direction:

$$\begin{aligned} \rho_1\xi_1 &= \rho_1 X_1 - \frac{dp_1}{dx} - A_{12}\rho_1\rho_2(u_1 - u_2)\,, \\ \rho_2\xi_2 &= \rho_2 X_2 - \frac{dp_2}{dx} - A_{21}\rho_1\rho_2(u_2 - u_1)\,, \end{aligned} \tag{3.7.1}$$

where ξ is the acceleration, ρ the density, X an external acceleration, p the pressure, u the velocity, and $A_{12} = A_{21}$ a material dependent constant. Stefan remarked, in addition, that the above two equations corresponded formally with those which were derived by Maxwell in his theory of gases.

Next, Stefan formulated the balance of mass for the first gas, excluding any mass exchange:

$$\frac{d\rho_1}{dt} + \frac{d(\rho_1 u_1)}{dx} + \frac{d(\rho_1 v_1)}{dy} + \frac{d(\rho_1 w_1)}{dz} = 0\,, \tag{3.7.2}$$

where u_1, v_1, and w_1 are the velocity components of the first gas in x–, y–, and z–directions, and he remarked that similar equations were valid for the second, third, etc., gas.

With the above relations, the general hydrodynamic equations, which are valid for a mixture of gases, were gained.

After some further remarks concerning the resistance force, Stefan (1871) stated that the relation for the resistance force was not only valid for gases, but also for liquids and for electrons in a conductor.

Then, Stefan (1871) applied his basic equations (3.7.1) and (3.7.2) to the diffusion of a mixture of two gases. He considered two gases with the same pressure which were separately situated in a pipe at the beginning of the experiment, and which then freely penetrated it.

He assumed that there were no body forces and that the pressure of both gases was field- and time-independent. Thus, the mixture as a whole was not in motion. He also neglected the accelerations of the individual constituents, because the mixing process happened very slowly.

In this case, Eqn. (3.7.1) reduced to:

$$\frac{dp_1}{dx} + A_{12}\,\rho_1\rho_2(u_1 - u_2) = 0\,,$$
$$\frac{dp_2}{dx} + A_{12}\,\rho_1\rho_2(u_2 - u_1) = 0\,. \tag{3.7.3}$$

To these relations, the balance equations of mass (3.7.2) had to be added:

$$\frac{d\rho_1}{dt} + \frac{d(\rho_1 u_1)}{dx} = 0\,,$$
$$\frac{d\rho_2}{dt} + \frac{d(\rho_2 u_2)}{dx} = 0\,. \tag{3.7.4}$$

It was assumed that the motions of the gases only occurred parallel to the axis of the pipe, which contained the x-axis.

For further development, it is more convenient to replace the densities by the pressures. Let the densities of both gases under the normal pressure p_0, and at $0°C$, be d_1 and d_2. Then let the absolute temperature of the freezing point ($0°C$) be denoted by T_0, and the absolute temperature of both gases by T. According to the law of Mariotte and Gay-Lussac:

$$\rho_1 = d_1\frac{T_0 p_1}{T p_0}\,, \quad \rho_2 = d_2\frac{T_0 p_2}{T p_0} \tag{3.7.5}$$

are then obtained. For further consideration, Stefan (1871) used the following abbreviations:

$$A_{12}\frac{d_1 d_2 T_0^2}{p_0 p_0 T^2} = b_{12}\,,$$
$$p_1 u_1 = q_1, \quad p_2 u_2 = q_2, \quad p_1 + p_2 = p\,. \tag{3.7.6}$$

With these abbreviations, the aforementioned relations can be slightly reformulated:

$$\frac{dp_1}{dx} + b_{12}(p_2 q_1 - p_1 q_2) = 0\,,$$
$$\frac{dp_2}{dx} + b_{12}(p_1 q_2 - p_2 q_1) = 0\,, \tag{3.7.7}$$

$$\frac{dp_1}{dt} + \frac{dq_1}{dx} = 0\,,$$
$$\frac{dp_2}{dt} + \frac{dq_2}{dx} = 0\,. \tag{3.7.8}$$

If one sums up the last two equations and considers:

$$\frac{d(p_1 + p_2)}{dt} = \frac{dp}{dt} = 0\,, \tag{3.7.9}$$

it follows that

$$\frac{d(q_1 + q_2)}{dx} = 0\,, \tag{3.7.10}$$

i.e., in the whole pipe, $q_1 + q_2$ is constant. However, since at the closed ends of the pipes no gas can escape, at the end, and also in the whole pipe, we have

$$q_1 + q_2 = 0\,. \tag{3.7.11}$$

Therefore, Eqns. (3.7.7) simplify to

$$\frac{dp_1}{dx} + b_{12} p q_1 = 0\,,$$
$$\frac{dp_2}{dx} + b_{12} p q_2 = 0\,. \tag{3.7.12}$$

If one eliminates q_1 and q_2 with the help of (3.7.8), one arrives at:

$$\frac{dp_1}{dt} = \frac{1}{b_{12} p} \frac{d^2 p_1}{dx^2} = k \frac{d^2 p_1}{dx^2}\,,$$
$$\frac{dp_2}{dt} = \frac{1}{b_{12} p} \frac{d^2 p_2}{dx^2} = k \frac{d^2 p_2}{dx^2}\,, \tag{3.7.13}$$

where k is the diffusion constant of the respective combination of both gases.

After the derivation of the differential equations (3.7.13), Stefan discussed solutions for special initial conditions and compared his theoretical results with test results from Professor Loschmidt. He stated that there was no remarkable difference. However, the correspondence between experience and theory cannot be considered as proof of the validity of the basic equations used because the differential equation (3.7.13) can also be derived in another way. Therefore, Stefan looked elsewhere for other problems in order to prove his theory. He remarked that the simplest problem would be to investigate the diffusion of three gas constituents. He solved this problem and again compared his theoretical results with those of test observations, concluding again that there was no essential difference between the theoretical and test results.

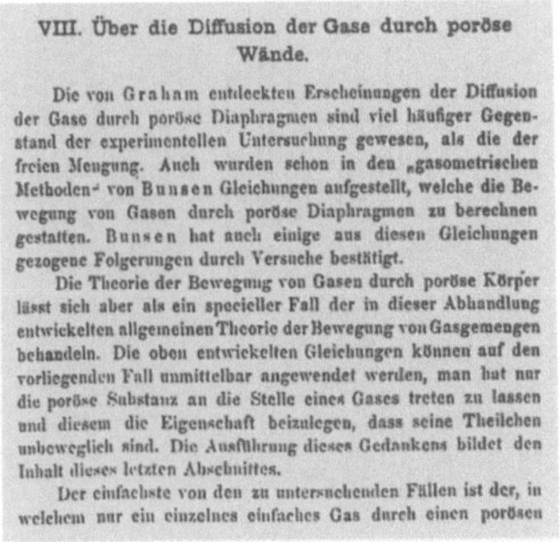

VIII. Über die Diffusion der Gase durch poröse Wände.

Die von Graham entdeckten Erscheinungen der Diffusion der Gase durch poröse Diaphragmen sind viel häufiger Gegenstand der experimentellen Untersuchung gewesen, als die der freien Mengung. Auch wurden schon in den „gasometrischen Methoden" von Bunsen Gleichungen aufgestellt, welche die Bewegung von Gasen durch poröse Diaphragmen zu berechnen gestatten. Bunsen hat auch einige aus diesen Gleichungen gezogene Folgerungen durch Versuche bestätigt.

Die Theorie der Bewegung von Gasen durch poröse Körper lässt sich aber als ein specieller Fall der in dieser Abhandlung entwickelten allgemeinen Theorie der Bewegung von Gasgemengen behandeln. Die oben entwickelten Gleichungen können auf den vorliegenden Fall unmittelbar angewendet werden, man hat nur die poröse Substanz an die Stelle eines Gases treten zu lassen und diesem die Eigenschaft beizulegen, dass seine Theilchen unbeweglich sind. Die Ausführung dieses Gedankens bildet den Inhalt dieses letzten Abschnittes.

Der einfachste von den zu untersuchenden Fällen ist der, in welchem nur ein einzelnes einfaches Gas durch einen porösen

Fig. 3.30. Stefan's treatment of a porous solid

In the eighth section of his valuable paper, Stefan (1871) treated a problem which is closely related to our subject, namely the diffusion of a gas through a porous diaphragm. He proceeded from his fundamental equations for the diffusion of two gases, and replaced one gas by the porous substance in such a way that he assumed that its particles were fixed. Stefan (1871) stated the ratio of the free gas pressure and the partial gas pressure inside the pores with the help of the porosity of the porous solid. Thus, he applied, for the first time, the mixture theory restricted by the volume fraction concept to a binary model within the framework of continuum mechanics.

In the time to follow, Stefan returned again to the problem of mixtures of gases (Stefan, 1872a, b). However, the later contributions did not reach the lucidity and importance of his work from 1871.

3.7.3 Biographical notes

Josef Stefan (1835–1893)

Josef Stefan was born as the son of very poor parents in St. Peter, close to Klagenfurt, Austria, on March 24, 1835. He went to the elementary school as well as the "Gymnasium" in Klagenfurt and enrolled at the University of Vienna in 1853 in order to study mathematics and physics. After passing the examinations, he found a position as a teacher at a Viennese "Realschule" in 1857. His excellent scientific works attracted the attention of experts and, by 1860, he had become a corresponding member of the "Kaiserliche Akademie" and, in 1865, a full member. In 1863, he was offered a full professorship in higher mathematics and physics at Vienna University. Josef Stefan worked mainly in the field of

Fig. 3.31. Josef Stefan (1835–1893)

kinetic gas theory. He, along with H. Helmholtz, made Maxwell's theory public in Europe. In 1879, he formulated the Stefan-Boltzmann-law together with his most famous student, L. Boltzmann. Josef Stefan died in Vienna on January 7, 1893.

3.8 The Foundation of Thermodynamics

Thermodynamics deals with the behavior of physical systems when heat energy, temperature changes, and/or forces are applied and taken away. The heat energy is of special interest because all macroscopic processes in nature are connected with heat-conversions (e.g., friction and reaction heat), and the largest part of the energy which is consumed in the world is gained in the form of heat energy, or appears during the production of energy. It was long ago recognized that heat is a manifestation of the motion of molecules of matter, which are in violent agitation.

There are therefore two basic approaches in thermodynamics: first, the phenomenological approach and, second, the molecular theory approach. In the following passage, only the phenomenological approach will be traced.

Phenomenological thermodynamics precedes from immediately measurable quantities and restricts itself to the macroscopic description of ideal processes. The state of a physical system is thereby completely laid out by a definite set of macroscopic, independent variables.

Thermodynamics is governed by two axioms. The first axiom concerns the balance of energy and states that the total energy of a closed system – for example the sum of mechanical, electrical, chemical, and thermal energy – remains constant. Heat energy can be produced, e.g., by mechanical and electrical work, and can be converted into mechanical and electrical work, respectively. The second axiom represents the entropy balance and states that the entropy of a closed thermodynamic system can change only through an exchange with the surroundings, or can only be increased by itself (entropy production, which is always zero or positive). This axiom also simultaneously expresses the direction of all heat processes: heat cannot flow by itself, i.e., without external work from a body with a low temperature to a body with a higher temperature. If the entropy production is equal to zero, the corresponding process is called a reversible process, otherwise it is an irreversible process.

The formulation of the fundamentals of thermodynamics represents a great achievement of the nineteenth century. Different topics concerning the development will be traced in the following sections (see R. Rühlmann, 1876). The early stages of the mechanical heat-theory dealt with the subsequent changes of the views on the idea of heat, and then with the development of the axioms of thermodynamics. For an extensive and broad overview of the historical development of thermodynamics, see the numerous valuable contributions by Truesdell (1969/84, 1971, 1975/76, 1978, 1980b, and 1985)

3.8.1 Development in the Early Days

The opinion that heat is a manifestation of the motion of the smallest parts of bodies can be found in antiquity. In the atomic-theory of Democritus, the view on the substance of heat seems to be the most developed. He assumed the existence of special heat-atoms of spherical shape and ascribed their motions to the heat phenomena. However, the naive view that heat was a form of matter (for Democritus, heat-atoms) was characteristic of all ancient philosophies of nature. During the Scholasticism, in the Middle-Ages, scientific thinking was solely determined in Europe by the Christian religion, and activities in the field of nature were considered godless. One tried to investigate the phenomena of nature not by observations, experiments, and logical deductions, but rather by studying the Bible, or at least the work of Aristotle. Also in the Renaissance period, there was not much progress concerning the heat phenomena. The first scientist to completely abandon the idea that heat was a kind of matter was Descartes (1677). He stated that a gas consisted of separated molecules which could move independently, "that hereby heat must be understood as nothing other than an acceleration of the motion of such molecules and by coldness a delay of those." However, Hobbes (1668) had also expressed some ideas in the direction of Descartes' view of heat. Locke (1722) also elucidated the molecular point of view very clearly.

Descartes (1677) formulated a first statement of the conversion of energy. He stated that the amount of motion in the universe remained constant. His fundamental statement suffers, however, from his incorrect use of the term "force."

Out of all the scientists in Newton's era, it was Hooke who expressed the clearest statements regarding the heat phenomena in the conversion of energy. He pointed out that the molecules of all bodies, in particular of solids, vibrate and that no other proof is necessary for this phenomenon than the fact that all bodies have a certain state of heat, and that never has an absolute cold body been found. The conversion principles he stated in a somewhat peculiar manner, saying namely that the totality of reality which stimulates our senses consists of matter and motion, and that these are unchangeable in totality.

Up until around 1700, the statements on heat phenomena and the conversion laws had been very vague, and merely had the value of speculation due to the lack of accurate and precise definitions and terms in mechanics, as well as the lack of principles. This changed in the eighteenth century. Hermann (1716), a former student of the famous Jacob Bernoulli, established a definite measure of heat. In his contribution, one can find the statement that the heat of a body is proportional to the density and to the square of the mean molecular speed. It is also remarkable that Johann Bernoulli (see M. Rühlmann, 1885) used the name energy for mechanical work. In a letter to Varignon he wrote:

"If any forces are applied in any manner and, of course, in such a way that they either directly or indirectly act, in this case equilibrium is on hand, if the sum of the positive energies is equal to the sum of the negative. The term energy is understood as the product of the force with the displacement in the projection on the direction of the force and this is to be taken negatively or positively, ..."

According to Truesdell (1975/76), it was Euler who came close to a real kinetic theory of gases. Daniel Bernoulli (1734) reverted to a model of Hermann's type in his book on hydromechanics. Thus, three members of the Basel school of mathematicians and physicists, working in Petrograd, "had laid out in mathematical form some of the elements of the modern molecular concept of gases and had derived equations of state."

The mathematicians and physicists of the first half of the eighteenth century had already rather clear views on heat phenomena, as has been stated above. It is astonishing that, in the second half of that century, the views became more and more murky until, finally, they were replaced by the erroneous idea that heat was a form of matter which remained conserved. In particular, the philosopher Wolf contributed to the propagation of this conception.

The idea of the existence of heat-matter was finally debunked by the experiments of von Rumford (1805). Rumford, originally known as Benjamin Thomson, was ennobled by the Elector Carl Theodor because of his contributions to Bavaria in Germany, and carried from then on the name Rumford after his birthplace, Rumford, New Hampshire, in the USA. He performed many tests on heat phenomena. At the end of the eighteenth century, he recognized that water had the same weight in both the liquid and the solid state. Moreover, he weighted iron and gold spheres in white-hot and in cold states, and always found approximately the same weight. Von Rumford reported on his observations, stating that all of these experiments had convinced him that the weight of a body was in no way changed by heat. However, he knew that with his experiments the dispute had not been completely solved, for the followers of the idea of heat matter could still argue that the weight of the heat matter was too light and had too small a density to establish its weight with a weighting machine. However, Rumford continued logically the chain of thoughts he had begun and finally succeeded in doing the right tests. He reported on these tests to the Royal Society in London on January 25, 1798, under the title: *An inquiry concerning the source of the heat which is excited by friction*. In his experiments, Rumford rotated a blunt drilling tool against the bottom of a cylindrical gun barrel with high pressure. The amount of heat which was produced by the continuous friction was extremely large. The apparatus produced heat as long as it was in action. Thus, the source of heat seemed to be inexhaustible.

Some scientists, notably Joule (1850), declared that Rumford had discovered the mechanical equivalent of heat. However, in all of Rumford's papers there is not one remark known which implies that Rumford had looked for a constant quantitative relation between the amount of heat and the amount of mechanical work which had been applied.

Not all researchers accepted Rumford's valuable results. In particular, some chemists and physicists adhered to the idea of heat-matter and doubted Rumford's test observations. However, Davy (1799) confirmed Rumfords's fundamental results by his own experiments and finally, the brilliant Thomas Young (1807) stated that there was no other alternative than to recognize that heat was produced by friction, and that if it would be produced out of nothing then it could not be matter.

3.8.2
The Achievements of Carnot (1796–1832) and Clapeyron (1799–1864)

Some completely new ideas were brought into the theory of heat by Sadi Carnot (1824). After years of extended studies, he published a treatise, *Réflexions sur la puissance motrice du feu et sur les machines propres à développer cette puissance*, which contained a peculiar consideration regarding processes occuring in machines, in which power is produced by heat. The contents of Carnot's paper can be sketched as follows: his considerations started from the observation that in all fields where power is produced by means of heat, simultaneously an exchange of heat from a hot to a less hot body occurs. In each machine operating by heat there is on the one hand a heat source and, on the other hand, a condenser. Furthermore, there is another body (steam) which comes into contact alternately with the heat source and the condenser and which produces, through its volume changes, the power on which the machine performs.

Carnot considered, at first, the question as to whether the power produced by heat depended on the quality of the third substance which caused the power

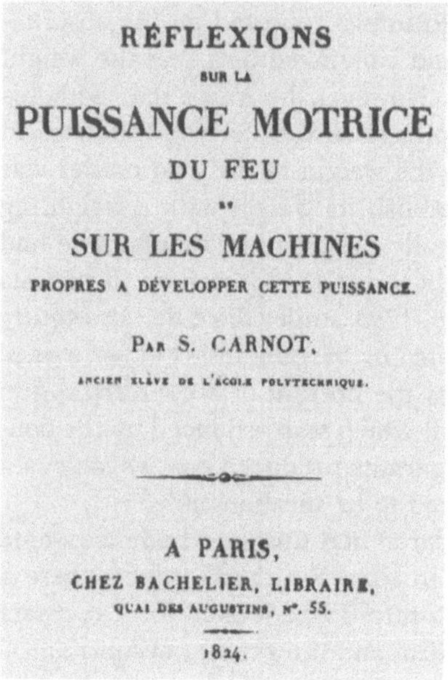
RÉFLEXIONS
SUR LA
PUISSANCE MOTRICE
DU FEU
ET
SUR LES MACHINES
PROPRES A DÉVELOPPER CETTE PUISSANCE.
PAR S. CARNOT,
ANCIEN ÉLÈVE DE L'ÉCOLE POLYTECHNIQUE.

A PARIS,
CHEZ BACHELIER, LIBRAIRE,
QUAI DES AUGUSTINS, N°. 55.
1824.

Fig. 3.32. Carnot's fundamental work

by its volume changes. He showed that produced power was independent of the quality of the steam and solely dependent on the temperature of both bodies. Furthermore, he showed that one must avoid every change in steam temperature which was not connected with a corresponding volume change, in order to obtain the maximum power. Each temperature change which is not connected with a simultaneous volume change must necessarily cause heat to pass over from a hot to a cold body. A harmful transition then occurs which must be avoided; e.g., it must be avoided that bodies of different temperatures come into contact with each other. At least, this should be avoided where possible. Sadi Carnot showed that it would be possible – with the help of envelopes, impermeable to heat – to construct machines which would completely fulfill this condition. He drew attention to the fact that such processes could be brought into the starting state at which the steam transfers the heat from the heat source to the condenser in the above-mentioned manner. In this case, the processes are reversible, i.e., they might also proceed in the opposite direction.

Proceeding from this, Sadi Carnot tried to determine the maximum power which could be produced by a certain amount of heat if this was transferred from a heat source to a colder body. He missed, however, the point that a part of the heat thereby disappeared and that only this part was transformed into power, due to his assumption that heat was a substance and that its quantity was therefore unchangeable. He adhered with his views to the mechanical comparison in which water flows from a higher to a lower level and thereby produces power. Sadi Carnot revealed, furthermore, that for a machine in which the steam passed through a completely reversible circular process, the ratio between the produced power and the amount of heat from the heat source was only a function of the temperature difference between the heat source and the substance which is in the process of cooling off. Carnot proved that for infinitesimally small cyclic processes, the produced amount of work was proportional to the temperature difference between the heat source and the cold body to which the heat was transferred, and an unknown temperature function. However, he was not able to completely develop the law for complete cyclic processes on the finite scale, because he was unable to determine the unknown temperature function, the so-called Carnot's function.

The drawback in Carnot's treatise is the following assumption: as already mentioned, he performed the theoretical foundation of the heat-machines as if the amount of heat which was carried by the steam appeared completely in the condenser, and as if only the temperature changed. Carnot's statement that the produced power is proportional to the temperature difference between the hot and the cold body is not, however, touched by this error.

Sadi Carnot's treatise ended with a discussion of different methods of using the amount of heat in machines in the best way. He arrived at the following points: in order to achieve the best success, the substance (steam) in the machine had to be heated up very high in the beginning and, finally, deeply cooled. One had to act in such a way that the transition from the highest to the lowest temperature be caused by the volume change of the steam. For the solution of this task, Carnot invented the already-mentioned cyclical processes.

Sadi Carnot found little recognition and understanding from his contemporaries. This may be based in the fact that at the time when he published his paper the assumption of heat-matter (caloric) was already doubted. Although Carnot also was of the opinion that this assumption was wrong, he was unable to change his results, for he died at a very young age.

Carnot's treatises went little-noticed in his own time. It was Benoit Pierre Emil Clapeyron (1834), a French mining engineer, who publicized the ideas of Sadi Carnot. He treated all of Carnot's principles and proofs graphically and analytically. In the beginning of his memoir, Clapeyron (1834) performed Carnot's simple cyclic process (see Mendoza, 1960):

"Let us imagine two bodies, one maintained at a temperature *T*, the other at a lower temperature *t*, such as for example the walls of a boiler in which the heat developed by combustion continually replaces that which the steam takes away; and the condenser of an ordinary heat-engine, in which a current of cold water all time removes the heat given up by the steam by condensation and that due to its own temperature. For simplicity we will call the first body *A*, the second *B*.

Then, let us take any gas whatever at temperature *T* and let us put it in contact with the source *A* of heat; let us represent its volume v_0 by the abscissa *AB* and its pressure by the ordinate *CB* (Fig. 3.33).

If the gas is enclosed in a deformable vessel and is allowed to expand in an empty space where it can lose heat neither by radiation nor by contact, the source *A* of heat will at all times provide the quantity of caloric which its increase of volume causes to become latent, and it will keep the same temperature *T*. Its pressure however will diminish following

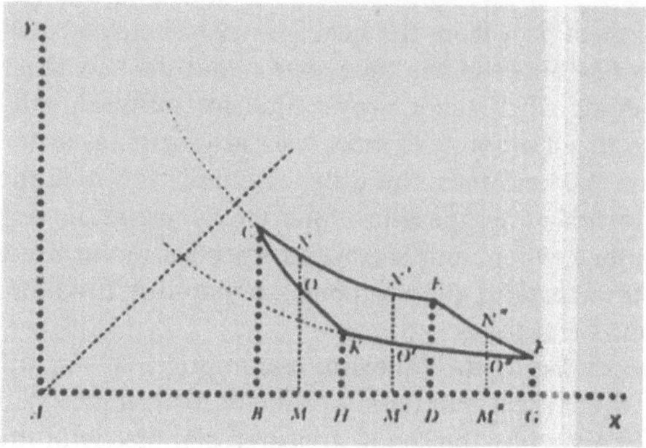

Fig. 3.33. Carnot's cyclic process

Mariotte's law. The law of this variation can be represented geometrically by the curve *CE* where the abscissae are the volumes, and the ordinates the corresponding pressures.

Let us suppose that the expansion of the gas is continued till the original volume *AB* has become *AD*; and let *DE* be the pressure corresponding to the new volume; the gas will have developed a quantity of mechanical action during its expansion given by the integral of the product of the pressure times the differential of the volume, represented geometrically by the area contained between the axis of abscissae, the two co-ordinates *CB*, *DE* and the portion *CE* of the hyperbola.

Let us suppose now that the body *A* is removed and that the expansion of the gas continues inside an envelope impermeable to heat; then since a part of its perceptible caloric becomes latent its temperature drops and its pressure decreases more rapidly according to an unknown law, which can be represented geometrically by a curve *EF* whose abscissae are the volume of the gas and whose ordinates are the corresponding pressures; we will suppose that the expansion of the gas is continued till the successive reductions of the perceptible caloric of the gas have brought it from the temperature *T* of the body *A* to the temperature *t* of the body *B*. Its volume is therefore *AG*, and the corresponding pressure *FG*.

It will be seen that the gas, during this second part of its expansion, develops a quantity of mechanical action represented by the area of the mixtilinear trapezium *DEFG*.

Now that the gas has been brought to the temperature *t* of the body *B*, let us bring the two into contact; if the gas is compressed in an envelope impermeable to heat, but in contact with the body *B*, the temperature of the gas tends to rise because of the release of latent caloric made perceptible by the compression, but as it is produced it is absorbed by the body *B* so that the temperature of the gas remains equal to *t*. As a result, the pressure increases according to Mariotte's law; it will be represented geometrically by the ordinates of a hyperbola *KF* and the corresponding abscissae will represent the volumes. Let us suppose that the compression is continued till the heat released by the compression of the gas and absorbed by the body *B* is exactly equal to the heat communicated by the source *A* to the gas, during its expansion in contact with it in the first part of the operation.

Then let *AH* be the volume of the gas and *HK* the corresponding pressure. In this state the gas possesses the same absolute quantity of heat as at the start of the operation, when it occupied the volume *AB* under pressure *CB*. If then the body *B* is removed and the gas is further compressed inside an envelope which is impermeable to heat until the volume *AH* becomes the volume *AB*, its temperature increases all the time by the release of latent caloric made perceptible by the compression. At the same time the pressure increases, and when the volume is reduced to *AB* the temperature returns to *T* and the pressure to *BC*. Now the different states in which a given mass of gas can exist are characterized by the volume, pressure, temperature, and the absolute quantity of heat which it contains; if two of these four quantities are known, the other two are determined; thus, in the case under discussion, since the absolute quantity of heat and the volume are the same as they were at the start of the operation, it is certain that the temperature and the pressure will also be what they were then. Consequently the unknown law, of how the pressure varies when the volume of the gas is reduced inside its impermeable envelope, is represented by a curve *KC* which passes through point *C*, and whose abscissae and ordinates always represent volumes and pressures.

However, the reduction of volume of the gas from *AC* to *AB* will have consumed a quantity of mechanical action which, by the same arguments which we have given above, will be represented by the two mixtilinear trapeziums *FGHK* and *KHBC*. If we subtract these two trapeziums from the first two *CBDE* and *EDFG* which represent the quantity of action developed during the expansion of the gas, the difference, which will be equal to the kind of curvilinear parallelogram *CEFK*, will represent the quantity of action developed by the cycle of operations just described, at the end of which the gas is in precisely its original state."

Clapeyron (1834) gave not only a graphical representation of Sadi Carnot's ideas, but rather he put them into an analytical framework in such an excellent way that his mathematical treatment became standard for a long time. He adhered, however, to Sadi Carnot's view that heat was a substance with an amount remaining constant. In his treatise, Clapeyron was not only the first scientist to speak of a reversible cyclic process, he also wrote the equation of state for gases:

"Let us first suppose that it is a gas which serves to transmit the caloric from the body *A* to the body *B*. Let v_0 be the volume of the gas at the pressure p_0 and at the temperature t_0; let p and v be the volume and the pressure of the same mass of the gas at the temperature t of the body *A*. Mariotte's law, combined with that of Gay-Lussac establishes between these different quantities the relation

$$pv = \frac{p_0 v_0}{267 + t_0}(267 + t) \tag{3.8.1}$$

or simply,

$$\frac{p_0 v_0}{267 + t_0} = R \tag{3.8.2}$$

$$pv = R(267 + t)\ ."$$

Clapeyron (1834) also used for the first time, for a small amount of heat, the expression:

$$dQ = \frac{dQ}{dv} \cdot dv + \frac{dQ}{dp} \cdot dp\,, \tag{3.8.3}$$

(the symbol *d*, for the small amount, is used for the total differential in mathematics; this description by Clapeyron and later by Clausius has led to much confusion in the literature) and introduced for the representation of the maximal efficiency of a differential reversible cyclic-process the quantity $\frac{dt}{C}$, where C is a smooth and positive function depending solely on the temperature.

Moreover, by applying Carnot's ideas to vapors, he developed a new relation between their latent heats, volumes, and pressures.

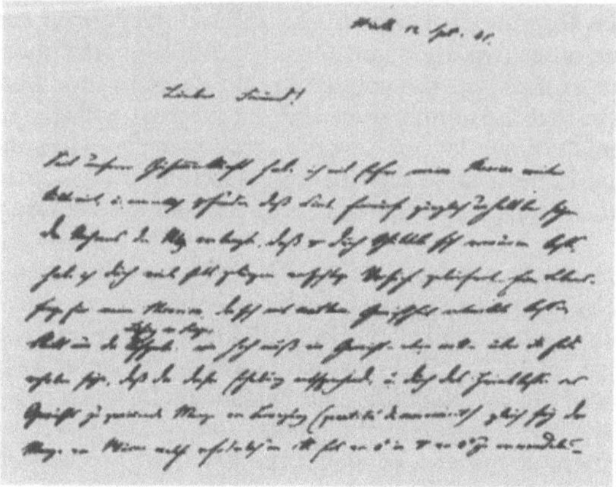

Fig. 3.34. Mayer's letter to his friend C. Bauer

In an essential part of his contribution, Clapeyron tried to determine the function *C*. Though he did not succeed, nevertheless he had already recognized the importance of this function:

"The function *C* .... is of great importance, as can be seen; it is independent of the nature of the gas and is a function of the temperature only; it is essentially positive and serves as a measure of the maximum quantity of action which heat can develop."

"The function *C* is, as we have seen, of great importance: it is the common link between the phenomena caused by heat in solid bodies, liquids and gases; it would be desirable for very exact experiments, such as researches on the propagation of sound at different temperatures, to be done to allow this function to be found with the desired accuracy; this would determine several other important things in the theory of heat, concerning which experiments have only led to poor approximations or have as yet told us nothing."

It was Clausius (1850) who identified the function *C* as being the absolute temperature divided by the mechanical equivalent of heat.

Although Clapeyron did not develop new ideas which exceeded Sadi Carnot's considerations, he clarified these by introducing the graphic representation and algebraic symbolism.

The ideas of Sadi Carnot, and the careful preparation of Clapeyron, later culminated in the formulation of the second law of thermodynamics. It is amazing to think that elements of the second law were already known before elements of the first law, and the first law itself, had been developed.

3.8.3
Robert Mayer, the Discoverer of the Mechanical Equivalent of Heat

The physician Robert Mayer is generally recognized as the discoverer of the mechanical equivalent of heat, i.e., that heat and mechanical work are mutually transformable according to a constant equivalent-ratio. Indeed, in 1841, he stated in a letter to his friend C. Bauer:

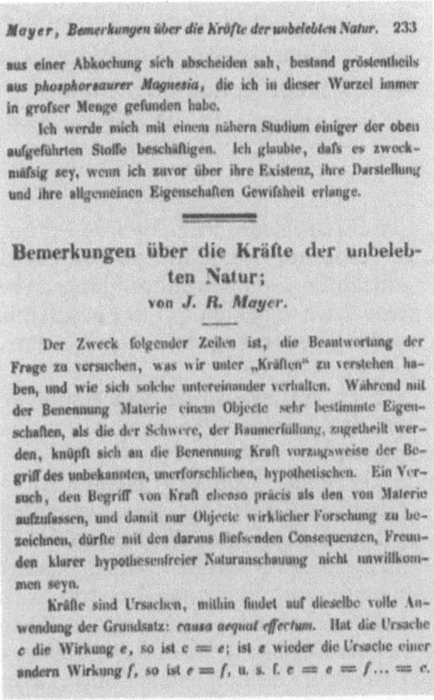

Mayer, Bemerkungen über die Kräfte der unbelebten Natur. 233

aus einer Abkochung sich abscheiden sah, bestand grösstentheils aus *phosphorsaurer Magnesia*, die ich in dieser Wurzel immer in grofser Menge gefunden habe.

Ich werde mich mit einem nähern Studium einiger der oben aufgeführten Stoffe beschäftigen. Ich glaubte, dafs es zweckmäfsig sey, wenn ich zuvor über ihre Existenz, ihre Darstellung und ihre allgemeinen Eigenschaften Gewifsheit erlange.

Bemerkungen über die Kräfte der unbelebten Natur;
von *J. R. Mayer.*

Der Zweck folgender Zeilen ist, die Beantwortung der Frage zu versuchen, was wir unter „Kräften" zu verstehen haben, und wie sich solche untereinander verhalten. Während mit der Benennung Materie einem Objecte sehr bestimmte Eigenschaften, als die der Schwere, der Raumerfüllung, zugetheilt werden, knüpft sich an die Benennung Kraft vorzugsweise der Begriff des unbekannten, unerforschlichen, hypothetischen. Ein Versuch, den Begriff von Kraft ebenso präcis als den von Materie aufzufassen, und damit nur Objecte wirklicher Forschung zu bezeichnen, dürfte mit den daraus fliefsenden Consequenzen, Freunden klarer hypothesenfreier Naturanschauung nicht unwillkommen seyn.

Kräfte sind Ursachen, mithin findet auf dieselbe volle Anwendung der Grundsatz: *causa aequat effectum*. Hat die Ursache c die Wirkung e, so ist $c = e$; ist e wieder die Ursache einer andern Wirkung f, so ist $e = f$, u. s. f. $c = e = f \ldots = c$.

Fig. 3.35. Mayer's paper *Remarks on the forces of the inanimate nature*

"Eine Lebensfrage für meine Theorien, die sich mit mathematischer Gewißheit entwickeln lassen, bleibt nun die Lösung der Frage: wie hoch muß ein Gewicht – etwa 100 *tt* – über die Erde gehoben sein, daß die dieser Erhebung entsprechende und durch das Herablassen des Gewichtes zu gewinnende Menge von Bewegung (quantité de mouvement) gleich sei der Menge von Wärme, welche erforderlich, um 1 *tt* Eis von 0° in Wasser von 0° zu verwandeln."

"A vital question for my theories, which can be developed with mathematical certainty, is now the solution of the question: how high must a weight – around 100 pounds – be lifted over the earth that the corresponding amount of motion, which is gained by letting down the weight, is equal to the amount of heat, which is necessary to convert 1 pound of ice of 0 degree into water of 0 degree."

He published his lucid discovery, finally, in a paper entitled *Bemerkungen über die Kräfte der unbelebten Natur* in Wöhler's and Liebig's Annalen (Mayer, 1842). His 1841 attempt to publish his finding in Poggendorf's Annalen had failed.

The discovery of the mechanical equivalent of heat was the decisive step towards developing the first law of thermodynamics. However, Mayer was not able to determine the exact value of this equivalent due to the lack of the required test observations – and probably also due to his lack of talent for doing the complicated experimental work needed in order to obtain sufficiently exact test results. This was done by the great James Prescott Joule.

Robert Mayer came upon his brilliant discovery while on duty as a ship's doctor on an East-India-steamer. In February, 1840, he embarked from Rotterdam on his sea voyage. During the long trip, he studied a handbook on physiology. This study, along with his profound knowledge of chemistry and of the process of combustion based on the theory of Lavoisier (a French chemist), enabled the young physician to draw the right conclusions from the occasional observation. Mayer (1851) reported on this:

"Die Erzeugung der Wärme durch die Reibung und durch andere mechanische Processe ist eine fundamentale Thatsache von so universaler Verbreitung, dass ihre wissenschaftliche Feststellung auch ohne eine vorausgeschickte Aufzählung von Nutzanwendungen dem Naturkundigen als werthvoll erscheinen wird, und es werden daher auch einige geschichtliche Bemerkungen über das Thatsächliche der Auffindung des vorliegenden Grundgesetzes hierwohl am Platze seyn.

Im Sommer 1840 machte ich bei Aderlässen, die ich auf Java an neuangekommenen Europäern vornahm, die Beobachtung, dass das aus der Armvene genommene Blut fast ohne Ausnahme eine überraschend hellrothe Färbung zeigte.

Diese Erscheinung fesselte meine volle Aufmerksamkeit. Von der Theorie *Lavoisier's* ausgehend, nach welcher die animalische Wärme das Resultat eines Verbrennungs-Processes ist, betrachtete ich die doppelte Farbenveränderung, welche das Blut in den Haargefässen des kleinen und großen Kreislaufes erleidet, als ein sinnlich wahrnehmbares Zeichen, als den sichtbaren Reflex einer mit dem Blute vor sich gehenden Oxydation. Zur Erhaltung einer gleichförmigen Temperatur des menschlichen Körpers muss die Wärme entwicklung in demselben mit seinem Wärme verluste, also auch mit der Temperatur des umgebenden Mediums nothwendig in einer Größenbeziehung stehen und es muss daher sowohl die Wärme-Produktion und der Oxydations-Process, als auch der Farbenunterschied beider Blutarten im Ganzen in der heißen Zone geringer seyn, als in kälteren Gegenden. ... Es bleibt also der Verbrennungs-Theorie, wenn sie sich nicht von vorn herein selbst aufgeben will, nichts übrig, als anzunehmen: dass die gesammte, theils unmittelbar, theils auf mechanischem Wege vom Organismus entwickelte Wärme dem Verbrennungs-Effecte quantitativ entspricht oder gleich ist.

Daraus folgt nun aber mit derselben Nothwendigkeit, dass die von lebendem Körper erzeugte mechanische Wärme mit der dazu verbrauchten Arbeit in einem unveränderlichen Größenverhältnisse stehen muss. Denn wenn, je nach der verschiedenen Construction der zur Wärmegewinnung dienenden mechanischen Vorrichtungen u. dgl., durch die nämliche Arbeit und bei gleichbleibendem organischen Verbrennungs-Processe ; verschieden grosse Wärmemenge erzielt werden könnten, so würde ja die producirte Wärme bei einem und demselben Material-Verbrauche bald kleiner, bald größer ausfallen können, was gegen die Annahme ist. Da aber ferner zwi- schen der mechanischen Leistung des Thierkörpers und zwischen anderen, anorganischen Arbeitsarten kein qualitativer Unterschied besteht,

so ist folglich eine unveränderliche Größenbeziehung zwischen der Wärme und der Arbeit ein Postulat der physiologischen Verbrennungs-Theorie.

Indem ich im Allgemeinen die angegebene Richtung einhielt, musste ich also nothwendig mein Hauptaugenmerk zuletzt auf den zwischen der Bewegung und der Wärme bestehenden physikalischen Zusammenhang richten, wo mir dann die Existenz des mechanischen Aequivalentes der Wärme nicht verborgen bleiben konnte. Wenn ich aber auch diese Entdeckung nur einem Zufall verdanke, so ist sie doch mein Eigenthum, und ich stehe nicht an, das Recht des Zuerstkommenden zu behaupten."

"The production of heat by friction and by other mechanical processes is a fundamental fact of such universal importance that its scientific statement will appear valuable for the nature-practitioner even without useful applications, and therefore also some historical remarks on the reality of the finding of the existing fundamental law may well be in order.

In the summer of 1840, I undertook blood-lettings on European newcomers to Java and made the observation that the blood taken from the arm-vein showed nearly without exception a surprising light red color.

This phenomenon fascinated me. Preceding from *Lavoisier's* theory, according to which animal heat is the result of a combustion-process, I considered the double color change, which the blood suffers in the capillary vessels of the small and large circulation, as a sensible noticeable sign, as the visible reflex of an oxydation of blood. For the conservation of a uniform temperature of the human body, the `heat development` in the body must be necessarily in one size relation with the `loss of heat`, i.e. also with the temperature of the surrounding medium. The heat production and the oxydation process must therefore, as well as the color difference of both kinds of blood in total, in the hot zone be less than in the colder areas. ... The combustion theory has therefore no choice but to assume, if it will not apriori surrender: that the whole heat corresponds quantitatively or is equal to the combustion-effects, in parts immediately, in parts in a mechanical way developed by the organism.

From this it now follows, however, with the same necessity `that the mechanical heat produced by the living body must be in a constant size-ratio with the work consumed for that purpose`. For if, according to the different construction of the mechanical devices for the production of heat etc., by the same work and at constant organic combustion processes, large amounts of `different` heat could be gained, the produced heat could turn out smaller as well as greater with the same consumption of material which contradicts the assumption. Because, however, no further qualitative difference exists between the mechanical power of the animal body and between other inorganic kinds of work, `a constant size-relation between the heat and the work is a postulate of the physiological combustion-theory`. While in general I kept the stated direction, I must necessarily direct my main attention at last to the physical relation existing between the motion and heat. Here the existence of the mechanical equivalent of heat could not remain concealed to me. Although I owe this discovery to pure chance, it is of course my property and I do not stand in a line to defend the right of the first-comer."

It was this point, namely the defense of the discovery of the mechanical equivalent, which again and again occupied him throughout the rest of his life. Robert Mayer, not well trained in physics and mathematics, with no scientific background, and without the understanding of formal scientific language, also missed no occasion to explain his standpoint in non-scientific circles. On May 14, 1849, for example, he announced his outstanding finding and its verification in the newspaper "Allgemeine Zeitung" of his hometown Heilbronn, Germany (see Fig. 3.36).

"Important Physical Finding"

" I succeeded in finding a simple procedure to state the equivalence of heat and the mechanical work or the conversion of the motion into heat and vice versa, (compare among others my treatise on the 'Kräfte der unbelebten Natur' in Wöhler und Liebig's Annalen, Maiheft 1842) by an easy experiment and in determining directly the corresponding equivalence-number with all desirable accurateness. The apparatus needed for these experiments which I have led, are manufactured by Mr. Mechanicus Wagner, and consist essentially of a metallic cylinder containing water, which is pressed through a narrow hole with the help of a pump piston and is heated up by this. If one now compares the amount of heat produced in this way with the simultaneously occurring consumption of work, one has solved the most natural scientific problem of the present time.

Forced by an opinion mentioned in an article written in the Journal des Débats on September 15 of last year, I maintain in public simultaneously my priority right for the

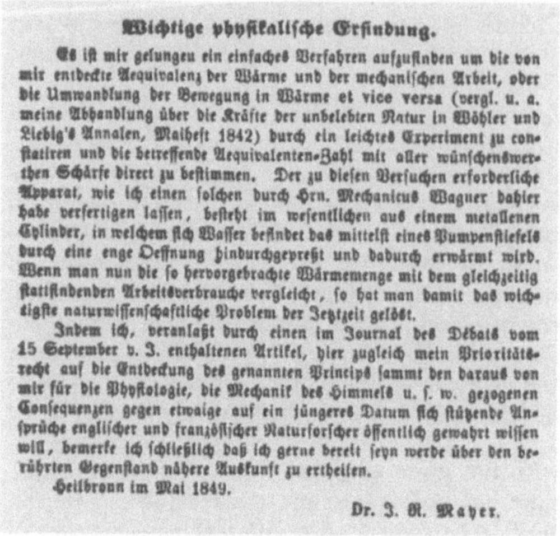

Wichtige physikalische Erfindung.

Es ist mir gelungeu ein einfaches Verfahren aufzufinden um die von mir entdeckte Aequivalenz der Wärme und der mechanischen Arbeit, oder die Umwandlung der Bewegung in Wärme et vice versa (vergl. u. a. meine Abhandlung über die Kräfte der unbelebten Natur in Wöhler und Liebig's Annalen, Maiheft 1842) durch ein leichtes Experiment zu constatiren und die betreffende Aequivalenten-Zahl mit aller wünschenswerthen Schärfe direct zu bestimmen. Der zu diesen Versuchen erforderliche Apparat, wie ich einen solchen durch Hrn. Mechanicus Wagner dahier habe verfertigen lassen, besteht im wesentlichen aus einem metallenen Cylinder, in welchem sich Wasser befindet das mittelst eines Pumpenstiefels durch eine enge Oeffnung hindurchgepreßt und dadurch erwärmt wird. Wenn man nun die so hervorgebrachte Wärmemenge mit dem gleichzeitig stattfindenden Arbeitsverbrauche vergleicht, so hat man damit das wichtigste naturwissenschaftliche Problem der Jetztzeit gelöst.

Indem ich, veranlaßt durch einen im Journal des Débats vom 15 September v. J. enthaltenen Artikel, hier zugleich mein Prioritätsrecht auf die Entdeckung des genannten Princips sammt den daraus von mir für die Physiologie, die Mechanik des Himmels u. s. w. gezogenen Consequenzen gegen etwaige auf ein jüngeres Datum sich stützende Ansprüche englischer und französischer Naturforscher öffentlich gewahrt wissen will, bemerke ich schließlich daß ich gerne bereit seyn werde über den berührten Gegenstand nähere Auskunft zu ertheilen.

Heilbronn im Mai 1849.

Dr. J. R. Mayer.

Fig. 3.36. Mayer's announcement of his outstanding findings in a newspaper

discovery of the mentioned principle as well as for the consequences drawn from the principle for physiology, the mechanics of the sky etc. against possible claims of English and French natural researchers which rest upon an earlier date. At last I mention that I am ready and will with pleasure give detailed information about the subject in question.
Heilbronn, May 1849.
Dr. J.R. Mayer"

Mayer also wrote a statement entitled "Wahrung literarischer Eigenthumsrechte" wherein he summarized those findings which he had published previously. In particular, he defended the finding of the mechanical equivalent of heat and pointed out that he had published this result in 1842, one year before Joule published his paper on the subject.

In his scientific career, Mayer also investigated other physical problems, for example, sunlight and the heat of the sun, the internal heat of the earth, and the creation of the sun's heat. However, almost all of his papers suffered from the lack of clear-defined terms and unique definitions and principles. Mayer (1851) was well aware of this disadvantage. In the preface to his paper *Bemerkungen über das mechanische Aequivalent der Wärme (Remarks on the mechanical equivalent of heat)*, he wrote:

"Warum ich mir bei meiner Stellung als praktischer Arzt in dieser wichtigen Sache mitzureden erlaubt habe, davon ist der Grund in der Schrift selbst angegeben. –

Mögen Sachverständige, welche die Schwierigkeiten kennen, mit denen man beim Bebauen eines neuen Feldes zu kämpfen hat, den Mängeln meiner Arbeiten eine nachsichtige Beurtheilung angedeihen lassen! Ars longa, vita brevis. – "

"The reason why I have taken the liberty in my position as a practical physician of putting in a word or two in this important matter, is given in the treatise. –

The experts who know the difficulties with which one has to fight during the construction of a new field, may judge the imperfections of my works with indulgence. Ars longa, vita brevis. – "

Although Mayer, as an outsider, did not succeed in all of his investigations into physical and physiological problems, due to the lack of a profound knowledge of physics and mathematics, he will maintain his place in the history of exact science as one of the great discoverers. He was highly gifted with a talent for recognizing important phenomena purely intuitively. The history of science has shown repeatedly that creative researchers, even those not confined by state-of-the-art knowledge, often intuitively find important effects. Further examples of such scientists are Fick (as already mentioned in Section 3.3) and von Terzaghi, the founder of modern soil mechanics (as will be seen in Sections 4.2 to 4.6).

3.8.4 The Contributions of Mohr, Séguin, Colding, Holtzmann, and Helmholtz

Before Mayer had developed his fundamental views on the mechanical equivalent of heat, Mohr (see R. Rühlmann, 1876) had already published some ideas on the nature of heat and the mechanical equivalent of heat in 1837. However, his treatise did not have any influence on the formulation of thermodynamics because it was published in a little-read journal and because his article did not contain truly new facts which surpassed, essentially, the findings of Rumford and Davy.

In 1839 R. Séguin also (see R. Rühlmann, 1876) reported, in a strange paper on railway systems, some ideas about the power of steam engines. One can read from his ideas the view that there existed a quantitative ratio between work and heat. Séguin's basic idea was as follows: if a body expanded it would lose heat, and the lost heat would be the equivalent of the mechanical work which was done during the expansion.

Moreover, Séguin tried to discover whether the heat which was carried by the steam from the furnace to the condenser in a steam engine was completely conserved. He did not come to an assured result due to the lack of sufficient devices and test methods. However, he recognized that the difference between the heat taken from the boiler and the heat in the condensor was larger in the case where the steam engine worked, than in the case where the steam flowed without working through the machine. Séguin was on the right track, but he did not attain a useful result.

An essential step further was made by the Danish engineer, Colding. In a treatise on power, reported to the Academy in Copenhagen in 1843, he expressed the law of the conservation of energy rather clearly and also communicated an approximate value for the mechanical equivalent of heat (350 mkg). In his paper, Colding proceeded from metaphysical considerations and arrived at the idea that power was everlasting and immortal, and that wherever power seemed to vanish – when mechanical, chemical, or some other work was done – only a conversion was occurring. The reason that Colding's contribution is not well known may lie in the fact that his terms lack the clarity of other authors and that he obviously did not recognize the full meaning of the principle.

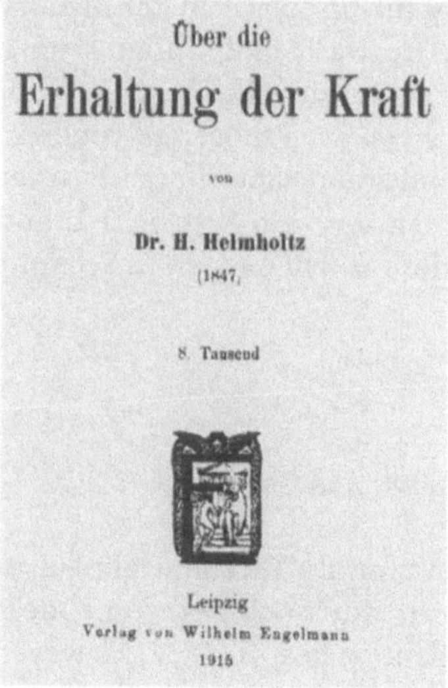
Über die
Erhaltung der Kraft
von
Dr. H. Helmholtz
(1847)
8. Tausend
Leipzig
Verlag von Wilhelm Engelmann
1915

Fig. 3.37. Helmholtz's treatise on thermodynamics

Further contributions to the foundation of thermodynamics came from Holtzmann (1845), who published a small booklet entitled *Ueber die Wärme und Elasticität der Gase und Dämpfe*. A translation of this booklet was published in the Philosophical Magazine and Journal of Science one year later (see Holtzmann, 1846). At first, Holtzmann pointed out that Clapeyron's results were only correct when the laws of Mariotte and Gay-Lussac remained valid. Then he showed that the unknown temperature function C, introduced by Clapeyron, could be determined under the assumption that the ratio of the specific heat was independent of temperature and had to have the following form:

$$\frac{A}{a+bt}, \tag{3.8.4}$$

where A is the mechanical equivalent of heat, a and b are constant numbers, and t is the temperature. Holtzmann did not prove that the reciprocal value of the function C was equal to the absolute temperature divided by the mechanical equivalent of heat. However, he was the first scientist to determine Carnot's function with the requirement that heat could be converted to work according to a determined fixed equivalent ratio.

In 1847, Helmholtz published his first treatise on thermodynamics (see Helmholtz, 1847), *Über die Erhaltung der Kraft*. He had reported on this topic at a meeting of the Physical Society in Berlin on July 23, 1847. His efforts to publish his talk in scientific journals failed. Thus, he edited his treatise by himself in a pamphlet. As indicated in the title, Helmholtz clearly stated the conservation of power (the German expression "Kraft" is in Helmholtz's sense a synonym for power. The use of this word for power was usual in the eighteenth and nineteenth century and does not fit with today's sense of the word). Thus, the next steps in his treatise could only consist of proving the equivalence of the different kinds of energy with the mechanical power. Afterwards, he treated the principle of the conservation of the motive power, in particular, in the case where central forces occurred. In the third section, he started with the sentence:

"Wir gehen jetzt zu den speciellen Anwendungen des Gesetzes von der Constanz der Kraft über."

"Now we proceed to the special applications of the law of the conservation of power."

Moreover, he remarked that, at first, such cases should be mentioned in which the principle of the conservation of the motive power had already been used and recognized. He listed all motions which occur under the influence of the gravitational force, the transmission of the motion by incompressible solids and liquids, and the motions of completely elastic solids and liquids. He then turned to the heat phenomenon and pointed out that there were two mechanical processes in which an absolute loss of power was generally assumed: the impact of inelastic bodies and friction. Following this, Helmholtz (1847) discussed these phenomena extensively:

"Man pflegt in der Mechanik die Reibung als eine Kraft darzustellen, welche der vorhandenen Bewegung entgegenwirkt, und deren Intensität eine Function der Geschwindigkeit ist. Offenbar ist diese Auffassung nur ein zum Behuf der Rechnungen gemachter, höchst unvollständiger Ausdruck des complicirten Vorgangs, bei welchem die verschiedensten Molecularkräfte in Wechselwirkung treten. Aus jener Auffassung folgte, dass bei der Reibung lebendige Kraft absolut verloren ginge, ebenso nahm man es beim elastischen Stosse an. Dabei ist aber nicht berücksichtigt worden, dass abgesehen von der Vermehrung der Spannkräfte durch die Compression der reibenden oder gestossenen Körper, uns sowohl die gewonnene Wärme eine Kraft repräsentirt, durch welche wir mechanische Wirkungen erzeugen können, als auch die meistentheils erzeugte Electricität entweder direct durch ihre anziehenden und abstossenden Kräfte, oder indirect dadurch dass sie Wärme entwickelt. Es bliebe also zu fragen übrig, ob die Summe dieser Kräfte immer der verlorenen mechanischen Kraft entspricht. In den Fällen, wo die molecularen Aenderungen und die Electricitätsentwicklung möglichst vermieden sind, würde sich diese Frage so stellen, ob für einen gewissen Verlust an mechanischer Kraft jedesmal eine bestimmte Quantität Wärme entsteht, und inwiefern eine Wärmequantität einem Aequivalent mechanischer Kraft entsprechen kann."

"In mechanics, one represents, friction as a force which counteracts the actual motion and whose intensity is a function of the velocity. Obviously, this interpretation is a most incomplete expression of the complicated process, made only for the purpose of the calculations, in which the most different molecular forces are in interaction. From those interpretations it followed that motive power would get absolutely lost in the case of friction, and one also assumed it for the elastic impact. However, what has not been considered is that, apart from the increase of stresses by the compression of the rubbing or pushing bodies, the gained heat represents a power by which we can produce mechanical agencies, as well as the electricity produced mainly either directly by the attracting and repulsing forces, or indirectly by the development of heat. The question therefore remains as to whether the sum of these powers always corresponds to the loss of mechanical power. In such cases where the molecular changes and the development of the electricity are possibly avoided, the question would arise as to whether, for a certain loss of mechanical power, a determined quantity of heat is created every time, and to what extent a heat-quantity can correspond to an equivalent of mechanical power."

In addition, Helmholtz pointed out that only a few tests were known for determining the mechanical equivalent of heat, and that Joule had performed some experiments with unsatisfactory results. Furthermore, he criticized the material theory of heat, as it was used by Carnot and Clapeyron, and remarked that the amount of heat could be fully increased by mechanical power and that therefore the heat phenomena could not be deduced from a substance but from changes of motions.

In the last two sections of his treatise, Helmholtz discussed the mechanical equivalent of the electrical processes and of magnetism and electromagnetism. Finally, he turned to a discussion on the heat development in plants and animals and came to some conclusions. He stated that nearly all information was missing for the comparison of the mechanical equivalents in the case of plants. In the case of animals, the situation was a little better. However, there still remained a lot to study. At the end, he drew some conclusions and stated that his law did not contradict any facts in the science. Moreover, he pointed out:

"Der Zweck dieser Untersuchung, der mich zugleich wegen der hypothetischen Theile derselben entschuldigen mag, war, den Physikern in möglichster Vollständigkeit die theoretische, practische und heuristische Wichtigkeit dieses Gesetzes darzulegen, dessen vollständige Bestätigung wohl als eine der Hauptaufgaben der nächsten Zukunft der Physik betrachtet werden muss."

"The purpose of this investigation, which I must also apologize for because of the hypothetical part of it, was to represent the theoretical, practical and heuristical importance of this law to the physicist, in the greatest possible completeness, and the complete verification must be considered as one of the main tasks in the near future of physics."

In 1881, Helmholtz added some footnotes to his treatise. In the fifth footnote, he made some remarks concerning the history of the discovery of the law of conservation of power:

"Zur Geschichte der Entdeckung des Gesetzes von der Erhaltung der Kraft wäre hier noch nachzutragen, dass R. Mayer 1842 seinen Aufsatz 'Ueber die Kräfte der unbelebten Natur', veröffentlicht hatte, und 1845 die Abhandlung über 'Die organische Bewegung in ihrem Zusamenhang mit dem Stoffwechsel'. Heilbronn. Schon in dem ersten Aufsatz ist die Ueberzeugung von der Aequivalenz der Wärme und Arbeit ausgesprochen und das Aequivalent der Wärme ... auf 365 meterkilogramm berechnet. Der zweite Aufsatz ist in seinem allgemeinen Ziele nach im wesentlichen zusammenfallend mit dem meinigen. Ich habe beide Aufsätze erst später kennengelernt, und seitdem ich sie kannte, nie unterlassen, wo ich öffentlich von der Aufstellung des hier besprochenen Gesetzes zu reden hatte, R. Mayer in erster Linie zu nennen, auch habe ich seine Ansprüche, so weit ich sie vertreten konnte, gegen die Freunde Joule's, welche dieselben gänzlich zu leugnen geneigt waren, in Schutz genommen."

"Concerning the history of discovery of the law of power conservation, it must be added here that in 1842 R. Mayer had published his article 'On the powers of the inanimate nature', and, in 1845, the treatise on 'The organic motion in its connection with the metabolism'. Heilbronn. Already in the first article, the conviction of the equivalence of heat and power is stated and the equivalence of heat ... is calculated at 365 meterkilogram. The general aim of the second article coincides, essentially, with that of mine. Both articles became known to me only later, and since I have known about them, I have not hesitated, wherever I have to talk in public of the establishment of the law discussed here, to mention R. Mayer in the first line; I have also defended, as far as I have been able to answer for, his claims against the friends of Joule who were inclined to deny those completely."

It seems that Helmholtz developed the mechanical equivalence of heat completely independently. For, in the small town of Potsdam, he had only the opportunity to use the library of the High School with its limited scientific literature. Helmholtz's style in investigating scientific problems was also totally different from Mayer's. Helmholtz, as a mathematically trained physician, wrote his treatises very clearly, with well-defined terms in strict scientific language; this talent also enabled him later to investigate further physical problems and to obtain fundamental results. In contradiction, Mayer, with his poor mathematical tools, used a colorful language with vague physical terms. Thus, it is not surprising that it was Helmholtz who made the new ideas on heat public in the scientific world.

There is no hint in Helmholtz's paper (1847) of the introduction of internal energy and, in no place, is the first law of thermodynamics formulated by Helmholtz, as is argued by many authors in the literature (see, e.g., Päsler, 1975).

3.8.5 The Decisive Investigations of Joule

James Prescott Joule is considered as one of the best scientists ever in the field of experimental research. He gained, in a strict, scientific way, important results through carefully performed tests and well-devised experimental methods. The certainty with which he was able to interpret the results of his observations led him to the discovery of some important principles.

At the beginning of his scientific career he was concerned with the influence of electrical and magnetic forces on mechanical work. After Faraday's discovery of the induction phenomena, Joule investigated the question as to whether mechanical work had to be applied to produce an electrical currrent of prescribed amperage by induction in a conductor with known resistance. He determined the necessary power needed to run a small electrical machine, let the produced electrical energy be converted directly to heat, and measured the heat produced in this way. He reported on his results in a talk given on August 21, 1843 (see R. Rühlmann, 1876). He stated that the mechanical equivalent of heat had the value 460 [mkg]. In an addendum, he wrote that he had already determined this important number earlier by measuring the amount of heat which was created by pressing water through narrow pipes, simultaneously with the mechanical work needed to do this. In this way, he had obtained for the mechanical equivalent of heat the value 438 [mkg].

During these investigations, Joule recognized that it did not matter whether mechanical work was transformed to heat directly, or via a roundabout way, for example, via electricity. In the time to follow, therefore, he turned to the exact experimental determination of the mechanical equivalent of heat, and to the question as to whether the process by which mechanical work was transformed to heat influenced the value of the mechanical equivalent of heat. He performed this through his friction experiments and refined his tests and measurements subsequently so that he obtained nearly the exact value.

On the basis of his experiments, he also came to the conclusion that the views of Carnot and Clapeyron were wrong: that in a cyclic process the total heat would be conserved.

Joule's investigations were a breakthrough for the theory of heat, at that time, due to their exactness and the ease with which they could be controlled.

3.8.6 The Foundation of Thermodynamics by Clausius, Rankine and Thomson

"Wenn auf Grund der gesammelten Erfahrungen und Erkenntnisse der Zeitpunkt heranrückt, in welchem die Wissenschaft reif wird, auf eine höhere Stufe der Vollkommenheit erhoben zu werden, von welcher aus eine grössere Zahl von Thatsachen unter einheitlichen Gesichtspunkten zusammengefasst werden können, so tritt gewöhnlich dieser Fortschritt nicht plötzlich ein, sondern man bemerkt, dass an verschiedenen Stellen wiederholt, anfänglich schüchtern und dann mit immer wachsender Energie, Versuche gemacht werden, diese höhere Stufe der Einsicht zu erklimmen, bis dann endlich, veranlasst durch die überraschenden Erfolge, welche solche erringen, die bereits den höheren Standpunkt einnehmen, die Gesammtheit zögernd sich entschliesst, den Bahnbrechern zu folgen." (R. Rühlmann, 1876)

"On the basis of collected experiences and findings the time is approaching, in which the science will become ripe to be raised to a higher step of perfection, from which a greater number of facts will be able to be summarized under common aspects. Usually this progress does not occur suddenly, but one notices that on different places repeated attempts are made, beginning timidly and then with ever increasing energy, to climb this higher step of insight. Then, finally, caused by the surprising successes which such scientists who have already taken the higher standpoint, achieve the other scientists in their entirety decide tentatively to follow the pioneers." (R. Rühlmann, 1876)

In the late 1840s Rudolf Clausius, a German professor, was concerned with a reformulation of the theory of heat, which had become necessary after the basic test observations of Joule, in particular, and in view of Carnot's fundamental assumption about the heat-substance. Clausius communicated the results of his investigations to the Academy in Berlin in February, 1850. He published his brilliant treatise *Ueber die bewegende Kraft der Wärme und die Gesetze, welche sich daraus für die Wärmelehre selbst ableiten lassen* (Clausius, 1850) (*On the motive power of heat, and on the laws which can be deduced from it for the theory of heat*) in April of the same year (Clausius' German statements were translated by Magie in Mendoza, 1960).

In his introduction, Clausius praised the investigations of Sardi Carnot and the skilfull formulation of Carnot's results by Clapeyron. He criticized, however, Carnot's view of heat and stated "that heat is not a substance, but consists in a motion of the least parts of bodies."

At the beginning of the first section of his paper, Clausius formulated a fundamental principle:

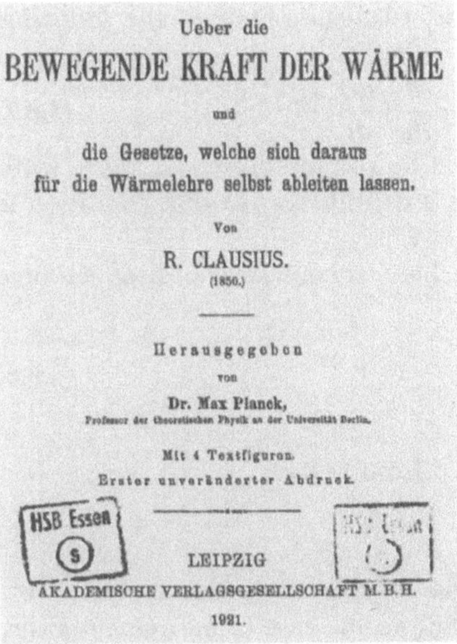
Ueber die
BEWEGENDE KRAFT DER WÄRME
und
die Gesetze, welche sich daraus
für die Wärmelehre selbst ableiten lassen.
Von
R. CLAUSIUS.
(1850.)
Herausgegeben
von
Dr. Max Planck,
Professor der theoretischen Physik an der Universität Berlin.
Mit 4 Textfiguren.
Erster unveränderter Abdruck.
HSB Essen
LEIPZIG
AKADEMISCHE VERLAGSGESELLSCHAFT M. B. H.
1921.

Fig. 3.38. Clausius' first fundamental treatise (1850)

"... dass in allen Fällen, wo durch Wärme Arbeit entstehe, eine der erzeugten Arbeit proportionale Wärmemenge verbraucht werde, und dass umgekehrt durch Verbrauch einer ebenso grossen Arbeit dieselbe Wärmemenge erzeugt werden könne."

"In all cases in which work is produced by the agency of heat, a quantity of heat is consumed which is proportional to the work done; and, conversely, by the expenditure of an equal quantity of work an equal quantity of heat is produced."

Then he drew some consequences from the principle of the equivalence of heat and work. After that he turned "to the mathematical discussion of the subject, in which we shall restrict ourselves to the consideration of the permanent gases and of vapors at their maximum density, since these cases, in consequence of the extensive knowledge we have of them, are most easily submitted to calculation, and besides that, are the most interesting." Clausius started with the combined laws of Mariotte and Gay-Lussac:

$$\text{(I.)} \qquad pv = R(a + t)\,, \tag{3.8.5}$$

and calculated the value $a = 273$. Then he discussed Carnot's cyclic process in the graphic representation by Clapeyron, and determined the work of the gas done in an infinitesimal cyclic process.

$$\text{The work done} = \frac{R\,dv\,dt}{v}\,, \tag{3.8.6}$$

where p, v, and t denote the pressure, volume, and temperature of the gas, and R is a constant. The next step was the determination of the heat consumed during the cyclic process. After some calculations, Clausius obtained the following result:

$$\text{The heat consumed} = \left[\frac{d}{dt}\left(\frac{dQ}{dv}\right) - \frac{d}{dv}\left(\frac{dQ}{dt}\right)\right] dv\, dt\,, \tag{3.8.7}$$

where Q is the amount of heat "which must be communicated to a gas, while it is brought from any former condition in a definite way to that condition in which its volume $= v$ and its temperature $= t$."

Thereupon, Clausius introduced the mechanical equivalent of heat, denoted by A:

$$\frac{\text{The heat consumed}}{\text{The work done}} = A\,. \tag{3.8.8}$$

With the Eqns. (3.8.6) and (3.8.7), he gained from (3.8.8):

$$\text{(II.)} \quad \frac{d}{dt}\left(\frac{dQ}{dv}\right) - \frac{d}{dv}\left(\frac{dQ}{dt}\right) = \frac{AR}{v}\,. \tag{3.8.9}$$

Clausius stated that this equation could be considered as the analytical expression of the fundamental principle applied to the case of permanent gases. Moreover, he showed that Q could not depend on v and t, "if these variables are independent of each other", and he brought Eqn. (3.8.9) "into the form of a *complete* differential equation":

$$\text{(II.a)} \quad dQ = dU + AR\frac{a+t}{v}dv\,, \tag{3.8.10}$$

where U is an arbitrary scalar depending on v and t (U is the internal energy, and it was Clausius who discovered and proved the existence of internal energy for the first time). The last term in (3.8.10) represents the mechanical work done during the change, multiplied by A. At first, Clausius did not elaborate on the above formula. Rather, he considered instead a permanent gas vapor, and developed the analytical expression for the fundamental principle for vapors at their maximum density. He obtained in this case, in a derivation similar for permanent gases:

$$\text{(III.)} \quad \frac{dr}{dt} + c - h = A(s-\sigma)\frac{dp}{dt}\,, \tag{3.8.11}$$

where s and σ represent the "volume of a unit weight of the vapor at its maximum density and the volume of the same quantity of liquid at the temperature t", respectively. The quantity denoted by r is the latent heat per mass element, h the quantity of heat per degree temperature, and c the specific heat of the liquid. It should be mentioned that a similar formula had already been developed by Clapeyron (1834). In the sequel to first section, Clausius developed the statement that "the difference of the specific heats referred to the unit of volume is therefore the same for all gases" and he drew further consequences from the fundamental principle.

In the second section of his valuable contribution, Clausius derived important consequences of Carnot's principle, in connection with his fundamental principle stated in the first section. At the beginning of this section, Clausius pointed out:

"*Carnot hat ... angenommen, dass der Erzeugung von Arbeit als Aequivalent ein blosser Übergang von Wärme aus einem warmen in einen kalten Körper entspreche, ohne dass die Quantität der Wärme dabei verringert werde.*

Der letzte Theil dieser Annahme, nämlich dass die Quantität der Wärme unverringert bleibe, widerspricht unserem früheren Grundsatze und muss daher, wenn wir diesen festhalten wollen, verworfen werden. Der erste Theil dagegen kann seinem Hauptinhalte nach fortbestehen."

"*Carnot assumed...that the equivalent of the work done by heat is found in the mere transfer of heat from a hotter to a colder body, while the quantity of heat remains undiminished.*

The latter part of this assumption - namely, that the quantity of heat remains undiminished - contradicts our former principle, and must therefore be rejected if we are to retain that principle. On the other hand, the first part may still obtain in all its essentials."

Moreover, he stated:

"Ein Uebergang von Wärme aus einem warmen in einen kalten Körper findet allerdings in solchen Fällen statt, wo Arbeit durch Wärme erzeugt, und zugleich die Bedingung erfüllt wird, dass der wirksame Stoff sich am Schlusse wieder in demselben Zustande befinde, wie zu Anfang.... Um jedoch diese übergeführte Wärme mit der Arbeit in Beziehung bringen zu können, ist noch eine Beschränkung nöthig. Da nämlich auch ein Wärmeübergang ohne mechanischen Effect stattfinden kann, wenn ein warmer und ein kalter Körper sich unmittelbar berühren, und die Wärme durch Leitung hinüberströmt, so muss, wenn man für den Uebergang einer bestimmten Wärmemenge zwischen zwei Körpern von bestimmten Temperaturen t und τ das Maximum der Arbeit erlangen will, der Vorgang so geleitet werden, wie es in den obigen Fällen geschehen ist, dass nie zwei Körper von verschiedener Temperatur in Berührung kommen.

Dieses *Maximum* der Arbeit nun ist es, welches mit dem Wärmeübergang verglichen werden muss, und dabei findet sich, dass man in der That Grund hat, mit *Carnot* anzunehmen, dass es nur von der Menge der übergeführten Wärme und von den Temperaturen t und τ der beiden Körper *A* und *B*, nicht aber von der Natur des vermittelnden Stoffes abhänge."

"A transfer of heat from a hotter to a colder body always occurs in those cases in which work is done by heat, and in which also the condition is fulfilled that the working substance is in the same state at the end as at the beginning of the operation ... Yet, in order to establish a relation between the heat transferred and the work done, a certain restriction is necessary. For since a transfer of heat can take place without mechanical effect if a hotter and a colder body are immediately in contact and heat passes from one to the other by conduction, the way in which the transfer of a certain quantity of heat between two bodies at the temperatures t and τ can be made to do the maximum of work is so to carry out the process, as was done in the above cases, that two bodies of different temperatures never come in contact.

It is this *maximum* of work which must be compared with the heat transferred. When this is done it appears that there is in fact ground for asserting, with Carnot, that it depends only on the quantity of the heat transferred and on the temperatures t and τ of the two bodies *A* and *B*, but not on the nature of the substance by means of which the work is done."

Clausius stated that it is justified to keep the first statement of Carnot's assumption.

Then, Clausius showed that the work may be expressed by the form $\frac{1}{C}dt$, where C is a function of t only. Moreover, he developed for permanent gases the analytical expressions of Carnot's principle as they were given by Clapeyron (1834) in a somewhat different form. Thus, he adhered here mainly to the view of Clapeyron (1834):

$$\text{(IV.)} \qquad \left(\frac{dQ}{dv}\right) = \frac{RC}{v}\,, \qquad (3.8.12)$$

where R is a constant. In the following investigations, Clausius developed corresponding expressions for vapor, in particular,

$$\text{(V.)} \qquad r = C(s-\sigma)\frac{dp}{dt}\,. \qquad (3.8.13)$$

Then, Clausius connected the last two equations with the result of the first principle and arrived at

$$\left(\frac{dQ}{dv}\right) = \frac{RA(a+t)}{v} \qquad (3.8.14)$$

and

$$\text{(Va.)} \qquad r = A(a+t)(s-\sigma)\frac{dp}{dt}\,, \qquad (3.8.15)$$

for which he also derived other forms. He performed some calculations on the basis of Regnault's experimental results in order to reveal the deviation from the Mariotte's and Gay-Lussac's law (see Clausius, 1850). At the end of his valuable paper, Clausius (1850) determined the numerical value of the mechanical equivalent of heat in different ways.

In a second paper entitled *Über eine veränderte Form des zweiten Hauptsatzes der mechanischen Wärmetheorie (On a varied form of the second law of the mechanical heat theory)*, Clausius (1854) came to some very deep insights on the nature of heat. At the beginning, he pointed out that he had left Carnot's statement unchanged in its essential form in his first treatise (Clausius, 1850), and that he had shown that the statement of the equivalence of heat and work and Carnot's statement did not contradict each other. Furthermore, he remarked that the form of Carnot's statement which he had derived in 1850 was imperfect in so far as it did not clearly reflect the essential nature of the statement and its relation to the first law, and he announced that he would communicate another form in the paper in question which would avoid the above-mentioned shortcoming and which would be very useful in application.

In the first part of this treatise, Clausius (1854) carefully discussed the first law; he paid particular attention to internal energy. He stated that in a process

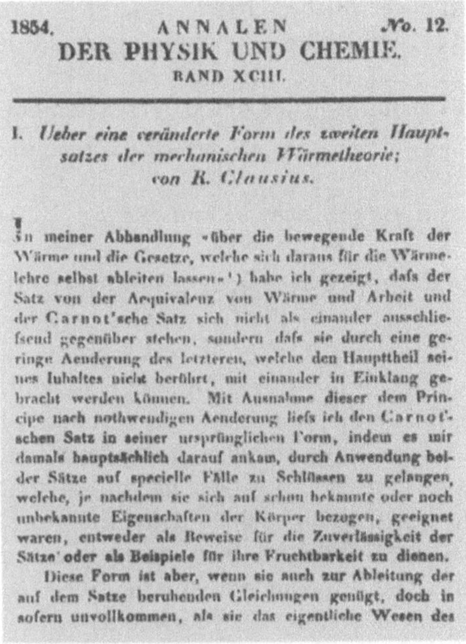

1854. ANNALEN No. 12.
DER PHYSIK UND CHEMIE.
BAND XCIII.

I. *Ueber eine veränderte Form des zweiten Hauptsatzes der mechanischen Wärmetheorie;*
von R. Clausius.

In meiner Abhandlung »über die bewegende Kraft der Wärme und die Gesetze, welche sich daraus für die Wärmelehre selbst ableiten lassen« [1]) habe ich gezeigt, dafs der Satz von der Aequivalenz von Wärme und Arbeit und der Carnot'sche Satz sich nicht als einander ausschließend gegenüber stehen, sondern dafs sie durch eine geringe Aenderung des letzteren, welche den Munpttheil seines Inhaltes nicht berührt, mit einander in Einklang gebracht werden können. Mit Ausnahme dieser dem Princip nach nothwendigen Aenderung liefs ich den Carnot'schen Satz in seiner ursprünglichen Form, indem es mir damals hauptsächlich darauf ankam, durch Anwendung beider Sätze auf specielle Fälle zu Schlüssen zu gelangen, welche, je nachdem sie sich auf schon bekannte oder noch unbekannte Eigenschaften der Körper bezogen, geeignet waren, entweder als Beweise für die Zuverlässigkeit der Sätze oder als Beispiele für ihre Fruchtbarkeit zu dienen.

Diese Form ist aber, wenn sie auch zur Ableitung der auf dem Satze beruhenden Gleichungen genügt, doch in sofern unvollkommen, als sie das eigentliche Wesen des

Fig. 3.39. Clausius' second fundamental treatise (1854)

cycle the internal energy is equal to zero and that the amount of heat is equal to the external work.

The second section in Clausius' (1854) paper has the title *Satz von der Aequivalenz der Verwandlungen (Statement of the equivalence of the conversions).* At the beginning, Clausius pointed out that Carnot's statement expressed a relation between two kinds of conversions, namely the conversion of heat into work and the transformation of heat from a hotter to a colder body, which could be denoted as a conversion of heat from a higher temperature to a lower temperature.

"In allen Fällen, wo eine Wärmemenge in Arbeit verwandelt wird, und der diese Verwandlung vermittelnde Körper sich schließlich wieder in seinem Anfangszustande befindet, muß zugleich eine andere Wärmemenge aus einem wärmeren in einen kälteren Körper übergehen, und die Größe der letzteren Wärmemenge im Verhältniß zur ersteren ist nur von den Temperaturen der beiden Körper, zwischen welchen sie übergeht, und nicht von der Art des vermittelnden Körpers abhängig."

"In all cases in which an amount of heat is converted into work and in which the body which has arranged this conversion is again in its initial state, another amount of heat must simultaneously pass over from a hotter to a colder body, and the quantity of the last amount of heat in ratio to the first is dependent only on the temperatures of both bodies between which it passes and not on the kind of the arranging bodies."

However, Clausius (1854) restricted his statement because, in the passage above, it was silently assumed that the heat converted to work came from one of the bodies between which the heat conversion occurred. In this way, one has already made *a priori* a determined assumption about the temperature of the heat converted into work. Thus, the influence which a change of temperature has on the ratio of both amounts of heat is revealed and the above statement is incomplete.

In order to determine this influence, Clausius returned to Carnot's statement:

"es kann nie Wärme aus einem kälteren in einen wärmeren Körper übergehen, wenn nicht gleichzeitig eine andere damit zusammenhängende Aenderung eintritt."

"heat can never pass over from a colder into a warmer body, without another change connected with it occurring simultaneously."

Then, Clausius (1854) again discussed Carnot's cyclic process, however, with the difference that this time he considered not only the two bodies between which the heat transformation occurred, but also a third body which produced the heat converted to work. He performed the cyclic process for simplicity with a permanent gas. He showed that during the cyclic process a certain amount of heat was converted into external work. Moreover, he stated that the described cyclic process could also be performed in a reverse way and he proved that both kinds of conversion were to be considered as procedures equal in nature. He called both conversions equivalent. In what followed, he tried to gain the corresponding mathematical description of this phenomenon and he called the mathematical value of a conversion the equivalent-value ("Aequivalenzwerth"). First, he introduced the positive sign of the conversion:

"Wir wollen im Folgenden *die Verwandlung aus Arbeit in Wärme, und demgemäß den Übergang von Wärme von höherer zu niederer Temperatur als positive Verwandlungen rechnen.*"

"In the following we will *denote the conversion from work into heat, and accordingly the transition of heat from a higher to a lower temperature as positive conversions.*"

Concerning the quantity of the equivalent-value, Clausius (1854) pointed out that it was obvious that the value of a conversion from work into heat had to be proportional to the amount of heat created in the process and could depend solely on the temperature. Therefore, Clausius assumed that the equivalent-value of the creation of the amount of heat Q with the temperature t from the external work could be represented, in general, by the expression:

$$Q\,f(t)\,, \tag{3.8.16}$$

wherein $f(t)$ is a temperature function equal for all cases. Furthermore, he stated:

"Wenn in dieser Formel Q negativ wird, so wird dadurch ausgedrückt, daß die Wärmemenge Q nicht aus Arbeit in Wärme sondern aus Wärme in Arbeit verwandelt wird."

"If in this formula Q becomes negative, so it is expressed by this that the amount of heat Q is converted not from work into heat but from heat into work."

After a long derivation, containing bad mathematics in parts and vague definitions which make it difficult to follow, Clausius arrived at the formulation of the following relation for a cyclic reversible process:

$$\int \frac{dQ}{T} = 0\,, \tag{3.8.17}$$

in which he had introduced

$$T = \frac{1}{f(t)} . \tag{3.8.18}$$

Here, T is an arbitrary function of the temperature and is neither the ideal-gas temperature nor the absolute temperature.

Clausius (1854) used the formula stated above to determine Carnot's function C:

$$\frac{\frac{dT}{dt}}{T} = \frac{A}{C} , \tag{3.8.19}$$

where A is the mechanical equivalent of heat. Furthermore, he considered an irreversible cyclic process and stated the second law in the form:

"Die algebraische Summe aller in einem Kreisprocesse vorkommenden Verwandlungen kann nur positiv seyn."

"The algebraic sum of all conversions occuring in a cyclic process can only be positive."

Finally, Clausius (1854) determined the time dependent function T for a permanent gas by using Mariotte's and Gay-Lussac's laws

$$T = (a + t) \cdot const. \tag{3.8.20}$$

where a denotes the reverse value of the expansion coefficient of permanent gases, which is nearly equal to 273, as Clausius remarked. Moreover, Clausius stated that the value of the constant in the above equation was not important, because by changing the value of the equivalence values, Eqn. (3.8.19) would not be influenced. He therefore used the simplest value, namely, unity:

$$T = a + t \tag{3.8.21}$$

and mentioned that T was simply the absolute temperature.

In his treatise *Über die Anwendung des Satzes von der Aequivalenz der Verwandlungen auf die innere Arbeit*, Clausius (1862) formulated the statement:

"*die algebraische Summe aller in einem Kreisprocesse vorkommenden Verwandlungen kann nur positiv oder als Gränzfall Null seyn.*

Der mathematische Ausdruck dieses Satzes ist folgender. Sey dQ das Element der von dem Körper während seiner Veränderungen an irgend ein Wärmereservoir abgegebene Wärme (wobei eine Wärmemenge, welche er einem Reservoir entzieht, negativ gerechnet wird) und T die absolute Temperatur, welche der Körper im Momente der Abgabe hat, so muß für jeden umkehrbaren Kreisproceß die Gleichung:

$$\text{(I)} \quad \int \frac{dQ}{T} = 0 , \tag{3.8.22}$$

und für jeden überhaupt möglichen Kreisproceß die Beziehung:

$$\text{(Ia)} \quad \int \frac{dQ}{T} \geq 0 \tag{3.8.23}$$

gelten."

"the algebraic sum of all conversions occuring in a cyclic process can only be positive or in the limit zero.

The mathematical expression of this statement is as follows. Be dQ the element of heat emitted to an arbitrary heat reservoir during its changes (whereby, the amount of heat removed from a reservoir, will be denoted as negative) and T the absolute temperature, which the body possesses at the moment of release, then for every reversible cyclic process the equation:

$$\text{(I)} \quad \int \frac{dQ}{T} = 0 , \tag{3.8.22}$$

and for every possible cyclic process, the relation

$$\text{(Ia)} \quad \int \frac{dQ}{T} \geq 0 \tag{3.8.23}$$

are valid."

Clausius published a series of other papers on thermodynamics, in particular in *Poggendorffs Annalen* and in *The London, Edingburg, and Dublin Philosophical Magazine and Journal of Science.* However, not all of his papers possess the lucidity of his first (Clausius, 1850).

Fifteen years later, Clausius (1865) published a paper that was interesting for several reasons, *Über verschiedene für die Anwendung bequeme Formen der Hauptgleichungen der mechanischen Wärmetheorie.* In Section 14 of this paper, he wrote:

"Alle vorstehenden Betrachtungen bezogen sich auf Veränderungen, welche in umkehrbarer Weise vor sich gehen. Wir wollen nun auch noch die *nicht umkehrbaren Veränderungen* in den Kreis der Betrachtungen ziehen, um wenigstens der Hauptsache nach kurz anzugeben, wie sie zu behandeln sind.

Bei mathematischen Untersuchungen über nicht umkehrbare Veränderungen handelt es sich vorzugweise um zwei Umstände, welche zu eigenthümlichen Größenbestimmungen Veranlassung geben. Erstens sind die Wärmemengen, welche man einem veränderlichen Körper mittheilen resp. entziehen muß, bei nicht umkehrbaren Veränderungen andere, als wenn dieselben Veränderungen in umkehrbarer Weise geschehen. Zweitens ist jede nicht umkehrbare Veränderung mit einer uncompensirten Verwandlung verbunden, deren Kenntniß bei gewissen Betrachtungen von Wichtigkeit ist."

"All aforementioned considerations, were related to changes, which occur in reversible form. Now, we will also consider the non-reversible changes in order to briefly state, at least, the main points, as to how they can be treated.

By mathematical investigation on non-reversible changes it is primarily a matter of two circumstances which gives rise to peculiar quantitative determinations. Firstly, the amounts of heat which must be transferred to or emitted from a changing body, respectively, are by non-reversible changes different from those in which the same changes occur in reversible form. Secondly, every non-reversible change is connected with a non-compensated conversion, whose understanding is of importance for certain considerations."

In the course of his further investigations, Clausius (1865) reintroduced the notion

$$dS = \frac{dQ}{T} \tag{3.8.24}$$

and after integration

$$S = S_0 + \int \frac{dQ}{T} , \tag{3.8.25}$$

where S_0 denotes S in its initial state. He explained:

"Sucht man für *S* einen bezeichnenden Namen, so könnte man, ähnlich wie von der Größe *U* gesagt ist, sie sey der *Wärme- und Werkinhalt des Körpers*, von der Größe *S* sagen, sie sey der *Verwandlungsinhalt* des Körpers. Da ich es aber für besser halte, die Namen derartiger für die Wissenschaft wichtiger Größen aus den alten Sprachen zu entnehmen, damit sie unverändert in allen neuen Sprachen angewandt werden können, so schlage ich vor, die Größe *S* nach dem griechischen Worte η'τροπη', die Verwandlung, die Entropie des Körpers zu nennen. Das Wort *Entropie* habe ich absichtlich dem Wort *Energie* möglichst ähnlich gebildet, denn die beiden Größen, welche durch diese Worte benannt werden sollen, sind ihren physikalischen Bedeutungen nach einander so nahe verwandt, daß eine gewisse Gleichartigkeit in der Benennung mir zweckmäßig zu seyn scheint."

"If one is looking for a significant name for *S*, so one could say, as it is said for the quantity *U*, that it be the *heat and work content* of the body, and for the quantity *S* the *conversion content* of the body. As I find it better to take the names for such important quantities in science, from the old languages, so that they can be applied in all new languages, therefore I suggest to denote the quantity *S* after the Greek word η'τροπη', the conversion, as the entropy of the body. I have intentionally formed the word *entropy* as similar as possible to the word *energy*, because both quantities, which shall be denoted by these words, are so closely related to each other according to the physical importance that a certain similarity in their naming seems for me to be advisable."

Finally, in the last section of his paper, Clausius (1865) wrote his famous statement on energy and entropy in the universe:

"Vorläufig will ich mich darauf beschränken, als ein Resultat anzuführen, daß, wenn man sich dieselbe Größe, welche ich in Bezug auf einen einzelnen Körper seine *Entropie* genannt habe, in consequenter Weise unter Berücksichtigung aller Umstände für das ganze Weltall gebildet denkt, und wenn man daneben zugleich den anderen seiner Bedeutung nach einfacheren Begriff der *Energie* anwendet, man die den beiden Hauptsätzen der mechanischen Wärmetheorie entsprechenden Grundgesetze des Weltalls in folgender einfacher Form ansprechen kann:

1) *Die Energie der Welt ist constant.*
2) *Die Entropie der Welt strebt einem Maximum zu.*"

"Temporarily, I will restrict myself to the statement of the following result: If one imaginarily forms the same quantity, which I have denoted, referred to a single body, as its *entropy*, in a consequent manner, considering all the circumstances for the whole universe, and if one besides that simultaneously applies the other, according to its importance, simpler notion *energy*, then both the statements of the mechanical heat theory can be expressed in the following simple form:

1) *The energy of the universe is constant.*
2) *The entropy of the universe goes to a maximum.*"

Clausius (1876) finally summarized all of his theoretical findings, as well as those of other authors, in the first volume of a book which represented the second completely rewritten edition of his pamphlet *Abhandlungen über die mechanische Wärmetheorie.*

During the time when Clausius (1850) was publicizing his fundamental findings, Rankine gave a talk to the Edingburg Royal Society: *On the mechanical action of heat, especially in gases and vapours*, which was first published in 1851, and again in 1854 (see Rankine, 1854). In his treatise, Rankine proceeded from an assumption which he had already founded earlier, that the heat phenomenon consisted of a vortical motion of the molecules. He supposed that the vortical material surrounds like the atmosphere dense kernels, whereby he did not specify these kernels. By heat "of a body", he understood the energy of the vortex-atoms contained in the body, and the absolute temperature was assumed to be the ratio of this energy and a coefficient characteristic for each substance. In an ideal gas, the elastic pressure should only change with the centrifugal force of the molecule vortices; Rankine concluded then, from well-known mechanical principles, that the pressure was proportional to the energy of the vortex-atoms and inversely proportional to the space which was occupied by these vortices. Moreover, Rankine stated that, in ordinary gases, the elasticity was additionally influenced by cohesive forces. However, in the case where the deviation of the ordinary gases from an ideal gas was small, Rankine calculated these forces approximately by a series expansion, and he applied his results to several different vapours. Rankine's particular merit lay in applying his peculiar methods of the mechanical theory of heat to several different technical problems, for example, showing the condensation of saturated vapour during an adiabatic expansion. He was also the first scientist to treat the different heat engines from a unique point of view. Later, Rankine summarized all his findings in a textbook entiteled *A Manual of the Steam-Engine and other Prime Movers* (see M. Rühlmann, 1885). In this textbook, Rankine stated (for the first time in a textbook) the first and second law of thermodynamics (so denoted by him):

First law: "Heat and mechanical energy are mutually convertible; and heat requires for its production, and produces by its disappearance, mechanical energy in the proportion of 772 foot-pounds for each British unit of heat."
Second law: "If the absolute temperature of any uniformly hot substance be divided into any number of equal parts, the effects of those parts in causing work to be performed are equal".

Rankine's textbook received much acceptance, ran into several editions and was for a long time commonly regarded by theorists and practical men alike as a classic work.

William Thomson (later Lord Kelvin) must be mentioned as one of the leading physicists of the nineteenth century. He began his investigations of the heat pheomenon in 1849, when he published a treatise called: *An account of Carnot's theory of the motive power of heat with numerical results deduced from Regnault's experiments on steam*, in which he drew his readers' attention to the special view of Carnot (see Thomson, 1849). In this work, he pointed to the contradiction which existed between Carnot's assumption of the conservation of the heat effective in a cyclic process and the results of Joule's experiments on the equivalence of mechanical power and heat, already known in England at that time. He pointed out that the difficulty could be avoided if one were to drop Carnot's fundamental assumption. However, Thomson could not make up his mind at that time whether to give up the false basis of Carnot's theory. Rather, he was of the opinion that other difficulties would appear which would not be

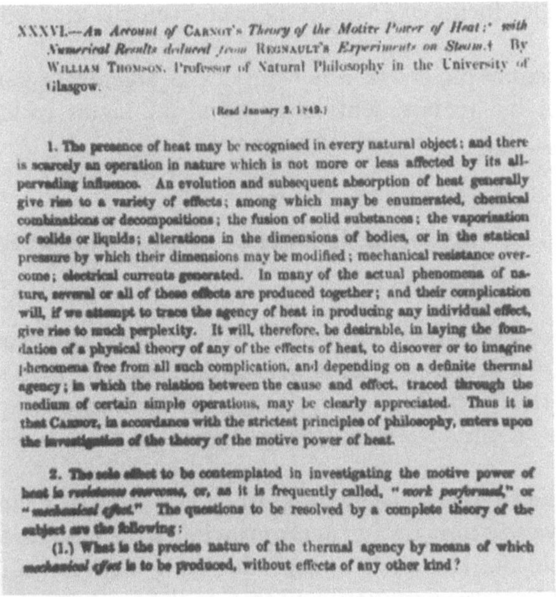

XXXVI.—*An Account of* CARNOT's *Theory of the Motive Power of Heat;* with Numerical Results deduced from* REGNAULT's *Experiments on Steam.*† By WILLIAM THOMSON, Professor of Natural Philosophy in the University of Glasgow.

(Read January 2, 1849.)

1. The presence of heat may be recognised in every natural object; and there is scarcely an operation in nature which is not more or less affected by its all-pervading influence. An evolution and subsequent absorption of heat generally give rise to a variety of effects; among which may be enumerated, chemical combinations or decompositions; the fusion of solid substances; the vaporisation of solids or liquids; alterations in the dimensions of bodies, or in the statical pressure by which their dimensions may be modified; mechanical resistance overcome; electrical currents generated. In many of the actual phenomena of nature, several or all of these effects are produced together; and their complication will, if we attempt to trace the agency of heat in producing any individual effect, give rise to much perplexity. It will, therefore, be desirable, in laying the foundation of a physical theory of any of the effects of heat, to discover or to imagine phenomena free from all such complication, and depending on a definite thermal agency; in which the relation between the cause and effect, traced through the medium of certain simple operations, may be clearly appreciated. Thus it is that CARNOT, in accordance with the strictest principles of philosophy, enters upon the investigation of the theory of the motive power of heat.

2. The sole effect to be contemplated in investigating the motive power of heat is *resistance overcome*, or, as it is frequently called, "*work performed*," or "*mechanical effect*." The questions to be resolved by a complete theory of the subject are the following:

(1.) What is the precise nature of the thermal agency by means of which *mechanical effect* is to be produced, without effects of any other kind?

Fig. 3.40. Thomson's basic paper of 1851

able to be solved without further experimental investigations and a complete reformulation of the theory of heat.

In 1851, Thomson's (1851) second paper on the heat problem was published: *On the dynamical theory of heat, with numerical results deduced from Joule's equivalent and Regnault's experiments on steam.* In this paper, he developed the second law of thermodynamics completely independently of Clausius' approach. His development was based on a mechanical principle, and he stated that it would be impossible to gain any mechanical power with the help of inanimate bodies by any substance if its temperature was lower than the lowest of all surrounding bodies. In his paper, Thomson gave credit to Clausius, remarking that Clausius had been the first scientist to found the second law of thermodynamics on the right principles, which he accomplished in May, 1850.

Another interesting consequence of the mechanical theory of heat was developed by Thomson (1852), namely the dissipation of mechanical energy. Thomson pointed out that with heat propagation and radiation, a scattering, and never a concentration of the heat, occurred, and that a larger amount of heat was always changed from a higher temperature to a lower temperature. For all irreversible cyclic processes, a dissipation of energy always occurs. Furthermore, he remarked that in such processes it was impossible to reconstruct the original state without adding new energies.

Thomson, like Clausius, drew the universal consequence that, according to the above statement, which is valid in the whole physical world, the universe will finally reach an equilibrium state in which all motive and living phenomena will be extinct, because all other forms of energy will have been changed into heat of the same temperature.

Thomson also wrote substantial papers on other physical problems. In particular, he became famous for his achievements concerning the laying of the first transatlantic cable, for which he was enobled by the Queen.

3.8.7
Discussions on the Correct Form of the Mechanical Theory of Heat and Further Developments

During the 1850s, the fundamental findings of Clausius, Rankine, and Thomson were discussed extensively in the literature and it was, in part, heavily argued as to whom one should give priority. It seems that, concerning the introduction of the internal energy and the second law of thermodynamics, Clausius (1850) made the first step.

As has already been pointed out in the previous sections, it was Carnot who explained the motive power of heat by a general principle which assumed that the quantity of heat would not change. However, this view was gradually replaced by the statement that heat was a motion, and that, in the production of mechanical work heat would be consumed. Finally, Mayer and Joule stated the equivalence of heat and mechanical work. Mayer and, in particular, Helmholtz formulated the conservation of energy, though not in the final form that we know today.

With these results, the departure-point for new investigations in the theory of heat was apparent.

The new era was pioneered by Thomson (1849), as aforementioned, who published a treatise in which he drew his colleagues' attention to the somewhat strange considerations of Carnot. He remarked:

> "It might appear, that the difficulty would be entirely avoided, by abandoning CARNOT's fundamental axiom, a view which is strongly urged by Mr. JOULE."

He added:

> "It is in reality to experiment that we must look – either for a verification of CARNOT's axiom, and an explanation of the difficulty we have been considering; or for an entirely new basis of the Theory of Heat."

Just at the time when Thomson's treatise appeared, Clausius was finishing his first paper on the mechanical theory of heat, which he presented to the Berlin Academy in February, 1850, and which was published in March/April, 1850. As already mentioned, he pointed out that the basic definitions and the entire mathematical treatment of heat had to be changed if one assumed the statement of the equivalence of heat and mechanical work. He went on to say that it was not necessary to reject Carnot's theory completely.

At the same time (February 1850) that Clausius was publicizing his epochal treatise, Rankine (1851) gave a valuable lecture. From his basic hypothesis that the heat phenomena resulted from a vortex of the molecules, he arrived at nearly the same results which Clausius had obtained from the first law of the mechanical theory of heat. In this paper, the second law was not mentioned. In a treatise published later, Rankine (1854) drew attention to the fact that this law could be derived from equations which were contained in the first section of his former article. Indeed, Rankine reported a new proof; however, according to Clausius (1876), this proof was neither consistent nor did it conform to the other ideas of Rankine.

In March, 1851, the second paper of Thomson also appeared (see Thomson, 1852), in which he left intact his former standpoint regarding Carnot's theory.

In this article,Thomson (1852) did not claim any priority. He wrote:

"The whole theory of the motive power of heat is founded on the two following propositions, due respectively to Joule, and to Carnot and Clausius."

After he had reported his own proof of the second law, he continued:

"It was not until the commencement of the present year that I found the demonstration given above, by which the truth of the proposition is established upon an axiom (§12) which I think will be generally admitted. It is with no wish to claim priority that I make these statements, as the merit of first establishing the proposition upon correct principles is entirely due to Clausius, who published his demonstration of it in the month of May last year, in the second part of his paper on the motive power of heat. I may be allowed to add, that I have given the demonstration exactly as it occurred to me before I knew that Clausius had either enunciated or demonstrated the proposition."

Attacks against the theory of Clausius came from Holtzmann, who criticized Clausius' mathematical treatment of the cyclic process and his determination of the specific heat, and in addition from Decker in 1858 (see Clausius, 1876). In particular, he slandered Clausius' mathematical development as a maltreatment of the analysis, a bad job, and nonsense. Obviously, Clausius was deeply hurt by Decker's attack and he rewrote the mathematical development of the theory of heat and used this extended development as an extra introductory chapter his book.

After the fundamental investigations and findings of the aforementioned scientists, the theory of heat has been further developed. We will call only a few names of researchers who have considerably contributed to this theory. It was Gibbs (1873a, b, 1875) who initiated a new line in the theory of heat. Although he spoke of thermodynamics, he treated thermostatics. Truesdell (1986) argumented:

"I see Gibbs as the one and only creator of *thermostatics*; in contrast, his writings seem to me to bear little on *thermodynamics*."

Gibbs' work is extensively discussed by Truesdell (1986).

The development of thermodynamics on the theoretical side has been, in a wide range, influenced by Planck's work – in particular, in Germany – which pioneered the field of homogeneous thermodynamic processes in bodies with irreversible changes. Planck (1897) summarized his findings in his famous textbook *Vorlesungen über Thermodynamik*, which appeared in Germany, already in its 11th edition. However, several translated editions have also been published in England.

The attempt of Carathéodory (1909) to axiomatically found thermodynamics had been heavily criticized by Truesdell (1980b, 1986), in particular, in the Appendix of Truesdell's (1980b) treatise with the title *Failure of Carathéodory's attempt to set the house in order.*

Important contributions, including continuum mechanical approaches, are due to Jaumann (1911) and Lohr (1917). It seems that their view on the balance of energy and the entropy principle are not evaluated yet.

In modern times, thermodynamic considerations have been incorporated in the mixture theory. In contrast to the balance of energy (first law of thermodynamics), the entropy principle (second law of thermodynamics) is an inequality for general irreversible processes. For a long time, it was unknown in the literature whether an entropy inequality had to be allowed for each constituent φ^α or whether one had to use only one entropy inequality for the mixture as a whole. Today, the opinion is that the postulate of separate entropy inequalities must be regarded as a sufficient, but too restrictive, requirement. The postulate of one common entropy is simultaneously a necessary and a sufficient condition for the existence of dissipation mechanisms within the mixture.

The entropy inequality for mixture bodies had been correctly developed, for the first time, by Bowen (1967), Müller (1968), and Truesdell (1968).

3.8.8 Biographical Notes

Sadi Carnot (1796–1832)

On June 1, 1796, a third son was born to Lazare Carnot, a member of the directorate of the first republic, and an excellent mathematician. He was named Sadi after a mediaeval Persian poet and moralist, whose poems had enjoyed something of a vogue. The boy spent his early childhood in S. Omer. A few years later, Lazare became Minister of War, and he often took his little son with him when he visited Napoleon. Sadi was the darling of Josephine, later the Empress, and he often spent half a day in the high society of this charming and intelligent woman. Sadi's father very carefully educated him himself. At the Lycée Charlemagne he prepared for the École Polytechnique and, in September 1812, the sixteen years old Sadi entered the famous school - during the year in which Napoleon's fortunes turned, the year of his retreat from Moscow.

In 1814 and 1815, he attended the School of War in Metz. Although he had already distinguished himself by excellent scientific achievements, the young officer found himself on garrison duty, far from Paris, doing the lowest routine jobs and quickly became sick of his duty. Therefore, in 1819, he transferred to the General Staff, but almost immediately retired on half-pay and moved to Paris; this was in 1820; he was 24 years of age.

Fig. 3.41. Sadi Carnot (1796–1832)

The period that followed was the most creative time of his life. He studied at the Sorbonne, the Collège de France, and the École des Mines, concentrating on physics and economics. He spent much of his time visiting factories and studying the organization and economics of various industries; he became an expert on the industry and trade in Europe.

In 1823, Lazare was in exile and Hippolyte, his younger son, returned to Paris. The two Carnot brothers set up house in a small apartment, and it was here that Sadi wrote *Réflexions suir la puissance motrice du feu et sur les machines*, which appeared in Paris in 1824. In the same year, the political scene worsened. For a time, Sadi was recalled to full-time military service, as a staff captain. But, in 1828, he resigned permanently and devoted himself to physics and economics. The incomplete manuscripts which have been preserved prove that Sadi had given up his opinion on the conservation of heat after the publication of his *Reflections*. Rather, he had come to full clarity about heat as a form of energy and had already approximately determined the mechanical equivalent of heat.

In 1832, Sadi caught scarlet fever, which turned into a brain fever; he recovered and was taken into the country to convalesce. Hippolyte and another friend went to nurse him. Some days later, he caught cholera and died within a few hours. He was just 36 years old (after Mendoza 1960, R. Rühlmann, 1876).

Benôit-Pierre-Émile Clapeyron (1799–1864)

Émile Clapeyron was born in Paris on February 26, 1799. He graduated from the École Polytechnique in 1818, and from the School of Mines in 1820. Together

Fig. 3.42. Benôit-Pierre-Émile Clapeyron (1799–1864)

with Lamé, he was recommended to the Russian government as a promising young engineer and he joined the Institute of Engineers of Ways of Communication in St. Petersburg. Here he did the same work as Lamé.

After his return to France in 1831, he was thereafter very active in practical work connected with the construction of French railroads. His main occupation was the application of thermodynamics to locomotive design. In 1834, he published his *Mémoire sur la puissance motrice de la chaleur* in the *Journal de l'École Polytechnique*. Clapeyron's paper was quite different from Carnot's concerning its manner of presentation, though it showed the same results. It was analytical throughout and made only cursory references to the problems of engine design and to the industrial applications which had been prominent in the original.

Starting in 1844, he gave a course on steam engines at the École des Pont et Chaussées and proved to be an excellent teacher. The combination of great theoretical knowledge and broad practical experience which he possessed especially served to attract students to his lectures.

In 1858, Clapeyron was elected a member of the Académie and remained at the École des Pont et Chaussées until his death on January 28, 1864 (after Timoshenko, 1953, and Mendoza, 1960).

Julius Robert Mayer (1814–1878)

Julius Robert Mayer was born as the son of a pharmacist in Heilbronn on November 25, 1814. As a boy, Robert was an absent-minded child. From his

Fig. 3.43. Julius Robert Mayer (1814–1878)

father, who was not only a clever businessman, but also a man familiar with the progress in chemistry, Robert learnt a lot of things in this field and became interested in natural science. Therefore, he decided to study medicine. During his studies at Tübingen University, from 1832 to 1837, he was not only busy with medicine, but also collected profound knowledge in the field of exact natural sciences. At the end of his study, he was taken into detention for being the leader of a forbidden fraternity. He was, however, of the opinion that he was not guilty and he went on hunger strike. Thereupon, his sentence was shortened and he was expelled from the university. Mayer finished his studies in medicine at the universities of Munich and Vienna and, after his final examination, attended the clinics and hospitals in Paris.

His father, who had quickly recognized his son's talent for natural sciences, allowed him to do duty as a ship's doctor on a steamer to East India and, in 1840, he started his trip, which proved to be rather boring due to a lack of work. During this time, he studied the *Handbuch der Physiologie* by Johannes Müller. This study, and Mayer's profound knowledge of chemistry, enabled the young doctor to draw far-reaching conclusions from his occasional observations. Mayer (1842) published his fundamentel finding on the equivalent of heat and mechanical work in Liebig's and Wöhler's *Annalen der Chemie und Pharmacie*. However, his work went unnoticed at that time.

In 1841, Mayer resettled in his hometown of Heilbronn, as a physician. He continued his scientific work with contributions in metabolism and astronomy.

Mayer suffered very much from the lack of acknowledgement and his experience concerning the fight for recognition of his discovery. In 1848, he was also deeply affected by the loss of two of his children, as well as his brother's participation in the revolution. He tried to commit suicide by jumping out of a window. The suicide attempt failed, but left his legs paralyzed. In 1851, he was put into a hospital for the mentally ill in Göppingen and Winnethal for a period of 15 months in which he was treated roughly and awfully.

In 1854, Mayer left the hospital. He suffered from depression for a long time before recommencing his publications with a treatise on fever in 1862. In 1867, he published a collection of all of his contributions to the mechanical theory of heat under the title *Die Mechanik der Wärme* in the publishing house Cotta. This work was widely enough recognized that within few years a new edition became necessary.

Apart from some public lectures, little is known concerning Mayer's further work. On March 10, 1878, he died of pneumonia.

James Prescott Joule (1818–1889)

In the history of science, Joule holds a special place. He was a scientist with a great ability to perform complicated experiments and to draw the right conclusions. This led him to the discovery of a considerable amount of important physical laws. His achievements are made all the more noteworthy when one considers that he was the owner of a large brewery who, as an autodidact, conducted research solely as a hobby; mathematics was not his field.

Joule was born close to Manchester on Christmas Eve, 1818. Already as a young man, he was concerned with physical problems. In the very beginning, he was interested in the utilization of electrical and magnetic forces for mechanical problems.

In the following years, he worked relentlessly to measure the production of heat caused by all possible proceedings. He stirred water and mercury, squeezed

Fig. 3.44. James Prescott Joule (1818–1889)

water through small holes in order to measure the frictional heat performed and he compressed gases and allowed them to expand again.

In all of these cases, he calculated the mechanical work which entered the system and the amount of heat which came out of the system. Every time he found that a certain amount of mechanical work produced a determined amount of heat.

In 1847, the first descriptions of his experiments appeared. However, there was no public response to his work. Finally, Thomson's allusions to Joule's works made him famous and established Joule's reputation.

At the end of his life, when Joule's economic situation became poor, Queen Victoria awarded him with a pension.

Joule died on October 11, 1889, in Sale (Cheshire).

Rudolf Clausius (1822–1888)

Rudolf Clausius was born on January 2, 1822 in Köslin, Pommern, in Germany. His father was a schoolmaster and later the headmaster of his private school. Clausius belonged to a large family; he had thirteen brothers and sisters. He entered the University of Berlin in 1840, and was trained by Berlin's leading mathematicians and physicists of the time. One of his teachers was Heinrich Dove, whose main field was meteorology but also lectured in theoretical physics. After passing the final examination (with honors) in 1844, he took over a position as a teacher at a "Gymnasium" in Berlin. As early as 1849, Clausius had already tried to get a professorship at Breslau. He wrote the minister that he

Fig. 3.45. Rudolf Clausius (1822–1888)

would like to take over a chair which was vacant at that time. On the third page, he mentioned that he would be enclosing his scientific papers, and, at the end of the page, he listed the professors who had trained him, names such as Dove, Magnus, Poggendorf, and Steiner. However, the minister was not able to fulfill his request.

In 1850, Clausius was appointed as a teacher of physics at the "Königliche Artillerie- und Ingenieurschule." Immediately thereafter, he finished his habilitation thesis and became "Privatdozent" at Berlin University.

When the Polytechnikum in Zürich was established, Clausius was offered a professorship at the Polytechnikum and at Zürich University (1855). For twelve years, he remained in Zürich, where he experienced a very fruitful scientific period; during this time he published numerous papers on the theory of heat. In 1867, he moved to Würzburg University and two years later, in 1869, he took over the professorship for physics at the University of Bonn, where he stayed for the rest of his life. He died in 1888.

William Thomson (Lord Kelvin) (1824–1907)

William Thomson was born of Scottish descent in Belfast in 1824. His father was a teacher at that time and later became a professor of mathematics at Glasgow University. In 1834, William Thomson enrolled at the University of Glasgow, where he studied classic languages, mathematics, and natural philosophy. In 1840, Thomson became acquainted with the famous book *Théorie analytique de la Chaleur* by Fourier. The contents of that book greatly influenced Thomson's early scientific work.

In order to help his children's study of foreign languages, Thomson's father took his family to Paris in the summer of 1839 and, in the summer of 1840, they travelled around Germany. During this time, he studied Fourier's book extensively.

In April of 1841, William Thomson left the University of Glasgow and entered St. Peter's College in Cambridge. Here, he continued to be interested in Fourier's work and, in November, 1841, Thomson's first scientific paper, dealing with Fourier's series, appeared in the *Cambridge Mathematical Journal*. Two more papers of a more advanced character appeared in the same journal during the course of the following year. In mathematical attitude and knowledge, Thomson was far ahead of other students in his class. It was expected that he would become senior wrangler for the year 1845; however, in the final examination he settled for the place of the second wrangler.

After graduating from Cambridge, Thomson decided to continue his studies, and for that purpose he went to Paris; while there, he met Liouville, Sturm, and Cauchy, the leading mathematicians of the day. He also met several French physicists. In order to acquire a better experimental method, he entered the physics laboratory of the Collége de France, where he worked with Professor Regnault, helping him in his famous experimental research into the laws of heat.

Fig. 3.46. William Thomson (Lord Kelvin) (1824–1907)

During this period, Thomson became acquainted with the famous paper by Clapeyron, *Mémoires sur la puissance motrice ...*, in which the writer explains Carnot's cycle. Thomson's early scientific work owes much to the writing of Carnot, Fourier, and Green.

After four months of study in Paris, he returned to Cambridge in May, 1846, and, in the fall of the same year, obtained a fellowship at St. Peter's College. He became a lecturer in mathematics at the college and began to coach pupils in mathematics. He also undertook the management of the *Cambridge and Dublin Mathematical Journal*.

In the fall of 1846, William Thomson was elected professor of natural philosophy at Glasgow University, where he taught this subject for fifty-three years.

In 1860, P. G. Tait was elected to the chair of natural philosophy in Edinburgh; very soon the two men, Thomson and Tait, were working together. Both felt a want of books which could be recommended to students interested in the various branches of theoretical physics. The two scientists decided that they would themselves write the necessary books and thus, in 1861, they started work on their *Treatise on Natural Philosophy*. The writing was significantly delayed by Thomson's participation in an important work on the Atlantic telegraph, which required plenty of theoretical and experimental investigations. It was not until 1867 that the first volume of the famous treatise was published. This volume dealt with the mechanics of rigid, elastic, and fluid bodies and contained much original work in the theory of elasticity due to Thomson. The broad outline of the *Treatise on Natural Philosophy*, which was to include all branches of theoretical physics, was never completely fulfilled.

In 1866, telegraphic connection was established between England and America. The success of this important enterprise was due to a great extent to Thomson's scientific advice and to his active and energetic cooperation. The honor of knighthood (Lord Kelvin) was conferred upon him in appreciation of his work and of his high position in the scientific world.

In 1899, Lord Kelvin retired from his professorship. On December 17, 1907, Lord Kelvin died.

Chapter 4: The Modern Era

The development of basic relations in the classical era: the stress concept, the elasticity law, the fundamental laws of Delesse, Fick, and Darcy, the foundation of the mixture theory by Maxwell and Stefan, as well as the foundation of thermodynamics, all these had provided enough background material in order to treat empty or fluid-saturated porous solids. Indeed, in this century, the theory of porous media has been at last firmly established based on the achievements of the nineteenth century.

Already from the second to the fourth decade, decisive progress was made towards creating a consistent porous media theory. There were two notable steps. First, during this time, important mechanical effects in a liquid-saturated *rigid* porous solid were discovered and/or investigated, namely the effects of uplift, friction, capillarity, and effective stresses. Second, in the 1920s and 1930s, the first attempts were carried out to investigate saturated *deformable* porous solids. The discovery of fundamental mechanical effects in saturated porous solids, and the formulation of the first porous media theories, are mainly due to two distinguished professors at the Technische Hochschule of Vienna, namely Paul Fillunger and Karl von Terzaghi. These two professors influenced the theory of porous media to a great extent in the first half of this century. It was Karl von Terzaghi who developed the one-dimensional consolidation theory, based on a variety of experimental datas, and also an analogous procedure to heat propagation. Moreover, he developed independently of Fillunger the concept of effective stresses, which is of great eminence in soil mechanics. He is responsible for introducing this concept to the minds of engineers. However, it was Paul Fillinger who founded the modern porous media theory on the basis of ensured mechanical principles. Unfortunately, the controversies regarding some scientific questions led to a deep hostility between the two outstanding Viennese professors and took a very tragic end. In the time that followed, Fillunger's masterpiece, as well as the valuable contributions of his successor, Gerhard Heinrich, were nearly completely forgotten and ignored. From the middle of the 1930s till the 1960s, porous media was governed by Maurice Biot who, in the main parts of his work, followed the scientific working of Karl von Terzaghi, namely to develop porous media theories more or less intuitively, only in parts founded on mechanical principles, but based on ensured experimental datas.

Biot's theory had an immense influence on the description of the mechanical behavior of saturated compressible media in the last decades. Even today, it plays a role in some parts of geomechanics.

After the creation of the modern mixture theory in the 1960s, the porous media theory got new impulses. At the beginning of this new development, saturated porous media were treated as pure mixtures, disregarding the volume fraction concept (As aforementioned, Fillunger's masterpiece was forgotten). However, this shortcoming was very soon discovered, and modern porous media theory (mixture theory restricted by the volume fraction concept) was created.

4.1 Discovery of Fundamental Effects of Liquid-Saturated Rigid Porous Solids

From 1913 to 1934, some important physical phenomena in liquid-saturated porous solids were described by two professors at the Technische Hochschule of Vienna, Paul Fillunger and Karl von Terzaghi. These concerned the effects of uplift, friction, capillarity, and effective stresses. It was Fillunger (1913) who pioneered the porous media theory of liquid-saturated porous solids. He investigated the *uplift* problem in such a saturated porous body. The study of

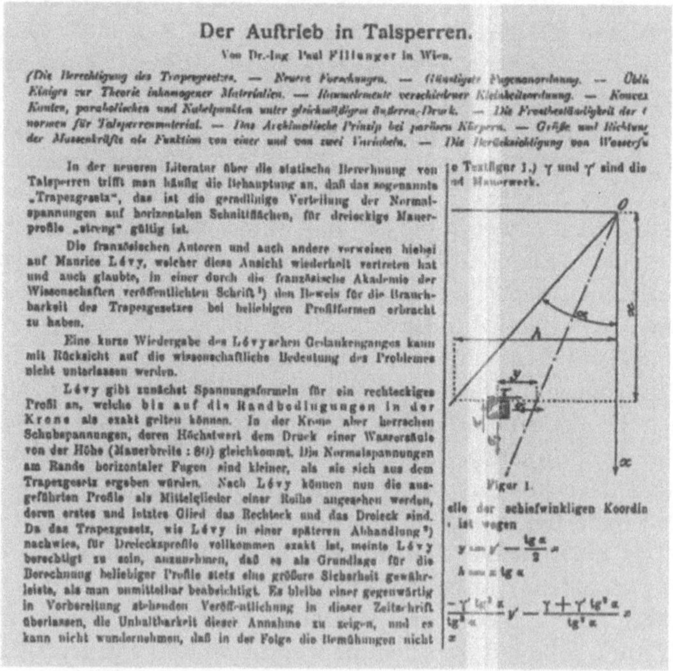

Der Auftrieb in Talsperren.

Von Dr.-Ing. Paul Fillunger in Wien.

In der neueren Literatur über die statische Berechnung von Talsperren trifft man häufig die Behauptung an, daß das sogenannte „Trapezgesetz", das ist die geradlinige Verteilung der Normalspannungen auf horizontalen Schnittflächen, für dreieckige Mauerprofile „streng" gültig ist.

Die französischen Autoren und auch andere verweisen hiebei auf Maurice Lévy, welcher diese Ansicht wiederholt vertreten hat und auch glaubte, in einer durch die französische Akademie der Wissenschaften veröffentlichten Schrift [1]) den Beweis für die Brauchbarkeit des Trapezgesetzes bei beliebigen Profilformen erbracht zu haben.

Eine kurze Wiedergabe des Lévyschen Gedankenganges kann mit Rücksicht auf die wissenschaftliche Bedeutung des Problemes nicht unterlassen werden.

Lévy gibt zunächst Spannungsformeln für ein rechteckiges Profil an, welche bis auf die Randbedingungen in der Krone als exakt gelten können. In der Krone aber herrschen Schubspannungen, deren Höchstwert dem Druck einer Wassersäule von der Höhe (Mauerbreite : 80) gleichkommt. Die Normalspannungen am Rande horizontaler Fugen sind kleiner, als sie sich aus dem Trapezgesetz ergeben würden. Nach Lévy können nun die ausgeführten Profile als Mittelglieder einer Reihe angesehen werden, deren erstes und letztes Glied das Rechteck und das Dreieck sind. Da das Trapezgesetz, wie Lévy in einer späteren Abhandlung [2]) nachwies, für Dreiecksprofile vollkommen exakt ist, meinte Lévy berechtigt zu sein, anzunehmen, daß es als Grundlage für die Berechnung beliebiger Profile stets eine größere Sicherheit gewährleiste, als man unmittelbar beabsichtigt. Es bleibe einer gegenwärtig in Vorbereitung stehenden Veröffentlichung in dieser Zeitschrift überlassen, die Unhaltbarkeit dieser Annahme zu zeigen, und es kann nicht wundernehmen, daß in der Folge die Bemühungen nicht

Fig. 4.1. Fillunger's first contributions to the porous media theory in the journal *Österr. Wochenzeitschrift f. d. öffentl. Baudienst* (1913)

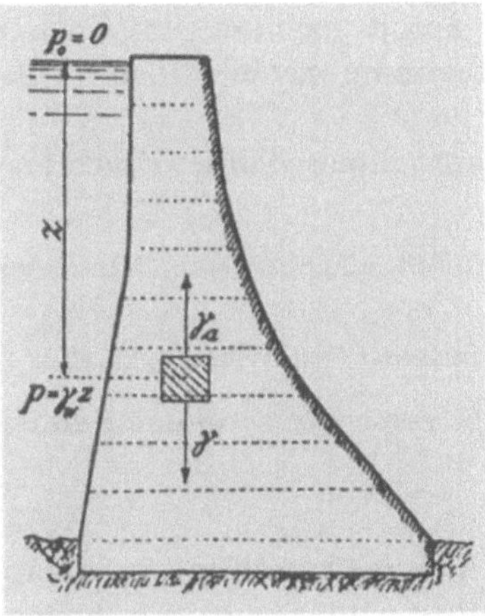

Fig. 4.2. Fillunger's consideration on the uplift in a barrage

the analysis of the forces acting in heavy-weight masonry dams had led him to this problem, and he specified the uplift force P_l specifically in respect to dam constructions, namely:

$$P_l = kV(\mu - \mu') \, . \tag{4.1.1}$$

In this equation, k defines the gradient of the liquid pressure, which Fillunger (1913) assumed to be constant, and V is the total volume of a liquid-saturated element. The quantities μ and μ' are the volume and surface porosity coefficients, respectively. In Fillunger (1913) and several subsequent publications (Fillunger, 1914, 1929, 1930a, b, 1934a, b, 1935), the various possibilities of (4.1.1), concerning the value of the uplift force, were discussed. Following this, the difference $\mu - \mu'$ can be positive, zero, or negative depending only on different intersecting techniques (see Fig. 4.3).

If the porous medium is cut (imaginarily, in an arbitrary manner) statistically (statistical cut), the Delessian law (see Section 3.3) yields $\mu = \mu'$ and thus the uplift force vanishes. On the other hand, if a granular porous medium, with point contacts between the grains, is exactly intersected only through these contact points (grain-to-grain cut), then $\mu' = 1$ and thus, the total uplift force

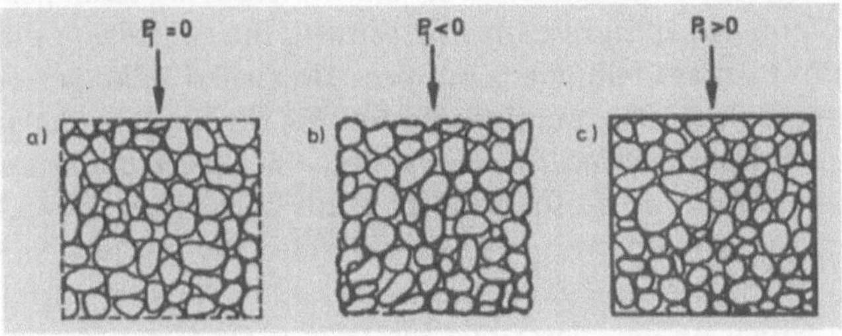

Fig. 4.3. Intersecting techniques: (a) statistical cut; (b) grain-to-grain cut; (c) closed solid surface

is effective. Finally, if $\mu' = 0$ – which implies a porous medium with a closed solid surface – then the "uplift force" acts in the same way as a load for the medium under discussion.

Following Fillunger (1913), $0 \leq \mu' \leq 1$ contains all possibilities which may occur in practice:

"Zwischen beiden Grenzen sind die praktisch möglichen Fälle eingeschlossen. 'Natura non facit saltus'."

"Between both limits, the practically possible cases are included. 'Natura non facit saltus'."

Analogous to (4.1.1), Fillunger's uplift force, with respect to a volume element of the porous medium, is (Fillunger, 1914):

$$\gamma_a = k(\mu - \mu') \,. \tag{4.1.2}$$

In 1929 and 1930, a polemical discussion over the uplift problem took place between Fillunger and the Italian engineer Hoffman without, however, further clarification of the problem. In contrast to Fillunger (1913, 1914), Hoffman's idea was (Hoffman, 1929) to always use that value of μ' which yielded the most unfavorable effect for the respective dam construction, thus increasing the stability of the building. This argument was strongly rejected by Fillunger (1929, 1930a, b).

Further considerations regarding the uplift problem were given by von Terzaghi (1933) and von Terzaghi and Rendulic (1934). It should, however, be mentioned, that von Terzaghi (1925a) had already considered the uplift problem in his famous book *Erdbaumechanik auf bodenphysikalischer Grundlage*, where he gave the correct formula for the uplift force in water-saturated natural sand. Nevertheless, the fundamental idea of von Terzaghi (1933) was to replace the surface porosity μ' in Fillunger's uplift formula (4.1.1) by a quantity n_W with which he called the effective surface porosity. In von Terzaghi and Rendulic (1934), n_W was determined by means of theoretical investigations into the strength of materials and by experiments with concrete specimens. The result was that $n_W \approx 1$ and therefore, the full value of the uplift force had to be effective.

Von Terzaghi and Rendulic (1934) found, approximately, the correct result. However, the proof of their formula must be rejected.

Karl von Terzaghi was led to a new investigation of the uplift problem during the preparation for his main lecture in the fall of 1932, while studying Fillunger's papers from 1913 to 1930 on the uplift in dams. He came to the conclusion that it was impossible for him to share Fillunger's opinion. He visited Fillunger on November 21 and December 14, 1932, and told him of his doubts in order to be sure that he had not misunderstood his views. Since Fillunger could not dispel these, von Terzaghi had to prove the validity of his own views through experiments. These were performed during 1933 and yielded the results which von Terzaghi expected. In the second half of September 1933, he showed Fillunger a manuscript in which he had described his results. On this occasion, von Terzaghi offered Fillunger a proposal that they should both present the central issue of the differences in their views in an oral discussion, and that they

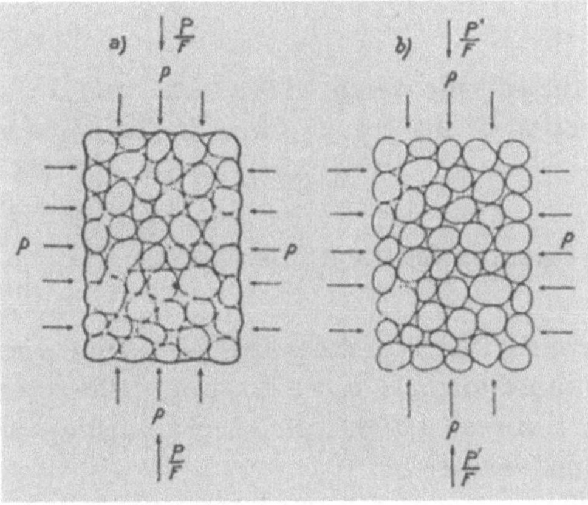

Fig. 4.4. Investigation of the uplift problem by von Terzaghi and Fröhlich (1934) (Determination of the effective surface porosity)

should publish the result of the investigations in an objective way, leaving any comment to the readers. In this way, the practical purpose of the dispute would be completely fulfilled without giving outsiders the sad spectacle of a polemic. Instead of agreeing to the proposal, Fillunger phoned Karl von Terzaghi on the morning of December 31, 1933, and asked him not to submit the manuscript for publication before he had talked with him. On this occasion, a meeting was arranged at 10:30 a.m. on the same day in the café Kuhnhof in Vienna. During this meeting, Fillunger ordered von Terzaghi, in a brusque way, to refrain from printing his manuscript, saying that otherwise von Terzaghi would be "pulled to pieces" by him (according to von Terzaghi).

Despite this heavy dispute, von Terzaghi offered a further proposal in a letter from February, 1934, in order to avoid the impression of personal differences, at least in the public eye. However, Fillunger refused this proposal in a telephone call. Finally, the manuscript was published in 1934.

The first investigations concerning the friction phenomenon, which occurs during the flow of liquid through a saturated porous solid, are again due to Fillunger (1914). In his paper, Fillunger (1914) pointed out:

"Bevor ich daran gehe zu zeigen, wie man den Auftrieb in die Berechnung von Talsperren einführen kann, müssen wir noch einer zweiten wenig bekannten Kraft nähertreten, nämlich der kapillaren Reibung des Wassers. Die Druckniveaulinien des ruhenden Wassers verlaufen horizontal und es entsprechen 10 m Höhendifferenz einem Druckunterschied von 1 *Atm*. Bei jeder anderen Lage der Niveaulinien muß man annehmen, daß das Wasser in Bewegung ist und daher Reibungskräfte hervorruft."

"Before I show how one can introduce the uplift into the calculation of barrages, we must still consider a second less known force, namely the capillary friction of the water. The level lines of the pressure of the water in repose run horizontally and 10 m of vertical interval corresponds to a pressure difference of 1 *Atm*. In every other location of the level lines one must guess that the water is in motion and therefore causes frictional forces."

Then, Fillunger stated a formula for the frictional force γ_ρ, which referred to the volume-element:

$$\gamma_P = \mu\sqrt{k^2 + \gamma_w^2 - 2k\gamma_w \cos\alpha}\,, \tag{4.1.3}$$

where k is the pressure gradient, γ_w the specific weight of the water, and α the angle between a z-axis and the direction of the gravity force. If this angle is zero, then we have:

$$\gamma_P = \mu(\gamma_w - k)\,. \tag{4.1.4}$$

This is a scalar version of the vector equation for the frictional force:

$$\gamma_r = \mu(\mathbf{k} - \operatorname{grad} p)\,, \tag{4.1.5}$$

which Fillunger (1929) developed in connection with the polemical dispute with the Italian engineer Hoffman. In the above formula, **k** is a downwards-directed vertical unit vector. In his response, Hoffman (1929) presented an analogous equation without, however, the volume porosity μ.

The theoretical treatment of *capillary* forces in saturated porous media was closely connected with the investigations of the uplift problem. Already in his first fundamental book on soil mechanics, von Terzaghi (1925a) had dealt with the capillarity problem. Later (von Terzaghi, 1933), he developed the formula:

$$p = \gamma\, m\, H_1\,, \tag{4.1.6}$$

where p is the hydrostatic "pressure" for the poreliquid caused by capillarity, γ is the specific weight of the liquid, m is an uplift coefficient similar to n_W mentioned above, and H_1 is the capillary rise. Von Terzaghi concluded that with respect to the vaporization process at the down stream face of masonry dams, the suction for the poreliquid caused an additional pressure on the masonry:

"... die in der Verdunstungszone wirksame Oberflächenspannung des Wassers erzeugt im Bereich der Luftseite der Mauer eine zusätzliche Druckbeanspruchung des Mauerwerks, die durch eine gleichgroße Zugspannung im strömenden Porenwasser ausgeglichen wird."

"... the surface stress of the water effective in the vaporization zone generates an additional pressure of the masonry in the region of the down stream face of the dam, which is balanced by an equal tension stress in the flowing porewater."

This point of view can only be explained if one considers the profile of a liquid interface, e.g., in circular tubes (see Fig. 4.5), where the surface stresses σ_S of the meniscus cause an axial pressure σ_A for the wall of the tube, and radial and tangential stresses are neglected. Since von Terzaghi found by experiment a value for the capillary rise of twenty meters, and thus a considerable pressure on the masonry, it was only natural for Fillunger (1934c) to attack von Terzaghi's result, which he again considered to be completely unrealistic. His article *Kapillardruck in Talsperren* was first submitted to the editor of the journal "Bautechnik". However, it was not accepted, as it would have only continued the scientific dispute, which had already started in the journal "Die Wasserwirtschaft". Fillunger's (1934c) article finally appeared in the journal "Die Wasserwirtschaft". In this article, he made some sarcastic remarks, e.g.,

"Zwar läßt die Natur sich jede Erklärung gefallen, allein sie hält sich nicht daran."

"Although nature puts up with any explanation, it does not stick to them."

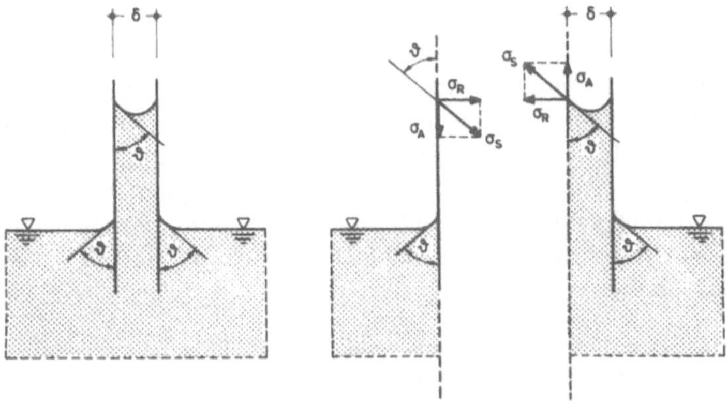

Fig. 4.5. The effect of capillarity in a cylindrical pore

Furthermore, Fillunger (1934c) criticized the common procedure for treating the capillarity problem:

"Die voranstehenden Betrachtungen sind nicht vollkommen befriedigend, da von den drei Hauptnormalspannungen des Spannungszustandes, der als Folge der Oberflächenspannungen des Wassers in einem durchtränkten porösen Körper entsteht, nur eine, nämlich der Kapillardruck σ, berechnet wurde. In einem Rohr herrschen außer dem Achsialdruck σ noch tangentiale und radiale Spannungen. Diesen entsprechen im durchtränkten porösen Körper ebenfalls gewisse Spannungen, doch kann über sie nur wenig Verläßliches ausgesagt werden."

"The preceding considerations are not completely satisfying, since from the three principal normal stresses of the stress state, which are produced as a consequence of the surface stresses of the water in a saturated porous body, only one was calculated, namely the capillary pressure σ. In a pipe there are, beside the axial pressure σ, tangential and radial stresses. These are equivalent to certain stresses in a saturated porous body, however, only few reliable remarks can be stated."

Although Fillunger (1934c) strongly criticized the formula given by von Terzaghi (1933), he also maintained the main statement, i.e., that the capillary suction for the liquid caused an additional pressure for the solid phase.

Von Terzaghi didn't respond to Fillunger's attack concerning the capillarity problem. He remarked in the questioning at the Technische Hochschule of Vienna on February 10, 1937:

"Der Grund hiefür war ein zweifacher. Der erste bestand in dem mir bereits bekannten, äusserst widerwärtigen Ton der Fillunger'schen Polemiken, der sich mit meinen Vorstellungen von wissenschaftlichen Auseinandersetzungen nicht verträgt; der zweite Grund bestand in der Sachunkenntnis, die der Inhalt des Artikels nach meiner Ansicht verriet."

"There were two reasons for this. The first lay in the awful form of Fillunger's polemics, which were already known to me and do not coincide with my views about scientific disputes; the second reason lay in the ignorance that the contents of the article in my opinion reveal."

According to Washburn (1921), the statical capillary problems had been investigated on both the theoretical and the experimental side, but the dynamical aspects of the subject did not appear to have received much attention up to the beginning of the 1920s. The attempt of Washburn (1921) failed because he did not proceed from the fundamental principles of mechanics. Much more evidence must be given to the contribution of Kozeny (1927). It is really amazing that both the brilliant professors, Fillunger and von Terzaghi, obviously overlooked the excellent paper on capillarity *Über kapillare Leitung des Wassers im Boden* (capillary motion of water in soil) by Josef Kozeny (1927), representative of hydraulics and a member of the Technische Hochschule of Vienna. In the introduction, Kozeny wrote:

"Wer die einschlägige Literatur überprüft, findet eine Menge schon vorhandener wertvoller Beobachtungen und Messungen. Doch fehlt meist die gesetzmäßige Beziehung, in der die einzelnen Größen zueinander stehen und die in einer mehr oder weniger einfachen Formel zum Ausdruck kommt."

"Who scrutinizes the literature relevant to the subject, finds a lot of already available valuable observations and measurements. However, in most cases the relations based on principles, which connect the individual quantities and which can be expressed in a more or less simple formula, are missing."

It seems that Kozeny (1927) was the first scientist formulating such a formula on the basis of ensured mechanical principles, see Cammerer (1963). After some micromechanical considerations, introducing among other things the filter velocity, Kozeny (1927) studied the capillary motion in saturated porous media, whereby he restricted his investigation to the one-dimensional instationary motion of water in a cylindrical pore of arbitrary form with the periphery u. Furthermore, he introduced the "effective capillarity diameter" d_w and the effective cross-section $n^F A$ with n^F as the volume fraction of the liquid. Then, he calculated the tension of a moistening liquid:

$$P = \sigma u = \frac{6\sigma}{d_w} n^S A \,, \tag{4.1.7}$$

where σ is the capillarity constant or the surface tension depending on the materials involved, A is the cross-section of the equivalent "flow tubes", and n^S is the volume fraction of the solid phase. The surface tension causes the water to rise in the pores of the sand mixture.

Kozeny introduced the parameter s and the angle β for the inclination of the "flowtube". Then, he calculated the friction force W and the weight of the water column $G \sin\beta$

$$W = \frac{\rho^{FR} g\, n^F A}{2k^F} \frac{ds^2}{dt} \,, \tag{4.1.8}$$

$$G \sin\beta = s\, n^F A\, \rho^{FR} g \sin\beta \,, \tag{4.1.9}$$

where k^F is the permeability coefficient. In order to describe the capillary motion of water, Kozeny proceeds from the equation of motion for the water column with the length s, where s is time dependent:

$$\frac{dB}{dt} = P - W - G\sin\beta\,, \tag{4.1.10}$$

where B is the momentum of the water column

$$B = \rho^{FR} n^F A\, s\frac{ds}{dt} = \frac{1}{2}\rho^{FR} n^F A\frac{ds^2}{dt^2}\,. \tag{4.1.11}$$

Considering (4.1.7), (4.1.8), and (4.1.9), Kozeny obtained from the equation of motion (4.1.10) with (4.1.11):

$$\begin{gathered}\frac{\rho^{FR} n^F A}{2}\frac{d^2 s^2}{dt^2} = \frac{6\sigma n^S}{d_w}A - \frac{\rho^{FR} g n^F A}{2k^F} - \rho^{FR} g\, s\, n^F A\sin\beta\,,\\ \frac{d^2 s^2}{dt^2} + m\frac{ds^2}{dt} + \nu s + n = 0\end{gathered} \tag{4.1.12}$$

with

$$\frac{g}{k^F} = m,\ 2g\sin\beta = \nu \quad \text{and} \quad -\frac{12\sigma}{\rho^{FR} d_w}\frac{n^S}{n^F} = n\,. \tag{4.1.13}$$

Kozeny gained analytical solutions of the fundamental differential equation for special problems, namely for the horizontal capillary motion and for the capillary rise. Moreover, he approximately solved (4.1.12) for the vertical motion. Kozeny, then, paid attention to the fact that the formula, just developed by him, could also be used to describe the filtration of water, by keeping the water level constant and changing the values of two coefficients. Furthermore, Kozeny (1927) pointed out that the filtration of water could also be described if the water level is changing. However, in this case he neglected the inertia effects. Moreover, Kozeny showed that his solutions can also be used for irrigation problems.

An essential mechanical effect in liquid-saturated porous solids had already been discovered at the beginning of this century, namely the effect of *effective stresses*. According to Skempton (1960), the first reflections on the concept of effective stresses were already made by the celebrated English geologist Sir Charles Lyell in 1871, by Boussinesq in 1876, and by Reynolds in 1886. Skempton (1960) also mentions some experimental investigations made by Föppl and Rudeloff in 1900 and 1912 and, in particular, by Fillunger in 1915:

"But more definite evidence was provided by Fillunger in 1915 when he published the results of tension tests on unjacketed specimens of Portland and slag cements, carried out under water in an apparatus in which the water pressure could be varied."

However, Fillunger had already considered the influence of the porewater pressure earlier. In his first paper about porous solids, Fillunger (1913) stated:

"Weniger die einzelnen angeführten Spannungsprobleme als die völlige Übereinstimmung der Endresultate führt geraden Weges zu der Überzeugung, daß eine das Mauerwerk durchdringende druckführende Flüssigkeit im Material gleichen Druck nach allen Richtungen erzeugt."

"It is less the actual individual stress problems, than the complete agreement of the final results, which lead straight to the conviction that a pressure-carrying liquid penetrating the masonry construction, creates a pressure in the material equal in all directions."

Einfluß des Wasserdruckes auf die Zugfestigkeit von Portlandzementmörtel 1:3.

Überdruck in Atm. beim Zugversuch	Anzahl der Versuche	Zugfestigkeit in kg/cm²			Angewendete Zerreißvorrichtung
		Höchstwert	Niedrigster Wert	Mittelwert	
0	12	40·4	29·8	36·1	Michaelis-Wage
100	12	39·2	25·7	32·4	Apparat
200	12	36·5	26·8	31·6	Apparat

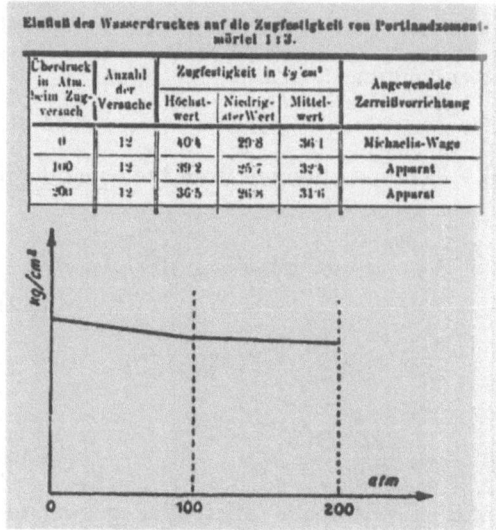

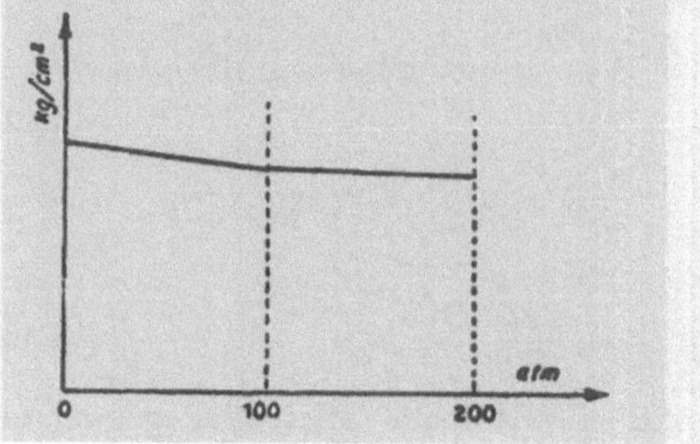

Fig. 4.6. Fillunger's test results from 1915 concerning the effective stress principle

He subsequently defined his statement more precisely regarding the causation of hydrostatic pressure by liquid in the pores of the masonry:

"man kann annehmen, daß der gleichmäßige Druck im Innern eine wesentliche Verminderung der Festigkeit nicht herbeiführen kann."

"one can assume that the uniform internal pressure cannot cause a significant reduction of the strength."

It seems that Fillunger (1913) was the first author to state that the poreliquid pressure did not have any influence upon the strength of the porous solid. One year later, Fillunger (1914) criticized the so-called Lichterfelder experiments described by Rudeloff and Panzerbieter (1912), which showed a strong influence of the poreliquid pressure upon the strength of the porous solid material. Considering the discrepancy between the Lichterfelder test results and his theoretical investigations, Fillunger (1915) decided to carry out his own experiments concerning the strength of liquid saturated porous solids. He communicated the test results that he had gained from tests on unjacketed specimens of cement under different water pressures (100 and 200 atmospheres). His main test result was:

"daß die Zugfestigkeit sich nicht mit dem Wasserdrucke ändert, ..."

"that the tension strength does not vary with the water pressure, ..."

Besides this comment that the tension strength did not change when the water pressure varied, he remarked that the porewater pressure did not affect the material behavior of the porous solid at all:

"Es zeigt sich auch, daß der in die Poren eindringende Wasserdruck keine sprengende Wirkung auszuüben vermag, wie von mancher Seite angenommen zu werden scheint, vom Verfasser aber bestritten wurde."

"It is also shown that the water pressure penetrating into the pores is not able to create any explosive effect, as seems to have been assumed by some, but has been disputed by the author."

Fillunger (1915) also alluded to experiments on salt specimens made by Voigt (1894, 1899), who had gained the same interesting result, i.e., that the tension strength did not depend on the hydrostatic pressure.

In Fillunger's opinion, the problem of the effect of the porewater pressure had obviously been solved because, in the following years, he did not work any further on this problem. Only in 1929 and 1930 did he again turn to this problem, and, during the polemical dispute with the Italian engineer Hoffman, Fillunger (1929, 1930a, b) repeated his view on the influence of the porewater pressure. Fillunger (1929) wrote:

"Ist der poröse Körper im unbelasteten Zustande nicht spannungslos, so überlagert sich der vom äußeren Druck überall erzeugte gleichmäßige Druckzustand seinen Eigenspannungen. Die Eigenspannungen können aus dem äußeren Druck niemals berechnet werden.

Gleicher Druck nach allen Richtungen kann bei der statischen Berechnung auch als ein anderen Spannungszuständen überlagerter Spannungszustand unberücksichtigt bleiben, weil er an der Bruchgefahr nichts ändert. Die genannten Versuche des Verfassers sollten hauptsächlich diesem Nachweis dienen."

"If the porous body is not stress-free in the unloading state, the uniform pressure-state caused by the external pressure superimposes its residual stresses. The residual stresses cannot be determined from the external pressure at any time.

Uniform pressure in all directions can also remain unconsidered in a static calculation, as a stress-state superimposing the other stress-states, because it does not influence the failure danger. The mentioned experiments of the author were mainly meant to be used as proof."

In the same paper he emphasized his view:

"... und man hätte an jeder Stelle dem durch andere Umstände hervorgerufenen Spannungszustand (Eigengewicht, Druck des Wassers im Becken gegen die Wasserseite) noch einen von Ort zu Ort wechselnden gleichen Druck nach allen Richtungen überlagert zu denken. Dieser überlagerte, 'hydrostatische' Druck bleibt ohne Einfluß auf die Festigkeit der Mauer."

"... and one would have to superimpose at each place the stress caused by other agencies (self-weight, pressure of the water in a pool against the water front) with a uniform pressure in all directions varying from place to place. This superimposed 'hydrostatic' pressure does not influence the strength of the masonry."

Hoffman (1929) obtained an interesting result in connection with the polemical dispute with Fillunger. He stated that in a saturated porous solid it is not the total porewater pressure which acted, but only that value which was diminished by the volume fraction of the solid:

"Nur wenn man annimmt, daß der Körper gleichförmig porös, d. h. μ' (surface porosity, the author) für jeden beliebig geführten Schnitt konstant sei, kann man behaupten, daß in ihm ein nach jeder Richtung gleicher Druck herrsche, der jedoch nicht dem Flüssigkeitsdrucke $-p$, sondern $-p(1-\mu')$ gleich ist."

"It is only when one assumes that the body is uniformly porous, i.e., μ' is constant for any cut made, that one can state, that a pressure equal in all directions exists, although this pressure is not equal to the fluid pressure $-p$ but $-p(1-\mu')$."

However, Hoffman (1929) expressed the opinion that the strength of the material of the saturated porous solid was influenced by the porewater pressure, a view which was strongly rejected by Fillunger, and which can also no longer be held in light of the modern theory of porous media.

Finally, Fillunger (1936) concluded, in a polemical paper, his investigations on the separation of the stress-state in connection with his fundamental treatment of the consolidation problem (see the next section). Fillunger (1936) added, to his basic equations of the description on the consolidation problem, an additional weighted pressure for the solid skeleton, for which a constitutive equation had to be formulated. He was the first author who clearly stated that a constitutive equation only had to be formulated for the excess over the weighted porewater pressure, and not for the total stress.

Von Terzaghi (1923) started the development of the idea of effective stress within the framework of the treatment of the consolidation problem for clay layers. He introduced the porewater pressure w in the form:

$$w = p_1 - p \,, \tag{4.1.14}$$

where p_1 marks a constant additional load of a clay layer, and where p is a field-dependent pressure increment. He connected this pressure increment with the change in the porosity. Hence, he formulated a constitutive relation for the effective stress, however, it seems that von Terzaghi gained his valuable results purely intuitively, rather than by mechanical principles, for in no sentence of his innovative paper did von Terzaghi (1923) mention the influence of the porewater pressure, or the idea of effective stress, nor did he do so in subsequent papers (von Terzaghi, 1924, 1925a, b, 1931, and 1933). Only in one publication, in connection with the vehement scientific dispute with Fillunger, did von Terzaghi (1934b) give a hint to the influence of the porewater pressure:

"Da außerdem die Ergebnisse der Fillungerschen Zugversuche in bester Übereinstimmung mit den meinigen zeigen, daß der hydrostatische Außendruck bei freiem Körper (im Gegensatz zum umhüllten) einen sehr geringen Einfluß auf die Größe der Bruchlast hat, kann sich der wirkliche Wert n_W (effective surface porosity, the author) nur um weniges von der Einheit unterscheiden."

"The results of Fillunger's tension tests also show, in the best agreement with mine, that the hydrostatic external pressure in a free body (contrary to a jacketed one) has very little influence on the magnitude of the failure load. Therefore, the true value n_W can only differ minimally from unity."

According to Skempton (1960), the idea of effective stress was formulated by von Terzaghi in 1936. Skempton stated in a paper entitled *Significance of Terzaghi's Concept of Effective Stress* (Skempton, 1960, slightly altered some notations of the original statement of von Terzaghi, 1936, in order to conform with the modern standard):

"The principle of effective stress has been stated by Terzaghi in the following terms.

The stresses in any point of a section through a mass of soil can be computed from the *total principal stresses* σ_1, σ_2, σ_3 which act in this point. If the voids of the soil are filled with water under a stress u, the total principal stresses consist of two parts. One part, u, acts in the water *and* in the solid in every direction with equal intensity. It is called the *neutral stress* (or the porewater pressure). The balance $\sigma_1' = \sigma_1 - u$, $\sigma_2' = \sigma_2 - u$ and $\sigma_3' = \sigma_3 - u$ represents an excess over the neutral stress u and it has its seat exclusively in the solid phase of the soil.

This fraction of the total principal stresses will be called the *effective principal stresses* ... A change in the neutral stress u produces practically no volume change and has practically no influence on the stress conditions for failure ... Porous materials (such as sand, clay and concrete) react to a change of u as if they were incompressible and as if their internal friction were equal to zero. All the measurable effects of a change of stress, such as compression, distortion and a change of shearing resistance are *exclusively* due to changes in the effective stresses σ_1', σ_2' and σ_3'. Hence every investigation of the stability of a saturated body of soil requires the knowledge of both the total and the neutral stresses.

This principle is of primary importance in soil mechanics. Its realization is entirely due to Terzaghi, and his earliest use, in 1923, of the equation $\sigma' = \sigma - u$ marks the beginning of the modern phase of our subject."

It seems, however, that von Terzaghi had used the concept of effective stresses purely intuitively in 1923, and that he first formulated the idea of effective stresses in the mid 1930s. The contribution of Rendulic (1936), a former associate of von Terzaghi, pointed out this possibility:

"Wir fassen unsere Erkenntnis über das Wesen der Spannungen in Ton zusammen und konstatieren, daß wir zwischen dreierlei Arten von Spannungen unterscheiden müssen. Es sind das:

1. Der Porenwasserdruck p. Er ist ein hydrostatischer, schubspannungsfreier Spannungszustand, welcher sich ungeachtet der Bezeichnung Porenwasserdruck, die sich dafür eingebürgert hat, gleichmäßig über Porenwasser und Festsubstanz verteilt. Wie bereits gezeigt, läßt sich dieser Porenwasserdruck auf natürliche Weise in den Porenwasserüberdruck w_0 und den natürlichen Druckzustand des Porenwassers zerlegen.
2. Der Spannungszustand, welcher allein durch die feste Phase des Tons übertragen wird. Wir betrachten dabei keine wirklichen, sondern mittlere oder durchschnittliche Spannungszustände, wie sie an der Begrenzungsfläche von Würfeln wirksam sind, deren Abmessungen groß sein sollen im Vergleich zu den Korndurchmessern. Für diesen Spannungszustand wollen wir mit Terzaghi die Bezeichnung 'Wirksamer Spannungszustand' gebrauchen.
3. Die Summe aus Porenwasserdruck und wirksamem Spannungszustand ergibt schließlich den 'totalen Spannungszustand'."

"We summarize our findings on the nature of the stresses in clay, and state that we must distinguish between three kinds of stresses. They are:

1. The porewater pressure p. It is a hydrostatic, shear stress-free stress state, which is, despite the notation porewater pressure usually used, uniformly distributed over the porewater and the solid material. As has already been shown, this porewater pressure can be decomposed in a natural way into the porewater overpressure w_0, and the natural pressure state of the porewater.
2. The stress state, which is transferred by the solid phase of clay. We consider no real but only mean or average stress states, as they are effective on the boundary surfaces of cubes whose dimensions should be large in comparison to the grain diameter. For this stress state we will use, after Terzaghi, the notation 'effective stress'.
3. The sum of the porewater pressure and effective stress state finally yields the 'total stress state'."

Furthermore, Rendulic (1936) remarked:

"Das mechanische Verhalten eines Tones (Formänderungen, Festigkeitsgrenzen usw.) hängt einzig und allein von den wirksamen Spannungen des Tons ab."

"The mechanical behavior of clay (deformations, limits of the strength etc.) depends solely on the effective stresses of the clay."

To conclude our discussion about the historical development of the idea of effective stress, Skempton (1960) is cited:

"Fillunger naturally concluded that the tensile strength does not vary with water pressure, at least within the pressure range and limits of accuracy of his experiments. This amounts to a corollary of the principle of effective stress, in the special case under consideration, yet neither he nor anyone else at the time realized the significance of these results. The same remark applies in the field of soil mechanics at this period for, as will be mentioned later, tests approximating to the undrained condition had been carried out by Bell in 1915 and Westerberg in 1921, which showed that the gain in strength in saturated clays under increasing external pressure was practically zero. This result is similarly a direct consequence of the principle of effective stress. Nevertheless it is clear that the physical meaning of these tests was in no way understood, and it required the genius of Terzaghi to clarify and enunciate this basic law of the mechanical properties of porous materials."

It is difficult to believe that von Terzaghi should be credited as the unique discoverer of the idea of effective stress, if one considers Fillunger's total work on the mechanical behavior of porous media; Fillunger (1913) had already stated, in his first study on the theory of water-filled porous media, that the porewater pressure had no influence on the strength of the solid skeleton.

Fillunger confirmed his conception, which he verified repeatedly through experiments, again and again in the following period. His valuable work on the consolidation problem (Fillunger, 1936), which contained the basic ideas of the porous media theory, shows that he understood fully the "principle of effective stress". Although he had gained his scientific findings on the influence of the porewater pressure essentially from brittle porous media, such as cement, concrete, and masonry, he was able to transfer the results obtained to the difficult consolidation problem of clay. As has already been pointed out, he formulated a constitutive equation only for the effective stress.

The introduction of effective stress by von Terzaghi (1936) can be understood only as a concept, not as a principle. It seems that von Terzaghi did not attach any importance to this concept in comparison with his other scientific findings. Neither in his autobiography (von Terzaghi, 1932), nor during the inquiry of 1936 and 1937, did he mention the discovery of the "principle of effective stress", although during the questioning he discussed his scientific investigations and findings in detail.

In conclusion, it seems that neither Fillunger nor von Terzaghi considered the mechanical effect of effective stresses to be as important as a general principle. Obviously, this mechanical effect was found independently. Nowhere in their polemics did they refer to this subject.

The discovery and description of fundamental mechanical effects in liquid-filled rigid porous solids in the first half of this century – *uplift, friction, cap-*

illarity, and *effective stresses* - by the Viennese professors Fillunger and von Terzaghi, and Dr. Kozeny, represents a brilliant scientific achievement in engineering. One should consider that, at that time, thermodynamics in the modern sense had not yet been developed, and the constitutive theory (with the procedure to gain restrictions from the entropy inequality and the incompressibility conditions) was still completely unknown. Therefore, it is not surprising that in the wake of the discoveries many errors and incorrect proofs appeared.

With these remarks, we will bring this section on the historical development of the theory of saturated *rigid* porous bodies to a close. The next section will follow the development of a general theory of saturated *deformable* porous media, a part of continuum mechanics which is still in the development process.

4.2 The Treatment of the Liquid-Saturated Deformable Porous Solid by von Terzaghi

Although much success had been gained with the model of a saturated rigid porous solid (see, e. g., the uplift problem), the theory of porous media remained incomplete because the description of the deformations and the determination of the stress state in saturated deformable porous bodies were at that time excluded from the research. It is true that the treatment of collapse states in soil bodies had reached a final stage in a certain sense; however, a theory of strength for fluid-filled porous solids was completely unknown before the beginning of the 1920s. From this time, however, comprehensive scientific activity began in this field accompanied, at times, by tough, polemical discussions. From this time on, one has been able to speak of a certain conclusion concerning the development of the material-independent relations of the theory of porous media. This statement is, however, not valid for the development of consistent constitutive equations for porous solids.

The first author to deal with the important problem of fluid-filled deformable porous solids was von Terzaghi (1923). At the beginning of this century, he recognized that a water-filled and deformable soil body, despite being of great relevance for the foundation of buildings, had still not been scientifically treated. During his activities as a building supervisor for the foundation of a bank building in St. Petersburg in 1910, he stated (von Terzaghi, 1932):

"dass auch anerkannte Autoritäten auf dem Gebiet des Bauwesens keine Anhaltspunkte für die Beurteilung der zu erwartenden Setzungen eines Bauwerkes haben. Die Meinung der Experten stützte sich lediglich auf theoretische Argumente ohne Beweiskraft."

"that also recognized authorities in the field of civil engineering have no basis for the assessment of the expected settlements of a building. The view of the experts relies only on theoretical arguments without conclusiveness."

After studies in the USA (1912–1914) and war service as a lieutenant (1914–1916), he received a lectureship for foundation and road construction at the Imperial Institute of Engineering in Constantinople.

"In den beiden Jahren, die nun folgten, hatte ich zum ersten Mal Gelegenheit, die für den Tiefbau in Betracht kommenden Theorien in Ruhe auf ihren Wahrheitsgehalt zu prüfen und mit den reichen Erfahrungen, die ich bereits gesammelt hatte, zu vergleichen. Im Laufe dieser Studien erkannte ich allmählich die Ursache des Misserfolges meiner bisherigen Bemühungen auf dem Gebiet der technischen Geologie. Die Ursache lag in der Unkenntnis der Beziehungen, die zwischen den Druckwirkungen und den ihnen entsprechenden Formänderungen der Böden bestehen und in dem Mangel an Versuchsmethoden, deren Anwendung unmittelbaren und ziffernmäßigen Aufschluss über die technisch wichtigen Eigenschaften der Böden liefert. Die Erddruck-Theorien waren unzulänglich, weil sie sich lediglich mit den Grenzzuständen des Gleichgewichtes der Böden befassen ohne auf die Formänderungen einzugehen, die dem Erreichen der Grenzzustände vorhergehen, und die rein geologische Betrachtungsweise musste versagen, weil zwei geologisch gleichwertige Schichten je nach ihrer Struktur sehr verschiedene Festigkeitseigenschaften aufweisen können. Durch diese Erkenntnis war die nächstliegende Aufgabe klar umschrieben. Sie bestand in der Schaffung einer Festigkeitslehre der Böden. Die Aufgabe konnte nur auf experimentellem Weg gelöst werden."

"In the following two years I had for the first time the opportunity to leisurely prove, the theories relevant to civil engineering and to compare them with the rich experiences which I had already collected. In the course of these studies, I gradually recognized the reason for the failure of my efforts hitherto in the field of technical geology. The reason lay in the ignorance of the relations which exist between the pressure effects and the corresponding deformations of the soils, and on the lack of test methods whose application yields an immediate and numerical explanation of the technically important properties of the soils. The earth-pressure theories were inappropriate because they only considered the collapse states of the equilibrium of the soils without going into the deformations, which precede the reaching of the collapse state, and the purely geological consideration must fail, because two geological equivalent layers can show very different strength properties according to their structures. Due to this finding, the next task was clearly defined. It consisted of the creation of a theory of strength for soils. The task could only be solved via experiments."

With these words, von Terzaghi (1932) described the research program which would occupy him intensively during the next years. During his employment at the American Robert-College in Constantinople, where he had moved in the meantime, he performed many experiments:

"Ich erkannte das Wesen der scheinbaren Kohäsion, den Zusammenhang zwischen den Festigkeitseigenschaften der Sande, der Tone und der festen Körper, und erfasste die Rolle, die der Strömungsdruck beim Zustandekommen des Grundbruches bei Stauwerken spielt."

"I recognized the nature of the apparent cohesion, the relation between the strength properties of the sands, the clays and the solids, and captured the role which the flow pressure played in the occurrence of the base failure at barrages."

Furthermore, von Terzaghi (1932) remarked:

"Im Jahre 1923 kamen meine Arbeiten neuerdings ins Stocken. Es handelte sich diesmal um die mathematische Erfassung des Vorganges bei der allmählichen Zusammendrückung, die der Ton unter dem Einfluß konstanten Druckes erfährt. Ich hatte mich derart in die Aufgabe verbissen, dass ich einen Monat lang alle meine Pflichten vernachlässigte und mich täglich bis tief in die Nacht hinein um die Lösung des Problems bemühte. Der Versuch misslang und ich entschloss mich zu der Veröffentlichung der Ergebnisse meiner bisherigen Untersuchungen, mit der Absicht, die Lösung dieses theoretischen Kernproblems der Bodenmechanik einem Nachfolger mit einer glücklicheren Hand zu überlassen.

Ein halbes Jahr später, während ich noch mit der Niederschrift der geplanten Abhandlung beschäftigt war, gelang mir die Lösung der Aufgabe mühelos nach halbstündigem

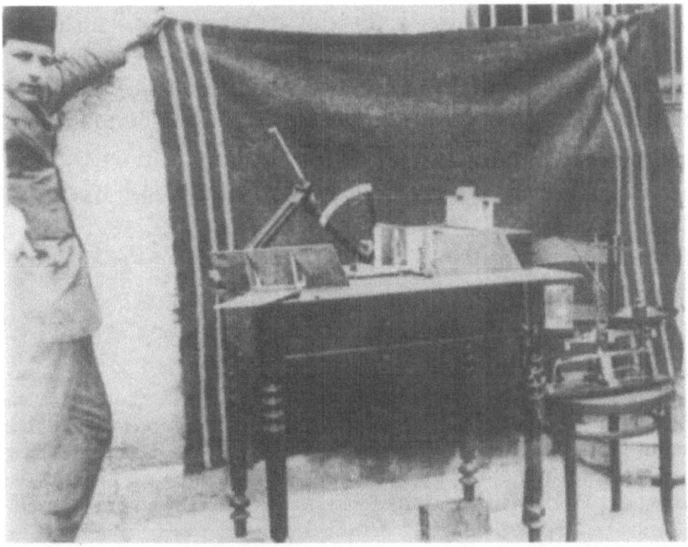

Fig. 4.7. Von Terzaghi's earth-pressure test in Constantinople (1919)

Nachdenken. Sie bestand in der Differentialgleichung des Verdichtungsvorganges, welche heute die Grundlage für sämtliche Untersuchungen betreffend die allmähliche Senkung von Bauwerken auf Tonböden bildet. Erst nach Aufstellung dieser Gleichung konnte der Versuch, die Festigkeitseigenschaften der Böden zu ergründen, als gelungen bezeichnet werden."

"In 1923 my work came to a standstill again. It was a question of the mathematical consideration of the process of gradual compression, which clay suffers under the influence of a constant pressure. I had grimly kept at this task to such an extent that I failed in my duties for about one month and I tried to solve the problem nearly every day until late into the night. The attempt failed and I decided to publish the results of my investigations up to then with the intention of leaving the solution of this central problem of soil mechanics to a successor with more luck.

Half a year later, while I was busy with the record of the planned paper, I succeeded in solving the task without trouble after half an hour of deep thought. It consisted of the differential equation of the consolidation process, which today is the basis for all investigations concerning the gradual settlements of buildings on clay. First, after the establishment of this equation, the attempt to reveal the strength properties of soils could be denoted as successful."

Von Terzaghi reported on the derivation of the partial differential equation of the consolidation process to the Academy of Sciences in Vienna on June 23, 1923. The derivation is contained in von Terzaghi's (1923) paper *Die Berechnung der Durchlässigkeitsziffer des Tones aus dem Verlauf der hydrodynamischen Spannungserscheinungen*. In this work, he was not particularly interested in settlement problems but – as is indicated by the title – in the calculation of the permeability coefficient of clay. The transfer of his ideas to settlement problems was done by von Terzaghi and Fröhlich (1936). At first, he stated that the permeability coefficient of clay was dependent on the water content, and the water content dependent on the pressure in the clay. Furthermore, he showed by the direct experimental proof method, the validity of Darcy's law for very small pores and lesser permeability. The direct proof method failed, however, for permeability coefficients $k \leq 0.06\,[cm/year]$. For this reason, he looked for a new proof method:

"Eine solche Methode wurde in der Beobachtung des zeitlichen Verlaufes der hydrodynamischen Spannungserscheinungen im Ton gefunden."

"Such a method was found in the observation of the time dependent course of the hydrodynamic stress phenomena in clay."

Von Terzaghi explained the term hydrodynamic stress phenomena as follows:

"Unter den hydrodynamischen Spannungserscheinungen versteht der Verfasser die Verzögerungen, welche die im Ton durch eine Außenkraft hervorgerufenen Spannungen durch die Widerstände gegen das Ausströmen des ausgequetschten Porenwassers erfahren. Die örtliche Änderung des im Ton herrschenden Druckes bedingt eine Änderung des Wassergehaltes am selben Ort. Nachdem die Änderung des Wassergehaltes ein Abströmen des Wassers und die Strömung des Wassers bei der sehr geringen Durchlässigkeit des Tones eines bedeutenden Gefälles bedarf, äußert sich die Wirkung einer Druckänderung zunächst in dem Auftreten örtlicher Differenzen in dem im Porenwasser herrschenden hydrostatischen Druck. Diese Druckdifferenzen verschwinden erst im Laufe der Zeit. Der zeitliche Verlauf des Spannungsausgleiches wird durch den Grad der Durchlässigkeit des Materials bedingt. Die Durchlässigkeitsziffer kann daher aus dem zeitlichen Verlauf des Ausgleiches berechnet werden, sofern alle anderen maßgebenden Faktoren bekannt sind."

"By hydrodynamic stress phenomena, the author means the delays which the stresses caused in the clay by an external force suffer by the resistances against the outflow of the squeezed-out porewater. The local variation of the stresses in the clay causes a change in the water content at the same place. Since the change of the water content requires a flow out of the water, and the flow of the water according to the very small permeability requires a considerable gradient, the effect of a pressure variation is firstly shown in the occurrence of local differences in the hydrostatic pressure in the porewater. These pressure differences disappear first in the course of time. The time-dependent process of the stress balance is caused by the stage of permeability. The permeability coefficient can therefore be determined from the time dependent course of the balance, if all other important factors are known."

The bases for the development of the partial differential equation for the description of the consolidation process were test results. These represented, for one-dimensional consolidation processes in a thin clay layer, on the one hand, the relation between the pressures acting on the surface of the clay and the pore number ϵ and, on the other hand, they yielded the relation between the permeability coefficient k and the pore number ϵ. These relations are depicted in Fig. 4.8. From Fig. 4.8 one can read that for increasing pressure p, the pore number ϵ and therefore also the permeability coefficient k decrease non-linearly. Von Terzaghi (1923) performed further considerations on a clay prism with the ground surface of $1cm^2$ and a mud content of $1cm^3$ drying substance. Furthermore, he restricted himself to small pressure variations so that, for the time dependent change of ϵ at a place z (see Fig. 4.8) of a clay layer, the following equation held (von Terzaghi's derivation, sometimes obscure, has been changed in order to conform with modern standards):

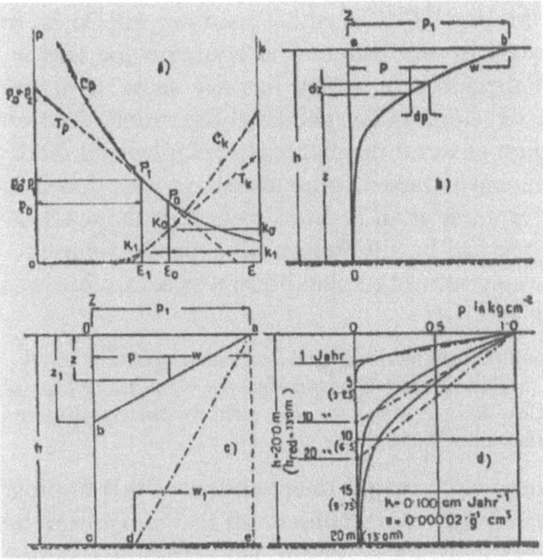

Fig. 4.8. Von Terzaghi's test results of loaded saturated clay (1923)

$$\frac{\partial \epsilon}{\partial t} = -\,a \frac{\partial p}{\partial t}\,, \tag{4.2.1}$$

whereby the coefficient a is denoted as a compression number. The variation of the pore number ϵ meant a change of the water content q. In order to gain the connection between the rates of q and ϵ, one can derive the following relation:

$$\frac{\partial q}{\partial z} = -\,\frac{\partial \epsilon}{\partial t}\,, \tag{4.2.2}$$

and with the above equation (4.2.1),

$$\frac{\partial q}{\partial z} = a \frac{\partial p}{\partial t}\,. \tag{4.2.3}$$

With the introduction of the porewater overpressure,

$$w = p_1 - p\,, \tag{4.2.4}$$

where p_1 represents the additional constant load in Fig. 4.8b, he obtained in the place of the aforementioned equation:

$$\frac{\partial q}{\partial z} = -\,a \frac{\partial w}{\partial t}\,. \tag{4.2.5}$$

With Darcy's law,

$$q = -\,k \frac{\partial w}{\partial z}\,, \tag{4.2.6}$$

he finally found the partial differential equation describing the problem of consolidation:

$$\frac{k}{a} \frac{\partial^2 w}{\partial z^2} = \frac{\partial w}{\partial t}\,. \tag{4.2.7}$$

It was this differential equation, among others, which made von Terzaghi famous.

Von Terzaghi's derivation of the partial differential equation (1923) is, in parts, obscure. Obviously influenced by the differential equation for the description of heat propagation (Fourier's law), which has the same structure mathematically, he succeeded in developing his partial differential equation more through intuition than through ensured mechanical principles and mathematical rules. Furthermore, in his basic paper, some terms are introduced in a very unscientific way, being in parts not at all defined, or first explained after having already been used. Thus, one can hardly follow his paper in some sections. Even so, the fact that the water-saturated clay body was a mixture was clearly known to him; he stated later:

"As a matter of fact, analysis of the phenomenon leads to new and very important conclusions. In order to perform it we have merely to keep in mind that the elementary laws of mechanics which apply to solids and liquids in general are also valid for the constituents of a mixture of clay and water" (von Terzaghi, 1925b).

However, in 1923 as well as later, von Terzaghi made no use of this finding. He was obviously not well-trained enough in mechanics and mathematics to be able to treat this difficult problem as a two-phase-system. The sometimes-sloppy work (von Terzaghi, 1923) is obviously founded in von Terzaghi's personality: he did not like to work out the final drafts of strenuous subjects. He was, however, an excellent and highly-gifted engineer, with great intuitive abilities, who was able to recognize difficult physical-technical problems very clearly, and could analyze and solve them. This ability was also revealed in his treatment of other problems, such as the solution of the uplift problem. Nevertheless, after he had solved a problem, he showed very little interest in the formulation of the scientific problem and the solution gained, as he declared in the questioning at the Technische Hochschule in 1936 and 1937.

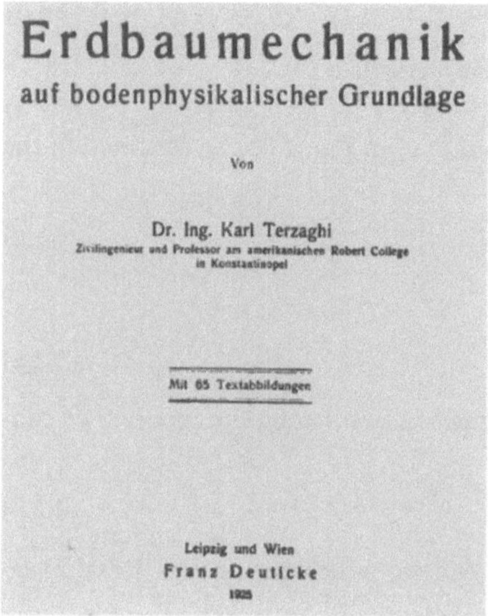
Erdbaumechanik
auf bodenphysikalischer Grundlage

Von

Dr. Ing. Karl Terzaghi
Zivilingenieur und Professor am amerikanischen Robert College in Konstantinopel

Mit 65 Textabbildungen

Leipzig und Wien
Franz Deuticke
1925

Fig. 4.9. Von Terzaghi's first book on soil mechanics

In the time to follow, von Terzaghi tried to publicize his ideas regarding the creation of a theory of strength for soils. After he had developed his fundamental differential equation for the description of the consolidation problem, von Terzaghi (1925a) published a book entitled *Erdbaumechanik auf bodenphysikalischer Grundlage*, which contained his basic views on soil mechanics. This book was recognized worldwide as the first substantial book on soil mechanics. In this book, von Terzaghi adhered to the description of the consolidation problem which he had formulated in 1923. Through a series of articles (von Terzaghi, 1925b), he also propagated his ideas in the English-speaking world.

In 1925, von Terzaghi was offered, and subsequently accepted, a visiting professorship at MIT in Cambridge (USA). It was at this time that he got in touch with the mathematican A. Ortenblad, who had just finished his studies at Harvard. Ortenblad had read von Terzaghi's (1925a) book *Erdbaumechanik*, and was not convinced about the analogy of heat propagation through an isotropic body with the consolidation problem. He started the development of his own theory in 1925, and presented it to the Board of Examiners as a thesis for a Doctorate in Science degree in May, 1926. In 1930, he published his thesis. In his thesis, Ortenblad (1930) formulated, for the first time, a consolidation theory for the three-dimensional stress and deformation states. However, he only extended von Terzaghi's differential equation.

Von Terzaghi's life remained busy during the 1920s. He gave lectures at the Technische Hochschule in Berlin and at MIT. Moreover, he became more and more involved in consulting, for example, with the United Fruit Company in Central America.

Then, in 1929, he was offered a chair at the Technische Hochschule of Vienna, which he accepted, beginning that same year. During his stay in Vienna, he threw himself into great activities in science as well as in consulting. His efforts to publicize his ideas culminated in the book *Theorie der Setzung von Tonschichten*, which he published with O.K. Fröhlich (von Terzaghi and Fröhlich, 1936), wherein the settlements of clay layers were calculated using the overpressure w gained from (4.2.7). They were able to perform this after von Terzaghi had recognized the concept of effective stresses. Von Terzaghi and Fröhlich stated:

"Vorliegende Schrift ist als Glied in einer zwangslosen Folge von elementaren Leitfäden gedacht, durch welche der Leser in die praktisch wichtigsten Teilgebiete der Bodenmechanik eingeführt werden soll."

"The present book is considered as a link in an informal series of elementary textbooks, by which the reader should be introduced to the most important branches of soil mechanics."

At the beginning of December, 1936, this book, which contained some imperfections in both the mathematical and mechanical representations of certain results, was heavily criticized in a polemic pamphlet by Fillunger (1936). The pamphlet, *Erdbaumechanik ?*, contained inexorable criticism of certain statements in the book, as well as personal, defamatory attacks directed mainly against von Terzaghi and, in some parts, against Fröhlich.

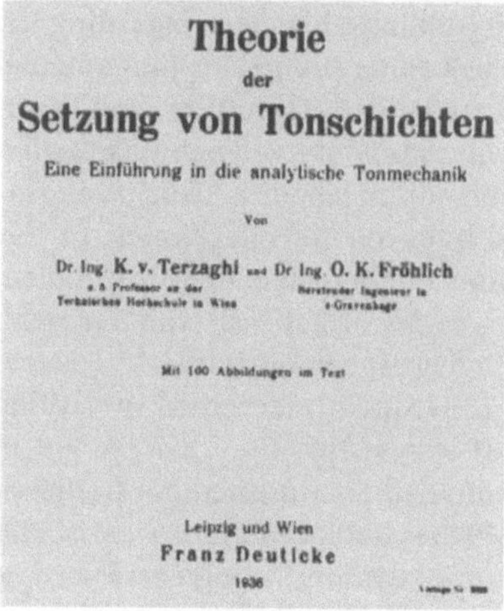
Theorie
der
Setzung von Tonschichten
Eine Einführung in die analytische Tonmechanik
Von
Dr. Ing. K. v. Terzaghi und Dr. Ing. O. K. Fröhlich
Mit 100 Abbildungen im Text
Leipzig und Wien
Franz Deuticke
1936

Fig. 4.10. The book *Theory of the settlement of clay layers* by K. von Terzaghi and O.K. Fröhlich

The reason for Fillunger's attack is obvious. The polemic disputes over the uplift problem had led to a deep hostility between the two distinguished professors at the Technische Hochschule of Vienna. Fillunger, who had a hunch that he was in error concerning his uplift formula, was not going to miss a good opportunity to attack von Terzaghi and to criticize his scientific work. Fillunger saw this opportunity in the aforementioned book by von Terzaghi and Fröhlich (1936). Fillunger's criticism, which culminated in the ruling that

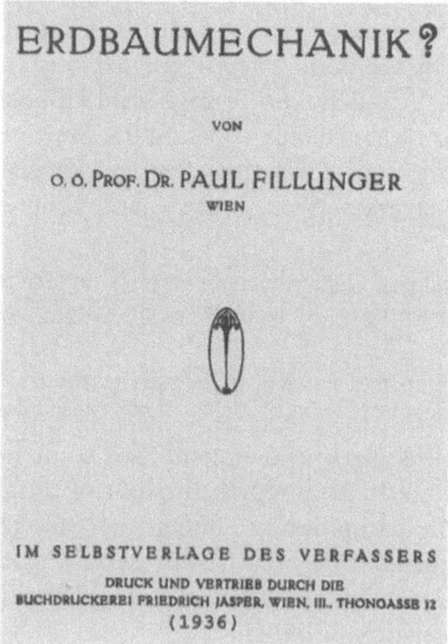
ERDBAUMECHANIK?
VON
O. Ö. PROF. DR. PAUL FILLUNGER
WIEN
IM SELBSTVERLAGE DES VERFASSERS
DRUCK UND VERTRIEB DURCH DIE
BUCHDRUCKEREI FRIEDRICH JASPER, WIEN, III., THONGASSE 12
(1936)

Fig. 4.11. The polemical brochure *Soil Mechanics?* by P. Fillunger

modern soil mechanics should be placed on the shelf because it was not scientifically founded, was not completely unfounded. As has already been mentioned, the partial differential equation describing the consolidation problem was developed by von Terzaghi more intuitively than on the basis of ensured mechanical principles. Although the derivation of the differential equation is in some parts obscure, one can state that it - together with some supplementary ideas - qualitatively describes the one-dimensional consolidation problem in an approximately correct way. From a scientific point of view, the derivation must, however, be refused. It was this point that encouraged Fillunger to work on his own ideas.

4.3 The Foundation of Modern Porous Media Theory by Fillunger

Fillunger (1936) studied the consolidation problem for six months. He then published his results in the second section of the brochure, *Erdbaumechanik?*. He proceeded from a two-phase-system, which he described with ensured mechanical axioms and principles:

"Man sieht leicht ein, daß bei der theoretischen Untersuchung der Setzung infolge von Grundwasserströmungen die Eulerschen Grundgleichungen der Hydrodynamik herangezogen werden müssen. Es würde keinerlei Schwierigkeiten bereiten, sie sofort unter möglichst allgemeinen Voraussetzungen aufzustellen, doch empfiehlt es sich, zunächst nur den einfachsten Fall zu behandeln, in dem alle Vorgänge nur von der Zeit t und einer einzigen Koordinaten z abhängen. Einerseits fällt es dem Leser dann wesentlich leichter, die Ableitung der Gleichungen auf ihre Richtigkeit zu prüfen, andererseits hatten die Herren v. Terzaghi und Fröhlich bei ihren Darlegungen auch nur diesen einfachsten Fall im Auge."

"One easily realizes that for the theoretical investigations of the settlement due to groundwater flows, the Eulerian fundamental equations of hydrodynamics must be applied. It would not cause any difficulties, to set them up immediately under possible general assumptions; however, it is recommended to treat only the simplest case at first, for which all processes depend only on the time t and on one single coordinate z. On the one hand it is then easier for the reader to examine the derivation of the equations concerning their correctness, and, on the other hand, also v. Terzaghi and Fröhlich had only treated this simple case."

Then, Fillunger started immediately to develop the fundamental equations of porous media theory. He stated that it was a matter of two coupled flows concerning the porewater flow and the settlement:

"Das Porenwasser (Körper 1) strömt nach aufwärts, der poröse Erdboden (Körper 2) mit der Setzungsgeschwindigkeit nach abwärts. Sehen wir ebenso wie die anderen Verfasser von der Wirkung des Eigengewichtes ab, so besteht die äußere Kraft für jeden Körper nur im Widerstand, den der andere Körper seiner Strömung entgegensetzt, und darauf beruht die Koppelung der zwei Bewegungen. Es empfiehlt sich ferner, diese äußere Kraft nicht mehr auf die Masseneinheit, sondern auf die Raumeinheit zu beziehen und sich vorzustellen, das Porenwasser erfülle stetig, aber mit veränderlicher Dichte den ganzen Raum und ebenso der Erdboden. Es ist dann so, als ob zwei Strömungen im gleichen Raume bestünden, die sich gegenseitig nur durch den Strömungswiderstand, nicht aber nach dem Gesetz der Raumverdrängung beeinflussen können."

$$
(7) \quad \ldots\ldots \quad \begin{cases} \dfrac{\partial v_1}{\partial t} + v_1 \dfrac{\partial v_1}{\partial z} = \dfrac{1}{\varrho_1}\left(-Z - \dfrac{\partial p_1}{\partial z}\right), \\ \dfrac{\partial v_2}{\partial t} + v_2 \dfrac{\partial v_2}{\partial z} = \dfrac{1}{\varrho_2}\left(Z - \dfrac{\partial p_2}{\partial z}\right), \\ \dfrac{\partial \varrho_1}{\partial t} + \dfrac{\partial (\varrho_1 v_1)}{\partial z} = 0, \\ \dfrac{\partial \varrho_2}{\partial t} + \dfrac{\partial (\varrho_2 v_2)}{\partial z} = 0. \end{cases}
$$

Fig. 4.12. Fillunger's basic equations

"The porewater (body 1) flows upwards, and the porous soil (body 2) flows downwards with the settlement rate. If we, like the other authors, disregard the effect of own weight, then, the external force for each body consists only of the resistance to this flow put up by the other body, and the coupling of the two motions is based on this. It is further recommended that this external force no longer be related to the mass unit but rather to the unit of volume, and that it be imagined that the porewater constantly, but with varying density, fills the total space as well as the soil. It is then as if two flows exist in the same space, two flows which can influence each other only through the flow resistance, but not according to the law of space-displacement."

After this statement, Fillunger (1936) wrote down the balance equations of momentum and mass for both bodies for the one-dimensional consolidation, excluding any mass exchange. These equations are represented in Fig. 4.12.

In these equations $v_{..}$ denotes the velocity parallel to the z-axis, $\rho_{..}$ the density, $p_{..}$ the pressure of the single constituents, and Z the interaction force between the porewater and the soil. In the following, Fillunger (1936) admitted only a hydrostatic pressure to both constituents and related this pressure, via the concept of volume fractions, to the real liquid pressure. Moreover, he considered only incompressible constituents and reformulated the equations in Fig. 4.12:

"In jedem Querschnitt z der Doppelströmung herrscht ein Druck p, der sich auf die beiden Körper oder Stoffe wie folgt verteilt. Ist n der Porenraum je Raumeinheit, so folgt aus dem Delesseschen Gesetz, wonach in einem gleichmäßigen Gemenge auf jeder Schnittfläche die Flächenanteile jedes Gemengeteiles gleich seinem Raumanteil sein müssen, daß auf das Porenwasser ein Partialdruck

$$p_1 = np\,, \tag{4.3.1}$$

auf die festen Stoffe im Ton ein Partialdruck

$$p_2 = (1-n)p \tag{4.3.2}$$

entfällt. In der Tat ist $p_1 + p_2 = p$. Bezeichnen wir die konstante Dichte des unzusammendrückbaren Porenwassers mit ρ_1', sein spezifisches Gewicht mit γ_1, ebenso die konstante Dichte der gleichfalls als unzusammendrückbar angesehenen festen Teilchen im Ton und ihr spezifisches Gewicht mit ρ_2' beziehungsweise γ_2, so ist (Erdbeschleunigung $= g$)

$$\rho_1' = \frac{\gamma_1}{g} \quad \text{und} \quad \rho_2' = \frac{\gamma_2}{g}\,. \tag{4.3.3}$$

Im verteilten Zustand, auf den ganzen Raum bezogen, haben wir daher

$$\rho_1 = n\rho_1' = n\frac{\gamma_1}{g} \quad \text{und} \quad \rho_2 = (1-n)\rho_2' = (1-n)\frac{\gamma_2}{g}\,. \tag{4.3.4}$$

Setzt man sowohl die Partialdrücke als auch die Dichten nach diesen Beziehungen in die Gleichungen (7) (siehe Fig. 4.12, the author) ein, wobei zu beachten ist, daß sowohl p als auch n Funktionen von t und z sind, so erhält man nach Kürzung des konstanten Faktors $\frac{\gamma_1}{g}$ beziehungsweise $\frac{\gamma_2}{g}$ in der dritten und vierten Gleichung:

$$
(8)\qquad
\begin{aligned}
&\frac{\partial v_1}{\partial t} + v_1 \frac{\partial v_1}{\partial z} = \frac{g}{n\gamma_1}\left(- Z - \frac{\partial (np)}{\partial z}\right), \\
&\frac{\partial v_2}{\partial t} + v_2 \frac{\partial v_2}{\partial z} = \frac{g}{(1-n)\gamma_2}\left(Z - \frac{\partial (1-n)p}{\partial z}\right), \\
&\frac{\partial n}{\partial t} + \frac{\partial (n v_1)}{\partial z} = 0\,, \\
&-\frac{\partial n}{\partial t} + \frac{\partial (1-n) v_2}{\partial z} = 0\,.\text{"}
\end{aligned}
\qquad (4.3.5)
$$

"In each cross-section z of the double-flow, there exists a pressure p which is distributed amongst the two bodies or materials as follows: If n is the pore space per unit of volume, then it follows from the Delessian law (where on each cut surface, in a uniform mixture, the surface ratios of each partial constituent must be equal to their volume ratios) that the partial pressure must be decomposed into p_1 (porewater)

$$p_1 = np\,, \qquad (4.3.1)$$

and into the partial pressure p_2 (clay phase)

$$p_2 = (1-n)p\,. \qquad (4.3.2)$$

In fact, $p_1 + p_2 = p$. If we denote the constant density of the incompressible porewater with ρ_1', its specific weight with γ_1, as well as the constant density of the solid particles in the clay (which are also considered as being incompressible) and its specific weight with ρ_2' and γ_2, then (acceleration due to gravity $= g$)

$$\rho_1' = \frac{\gamma_1}{g} \quad \text{and} \quad \rho_2' = \frac{\gamma_2}{g}\,. \qquad (4.3.3)$$

In the distributed state, related to the total space, we then have

$$\rho_1 = n\rho_1' = n\frac{\gamma_1}{g} \quad \text{and} \quad \rho_2 = (1-n)\rho_2' = (1-n)\frac{\gamma_2}{g}\,. \qquad (4.3.4)$$

If one introduces the partial pressures as well as the densities according to these relations into Eqn. (7), (see Fig. 4.12, the author) taking into consideration that both p and n are functions of t and z, then after dividing by the constant factor $\frac{\gamma_1}{g}$ or $\frac{\gamma_2}{g}$, respectively, one obtains in the third and fourth equations:

$$
(8)\qquad
\begin{aligned}
&\frac{\partial v_1}{\partial t} + v_1 \frac{\partial v_1}{\partial z} = \frac{g}{n\gamma_1}\left(- Z - \frac{\partial (np)}{\partial z}\right), \\
&\frac{\partial v_2}{\partial t} + v_2 \frac{\partial v_2}{\partial z} = \frac{g}{(1-n)\gamma_2}\left(Z - \frac{\partial (1-n)p}{\partial z}\right), \\
&\frac{\partial n}{\partial t} + \frac{\partial (n v_1)}{\partial z} = 0\,, \\
&-\frac{\partial n}{\partial t} + \frac{\partial (1-n) v_2}{\partial z} = 0\,.\text{"}
\end{aligned}
\qquad (4.3.5)
$$

These are the material-independent fundamental equations – within the framework of a purely mechanical theory for a one-dimensional motion – of a binary model consisting of an incompressible liquid and an incompressible solid skeleton. Fillunger (1936) consequently used the balance equations of momentum and mass excluding any mass exchange, separately for both constituents, whereby he included, in the balance of momentum, interaction forces Z as volume forces. Subsequently, he determined the interaction force, with the help of Darcy's law:

$$Z = \frac{v_1 n}{k}\,, \tag{4.3.6}$$

where k is the coefficient of permeability.

The drawback of this derivation was entirely known to him, for he later indicated that the above equation had been derived with the assumption of a stationary and a homogeneous flow and a rigid solid skeleton with the velocity v_2 equal to zero and the porosity n constant:

"Es ist selbstverständlich, daß bei der Behandlung der Porenwasserströmung, die mit der Setzung gekoppelt ist, an Stelle von v_1 die Relativgeschwindigkeit $(v_1 - v_2)$ zu verwenden ist. Man hat aber auch zu beachten, daß beim Durchlässigkeitsversuch $n = konst.$ ist."

"It is obvious that in the treatment of the porewater flow which is coupled with the settlement, v_1 must be replaced by the relative velocity $(v_1 - v_2)$. One has, however, also to consider that $n = const.$ in the filtration test."

Even if one cannot agree with all of Fillunger's further arguments (1936), four remarks in view of "desirable generalisations" are of interest:

"Man sieht nun leicht ein, welche Verallgemeinerungen noch wünschenswert, ja zum Teil notwendig wären, um ein für die Baupraxis verwendbares theoretisches Rüstzeug daraus zu bilden. Sie mögen hier in Kürze aufgezählt werden.

1. Einführung von x, y, z als Koordinaten statt z allein, da Bauwerke immer nur einen kleinen Teil der Erdoberfläche unmittelbar belasten. Es dürfte dann unvermeidbar werden, die innere Reibung zäher Flüssigkeiten zu berücksichtigen und Zusatzglieder in die Bewegungswiderstände aufzunehmen, die das Geschwindigkeitsgefälle und eine besondere Reibungskonstante enthalten müßten.
2. Einführung weiterer Durchlässigkeitskoeffizienten, etwa neben k' für lotrechte Strömung, k'' für waagrechte Strömung, da die Schichtung des Erdbodens verschiedene Durchlässigkeiten in diesen zwei Richtungen vermuten läßt.
3. Einführung des Eigengewichtes von Wasser und festen Tonpartikeln als Zusatzglieder zu $-Z$ beziehungsweise $+Z$ in der ersten beziehungsweise zweiten Gl. (8). Dann könnte auch die sogenannte 'Sedimentation', das heißt die natürliche Ablagerung und Verfestigung von Schlammassen, erfaßt werden.
4. Besitzen die Tonteilchen einen Verdichtungswiderstand ähnlich dem elastischer Körper, so wäre in der zweiten Gl.(8) (equation of motion for the partial solid, the author), und zwar nur dort, ein Zusatzglied zu Z (interaction force, the author) von der Form $\frac{\partial f(n,k''')}{\partial z}$ anzufügen (n is the porosity, the author). Es bedeutet dann $p = f(n, k''')$ eine Zustandsgleichung, p einen Druck (kg/cm^2), k''' einen neuen Stoffwert. Die Gestalt der Funktion $f(n, k''')$ und die Konstante k''' müßten aus Versuchen bestimmt werden, die wir getrost als Ödometerversuche bezeichnen können."

"One easily realizes which generalizations are desirable even partly necessary in order to form a theoretical tool applicable in the field of civil engineering. They may be shortly listed here:

1. Introduction of x, y, z as coordinates instead of z alone, since buildings immediately load only a small region of the earth surface. Then, it may become unavoidable to

consider the internal friction of viscous liquids and to incorporate additional terms in the motion-resistances which must contain the velocity gradients and a special friction constant.

2. Introduction of further permeability coefficients, e.g., besides k' for vertical flow, k'' for horizontal flow because the arrangement in layers can suppose different permeabilities in these two directions.
3. Introduction of the own weight of water and solid clay particles as additional terms to $-Z$ and $+Z$, respectively, in the first and second, respectively, Equation (8). Then, also the so-called 'sedimentation' could be incorporated, i.e., the natural deposition and hardening of mud.
4. If the clay particles have a compression resistance similar to the elastic body, then in the second of Eqns.(8), and only there, an additional term to Z of the form $\frac{\partial f(n,k''')}{\partial z}$ should be added. This means that $p = f(n, k''')$ is an equation of state, p is a pressure (kg/cm^2) and k''' is a new material-dependent value. The form of the function $f(n, k''')$ and the constant k''' would have to be determined from tests, which we could confidently call oedometer tests."

In a footnote, Fillunger (1936) pointed out that von Terzaghi and Fröhlich had obviously not needed any extension of their relations to calculate the sedimentation. They had referred to the partial differential equation in Section 4.2.

Contrary to von Terzaghi's (1923) procedure, Fillunger's (1936) approach corresponds to the modern concept of the treatment of porous media, namely to the mixture theory restricted by the volume fraction concept. It represents an extraordinary scientific achievement of high originality when one considers that, at that time, the modern mixture theory had not yet been developed, much less a porous media theory.

4.4 The Tragic Controversy Between the Viennese Professors Fillunger and von Terzaghi in 1936/37

Up to the end of the 1920s, the academic career of Professor P. Fillunger had taken a normal course, without any particularly noteworthy events. He was a well-known scientist in the field of mechanics, especially in porous media and elasticity theories, a strict teacher and an inexorable examiner. Moreover, in his neighborhood, he was known as a person possessing engaging manners and, within his circle of friends, he enjoyed decidedly great popularity. However, some colleagues reported that he always stood firm in his view concerning the fundamental principles of mechanics.

In 1928, he published a polemical statement against the civil engineer Ottokar Stern (Fillunger, 1928), whose work on the foundation of concrete piles was strongly attacked. In the following period, the polemics continued. In his last statement on the aforementioned subject, he revealed his motivation:

"Eine sachliche Kritik ist auch dann nützlich und notwendig, wenn sie nur den Zweck verfolgt, ein ganz unbegründetes Zutrauen zu einer wissenschaftlich unhaltbaren Theorie zu zerstören. Ich werde mir das Recht hiezu auch in Zukunft nicht streitig machen lassen, wenn es sich um eine Angelegenheit handelt wie diese, wo durch eine übermäßige Anpreisung einer technischen Neuerung nicht nur großer Sachschaden herbeigeführt werden kann, sondern unter Umständen auch Menschenleben gefährdet werden können."

"An objective criticism is also in that case useful and necessary, if it pursues only that purpose to destroy a totally unfounded confidence in a scientifically indefensible theory. I take it upon myself to do it also in the future if it refers to a matter like this whereby excessive praising of a technical innovation could not only cause great material damage but, under certain circumstances, human life could also be jeopardized."

This statement is without a doubt the main strength in Fillunger's polemical criticism. However, this was in all cases, as we will see in the following events, connected with over-sensitive reactions to the criticism of his own work and statements. As Stern (1928) complained, the polemical discussion was caused by a personal hostility which originated in a confrontation in a standards committee.

In 1929, Fillunger was obviously deeply hurt by the already-mentioned article by Hoffman (1929), in which his uplift formula of 1913 had been criticized. He answered this with heavy polemics because he was convinced that Hoffman's (1929) contribution was incomplete and, in the main parts, incorrect. Although the polemics did not lead to a clarification of the scientific issues, the dispute obviously encouraged Fillunger from then on to fight against errors and inadequate representations in scientific publications.

Then, in 1933 and 1934, the previously-described dispute (Section 4.2) with von Terzaghi over the uplift and capillary problems arose and, on February 19, 1935, Fillunger heavily criticized a report by Lehr (1934) in a speech to the "Österreichische Ingenieur- and Architektenverein". This speech was filled with many polemical statements and caused such a sensation that a discussion was arranged to take place on April 5, 1935. The discussion evening ended with a final remark by Fillunger in which he concluded that his 25 years of teaching about the strength of materials qualified him to criticize Lehr's contribution (1934). Moreover, in controversy, the rector of the Technische Hochschule was obliged to intervene. The "Österreichische Ingenieur- and Architektenverein" refused the publication of Fillunger's talk in its own journal, as was usual, with the excuse of lack of space. Therefore, Fillunger edited the talk by himself (Fillunger, 1935). In seeking to clarify the truth, he continued to attack Lehr's report (1934). In December, 1935, Lehr's report (1934) was advertised in the journal, "Zeitschrift für Angewandte Mathematik und Mechanik", which was, at that time, edited by Professor Trefftz. In two letters of January 7 and 21, 1936, to the editor, Fillunger intervened against the advertisement. In the last letter, he mentioned that not only the "realistic strength of materials" of Lehr (1934) had to be criticized extensively, but also "a certain soil mechanics". It then also seems that in this letter Fillunger announced for the first time his objective to attack von Terzaghi's work.

The two publications of von Linsemann (see Fillunger, 1935) and Foerster (see Fillunger, 1936) finally convinced Fillunger to act as a rigorous reviewer in the new fields of technical science.

"Wer aber mit frechem Schritte einhertritt, sich mit fremden Federn schmückt, ohne denen zu danken, von denen er gelernt oder entlehnt, wer selbst seine Krallen überall einschlägt, nichts neben sich gelten läßt, jeden Gegner oder Vorgänger verkleinert, wer vollends mit hervorstechender Arroganz eigene Unfähigkeit verbindet, der möge nur einem mutigen Zensor in die Hände fallen, welcher eine kräftige Geißel führt und die friedlichen Tauben gegen den krächzenden Geier schützen kann. Es ist genug, daß überhaupt das *oberflächliche Wissen*, die fade Nichtigkeit der modernen Weltbildung einen so breiten Raum beherrscht und man genötigt ist, in das stagnierende Sumpfwasser von Zeit zu Zeit ein paar kräftige Steine zu werfen.

Wenn aber dann die gleißende Nichtigkeit auf den Thron gehoben wird und der Literatur Regeln und den Sitten Befehle geben will, da bedarf es eines Gegenstoßes, der stark genug ist und laut genug, um sich den Besseren in der menschlichen Gesellschaft vernehmlich zu machen, und da bedarf es auch starker und strenger Worte. ...

Der Kritiker hat nicht bloß einen offenen und ehrlichen Kampf mit dem Autor zu bestehen, sondern auch noch einen anderen mit unsichtbaren Mächten, welche das Urteil irreführen möchten."

"Wer kennt sie nicht, die Virtuosen und Schwindler des Wortes, gegen die der Mensch, der bewußt einmal die Unwahrheit sagt, primitiv und harmlos wirkt? Die sogenannte, 'schöngeistige', aber auch die wissenschaftliche Literatur Deutschlands war nicht gerade arm an ihnen; und sie ist es heute auch noch nicht. Bei außerordentlich geringfügiger eigener Substanz an Geist, Seele, Wesen, Gefühl, Glauben, unmittelbarem Wissen und Erleben setzen diese Lügner einen erstaunlichen Schatz an Vokabeln in Betrieb. Die Sprache hat scheinbar ihre Wurzeln eingebüßt und führt auf eine gespenstige Art ein eigenes Leben; sie ist angewandte Technik des reinen Redens, ist Gedächtnis, Assoziations- und Reproduktionsfunktion. Ein grauenvoller Leerlauf der Worte stellt sich ein: reich, schillernd, blendend, pompös, frappierend - und hohl."

"Whoever walks along with insolent steps, dresses up with strange feathers without thanking those from whom he learned or borrowed, who digs his claws into everything, who accepts nothing other than himself, who belittles every opponent or predecessor, who combines his predominant arrogance with incompetence, falls into the hands of only a courageous censor who carries a strong scourge and can protect the peaceful doves against the croaking vulture. It is sufficient that in all the *superfluous knowledge*, the stale vanity of modern education governs such a wide region and one is obliged to throw some stones into the stagnating, swampy water. ...

If then, however, the glistening emptiness is enthroned in order to give the literature rules and the morals commands, a counterthrust is necessary which is strong enough and loud enough to be noticed by the better people in human society, and also strong and strict words are necessary. ...

The reviewer has not only to fight an open and honest battle with the author, but rather also another one with invisible powers which intend to mislead opinion."

"Who does not know the virtuosos and swindlers of the word against whom the human being, who deliberately once says an untruth, appears primitive and harmless? The so-called 'aesthetic', but also the scientific literature of Germany, was not exactly poor in those; and is also still not so today. With extremely little own substance of intellect, spirit, mind, emotion, belief, immediate knowledge and experience, these liars initiate an amazingly wealthy vocabulary. The language has apparently lost its roots and performs in a ghostlike manner an independent life; it is the applied technique of pure talk, is recollection, an association and reproduction function. A horrible, idle motion of the words sets in: rich, glittering, dazzling, pompous, astonishing - and hollow."

In the background of the aforementioned polemical discussions with the consequences drawn by Fillunger, an occurrence which is unique in European scientific history must be evaluated. Obviously, Fillunger was so deeply hurt by

Stern's, Hoffman's, and, in particular, von Terzaghi's attacks, that he himself became the aggressor. As already indicated in a letter to Professor Trefftz in Dresden, Germany, of January 21, 1936, he intended to criticize extensively "a certain soil mechanics". The opportunity to do this came shortly afterwards. In the spring of 1936, Fillunger made an effort to obtain permission to publish a book on material strength with the publishing house Deuticke - an effort which subsequently failed. On this occasion, the book *Theorie der Setzung von Tonschichten* by von Terzaghi and Fröhlich (1936) was shown to him for the first time. Fillunger bought the book and studied it extensively during the summer semester. He came, finally, to the conclusion that he must refuse the book on scientific grounds. In the time to follow, he developed the fundamental equations of the purely mechanical porous media theory; he had succeeded in this by the end of September, 1936. In October, 1936, he began to write down his criticism of von Terzaghi's work, especially of the book by von Terzaghi and Fröhlich (1936), and he represented his own view of the description of settlement problems.

He decided to publish his criticism and his investigations in a pamphlet which he edited himself, due to the bad experiences he had with the publication of his polemics against von Terzaghi. In the course of these polemics, some of Fillunger's contributions had been refused previously by the editors, as had already been mentioned. Thus, in editing the brochure by himself, he saw the only possibility to publicize his ideas.

The public had first learned of the objections through rumors during the summer of 1936. Then, at the end of the summer semester, Fillunger visited his colleague Professor Lechner, who had formerly been a co-worker of the professors Jaumann and Hamel at the Deutsche Technische Hochschule in Brno, and declared that he had discovered severe mistakes in the book by von Terzaghi and Fröhlich. Fillunger also repeated his objections to the "Dozent", Dr. Magyar (fluid mechanics). Finally, in October, 1936, Fillunger announced to Professor Lechner, during a bus ride, that he would hold a talk on soil mechanics in the "Ingenieur- und Architektenverein, Wien". On this occasion, Fillunger stated that one had to maintain order at the Technische Hochschule of Vienna or one would be driven directly to Bolshevism. Moreover, it would also be necessary to attack other professors. However, since at this time he had been too involved in the affair with von Terzaghi, he was forced to hold back on this subject.

Fillunger started his last attempt to represent his view on soil mechanics in a talk in the "Österr. Ingenieur- and Architektenverein" on October 23, 1936. The talk, entitled *Erdbaumechanik?*, was at once declared unwelcome by the general manager, Ing. Willfort. However, it was decided that the governing board should be consulted. A spontaneous counterproposal from the general manager was accepted by Fillunger. Ing. Willfort proposed that a talk should be held in which Fillunger would speak for the first 45 minutes, and then von Terzaghi would have his turn. In addition Fillunger was to hand his explanation over in written form to von Terzaghi. Fillunger agreed to this proposal. However, in a later discussion, Ing. Willfort did not uphold his proposal. Finally, the governing board declined to hold the talks on November 19, 1936. Fillunger was informed orally and then in writing (November 25, 1936).

Thereupon Fillunger decided to distribute his pamphlet *Erdbaumechanik?*, which had been printed in the meantime. The distribution took place on December 2, 1936. The brochure was sent to more than 100 different people and institutions all over the world but, in particular, to those in Austria and Germany. He explained his procedure to the president of the "Österr. Ingenieur- und Architektenverein":

"Es handelt sich hier wie auch schon in meinem Vortrag am 15. Februar 1935 um eine Lebensfrage der technischen Wissenschaft. Es hätte keinen Sinn mehr, wirklich technisch-wissenschaftlichen Unterricht zu erteilen, wenn auch weiterhin der Boden, den wir bebauen, durch solche Lehren verseucht werden dürfte. Professor heißt aber 'Bekenner', und darum muss zu entschiedener und rascher Abwehr gegriffen werden, sobald so grandiose Irrtümer (milde gesagt) als soche erkannt worden sind. ...

Gestatten Sie mir, Herr Präsident, zum Schlusse noch eine Bemerkung. Es wird bestimmt versucht werden, mir den Vorwurf der Hinterhältigkeit zu machen, weil ich die schriftliche Erledigung meines Vortragsbegehrens urgiert habe, ohne gleichzeitig von meinen Absichten und in welchem Stadium der Durchführung sie sich bereits befinden weder Ihnen selbst noch Herrn Gen.-Sekr. Willfort etwas mitzuteilen. Ich glaube nicht, daß ich dazu verpflichtet gewesen wäre. Die Schwere des Kampfes, den ich mit sehr ungleichen Waffen auszukämpfen gezwungen bin, rechtfertigt wohl, dass ich den kleinen Vorteil nicht auch noch aus der Hand geben mochte, einen nicht ganz vorbereiteten Gegner entgegentreten zu können. Auch habe ich ja alles getan, was in meiner Macht stand, ihn über meine Absicht überhaupt einen Angriff zu unternehmen nicht im Unklaren zu lassen."

"It is here a matter of a vital question concerning technical science, as I have already stated in my talk of February 15, 1935. It would not make any sense to give real technical-scientific teaching if the areas which we develop further are contaminated by such theories. Professor means, however, 'confessor', and for this reason it must be reached for decisive and prompt defense as soon as such great errors (to put it mildly) have been recognized....

Allow me, Mr. President, to make a remark in conclusion. It will surely be attempted to accuse me of being perfidious because I have urgently written my wish to give a talk, without simultaneously informing either you or general manager Mr. Willfort, of my objectives and at which stage of performance they are already positioned. I did not believe that I would be obliged to do this. The severity of the battle, which I am forced to fight with very unequal weapons, doubtlessly justifies the fact that I might not give away the little advantage of being able to face a not so well-prepared opponent. Also, I have done everything in my power to not leave him in the dark about my objective to attack him."

On Thursday, December 3, 1936, von Terzaghi was surprised by the news that Fillunger had started an attack against him and against soil mechanics as a science. Von Terzaghi's friends and associates were in a great panic. In the evening of the same day, he received a copy of Fillunger's brochure.

Fillunger's (1936) pamphlet, *Erdbaumechanik?*, contained six main parts. In his brief introduction, Fillunger stated that the consolidation theory of von Terzaghi and Fröhlich (1936) was in no way satisfactory, and had to be declined. Due to the eminent importance of the theory of settlement for civil engineering, however, Fillunger continued, a scientific-critical treatment of this subject was justified by another author. Then, Fillunger started his comments, which were a mix between technical criticism and personal defamations.

In the first section of his pamphlet, he showed briefly the derivation of von Terzaghi's partial differential equation of the consolidation problem from 1923. He accompanied nearly each step of the derivation with comments, and remarked finally that the acceleration was missing in von Terzaghi's investigations. Moreover, he criticized von Terzaghi's "thermodynamisches Gleichnis" (thermodynamic parable).

In the second section, he developed his own ideas regarding the consolidation problem, in which he founded the pure mechanical theory of porous media (see the explainations in Section 4.3).

Then, in the third part of his pamphlet, Fillunger vehemently criticized the integration of von Terzaghi's differential equation performed by von Terzaghi and Fröhlich (1936). In particular, he criticized single calculation steps and accompanied them with sarcastic remarks.

Soil-mechanical tests in the laboratory were the target of Fillunger's criticism in the fourth passage of his brochure. After some general critical remarks about experimental research, he tried to prove that von Terzaghi and Fröhlich (1936) had used a particular form of a total differential which was denoted by Fillunger as the "Terzaghi-Differential". In this specific point, Fillunger was in error, as will be seen later. He also criticized single experiments, in particular the oedometer test with disturbed soil samples. Finally, he attacked the tests on the shear strength and on the internal friction of soils.

In the fifth section, concerning applications in constructional engineering, Fillunger's criticism culminated with the statement:

"Bautechnische Anwendungen! Es gibt deren keine, wäre man versucht, kurz angebunden zu sagen und mit dieser Feststellung das unerfreuliche Thema ein für allemal zu beschließen. Man könnte bloß noch hinzufügen: es wird auch keine geben, wenigstens so lange nicht, als das Bauen von Häusern, Straßen und Brücken in der bisher üblichen Weise auf der festen Oberfläche der Kontinente erfolgt."

"Applications in constructional engineering! There are not any if one attempts by this statement to conclude the unpleasant topic once and for all. One could only simply add: there will never be any applications, at least not as long as the construction of houses, roads and bridges takes place in the hitherto usual way on the solid surface of the continents."

The last section, finally, contained some very general remarks and statements, and it was particularly in this section that Fillunger justified his heavy criticism.

As already-mentioned, the technical comments were accompanied by personal defamation. The main attacks are summarized in the following:

"Schwer ist der Gedanke zu ertragen, daß ein Prüfungskandidat, der dieses mathematisch-geometrische Problem vielleicht besser zu bewältigen verstand als die Verfasser, deshalb eine Verzögerung in der Vollendung seiner Studien oder gar ihr vorzeitiges Ende erlitten haben könnte!..."

"It is difficult to bear the idea that an examination candidate, who was probably better able to manage this mathematical-geometrical problem than the authors, could therefore have suffered a delay in completion of his studies or even their premature failure!..."

"Das Herumwerfen mit aufgeschnappten wissenschaftlichen Ausdrücken wie dieser muß natürlich bei Leuten von gleicher wissenschaftlicher Scheinbildung oberflächlichster Art einen außerordentlich günstigen Eindruck hinterlassen."

"The throwing around of scientific terms like this must of course leave an extraordinarily auspicious impression on people of the same fictitious scientific education of the most superficial kind."

"Taschenspieler müssen, wollen sie Erfolg haben, die Aufmerksamkeit ihrer Zuschauer ermüden und ablenken. Bei solchen wissenschaftlichen Taschenspielereien wie diese Fröhliche Terzaghi-Mathematik eine ist, kommt noch als weiterer den Erfolg der Eskamotagen begünstigender Umstand die Scheu des Lesers und des Hochschulhörers hinzu, sie könnten durch unberechtigte Kritik, ja schon durch bloßes Einbekennen: Das begreife ich nicht! schwere Mängel ihres Wissen und ihrer Geistesschärfe aufdecken. Allerdings, den wissenschaftlichen Kapazitäten, den Hochschullehrern, die den Dingen fachlich nicht vollständig fernstanden, ihnen vom Anfang an zugesehen, ohne lauten Widerspruch vernehmen zu lassen, sogar fördernd oder sonstwie daran teilgenommen haben, wird man diese Entschuldigung nicht gelten lassen können. ..."
"Jugglers must, if they want success, exhaust and divert the attention of their audience. With such scientific juggling, as this Fröhlich- [10] Terzaghi-mathematics is, readers and students shy away from what they could show by unfounded criticism or already by pure confession: This is what I do not understand! A severe lack of knowledge and lack of keen intellect is added in, which is favorable to the success of the jugglers. However, one cannot excuse the leading scientific authorities, the professors, who were technically not complete strangers to these things, who observed these from the very beginning without protesting even promoting or in some other way participating ..."

"Und von dem schönen Freimut, mit dem die 'offenkundigen Mängel' einbekannt, in der geschwollenen Vorrede die, 'wichtigsten Fehlerquellen' aufgezählt werden, wobei die allerwichtigsten, nämlich die in offenkundigen Mängeln im Wissen und Können der Verfasser liegenden, natürlich nicht erwähnt sind, und daß das Ganze nur einen 'Erstlingsversuch' bedeute, darf man auch nicht allzuviel halten. Das kennt man schon!"

"One should think little of the frankness with which the 'obvious faults' are confessed, the 'most important sources for mistakes' are listed in the bombastic preface, whereby the most important lack of knowledge and the ability of the authors are, of course, not mentioned and that the entirety only means a 'first approach'. This is already well-known."

The statements (not all are cited here) that Fillunger printed in his pamphlet represented in totality the scandalous accusation that von Terzaghi had invented a dummy science and had betrayed all professional circles through a clever swindle, all for the purpose of self-enrichment. This was implied to have begun in his teaching program at the universities and subsequently ending with the "International Congress for Soil Mechanics" at Harvard in 1936.

Immediately after the appearance of the pamphlet *Erdbaumechanik?*, von Terzaghi embarked upon great activities. Already on December 4, 1936, he had had discussions with his colleagues and friends Professor Orley and Schaffernak, specialists for road, railway, and tunnel construction, as well as hydraulics. Moreover, he was accepted by the Rector of the Technische Hochschule of Vienna, Prof. Dr. techn. Friedrich Böck, and he received several telephone

[10] Fillunger used O.K. Fröhlich's name as an adjective, punning on the German word for "happy", the author.

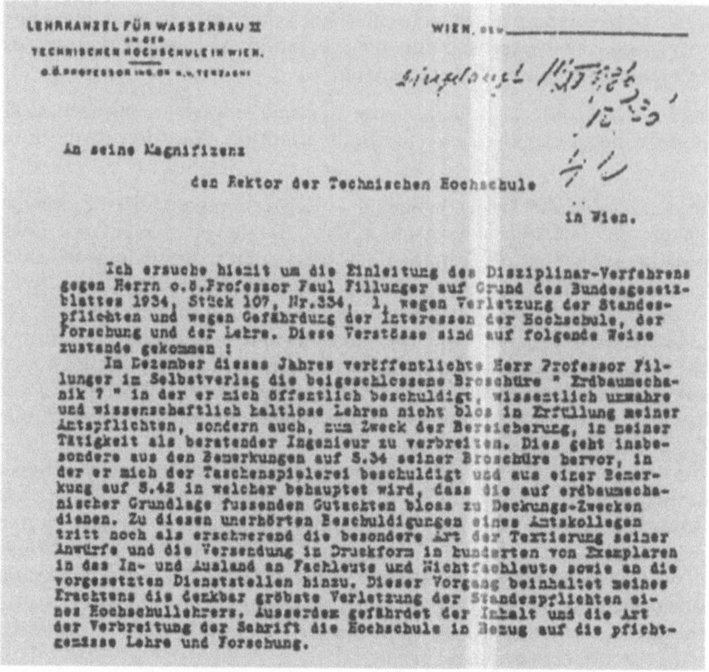

LEHRKANZEL FÜR WASSERBAU II
AN DER
TECHNISCHEN HOCHSCHULE IN WIEN.
O.Ö.PROFESSOR ING.DR K.v.TERZAGHI

WIEN, den

An seine Magnifizenz

den Rektor der Technischen Hochschule

in Wien.

Ich ersuche hiemit um die Einleitung des Disziplinar-Verfahrens gegen Herrn o.ö.Professor Paul Fillunger auf Grund des Bundesgesetzblattes 1934, Stück 107, Nr.334, 1, wegen Verletzung der Standespflichten und wegen Gefährdung der Interessen der Hochschule, der Forschung und der Lehre. Diese Verstösse sind auf folgende Weise zustande gekommen :

Im Dezember dieses Jahres veröffentlichte Herr Professor Fillunger im Selbstverlag die beigeschlossene Broschüre " Erdbaumechanik ? " in der er mich öffentlich beschuldigt, wissentlich unwahre und wissenschaftlich haltlose Lehren nicht blos in Erfüllung meiner Amtspflichten, sondern auch, zum Zweck der Bereicherung, in meiner Tätigkeit als beratender Ingenieur zu verbreiten. Dies geht insbesondere aus den Bemerkungen auf S.34 seiner Broschüre hervor, in der er mich der Taschenspielerei beschuldigt und aus einer Bemerkung auf S.42 in welcher behauptet wird, dass die auf erdbaumechanischer Grundlage fussenden Gutachten bloss zu Deckungs-Zwecken dienen. Zu diesen unerhörten Beschuldigungen eines Amtskollegen tritt noch als erschwerend die besondere Art der Textierung seiner Anwürfe und die Versendung in Druckform in hunderten von Exemplaren in das In- und Ausland an Fachleute und Nichtfachleute sowie an die vorgesetzten Dienststellen hinzu. Dieser Vorgang beinhaltet meines Erachtens die denkbar gröbste Verletzung der Standespflichten eines Hochschullehrers. Ausserdem gefährdet der Inhalt und die Art der Verbreitung der Schrift die Hochschule in Bezug auf die pflichtgemässe Lehre und Forschung.

Fig. 4.13. K. von Terzaghi's request to open a disciplinary action

calls concerning Fillunger's attack, including one from the general manager of the "Österreichische Ingenieur- und Architekten Verein", Willfort. In the afternoon of December 5, 1936, Dr. W. Herold visited von Terzaghi. Dr. Herold had written a book concerning some strength problems which had been published by Julius Springer in 1934. This book was strongly and sarcastically criticized by Fillunger in his talk on "Alte und neue Probleme der Festigkeitslehre" in the "Österreichische Ingenieur- und Architektenverein". Von Terzaghi and W. Herold arranged that Herold should write a memorandum against Fillunger in order to weaken Fillunger's position. This was written the next day and then typed on December 12, 1936. In this memorandum, Herold inexorably criticized Fillunger's scientific work. In the time between December 5 and December 11, von Terzaghi studied Fillunger's pamphlet and worked, together with Fröhlich, on a rebuttal. The publisher, Deuticke, acquainted with von Terzaghi for a long time, promised to publish the reply in his publishing house.

Then, on December 10, 1936, von Terzaghi formulated an urgent request to open a disciplinary action against Fillunger on the basis of a paragraph in the "Bundesgesetzblatt of 1934", because of an alleged violation of Fillunger's class-duties and his jeopardizing of the interests of the Technische Hochschule. He included his request in a two-page letter to the Rector of the Technische Hochschule of Vienna. The Rector, Prof. Dr. techn. Friedrich Böck, reacted im-

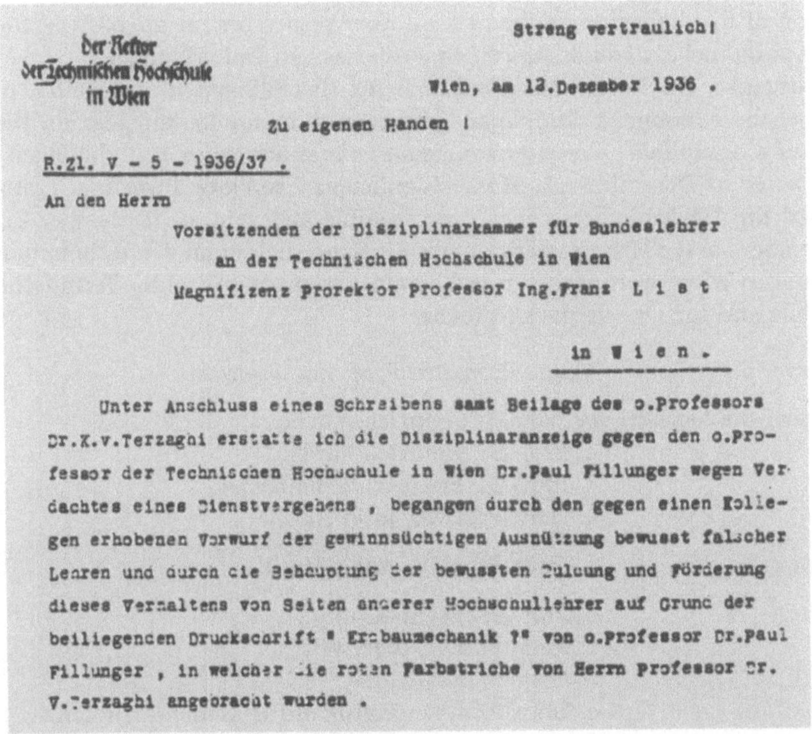

Streng vertraulich!

Der Rektor
der Technischen Hochschule
in Wien

Wien, am 12.Dezember 1936 .

Zu eigenen Handen !

R.Zl. V - 5 - 1936/37 .

An den Herrn

Vorsitzenden der Disziplinarkammer für Bundeslehrer
an der Technischen Hochschule in Wien
Magnifizenz Prorektor Professor Ing.Franz L i s t

in W i e n .

Unter Anschluss eines Schreibens samt Beilage des o.Professors Dr.K.v.Terzaghi erstatte ich die Disziplinaranzeige gegen den o.Professor der Technischen Hochschule in Wien Dr.Paul Fillunger wegen Verdachtes eines Dienstvergehens , begangen durch den gegen einen Kollegen erhobenen Vorwurf der gewinnsüchtigen Ausnützung bewusst falscher Lehren und durch die Behauptung der bewussten Duldung und Förderung dieses Verhaltens von Seiten anderer Hochschullehrer auf Grund der beiliegenden Druckschrift " Erdbaumechanik ?" von o.Professor Dr.Paul Fillunger , in welcher die roten Farbstriche von Herrn Professor Dr. v.Terzaghi angebracht wurden .

Fig. 4.14. Disciplinary action of the Rector

mediately. He had already written, on December 9, to professor Dr. Merkl, the "Disziplinaranwalt" of the Technische Hochschule of Vienna, asking him whether an offense against regulations had been committed concerning the content of some statements in the brochure *Erdbaumechanik ?*. Prof. Dr. Merkl answered the same day at 11 p.m., declaring that Professor Fillunger's comments in his brochure exceeded the approved special scientific criticism allowed under constitutional law and gave reasons for the suspicion of an offense against regulations. Prof. Merkl left it to the Rector's discretion whether or not to lay a disciplinary action against Professor Fillunger at the "Disziplinarkammer" (Disciplinary Court). This was done by the Rector in a letter to the president of the "Disziplinarkammer", Prof. Dr. List, on December 12, 1936. In a strictly confidential letter to the president of the "Disziplinarkammer" at the Technische Hochschule of Vienna, Magnifizenz Prorektor Professor Ing. Franz List (nearly all documents in this affair were sent strictly confidentially), the Rector opened the disciplinary action, sighting the suspicion of an offense against regulations, in the form of an accusation against a colleague, of profit-seeking utilization of deliberately falsified theories, further sighting the deliberate toleration and promotion of this behavior on the part of other professors on the basis of the enclosed pamphlet *Erdbaumechanik?* by Professor Dr. Paul Fillunger.

Furthermore, the Rector remarked that the disciplinary lawyer would, in the case of the announced disciplinary information by the Rektor, seek for the opening of a disciplinary investigation against the o. Professor Paul Fillunger.

In a letter of December 15, 1936, Magnifizenz Prorektor Prof. Ing. Franz List called the "Disziplinarkammer" to assemble at 6 p.m. on December 21, 1936. Members of the "Disziplinarkammer für Bundeslehrer an der Technischen Hochschule in Wien" were at that time the following professors of the Technische Hochschule and the Universität of Vienna:

President:	Magn. Prorektor Prof. Ing. Franz List
Members of the Senate:	o. Prof. Dr. Heinrich Mache o. Prof. Dr. Heinrich Pawek o. Prof. Dr. Friedrich Schaffernak o. Univ. Prof. Dr. Josef Hupka
Disciplinary lawyer:	o. Univ. Prof. Dr. Adolf Merkl
Substitutes:	o. Prof. Dr. Erwin Kruppa o. Prof. Dr. Engelbert Wist o. Prof. Dr. Karl Holey o. Univ. Prof. Dr. Ferdinand Degenfeld-Schonburg

In the meantime, the first protest letters arrived at the "Rektorat" of the Technische Hochschule of Vienna. In a letter of December 12, 1936, the president of the "Österreichischer Ingenieur- und Architekten-Verein", Ing. Brabbée, expressed his urgent request to the collegium of professors to do everything possible in order to clean up the whole affair. This was in the interest of the Technische Hochschule, as well as in the interest of the entire engineering industry.

The next protest letter was from Dr.-Ing. Otto Karl Fröhlich, the co-author of the book *Theorie der Setzung von Tonschichten* which had been strongly attacked by Fillunger (1936). Fröhlich mentioned in his letter that in the mathematical part of the pamphlet, Fillunger had defamed him in a severe way, and that this had caused him substantial economic losses in his consulting work. Moreover, he stated that he was able to refute Fillunger in the application of the higher analysis of physical problems, and he requested the establishment of a committee of experts.

The Rector of the Technische Hochschule also received protest letters from abroad, namely from Mr. Cummings of the Raymond Concrete Pile Company (USA), from Mr. Proctor of the American Society of Civil Engineers (Soil Mechanics and Foundations Division), and from Mr. A. Casagrande, a professor at Harvard University, Cambridge, Massachusetts. The most severe letter was written by Casagrande, a disciple of von Terzaghi. He threatened the Technische

Hochschule of Vienna that the International Conference on Soil Mechanics and the American Society of Civil Engineers were ready to intervene in the affair if the Technische Hochschule did not take action against Fillunger's defamations. Finally, A. Casagrande pointed out that von Terzaghi had planned to organize the next Second International Conference of Soil Mechanics in Austria. Undoubtedly, the decision on this proposal would depend strongly on the reactions of the Technische Hochschule and the Austrian engineers in the affair.

The activities of both von Terzaghi and Fröhlich to overcome Fillunger's brutal attack continued. After von Terzaghi had cancelled some lectures for the students and had handed others over to his co-worker Steinbrenner, he delivered a small address to his students at 9.45 a.m. on December 15, 1936. He began by explaining the conflict with Fillunger and then went on to make some general statements: "Successful teaching is based upon the students' respect for their teachers and upon the confidence of the students in the expertise of their professors. The aggressor has undertaken a large-scale attempt to sap your respect in my person and to shake your confidence in my expertise". Von Terzaghi announced that he would comment on the contents of the attack at a later date. He stated that until then he would have to turn his lectures over to his co-workers because he needed every minute of his time to defend himself from the attacks.

Indeed, von Terzaghi and Fröhlich worked without rest on their reply to Fillunger's brochure. They were also in contact with Dr. Leo Rendulic, a respected former collaborator of von Terzaghi, who had in the meantime moved to Berlin. Fröhlich wrote to Rendulic that he and von Terzaghi wished to incorporate him into fighting the attack. He further remarked that the publisher Deuticke had offered to publish the report. The title of the brochure came to von Terzaghi's mind in a flash:

ERDBAUMECHANIK?	SOIL MECHANICS?
und	and
ERDBAUMECHANIK!	SOIL MECHANICS!
Eine	A
Entgegnung	Reply
von	by
K. V. TERZAGHI	K. V. TERZAGHI
Unter Mitarbeit von	In Cooperation with
O. K. Fröhlich	O. K. Fröhlich
und	and
L. Rendulic	L. Rendulic

In the final form of the reply the title was changed, as well as the fact that Rendulic did not appear as a co-author.

On December 16, 1936, the council of professors held a meeting. The extent to which life at the Technische Hochschule had been effected by Fillunger's attack became evident in Rector Dr. Friedrich Böck's statement:

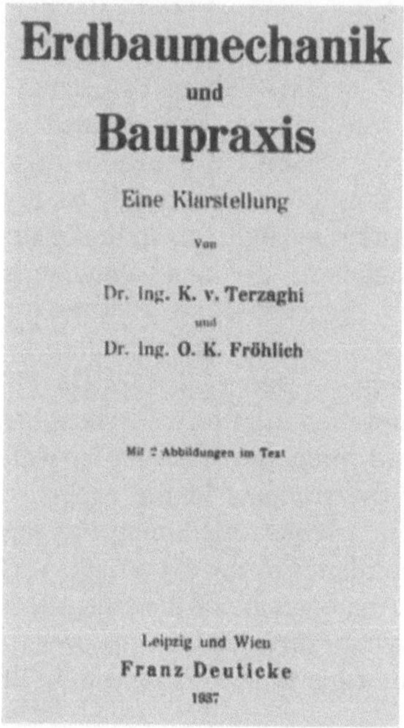

Fig. 4.15. Defense brochure of K. von Terzaghi and O.K. Fröhlich

"Ich komme nunmehr zum Schlußteil meiner mündlichen Mitteilungen, der eine ganze Anzahl von Punkten umfasst, die zweifellos untrennbar miteinander verknüpft sind und ein Ereignis betreffen, das, wie ich wohl sagen kann, uns alle seit 14 Tagen mit schwerster Sorge belastet."

"Now I come to the final part of my speech, which contains a large number of points, which are without any doubt inseparable and concern an occurence that, I can say, has burdened us all seriously for 14 days."

Then, the Rector listed the aforementioned points chronologically, from the appearance of the pamphlet to the letter from Dr. Fröhlich. Furthermore, he mentioned that von Terzaghi had already expressed the urgent wish to comment on Professor Fillunger's attack in the current meeting. The Rector had convinced von Terzaghi to hold off temporarily and to save his defence for later. In the same meeting, Professor Fillunger had applied for a discussion about von Terzaghi's teaching program. However, the application had been refused by the council against the vote of Professor Fillunger.

After having studied Fillunger's pamphlet very carefully between December 3 and 6, 1936, von Terzaghi and Fröhlich began immediately with the formulation of their reply on Monday, December 7, 1936.

They worked almost every day on their brochure, including Christmas, 1936, finally finishing the manuscript at the very beginning of January, 1937.

In their preface to the rebuttal paper *Erdbaumechanik und Baupraxis: Eine Klarstellung*, von Terzaghi and Fröhlich (1937) pointed out that soil mechanics was a part of applied hydraulics (in a certain sense, a strange statement).

Moreover, they emphasized again and again that soil mechanics served to solve practical problems. The preface contained the term "practical" nine times. In no sentence did the term "scientific" appear. It may be that Fillunger's attack had made both authors feel insecure due to the fact that Fillunger was recognized in the literature as a theorist with a profound knowledge in the field of mechanics. Therefore, the authors of the rebuttal restricted themselves mainly to practical problems.

The reply by von Terzaghi and Fröhlich consisted of three main parts: part A treated the history of the creation and the practical applications of soil mechanics; part B contained the rejection of different criticisms on the theory of the settlement of clay layers and, finally, part C contained a brief review of laboratory tests in the field of soil mechanics.

Part A "Die Erdbaumechanik als technische Wissenschaft" was written by von Terzaghi. He first repeated the main points of Fillunger's attack. Then he described extensively the creation of modern soil mechanics in Sweden, in the United States, in Germany, and in Turkey, where he had begun his fundamental work in 1917. He reported that the great success of his book from 1925 was based partially on the fact that it contained the solutions to many important practical problems. Later, von Terzaghi gave a brief overview on the historical development of foundation engineering and also discussed the importance of modern soil mechanics.

In the next section of part A, von Terzaghi considered the practical application of soil mechanics. The practical problems of soil mechanics can be summarized as follows:

a) Numerical description of the technically important properties of soil as, e.g., the compression, permeability, shear resistance etc.
b) Avoidance of the role of pure chance during the construction of artificial soil bodies.
c) The setting up of rules for the adjustment of road constructions for artificial roads based on the consistency of the ground.
d) Elaboration of reliable methods for the measurement of motions and pressures on finished buildings in order to gain numerical experience.
e) Revealing the qualitative relations between cause and effect in underground engineering.
f) Elaboration of roughly approximate methods for the numerical prediction of effects in underground engineering.

Subsequently, von Terzaghi discussed points a) through f) in detail, and he refused Fillunger's criticism that an application of the new soil mechanics to civil engineering did not exist and that there would never be such an application.

In the final section of part A, von Terzaghi considered the theories of soil mechanics. These dealt with, according to von Terzaghi, stability of slopes, the earth pressure on retaining walls, the methods of dynamical underground investigations, the stress distribution in loaded foundation soil, and the settlement of clay layers. He discussed the settlement of clay layers in detail because only this one had been attacked by Fillunger. Karl von Terzaghi repeated all the basic assumptions of his theory, developed his partial differential equation, and stated that the main predictions of his consolidation theory had been proven again and again by tests and by experience, although some problems had yet to be solved. Nevertheless, there had been a great amount of progress in contrast to the state of the science in 1920.

Part B of the reply "Erdbaumechanik und Baupraxis" was formulated by Fröhlich. He proved that many attacks against the consolidation theory were based on improper statements by Fillunger. In particular, Fröhlich included the acceleration in the derivation of von Terzaghi's differential equation - a main point of Fillunger's approach to the consolidation problem. However, he had in no way understood Fillunger's completely new idea for treating saturated porous bodies. He criticized Fillunger for not having included a constitutive equation for the effective stresses of the porous solid, although Fillunger listed this problem as a desirable generalization of his theory under point 4. Moreover, Fröhlich pointed out that an extensive discussion of Fillunger's theory would be published elsewhere. This did not happen. Instead, Fillunger's theory was completely ignored by the von Terzaghi school in the time to follow.

Fröhlich then refused Fillunger's criticism of the approximate solution of von Terzaghi's differential equation and the theoretical treatment of the test results in the laboratory. He showed that there were some considerable mistakes in Fillunger's attacks. Indeed, this was the case. This was, in particular, evident in Fillunger's criticism of the use of differential calculus on the part of von Terzaghi and Fröhlich. Fillunger had assumed that both authors had applied a special calculus in order to gain the total differential of the pore number, and he spoke of the "Terzaghi-Differential". However, Fillunger made an unimaginable mistake when he introduced the constant value 1 for the variable length l of a clay sample. It is hard to understand how such a conscientious scientist could overlook such a basic fault. It may be that he was blinded by his extraordinary hate for von Terzaghi.

Finally, Fröhlich summarized his view by stating that Fillunger's criticisms were directed against a theory which described the consolidation problem sufficiently for practical purposes. He showed that this criticism was based in part on improper mathematical rules, and that the relations developed by Fillunger to describe the consolidation problem were practically useless because their solution was not known.

Part C, "Versuchstechnik und praktische Anwendungen der Erdbaumechanik", was again written by von Terzaghi. This part dealt with the objections raised by Fillunger against the test technique and the applications of soil mechanics. Whereas von Terzaghi had described the test technique and practical application of soil mechanics very generally in part A, he now discussed these points in detail in part C, refuting Fillunger's objections.

The reply was finished in the very beginning of January, 1937, and was sent to colleagues, friends, and other individuals in the field of soil mechanics. Enclosed was a handbill:

"Zur gefl. Kenntnisnahme:

Im Dezember 1936 hat Herr Dr. Fillunger, o.ö. Professor an der Technischen Hochschule in Wien und Mitglied des Österreichischen Ingenieur- und Architektenvereines im Selbstverlag ein Druckwerk auf Erdbaumechanik veröffentlicht. Trotzdem Herr Fillunger als ein dem Fachgebiet ferne Stehender keinen Einblick in die praktischen Forderungen des Tiefbaus haben kann und niemals die persönliche Verantwortung für das Gelingen einer Fundierung trug, hielt er sich für kompetent, in seiner Schrift ein Urteil über die Erdbaumechanik als technische Hilfswissenschaft zu fällen. Zur Entkräftung der praktischen Einwändungen des Herrn Fillunger gegen die Erdbaumechanik genügt ein Hinweis auf die zahlreichen praktischen wertvollen Leistungen dieser jüngsten technischen Hilfswissenschaft. Diese Leistungen wurden wiederholt im technischen Schrifttum behandelt. Eine Übersicht über dieselben findet sich in der beigeschlossenen zum Zwecke der Klarstellung des Sachverhaltes verfaßten Broschüre 'Erdbaumechanik und Baupraxis: eine Klarstellung'. Die Broschüre enthält auch eine Erörterung des mathematischen Teils der Fillungerschen Kritik. Trotzdem sich Herr Fillunger in diesem Teil seiner Kritik auf seinem engsten Fachgebiet befindet, verstieß er nicht nur gegen die Regel sondern auch gegen den Geist der angewandten Mathematik und seine Kritik enthielt leider nicht einen einzigen aufbauenden Gedanken. Die von Herrn Fillunger verfaßte Druckschrift 'Erdbaumechanik?' wurde nicht bloß an Fachleute sondern auch an Juristen, Verleger, Verwaltungsbeamte, Institute und sogar an die Schriftleitungen von Tageszeitungen versendet. Sie kam dadurch in die Hände von zahlreichen Personen, deren Interesse an dem behandelten Gegenstand lediglich durch die in der Schrift enthaltenen Unterstellungen geweckt wird. Außerdem ist die Schrift derart verfaßt, daß sie sich nicht bloß auf meine eigene Person bezieht sondern auch den Ruf und die wirtschaftliche Existenz meiner früheren und jetzigen Mitarbeiter und meines Erachtens auch das Ansehen der Wiener Technischen Hochschule auf das schwerste gefährdet.

Schließlich sei hervorgehoben, daß es mir nur gelungen ist, ein äußerst lückenhaftes Verzeichnis der Adressen, der mit der Fillungerschen Broschüre beteiligten Personen zu bekommen. Durch diese Umstände bin ich genötigt, an die Empfänger dieses Flugblattes die Bitte zu richten, seinen Inhalt in ihrer Wirkungsweise nach Möglichkeit bekannt zu machen und die beigeschlossenen Exemplare meiner Broschüre an die am Gegenstand derselben interessierten Fachleute ihres Bekanntenkreises oder an jene Stellen und Personen weiterzuleiten, von denen angenommen werden kann, daß sie die Fillungersche Druckschrift 'Erdbaumechanik?' bekommen haben. Die Spötteleien und Verdächtigungen, mit denen die Fillungersche Schrift durchsetzt ist, haben dem sachlichen Zweck seiner Schrift in keiner Weise gedient. Es versteht sich von selbst, daß sich die akademische Behörde der Hochschule, an welcher Herr Fillunger wirkt, mit dieser Angelegenheit beschäftigen wird. Die Versendung dieses Flugblattes erfolgt an Institute und Personen in allen Ländern, in denen Herr Fillunger nach eigenen Angaben seine Broschüre verbreitet hat. Diese Länder sind: Österreich, Deutschland, Danzig, Schweiz, Tschechoslowakei, Holland, Polen, Norwegen, Italien, Letland, Rußland, Türkei und die Vereinigten Staaten von Nord-Amerika.

Wien, im Jänner 1937
Dr.-Ing. Karl von Terzaghi
o. ö. Professor an der
Technischen Hochschule Wien."

"To whom it may concern:

In December, 1936, Dr. Fillunger, o. ö. professor at the Technische Hochschule of Vienna and member of the Österreichischen Ingenieur- und Architektenvereines published by himself a pamphlet on soil mechanics. Although Mr. Fillunger, as a person far away from the subject, can not have any insight into the practical requirements of underground engineering and has never carried personal responsibility for the success of a foundation, he believes himself to be competent to give his opinion on soil mechanics as a technical auxiliary science. In order to weaken the practical objections of Mr. Fillunger against soil mechanics, a reference to the numerous valuable practical achievements of this recently developed technical auxiliary science is sufficient. These achievements have been treated again and again in the technical literature. A survey of these can be found in the enclosed brochure 'Erdbaumechanik und Baupraxis: eine Klarstellung', formulated for the purpose of clarifying the facts. The brochure also contains a discussion of the mathematical portion of Fillunger's criticism. Although Mr. Fillunger is in his own special field in this part of his criticism, he violates not only the rules but also the spirit of applied mathematics and his criticism unfortunately does not contain even one constructive idea. The publication 'Erdbaumechanik?' formulated by Mr. Fillunger was sent not only to experts but also to lawyers, editors, administrative officers, institutes and even to the editorial staff of newspapers. It came thereby to numerous persons whose interest in the treated subject was created exclusively by the allegations contained in the publication. In addition, the publication is composed in such a way that it is not only related to my person, but it also heavily jeopardizes the reputation and the economic existence of my former and present co-workers, and, in my opinion, the reputation of the Viennese Technische Hochschule as well.

Finally, it must be emphasized that I have only succeeded in obtaining an extremely incomplete list of the addresses of the individuals who have received Fillunger's brochure. Due to these circumstances, I am forced to direct the request to the receivers of this handbill, that they make the contents public if possible and that they pass on the enclosed copies of my brochure to experts who are interested in the subject or to those institutions and individuals that have probably received Fillunger's publication 'Erdbaumechanik?'. The sarcastic and suspicious remarks with which Fillunger's brochure is filled have in no way supported the factual purpose. It is natural that the academic staff of the Hochschule where Fillunger is appointed will deal with this affair. The handbill will be sent to institutes and individuals in all countries where Fillunger, according to his own statement, has distributed his brochure. These countries are: Austria, Germany, Danzig, Switzerland, Czechoslovakia, The Netherlands, Poland, Norway, Italy, Latvia, Russia, Turkey and the United States of America.

Vienna, January, 1937
Dr.-Ing. Karl von Terzaghi
o. ö. Professor at the
Technische Hochschule of Vienna."

K. von Terzaghi and O.K. Fröhlich received a considerable amount of positive replies. These were appreciative of the purely factual tone of the rebuttal.

The disciplinary committee ("Disziplinarkammer") met on December 21, 1936, in the Rektorat of the Technische Hochschule at 10.15 a.m., and not at 6 p.m., as scheduled. Present were:

President:	Magnifizenz Pror. Prof. Ing. List
Members:	o. Prof. Dr. Mache o. Univ. Prof. Dr. Josef Hupka
Alternates:	o. Prof. Dr. Kruppa o. Prof. Dr. Richter
Disciplinary lawyer:	o. Univ. Prof. Dr. Merkl
Secretary:	Dr. Stein

The president, Prof. List, opened the session by welcoming the members and thanked, in particular, the professors of the University. He then read out a letter from Magn. Böck to the disciplinary lawyer, Prof. Dr. Merkl, and Dr. Merkl's response of December 10, 1936, as well as a letter from Magn. Böck to the president of the disciplinary committee containing the disciplinary information against Prof. Dr. Fillunger, a letter from Prof. Dr. von Terzaghi to Magn. Böck with the request to open a disciplinary action against Prof. Dr. Fillunger, Dr. Fröhlich's letter, and a letter written by Prof. Dr. von Terzaghi (without enclosures) to Magn. Böck from December 21, 1936.

The disciplinary lawyer Prof. Dr. Merkl pointed out: "The problem on hand touches the fundamentals of the freedom of science and teaching. The opportunity to criticize must be given. However, the limits of the discipline must be guaranteed. Prof. Dr. Fillunger raises defamatory attacks. Thus, an order to open a disciplinary action is necessary:

1. Prof. Dr. Fillunger must be called as an accused person. He must factually justify his attacks and must say whether personal conflicts existed before this publication.
2. Prof. Dr. von Terzaghi and Dr. Fröhlich are to be called into the witness-stand. They are also to be examined as to whether the pamphlet has personal backgrounds.

The main issue, however, is the need for an expert opinion; this must be written by a commitee of experts. In order to write the expert opinion completely independently, reviewers from outside are to be incorporated, e.g., from other Austrian Technische Hochschulen (or even from abroad). Prof. Dr. Fillunger starts attacks against his colleagues and, moreover, an attack against the scientific image of the staff of professors. These accusations are, however, too vague, so that I will not yet apply for the suspension of Prof. Dr. Fillunger from his duties. The question of suspension will become ripe first after the expert opinion has been finished."

Professor Dr. Hupka was of the opinion that the disciplinary action could be commenced after the decision of the experts.

This opinion was strictly refused by the disciplinary lawyer Prof. Dr. Merkl, who remarked in a severe tone that the opening of the disciplinary action was necessary because the accusation had become widely public and therefore the involved groups would be able to see that the Hochschule was already officially examining the matter. The expert opinion would only be able to state the gravity of the offense.

Professor Dr. Hupka replied: "Without doubt, the opening of the disciplinary investigation should be voted. However, within the framework of the disciplinary action, expert opinion must be brought in by the professors' council."

In contrast to this professor's view, Dr. Merkl answered: "The expert opinion must be objective; therefore, no one depending on Prof. Dr. Terzaghi or one of his disciples, may be a member of the expert committee."

Then, the president, Prorektor Prof. Dr. List, recommended that the Rector should nominate the experts. Moreover, he stated that the disciplinary lawyer should request the opening of the disciplinary investigation.

Professor Dr. Merkl formulated his motion:

"Die Konkretisierung der Beschuldigung durch Zitierung der diesbezüglichen Stellen aus der Schrift 'Erdbaumechanik?' des Prof. Dr. Fillunger hat erst im Verweisungsbeschluss zu erfolgen. Ich beantrage also, gemäß §113 D.P. die Einleitung einer Disziplinaruntersuchung gegen Herrn Professor Dr. Paul Fillunger zu beschlissen, wegen des durch Veröffentlichung und Verbreitung der Druckschrift 'Erdbaumechanik?' begründeten Verdachtes eines Dienstvergehens."

"Making the accusation concrete, by citing the corresponding parts of the brochure 'Erdbaumechanik?' of Prof. Dr. Fillunger, has to take place first in the resolution of reference. I therefore move according to §113 D.P. to pass the resolution of the opening of a disciplinary investigation against Professor Dr. Paul Fillunger due to a suspicion of offense of regulations founded by the publication and distribution of the printed book 'Erdbaumechanik?' "

In the ensuing discussion, Professor Dr. Kruppa declared that members of the expert committee should also be nominated by the professors, Dr. Terzaghi and Dr. Fillunger. Professor Dr. Hupka remarked that the expert committee should be able to examine both sides. Prorektor Prof. Dr. List proposed that the Rector set up a mixed commission, and that this commission determine further experts who would then be delegated to an expert committee.

Finally, the president Prorektor Prof. Dr. List put the motion (of the disciplinary lawyer) concerning the opening of a disciplinary investigation to a vote. The motion was universally accepted. The president stated: "Professor Dr. Fillunger is now to be examined as an accused person; he has to prove the truth of his accusations. Moreover, it must be proven whether there are personal motives. In addition, an expert opinion must be brought in. The composition of the expert committee should be arranged by the Rector on the basis of the nominations by the restricted commission, after the hearing of the accused and attacked persons. The task of the expert committee will be the special scientific review of the technical explanations in Professor Dr. Fillunger's brochure."

In addition, Professor Dr. Merkl pointed out that the decision to open a disciplinary investigation had to be served to Professor Dr. Fillunger and accompanied by information on his right of appeal. The hearing ended at 11.15 a.m.

The Rector of the Technische Hochschule of Vienna and Professor Dr. Fillunger were informed of the decisions of the disciplinary senate on the same day. Professor Dr. Fillunger was simultaneously informed that no right of appeal was admissable. The Rector appointed the Administrationsrat Regierungsrat Dr. Josef Goldberg on the same day to be the investigation-commissioner in the disciplinary investigation against Professor Dr. Paul Fillunger.

In addition, the Rector appointed the professors, Dr. Franz Jung (mechanics), Dr. Ludwig Flamm (physics), Dr. Lothar Schrutka-Rechtenstamm (mathematics), and Dr. Josef Kozeny (hydraulics) to form a commission. He requested that the commission be formed by Monday, January 4, 1937, and entrusted the commission with the task of nominating experts for the investigation committee. This committee was set up on January 7, 1937, and consisted of the professors, Dekan Dr. Karl Wolf (mechanics), Dr. Rudolf Saliger (concrete construction), Dr. Friedrich Schaffernak (hydraulic engineering), Dr. Franz Jung (mechanics), Dr. Ludwig Flamm (physics), Dr. Schrutka-Rechtenstamm (mathematics), Dr. Franz Aigner (physics), Dr. Alfred Lechner (mechanics), and Dr. Josef Kozeny (hydraulics).

The Rector remarked in his letter that, in the case of inviting an out-of-town expert, the "Rektorat" had to be informed for the reason of covering the expenses. Moreover, he made an appeal to begin immediately with the investigations and to finish them, if possible, by January 23, 1937.

Referring to statements of von Terzaghi, the professors, Dr. Jung, Dr. Lechner, Dr. Wolf, and Dr. Flamm were recommended by Prof. Dr. Fillunger, whereas the professors, Dr. Schaffernak, Dr. Aigner, and Dr. Schrutka-Rechtenstamm were recommended by von Terzaghi. Von Terzaghi ignored Prof. Dr. Salinger because of factual differences. (It should be mentioned, however, that in a letter of December 28, 1936, von Terzaghi requested Professor Dr. Flamm to act in the commission).

The appointed investigation-commissioner, Dr. Goldberg, began immediately with his questioning. He started questioning Professor Fillunger just one day after his appointment, on December 22, 1936. First, Fillunger discussed in detail the whole previous history of the affair, in particular the dispute over the uplift problem. In further questioning on February 17, 18, and 19, 1937, Fillunger defended the personal attacks against von Terzaghi, Fröhlich, and his colleagues from the Technische Hochschule. He stated that the main motive for his sharp criticism had been the fact that the theories of von Terzaghi and Fröhlich were liable to cause great damage:

1. In the practice, due to an unjustified confidence in these theories, it could happen that false foundations could be constructed which would weaken the reputation of the engineers. This would be very critical in view of the political situation (Bolshevism).
2. In the teaching program of the Technische Hochschule the theories are taught in the higher semesters where the students do not have relations to the theoretical subjects and therefore can not reflect critically on the theories.

3. In scientific circles it can happen that real science, complicated by great intellectuals, will be refused, whereas superficial simple statements spread rapidly.

The question as to whether he had violated the reputation of the Technische Hochschule by his personal attacks was denied by Fillunger.

The questioning of von Terzaghi began on December 23, 1936, from 9.30 a.m. to 12.30 p.m. The topics were the same as in the questioning of Fillunger. Further questioning was performed on February 10, 1937, and March 3 to 5, 1937. In these last hearings, von Terzaghi gave an interesting inside-look into his method of scientific working. He pointed out that, particularly in the first decade of his research, his method was determined by the fact that his field had in no way been scientifically investigated. In such a situation, the success of his efforts would depend exclusively on a talent which he called 'physical scent' and which could not be learned. His method of working could be described as follows: as soon as a phenomenon in nature excited his interest, he tried at first to understand purely intuitively the progress of the phenomenon due to the properties of the substances which were involved in the progress. He therefore repeated the experiment again and again over longer time frames. When, in 1923, he made efforts to describe analytically the nature of the settlement of clay layers, he gave up the attempt in an exhausted state after several weeks of hard work. Three months later, he succeeded easily within two hours, and after five further hours, the derivation of the fundamental equation of the consolidation problem was completed. This method is sometimes called intuition. The next step consisted of the mathematical description of the conception of the nature of the problem, which he had already formed in his mind, in which he neglected apriori all such factors which he considered intuitively as unimportant. Thereafter, he invented such experiments whose results allowed for the proof of the theoretical ideas. The completion of the investigation consisted of performing the experiments. If, then, the experiment confirmed the correctness of the basic idea, the further development of the theory could be performed with the help of recognized and generally used calculation rules. Because he did not have, however, any interest in this part of the work, he encouraged qualified colleagues to continue the work. The result of such an encouragement becomes evident in the book *Theorie der Setzung von Tonschichten*, in which Fröhlich did the strenuous work of the evaluation of the partial differential equation. In this connection, von Terzaghi declared that he had never read and would never read the sections worked out by Fröhlich, since he was satisfied if he knew the assumptions and the results. The role of the 'physical scent' becomes very clear in the fact that most of his experiments were not new. The difference between his own work and that of his predecessors consists only in the manner of observation and in the apparent slight changes of the test arrangements. In order to discover a new relation through an experiment, one must already suppose its existence beforehand, and must perform the test-program in such a way that the validity of the supposition becomes evident in the test results. Von Terzaghi became convinced, over the course of several years, that the talent for doing this kind of work could not be taught. Despite severe efforts, he succeeded only occasionally in discovering, among his students and co-workers, young people who had at least a point of attachment to this ability. More widespread, even in mathematically trained circles, are such minds that cannot understand a physical process even if the description is already completed. According to the aforementioned way of working, von Terzaghi started almost from a more or less intuitively gained set of conceptions; he proved, later on, whether these conceptions were correct. In his papers, however, he had to choose the inverse manner. From this inversion of his thinking process, there resulted not infrequently inconsistent jumping from one idea to another.

With this glimpse inside von Terzaghi's working-style, the questioning on March 5, 1937, was closed.

Not only Fillunger and von Terzaghi were questioned by Dr. Goldberg in 1936 and 1937, but also Dr. Fröhlich and other professors of the Technische Hochschule of Vienna. As early as December 23, 1936, the questioning of Fröhlich concerning his relations to von Terzaghi and Fillunger had started, lasting from 12.30 p.m. to 1.25 p.m. A further hearing with Fröhlich took place on February 24, 1937. He expressed his admiration for von Terzaghi's talent for intuitively recognizing important physical phenomena. Fröhlich recalled a related statement of the famous hydraulician Professor Forchheimer: "Professor von Terzaghi has a distinct ingenius talent for technical-physical problems." Furthermore, Fröhlich pointed out that von Terzaghi's papers contained, in general, only the main ideas, so that it was not surprising that they had not been understood for a long time.

Professors of the Technische Hochschule of Vienna were questioned in January, February, and March, 1937, mainly concerning the jeopardization of the reputation of the Technische Hochschule by Fillunger's attack. Professors Urbanek, Lösel, Abel, and Theis were questioned in this regard as well.

The commission to set up an investigation committee had already met in December, 1936. On Tuesday, December 29, 1936, Professor Flamm declared that one should not construct a new theory but should keep the model of von Terzaghi. Moreover, he doubted the validity of the incompressibility condition for the partial water body.

As already mentioned, the investigation committee began its work at the very beginning of January, 1937. It held eighteen meetings. The leading member of the committee was Professor Flamm, a son-in-law of the famous physicist Professor Ludwig Boltzmann. Professor Flamm had taken over a professorship for theoretical physics at the Technische Hochschule of Vienna in the 1920s. He formed a strong group with the professors Schaffernak (hydraulics, and a friend of von Terzaghi) and Aigner (physicist), who had no opinion of his own in most cases, but agreed with Flamm and Schaffernak. There was obviously some animosity between Flamm and the professors of mechanics and, in a meeting in January, 1937, Professor Flamm asked his colleagues whether they had ever heard anything about mechanics. Upon hearing this, Professor Jung severely denounced Flamm's arrogant tone. The professors of mechanics investigated the scientific side of the affair very carefully. In particular, Professor Lechner, together with his co-worker Dr. Heinrich, showed that von Terzaghi's partial differential could be gained from Fillunger's basic system of differential equations using, however, strongly simplified assumptions. Lechner's and Heinrich's investigations later became part of the body of expert opinion. They were also later published by Heinrich (1938a). For Flamm and Schaffernak, the investigations of the mechanics-group were obviously a little too boring. In order to convince his colleagues that von Terzaghi's work was sound, at the end of January Schaffernak constructed a simple mechanical model of the consolidation problem. However, the members of the mechanics group continued their work to develop von Terzaghi's differential equations based on Fillunger's fundamental equations.

A decisive meeting of the investigation committee was held on January 30, 1937, between 11 a.m. and 1 p.m. During the discussions, it became evident that Fillunger had made serious mistakes in Sections III and IV of his brochure *Erdbaumechanik?*. This statement was devastating for Fillunger's position. The committee also visited von Terzaghi's laboratory, and the members saw for themselves that von Terzaghi and his staff worked with scientific methods.

Obviously, Professor Salinger reported the results of the meeting to Fillunger in the afternoon, to which Fillunger expressed his feelings in a very loud voice, leading to a gathering of the students on the floor.

In a letter of February 5, 1937, addressed to Professor Wolf, the chairman of the investigation committee, Fillunger requested that the committee allow him to present a statement to the reply *Erdbaumechanik und Baupraxis: Eine Klarstellung* by von Terzaghi and Fröhlich. In a letter of February 8, 1937, Professor Wolf answered that the members were only prepared to receive Fillunger's statement privately, as they had done with von Terzaghi's and Fröhlich's defending report.

Fillunger finished his statement in a great hurry, and in a short letter to the members of the committee he apologized for possible mistakes in his statement, *Vorläufige Erwiderung auf die Verteidigungsschrift der Herren v. Terzaghi und Fröhlich*, due to his haste. In the very beginning, he repeated his main point of criticism:

"Sowohl die 'Theorie der Setzung von Tonschichten' der Herren v. Terzaghi und Fröhlich als auch die Berechtigung zur vollständigen Ablehnung dieser Theorie aus wissenschaftlichen Gründen durch den Verfasser dieser Erwiderung *stehen und fallen* mit der Beantwortung der Frage, ob die Differentialgleichung der Porenwasserströmung, wie sie durch v. Terzaghi aufgestellt und ausser von diesem auch von Herrn Dr. Fröhlich vertreten wird, nämlich die Gleichung

$$c \frac{\partial^2 w}{\partial z^2} = \frac{\partial w}{\partial t} , \tag{4.4.1}$$

ERDBAUMECHANIK
UND WISSENSCHAFT

EINE ERWIDERUNG

VON

BAURAT PROF. DR PAUL FILLUNGER
WIEN

ALS MANUSKRIPT GEDRUCKT

DRUCK UND VERLAG DER
BUCHDRUCKEREI FRIEDRICH JASPER, WIEN, III., THONGASSE 12

Fig. 4.16. P. Fillunger's reply to the defense brochure of K. von Terzaghi and O.K. Fröhlich, edited by his son

wissenschaftlich begründet werden kann. Daher soll diese Frage hier in erster Linie erörtert werden. Hinter dieser Frage treten andere, wie z.B. die nach den experimentellen Grundlagen, nach der praktischen Anwendbarkeit und der im Buche 'Theorie der Setzung von Tonschichten' gewählten Darstellungsform an Bedeutung weit zurück ..."

"The 'Theorie der Setzung von Tonschichten' of Mr. v. Terzaghi and Mr. Fröhlich as well as the right to completely reject this theory from a scientific point of view by the author of this reply, *stand and fall* with the answer of the question as to whether the differential equation of the porewater-flow can be scientifically founded, as it has been developed by v. Terzaghi and apart from this has been supported by him and Mr. Dr. Fröhlich, namely the equation

$$c\,\frac{\partial^2 w}{\partial z^2} = \frac{\partial w}{\partial t}\,. \tag{4.4.1}$$

Therefore, this question should be discussed here first of all. Compared to this question, other questions, for example, those referring to the experimental basis, to the practical application and to the representation form in the book 'Theorie der Setzung von Tonschichten', are unimportant ..."

In the first part of Fillunger's treatise, the theoretical part, he remarked that the formal correspondence of von Terzaghi's differential equation with the Fourier-equation for non-stationary heat propagation could not be sufficient in scientifically founding the differential equation. He then discussed in detail all terms in the derivation of von Terzaghi's differential equation and criticized the analogous conclusion, as well as the attempt on the part of Dr. Fröhlich to include the acceleration in the differential equation. He also tried to prove that Dr. Fröhlich's approximate solution was false. In the second and third section of his reply, Fillunger treated the consolidometer test and justified his criticism.

After a very intense committee meeting on February 13, 1937, Professor Schaffernak became ill and was not able to attend the meetings in the following weeks. Thus, von Terzaghi lost a personal friend on the investigation committee. However, the affair was already decided, and the subsequent meetings served only to formulate the details of the result.

Karl von Terzaghi was deeply hurt by the whole affair. There were also at the time several ugly rumors about his role in connection with the construction of the Reichsbrücke in Vienna. In 1935, the news had already been spread in Germany, by a former co-worker of von Terzaghi, that his position as a specialist in Austria had been severely shaken. Among the teaching staff of the Technische Hochschule of Prague, it was an open secret that von Terzaghi had at first proposed a suspension bridge in Vienna, and thereafter recognized that the soil was not suitable for the anchoring of such a construction. In Vienna, the newspapers reported that the high costs of the Reichsbrücke were due to the excessive caution of von Terzaghi. Although the responsible persons in the government declared that the news had been invented by reporters, it was clear to von Terzaghi that the information had come from certain circles in the government. Von Terzaghi also had some "enemies" in Berlin.

After Fillunger's attack, and the rumors cited above, von Terzaghi believed that there was a "hintermann" (a man operating behind the scene). Obviously, he suspected his colleague Professor Dr. Saliger (professor for concrete construction) as a possible source of such talk. At the very beginning of January, 1937, von Terzaghi called Saliger by phone and then subsequently met with him on January 8, 1937. Saliger promised to send von Terzaghi a letter. This was done on January 9, 1937. In the letter, Saliger assured von Terzaghi that he himself had never spread rumors about him and that he was upset by Fillunger's attack.

On January 11, 1937, the dean of the faculty, Professor Wolf, informed the Rector that von Terzaghi had cancelled his lectures. An urgent request from the Rector convinced von Terzaghi to take up his lectures again.

As has already been pointed out, von Terzaghi and Fröhlich worked hard to fend off Fillunger's attack, along with certain rumors. After completing the reply (von Terzaghi and Fröhlich, 1937), von Terzaghi worked on the opinion paper of the Reichsbrücke, and met with some members of the committee. In February, 1937, von Terzaghi formulated a memorandum, consisting of three parts. The first part explained extensively the previous history and the factual background behind the attack. He repeated the story of the disagreement over the uplift problem, and formulated a first statement regarding the factual side of Fillunger's attack. The essential points had already been included in the reply from von Terzaghi and Fröhlich (1937). In the appendix of his memorandum (part one), von Terzaghi denied the accusation that he had spread his view on soil mechanics solely for profit-seeking reasons. He stated that his consulting work was only dedicated to those technical problems which were interesting from a scientific point of view. Moreover, he declared that his financial gains had always been very modest, and he proved this by various examples.

In the second part of his memorandum, von Terzaghi gave some advice to the members of the investigation committee, which consisted mostly of mathematicians, physicists, and mechanicians. He defined soil mechanics as a branch of applied hydraulics, and stated that soil mechanics served exclusively for practical purposes. He further stated that the essential and seriously offensive remarks by Fillunger could be judged only by representatives from the applied fields or by experts in civil engineering. The task of the mathematicians, physicists, and mechanicians could only consist in discovering the errors in Fillunger's reasoning, which he must have made in order to declare a theory to be absurd which had been proven in the laboratory as well as on the building site. Then, von Terzaghi suggested a detailed schedule to the committee in order to accelerate the passing of a resolution. Von Terzaghi was very much interested in the immediate clarification of the attack. Besides the detailed schedule, he also developed a questionnaire to be answered by the members of the committee, via the multiple-choice-method. Obviously, this memorandum never came to the knowledge of the committee; this was certainly good for von Terzaghi's position. One can imagine how the members of the committee, all professors in their own right with their own views on the whole affair, would have reacted to such a strong attempt to influence their opinion.

In the third part of his memorandum, von Terzaghi presented a broad and deep exposé of Fillunger's scientific talent. First, he discussed the papers of Fillunger from 1912 on. He accepted the papers on the elasticity theory, but rejected the papers which dealt with new physical phenomena, for example, with the uplift problem. He heavily criticized Fillunger's first paper on the uplift problem, published in 1913, and the subsequent papers, as well as Fillunger's polemics against Hoffman and Lehr. Von Terzaghi stated that Fillunger had a particular talent for reacting when he recognized that he was in error. In this case, he would mask his results again and again by going far back into his mathematical considerations and drawing the readers' attention away from the weakest point of his theory. Moreover, he would invent new mathematical proofs in order to show that his formulas were correct, and would present this for the most part in a rather confused manner. On the basis of all these experiences, von Terzaghi finally gained (and he was not alone) the following impression of Fillunger's intellectual personality: "Fillunger connects good mathematical training with a typical defect in physical talent, and with a distinct defect in scientific self-criticism. He does not have any imagination of the possible differences between the physical reality and the abstract assumptions on which the mathematical description of the problem is based. With this, the apparent contrast between Fillunger's success in the field of the elasticity theory and his failure considering physical problems is explained. If the starting point for a thinking process is uniquely given as, for example, in the elasticity theory, every mathematically trained person is able to calculate further in a correct way. Fillunger was in this position when he developed his theoretical papers in the fields of the elasticity theory and the strength of materials. If, however, the formulation of the approach requires physical understanding, he moves away from the reality caused by his purely abstract ideas, and does not return to the reality. This was the case in the development of his uplift theory and was repeated in his criticism of the consolidation theory. Moreover, if Fillunger does not understand the ideas of another researcher, he declares them as false and then invents, in the purely abstract field, complicated proofs for his unfavorable judgement and he does not change his view, notwithstanding the most obvious experimental proof to the contrary. In the course of the last decade these described oddities of Fillunger's scientific activites have raised another point, namely the emphasis of his own importance by mockery and disparagement of persons who express other views. This point is the source of his attacks against other scientists 'to free science from harmful elements'."

This deep analysis of Fillunger's scientific personality was the main point in von Terzaghi's memorandum.

While von Terzaghi's life during the affair, from December, 1936, through February, 1937, can easily be followed by reading von Terzaghi's diary, little is known regarding the course of Fillunger's life during this time. Of course, his daily life too was filled with efforts to defend his polemical pamphlet and further attacks. Already in the council of professors, as has been mentioned, he had applied for a discussion over von Terzaghi's teaching program. This request was, however, refused. Also, in the questioning, he heavily defended his own polemical statements and attacks on von Terzaghi's scientific work. After the appearance of the defending treatise from von Terzaghi and Fröhlich (1937), Fillunger worked intensively on the brochure in question. The Dean Professor Wolf, chairman of the investigation committee, told Fröhlich that Fillunger had, by January 21, 1937, realized his fault concerning his attacks against the mathematical treatment ("Terzaghi-differential" and approximative procedure) adopted in the book by von Terzaghi and Fröhlich (1936). Moreover, Wolf reported that Fillunger had been totally bewildered. As has already been mentioned, after the decisive meeting of the investigation committee on January 30, 1937, Fillunger made such a noise in Salinger's office that students gathered on the floor. In the evening, Fillunger formulated a letter to the Rector which, however, he did not mail:

"

30. Januar 1937

Euer Magnifizenz!

Die Schrift von Dr. Fröhlich belehrte mich, daß meine Anwürfe unbegründet waren. Ich bitte alle die ich beleidigt habe um Verzeihung.

Unterschrift"

"

January 30, 1937

Euer Magnifizenz!

The paper by Dr. Fröhlich taught me that my offensive remarks were unfounded. I beg all those whom I have hurt to pardon me.

signature"

On the same evening, he destroyed all those letters which supported him in the affair.

During the next weeks, Fillunger nevertheless worked on the aforementioned preliminary reply to the defending treatise by von Terzaghi and Fröhlich. After he had handed it over to the members of the investigation committee, he revised the manuscript in February. Obviously, he was gradually realizing that he had been wrong in some parts of his polemical pamphlet *Erdbaumechanik?*. The strain on his nerves became too much and he wasted away. Gradually, he came to believe that the only means of escaping the affair lay in committing suicide. He would otherwise be suspended from his duty at the Technische Hochschule and would lose his civil rights. However, there were obviously days when he believed he could win the fight, and on those days he worked hard on his preleminary statement, changing some parts completely.

On Saturday, March 6, 1937, Fillunger worked in his laboratory at the Technische Hochschule, as usual. At 11 a.m., he phoned his wife, Margarete, at their apartment in the Messerschmidtgasse 28 in Vienna. One can only guess that someone had informed him that the investigation committee had come to a final decision, voting to support von Terzaghi on all points. After the telephone call, Mrs. Fillunger rushed out of the apartment, completely dazed, returning only after a while.

Around noon, Fillunger came home from the Technische Hochschule, changed his clothes, and left the apartment again at 2 pm, in great agitation. His wife followed some minutes later, herself also very excited, shouting: "I must hold him back".

In the late afternoon, the couple came back to their apartment, which was located in an apartment building with four other apartments. The Fillungers had lived in a very quiet residential area, characterized by two-story buildings with small front gardens, for three years. They lived in great harmony, enjoying a happy marriage. Mrs. Fillunger had shared all of the professor's scientific work and was also intensively involved in the polemic conflict. Both made the grave decision to commit suicide.

Both prepared the suicide very carefully. Mrs. Fillunger cancelled the daily milk delivery from the shop. This was not surprising, as the couple had often spent weekends outside Vienna. Moreover, Mrs. Fillunger deposited the key for the apartment with the caretaker. That evening, they wrote ten farewell letters and, in addition, Professor Fillunger wrote his last will and testament. Among others, there were farewell letters to the Rector, to his co-worker Dr. Ježek, and to the police. In the farewell letter to the Rector, he included a letter which he had already written on January 30, 1937, and which has been cited earlier:

" 6. III. 1937

Die wieder erwachte Hoffnung schwand neuerlich, immer stärker befällt mich das Gefühl meiner Schuld, die ich nur damit zu mildern vermag, daß ich versichere, schon am 30. I. alle Briefe, die mir zuzustimmen scheinen, vernichtet zu haben."

"The again awakened hope has disappeared recently, more and more am I struck by the sense of my guilt, which I might be able to weaken only by the assurance that I had already destroyed all letters which seem to agree with me on January 30."

In the farewell letter to his co-worker of many years, Dr. Ježek, he thanked him for his cooperation:

" Lieber Doktor Ježek!
Denken Sie nicht allzu unbillig an ihren langjährigen Vorgesetzten, dem Sie immer ein treuer Helfer waren. Ich danke Ihnen für alle Freundlichkeit
Fillunger.
Leider war ich taub gegen Ihr Abraten."

" Dear Doktor Ježek!
Do not think too unfairly of your superior of many years, to whom you were always a true co-worker. I thank you for all your kindness
Fillunger.
Unfortunately, I was deaf to your warning."

The letters to the police by both Paul and Margarete Fillunger are particularly deeply moving:

"Meine Verblendung ist von mir gewichen – der gute Glaube, in dem ich noch vor nicht allzu langer Zeit zu sein glaubte, besteht nicht mehr. Schwere Angriffe auf einen anderen fordern Sühne, die ich nur selbst geben kann. Meine überaus brave Gattin will mich nicht allein sühnen lassen.
Wir bitten uns vor der Beerdigung den Herzstich geben zu lassen."

"Die große Hingabe und Liebe zur Wissenschaft die mein armer Mann stets nur als sein Höchstes hielt führte uns durch einen wissenschaftlichen Irrtum in den Tod. Unsere Ehre kann nur so gerettet werden und uns der Glaube geschenkt werden, daß nur der edelste Gedanke und die ehrlichste feste Überzeugung der Wahrheit zu dienen zu diesem wissenschaftlichen Schritt Berechtigung gab."

"My blindness has left me – the good faith, in which I believed myself to be in not too long ago, does not exist. Heavy attacks upon another person require penance which I can only give to myself. My extremely brave wife will not let me expiate alone.
We ask for the heart-stab before the funeral."

"The great devotion and love for science which my husband always held as his most precious possession, led us, due to a scientific error, into death. Our honor can only be saved in this way, and the belief be credited to us that only the noblest idea and the most honest strict conviction to serve the truth justified to this scientific step."

Mr. and Mrs. Fillunger put the letters on the desk in Professor Fillunger's study room. They put their door bell out of use by winding fabric around it and prepared their bathroom. They brought in an arm-chair and sealed the bathroom door with newspapers. In addition, they fixed a warning outside on the bathroom door: "Bitte kein Licht machen!" ("Please do not turn on the light!"). Both changed their clothes, dressing themselves totally in black. They took soporifics and opened the gas tap.

On Monday, March 8, 1937, at 8:30 a.m., a carrier called to deliver pictures which had been sent by Fillunger's son, Erwin, who lived in Munich. Because no one opened the housedoor in the Messerschmidtgasse 28, the caretaker, who had the key, let the carrier into the apartment. The caretaker, who was very well acquainted with Fillunger's habits, noticed at once that the newspaper was lying unread at the door in the same shape as when she had put it into the door slot.

1. Auflage

Das Echo

Zehn Groschen

Nr. 55 Montag, 8. März 1937 Jahrgang 4

Wissenschaftlicher Streit als Selbstmordmotiv

Die Tragödie des Hochschulprofessors Fillunger

Fig. 4.17. Report on the suicide of P. and M. Fillunger

Moreover, the door to one of the four rooms was open and on the desk lay, in great disorder, numerous galley proofs, money and several letters. The paralysing silence was ominious, and when the caretaker and her accompanier saw the sealed bathroom door, they had a hunch that dreadful things must have happened in the apartment. When they courageously opened the door they saw the dead Margarete Fillunger sitting in the arm-chair, and a lifeless Paul Fillunger lying collapsed on the floor. He had obviously fallen from his stool, and he lay with his head close to the feet of his wife.

The suicides of Margarete and Paul Fillunger were a great scandal in Vienna. All of the newspapers reported on this event during the following days, describing the facts and searching for the motives. Professor Karl von Terzaghi was upset when he was informed of the suicides, as he declared in the "Neues Wiener Journal".

The expert opinion paper was completed on April 9, 1937, and consisted of three sections: I. Befund, II. Gutachten, III. Begründung des Gutachtens (I. Findings, II. Opinion, III. Foundation of the opinion), also containing four appendixes. In the first section, the history of the affair was briefly sketched and the main points of the accusation were stated:

"Herr Fillunger nennt in seiner Streitschrift die Theorie der Herren Terzaghi und Fröhlich einen Unsinn und stellt die zu ihrer bautechnischen Anwendung erforderlichen Laboratoriumsversuche als eine Unmöglichkeit hin. Er stellt einen eigenen theoretischen Ansatz auf, dem er aber keine praktische Lösungsmöglichkeit zuerkennt."

"Mr. Fillunger, in his polemic pamphlet, states that the theory of Mr. Terzaghi and Mr. Fröhlich is nonsense and denotes the laboratory tests as an impossibility. He introduces his own theoretical approach, which he cannot prove in practice."

In the second section, the expert opinion was formulated. At the very beginning of this section, it was stated that Fillunger was competent as a critic only in the theoretical field. Moreover, it reproached him for overlooking, in his treatise *Erdbaumechanik ?*, the theoretically important fact that, for the special dynamic problem which plays a decisive role for the time dependent settlement of clay layers, an initial value-problem had to be solved. The committee was of the opinion that this could only be performed generally with linear differential equations. Concerning the kind of representation in von Terzaghi's and Fröhlich's book, the committee came to the conclusion that Fillunger's criticisms were partly justified. However, the poor representation was an unimportant factor and did not play any important role in the further investigation.

The ideas of von Terzaghi and Fröhlich were often misunderstood by Fillunger. This, one can suppose, was partially the reason for the factual mistakes and contradictions in Fillunger's brochure.

Even in the theoretical field, the committee stated that Mr. Fillunger was not right in view of the statements which referred to the subject. Even worse were his attacks against the laboratory tests and the applications in foundation engineering, which did not hold either.

The investigation committee remarked, furthermore, concerning the pamphlet *Erdbaumechanik ?*: "This treatise does not consider the usual fundamental rules for scientific publications which are characterized by the setting up of clear statements and by subsequent proofs. Moreover, the literature is only partially considered; also imperfections concerning the rendering of ideas and texts can be found." Although these points of criticism are also valid for the book by von Terzaghi and Fröhlich (see Appendix C of the expert opinion) and, in particular, for other papers written by von Terzaghi, the investigation committee did not mention this fact in the opinion.

In the third section on the foundation of the opinion, the committee discussed the first and second sections in Fillunger's brochure. It was recognized by the committee that, in the second section of Fillunger's treatise, new contributions were contained which could be used to treat the same problem in another way. However, with the assumption of quasi-stationary flow, one essentially obtained no other differential equation than that of von Terzaghi and Fröhlich (The calculation was contained in Appendix A). Appendix A was based on the calculations by Professor Flamm, who had used in his contribution important investigations by Professor Lechner and Dr. Heinrich, a co-worker of Professor Lechner. Flamm pointed out that those terms, in Fillunger's general system of field equations, which contain the porewater pressure must be set right. However, Flamm's "corrections" consisted only of a simplification of Fillunger's general balance equations of momentum and mass towards the geometrically-linear theory.

The committee then considered the third section of Fillunger's pamphlet, which contained the integration of von Terzaghi's differential equation and some conclusions. The committee recognized that one could derive the integral formula without any neglect, in a shorter way than Fröhlich had done. Appendix B yielded the exact derivation. Moreover, Appendix B weakened all attacks from the third section of Fillunger's treatise. However, in Appendix B the committee criticized the kind of representation of the integration process contained in von Terzaghi and Fröhlich's book.

In the fourth section of his brochure *Erdbaumechanik?*, Fillunger made the most severe errors. The committee showed this in detail in Appendix C of the opinion paper. In his pamphlet, Fillunger had given the impression via a self-invented and false formula which he denoted as the "Terzaghi-differential", that differential calculus was violated in the book by von Terzaghi and Fröhlich. Fillunger stated that the pore number ϵ used by von Terzaghi and Fröhlich was a useless quantity. This quantity is defined as:

$$\epsilon = \frac{e}{c}\,, \tag{4.4.2}$$

where c means the volume of the solid substance in a clay prism of the length l and the cross-section "unity", and e the corresponding pore volume. Thus,

$$l = c + e \tag{4.4.3}$$

or

$$e = l - c\,. \tag{4.4.4}$$

From (4.4.2) and (4.4.4),

$$\epsilon = \frac{1}{c}l - 1 \tag{4.4.5}$$

is obtained. For the compression of a clay cylinder in the oedometer, it is assumed that

$$c = \text{const.} \tag{4.4.6}$$

Thus, the pore number ϵ is connected with the direct measurable quantity l in a very simple way, see (4.4.5) and (4.4.6). The volume fraction n of the pores, which is defined by:

$$n = \frac{e}{l} \tag{4.4.7}$$

leads with (4.4.4) to

$$n = 1 - \frac{c}{l}\,. \tag{4.4.8}$$

From (4.4.2), in consideration of (4.4.6),

$$d\epsilon = \frac{de}{c} \tag{4.4.9}$$

is gained. Fillunger meant, in his brochure, that this formula was identical with the Terzaghi-differential which was constructed by him,

$$d\epsilon = \frac{dn}{1-n}\,. \tag{4.4.10}$$

One can reformulate (4.4.9) by setting, according to (4.4.7),

$$e = nl \tag{4.4.11}$$

and (4.4.4)

$$c = l - e$$

$$c = (1 - n)l \,. \tag{4.4.12}$$

Thus, Eqn. (4.4.9) can also be written as

$$d\epsilon = \frac{d(nl)}{(1 - n)l} \,. \tag{4.4.13}$$

In his brochure, Fillunger stated that, for $l = 1$, Eqns. (4.4.9) and (4.4.10) coincided. As one can see from the reformulation (4.4.13), such a substitution is only permitted for

$$l = const. \tag{4.4.14}$$

In this case, the quantity l can be eliminated and it is not necessary to set l equal to "one". However, l is a variable quantity, as was also mentioned by Fillunger in another place in his brochure. Then, from (4.4.12), in consideration of (4.4.6),

$$dl = d(nl) \tag{4.4.15}$$

is obtained. Therefore, Eqn. (4.4.13) can also be written as follows:

$$d\epsilon = \frac{dl}{(1 - n)l} \,. \tag{4.4.16}$$

Moreover, Eqn. (4.4.15) yields

$$(1 - n)dl = ldn \tag{4.4.17}$$

or

$$\frac{dl}{l} = \frac{dn}{1 - n} \,. \tag{4.4.18}$$

With this relation, from (4.4.16)

$$d\epsilon = \frac{dn}{(1 - n)^2} \tag{4.4.19}$$

is obtained.

From (4.4.2), (4.4.11), and (4.4.12), it follows

$$\epsilon = \frac{n}{1 - n} \,. \tag{4.4.20}$$

From this formula, Eqn. (4.4.19) is immediately gained by differentiation. Since the relation (4.4.9) leads with the correct calculation to the correct formula (4.4.19), the labelling of (4.4.9) as the "Terzaghi-differential" is unfounded.

In Appendix C, the investigation committee completely adopted the arguments which Dr. Fröhlich developed in the rebuttal, *Erdbaumechanik und Baupraxis: Eine Klarstellung*. The committee severely criticized, however, also in Appendix C, the representation of the theory in the book by von Terzaghi and Fröhlich (1936):

"Allerdings muß auch gesagt werden, daß in dem Buch 'Theorie der Setzung von Tonschichten', trotzdem es den Untertitel 'Eine Einführung in die analytische Tonmechanik' führt, der Leser mit der Porenziffer ϵ das erste Mal durch den folgenden Satz auf S. 24 bekannt gemacht wird: 'Unter der Porenziffer ϵ versteht man bekanntlich das Verhältnis des Porenvolumens n zum Volumen des festen Endstoffes $(1 - n)$ einer gegebenen Bodenprobe'. Gemeint ist die durch Formel (1) dieser Beilage gegebenen Definition, die sich auch in der Form (16) schreiben läßt. Der angeführte Satz ist eine mißglückte Formulierung dieses Gedankens und führt leicht dazu, den Leser zu verwirren."

"It must, however, also be said that although the book 'Theorie der Setzung von Tonschichten', carries the subtitle 'An introduction to analytical clay mechanics', the reader becomes acquainted for the first time with the pore number ϵ through the following sentence on p. 24: 'By the porenumber ϵ one understands, as everybody knows, the fraction of the pore volume n to the volume of the solid final matter $(1 - n)$ of a given soil sample'. However, the definition given by the formula (1) in this Appendix is meant, and which can also be written in the form (16). The aforementioned sentence is an unsuccessful formulation of this idea and easily confuses the reader."

Concerning Fillunger's criticism of the experimental investigations of von Terzaghi and Fröhlich, the committee refused all points of the attack and stated that the tests methods of von Terzaghi's laboratory were correctly developed. Moreover, the time dependent settlement curve, based upon the relationship between the pore number and the grain-to-grain pressure, was in agreement with calculation and observation, as had been recognized during the committee members' visit to von Terzaghi's laboratory.

Section V of Fillunger's polemical pamphlet treated the construction technique of soil mechanics and, in particular, that of the theory of consolidation. The investigation committee stated: "Since Fillunger considers the theory as false, he also declares the application of the theoretical results in practice as impossible." The committee showed that, with the results of von Terzaghi's and Fröhlich's theory, an engineer was able to draw the right conclusions from bore profiles in order to recognize instinctively the possibility of settlements, and

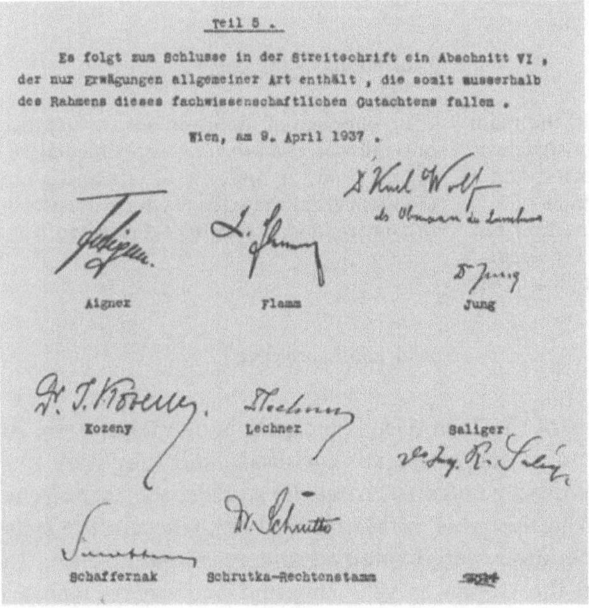
Teil 5 .

Es folgt zum Schlusse in der Streitschrift ein Abschnitt VI , der nur Erwägungen allgemeiner Art enthält , die somit ausserhalb des Rahmens dieses fachwissenschaftlichen Gutachtens fallen .

Wien, am 9. April 1937 .

Aigner Flamm Jung

Kozeny Lechner Saliger

Schaffernak Schrutka-Rechtenstamm

Fig. 4.18. The expert opinion signed by the members of the investigation committee

to prevent later surprises. Moreover, the committee stated that this remarkable advance had not been considered by Mr. Fillunger. In Appendix D of the opinion paper, a list of research institutes from all over the world working in the field of soil mechanics was included, which was enough to reveal that Fillunger's criticism concerning construction techniques was unjustified.

The final section of Fillunger's brochure was not discussed by the committee due to the fact that it contained only general reflections which lay outside the framework of the expert opinion.

The opinion was signed by the members of the investigation committee on April 9, 1937 (see Fig. 4.18), and subsequently put under seal. Only individuals having an authorized interest in the entire case were permitted to read the opinion.

With the suicide of Paul Fillunger and the signed experts opinion, the fight with Karl von Terzaghi was essentially decided. How far this conflict went on to influence the further development of the porous media theory will be discussed in the next section.

4.5 The Further Development of the Viennese Affair and in Soil Mechanics

"...; die reine Theorie hat auf dem Gebiet der Erdbautechnik trotz zweihundertjähriger Bemühungen der Hauptsache nach versagt und eine gesunde Entwicklung dieses Wissenzweiges kann nur in empirisch-wissenschaftlicher Richtung, durch eine Darlegung der vielseitigen Bedingtheit der Erscheinungen angebahnt werden."

Karl von Terzaghi (1925a)

"Das Gegenteil von Theorie ist nicht die Praxis, sondern die Beobachtung, die Experimentalforschung. Und das Gegenteil von Praxis ist nicht die Theorie, sondern die Wissenschaft, die aus theoretischer und experimenteller Forschung zusammengesetzt ist."

Paul Fillunger (1935)

"...; the pure theory has failed in the main point, in the field of earth construction, despite the efforts of two centuries, and a sound development of this branch of science can only be initiated in an empirical-scientific direction through the representation of the many-sided conditionality of the phenomena."

Karl von Terzaghi (1925a)

"The opposite of theory is not the practice but the observation, the experimental research. And the opposite of practice is not the theory, but the science which consists of theoretical and experimental research."

Paul Fillunger(1935)

Nearly all of the members of the Technische Hochschule of Vienna were in a state of shock due to the tragic end of the strange affair, and they were unable to understand the tragic deadly outcome. After the suicide, the Technische Hochschule gave a statement to the press on March 12, 1937, wherein the facts concerning the history of the affair were explained and commented upon. In particular, it was cited from the – then as yet – unpublished expert opinion paper, that the factual attacks of Professor Fillunger had had no any qualified basis.

Professor Dr. Fillunger and his wife Margarete were buried in the Schwechater cemetery on March 12, 1937. At the grave, the Rector of the Technische Hochschule spoke the farewell words, in the name of the council of professors. The funeral was attended by delegations from the Ingenieur- und Architektenverein, by the council of professors, and by Fillunger's students.

The disciplinary action was put to an end during the summer of 1937. The scientific conflict, however, continued.

In July, 1937, the revised version of *Vorläufige Erwiderung auf die Verteidigungsschrift der Herren v. Terzaghi und Fröhlich* by Paul Fillunger was edited by his son Erwin Fillunger, with the new title, *Erdbaumechanik und Wissenschaft – eine Erwiderung* (*Soil mechanics and science – a reply*). In the preface Erwin Fillunger pointed out:

"Die Schrift befand sich in der vorliegenden Form, als mein guter Vater leider zu der furchtbaren Ansicht kam, selbst im Irrtum zu sein. Durch seinen sofortigen unerbittlichen Entschluß unterblieb die Drucklegung.

Ich gebe die Schrift nun heraus, in dem Bewußtsein, damit eine ernste Kindespflicht der Dankbarkeit gegenüber dem Andenken an meine lieben Eltern zu erfüllen. Die Herausgabe erfolgt in kleinster Auflage und nur zu dem Zweck, es den Freunden meines Vaters zu ermoeglichen, auch seinen letzten Standpunkt noch kennenzulernen."

"The paper was in the existing form when my good father unfortunately came to the terrible conviction that he was wrong. Due to his immediate inexorable decision, the printing remained unfinished.

I am editing the paper now in the responsibility to fulfil with this a serious filial duty of thankfulness towards the memory of my beloved parents. The publication takes place in the smallest edition and only for the purpose of making it possible for the friends of my father to also become acquainted with his last standpoint."

This last paper of Fillunger did not reach the public and has not been cited in the literature.

At the very beginning of January, 1938, Dr. Heinrich (1938a), later also professor at the Technische Hochschule of Vienna, published a paper entitled *Wissenschaftliche Grundlagen der Setzung von Tonschichten* (*Scientific fundamentals of the theory of the settlement of clay layers*), in which he presented the findings of Professor Lechner and himself, which had been discovered within the framework of their co-operation in the investigation committee in the first months of 1937. Heinrich (1938a) pointed out in the introduction:

"Die von Herrn Professor v. Terzaghi abgeleitete der Theorie der Setzungen von Tonschichten zu Grunde liegende Differentialgleichung trägt bewußt den Charakter einer ersten Annäherung an das gestellte Problem. Die Annäherung mag in den meisten Fällen für die Bedürfnisse der praktischen Erdbaumechanik hinreichen, doch bleibt, in Anbetracht der grundlegenden Bedeutung einer den Setzungsvorgang beschreibenden Differentialgleichung, der Wunsch nach einer exakteren Behandlung offen.

Es soll im folgenden versucht werden, die Grundlagen für eine strengere Theorie der Setzung zu geben, wobei insbesondere darauf Wert gelegt werden soll, nach Möglichkeit über die gemachten Voraussetzungen genau buchzuführen. Auch soll angestrebt werden,

Wissenschaftliche Grundlagen der Theorie der Setzung von Tonschichten.

Von Dr.-Ing. Gerhard Heinrich, Wien.

Die von Herrn Professor v. Terzaghi abgeleitete, der Theorie der Setzung von Tonschichten zu Grunde liegende Differentialgleichung[1]) trägt bewußt den Charakter einer ersten Annäherung an das gestellte Problem. Die Annäherung mag in den meisten Fällen für die Bedürfnisse der praktischen Erdbaumechanik hinreichen, doch bleibt, in Anbetracht der grundlegenden Bedeutung einer den Setzungsvorgang beschreibenden Differentialgleichung, der Wunsch nach einer exakteren Behandlung offen.

Es soll im folgenden versucht werden, die Grundlagen für eine strengere Theorie der Setzung zu geben, wobei insbesondere darauf Wert gelegt werden soll, nach Möglichkeit über die gemachten Voraussetzungen genau buchzuführen. Auch soll angestrebt werden, die sich als notwendig erweisenden Vernachlässigungen womöglich einer numerischen Abschätzung zugänglich zu machen.

Schon Professor Dr. P. Fillunger hat in seiner Schrift „Erdbaumechanik?"[2]) darauf hingewiesen, daß es sich bei der Setzung in Verbindung mit der Porenwasserströmung um eine gekoppelte Bewegung zweier Phasen, einer festen und einer flüssigen, handelt. Dies setzt allerdings voraus, daß der zwischen der Festsubstanz des Tones befindliche Porenraum ganz mit Wasser ausgefüllt sei. Die hier gewählte Behandlungsweise schließt sich, was die grundsätzliche Form der aufgestellten Gleichungen betrifft, an die in der erwähnten Schrift enthaltenen Darstellung an.

Wie bei Prof. Fillunger so sollen auch hier die Bewegungsgleichungen für jede der beiden Phasen getrennt aufgestellt werden. Die tatsächlich auftretenden Bewegungen sowohl der festen als auch der flüssigen Phase sind äußerst verwickelt und von der zufälligen Konfiguration der Körner (resp. Blättchen oder Stäbchen) abhängig, so daß die vorliegende Aufgabe den Charakter eines statistischen Problems besitzt und als solches behandelt werden muß. Die eingeführten Größen sind also grundsätzlich statistische Mittelwerte. Dies setzt voraus, daß die gewählten Elementargebiete (Volumselemente) eine gewisse Größe nicht unterschreiten, so daß auch in diesen Elementen noch mit Mittelwerten gerechnet werden darf. Andrerseits müssen sie natürlich klein sein in bezug auf die Gesamtabmessungen der betrachteten Tonschichte.

[1]) Siehe „Theorie der Setzung von Tonschichten" von Dr. K. v. Terzaghi und Dr. O. K. Fröhlich, erschienen bei Franz Deuticke Leipzig und Wien.
[2]) Erschienen im Selbstverlag des Verfassers.

Fig. 4.19. Heinrich's first paper on the porous media theory

die sich als notwendig erweisenden Vernachlässigungen womöglich einer numerischen Abschätzung zugänglich zu machen.

Schon Professor Dr. P. Fillunger hat in seiner Schrift, 'Erdbaumechanik?' darauf hingewiesen, daß es sich bei der Setzung in Verbindung mit der Porenwasserströmung um eine gekoppelte Bewegung zweier Phasen, einer festen und einer flüssigen, handelt. Dies setzt allerdings voraus, daß der zwischen der Festsubstanz des Tones befindliche Porenraum ganz mit Wasser ausgefüllt sei. Die hier gewählte Behandlungsweise schließt sich, was die grundsätzliche Form der aufgestellten Gleichungen betrifft, an die in der erwähnten Schrift enthaltenen Darstellung an."

"The differential equation developed by Professor v. Terzaghi, which is the basis for the theory of the settlement of clay layers, carries consciously the character of a first approach to the problem raised. The approach may be sufficient in many cases for the calculation of practical soil mechanics. Due to the fundamental importance of a differential equation describing the settlement process, the demand for a more exact treatment remains, however, open.

In the following it will be attempted to give the fundamentals for a stricter theory of the settlement, whereby in particular a high value shall be set on bookkeeping exactly over the introduced assumptions if it is possible. It shall also aim to make the necessary approximations available to a numerical estimation.

Professor Dr. P. Fillunger has already pointed out in his paper 'Erdbaumechanik?' that the settlement in connection with the porewater flow is a matter of a coupled motion of two phases, a solid and a liquid. This assumes, however, that the pore space between the solid substance of the clay is totally filled with water. The treatment chosen here follows the representation in the mentioned paper, regarding the fundamental form of the advanced equations."

Moreover, in the conclusion to his paper, Heinrich (1938a) stated:

"Es wird versucht, eine Behandlung des Problems der Setzung von Tonschichten auf wissenschaftlicher Grundlage durchzuführen, ..."

"It is attempted to realize a treatment of the problem of the settlement of clay layers on a scientific basis, ..."

This statement corresponds to Fillunger's question (see Fillunger, 1937) in his response to the defending paper from von Terzaghi and Fröhlich (1937), namely whether von Terzaghi's differential equation could be scientifically founded. Professor Flamm was very indignant about Heinrich's comments. Flamm had already visited von Terzaghi on January 24, 1938, in order to discuss Heinrich's paper. In the next days, Flamm (1938) wrote a treatise which he called *Beitrag zur Theorie der Setzung von Tonschichten* (*Contribution to the theory of the settlement of clay layers*). In the second half of February, 1938, Flamm discussed the manuscript with von Terzaghi. The manuscript appeared in May, 1938, in the same journal as Heinrich's paper. In the introduction, Flamm (1938) cited Heinrich's aforementioned statements and commented angrily with:

"Beim Leser könnte dies leicht den Eindruck hervorrufen, als wären die Rechnungen v. Terzaghis wissenschaftlich nicht einwandfrei. Die folgenden Zeilen sollen dazu dienen, eine solche Auffassung als unbegründet zu widerlegen und auch zu anderen Ausführungen des Herrn Dr.-Ing. Heinrich klärend Stellung zu nehmen."

"For the reader this could easily give the impression that the calculations by v. Terzaghi would not be scientifically correct. The following lines shall determine that such an opinion is unfounded, and will clarify also the other statements from Mr. Dr.-Ing. Heinrich."

With Flamm's (1938) paper, the scientific conflict (Fillunger-von Terzaghi) was finished.

The reader may put the question as to whether or not Fillunger had a chance to overcome the disciplinary action as an innocent person and to win the fight against von Terzaghi. The answer is an absolute "no".

Fillunger lived in a very small world of his own; his whole life - schooling, studies, as well as his occupational career - was confined to the city of Vienna. He was very deeply involved in his scientific research, which he very much overrated, and had a certain arrogance, in short, he was a know-it-all who was always busy teaching others and all too often did not accept any other opinion.

In contrast to Fillunger, von Terzaghi was an internationally known personality, a cosmopolitan, having made international trips to many countries, e.g., Russia, USA, and Turkey. Science was not the only field which occupied him; he was very busy as a consultant, which in turn gave him the opportunity to interact with more people. He was a very perceptive person, having the ability to sharply analyze problems as well as to recognize the important issues. Furthermore, he was very tolerant, charming, and easy-going in his dealings with people which, in turn, gave him an attractive personality.

Von Terzaghi was much better equipped for the controversy compared to Fillunger. He used his aforementioned capabilities to build up his defense in which he won over the "right" people, who later promoted his ideas in the committee.

Fillunger took the technical criticism personally and was not able to further advocate his ideas to the committee through his representatives. As has already been discussed, von Terzaghi's supporters dominated the inquiry committee; the leading person there was Professor Flamm. He pinned the members down to concentrate on Fillunger's factual offensives and to discuss only von Terzaghi's consolidation model and not other models, such as Fillunger's. At no stage of the entire investigation was the question raised as to whether von Terzaghi's differential equation could be scientifically proved, as suggested by Fillunger. Instead of concentrating on this main point, the committee was more involved in petty matters such as Fillunger's criticism of von Terzaghi's and Fröhlich's statements and mathematical details, which were not of as great importance. Surprisingly, none of the members of the committee mentioned that von Terzaghi's differential equation was valid only for one-dimensional consolidation and that the derivation of this equation was inadequate. Likewise, it is also very surprising that no one perceived (or wanted to perceive) that Fillunger had founded a general mechanical theory for saturated porous media. It appears, in some respect, perfidious that Fillunger's basic ideas and his field equations served to show that von Terzaghi's differential equation was equal to Fillunger's field equations. In the expert opinion, no hint appeared that von Terzaghi had derived his differential equation by an analogous conclusion and not from the fundamental balance equations of mechanics. Moreover, in order to reveal the "equality" of both approaches, Flamm manipulated Fillunger's field equations by setting the volume fractions constant, which is only approximately valid within the geometrically-linear theory. He maintained, in the Appendix A of the expert opinion:

"Die in (22) and (23) (the two equations of motion for the fluid and solid phases, the author) den Porenwasserdruck w enthaltenden Glieder müssen aber richtig gestellt werden; die Gleichungen haben zu lauten:"

"The terms containing the porewater pressure w in (22) and (23) must, however, be set right; the equations have to read":

In Appendix A, the manipulated equations of motion followed after this remark. With this scandalous manipulation, Flamm obviously pursued two purposes: first, only with his manipulated equations could he derive von Terzaghi's differential equation; second, with his manipulation he could bring Fillunger into disrepute. It is not known how far Flamm's procedure was arranged with von Terzaghi.

Clearly, it seems as if the representative from the field of the mechanics in the inquiry committee had not paid attention to Flamm's procedure and let themselves be misled by Flamm. It seems as if they had a guilty conscience, otherwise the prompt publication of Heinrich's (1938a) paper at the very beginning of 1938, in which he criticized, in a somewhat hidden form, von Terzaghi's scientific working, cannot be explained. In his paper, Heinrich simplified, in a very correct form, Fillunger's field equations to the geometrically-linear case, thus arriving at the manipulated formulas of Flamm. Flamm (1938) wrote, in his rebuttal, that Heinrich was wrong on many points and that Heinrich's merit would consist of setting right Fillunger's balance equation. Thus, he used the same words as in the Appendix A of the expert opinion in 1937; from this, we can conclude that Flamm's manipulation of Fillunger's equations does not seem to be a slip. This can be stressed by the point that Flamm visited von Terzaghi at least two times to discuss his reply to Heinrich's paper.

After the bizzare affair, there was a rumor, according to the relatives of Fillunger, that von Terzaghi was requested by Fillunger to stop the disciplinary action. Von Terzaghi, however, completely rejected Fillunger's request.

How the strange Fillunger-von-Terzaghi-affair was seen by the late Ruth D. Terzaghi, the second wife of Karl von Terzaghi, can be read from a statement in a letter to the author from December 13, 1989:

"I believe that Arthur Casagrande gave the matter (the Fillunger-von-Terzaghi-affair, the author) such slight attention because no one in this country took the matter seriously. Fillunger was regarded as, at best, a freak and, at worst, as a psychotic."

As has already been pointed out, Fillunger's brilliant findings were nearly totally ignored in the literature, a sad fact which was partially enforced by the disciples of von Terzaghi. For example, within the framework of the edition of the book, *From theory to practice in soil mechanics* (Bjerrum, Casagrande, Peck, and Skempton, 1960), Professor Skempton sent a first draft of his article *Significance of Terzaghi's concept of effective stress* to Professor Arthur Casagrande, where he mentioned Fillunger's contribution, referring to Fillunger's paper from 1915 (see Section 4.1) on the discovery of effective stresses, with the words:

"... yet neither he nor anyone else at the time realized the full significance of these results."

A. Casagrande answered Skempton very angrily in a letter of June 9, 1959:

"I am still not happy about the manner in which you bring Fillunger into the picture. I don't know whether you have ever read Fillunger's booklet which he published for the purpose of discrediting all that Terzaghi has done. Not only was he scientifically completely wrong but it was such a viscous and uncalled for attack in which he ridiculed the principles of soil mechanics that were established by Terzaghi. You are still giving him credit where no credit is due, by stating: 'This amounts to a corollary of the principle of effective stress', even though you go on to say that 'neither he nor anyone else at the time realized the full significance of these results'. A reader may gain the impression that in a vague way Fillunger had understood the mechanics but had not yet realized the full significance. The fact is that he denounced as nonsense Terzaghi's concept of pore pressures and effective stresses. The least I would ask you to do is to delete the word 'full' in the sentence '.. yet neither he nor anyone else at the time realized the full significance of these results'."

In a letter to A. Casagrande of June 22, 1959, Skempton agreed:

"Regarding your comments on my revisions, I will certainly omit the word 'full' in the sentence to which you refer..."

Indeed, in the final form of the book in question, the word "full" was omitted.

The Fillunger-von-Terzaghi-affair greatly influenced the further development of porous media theories and, in particular, the development of soil mechanics.

Fillunger's work was continued by Heinrich. In the aforementioned paper (Heinrich, 1938a), Fillunger's concept (1936) was consequently used for the investigation of the one-dimensional mechanical behavior of a liquid saturated porous solid. Moreover, in remarkable contributions during the 1950s and 1960s,

Heinrich extended these investigations, on the basis of Fillunger's fundamental equations, to ground-water flow and three-dimensional consolidation problems (Heinrich and Desoyer, 1955, 1956, and 1961). Heinrich's work will be discussed in Section 4.7.2. Unfortunately, Heinrich's work has only been mentioned by a few authors in the literature and has not had a decisive influence on the development of porous media theories.

Von Terzaghi, who always had, from the beginning of his scientific career, a skeptical view of any theory (see the remark at the beginnings of this section) obviously completely stopped his theoretical research in the field of porous media. After his brilliant success in discovering and explaining fundamental mechanical effects in saturated porous solids, as for example, capillarity, uplift, and the effect of effective stresses (see Section 4.1), he proceeded essentially to practical problems in the field of soil mechanics and foundation and dam constructions. Although von Terzaghi and Fröhlich announced, in their defending paper (von Terzaghi and Fröhlich, 1937), that a discussion of Fillunger's theory would be published elsewhere at the right time, and although von Terzaghi worked again on the uplift problem during March, 1937, no contributions to these theoretical problems were published. One can only surmise that von Terzaghi was tired of doing theoretical research work.

Obviously, von Terzaghi, who was not very well-trained in either mechanics or mathematics, did not dare to fight on the theoretical field against Fillunger, who was known in the German-speaking countries as a theoretical scientist with well-founded knowledge in mechanics. He left it to Fröhlich to refute those of Fillunger's attacks which had been made on theoretical ground. It may be that the confrontation with attacks from a theorist, had convinced him to completely stop scientific work in the theoretical field. In any case, after leaving Vienna in 1938, he promoted only one student at Harvard University, where he had taken over a visiting lectureship. From that point on, no substantial contribution by von Terzaghi to the theory of the multicomponent soil body is known. All of his further contributions to soil mechanics were devoted to the practical, engineering-scientific side of soil mechanics. He was also of the opinion that laboratory work and the theoretical aspects of soil mechanics had already been brought to a close, in particular by himself. In the opening address at the Second International Conference on Soil Mechanics and Foundation Engineering (Rotterdam) in 1948, von Terzaghi pointed out:

"Every science, pure or applied, is based on what is known as fundamental research. Prior to 1936 fundamental research in soil mechanics consisted chiefly in the investigation of the significant soil properties by laboratory tests, and in the development of theories of earth pressure, stability and settlement. In 1936, this pioneering stage of soil mechanics research was already completed. It had created what may be called an ideal pattern of soil behavior and it had placed at the disposal of the practicing engineer a set of theoretical concepts covering every important field of soil behavior. The concepts were based on the laws of applied mechanics and on the results of laboratory tests performed under rigidly controlled conditions on soil samples which were believed to be almost undisturbed".

Moreover, in the outlook of his opening address, he stressed his standpoint:

"The days in which significant discoveries could be made in the laboratory, or at the writing desk, appear to be gone forever".

His view, and his criticism on the theoretical penetration of soil mechanics, were summarized in a somewhat ironic tone in a letter to a Mr. Harding (England) in 1952:

"I greatly enjoyed your occasional reference to the mathematical mind. I would define a mathematician as a man who can solve every equation but he may fail to notice it when he displaces a decimal point. Once I assigned to a mathematically minded engineer, of high academic standing, the following task. He should estimate the flow of water into an open excavation for the different stages of excavation to bedrock at a depth of about 100 ft. He promptly set up the differential equations which were rather involved and one week later, full of pride, he presented the results. According to the results of his computations – which were correct provided you endorsed the assumptions – the inflow is a maximum at a depth of about 60 ft. As excavation proceeds beyond this depth the inflow decreases and becomes very moderate at the final stage. 'Fine', I said. 'Now you work out a procedure for excavating down to 100 ft without passing through that confounded intermediate stage in the proximity of 60 ft'. Grand fellows: I have suggested repeatedly that we should keep them in cages and feed them their problems through the bars. If there is no danger that they may get out they can be quite useful.

Once every few months I get a manuscript full of differential equations describing a revolutionary discovery in the realm of theoretical soil mechanics, for review and comments. I have composed a sort of standard letter for my reply and the manuscripts are tucked away in a fat file labeled 'Nut House'."

If one reads these statements carefully, it is not surprising that, for the creation of a theory of saturated and non-saturated porous solids, scientists in the field of soil mechanics have contributed little – with only few exceptions – due to the fact that von Terzaghi had an eminent influence on the development of soil mechanics. In particular in the USA, von Terzaghi was, and still is in certain circles, a legend. Von Terzaghi's view is in some ways hard to understand because it was completely clear to him that a saturated porous solid was in a certain sense a mixture; this view coincides with modern standards (von Terzaghi, 1925b). However, von Terzaghi did nothing to transform his ideas into theory. Moreover, his disciples – in particular, Arthur Casagrande – adhered to von Terzaghi's viewpoint. Thus, one can understand that, in the time that has followed, many branches of soil mechanics have drifted into a direction of a purely experience-science, with no scientific foundation of its results. Of course, it should be emphasized that soil, as a natural material, is very complex, and thus in general is not open for an exact mechanical-mathematical treatment. However, that is not the point. Rather, a theory always deals with ideal materials and serves to recognize the assumptions and limits of any calculation of practical problems. This criticism is directed against the attempt on the part of representatives of soil mechanics to hinder the scientific investigation on the basis of well-founded mechanical axioms and mathematical rules.

The main efforts, after the fundamental findings of von Terzaghi, were directed towards carrying out the ideas on soil mechanics which had been developed by von Terzaghi and his followers. At the beginning of the 1930s, the idea was born to establish National Committees in soil mechanics and foundations. In 1935, von Terzaghi stated in a letter to A. Casagrande:

"Ich machte mich sofort an die Durchsicht Ihres Entwurfes betreffend die Boden-Konferenz in Cambridge ... Bei der ersten Lesung vermisse ich nur den Schritt vom Individuum zur Organisation. Der gegenwärtige Entwurf enthält bloss eine 'Application Form' für einzelne Teilnehmer. Wenn ich recht verstehe, so ist der Vorgang wie folgt: Zunächst werden die National-Kommittees ernannt und dann erst wird die Flugschrift ausgesendet, so dass die Namen der Mitglieder der National-Kommittees bereits in der Flugschrift enthalten sind. Es frägt sich nun auf welchem Weg die National-Kommittees ins Leben gerufen werden sollen. Ich glaube in verschiedenen Ländern wäre ein verschiedener Vorgang am Platz. "

"I immediately took a look at your draft concerning the soil conference in Cambridge ... At first glance I only miss the step from the individual to the organization. The current draft contains only an 'Application Form' for the individual participants. If I understand it in the right way, the procedure is as follows: at first the national-committees are nominated and only then will the pamphlet be mailed so that the names of the members of the national-committees are already contained in the pamphlet. It would be doubtful as to which way the national-committees should be called to commence. I believe in different countries there should be different procedures."

On October 9, 1935, A. Casagrande answered. He proposed that von Terzaghi should decide the way in which to form the national-committees in Austria, Sweden, France, and Russia. For the remaining countries – in particular, Germany, Italy, Egypt, etc. – he himself would arrange the formations. The efforts to establish national-committees continued with much success in the years following. Finally, A. Casagrande stated in a letter to von Terzaghi of February 1, 1938:

"Ich moechte Sie zu dem ausgezeichneten Erfolg Ihrer Bemuehungen zur Organisation von National-Komittees begluecktwuenschen. – Somit koennen wir die folgenden Laender als geloest betrachten: England, Frankreich, Italien, Schweiz, Oesterreich, U.S.A., Mexico, China, Aegypten, Russland. In den folgenden Laendern koennen wir uns darauf verlassen, dass, so bald wir uns etwas darum bemuehen, ohne weiteres die Organisation zustande kommt, da geeignete Maenner vorhanden sind: Holland, Daenemark, Schweden, Finnland, Estland (vielleicht koennte man ein nordisches Kommittee gruenden in welchen Daenemark, Schweden, Norwegen, Estland und Finnland vertreten sind?), Ungarn, Niederlaendisch-Indien. Dazu wahrscheinlich Canada, Brazilien, Australien und Polen. Es bleiben somit nur noch der komplizierte Fall Deutschland uebrig, und dann die Tschechoslovakai und einige Laender wie Tuerkei, Sued-Afrika, Neu-Seeland, Palaestina, etc.

Der naechste notwendige Schritt waere, dass Sie das Executive Committee der Permanent International Organization ernennen. Laut Resolution No. 3 soll dieses Komittee ausser Ihnen und mir noch aus drei Mitgliedern bestehen. Fuer diese drei Mitglieder schlage ich vor: Proctor, Bretting, Mayer-Peter oder Buisman. Deutschland ist indirekt durch Sie und mich vertreten, offiziell aber nicht, so dass Frankreich und England sich darueber nicht aufregen koennen.

Das Internationale Komittee (Punkt 'd', Resolution No. 3) soll von den einzelnen National-Komittees erwaehlt werden und es ergibt sich die Frage wie viel Mitglieder jedes Land waehlen darf oder soll."

"I would like to congratulate you on the outstanding success of your efforts for the organization of national-committees. – Thus we can consider the following countries as settled: England, France, Italy, Switzerland, Austria, U.S.A., Mexico, China, Egypt, Russia. In the following countries we can be assured that, if we make an effort soon, an organization could be set up without difficulties because there are qualified men: The Netherlands, Denmark, Sweden, Finland, Estonia (perhaps one could found a Nordic committee in which Denmark, Sweden, Norway, Estonia and Finland act?), Hungary, The Dutch Indies. In addition, probably Canada, Brazil, Australia, Poland. Thus only the complicated cases remain, Germany and Czechoslovakia, and some countries like Turkey, South-Africa, New-Zealand, Palastine etc.

The next necessary step will be to nominate the Executive Committee of the Permanent International Organization. According to the Resolution No.3, this Committee should consist of three members, besides you and me. For these three members I suggest: Proctor, Bretting, Meyer-Peter or Buisman. Germany is represented indirectly by you and I, however, not officially so that France and England cannot be upset about this.

The international committee (point 'd', Resolution No.3) should be elected by the individual national-committees, and the question arises as to how many members each country is allowed to elect or should elect."

The above-mentioned resolution, No. 3, was adopted by the participants of the Harvard conference after some discussions at the end of the last meeting on June 25, 1936.

In the letter to von Terzaghi from February 1, 1938, A. Casagrande announced further steps:

"Der naechste Schritt wuerde dann die Gruendung einer Internationalen Gesellschaft fuer Erdbaumechanik sein. Die Mitgliedsbeitraege sollte man gering halten um Mitgliedschaft allen ernstlich interessierten zu ermoeglichen. Die geeignetste Verwendung der Mitgliedsbeitraege waere meiner Meinung nach fuer eine vierteljaehrliche 'Soil Mechanics Review' in mehreren Sprachen. Ich stelle mir die Sache so vor, dass entweder Sie in Wien, oder ich hier in Cambridge, die Schriftleitung inne haben und man fuer jede der wichtigsten Sprachen mindestens einen Mitarbeiter gewinnt der eine gute Kenntnis der Erdbaumechanik hat und dem die entsprechende Literatur leicht zugaenglich ist. Er wuerde die wichtigsten Veroeffentlichungen in dieser Sprache kurz zusammenfassen oder die entsprechenden Verfasser dazu veranlassen so etwas selbst zu schreiben. Diese Beitraege wuerden an die Schriftleitung gesandt werden, wenn notwendig noch weiter gekuerzt, geordnet und fuer eine Nummer zusammengestellt, und dann an die einzelnen Schriftleitungen fuer die verschiednen Sprachen gesandt, die fuer die Uebersetzung und den Druck zu sorgen haben. Fuer den Druck sollte man einen billigen und raschen Prozess (so wie unsere Proceedings, fuer kleine Auflagen selbst mimeographirt) waehlen. Im Anfang wird man sich mit zwei oder drei Sprachen begnuegen. Im Laufe der Zeit glaube ich, dass englische, franz., deutsche, spanische und schwedische (oder daenische) Auflagen erscheinen werden. Die Veroeffentlichung der englischen Auflage koenn- te von der hiesigen Zentrale der Internationalen Gesellschaft uebernommen werden und die uebrigen Auflagen muessten von den National-Komitees verwaltet werden. Die Finanzierung stelle ich mir so vor, dass jedes Mitglied der Internationalen Gesellschaft das Recht hat die Review in einer der Sprachen fuer einen Bezugspreis von, z. B. $ 3.00 per Jahr zu beziehen, waehrend fuer Nichtmitglieder der Preis beispielsweise das Doppelte betragen wuerde. Die Auflage in einer neuen Sprache wuerde erst dann unternommen werden, wenn genuegend Mitglieder sich bereit finden sie zu abonnieren. Wenn notwendig, kann der Bezugspreis fuer verschiedene Sprachen verschieden hoch sein. Den Mitgliedsbeitrag selbst koennte man mit $ 1.00 festlegen. Was man von diesen Beitraegen erspart kann man als Zuschuss zur Veroeffentlichung der Proceedings der naechsten Konferenz verwenden.

Ich wuerde Ihnen dankbar sein wenn Sie alle diese Fragen ueberlegen wuerden und mir Vorschlaege fuer Aenderungen und Ergaenzungen mitteilen wuerden. Sobald wir uns geeinigt haben wuerde ich ein ausfuehrliches gedrucktes Rundschreiben an alle Mitglieder der ersten Konferenz senden in dem all das enthalten ist, sowohl auch die Statuten schon bestehender National-Komittees und Mitgliedschaft (als Vorbild und zur Aneiferung andrer

Laender). In dieser Schrift wuerde ich die Mitglieder bitten zu den Vorschlaegen Stellung zu nehmen, um dem Executive Committee die Moeglichkeit zu geben, die Anschauungen der Mitglieder zu beruecksichtigen, bevor die endgueltige Organisation der Internationalen Gesellschaft durchgefuehrt wird."

"The next step will be the formation of an International Society for Soil Mechanics. The membership fee should be kept low in order to allow all seriously interested people the membership. The most suitable use of the fees is in my opinion for a quarterly 'Soil Mechanics Review' in several languages. I suggest that either you in Vienna, or I here in Cambridge should do the editorial work and that one should gain, for each of the most important languages, one co-worker who has a good knowledge of soil mechanics and to whom the corresponding literature is easily accessible. He would briefly summarize the most important papers in this language or oblige the corresponding author to write the summary by himself. These contributions would be sent to the editorial board, further shortened if necessary, sorted and put together for one issue and then sent to the individual editorial boards, which would take care of the translation and the print. For printing, one should choose a cheap and quick process (like our Proceedings, for small numbers of copies "mimeographirt"). At the beginning, two or three languages will be sufficient. Over the course of time I believe that English, French, German, Spanish and Swedish (or Danish) editions will appear. The publication of the English edition could be taken over by the local center of the International Society, and the remaining editions will have to be managed by the national-committees. I suggest financing this in such a way that each member of the International Society would have the right to receive the Review in one of the languages for a price of, e. g., $ 3.00 per year, whereas for non-members the price, would for example be the double. The edition in a new language would only then be undertaken when a sufficient number of members had declared their intention to subscribe to it. If necessary, the price could be at different levels for different languages. One could fix the membership fee at $ 1.00. The amount which can be saved can be used as a contribution to the publication of the proceedings of the next conference.

I would be grateful to you if you would think over all of these problems and would report to me proposals for changes and supplements. As we have agreed I will send an extended printed circular letter to all members of the first conference which will contain all this, as well as the regulations of already existing national-committees and membership (as a model and stimulation for other countries). In this treatise, I will ask the members to comment on the proposals in order to give the Executive Committee the opportunity to consider the opinions of the members before the final organization of the International Society is carried out."

In order to publicize and make known their standpoint on soil mechanics, the von Terzaghi-school came up with another idea, to write books on the subject. In a letter of November 3, 1936, written to the von Terzaghi family, A. Casagrande stated:

"Mr. Hamilton war vor einigen Tagen hier und erzaehlte mir von den Abschiedsstunden in New York. U.a. habe ich ihm ueber das neue Schmarotzertum in der Bodenmechanik geklagt, worauf er richtig bemerkte: 'Umso wichtiger, dass wir unsere Buecher herausbringen'. Das ist wohl die einzige Loesung."

"Some days ago, Mr. Hamilton was here and told me of the 'goodbyes' in New York. Among others, I have complained to him about the new parasitism in soil mechanics, whereupon he correctly remarked: 'It is important that we edit our books'. This is probably the only solution."

Indeed, as is well-known in soil mechanics, von Terzaghi wrote two remarkable books on theoretical and practical soil mechanics in 1943 and 1948, the last one with Professor Peck, a former co-worker of his. Moreover, the von Terzaghi-school did not hesitate to revile other scientists who did not share their views on soil mechanics. For example, in 1936, Professor Housel from the University of Michigan, an already-recognized scientist in the field of soil mechanics, had his own ideas which did not coincide with those of the von Terzaghi-school. At the 1936 conference, the reports over his large work were restricted by the organizing committee to a summary of less than five pages, and his talk to ten minutes. Housel complained bitterly about this situation:

"Ten minutes is far too short a time to present completely one's position on one or several controversial points which have been the subject of discussion during the meetings of this Conference. Consequently I have attempted to formulate as briefly as possible in written form some of my own reactions which have been accumulating during the discussion of the past several days and which are clamoring for expression too insistently to be ignored for my own peace of mind."

Housel spent much of his time after the 1936 conference attempting to discredit various of von Terzaghi's concepts.

Von Terzaghi did not react very much to these attacks. It was A. Casagrande who carried the banner higher than the leader. With ironic and sometimes rough words, he attacked several researchers: In 1936, for example he wrote:

"A propos, Mr. Knappen hat mich auch besucht und mir voller Stolz erzaehlt, dass er Secretary des Executive Committee der neuen Division der Am. Soc. Civ. Eng. ist. Wenn man ihn reden hoert, wuerde man annehmen dass er die Bodenmechanik mit Schoefloeffeln gefressen hat. Er und Philippe schreiben zusammen an einem Buch! Die neue Division funktioniert auch wie ich erwartet habe. Nun haben sie auch gluecklich ein Committee on the 'Classification of Soils, Terms and Definitions in Soil Mechanics' mit W.P. Kimball, unserem gemeinsamen guten Freund, ausgerechnet Kimball, als Chairman. Er eignet sich natuerlich sehr gut dazu, denn er ist ja einer von jenen der in seiner Beschraenktheit alles standardisieren will. Ich koennte mir meine letzten Haare ausraufen, dass ich mich doch dazu ueberreden liess Mitglied in der Society zu werden."

"A propos, Mr. Knappen has also visited me and has told me with pride that he is the Secretary of the Executive Committee of the new division of the Am. Soc. Civ. Eng. If one hears him talk one would assume that he has eaten soil mechanics with a ladle. He and Philippe are writing a book together! The new division works as I have expected. Now they fortunately also have a committee on the 'Classification of Soils, Terms and Definitions in Soil Mechanics' with W.P. Kimball, our common good friend, as the only chairman. He is of course very much qualified for that because he is one of those who will standardize everything in his stupidity. I could kick myself that I allowed myself to be persuaded to be a member of this Society."

Professor Burmister, a famous professor at Columbia, was another person who was adversely affected. In a letter to von Terzaghi from May 31, 1937, A. Casagrande pointed out:

"Ich glaube, dass weder Hogentogler's noch Plummer's Buch viel Schaden anrichten werden, weil sie in einer Art geschrieben sind, dass jeder der etwas Verstand hat, sehen muss wie wertlos diese Lektuere ist. Umso gefaehrlicher ist meiner Meinung das Buch, das Burmister bei Wiley (!) veroeffentlichen wird. Ich habe Hamilton rueckhaltlos meine Meinung ueber Burmister ausgedrueckt. Burmister hat aber sehr viel 'pull' in New York durch seine Kollegen an Columbia, die ihn sehr unterstuetzen und ihn fuer ein Genie halten, sowie auch bei den New York Consulting Engineers fuer die er viel arbeitet. Es ist wirklich ein Jammer dass so ein grossmäuliger Prahler sich in New York so breit gemacht hat. Solche wertlose Arbeiten wie sie Burmister und Krynine in den letzten Monaten in den Proceedings A.S.C.E. (American Society of Civil Engineers, the author) veroeffentlicht haben, werden ohne weiteres vom publication committee akzeptiert, waehrend eine ausgezeichnete Arbeit, die Taylor von M.I.T. eingereicht hat, abgewiesen wurde. Es wuerde mich wirklich interessieren zu wissen, wer die Spezialisten sind, an die sich das Publications-Committee um Rat wendet. Unter solchen Umstaenden wird es mir natuerlich nie einfallen, in den Proceedings etwas zu veroeffentlichen. Man kann es ja nicht einmal wagen, eine ehrliche Kritik von Krynine's Arbeit zu schreiben, weil sie es nie veroeffentlichen wuerden und damit selbst eingestehen, was fuer einen Bloedsinn von Ihnen angenommen wurde."

"I believe that neither Hogentogler's nor Plummer's book will cause much damage because they are written in such a manner that anyone who has some intellect must see how useless these books are. The more dangerous is in my opinion the book which Burmister will publish with Wiley (!). I have expressed without reserve my opinion of Burmister to Hamilton. However, Burmister has a lot of 'pull' in New York through his colleagues at Columbia, who support him greatly and who consider him as a man of genius, as well as from the New York Consulting Engineers for whom he works so much. It is really a pity that such a big-mouthed braggart has spread himself so far in New York. Such valueless works as Burmister and Krynine have published in the Proceedings A.S.C.E. in the last months have been accepted without further ceremony by the publication committee, whereas an outstanding work which Taylor of M.I.T. submitted, was refused. It would be really interesting to know who the specialists are that the publication committee seeks for advice. Under such circumstances it will, of course, never occur to me to publish something in the Proceedings. One cannot dare by any means to write an honest criticism of Krynine's work, because they would never publish it, and thus confess what an idiocy they have taken on."

A. Casagrande was not the only follower who tried to carry the Harvard point, although he was the leading person among the other three important adherents, namely, the professors R. Peck from the University of Illinois, Urbane, Illinois, A.W. Skempton from the Imperial College of Science and Technology, London, and L. Bjerrum, Director of the Norwegian Geotechnical Institute, Oslo. Their efforts to reveal von Terzaghi's dominance in the field of soil mechanics culminated in the publication of the book *From Theory to Practice in Soil Mechanics* (Bjerrum, Casagrande, Peck, Skempton, 1960). In the preface, the four disciples described the purpose of the book:

"Its purpose is to present a detailed account of Terzaghi's personality, his professional achievements, and his method of working. To the greatest possible extent the essence of his contributions is presented in his own words. When set forth in orderly sequence and supplemented by explanatory comments, Terzaghi's writings and reports show clearly how he proceeded, step by step, first to establish the fundamental principles of soil mechanics and then to use them as powerful tools in his engineering practice. Hence the contents of this book provide an ideal means to demonstrate, especially to young engineers, the prerequisites and techniques for successfully practicing soil mechanics. This belief was not only one of the incentives for the preparation of the book but it was also responsible for the selection of the title."

In the late 1940s and early 1950s, it was recognized by several scientists that the Harvard-school was dominating soil mechanics and that the views of other researchers were put down by von Terzaghi and his disciples, and above all by A.

Casagrande. These scientists were concerned that the Harvard-school was dominating the soil mechanics profession, especially A.S.C.E. (American Society of Civil Engineers), and controlling the publication of papers. Moreover, it was rumored that Professor Taylor of M.I.T. was never promoted to full professor allegedly due to von Terzaghi's intervention. It was also rumored that von Terzaghi tried to have Professor Tschebotarioff fired from Princeton. The story was always the same. In order to be successful, one had to have attended Harvard and had to be obedient to von Terzaghi's views. It was then believed that every paper in soil mechanics submitted to the A.S.C.E. had to have von Terzaghi's approval to be published. However, one must state that there is no evidence that von Terzaghi personally prevented publication. On the other hand, the late Professor A. Casagrande was instigating the censorship in von Terzaghi's name. He suppressed several outstanding pieces of work.

The Terzaghi-school, by no means, allowed anyone in the field of soil mechanics to dispute von Terzaghi's right to stand as the founder of soil mechanics. In the beginning of 1962, an article appeared in the "Oesterreichische Ingenieur-Zeitschrift" in which a statement was quoted to the effect that O.K. Fröhlich was, together with von Terzaghi, the founder of soil mechanics. A. Casagrande reacted immediately. He persuaded Dr. Golder from Toronto, Canada, who was known as a brilliant writer with a special sense of humor, to write a letter to the leading journal in soil mechanics, *Geotechnique*, because he was able to "give to such a letter just the right touch of sting". Dr. Golder wrote such a letter to the Editor of Geotechnique, with the title "Claims of Fatherland", not mentioning, however, the name Fröhlich. In the last two passages he stated:

> "For the sake of future historians of the subject, and as there will assuredly arise many kings who knew not David, cannot we who know him agree that there is no ambiguity of paternity in this case, although the identity of the mother remains in doubt unless this be fecund mother earth.
>
> To continue my metaphor, the *conception* of soil mechanics was, in my opinion, the paper 'Die Berechnung der Durchlässigkeitsziffer des Tones aus dem Verlauf der hydrodynamischen Spannungserscheinungen' in 1923, the *birth* was 'Erdbaumechanik auf bodenphysikalischer Grundlage' in 1925 and the *coming of age* was 'From Theory to Practice in Soil Mechanics' in 1960. One man alone was responsible for this achievement, and no one could be happy who claimed to share it with him."

Furthermore, after the death of von Terzaghi, his disciples did not hesitate to turn him into a legend. On the occasion of the 150th-anniversary of the Technische Hochschule of Vienna in 1965, A. Casagrande donated a bust of von Terzaghi to the Technische Hochschule. This bust was not unveiled, however, before the renaming of a lecture-hall as the "Terzaghi Lecture-Hall" on March 31, 1967. This was a big event at the Technische Hochschule and, in the evening, the Rector gave a reception for von Terzaghi's closest relatives living in Europe at that time.

A. Casagrande also made great efforts to establish the "Terzaghi Lecture". A street was even rechristened "von Terzaghigasse".

In 1969, a memorial tablet was unveiled at von Terzaghi's birthplace, and further, at the house in Graz, a plaque reminds all to this day of the founder of modern soil mechanics.

In 1975, there was a conference in Vienna. At this conference, the efforts to praise von Terzaghi culminated in this speech by his wife:

"I believe that Karl's personality was accurately characterized by our good old friend of the thirties Professor Fritz Haas who said that Karl was a renaissance man. It was in the renaissance period that men first saw earth and sky in all their fresh and unexplored splendor, unclouded by tradition, myths or superstition. It was during those years of change and development that they first dared to give more credence to the witness of their own eyes than to the believes of past generations. Instead of consulting traditional authority they began to ask questions of nature. Although this liberation spread only slowly throughout the fabric of intellectual life and indeed is not yet complete, its impact was revolutionary, wherever it took place and men responded with excitement and with joy, as the domination of the unseen world began to fade and the appreciation of the pleasures of the real world increased. For the first time in perhaps a thousand years at least a select few could expect to enjoy life for its own sake. In retrospect it appears that renaissance man stood on a divide, liberated from the scholasticism of the past but not yet dominated by the machines to which later generations would become addicted. Perhaps it was this freedom both from the burdens of the past and from those of the future that gave such an exceptionally large number of individuals the leisure as well as the creative urge to develop a broad spectrum of talents and interests. This was I believe an ambience that Karl would have found very congenial. He too occupied that divide to which man had ascended from the foggy valleys dominated by myth and from which they had not yet descended to a slavish addiction to gadgets. Karl had neither time nor interest for such distractions as radio, TV or automobiles. But throughout much of his life he found the leisure for capturing his delightful impressions of a landscape or a seascape with pencil and water color and he never failed to respond with appreciation and enthusiasm to the attainments of others in a wide variety of fields including literature, painting, music and last but by no means least the culinary and the wine making arts. He was born to doubt the validity of tradition and authority and he began by rejecting both the military school and the military career to which tradition assigned an officer's son. He felt this was a condemnation rather than a good tradition. His interest in Darwin's theory of evolution, which was then still anathema to many of his elders nearly led to his expulsion from high-school. Still later he was classified as a dangerous rebel who actually wished to substitute fact for fancy when he firmly rejected most of the rules of thumb which passed for engineering science throughout much of his career. In all these activities he shared with his renaissance forbears a zest for the world around him and a boundless joy in discovery."

4.6 Biographical Notes

Paul Fillunger (1883–1937)

Paul Fillunger, born in Vienna on June 25, 1883, came from a family where several members had already worked in the field of engineering. His father was involved in the development of the so-called "chain-bridge". One of his brothers worked as an engineer, while another worked as a consular employee

Fig. 4.20. Paul Fillunger (1883–1937)

in Germany. Only distant relatives lived in Vienna later, with the exception of one cousin. The family belonged to the Roman Catholic church.

Fillunger successfully completed the four classes of the junior high school in Vienna from 1894/95 until 1897/98, and the three classes of the upper high school in the k.k. (kaiserlich, koeniglich) state-upper-high school in Linz from 1888/89 until 1900/01. He passed his final examination with honors. In 1901, he enrolled at the faculty of mechanical engineering of the k.k. Technische Hochschule of Vienna. After studying for four semesters, he did some practical studies in 1903. In the same year, he passed the 1st examination (Staatsprüfung) with the degree "capable" and, after doing further practical studies, he passed the 2nd examination in 1904 (Staatsprüfung) with the degree "very capable".

In 1906, Fillunger entered the private Austria-Hungarian State Railway Company, where he stayed until 1910. During this period, he had some free-time for his hobbies. For example, in 1907 he passed the official examination for becoming an engine-driver and a tank-attendant. Moreover, he taught himself differential and integral calculus with enthusiasm, using E. Czuber's book. In addition, he widened his knowledge in the field of strength of materials, guided by books by C. Bach, *Elastizität und Festigkeitslehre*, by L. v. Tetmayers' textbook with the same title, and by G. Kaiser's textbook, *Construction der gezogenen Geschützrohre*. In 1910, Fillunger took over a teaching position at the Technologisches Gewerbemuseum in Vienna and, in 1913, he was awarded with the title k.k. Professor; his teaching areas were mathematics, mechanics, and mechanical engineering. His career was interrupted in 1915 when he was

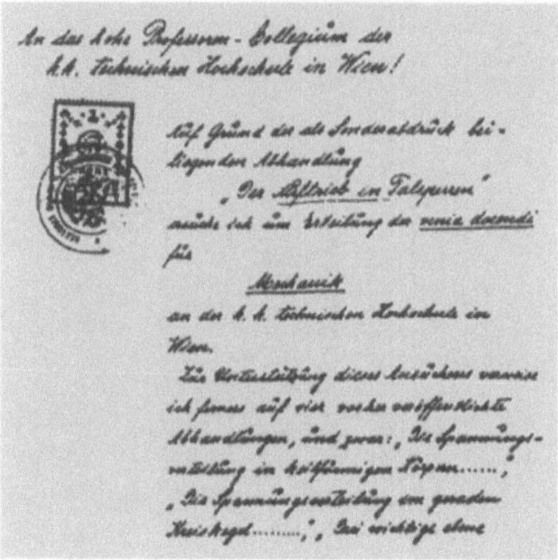
An das löbl. Professoren-Collegium der k.k. Technischen Hochschule in Wien!

Auf Grund der als Sonderabdruck beiliegenden Abhandlung „Der Auftrieb in Talsperren" ersuche ich um Erteilung der venia docendi für

Mechanik

an der k.k. Technischen Hochschule in Wien.

Zur Unterstützung dieses Ansuchens verweise ich ferner auf vier von mir veröffentlichte Abhandlungen, und zwar: „Die Spannungsverteilung in keilförmigen Körpern......", „Die Spannungsverteilung im geraden Kreiskegel.........",

Fig. 4.21. Fillunger's application for his habilitation

called up as a "Landsturmingenieur" (a position in the army). In the last years of World War I, he did work which provided to be useful at the "Fliegerarsenal" in Aspern. At that time, Fröhlich and von Mises (a famous mathematician) were also engaged at the "Fliegerarsenal". After World War I, Fillunger was appointed chairman of the laboratory of the Technologisches Gewerbemuseum. Fillunger worked successfully there and, in September 9, 1920, he was honoured with the title "Baurat". In 1923, he was offered the chair of Technical Mechanics at the Technische Hochschule of Vienna, which he subsequently accepted; he received the title o. ö. professor (full professor). However, he was only able to take over the chair, and not the laboratory, which was incorporated in the chair of mechanical technology (Prof. Dr. Ludwik).

Fillunger started his scientific career in 1908, when he received his doctorate degree at the Technische Hochschule of Vienna. The title of his dissertation work was: "Ein Versuch, die Spannungsverteilung in keilfoermigen Koerpern auf theoretischem Wege zu finden" ("An attempt to find the stress distribution in wedge-shaped bodies in a theoretical way"). Fillunger's achievements were honoured with "excellence". In 1914, Fillunger handed in some papers to the Technische Hochschule of Vienna in order to obtain the venia legendi, namely papers on the stress distribution in wedge-shaped bodies and, as a habilitation work, the already-mentioned treatise on the uplift problem from 1913. The expert Prof. Dr. Wighardt pointed out that Fillunger was correct to declare his paper on the uplift problem as his habilitation work because this treatise was the most important. Moreover Wighardt wrote: "... he (Fillunger) has enough of his own ideas, ... Occasional unskillfulness and small mistakes come certainly from the fact that the general mathematical training of the author did not occur on a broad basis". Furthermore, Wighardt remarked that the wide field of mechanics was open only to those persons who had had a solid mathematical training. Therefore, Fillunger only received the venia legendi for the narrower field of elasticity theory and strength of materials, and not for the intended field of mechanics. The colloquium took place on December 7, 1914, at 4 p.m., and referred to Airy's stress function, to some important questions regarding the strength of materials (in particular, to the elastic hysterese), to the relation of this to the fading of metals, and to purely elastic occurrences. The habilitation committee voted unanimously to allow Fillunger to give a test presentation. Fillunger made three proposals: "Theorie der Festigkeit von Zughaken" ("Theory of the strength of tension hooks"), "Die ebenen Probleme der Elastizitäts-Theorie" ("The plane problems of the theory of elasticity"), and "Kurzer Abriß der Potentialtheorie und das Problem der Ausbreitung der Kräfte" ("Brief survey of the potential theory and the problem of the propagation of forces"). The committee chose the last mentioned title and, at 4 p.m. on January 15, 1915, Fillunger delivered his presentation. The committee was satisfied and the habilitation procedure was concluded. In the next years, Fillunger published several remarkable papers and his reputation grew steadily. Thus, it was quite natural that he was offered a professorship at the Deutsche Technische Hochschule of Brno. However, he refused this offer and took over the aforementioned o. ö. professorship at the Technische Hochschule of Vienna. In the laudatory report, the search committee pointed out that Fillunger connected theory and experiment in an excellent way. In the time following, Fillunger was quite busy with teaching and with the organization of his chair. Thus, it was quite natural that his scientific output in the form of papers was a little poor. This changed, as has been pointed out in the foregoing sections, when his polemical disputes started with several scientists at the end of the 1920s and the beginning of the 1930s. Moreover, in 1932, he was appointed as a dean.

Paul Fillunger married Margarete Gregoritsch, a young woman from Graz, on May 31, 1908. Their only son, Erwin Fillunger, was born on February 22, 1909. It seems that Paul and Margarete Fillunger lived in happy matrimony. That Margarete Fillunger decided to commit suicide simultaneously with her husband indicates that the Fillunger's were not only deeply connected emotionally, but also that Mrs. Fillunger was also intensively interested in her husband's profession as a professor, as well as in the scientific problems with which he was concerned. Everyone who was closely related to the couple stated that there could not be a more ideal marriage than that of Professor Fillunger and his wife. One of Fillunger's cousins pointed out that Mrs. Fillunger's husband was always her highest priority and that she always believed deeply in him. The partners were never separated, except during the time that Professor Fillunger was on duty during World War I. The dean of the Technische Hochschule of Vienna also spoke, in his farewell address, of their ideal matrimony during their last thirty years together. The couple were always tenderly connected, so much so that it seems unlikely that Mrs. Fillunger would have survived the death of her husband.

The residents of the quiet, two-floor house in the "Messerschmidtgasse im achtzehnten Bezirk" in Vienna, in which the couple lived in a middle-class, simply furnished, four-room apartment, spoke of the kind and discreet character of the couple.

However, several relatives reported that at home Fillunger was at times a "know-it-all", as in his scientific life. Towards his son, he was correct and stern.

On March 7, 1937, Professor Fillunger and his wife committed suicide.

Karl von Terzaghi (1883–1963)

Karl von Terzaghi, whose full name was Karl Anton Terzaghi Edler von Pontenuovo, was born on October 2, 1883, in Prague, at that time the capital of the Austrian Province of Bohemia. He was the descendant of a long line of Austrian professional men and army officers who lived in Lodi, Lombardy. His father, Antonius de Padua Petrus Terzaghi (1839–1890), was stationed as a commander of an infantry battalion in Prague. He had earned great merits in the terrible battle of Solferino (1859). Therefore, he was rewarded with the title "Edler von Pontenuovo". In 1882, Antonius Terzaghi, at that time in the position of "k. k. Major", married Amalie Philippine Eberle (1853 - 1942) in Graz. Her father, Karl Andreas Eberle (1823 - 1916), a tobacco factory owner, later became Karl von Terzaghi's guardian. In 1885, his sister, Gabriele Anna Johanna Nep. Terzaghi von Pontennovo, known later as Ella, was born in Prague.

Karl von Terzaghi's father retired in 1887 and moved to Graz, beside Meran, the preferred place for retired officers. Unfortunately, he did not enjoy his retirement for very long, for he died in 1890. Karl von Terzaghi's maternal grandfather, Karl Eberle, seems to have been happy to act in loco parentis. He was a wealthy man. In 1846, he had finished his studies at the Technische Hochschule of Vienna and, until the 1870s, he was in the duty of the Austrian tobacco administration. Then, he organized, as a general manager of an international bank syndicate, the production of tobacco in Romania.

As a child, von Terzaghi might perhaps have been described as either "mischievous" or "troublesome". According to the custom of the time, he was destined for a career in the Austrian army. When he was ten, he became a pupil at the military-Unterrealschule in Güns, with the objective of applying later for admission to the k. k. navy. When his application was refused due to a slight eye defect, he left the military school in 1898, graduated from the Landes-Oberrealschule in Graz in 1900, and enrolled at the faculty of mechanical engineering at the Technische Hochschule of Graz. In the course of his first two years, he enjoyed the academic liberties in a rather dissolute manner. He joined a bellicose fraternity, spent most of his time drinking, rioting, and duelling, and was only occasionally seen in the lecture halls of the Technische Hochschule. After he had sowed his wild oats, he spent most of his time at the Technische Hochschule, attending not only his mechanical engineering courses, but also courses in philosophy, geology, and petrography. He graduated in 1904. He had a special method for passing examinations. In the last weeks before examinations, he spent the whole time alone in his room in his grandfather's house,

Fig. 4.22. K. von Terzaghi with sister, mother and grandparents around 1897

which was located opposite the Technische Hochschule, preparing nearly day and night.

After his graduation, he worked for some months as an unpaid trainee in a machine factory in Andritz and soon realized that he would be unhappy working in the profession of a mechanical engineer. He then served one year in the army. During his watch duties, and also during the completion of his (in parts) long confinements, he translated a book on geology from English into German. Also during this time, it became increasingly clear to him that he could not spend his whole life as a mechanical engineer. His grandfather gave him permission to continue his studies for one year at the University and the Technische Hochschule in Graz, where he concentrated on geological subjects, as well as bridge and railway constructions. In the summer of 1906, he was severely injured in a mountain climbing accident in the Alps, so that he was not able to take part in a planned expedition across Greenland nor in any future expeditions. Therefore, he decided to work in the field of civil engineering, where he had the opportunity to keep in touch with his favorite field, geology.

In the fall of 1906, von Terzaghi accepted a position as a junior engineer in the concrete construction firm of Adolph Baron Pittel in Vienna. One year after his entrance into the Viennese firm, he was sent to Romania to act as an independent building supervisor in constructing plaster-silos, despite the fact that he had never seen a large building site. There he learned the elements of practical civil engineering in a very short time. After finishing the construction of the silos in Romania, he was entrusted with the construction of a hotel building in Semmering, along with the conception and construction of a small but interesting hydroelectric power development in Lower Austria. However, he was not satisfied because he was not making any progress in the field of technical geology. Therefore, after three years of affiliation with the construction firm, he applied for a position as an officer in the Dutch colonial troops. Before the negotiations came to an end, von Terzaghi was put in charge of the exploration of the preliminary hydrographic and geologic studies for a proposed large hydroelectric power development on the Gacka river in Croatia, in the hinterland of the Adriatic coast, beginning in the fall of 1909. During the two years that he worked in Croatia, he became more and more interested in the application of geology to engineering problems. In the fall of 1910, von Terzaghi completed his assignment and returned to Austria to concentrate on writing a paper on the geology and hydrology of the region which he had explored. While working on the final version of the manuscript, he learned, via a letter from a friend, of the difficulties which had been encountered in the foundation of a large bank building in St. Petersburg. Von Terzaghi offered his services, was engaged by the construction firm, and arrived three days later in St. Petersburg. Within four weeks, he had the situation under control and the project was completed in time.

His tasks in the time to follow were the organization of a branch of the firm in Riga, on the Baltic Sea, and the establishment of the construction sites there. Although life in St. Petersburg was pleasant, there was no compelling challenge in sight for von Terzaghi. Finally, his restlessness motivated him to deepen his technical-geological knowledge through personal experience with the American civil engineering practices. In December, 1910, he transferred his duties to Dr. O. K. Fröhlich, who had worked with him in Croatia. Von Terzaghi returned to Graz via Finnland and Sweden and, in January, 1912, he received his doctorate degree at the Technische Hochschule in Graz. In February, 1912, he embarked for the United States of America on the steamer "Amerika" of the Hamburg-Amerika-Linie.

Several letters of recommendation from his Austrian friends and the administration introduced him to leading American experts. In the United States, a large-scale irrigation program was about to commence, initiated by President Roosevelt in the "Reclamation Act" of 1902. It included the construction of sixty dams. The United States Reclamation Service was entrusted with the realization of the project. On the dam sites, the engineers had to deal with nearly all technical-geological difficulties which could occur in the field of civil engineering. Thus, if anywhere in the world there was an opportunity to create a broad basis for the development of technical geology, it was provided by the systematic working out of the experiences which were being collected on the building sites there. In his pursuit, von Terzaghi was kindly supported by Mr. F. H. Newell, at that time Director of the Service. For nearly two years, von Terzaghi travelled from site to site (New Orleans, New Mexico, Idaho, Arizona, California, Nevada, and Washington). Following his work with the United States Reclamation Service, he spent some time in the southwest of the United States, where he even briefly considered setting up business as a date farmer. Later, he went to Mexico, put money into a construction business and then lost everything as a result of the Mexican revolution.

He probably received financial support from his wealthy maternal grandfather and from the "Österr. Ingenieur- und Architektenverein", to whom he had applied for funds in a research proposal of March 31, 1912. Because he could not find a position as an engineer, he worked temporarily as a driller on the rapids of Big Eddy. At the end of winter 1912/13, he was injured by an explosion. He was admitted to the hospital in Portland. Here, he had time to think over his objectives and his aims to develop a rational approach to earth-work and foundation engineering. This attempt proved a discouraging failure, and he was disappointed and disheartened. An impression of his disappointment can be obtained from a letter to Professor Wittenbauer (von Terzaghi's mechanics professor in Graz) of March 6 (see Bjerrum, 1969):

> "I am experiencing the most miserable period of my entire life so far; unemployed and my mind divided and tortured by doubts. Sadly I have to confess that thoughts of home fill me with indignation, for at home they grieve over my existence in a small-minded way, and my communications with them are limited to dry and factual reports. How pitifully unimportant my security is when I have lost what I most valued, my belief that engineering is the most blessed profession, and the prevailing hypocrisy fills me with disgust ... I have broken off practically all my correspondence, because I do not wish to involve others in my depressed state of mind, and I also wish to avoid the pain inflicted by an indifferent shrug of the shoulders, with which even an old friend is brushed off under such circumstances. I have no one here, only the stubborn decision to keep going, and not to leave the United States until I have regained my old confidence in a new form."

After leaving the hospital in Portland, he became acquainted in a pub with an architect who was having difficulties with some constructions. Von Terzaghi took over the job as a designer. He used the evening hours to work up the rich material which he had collected during the last year. However, the result was still discouraging.

In December, 1913, von Terzaghi returned to Europe. Under the influence of his miserable failure, he wrote to a friend from Graz, who was working as a contractor in the development of an irrigation system in the northern part of Argentina, with the objective of applying his collected experiences in South America. However, due to a serious economic crisis in Argentina, he cancelled his plans for departure. After this, von Terzaghi was trying with two friends to start a small construction company in Vienna; then the First World War broke out. He joined the army as a reserve officer and served as an infantry officer in Croatia and Serbia. In a letter to his son, he reported later that the siege of Belgrade, and the capture of the city on October 9, 1915, had been spectacular military events which he would never forget. After the collapse of the Serbian resistance, von Terzaghi applied for a transfer to the more eventful Italian theater of war. After his request was refused, he was incorporated, with the help of an old military school friend, into the air force and soon after became the

Fig. 4.23. K. von Terzaghi in Constantinople with his students

commander of the Flugfeld in Aspern, where captured Allied military planes were patched up and tested, with the purpose of learning about any advances which the Allies might have made over their Central Powers counterparts. It was a hazardous occupation, and one of his former professors (Professor Dr. Forchheimer), deciding that it was a shame to waste a talented man in such a way, arranged to have him appointed to the faculty of the Kaiserlich Ottomanische Ingenieur-Hochschule in Constantinople. In the fall of 1916, von Terzaghi received an order from the Department of Foreign affairs in Vienna to take over a position as a professor for foundations and roads. In accordance with the order, he went to Constantinople in September, 1916.

In Constantinople, he had the time and leisure to prepare his lecture notes. This was, however, a lot of work because he had to give the lectures in French. The appointment to a teaching position gave von Terzaghi a welcome opportunity during the next two years to resume his attempts to develop a rational approach to earthwork engineering. At the end of the First World War, all members of the teaching staff affiliated with the defeated nations were dismissed, and von Terzaghi accepted a poorly paid teaching job at the American Robert College, where he took over the position of a teacher for thermodynamics and gas engines who had left the college suddenly. Although he was not well-trained in these fields, he did not hesitate to take the position. Apart from the teaching tasks, he established a small soils-laboratory and performed a long series of tests on soils.

Fig. 4.24. Karl von Terzaghi around 1916

In 1924, the Dean of the School of Engineering at Purdue University, Lafayette, Indiana (USA), Professor A. Potter, received a letter from the Dean of the American Roberts College, Istanbul (Turkey), Professor A. Scipio, describing a scientist in the department of civil engineering, von Terzaghi, who was advancing some new ideas and philosophies concerning foundation engineering. Professor Scipio wrote that von Terzaghi should come to the United States and secure a position which had been offered for research in soil mechanics and foundation engineering. The Dean of Robert College thought that von Terzaghi would accept a salary of one hundred dollars per month. However, the head of the civil engineering department at Purdue University expressed a negative attitude regarding the venture. Professor Potter transmitted the information to Dr. Stratton, the President of the Massachusetts Institute of Technology at that time, with the suggestion that M.I.T. might be interested in von Terzaghi. Dr. Stratton agreed, offered him a salary of two thousand five hundred dollars per year and, in 1925, engaged von Terzaghi as a visiting lecturer.

Von Terzaghi immediately commenced with great activities at M.I.T. One week after his arrival, he started his consulting work for the Bureaus of Public Roads. His next task consisted of authoring a popular representation of the contents of his book "Erdbaumechanik" (von Terzaghi, 1925a), which he had published right before his departure for the USA. His consulting work grew and grew and, in 1928, he was sent to Central America by the United Fruit Company, in order to investigate the permeability of primeval forest soils. After some months, he began to suffer from a painful tropical illness and, in the late fall, he returned from Quirgua, where he had been in a hospital for one month, to Cambridge via New Orleans. Two months later, von Terzaghi was requested to prove the calculation documents concerning a retaining wall of fifty five meters height on the Connecticut River. Von Terzaghi accepted, despite his weak health, because it was a chance for him to work with an earth pressure apparatus equipped with modern measurement techniques. Apart from his consulting and research activities, von Terzaghi gave lectures on soil mechanics at M.I.T. However, he started with subject matter well beyond an engineer's level, which proved to be too taxing for his students. Furthermore, he talked too fast. The dean requested that Mr. A. Ortenblad (later a student of von Terzaghi), who had studied mathematics at Harvard and was enrolled at M.I.T., to observe von Terzaghi's course and report back to him. One day the dean visited the course and asked von Terzaghi to start the course again, due to the fact that no one was able to understand the subject, and also asked him to teach more slowly, explaining in more detail. Von Terzaghi became furious and said that there must have been a "spy" in class. Finally, von Terzaghi agreed to start the course again. From then on, he became a more communicative man.

In 1928, a committee at the Technische Hochschule of Vienna for the re-appointment of the chair for Hydraulic Engineering II, demanded the establishment of a chair for soil mechanics. In a preliminary report, the committee stated:

"The relations in the field of earth work mechanics are something different. This field should be carried out, according to the opinion of the committee, at a far higher degree than up to now, and, of course, in this sense mechanics would be placed into the foreground and would be brought to the experimental-computing side, in connection with the tasks of ground construction ... The committee recommends therefore that the admirable Professoren-Kollegium may decide:

that the lecture subjects of hydraulic engineering be split into two chairs and one Honorardozentur on the occasion of the new appointment of the chair for Hydraulic Engineering II, and indeed that the chair I hydrology should consist of river construction and water power plants, and the chair II should consist of earth work mechanics, ground construction, construction realization in civil engineering and traffic water construction, ..."

In a second report, the committee listed nine scientists who were qualified for the position of the second chair. Amongst the names on this list was that of von Terzaghi. The committee informed the Professoren-Kollegium that, after a careful discussion, the committee wished to propose:

"Primo loco: Ing. Dr. Karl Terzaghi"
"Secundo loco: Ing. Dr. Ludwig Mühlenhofer"

In the recommendation, the committee mentioned that von Terzaghi was a noted authority in the field of earthwork mechanics and foundations. In Constantinople, he had already begun his basic, pioneering works in the fields of the earth pressure theory and soil physics, whose results he had summarized in his book *Erdbaumechanik auf bodenphysikalischer Grundlage* (see von Terzaghi, 1925a). The Professoren-Kollegium agreed to the committee's proposal and von Terzaghi was offered the professorship for Water Construction II.

At the beginning of October, 1929, von Terzaghi embarked from Boston on the steamer "Bremen", bound for Europe. Early in the morning of October 11, the "Bremen" arrived at Bremerhaven. Just before his departure, he was asked by the Soviet Government to prepare a report concerning the foundations for the locks of the Volga-Don Canal. In Berlin, he met the trade-attaché of the Soviet Embassy and then flew with him on Lufthansa to Moscow, where he visited the building site on October 14th.

In October, 1929, von Terzaghi was appointed as a Professor of Water Construction at the Technische Hochschule of Vienna. He was successful in his negotiations with the administration, in particular concerning his personal salary. Moreover, a committee of Austrian engineers, under the chair of Hofrat Dr. W. Exner, at that time a recognized authority in the field of engineering in Austria, agreed to bear a portion of the costs for the establishment of a soil mechanics laboratory.

Von Terzaghi began teaching in Vienna in the beginning of 1930. In the summer of 1930, he was invited by a Swedish firm to return to Russia, where he did extensive consulting work. He established a laboratory at the Technische Hochschule of Vienna. However, the economic plight of Austria hindered the completion of this project. Despite this difficulty, von Terzaghi's chair soon became the Mecca for engineers interested in earthwork engineering. Young engineers (sometimes up to ten of them) came not only from Austria, but also from Germany, Sweden, Norway, Australia, and the USA. During his first years in Vienna, von Terzaghi and his students collected numerous field observations which they had acquired at various building sites. In the summer of 1931, von Terzaghi was invited by an Italian firm to study the methods for making the foundation of a large dam in North Africa stable.

Von Terzaghi's reputation grew steadily. It was quite natural that, in December, 1932, he was asked by the "österr. Bundesministerium für Handel und Verkehr" to investigate whether the existing Reichsbrücke could be widened to double its existing width without strengthening the foundation. In his May, 1933, opinion paper, he pointed out that he could only carry the responsibility for the appointed task if certain conditions were fulfilled. Von Terzaghi's objections were not considered by the Federal Administerium and therefore the costs for the project were exceeded considerably, which led to a scandal in Austria.

After his clarification of the uplift problem, von Terzaghi again turned to some tasks of a more practical nature. In June, 1934, he got the chance to carry out a large-scale test on the capacity load of an artificially condensed sand layer, namely in connection with the foundation of the National Museum in Kracow.

In the following time, von Terzaghi continued his consulting work. In the spring of 1935, he was requested by Swedish consultants to go to Leningrad, via Stockholm, in order to investigate the failure of a part of a dam at a power

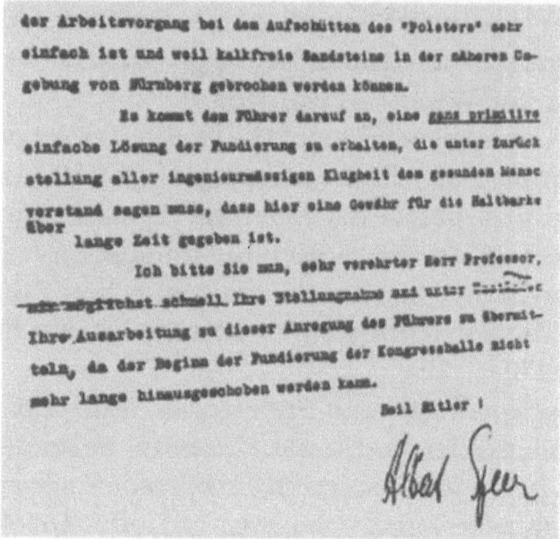

der Arbeitsvorgang bei dem Aufschütten des "Polsters" sehr einfach ist und weil kalkfreie Sandsteine in der näheren Gegebung von Nürnberg gebrochen werden können.

Es kommt dem Führer darauf an, eine ganz primitive einfache Lösung der Fundierung zu erhalten, die unter Zurückstellung aller ingenieurmässigen Klugheit dem gesunden Menschverstand sagen muss, dass hier eine Gewähr für die Haltbarkeit über lange Zeit gegeben ist.

Ich bitte Sie nun, sehr verehrter Herr Professor, mir möglichst schnell Ihre Stellungnahme und unter [illegible] Ihre Ausarbeitung zu dieser Anregung des Führers zu übermitteln, da der Beginn der Fundierung der Kongresshalle nicht mehr lange hinausgeschoben werden kann.

Heil Hitler !

Albert Speer

Fig. 4.25. Letter of Hitler's architect Albert Speer to K. von Terzaghi concerning the foundation of the Kongresshalle

plant. Von Terzaghi solved the problem in a relatively short time. In October, 1935, he was telephoned by Dr. Fritz Todt, then General Inspector of German road affairs, and asked to salvage a section of the Autobahn between Munich and Salzburg which was severely damaged by foundation failures in parts.

Some weeks later, he was again called back to Germany. This time, it was a question concerning the foundation of the Kongreßhalle for the Reichsparteitag in Nürnberg. This building was a special project of the Führer, Adolf Hitler, and it is therefore not surprising that Hitler invited von Terzaghi to Munich to report on his observations regarding the building site. Von Terzaghi was deeply impressed by Hitler's knowledge of soil mechanics and foundations.

In 1935, he applied to take leave from the Technische Hochschule for one year without salary, and he requested Dr. O. K. Fröhlich to take over his lectures during the winter semester of 1935/36 and the summer semester of 1936. The reasons for his leave-taking were invitations to give guest lectures at the Technische Hochschule of Berlin and at Harvard and an invitation to serve as President of the First International Conference on Soil Mechanics and Foundation Engineering. In the middle of November, 1935, he moved to Berlin with his second wife, whom he had married in 1930, and lectured for three months on the practical applications of soil mechanics in the presence of Dr. Todt, who thanked von Terzaghi at the conclusion of his lectures with warm words and handed over a gift of appreciation with a personal dedication.

From January 16 until January 24, 1936, von Terzaghi was on his way to the United States of America via Bremen, Germany. During the journey, he began to work out an opinion paper regarding a failure accident at a subway site in Berlin. After landing in the United States, he continued his work on the paper, first in New York and then in Boston. During the spring term, he lectured at Harvard University. His lectures were attended by an audience of around 100 people which consisted of, as in Berlin, mostly members of the building administration and the building industry. On March 28, he left Boston by plane in order to lecture at the University of Illinois, Urbana-Champaign, where he was welcomed by many old friends (he had given lectures there several times before moving to Vienna in 1929). He returned to Boston via Washington D.C., where he again visited the laboratories of the Bureau of Public Roads. In the beginning of June, he finished his talks at Harvard University and, on June 19, 1936, the First International Conference on Soil Mechanics and Foundation Engineering, which had been organized by his disciple A. Casagrande, started at the Rockfeller Center in New York. Von Terzaghi, as the president of this conference, welcomed members from nearly twenty different countries. The conference continued at Harvard University from June 22 until June 27. On September 10, von Terzaghi left New England in order to lecture in the western part of the United States. His schedule included talks at the headquarters of the U.S. Reclamation Service in Denver, Colorado, at universities in Washington, Idaho, and California, the California Institute of Technology in Pasadena, the headquarters of Californian Road Construction in Sacramento, and the American Societies of Civil Engineering in San Francisco and Los Angeles.

When he landed in Cuxhaven, Germany, on October 30, 1936, he could not foresee that he would soon be violently attacked by his colleague Professor Fillunger (see Section 4.4.), nor that this attack would paralyze his ability to work for half a year.

After the strange affair with Professor Fillunger (see Section 4.4.), von Terzaghi had to defend himself against ugly rumors concerning his role in the Reichsbrücke affair. He did this in a speech to the "Österr. Ing. und Arch. Verein" on April 30, 1937. The reaction on the part of the audience was very cool; people had not yet forgotten the tragic end of the Fillunger affair. He repeated the talk before a friendlier audience on May 10 and the next day he went on vacation to the Dalmatian coast. After his vacation, he travelled to the middle part of Sweden in order to deliver an opinion paper on a conflict between the city of Stockholm and the Swedish Government. In the middle of June, he returned to Vienna and the von Terzaghi family moved from their apartment in the center of Vienna, at Dollfußplatz 2, to Grinzing, Kahlenbergerstraße 59. His work at the Technische Hochschule was pleasantly interrupted several times by visitors from abroad.

In July, 1937, several professors, including von Terzaghi and a group of civil engineering students, travelled for one week through the southern part of Germany under the leadership of Professor Salinger, where they visited many building sites in the area of road and water construction. The excursion was generously supported by the Generalinspektor Dr. Todt.

Following the excursion, von Terzaghi's consulting activities spread steadily. He examined several landslides at the Aachensee, became a consultant of the French firm "Les Travaux Souterrains", and gave his opinion on the foundation of a power plant in Latvia in September, 1937.

Fig. 4.26. K. von Terzaghi on an excursion through the southern part of Germany

In October, 1937, von Terzaghi travelled with his wife to Paris in order to hold a speech there and to give advice to "Les Travaux Souterrains" concerning the stability of a barrage at Béni Bahdel in Algeria. At the end of October, the von Terzaghis returned to Vienna. In the two months to follow, he had enough leisure time to work out his lecture notes on soil mechanics. However, by the end of December, he was on his way to London and, in January, 1938, he was studying the first results concerning the observations of the barrage at Ghrib, a city near Algier. After his return to Vienna, von Terzaghi wrote up the final results and soon was able to deliver his opinion paper to the directorate of "Travaux Souterrains".

On March 3, 1938, von Terzaghi took over the honorable function of outlining the merits of the former President of the United States of America, Herbert Hoover, in economics and engineering in a ceremonial act at the Technische Hochschule of Vienna, and of conferring the honorary doctorate-degree on him. On this occasion, the high officials of the Schuschnigg-Austria gathered for the last time and the festival hall was completely occupied.

By March, 1938, von Terzaghi was asked again to visit the barrage in Algeria, travelling this time via Paris. He wrote the Rector of the Technische Hochschule that he would need a visa for France. The temporary Rector forwarded von Terzaghi's request by letter to the police headquarters of Vienna on March 28, 1938. Obviously, von Terzaghi must have received his visa immediately, for he was on his way to Paris on March 29, arriving in Paris on March 30. In the

Fig. 4.27. Honorary Doctor degree for President H. Hoover

beginning of April, he went via Marseille to Ghrib. The next days were filled with studies and discussions over the foundation of the barrage. In the middle of March, he returned to Paris where he worked hard during the following two weeks to complete his opinion paper on the barrage in North Africa. However, he also took time to enjoy Parisian life (the Casino de Paris). His stay in Paris was interruped only by a visit to London (April 25 to 27) in order to give some advice in connection with difficulties encountered in the construction of an earth dam in England. He was back at the Technische Hochschule on April 29, 1938. In his absence, he had missed a great part of the troubles and the ceremonies in Vienna in connection with Austria's "Anschluß" in April, 1938. In Vienna, von Terzaghi again took up lecturing students and worked on the opinion paper for the barrage at Ghrib. On May 23, he was again in Paris. In a letter dated the same day, he informed the Rector that he would go to Paris and London for some consulting work, and that his co-worker Kienzle would substitute for him in his lectures. He returned to Vienna on the Orient Express on June 9, 1938. At the beginning of June, Mrs. von Terzaghi settled in Paris and the von Terzaghis gave up their apartment in Vienna. In the middle of June, von Terzaghi finished his work on the opinion paper and, on June 27, he gave his last lecture to his Viennese students. In the next days, he sorted out his files and on July 9, 1938, he was, via Mannheim and Heidelberg, on his way to Paris. In his diary, he drew a line. Obviously, it was clear to him that he would never again return to Vienna. His co-workers, his colleagues, and even his relatives in Graz had not been informed; nobody knew exactly where von Terzaghi had gone.

In July and August, 1938, von Terzaghi became seriously ill, but he was able to recover by the sea. After that, he travelled to London, returning to Paris at the end of August. In a letter to the Rector of the Viennese Technische Hochschule dated August 31, 1938, von Terzaghi urgently requested that he be dismissed at the end of the winter semester 1938/1939, and that he be suspended from office immediately, without salary, due to his bad state of health after twenty years of continuous teaching and research work. He pointed out that, in 1929, he had accepted the professorship at the Viennese Technische Hochschule despite a loss in personal financial earnings, although he could have accepted American citizenship after working such a long time in American duty. Moreover, in this connection, he also spoke of the fact that certain promises in the negotiations with the administration had not been fulfilled so that he had been obliged to cover the expenses of his chair partially with his own funds. Furthermore, he stated: "In 1936, my colleague Professor Fillunger started the attempt to destroy my good reputation as a human being and a researcher. And the prevention of this attack hindered me for half a year in continuing my scientific work." He concluded that the work conditions at the Technische Hochschule had been for him by no means ideal. Von Terzaghi remarked:

"Ich hatte jedoch die Genugtuung durch die Fernhaltung volksfremder Elemente von meinem Lehrkanzelbetrieb und durch gewissenhafte Schulung einer größeren Zahl von begabten Hörern meinem Volk nennenswerte Dienste zu leisten. Infolge meiner offenen Bevorzugung des völkisch eingestellten Teils der akademischen Jugend mußte ich jahrelang auf eine Förderung der Lehrkanzel durch die alte Regierung verzichten. Die aus meiner Schule herausgegangenen Arbeitskräfte haben im deutschen Reich hochverdiente Anerkennung gefunden."

"I had, however, the satisfaction of holding off the non-Aryan elements from my chair and my conscientious training of a larger number of talented students to provide my nation with remarkable services. Due to my open preference for the national-minded part of the academic youth, I had to renounce support for my chair from the old government for years. The employees who have come from my school have found in the German Reich a highly-deserved appreciation."

Moreover, von Terzaghi pointed out that, due to the "Anschluß" in March, Mrs. von Terzaghi was forced either to renounce her American citizenship and her freedom of movement, or leave the country. In view of the premises under which she had decided at that time to move to Europe, it was understandable that she refused to stay in Germany. She was forced to take the necessary steps that the new political situation demanded. Von Terzaghi closed his letter with a request:

"Ich bitte mir im Hinblick auf die geschilderten Umstände einen solchen Abgang von der Technischen Hochschule Wien zu erwirken, daß meine menschlichen und wissenschaftlichen Beziehungen zu meinen alten Mitarbeitern keine Einbußen erleiden."

"I request in view of the described circumstances to take out such a resignation from the Technische Hochschule of Vienna that my human and scientific relations to my old co-workers suffer no damages."

The Rector answered von Terzaghi in a letter dated September 6, in which he expressed his disappointment over von Terzaghi's request; he reported to the Minister on September 7. In this letter, the Rector asked the Minister to refuse von Terzaghi's request concerning his suspension, and to dismiss him immediately.

In the middle of September, von Terzaghi returned to London and, on the evening of September 16, 1938, he left Europe again, bound for the United States.

In order to clean up the affair after von Terzaghi's departure, the Rector of the Technische Hochschule established a committee which consisted, among others, of the professors Saliger, Hartmann, Schaffernak, Kozeny, and "Dozentenführer" Schober (the Nazis had installed someone in the position of "Dozentenführer" at every major university). In a letter, the committee summarized the results of a meeting held on September 28, stating that the chair (Water Construction II) had not been handed over in an orderly fashion by von Terzaghi. Furthermore, it was stated that von Terzaghi had moved to Paris in July, 1938.

Moreover, the idea of suspending him from his duties during the winter semester 1938/39 and replacing him in his seminars with his co-worker Dr. Kienzle was refused. The committee proposed that Professor Schoklitsch, a former professor at Brno, should replace von Terzaghi. Finally, the committee painfully regretted von Terzaghi's resignation and asked him to reconsider his decision because, at that time, huge tasks in the building field which necessitated his cooperation were forthcoming. In a letter dated October 27, 1938, the Minister of Education refused von Terzaghi's request concerning suspension. This decision was conferred to von Terzaghi by the Rector of the Technische Hochschule. In a telegram from Cambridge, USA, of November 25, von Terzaghi requested for immediate dismissal from his duties.

In the time to follow, some strange things happened in the wake of von Terzaghi's dismissal. Obviously, von Terzaghi was very much interested in weakening the negative image surrounding his obscure disappearance from Vienna, for he was also very much interested in receiving the thanks of the "Führer" in his letter of dismissal. In two nearly identically formulated letters written on April 3, 1939, one to von Terzaghi's daughter, Vera, and the other to his wife, the Rector stated that the thanks of the Führer and Reichskanzler for von Terzaghi's achievements could only be included in the letter of dismissal if the "Arier-Nachweis" (evidence of pure Aryan heritage) for von Terzaghi and his wife was successful. Mrs. von Terzaghi replied in a letter of April 12, 1939, pointing out that it would be extremely difficult to obtain the required documents due to the fact that in the United States it was not usual to keep all the documents. Von Terzaghi himself wrote a letter to the Rektor on April 24, 1939, in which he pointed out that he had received a copy of the letter to his wife. He stated that it would be easy for his brother-in-law, Professor Dr. Byloff (Graz, Austria) to completely document his descent. Before his marriage he had naturally been interested in his wife's family tree. However, from memory he was only able to give a little information regarding the history of his wife's family. Finally, he made some general remarks referring to his wife's family history: "In the United States the social separation between Aryan and non-Aryan elements is essentially sharper than in Germany and Austria. The circle of friends of my wife's family consists solely of Aryans, and the whole personality of my wife is to such an extent Aryan, that the desired proof concerning the descent of my wife seems to be a mere formality. I do not have any doubt concerning the result".

On April 26, 1939, the Rector of the Technische Hochschule requested, in a letter to Professor Dr. Byloff, that he send him the descent documents. At the end of the letter he stated: "The proof of Aryan descent is necessary for the dismissal-degree of Professor Terzaghi". In letters dated April 27 and August 8, 1939, the Rector reported to the minister on his efforts concerning the proof of Aryan descent for von Terzaghi and his wife, saying that his efforts had not been successful. Finally, he applied for von Terzaghi's dismissal without the thanks of the "Führer". In a letter of November 18, 1939, the Rector declared to von Terzaghi that a dismissal could only be effected if he would renounce his salary and his pensions. Von Terzaghi agreed to this proposal on December 8, 1939. Moreover, he requested that the letter of dismissal be sent to his address in the United States.

The "Ministerium für innere und kulturelle Angelegenheiten Abtl. IV, Erziehung, Kultur und Volksbildung" in Berlin wrote to the Rector on November 4, 1939, to dismiss von Terzaghi. The dismissal was valid until the end of March 1939. Moreover, in the same letter, it was left to the Rector as to whether to express thanks.

The letter of dismissal (dated September 22, 1939), as well as the corresponding document, did not contain any words of thanks.

After this, several rumors circulated about von Terzaghi's departure from Vienna. In a letter of August 31, 1943, the Rector reported to the "Reichsminister für Wissenschaft, Erziehung und Volksbildung" in Berlin, saying that von Terzaghi had moved to the United States without an understandable reason. Therefore, he continued, the name von Terzaghi should not be mentioned in the correspondence of the Hochschule. Furthermore, the Rector declined a personal ceremony on the occasion of von Terzaghi's sixtieth birthday. Some of von Terzaghi's colleagues were also deeply disappointed by his sudden and somewhat strange disappearance. In his memoir, Professor Salinger wrote: "He (von Terzaghi) has stealthily left Vienna and Austria soon after their incorporation into the Reich and our trust has been coarsely deceived".

Von Terzaghi arrived in America on a clear day on September 25, 1938, at 7.30 a.m. He was welcomed by A. Casagrande. His wife and son had remained in Paris. The next day, von Terzaghi had a meeting with Dean Westergaard of Harvard University, obviously in order to discuss his future position at Harvard. A. Casagrande had already made great efforts to find a position for von Terzaghi at a university in the United States in the beginning of 1938. He had written, for example, to Dean S.B. Moris at the School of Engineering, Stanford University,

California, asking for a half-time professorship for von Terzaghi. However, Dean Morris replied by letter: "Our budgetary situation does not permit me to make any encouraging statement at this time ..." Finally, von Terzaghi acquired a position as a visiting lecturer at Harvard University, though with a smaller salary and, in the semesters to follow, he lectured on Engineering Geology and Applied Soil Mechanics.

By the end of 1938, von Terzaghi had started his consulting work in the United States and this was to intensify in the following years. He began as a consultant for the Department of Subways and Fraction of the city of Chicago, in connection with the construction of a subway system and, in the next two years, his activities were focused mainly on this project. In April, 1939, he returned to London to lecture at Imperial College. Before leaving for London, he was asked to investigate the considerably large settlement of the Charity Hospital in New Orleans.

In the spring of 1939, von Terzaghi's wife and son moved from Paris to Cambridge and, in June, 1939, von Terzaghi and his family settled in Winchester, Massachusetts, a small town to the north of Cambridge.

As has already been mentioned, from December 1938 on von Terzaghi was involved primarily in the Chicago subway project. Although most of his previous practical application of soil mechanics had been in connection with foundations and dams, he did not hesitate to take on the new task. In January, 1939, von Terzaghi arrived for the first time in Chicago in connection with the subway project, and then followed this during the next three years with many extensive advisory visits, which became less frequent during the last year. In 1942, he was asked to investigate the stability of the Tecaxa Dam in eastern Mexico. During the following years, von Terzaghi acted as a consultant to the Mexican Government, giving advice concerning the subsidence of the City of Mexico due to the lowering of the water table. Apart from these projects, he also found time to write two remarkable books, which attracted much interest throughout the world and were translated into many different languages. In the years to follow, von Terzaghi extended his consulting activities. In particular, he gave advice on foundation problems in British Columbia, Canada. However, he was also a welcomed consultant in Sweden, Brazil, Turkey, and France.

How extensive his consulting and lecturing work was can be read from a letter to his daughter Vera in Austria, which he had begun in the Westward Hotel in Anchora, Alaska. In this letter, he described his activities between July and November, 1952, at which time he was nearly 69 years old. At the beginning of July, 1952, he started his third trip to the northwest of the United States. First, he went to Urbana, south of Chicago, to lecture at the university there. From Urbana, he proceeded to Spokane, Washington, from which he visited, by ship and plane, the tremendous landslides which had occurred on the northern part of the Roosevelt Sea, destroying the roads on the plateau. In Trail, on the Columbia River, north of the Canadian border, he visited the drainage-tunnel which he had designed. In Vancouver, British Columbia, he conferred

Fig. 4.28. K. von Terzaghi on a building site

with the directors of the electric plants for three days and gave a talk to the Canadian Engineering Society. From Vancouver, von Terzaghi flew with a small plane to the Nechako River, west of Prince George, in order to see the building site of a rock-filled dam, measuring one hundred meters high, which he had designed. One week later, he visited the building sites of other small dams by plane. Again by plane, he flew to the west, across the coastal mountain range, to the building site of a power station, and visited the line of construction for a cable in the glacial coastal mountain range. The terrain was so inaccessible that the various points could only be reached by helicopter. He returned to Vancouver and admired the grandious glacial fjord landscape. In Vancouver, two telegrams awaited him, the first one calling him back to Trail, the second one requesting him to fly to Southern California. He travelled back to Trail, on the Columbia River, and then went by plane to Los Angeles, then to friends in La Jolla, where he spent a nice weekend. From California, he returned to the east, to Hartford, Connecticut, where he visited a dam. After spending one week on vacation in Maine with his family, von Terzaghi was again on his way to Chicago, where he had to give the opening speech on the occasion of the one-hundredth anniversary of the American Society of Engineers. In the south of Chicago he also had to review an underground storeroom. He returned from Michigan to Winchester, hoping to be able to relax at his home. However, three days later, a telegram was calling him to Alaska. On his birthday, he flew again to Vancouver and Trail, and then back to Winchester. At the end of October, he lectured again in Urbana and, at the end of November, he was on his way

Fig. 4.29. K. von Terzaghi at home

to Mexico, where he received an Honoray Doctoral Degree at the University of Mexico.

It is amazing that von Terzaghi could still physically and mentally stand such exhausting trips at the age of 69.

In 1953, von Terzaghi suffered a heart attack. After a stay in the hospital, he recovered fully. A few months later, he studied the Sasumusa dam site in the Kenya Colony in Central Africa. In 1954, von Terzaghi was asked to take over the position of the Chairman of the Board of Consultants for the planned High Aswan Dam in Aswan, Egypt. From then on, he took the chair in numerous meetings of this board. In 1959, the Egyptian government decided, due to disputes with the western countries, to carry out the project with the financial and technical assistance of the Soviet Union. Von Terzaghi resigned immediately because there was now no possibility to make any influence on the design and the realization of this enormous engineering project.

His last years were overshadowed by a serious operation in 1960 and the subsequent loss of one leg. However, with his iron willpower, he was able to continue his work until he died on December 25, 1963.

Karl von Terzaghi's character was likely formed by his Italian and Bohemian heritage. As a child he liked to fool around. At the Technische Hochschule of Graz he joined, as mentioned earlier, a fraternity and in the course of his first two years enjoyed his academic liberties in a rather dissolute manner. His days were filled with dueling and drinking, which sometimes culminated in being so drunk that he passed out. Moreover, on several occasions he came into conflict

Fig. 4.30. Olga Byloff (1883–1961)

with the university officials and the police. His study habits were so disordered that he could not follow the teaching program. In order to pass the examinations, he locked himself in his room, only leaving to eat meals, and stayed there for weeks to learn the class material. His ability to concentrate on the main issues led him to pass the examinations. These characteristics of his person, namely the restless manner and ability to focus on main issues, became an important part of his personality and were the motivating force in his life.

Later on, von Terzaghi was recognized in society as a charming, handsome man with good manners and a special sense of humor. He was a master at entertaining people. These abilities were also demonstrated in the scientific world. Karl von Terzaghi was a person with stage presence. In his talks, he always used pauses between the words so that everybody could follow his lectures. He also wrote very well and thus was able to make his ideas public. He also had a very dominating personality, when he was bothered by competitors he tried to put them down.

Women always played an important role in von Terzaghi's life. In his youthful years, his relationships were often pleasant, but sometimes turbulent and passionate and, for the most part, short-lived. There were only two women that played decisive roles in his life. The first was Olga Amalia Katarina Byloff, born on September 27, 1883, a descendant of a wealthy university-educated family.

Fig. 4.31. Ruth Doggett (1903–1993)

Von Terzaghi had met Olga Byloff during the wedding celebration of Karl von Terzaghi's sister Ella with Friedrich Ottokar Byloff in 1911. In the time to follow, they saw each other very often and their meetings were not only restricted to Graz. In 1912, for example, Olga visited Karl von Terzaghi in New York for four weeks. Then, Olga moved to Mexico in order to take over the position of a governess in a very wealthy Mexican family. Von Terzaghi rejoined Olga Byloff again in Mexico, as well as in other places of the world, as von Terzaghi indicated in a letter to Professor Wittenbauer, his former professor in mechanics, of November 17, 1916. Their love for each other grew and, in June, 1916, they married. In a letter to his son, von Terzaghi later explained: "In May 1916 my daughter Vera was born and a few months later I decided to marry her mother. I knew it was a daring experiment, but I wanted to do my best to provide the child with a home." After many happy years in Constantinople, the time came when Olga could not adjust to Karl von Terzaghi's restless life and, in 1922, they decided to spend some time apart. The separation caused a lot of suffering for both of them; during this time, von Terzaghi invited his sister Ella to spend some time with him. In 1926, von Terzaghi asked Olga to go with him to the United States, but she refused his invitation. That same year, they divorced and, on September 1, Olga married the pharmacist Franz Ruchty in Graz.

Another woman who had a lasting impact on von Terzaghi's life was Ruth Doggett. She was a descendant of an old family of European immigrants with some roots going back to the first immigrants in Boston. Because there are only ew personal notes of von Terzaghi available concerning this period in his life, Casagrande (1960) is cited: "In 1928 he (Terzaghi) met Ruth Doggett who was hen engaged in doctoral research in geology at Radcliffe College. They were narried in 1930, after she received her doctor's degree. In Ruth he found a harming and gifted wife who provided him with a happy family life, and not nly shared his interest in geology, but became a highly competent associate in is work including the teaching of his courses." They had two children, a son nd a daughter.

Although von Terzaghi spent influential years of his life in the heavy political urroundings of Europe, he remained basically politically uninvolved. Among is circle of friends in Vienna were some "Nazis". It seems, however, that von Terzaghi considered these friends only as acquaintances. His departure from Vienna, giving up his position and rights at the Technische Hochschule, was lso not politically motivated, as some rumors had mentioned. It was his restless ature that led him to appropriate new tasks. Von Terzaghi's flattering response o the new leadership of the Technische Hochschule, after the "Anschluss" in 938, is obviously founded in his great vanity, which he also showed in situations ther than his political view.

In his scientific and engineering career, he received numerous honors. mong others he was a member of the Österr. Betonverein, the Academy of cience in Vienna, the American Society of Civil Engineers, and the Boston ociety of Civil Engineers. He received the Norman Medal of the American ociety of Civil Engineers four times. As well as other awards, he was also honred by the "Goldene Ehrenmünze" of the "Österreichischer Ingenieur- und rchitekten-Verein". His outstanding achievements were recognized worldwide s he was bestowed with several honorary doctor degrees. At the age of 65 years, e received his first "Dr. sc.h.c." by Trinity College of Dublin. In the years to ollow, he was awarded with additional honorary doctor degrees from universiies in Istanbul, Mexico D.F., Zürich, Bethlehem (USA), West Berlin, Trondheim, Graz, and Columbus, Ohio (USA).

Without any doubt, there is no man in the field of civil engineering in this entury who had such an impact on the development of an important branch f engineering, namely soil mechanics.

.7 he Followers of von Terzaghi and Fillunger: Biot, Heinrich and Frenkel

fter the brilliant findings of von Terzaghi and Fillunger in the first half of this entury, this work was continued by two excellent scientists who worked on orous media theory, Maurice Biot and Gerhard Heinrich. At the very beginning f Biot's career, he mostly followed von Terzaghi in his scientific work, whereas Heinrich exclusively used Fillunger's concept in his investigations of porous nedia.

Frenkel, in his studies of seismic and seismoelectric phenomena in moist soil, used the volume fraction concept and parts of the mixture theory, thus also using Fillunger's ideas. However, it is not known if Frenkel was familiar with Fillunger's concept.

In this section, the important contributions of the aforementioned scientists will be reviewed.

4.7.1 Biot's Theory

Biot (1935, 1941a) generalized von Terzaghi's theory of consolidation by extending it to the three-dimensional case, and by establishing equations valid for any arbitrary load varied with time. In his work, he discussed the number of physical constants necessary to determine the properties of soil and developed the general equations for the prediction of settlements and stresses for three-dimensional problems.

Numerous developments in porous media theory, concerning its basic equations, have referred to the work of Biot. Therefore, it seems advantageous to review Biot's prevailing publications, so some parts of the following section rely heavily on the original papers. Biot (1941a) assumed the following properties of soil:

"(1) isotropy of the material,
(2) reversibility of stress-strain relations under final equilibrium conditions,
(3) linearity of stress-strain relations,
(4) small strains,
(5) the water contained in the pores is incompressible,
(6) the water may contain air bubbles,
(7) the water flows through the porous skeleton according to Darcy's law."

Furthermore, Biot investigated the soil stresses as follows:

"Consider a small cubic element of the consolidating soil, its sides being parallel with the coordinate axes. This element is taken to be large enough compared to the size of the pores so that it may be treated as homogeneous, and at the same time small enough, compared to the scale of the macroscopic phenomena in which we are interested, so that it may be considered as infinitesimal in the mathematical treatment.

The average stress condition in the soil is then represented by forces distributed uniformly on the faces of this cubic element. Physically we may think of these stresses as composed of two parts; one which is caused by the hydrostatic pressure of the water filling the pores, the other caused by the average stress in the skeleton. In this sense, the stresses in the soil, (denoted by $\mathbf{T}^S$, the author), are said to be carried partly by the water and partly by the solid constituent."

Assuming the strain in the soil to be small, the strain tensor can be identified by the well-known linearized Green strain tensor $\overset{L}{\mathbf{E}}$:

$$\overset{L}{\mathbf{E}} = \frac{1}{2}\left[\operatorname{grad}\mathbf{u} + (\operatorname{grad}\mathbf{u})^T\right] , \tag{4.7.1}$$

where $\mathbf{u}$ denotes the displacement vector of the soil.

"In order to describe completely the macroscopic condition of the soil we must consider an additional variable giving the amount of water in the pores. We therefore denote by θ the increment of water volume per unit volume of soil and call this quantity the *variation in water content*. The *increment of water pressure* will be denoted by σ....

It is clear that if we assume the changes in the soil to occur by reversible processes the macroscopic condition of the soil must be a definite function of the stresses and the water pressure... Furthermore if we assume the strains and the variations in water content to be small quantities, the relation between these two sets of variables may be taken as linear in first approximation."

For the particular case of an isotropic soil with $\sigma = 0$, the relation between strains and stresses must reduce to Hooke's law for an isotropic elastic body:

$$\overset{L}{\mathbf{E}} = \frac{1}{2G}\mathbf{T}^S - \frac{\nu}{E}(\mathbf{T}^S \cdot \mathbf{I})\mathbf{I}\,, \tag{4.7.2}$$

where the constants E, G, and ν may be interpreted in the sense of bulk material constants, respectively, as Young's modulus, the shear modulus, and Poisson's ratio of the solid skeleton; $\mathbf{I}$ represents the identity tensor.

Suppose that the effect of the water pressure σ is introduced. It cannot produce any shearing strain because of the assumed isotropy of the soil and therefore the relation (4.7.2), taking into account the influence of σ, becomes:

$$\overset{L}{\mathbf{E}} = \frac{1}{2G}\mathbf{T}^S - (\frac{\nu}{E}\mathbf{T}^S \cdot \mathbf{I} - \frac{\sigma}{3H})\mathbf{I}\,, \tag{4.7.3}$$

where H is an additional physical constant. Relation (4.7.3) expresses the six strain components of the soil as functions of the stresses in the soil and the pressure of the water in the pores. The increment of the water content θ must depend on the same variables. The most general relation is given by:

$$\theta = a^i T_i + a\sigma\,, \quad i = 1, \ldots, 6\,, \tag{4.7.4}$$

where T_i denotes the six stress components (T_1, T_2, T_3: normal stresses; T_4, T_5, T_6: shear stresses) and a^i as well as a represent the seven possible unknown physical constants. The number of the constants can be reduced by the following assumptions: first, because of the isotropy of the material, a change in sign of the shear stresses cannot affect the water content, therefore, the constants a^4, a^5, a^6 are equal to zero; second, the normal stress components must be equal to a hydrostatic stress state for a change in water content, thus, $a^1 = a^2 = a^3$. Therefore, relation (4.7.4) may be written in the form:

$$\theta = \frac{1}{3H_1}\mathbf{T}^S \cdot \mathbf{I} + \frac{\sigma}{R}\,. \tag{4.7.5}$$

In equation (4.7.5), H_1 and R represent two additional physical constants.

The relations (4.7.3) and (4.7.5) contain five distinct physical constants, but this number can be reduced to four. Biot established the equivalence of the constants H_1 and H, and thus equation (4.7.5) results in:

$$\theta = \frac{1}{3H}\mathbf{T}^S \cdot \mathbf{I} + \frac{\sigma}{R}\,. \tag{4.7.6}$$

For further considerations, it is convenient to express the stresses as functions of the strains and the water pressure. Solving equation (4.7.3), with respect to the stresses, yields:

$$\mathbf{T}^S = 2\mu\mathbf{E}^L + (\lambda\mathbf{E}^L \cdot \mathbf{I} - \alpha\sigma)\mathbf{I}\,, \tag{4.7.7}$$

where μ and λ are the Lamé constants and α is defined by

$$\alpha = \frac{K}{H}\,, \tag{4.7.8}$$

with K as the compression modulus. The water content, expressed in dependence upon the strains and the water pressure, is given in the following form:

$$\theta = \alpha \overset{L}{\mathbf{E}} \cdot \mathbf{I} + \frac{\sigma}{Q} \tag{4.7.9}$$

with

$$\frac{1}{Q} = \frac{1}{R} - \frac{\alpha}{H}\,. \tag{4.7.10}$$

The two well-known elastic coefficients and the constants R and H completely define the physical proportions of an isotropic soil in the equilibrium condition. The inverse values of R and H can be interpreted as follows:

1/H is a measure of the compressibility of the soil for a change in water pressure;
1/R is a measure of the change in water content for a given change in water pressure.

The other coefficients were derived from the constants defined above.

"For instance α... measures the ratio of the water volume squeezed out to the volume change of the soil if the latter is compressed while allowing the water to escape ($\sigma = 0$). The coefficient $1/Q$ ... is a measure of the amount of water which can be forced into the soil under pressure while the volume of the soil is kept constant"

More details with respect to the elastic coefficients of the theory of consolidation can be found in papers by Biot and Willis (1957) and Fatt (1959).

Biot (1941a) continued:

"We now proceed to establish the differential equations for the transient phenomenon of consolidation, i.e., those equations governing the distribution of stress, water content and settlement as a function of time in a soil under given loads."

The corresponding stresses must satisfy the equilibrium conditions of a stress field (inertia forces and body forces are neglected):

$$\operatorname{div} \mathbf{T}^S = \mathbf{0}\,. \tag{4.7.11}$$

Insertion of the physical relations (4.7.7) and (4.7.9) into Eqn. (4.7.11) and consideration of the linearized strain tensor in the form (4.7.1) result in:

$$\mu \operatorname{div}\operatorname{grad}\mathbf{u} + (\lambda + \mu)\operatorname{grad}\operatorname{div}\mathbf{u} - \alpha \operatorname{grad}\sigma = \mathbf{0}\,. \tag{4.7.12}$$

In order to have a complete set of equations for determining the displacement vector $\mathbf{u}$ and the water pressure σ, one more equation is necessary. This equation can be obtained by introducing Darcy's law governing the flow of water in a porous medium:

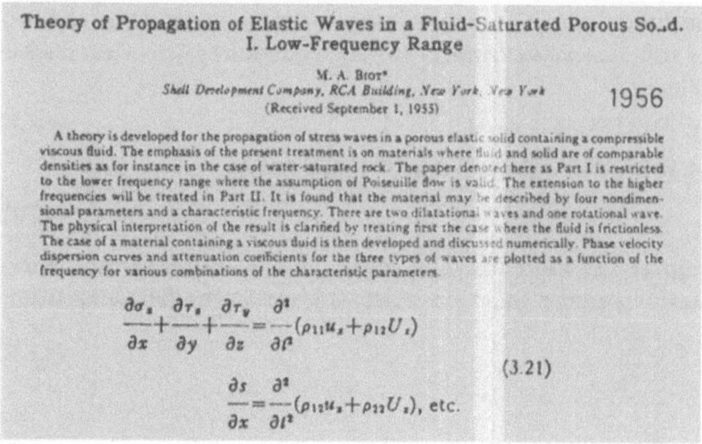

Theory of Propagation of Elastic Waves in a Fluid-Saturated Porous Solid. I. Low-Frequency Range

M. A. Biot*
Shell Development Company, RCA Building, New York, New York

1956

(Received September 1, 1955)

A theory is developed for the propagation of stress waves in a porous elastic solid containing a compressible viscous fluid. The emphasis of the present treatment is on materials where fluid and solid are of comparable densities as for instance in the case of water-saturated rock. The paper denoted here as Part I is restricted to the lower frequency range where the assumption of Poiseuille flow is valid. The extension to the higher frequencies will be treated in Part II. It is found that the material may be described by four nondimensional parameters and a characteristic frequency. There are two dilatational waves and one rotational wave. The physical interpretation of the result is clarified by treating first the case where the fluid is frictionless. The case of a material containing a viscous fluid is then developed and discussed numerically. Phase velocity dispersion curves and attenuation coefficients for the three types of waves are plotted as a function of the frequency for various combinations of the characteristic parameters.

$$\frac{\partial\sigma_x}{\partial x}+\frac{\partial\tau_z}{\partial y}+\frac{\partial\tau_y}{\partial z}=\frac{\partial^2}{\partial t^2}(\rho_{11}u_x+\rho_{12}U_x)$$

$$\frac{\partial s}{\partial x}=\frac{\partial^2}{\partial t^2}(\rho_{12}u_x+\rho_{22}U_x),\ \text{etc.} \qquad (3.21)$$

Fig. 4.32. M. Biot's work on wave propagation

$$\mathbf{w}^F = -\,k^F \operatorname{grad}\sigma\,, \qquad (4.7.13)$$

where $\mathbf{w}^F$ represents the volume of water flowing per second and unit area through the faces of an elementary cube and k^F denotes the permeability coefficient of the soil. Furthermore, under the assumption of the incompressibility of water, the rate of water content of an element of soil must be equal to the volume of water entering per second through the surface of the elements. Hence:

$$\frac{\partial\theta}{\partial t} = -\operatorname{div}\mathbf{w}^F\,. \qquad (4.7.14)$$

Combining the equations (4.7.9), (4.7.13), and (4.7.14) results in:

$$k^F \operatorname{div}\operatorname{grad}\sigma = \alpha\frac{\partial(\overset{L}{\mathbf{E}}\cdot\mathbf{I})}{\partial t} + \frac{1}{Q}\frac{\partial\sigma}{\partial t}\,. \qquad (4.7.15)$$

Equations (4.7.12) and (4.7.15) are the basic relations of Biot's theory of consolidation, where equation (4.7.12) describes the settlement of the soil, and equation (4.7.15) describes the change in the water pressure. Furthermore, it is important to note that these two relations are coupled.

Biot's following investigations considered the development of a theory for the propagation of stress waves in a porous elastic solid containing a compressible viscous fluid (see Biot, 1956a, b). Heinrich and Desoyer (1961) stated that Biot developed his theory without a detailed analysis of mass continuity equations, nor of the forces acting on the individual constituents (skeleton and fluid), so that the theory was found not to be derived from proper fundamentals. However, we will demonstrate that, in special cases, Biot's equations cover the equations developed by other authors.

Biot's theory of dynamic consolidation was successfully employed by many authors (see, e.g., references in Dziecielak, 1986) and, for this reason, it will be presented here.

Biot's theory of consolidation proceeds differently from those chosen by other authors. The constitutive equations in his theory were obtained by extending the theory of elasticity for a two-phase system through the introduction of some additional coupling parameters.

Biot considered a porous isotropic elastic solid saturated with a viscous fluid, and he assumed that the fluid was compressible and could flow relative to the solid, causing friction to arise. The changes in the mass densities of both constituents and in the porosity ratio were assumed to be small. In other words, Biot considered the partial mass densities as constant quantities.

Analogous to his previous findings (see Biot, 1935), Biot defined the fluid pressure, with respect to the real fluid pressure p, using the volume porosity n, thus:

$$\sigma = -\, np \,. \tag{4.7.16}$$

In order to describe the two-phase system with different states of motion for the constituents, Biot assumed different displacement vectors. Thereby the displacement vector of the skeleton $\mathbf{u}_S$ is defined as the displacement of the material considered to be uniform and averaged over the cubic element. The average fluid displacement vector $\mathbf{u}_F$ is defined in such a way that the product of this displacement multiplied by the cross-sectional fluid area represents the volume flow. The strain in the fluid (increment of fluid volume) is defined by the dilatation:

$$\theta = \operatorname{div} \mathbf{u}_F \,. \tag{4.7.17}$$

The strain tensor of the solid is introduced similarly to Eqn. (4.7.1) by:

$$\overset{L}{\mathbf{E}} = \frac{1}{2}\left[\operatorname{grad} \mathbf{u}_S + (\operatorname{grad} \mathbf{u}_S)^T\right] . \tag{4.7.18}$$

The constitutive relations (4.7.7) and (4.7.9) will be used in the following modified forms:

$$\sigma = Q\theta - \alpha Q \overset{L}{\mathbf{E}} \cdot \mathbf{I} \,, \tag{4.7.19}$$

and

$$\mathbf{T}^S = 2\mu \overset{L}{\mathbf{E}} + \lambda(\overset{L}{\mathbf{E}} \cdot \mathbf{I})\mathbf{I} - \alpha Q\theta \mathbf{I} + \alpha^2 Q(\overset{L}{\mathbf{E}} \cdot \mathbf{I})\mathbf{I} \,, \tag{4.7.20}$$

where $\overset{L}{\mathbf{E}}$ is defined by relation (4.7.18).

For the derivation of the equations of motion for the two-phase system, Biot used the Lagrangian viewpoint and the concept of generalized coordinates. The generalized coordinates were chosen as the six average displacement components of the solid and the fluid, respectively. The kinetic energy of the system per unit volume was assumed via:

$$K = \frac{1}{2}\left(\rho^{11}\dot{\mathbf{u}}_S \cdot \dot{\mathbf{u}}_S + 2\rho^{12}\dot{\mathbf{u}}_S \cdot \dot{\mathbf{u}}_F + \rho^{22}\dot{\mathbf{u}}_F \cdot \dot{\mathbf{u}}_F\right) , \tag{4.7.21}$$

where a dot over symbols denotes the partial time derivative. This expression is based on the assumption that the material is statistically isotropic, and hence the directions in space are equivalent and uncoupled dynamically.

"The coefficients $\rho^{11}, \rho^{12}, \rho^{22}$ are mass coefficients which take into account the fact that the relative fluid flow through the pores is not uniform."

These mass coefficients will be discussed later. Furthermore, Biot introduced a dissipation function as follows:

"It will be assumed that the flow of the fluid relative to the solid through the pores is of the Poiseuille type. That this assumption is not always valid is well known, e. g., when the Reynolds number of the relative flow exceeds a certain critical value. If we accept the assumption of Poiseuille flow the microscopic flow pattern inside the pores is uniquely determined by the generalized velocities $\dot{\mathbf{u}}_S$ and $\dot{\mathbf{u}}_F$. Dissipation depends only on the relative motion between the fluid and the solid. Introducing the concept of dissipation function, we may write this function as a homogeneous quadratic form with the foregoing six generalized velocities... The dissipation also vanishes when there is no relative motion of fluid and solid; hence, when $\dot{\mathbf{u}}_S = \dot{\mathbf{u}}_F$."

Therefore, the dissipation function D is:

$$D = \frac{1}{2} b \, (\dot{\mathbf{u}}_S - \dot{\mathbf{u}}_F) \cdot (\dot{\mathbf{u}}_S - \dot{\mathbf{u}}_F) \, . \tag{4.7.22}$$

The Lagrange equations concerning the dissipation function (4.7.22), can be written as:

$$\begin{aligned} \frac{\partial}{\partial t} \frac{\partial K}{\partial \dot{\mathbf{u}}_S} + \frac{\partial D}{\partial \dot{\mathbf{u}}_S} &= \mathbf{f}^S \, , \\ \frac{\partial}{\partial t} \frac{\partial K}{\partial \dot{\mathbf{u}}_F} + \frac{\partial D}{\partial \dot{\mathbf{u}}_F} &= \mathbf{f}^F \, , \end{aligned} \tag{4.7.23}$$

where $\mathbf{f}^S$ and $\mathbf{f}^F$ denote the total generalized forces per unit volume acting on the solid and the fluid, respectively. Biot expressed these forces by the gradients of the solid stresses and of the fluid pressure, neglecting the body forces. In general, the body forces can be added, thus giving $\mathbf{f}^S$ and $\mathbf{f}^F$ the forms:

$$\begin{aligned} \mathbf{f}^S &= \operatorname{div} \mathbf{T}^S + \rho^S \mathbf{b} \, , \\ \mathbf{f}^F &= \operatorname{div} \mathbf{T}^F + \rho^F \mathbf{b} \, . \end{aligned} \tag{4.7.24}$$

Inserting the relations (4.7.21), (4.7.22), and (4.7.24) into (4.7.23), we obtain the dynamic equations for the theory of consolidation:

$$\begin{aligned} \rho^{11} \ddot{\mathbf{u}}_S + \rho^{12} \ddot{\mathbf{u}}_F + b \, (\dot{\mathbf{u}}_S - \dot{\mathbf{u}}_F) &= \operatorname{div} \mathbf{T}^S + \rho^S \mathbf{b} \, , \\ \rho^{12} \ddot{\mathbf{u}}_S + \rho^{22} \ddot{\mathbf{u}}_F + b \, (\dot{\mathbf{u}}_F - \dot{\mathbf{u}}_S) &= \operatorname{grad} \sigma + \rho^F \mathbf{b} \, . \end{aligned} \tag{4.7.25}$$

Through Poiseuille's type of fluid flow, the coefficient b is related to Darcy's permeability coefficient k by:

$$b = \frac{\mu^F n^2}{k} \, , \tag{4.7.26}$$

where μ^F is the fluid dynamic viscosity and n the porosity.

Furthermore, let us discuss the nature of the coefficients ρ^{11}, ρ^{12}, and ρ^{22}. To clarify the meaning of these coefficients, Biot made certain assumptions.

First, he investigated a fluid-soil system without relative motion between fluid and solid. In this case:

$$\dot{\mathbf{u}}_S = \dot{\mathbf{u}}_F \,, \tag{4.7.27}$$

and thus, instead of equation (4.7.21),

$$K = \frac{1}{2}(\rho^{11} + 2\rho^{12} + \rho^{22})\dot{\mathbf{u}}_S \cdot \dot{\mathbf{u}}_S \,, \tag{4.7.28}$$

with the consequence that the sum of the mass coupling parameters corresponds to the total mass of the fluid-solid system per unit volume, i.e.,

$$\rho^{11} + 2\rho^{12} + \rho^{22} = \rho \,. \tag{4.7.29}$$

Insertion of the assumption (4.7.27) into the equations of motion (4.7.25) results in:

$$\begin{aligned} (\rho^{11} + \rho^{12})\,\ddot{\mathbf{u}}_S &= \operatorname{div} \mathbf{T}^S + \rho^S \mathbf{b} \,, \\ (\rho^{12} + \rho^{22})\,\ddot{\mathbf{u}}_F &= \operatorname{grad} \sigma + \rho^F \mathbf{b} \,. \end{aligned} \tag{4.7.30}$$

Thus, the partial densities of the solid and fluid constituents, ρ^S and ρ^F, can be identified by:

$$\begin{aligned} \rho^{11} + \rho^{12} &= \rho^S \,, \\ \rho^{12} + \rho^{22} &= \rho^F \,. \end{aligned} \tag{4.7.31}$$

With respect to the porosity n, Biot formulated the following relation between the partial densities and the real partial densities:

$$\begin{aligned} \rho^S &= (1 - n)\rho^{SR} \,, \\ \rho^F &= n\rho^{FR} \,. \end{aligned} \tag{4.7.32}$$

A second assumption on the part of Biot leads to an interpretation of the parameter ρ^{12} as a mass coupling coefficient. By assuming that the average displacement of the fluid is zero, i.e., $\mathbf{u}_F = \mathbf{0}$, the following equations are obtained with the aid of (4.7.24) and (4.7.25):

$$\begin{aligned} \rho^{11}\ddot{\mathbf{u}}_S &= \mathbf{f}^S \,, \\ \rho^{12}\ddot{\mathbf{u}}_S &= \mathbf{f}^F \,. \end{aligned} \tag{4.7.33}$$

Biot concluded from (4.7.33) that

"when the solid is accelerated, a force $\mathbf{f}^F$ must be exerted on the fluid to prevent an average displacement of the latter. This effect is measured by the 'coupling' coefficient ρ^{12}. The force $\mathbf{f}^F$ necessary to prevent the fluid displacement is obviously in a direction opposite to the acceleration of the solid:"

$$\rho^{12} < 0 \,; \tag{4.7.34}$$

thus ρ^{12} was understood as an additional apparent mass with a change in sign.

The same conclusion is reached after considering $(4.7.33)_1$, where ρ^{11} represents the total effective mass of the solid moving in the fluid. This total mass must be equal to the proper mass of the solid ρ^S, plus an additional mass ρ^a due to the fluid:

$$\rho^{11} = \rho^S + \rho^a \, . \tag{4.7.35}$$

From equations $(4.7.31)_1$ and (4.7.35), it is evident that

$$\rho^{12} = -\rho^a \tag{4.7.36}$$

holds. Hence, ρ^{12} is the additional apparent mass with a change in sign. Therefore, the dynamic coefficients can be written as:

$$\begin{aligned} \rho^{11} &= \rho^S + \rho^a \, , \\ \rho^{22} &= \rho^F + \rho^a \, , \\ \rho^{12} &= -\rho^a \, . \end{aligned} \tag{4.7.37}$$

Further conditions must be satisfied by these dynamic coefficients to assure the kinetic energy to be of a positive definite quadratic form. Thus, the coefficients ρ^{11} and ρ^{22} must be positive, i.e.:

$$\rho^{11} > 0 \, , \; \rho^{22} > 0 \, , \tag{4.7.38}$$

and

$$\rho^{11}\rho^{22} - (\rho^{12})^2 > 0 \, . \tag{4.7.39}$$

Biot (1956a) pointed out concerning the relations (4.7.38) and (4.7.39):

"These inequalities are always satisfied if the coefficients are given by the relations (4.7.37) where ρ^S, ρ^F and ρ^a are positive by the physical nature."

Finally, Biot (1956a) developed the basic equations for the description of the elastic wave propagation in saturated porous bodies.

It can be shown that Biot's theory from 1956, with the mass coupling terms, describes a ternary model consisting of a porous solid and free and trapped fluids (see de Boer *et al.*, 1991).

Biot and Willis (1957) reviewed Biot's poroelasticity theory and stated

"methods of measurement are described for the determination of the elastic coefficients of the theory."

Moreover, they discussed

"the physical interpretation of the coefficients in various alternate forms."

For an isotropic system, in which there are four coefficients, the four measurements of shear modulus, jacketed and unjacketed compressibility, and coefficient of fluid content, together with a measurement of porosity, appear to be the most convenient. The porosity is not required if the variables and coefficients are expressed in the proper way. The coefficient of fluid content is a measure of the volume of fluid entering the pores of a solid sample during an unjacketed compressibility test.

After the introduction, the authors summarized the stress-strain relations for the isotropic case:

$$\begin{aligned} \mathbf{T}^S &= 2N \, \mathbf{E}_S + A \, e_S \, \mathbf{I} + Q \, e_F \mathbf{I} \, , \\ p^F &= Q \, e_S + R \, e_F \, , \end{aligned} \tag{4.7.40}$$

"in which the $\mathbf{T}^S$ (The original notations for $\mathbf{T}^S$ and p^F have been changed by the author in order to correspond to modern standards) are the forces acting on the solid portions of the faces of a unit cube of porous material, and p^F is the force acting on the fluid portions."

Although Biot and Willis (1957) used the term forces for $\mathbf{T}^S$ and p^F, the term stress is meant. The average strains $\mathbf{E}_S$, and the average volume strains e_S and e_F, are connected with the components of the average displacement of the solid phase u_i and of the fluid phase U_i in rectangular coordinate systems, namely

$$\begin{aligned} e_{ik} &= \frac{\partial u_i}{\partial X_k} + \frac{\partial u_k}{\partial X_i}\,, \\ e_S &= \frac{\partial u_i}{\partial X_i}\,, \quad e_F = \frac{\partial U_i}{\partial X_i}\,, \end{aligned} \tag{4.7.41}$$

where X_i are the components of the position vector. Biot and Willis (1957) stated:

"It should be pointed out that this expression (the expression for the volume strain e_F, the author) is not the actual strain in the fluid but simply the divergence of the fluid-displacement field which itself is derived from the average volume flow through the pores."

Furthermore they wrote:

"Four independent measurements, in addition to the porosity n^F (Biot and Willis denoted the porosity by f, the author), are required to fix the four elastic coefficients A, N, Q, and R. Satisfactory combinations of measurements may be made in a variety of different ways, but the most convenient appears to be the combination of measurement of shear modulus, jacketed and unjacketed compressibility of the porous solid, and an unjacketed coefficient of fluid content."

They remarked that the shear modulus of the skeleton is equivalent to N and that a coefficient of jacketed compressibility κ is determined by

$$\kappa = -\frac{e}{p'}\,, \tag{4.7.42}$$

where e is the dilatation of the specimen and p' the external pressure at the jacket. In this test, the partial pressure of the fluid p^F is equal to zero and

$$T^{11} = T^{22} = T^{33} = -\,p' \tag{4.7.43}$$

with T^{11}, T^{22}, and T^{33} as principal stresses.

The reciprocal value of the coefficient of the jacketed compressibility κ is connected with the response parameters in (4.7.40) by

$$\frac{1}{\kappa} = \frac{2}{3}N + A - \frac{Q^2}{R}\,, \tag{4.7.44}$$

and it is easily recognized that (4.7.44) can be written as

$$\frac{1}{\kappa} = \frac{2}{3}\mu + \lambda\,, \tag{4.7.45}$$

whereby $\mu = N$ is the shear modulus and

$$\lambda = A - \frac{Q^2}{R} \tag{4.7.46}$$

the Lamé-coefficient of the skeleton, if the partial fluid pressure is equal to zero.

In the unjacketed test, the compressibility coefficient δ is determined by

$$\delta = -\frac{e_S}{p'} \,. \tag{4.7.47}$$

Without any foundation, Biot and Willis (1957) distribute the fluid pressure p', which is equal to the porefluid pressure p, over the partial solid and fluid phases

$$\begin{aligned} T^{11} &= T^{22} = T^{33} = -\,n^S p' \,, \\ p^F &= -\,n^F p' \,. \end{aligned} \tag{4.7.48}$$

From (4.7.40), considering (4.7.46) and (4.7.47), the relations between the response functions in (4.7.40) and the coefficients of the jacketed and unjacketed compressibilities κ and δ are obtained, considering (4.7.43):

$$1 - \frac{\delta}{\kappa} = (\frac{Q+R}{R}) n^F \,. \tag{4.7.49}$$

This relation can be reformulated if one considers the quantity

$$\xi = n^F (e_S - e_F) \,, \tag{4.7.50}$$

which

"represents the volume of fluid which enters the pores of a unit volume of bulk material. If, during a jacketed compressibility test the interior of the jacket is connected to the atmosphere by a tube the fluid volume passing through this tube is equal to ξ."

Moreover, the last equation of (4.7.40) yields, considering p^F equal to zero in the jacketed test,

$$\frac{\xi}{e_S} = (\frac{Q+R}{R}) n^F \,. \tag{4.7.51}$$

Thus, with (4.7.49)

$$\xi = e_S (1 - \frac{\delta}{\kappa}) \tag{4.7.52}$$

is obtained.

There is still one additional measurement required

"...which must involve the fluid strain... Again we consider the volume of fluid which enters the pores of a unit volume of porous material ... but in this case with reference to an unjacketed compressibility test. We may define a coefficient γ of fluid content by

$$\gamma = \frac{n^F \left(e_S - e_F\right)}{p'} \tag{4.7.53}$$

for an unjacketed compressibility test. This gives for the fluid strain."

$$e_F = -\,\frac{\gamma}{n^F} p' + e_S \tag{4.7.54}$$

or considering (4.7.47), we get:

$$e_F = -\frac{\gamma}{n^F} p' - \delta p' . \tag{4.7.55}$$

From the last equation in (4.7.40), considering (4.7.47), (4.7.48), and (4.7.54),

$$n^F = (Q + R)\delta + R\frac{\gamma}{n^F} \tag{4.7.56}$$

is obtained.

In the remaining part of the section on isotropic porous solids, Biot and Willis (1957) summarized their results and finally they drew attention to the fact that

"....it is possible to determine the coefficient of fluid content γ directly from the fluid compressibility c. Considering the unjacketed test, the pore space in this special case will undergo the same strain as the solid matrix. Therefore the porosity n^F of the material will not undergo any strain and, if the fluid completely saturates the pores, the fluid dilatation will be given by

$$e_F = -cp' .\text{"} \tag{4.7.57}$$

Thus from (4.7.53), considering (4.7.47) and (4.7.57),

$$\gamma = n^F(c - \delta) \tag{4.7.58}$$

is derived.

"This relation will be strictly valid only for materials such that the pore volume and the bulk volume remain in constant ratio; i.e., the porosity does not vary when the specimen is subjected to fluid pressure in an unjacketed test. This of course will be true if the material of the porous matrix is homogeneous, isotropic, and elastically linear."

The remaining part of Biot's and Willis' (1957) article is concerned with the introduction of alternative variables, transverse isotropy, and elastic coefficients for incremental deformations of a prestressed material. In Section 3 on elastic coefficients with alternative variables, there is another interesting hint to the mechanical meaning of ξ, one should regard (4.7.52) and the following remarks. In this section, Biot and Willis (1957) stated:

"This again results from the existence of an elastic potential energy with p and ξ acting as conjugate variables."

This statement is also repeated in the Detournay's and Cheng's (1993) review article.

Moreover, in Section 3 (Biot and Willis, 1957), the constitutive equation for the total stress is given by

$$\mathbf{T}^S + \mathbf{T}^F = -p(1 - \frac{\delta}{\kappa})\mathbf{I} + 2\mu\mathbf{E}_S + \lambda(\mathbf{E}_S \cdot \mathbf{I})\mathbf{I} , \tag{4.7.59}$$

where δ and κ characterize the compressibility of the real solid material and that of the solid skeleton. Furthermore, μ and λ are the Lamé constants of the solid material.

In what follows, the results of Biot's poroelasticity will be compared with those of modern standards.

At the beginning of Biot's and Willis' (1957) treatise, the stress and strain states are defined as being partial quantities, see (4.7.40), (4.7.41), and the corresponding remarks. However, in the sequel, these clear definitions do not always seem to be considered in the described experiments. Whereas the jacketed test can be duplicated with the aforementioned notions, the treatment of the unjacketed test raises some doubts on the correct correspondence of the mechanical quantities. In (4.7.47), for example, the partial strain is combined with the pore-fluid pressure p. Indeed, it is legitimate to measure real and partial quantities in a test. However, it is doubtful as to whether it makes sense to integrate such test results within a general theory. This comment is also valid for Eqn. (4.7.57). In this relation, the pore fluid is connected with the partial volume strain e_F of the fluid via the real compressibility of the fluid. Although Biot and Willis (1957) restrict this correspondence to constant volume fractions, their method to include the compressibility into the determination of response parameters remains partly obscure.

This is also valid for the introduction of the notion $\xi = n^F(e_F - e_S)$ as a conjugate variable to the pore-fluid pressure p. It is hard to follow Biot's view because the only conjugate variable to the pore-fluid pressure p is the real volume strain e_{FR}, e.g., see Macvean (1968). But the real volume strain e_{FR} is nowhere defined in Biot's theory. The weighted difference of the partial volume strains of the solid and the fluid phases never yields the real volume strain of the fluid (see de Boer, 1996), where the corresponding relations are developed.

Without any doubt, Biot's poroelastic theory, within the framework of the geometrically- and physically-linear theory, gives good results for a wide range of practical problems – in particular for dynamic problems – and represents an enormous achievement. Until recently, it was the only theory which included the compressibility of the porous solid phase nearly correctly (see Lade and de Boer, 1997). However, the incorporation of the compressibility of the fluid phase seems to have failed. Therefore, one should be careful if an extension to geometrically and physically non-linear behavior of saturated porous solids is required, for example, to describe the mechanical behavior of porous solids in the viscous and plastic range saturated by compressible viscous and non-viscous fluids. These problems need clear and distinct notions and definitions which are, in some parts, not contained in Biot's theory.

In the time to follow, the objectives of Biot's papers were mainly directed towards completing his theory and throwing light upon some open problems. In two papers (see Biot, 1962a, b), he investigated the acoustic wave propagation in a porous solid, whereby the last considered the influence of anisotropy, of viscoelasticity, and the dissipation in solids.

Finally Biot (1972) developed, within the framework of quasi-static and isothermal deformations, a theory of finite deformations of porous media, stating in this paper:

"The mechanics of porous media is thus brought to the same level of development of the classical theory of finite deformations in elasticity."

It should be remarked, in conclusion, that Biot's treatment of the saturated porous body is, in some respects, unsatisfactory. Clear definitions and terms are missing in certain parts. In addition, terms are omitted in the balance equations, and an unnecessary approximation is used which greatly restricts the theory. In Darcy's law, for example, the convective expressions are neglected so that the developed equations are only valid for a slow flow of the liquid in a porous body. The essential disadvantage of Biot's models lies, however, in the fact that the corresponding theory is not developed from the fundamental axioms and principles of mechanics and thermodynamics. Thus, some derivations are very obscure.

These statements should not be understood as pedantic criticism. Without Biot's works, in which several effects had been described in a correct way through pure intuition, the theory of porous media would not have been developed so extensively in the past.

4.7.2 Heinrich's Theory

After the publication of the fundamental work *Wissenschaftliche Grundlagen der Theorie der Setzung von Tonschichten* in 1938, Heinrich published several papers on different problems in mechanics, in particular on some problems in the mixture theory. It was just before 1955 and 1956 that he returned to the porous medium problem. In these years, he wrote, together with Desoyer, two remarkable papers on stationary and instationary ground-water flow (see Heinrich and Desoyer, 1955, 1956) through reposed anisotropic skeletons. In the first paper, it was assumed that the grains of the porous body were only in contact at single points, while in the second paper, the considerations were extended to general solid matrices. In the paper from 1955, Heinrich and Desoyer used Euler's cut principle in the sense that they introduced a grain-to-grain cut, instead of a statistical one. For such a finite volume of the mixture, consisting of the porous solid and the liquid, they performed a careful analysis of the forces acting on this volume. They stated that for the macro-flow on the surface of the volume, only pure pressure was acting according to the assumption of point-to-point-contacts. Further, they pointed out that the forces in the micro flow caused by the grains were distributed as quasi-volume forces over the whole volume in the macro-flow.

The authors introduced the following quantities: the surface vector of the surface element was denoted by $\mathbf{do}$, a volume element of the mixture by dV, the pore volume by n, the acceleration of gravity by g, the specific weight by γ_w, the overpressure in the liquid by p, the potential of the gravity force by U, and the friction force $\mathbf{R}_w$. The sign ∇ is the well-known differential gradient operator. With these notations, Heinrich and Desoyer introduced the resultant of the surface forces as $-\int p\mathbf{do}$, and that of the gravity force as $-\int \frac{\gamma_w}{g} n\nabla U dV$. Moreover, they computed the friction force as $\int \mathbf{R}_w dV$, and the quasi-volume force on the liquid caused by the pressures of the grains as $\int (1-n)\nabla p dV$. They

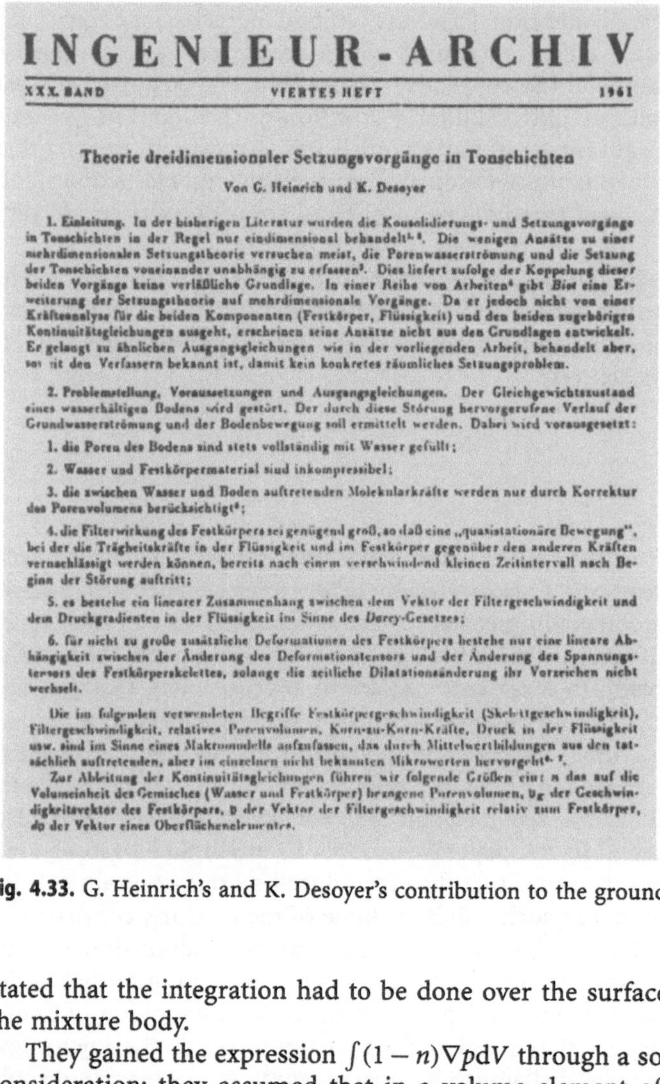

INGENIEUR-ARCHIV

XXX. BAND VIERTES HEFT 1961

Theorie dreidimensionaler Setzungsvorgänge in Tonschichten

Von G. Heinrich und K. Desoyer

1. Einleitung. In der bisherigen Literatur wurden die Konsolidierungs- und Setzungsvorgänge in Tonschichten in der Regel nur eindimensional behandelt[1, 2]. Die wenigen Ansätze zu einer mehrdimensionalen Setzungstheorie versuchen meist, die Porenwasserströmung und die Setzung der Tonschichten voneinander unabhängig zu erfassen[3]. Dies liefert zufolge der Koppelung dieser beiden Vorgänge keine verläßliche Grundlage. In einer Reihe von Arbeiten[4] gibt *Biot* eine Erweiterung der Setzungstheorie auf mehrdimensionale Vorgänge. Da er jedoch nicht von einer Kräfteanalyse für die beiden Komponenten (Festkörper, Flüssigkeit) und den beiden zugehörigen Kontinuitätsgleichungen ausgeht, erscheinen seine Ansätze nicht aus den Grundlagen entwickelt. Er gelangt zu ähnlichen Ausgangsgleichungen wie in der vorliegenden Arbeit, behandelt aber, soweit den Verfassern bekannt ist, damit kein konkretes räumliches Setzungsproblem.

2. Problemstellung, Voraussetzungen und Ausgangsgleichungen. Der Gleichgewichtszustand eines wasserhältigen Bodens wird gestört. Der durch diese Störung hervorgerufene Verlauf der Grundwasserströmung und der Bodenbewegung soll ermittelt werden. Dabei wird vorausgesetzt:

1. die Poren des Bodens sind stets vollständig mit Wasser gefüllt;

2. Wasser und Festkörpermaterial sind inkompressibel;

3. die zwischen Wasser und Boden auftretenden Molekularkräfte werden nur durch Korrektur des Porenvolumens berücksichtigt[5];

4. die Filterwirkung des Festkörpers sei genügend groß, so daß eine „quasistationäre Bewegung", bei der die Trägheitskräfte in der Flüssigkeit und im Festkörper gegenüber den anderen Kräften vernachlässigt werden können, bereits nach einem verschwindend kleinen Zeitintervall nach Beginn der Störung auftritt;

5. es bestehe ein linearer Zusammenhang zwischen dem Vektor der Filtergeschwindigkeit und dem Druckgradienten in der Flüssigkeit im Sinne des *Darcy*-Gesetzes;

6. für nicht zu große zusätzliche Deformationen des Festkörpers bestehe nur eine lineare Abhängigkeit zwischen der Änderung des Deformationstensors und der Änderung des Spannungstensors des Festkörperskelettes, solange die zeitliche Dilatationsänderung ihr Vorzeichen nicht wechselt.

Die im folgenden verwendeten Begriffe Festkörpergeschwindigkeit (Skelettgeschwindigkeit), Filtergeschwindigkeit, relatives Porenvolumen, Korn-zu-Korn-Kräfte, Druck in der Flüssigkeit usw. sind im Sinne eines Makromodells aufzufassen, das durch Mittelwertbildungen aus den tatsächlich auftretenden, aber im einzelnen nicht bekannten Mikrowerten hervorgeht[6, 7].

Zur Ableitung der Kontinuitätsgleichungen führen wir folgende Größen ein: n das auf die Volumeinheit des Gemisches (Wasser und Festkörper) bezogene Porenvolumen, $\mathfrak{v}_g$ der Geschwindigkeitsvektor des Festkörpers, $\mathfrak{v}$ der Vektor der Filtergeschwindigkeit relativ zum Festkörper, $d\mathfrak{o}$ der Vektor eines Oberflächenelementes.

Fig. 4.33. G. Heinrich's and K. Desoyer's contribution to the ground-water flow

stated that the integration had to be done over the surface and the volume of the mixture body.

They gained the expression $\int(1-n)\nabla p dV$ through a somewhat semi-micro consideration: they assumed that in a volume element of the micro flow dV there may be on an average about N grains. The resultant of the pressure forces acting on the liquid caused by the granules is, then, $\sum\limits_{j=1}^{N}\oint p\, d\mathbf{o}_j$, where $d\mathbf{o}_j$ marks the surface vector of the surface element of the j-grain. Applying Gauß's theorem, they obtained:

$$\sum_{j=1}^{N}\int \nabla p\, dV_{kj} = \nabla p \sum_{j=1}^{N} V_{kj}\,, \tag{4.7.60}$$

where dV_{kj} means the volume element of the j-grain and V_{kj} the whole volume of the j-grain. Moreover, the relation:

$$\sum_{j=1}^{N} V_{kj} = (1-n)dV \tag{4.7.61}$$

is valid. From these equations, it follows

$$\sum_{j=1}^{N} \oint p\, d\mathbf{o}_j = (1-n)\nabla p dV \,. \tag{4.7.62}$$

Neglecting the inertia forces, the authors gained the equation of balance of momentum for the liquid:

$$\mathbf{R}_w = n\nabla\left(\frac{\gamma_w}{g}U + p\right) . \tag{4.7.63}$$

For the porous solid, Heinrich and Desoyer developed, with the help of the equation of balance of momentum,

$$\operatorname{Div} \mathbf{S}_K = \frac{(1-n)\gamma_k}{g}\nabla U + \mathbf{R}_w + (1-n)\nabla p \,, \tag{4.7.64}$$

where $\mathbf{S}_K$ is the stress tensor of the porous solid, and Div the well-known differential operator. Substituting $\mathbf{R}_w$ in (4.7.64) from (4.7.63), the authors obtained:

$$\operatorname{Div} \mathbf{S}_K = \frac{\gamma_k - n(\gamma_k - \gamma_w)}{g}\nabla U + \nabla p \,. \tag{4.7.65}$$

In the special case where γ_k and n do not depend on the place, Eqn. (4.7.65) can be reformulated as:

$$\operatorname{Div}\left\{\mathbf{S}_K - \left[\frac{\gamma_k - n(\gamma_k - \gamma_w)}{g}U + p\right]\mathbf{I}\right\} = \mathbf{0} \,. \tag{4.7.66}$$

In the remaining part of their treatise, the authors described in detail the water flow through the porous solid consisting of grains.

In their second paper on the stationary and instationary ground-water flow, Heinrich and Desoyer (1956) generalized their investigations by dropping the strong restriction of point-to-point contact for the grains. They extended the problem to general solid matrices. First, they made some remarks concerning the Delessian law. Then, they again analyzed the forces acting on the volume element on the macroscopic scale. From the equation for balance of momentum they gained the same relation for the friction force as in Eqn. (4.7.63), and they stated that this did not play any role at all in the assumption of point contacts for the grains. For the skeleton with the stress tensor $\mathbf{S}$, under consideration of some semi-micro investigations, the authors gained the following equilibrium equation using the statistical cut:

$$\operatorname{Div} \mathbf{S} = n\nabla p + \frac{\gamma_k - n(\gamma_k - \gamma_w)}{g}\nabla U + p\nabla n \,. \tag{4.7.67}$$

In what followed, Heinrich and Desoyer drew some obscure conclusions. Strongly influenced by Fillunger, they mixed up the statistical cut (the only possible one for developing a continuum-mechanical approach to the problem of saturated porous media) and the grain-to-grain cut (in this case the porosity n is equal to one) which yields a special boundary problem. They stated:

"Damit ist eine alte Streitfrage aufgeklärt, die in der hydraulischen Literatur bis heute noch nicht bereinigt werden konnte, die Frage nämlich, ob bei der Wirkung der strömen- den Flüssigkeit auf den Festkörper der Druckgradient in voller Stärke oder mit dem Faktor *n* versehen einzuführen ist. Es ist dies also lediglich davon abhängig, ob der betrachtete Schnitt statistisch geführt oder durch die Berührungszonen der Körner hindurchgelegt wird."

"With this an old point of controversy has been clarified which has not yet been settled in the hydraulic literature, namely the question as to whether with the action of the flowing liquid on the solid the full pressure gradient or the pressure gradient provided with the factor *n*, must be introduced. It depends thus only on the fact of whether the considered cut has been statistically conducted or has been put through the contact zones of the grains."

Furthermore, Heinrich and Desoyer pointed out:

"Im Falle der Punktberührung ist für ausgewählte Schnitte durch die Berührungspunkte... der volle Auftrieb wirksam. Für statistische Schnitte... verschwindet der Auftrieb zur Gänze, unabhängig davon, ob Punktberührung vorliegt oder nicht

In allen Fällen also, in denen es auf statistische Mittelwerte der Spannungen im Festkörper ankommt (z.B. wenn man Spannungs-Dehnungs-Beziehungen verwendet, um das elastische oder plastische Verhalten des Festkörpers zu untersuchen) darf... nicht mit einem Auftrieb gerechnet werden, da in diesem Fall der Festkörper statistisch zu schneiden ist..."

"In the case of the point contact, the full uplift is effective for selected cuts through the contact points. For statistical cuts the uplift disappears completely, independently of the fact whether point-contact occurs or not.

In all cases in which the statistical averages of the stresses in the solid are the point (e.g. if one uses stress-strain-relations in order to investigate the elastic or plastic behavior of the solid body) an uplift cannot be expected, because in this case the solid body must be cut statistically."

It is not easy to understand how Heinrich and Desoyer could have overlooked the fact that, for a continuum-mechanical approach, only the statistical cut is possible and that physical effects, like the effect of the flowing liquid on the solid and the uplift, cannot depend on special cuts.

In 1961, Heinrich and Desoyer (1961) developed a consolidation theory which was based mostly on the same fundamentals as the preceding works. In this paper, they criticized Biot's method:

"In einer Reihe von Arbeiten gibt *Biot* eine Erweiterung der Setzungstheorie auf mehrdimensionale Vorgänge. Da er jedoch nicht von einer Kräfteanalyse für die beiden Komponenten (Festkörper, Flüssigkeit) und den beiden zugehörigen Kontinuitätsgleichungen ausgeht, erscheinen seine Ansätze nicht aus den Grundlagen entwickelt."

"In a series of papers, *Biot* gives an extension of the settlement theory to multi-dimensional processes. Because he does not preceed, however, from an analysis of the forces of both components (solid, liquid) and the corresponding continuity equations, his relations do not seem to be developed from the fundamentals."

In the 1961 paper, Heinrich and Desoyer used a stress-strain-relation which is closely connected to Hooke's law. However, they formulated this constitutive relation in the increments and distinguished carefully between consolidation and swelling. With this constitutive relation, they developed the partial differential equations which describe the consolidation problem. They solved these equations, for some boundary value problems, with the help of Laplace transformations.

Heinrich (1962) later returned to the consolidation problem with a simplified theory. There are, however, no papers known dating from later than 1962.

4.7.3
Frenkel's Description of Moist Soil

Frenkel's investigations concerning moist soil resulted in a quantitative determination of the electric effect associated with the propagation of elastic vibrations in the soil. In relation to porous media theories, only his basic equations of the two-phase system are of interest. Thus, the derivation of the equations of motion, with respect to the liquid and soil phases, will be reviewed in the following (see Frenkel, 1944).

Frenkel stated that any theory of the motion of water in a soil based upon Darcy's law did not take into account the fact that the particles of the soil could be elastically compressed and extended. He assumed that the external forces and the hydrostatic pressure acted only on the liquid filling these pores. This simplifying assumption necessitates a correction, even in the case of problems on the steady flow of porewater under the influence of given external forces. The soil, as a two-phase system, is characterized by a partial independence of the two components from each other.

The macroscopic theory of soil only considers such distances that are large compared with the dimensions of the solid particles or of the pores, and such elements of volume that contain a large number of these particles and pores. The degree of porosity is taken into account by a coefficient n equal to the ratio of the volume of the liquid-saturated pores v^F to the total (macroscopic) volume occupied by the soil, $\mathrm{v} = \mathrm{v}^S + \mathrm{v}^F$. Thus,

$$n = \frac{\mathrm{v}^F}{\mathrm{v}^S + \mathrm{v}^F} \tag{4.7.68}$$

holds.

Frenkel defined the partial density ρ^i and the effective density ρ^{iR} of the phases as follows. The actual density will be denoted by ρ^{iR}, and the mean (macroscopic) density by ρ^i. Referring the volumes v^i and v to unit mass, 1, we have:

$$\rho^{iR} = \frac{1}{\mathrm{v}^i}\,, \quad \rho^i = \frac{1}{\mathrm{v}}\,, \tag{4.7.69}$$

and consequently, with respect to the solid:

$$\rho^S = \rho^{SR}\,\frac{\mathrm{v}^S}{\mathrm{v}} = \rho^{SR}\,(1 - \frac{\mathrm{v}^F}{\mathrm{v}})\,, \tag{4.7.70}$$

i.e., according to the definition (4.7.68):

$$\rho^S = \rho^{SR}\,(1 - n)\,. \tag{4.7.71}$$

If all pores of the soil are completely filled with a liquid that can flow freely in and out of them, in order to remain in equilibrium and in the absence of external forces the liquid phase must be subjected to the same hydrostatic pressure p, at all points of the multiply connected space formed by the pores. This pressure p must be exerted on the solid skeleton of the soil.

Under the assumption of an absolutely rigid solid skeleton, the flow of the liquid phase is determined by Darcy's equation:

$$\mathbf{v}_F = \frac{K^S}{\mu^F}(-\operatorname{grad} p + \rho^F \mathbf{b}) , \tag{4.7.72}$$

where $\rho^F \mathbf{b}$ denotes the external force acting on the liquid contained in a unit volume of the soil, μ^F represents the viscosity coefficient of the liquid, and K^S the intrinsic permeability of the soil. The latter is proportional to the degree of the porosity n, and to the mean value of the cross-section of the pores, i.e., to the square of their linear dimensions d:

$$K^S = \text{const.} \cdot n\, d^2 . \tag{4.7.73}$$

Eqn. (4.7.72) must be completed by the continuity equation:

$$\frac{\partial \rho^F}{\partial t} + \operatorname{div}(\rho^F \mathbf{v}_F) = 0 . \tag{4.7.74}$$

Darcy's equation (4.7.72) refers to the case of a steady flow. In the case of a variable flow, it is replaced by:

$$\rho^F \frac{\partial \mathbf{v}_F}{\partial t} = -\operatorname{grad} p + \rho^F \mathbf{b} - \frac{\mu^F}{K^S}\mathbf{v}_F . \tag{4.7.75}$$

The convective part of the material time derivative of $\mathbf{v}_F$ can be neglected. However, Frenkel stated that this equation was inexact, because of the absence of the factor n at the gradient of the pressure p. Introducing this factor, he obtained the correct equation for the motion of the liquid:

$$\rho^F \frac{\partial \mathbf{v}_F}{\partial t} = -\, n \operatorname{grad} p + \rho^F \mathbf{b} - \frac{\mu^F}{\kappa}\mathbf{v}_F , \tag{4.7.76}$$

where the ordinary filtration coefficient K^S is replaced by the coefficient,

$$\kappa = \frac{K^S}{n} = \text{const.} \cdot d^2 , \tag{4.7.77}$$

which ensures the validity of Darcy's law in its usual statement (4.7.72), for the special case of a steady flow of the liquid.

Equation (4.7.76) is easily generalized to the case where the deformability of the solid skeleton becomes important, i.e., the absolute velocity of the liquid $\mathbf{v}_F$ must be replaced by the relative velocity with respect to the solid phase at the corresponding point. This relative velocity is connected with the friction force acting on the liquid in a unit volume of the soil by the relation:

$$\mathbf{f}^{FS} = -\frac{\mu^F}{\kappa}(\mathbf{v}_F - \mathbf{v}_S) . \tag{4.7.78}$$

With respect to the solid phase, the friction force acts with opposite sign, i.e., $\mathbf{f}^{SF} = -\mathbf{f}^{FS}$. Replacing $\mathbf{v}_F$ in Eqn. (4.7.76) by $\mathbf{v}_F - \mathbf{v}_S$, concerning the friction force, we obtain the final form of the equation of motion for the liquid phase:

$$\rho^F \frac{\partial \mathbf{v}_F}{\partial t} = -\, n \,\text{grad}\, p + \rho^F \mathbf{b} - \frac{\mu^F}{\kappa}(\mathbf{v}_F - \mathbf{v}_S)\,. \tag{4.7.79}$$

The equation of motion for the solid phase, in the general case of a relative motion of the liquid, can be written as follows:

$$\rho^S \frac{\partial \mathbf{v}_S}{\partial t} = \text{div}\, \mathbf{T}^S - (1-n)\,\text{grad}\, p + \rho^S \mathbf{b} + \frac{\mu^F}{\kappa}(\mathbf{v}_F - \mathbf{v}_S)\,. \tag{4.7.80}$$

Unfortunately, Frenkel's valuable paper was nearly completely forgotten and seldom cited in the literature.

4.7.4 Further Developments

In 1978, Derski published a theory concerning a porous elastic solid containing a fluid in which a mass coupling between fluid and solid was neglected, and in which the mass coefficients of the kinetic energy took on a new meaning, in relation to the coefficients introduced by Biot 1956 (see Derski, 1978).

Derski assumed that one part of the mass of the system (skeleton and fluid), denoted by ρ^1, moved with the velocity of the skeleton $\mathbf{v}_S$, and that the remaining part, denoted by ρ^2, moved with the velocity of the fluid $\mathbf{v}_F$. In this case, the specific kinetic energy K is given by:

$$K = \frac{1}{2}(\rho^1 \mathbf{v}_S \cdot \mathbf{v}_S + \rho^2 \mathbf{v}_F \cdot \mathbf{v}_F)\,. \tag{4.7.81}$$

Using the dissipation function and the generalized forces of Biot, cf., equations (4.7.22) and (4.7.24), Derski obtained, from the Lagrange equations, the following relations which neglect body forces:

$$\begin{aligned} \text{div}\, \mathbf{T}^S &= \rho^1 \dot{\mathbf{v}}_S + b(\mathbf{v}_S - \mathbf{v}_F)\,, \\ \text{grad}\, \sigma &= \rho^2 \dot{\mathbf{v}}_S + b(\mathbf{v}_F - \mathbf{v}_S)\,. \end{aligned} \tag{4.7.82}$$

The exact derivation of the equations (4.7.82), combined with a physical interpretation of the coefficients ρ^1 and ρ^2, is based on the general principles of continuum mechanics. Derski developed the mass continuity equations and the balance of momentum for the whole system and for the free fluid. He introduced the free fluid as a part of the fluid which moves with its own velocity, i.e., the velocity of the fluid. The other part of the fluid (trapped fluid) moves with the velocity of the skeleton.

The mass densities per unit volume, for the solid and the fluid, respectively, are given as ρ^S and ρ^F, and the mass density of the system ρ is equal to the sum of the partial densities, i.e., $\rho = \rho^S + \rho^F$. The density of the fluid can be divided into the density of the free fluid, ρ^{Ff}, moving with the velocity $\mathbf{v}_F$ of the fluid, and the density of the trapped fluid, ρ^{Fg}, moving with the velocity of the skeleton $\mathbf{v}_S$. Thus, using the above notation, two different mass continuity equations are obtained.

The local mass balance equation for the system is:

$$\frac{\partial \rho}{\partial t} + \operatorname{grad} \rho \cdot \mathbf{v}_S + \rho \operatorname{div} \mathbf{v}_S = -\operatorname{div}[\rho^{Ff}(\mathbf{v}_F - \mathbf{v}_S)] . \tag{4.7.83}$$

The local mass balance equation for the free fluid is:

$$\frac{\partial \rho^{Ff}}{\partial t} + \operatorname{grad} \rho^{Ff} \cdot \mathbf{v}_F + \rho^{Ff} \operatorname{div} \mathbf{v}_F = 0 . \tag{4.7.84}$$

The derivation of the equations of balance of momentum results in the following:

The balance of momentum of the skeleton with the trapped fluid is:

$$\operatorname{div} \mathbf{T}^S + (\rho - \rho^{Ff})\mathbf{b} = (\rho - \rho^{Ff})[\frac{\partial \mathbf{v}_S}{\partial t} + (\operatorname{grad} \mathbf{v}_S)\mathbf{v}_S] - b(\mathbf{v}_F - \mathbf{v}_S) . \tag{4.7.85}$$

The balance of momentum of the free fluid is:

$$\operatorname{grad} \sigma + \rho^{Ff}\mathbf{b} = \rho^{Ff}[\frac{\partial \mathbf{v}_F}{\partial t} + (\operatorname{grad} \mathbf{v}_F)\mathbf{v}_F] + b(\mathbf{v}_F - \mathbf{v}_S) . \tag{4.7.86}$$

Neglecting the convective part of the time derivative, as well as the body forces in the equations (4.7.85) and (4.7.86), the coefficients ρ^1 and ρ^2 of the relations (4.7.82) can be identified by the mass density of the skeleton connected with the mass density of the trapped fluid, and by the mass density of the free fluid, viz.:

$$\rho^1 = \rho - \rho^{Ff} , \; \rho^2 = \rho^{Ff} . \tag{4.7.87}$$

Kowalski (1979) compared the equations of motion for a fluid-saturated porous solid given by Biot (1956a) with those given by Derski (1978). Kowalski stated that the transition from one of the systems of equations to the other was possible, and thus Biot's equations were found to be equivalent to those given by Derski. They differ from each other only in the interpretation of the introduced quantities and coefficients.

Biot defined the velocity of the fluid $\dot{\mathbf{u}}_F$ as the time derivative of the displacement vector of the fluid $\mathbf{u}_F$, while Derski distinguished between the trapped fluid moving with the velocity of the skeleton $\mathbf{v}_S$, and the free fluid moving with the velocity of the fluid, $\mathbf{v}_F$. Combining the trapped and free fluids, Kowalski defined the mass-centre velocity of the whole fluid $\dot{\mathbf{u}}_F$ as follows:

$$\rho^F \dot{\mathbf{u}}_F = \rho^{Ff}\mathbf{v}_F + \rho^{Fg}\mathbf{v}_S . \tag{4.7.88}$$

Bearing in mind that the velocity of the solid $\dot{\mathbf{u}}_S$, given by Biot, is equal to the velocity of the skeleton $\mathbf{v}_S$, given by Derski, the following linear transformations between the different velocities may be obtained:

$$\begin{aligned} \mathbf{v}_S &= \dot{\mathbf{u}}_S , \\ \mathbf{v}_F &= \frac{\rho^F}{\rho^{Ff}}\dot{\mathbf{u}}_F - \frac{\rho^{Fg}}{\rho^{Ff}}\mathbf{v}_S . \end{aligned} \tag{4.7.89}$$

The application of equation (4.7.87) and of the relations (4.7.89) to Derski's form of the kinetic energy, (4.7.81), results in:

$$K = \frac{1}{2}\left[(\rho^S + \rho^{Fg} + \frac{(\rho^{Fg})^2}{\rho^{Ff}})\dot{\mathbf{u}}_S \cdot \dot{\mathbf{u}}_S - 2\rho^F \frac{\rho^{Fg}}{\rho^{Ff}}\dot{\mathbf{u}}_S \cdot \dot{\mathbf{u}}_F + \frac{(\rho^F)^2}{\rho^{Ff}}\dot{\mathbf{u}}_F \cdot \dot{\mathbf{u}}_F\right], \quad (4.7.90)$$

where Biot's mass coefficients can be identified with Derski's mass densities via:

$$\begin{aligned} \rho^S + \rho^{Fg} + \frac{(\rho^{Fg})^2}{\rho^{Ff}} &= \rho^{11}, \\ \rho^F \frac{\rho^{Fg}}{\rho^{Ff}} &= -\rho^{12}, \\ \frac{(\rho^F)^2}{\rho^{Ff}} &= \rho^{22}. \end{aligned} \quad (4.7.91)$$

The kinetic energy K, given in equation (4.7.90) with respect to the relations (4.7.91), is the same energy equation set up by the Biot equation (4.7.21). Thus, the equations of motion resulting from the Lagrange equations describe the same physical phenomena. It seems that, in some respects, Derski's equations are better than Biot's. The velocities $\mathbf{v}_S$ and $\mathbf{v}_F$ are just those which can be directly measured in experiments. It would be much more difficult to measure the velocity $\dot{\mathbf{u}}_F$ because, as was shown by relation (4.7.88), it is a barycentric velocity of the whole fluid, i.e., of the free and the trapped fluid.

4.7.5 Biographical Notes

Gerhard Heinrich (1902–1983)

Gerhard Heinrich was born in Fürstenfeld, Steiermark, on April 18, 1902. He visited the local highschool and passed the final examination with distinction.

Gerhard Heinrich moved, after having studied mechanical engineering at the Technische Hochschule in Graz, in 1922 to the Technische Hochschule of Vienna. Here, he was also awarded with a Diploma in electrical engineering in 1926. In the following time, he was employed as an unpaid trainee in Vienna, for example, with the firm Siemens Schuchart, until 1927. After this he returned to the Technische Hochschule in Vienna where he was engaged, at first, from May 1, 1927 to May 31, 1932, as a temporary auxiliary assistant for the chair for generator-construction. During this time he passed the examination to be a teacher for physics and mathematics at the University of Vienna (July 13, 1930). From 1932 on, Heinrich turned more and more to mechanics. At first, he took over the duties of an auxiliary assistant (June 1, 1932) for the chair for general and analytical mechanics at the Technische Hochschule of Vienna (Professor F. Jung) and then from November 1, 1937, the position of an assistant for the same

Fig. 4.34. Gerhard Heinrich (1902–1983)

chair. During this time his doctorate examination (April 6, 1935) was submitted. Later, he completed his habilitation thesis (July 11, 1938).

War duties led Heinrich in 1943 to the German naval observatory at Hamburg. After the war, he was unable to return to his old position at the Technische Hochschule of Vienna. However, in 1950 he was offered a professorship for general mechanics at the Technische Hochschule (as a successor to Professor K. Wolf), in recognition of his long and profound education as an academic teacher, and his outstanding achievements in the complete field of mechanics (hydromechanics, gyro-theory, statics of ropes, application of elasticity theory in the theory of friction, vibration, and application of mechanics in physical chemistry). Also in recognition of his scientific achievements, the Austrian Academy of Sciences chose him as a corresponding member in 1962, and subsequently as a true member in 1964.

Heinrich was an ideal scientist with a deep knowledge of the different branches of mechanics and he was, to his further credit, an enthusiastic professor.

Maurice A. Biot (1905–1985)

Maurice A. Biot was born in Belgium in 1905 and studied electrical engineering, mining engineering, and philosophy in Belgium. In Pasadena (Caltech) he wrote his doctorate thesis and became a co-worker of Theodor von Kármán. Later, he taught at the universities of Louvain, Harvard, and Columbia where he was a

Fig. 4.35. Maurice A. Biot (1905–1985)

member of the faculty of physics. At the beginning of World War II, he entered the US Navy as a volunteer. After the war, he belonged to the Military Intelligence Unit in Europe which questioned German scientists and engineers regarding their war-time research on jets and missiles. After a short stay at Brown University, he was busy as a consultant for the Shell Company and other private firms.

Biot published nearly two hundred papers on elasticity theory, thermodynamics, applied mathematics, soil mechanics, wave propagation, and other fields of mechanics.

4.8
Further Development of the Elasticity and Plasticity Theories

In this century the theory of elasticity, as well as the plasticity theory, have made great progress. This concerns the development of theories with finite deformations for solids with elastic and plastic behaviors. Moreover, in the plasticity theory the behavior of geomaterial has been included, which shows a distinctly different mechanical behavior from ductile materials. In the following, several important developments will be discussed.

4.8.1
Elasticity Theory

As has been discussed in Section 3.2, the linear elasticity theory – within the geometrically- and physically-linear theories – was a creation of the nineteenth century. Some results had already been transferred to porous media theory in the 1920s and 1930s. The investigations by von Terzaghi (1923) and Fillunger (1936), and also the first investigations by Heinrich (1938a, b) were, however, restricted mainly to the development of the relation between the increments of the hydrostatic pressure and the volume change. Biot (1935) treated a saturated porous solid like an ordinary isotropic elastic solid. Later (Biot, 1941a, b, Biot and Clingan 1941, 1942), he extended Hooke's law by adding the increment of the porewater pressure provided with a material-dependent constant. Heinrich and Desoyer (1961) brought some new ideas into the development of their constitutive equations in order to describe three-dimensional consolidation problems. They postulated the following relation:

(7) $$\mathbf{D}' = \alpha\left[(1+\nu)\mathbf{S}'_1 - \nu S'_1 \mathbf{I}\right] , \tag{4.8.1}$$

where $\mathbf{D}'$ is an additional change of the deformation tensor $\mathbf{D}$ of the solid-skeleton, and $\mathbf{S}'$, the corresponding change of the stress tensor $\mathbf{S}$. The material dependent quantities are denoted by α and ν. Moreover S'_1 is the first invariant of the stress increments and $\mathbf{I}$ is the identity tensor. Then, Heinrich and Desoyer (1961) determined the volume change ε':

(9) $$\varepsilon' = D'_1 = \mathbf{D}' \cdot \mathbf{I} = \alpha(1-2\nu)S'_1 \tag{4.8.2}$$

and they stated:

"In sinngemäßer Erweiterung der bisherigen Ansätze wird angenommen, daß die positiven Beiwerte α und ν nur von $sgn\ D\varepsilon'/dt$ abhängen, wenn D/dt die substantielle Ableitung bezüglich der Skelettbewegung bedeutet. Da $\nu < \frac{1}{2}$ sein muß, gilt $sgn\ D\varepsilon'/dt = sgn\ DS_1/dt$. Ist $D\varepsilon'/dt$ negativ (Konsolidierung), dann sollen die Beiwerte mit α_C und ν_C; ist $D\varepsilon'/dt$ positiv (Schwellung) mit α_S und ν_S bezeichnet werden. Es ist anzunehmen, daß die Beiwerte α und ν in geringem Maße auch noch in anderer Weise von den Invarianten des Deformationsaffinors und des Affinors der Deformationsgeschwindigkeiten abhängen werden, doch sollen im folgenden $\alpha_C, \nu_C, \alpha_S, \nu_S$ als Konstante angesehen werden. Ferner soll angenommen werden, daß der Sprung der Werte α und ν nur bei Vorzeichenänderung von $D\varepsilon'/dt$ auftritt.

Aus (7) folgt mit Verwendung von (9)

(10) $$\mathbf{S}' = \frac{1}{\alpha(1+\nu)}\left(\mathbf{D}' + \frac{\nu}{1-\nu}\varepsilon'\mathbf{I}\right) . \tag{4.8.3}$$

Um einen Zusammenhang zwischen der Änderung n' des relativen Porenvolumens n und der Dilatationsänderung ε' herzustellen, denken wir uns ein Volumenelement dV, das sich aus den Anteilen dV_K (Kornmaterial) und dV_W (Wasser) zusammensetzt:

(11) $$dV = dV_K + dV_W . \tag{4.8.4}$$

Nach Definition ist

(12) $$n = \frac{dV_W}{dV} . \tag{4.8.5}$$

Bedeutet D die substantielle Änderung, dann ist

(13) $$D\varepsilon = \frac{D(dV)}{dV} . \tag{4.8.6}$$

Wegen der Erhaltung des Kornvolumens folgt aus (11)

(14) $$D(dV) = D(dV_W) \tag{4.8.7}$$

und damit aus (12)

$$Dn\, dV + nD(dV) = D(dV_W)\,. \qquad (4.8.8)$$

Daraus ergibt sich bei Verwendung von (13) und (14)

$$\frac{Dn}{1-n} = D\varepsilon\,. \qquad (4.8.9)$$

Die Integration liefert, wenn zum Anfangswert ε_0 das relative Porenvolumen n_0 gehört,

$$-\,ln\left(1 - \frac{n-n_0}{1-n_0}\right) = \varepsilon - \varepsilon_0\,. \qquad (4.8.10)$$

Setzt man

(15) $$n - n_0 = n' \qquad (4.8.11)$$
$$\varepsilon - \varepsilon_0 = \varepsilon'\,,$$

so ergibt sich

(16) $$ln\left(1 - \frac{n'}{1-n_0}\right) = -\,\varepsilon'\,. \qquad (4.8.12)$$

Da die Gleichungen (7) und (9) nur für kleine Änderungen Gültigkeit haben, wird man im Rahmen dieser Größenordnung statt (16) die Näherung

(17) $$\frac{n'}{1-n_0} = \varepsilon' \qquad (4.8.13)$$

verwenden."

"Within the corresponding extension of the previous approaches it is assumed that the positive coefficients α and ν depend only on $sgn\, D\varepsilon/dt$, where D/dt means the material time derivative with respect to the motion of the skeleton. Since $\nu < \frac{1}{2}$ must be, $sgn\, D\varepsilon'/dt = sgn\, DS_1'/dt$ is valid. When $D\varepsilon'/dt$ is negative (consolidation) the coefficients shall be denoted by α_C and ν_C; when $D\varepsilon'/dt$ is positive (swelling) the coefficients shall be denoted by α_S and ν_S. It is to be assumed that the coefficients α and ν also depend to a small degree in another way on the invariants of the deformation tensor and the tensor of the deformation velocities. However, in the following, $\alpha_C, \nu_C, \alpha_S, \nu_S$ are considered as constants. Moreover it shall be assumed that the jump of the values α and ν only occurs if the sign of $D\varepsilon'/dt$ changes.

From (7), in consideration of (9), it follows that

(10) $$\mathbf{S}' = \frac{1}{\alpha(1+\nu)}\left(\mathbf{D}' + \frac{\nu}{1-\nu}\varepsilon'\mathbf{I}\right). \qquad (4.8.3)$$

In order to construct a relation between the change n' of the relative pore volume n and the change of the dilatation ε', we consider a volume element dV which is composed of the constituents dV_K (grain material) and dV_W (water):

(11) $$dV = dV_K + dV_W\,. \qquad (4.8.4)$$

By definition it is:

(12) $$n = \frac{dV_W}{dV}\,. \qquad (4.8.5)$$

When D means the substantial change, then it is

(13) $$D\varepsilon = \frac{D(dV)}{dV}\,. \qquad (4.8.6)$$

Because of the conservation of the grain volume it follows from (11) that

(14) $$D(dV) = D(dV_W) \qquad (4.8.7)$$

and thus from (12) that

(15) $$Dn\, dV + nD(dV) = D(dV_W)\,. \tag{4.8.8}$$

From this, utilizing also (13) and (14)

$$\frac{Dn}{1-n} = D\varepsilon \tag{4.8.9}$$

results. When the relative pore volume n_0 belongs to the initial value ε_0, the integration yields,

$$-\log\left(1 - \frac{n - n_0}{1 - n_0}\right) = \varepsilon - \varepsilon_0\,. \tag{4.8.10}$$

If we introduce

(15) $$\begin{aligned} n - n_0 &= n' \\ \varepsilon - \varepsilon_0 &= \varepsilon'\,, \end{aligned} \tag{4.8.11}$$

we obtain

(16) $$\log\left(1 - \frac{n'}{1 - n_0}\right) = -\,\varepsilon'\,. \tag{4.8.12}$$

Since the equations (7) and (9) are only valid for small changes, one will use within the framework of this order, instead of (16), the approximation

(17) $$\frac{n'}{1 - n_0} = \varepsilon'\,.\text{"} \tag{4.8.13}$$

While it is true that further important papers on the consolidation theory have been published, there are hardly any constitutive equations to be found which have gone beyond the approach of Heinrich and Desoyer (1961). It should, however, be recalled that Rendulic (1936) had already clearly distinguished between loading and unloading.

General non-linear constitutive equations for describing finite deformations in one-component elastic materials have been developed in this century. Murnaghan's treatise (1937) on finite deformations of an elastic solid is incomplete. Truesdell and Noll (1965) criticized:

"The exposition most widely read in recent years is that of Murnaghan, which is brief, easy, and so incomplete as scarcely to be representative."

The efforts of Mooney (1940), Rivlin (1948), as well as those of Rivlin and Saunders (1951) were, on the contrary, much more successful. An extended survey on the finite elasticity theory is contained in the book by Truesdell and Noll (1965). Further fundamental scientific findings can be found in the contributions by Doyle and Ericksen (1956), by Wang and Truesdell (1973) as well as in those by Marsden and Hughes (1983).

Based on Rivlin's Neo-Hooke model (1948), Simo and Pister (1984) developed a new elasticity law which contained two material-dependent coefficients which allowed for the description of finite volume strains. The elasticity laws discussed refer exclusively to the description of the elastic behavior of non-porous solids. Publications over elasticity laws for porous solids are rare. Ehlers (1989a) and Bluhm and de Boer (1994) have discussed new laws on the basis of the constitutive theory of elastic non-porous material, following the investigations of Simo and Pister (1984).

4.8.2 Plasticity Theory

As has already been pointed out in Section 3.5, basic investigations into the plastic behavior of solids had been performed in the second half of the last century and the early part of this century. These concerned, however, the treatment of the failure and the ideal-plastic behavior.

After von Mises's fundamental work of 1913, it was Schleicher (1926) who presented a careful paper on the mechanical behavior of frictional materials at the limit of the elastic range, which remained at that time widely unknown. In this paper, he reviewed all of the primary existing yield conditions and cited various experimental observations which had been gained by von Kármán (1911) and Böker (1915) through tests for different stress paths on marble, sandstone, and zinc. Schleicher (1926) defined the limit of the elastic range by the hypothesis:

"Das Maß für die Höhe der Beanspruchung ist die gesamte in der Raumeinheit aufgespeicherte Formänderungsarbeit."

"The measure for the amount of the strength is the entire mechanical work stored in the space unit."

Moreover, he postulated that the quantity which limited the mechanical work of a linear-elastic solid had to be a function of the hydrostatic pressure. He showed that the new yield condition was able to predict the experimental results satisfactorily; however, he missed the comparison in the octahedral plane. For this reason, he could not recognize that his yield (failure) condition had to fail, since it did not contain the necessary information in order to describe compression and extension states, namely the third invariant of the deviator stresses. Incidentally, a vehement dispute over priority questions arose between Schleicher and von Mises (Schleicher, 1926, 1928, von Mises, 1928) concerning the new yield condition.

Von Mises' (1928) subsequent paper has caused, in some respects, some misunderstandings in the secondary literature. In this article, von Mises developed the "concept of plastic potential", also known as the associated flow rule:

"Wenn wir die gleichen Überlegungen auf Kristalle übertragen und fordern, daß die sechs Größen $\varepsilon_x, \ldots, \gamma_x, \ldots$ (mit $\varepsilon_x + \varepsilon_y + \varepsilon_z = 0$) solche lineare Funktionen der fünf Spannungsgrößen $\sigma_x - \sigma_z$, $\sigma_y - \sigma_z$, τ_x, τ_y, τ_z sind, die den Invarianzbedingungen des betreffenden Kristallsystems genügen, so gelangt man zu einem sehr klaren und einfachen Ergebnis. Die sechs linearen Funktionen müssen die Ableitungen einer quadratischen Form F der Spannungskomponenten sein:

$$\begin{aligned} \varepsilon_x &= \frac{\partial F}{\partial \sigma_x}, \quad \varepsilon_y = \frac{\partial F}{\partial \sigma_y}, \quad \varepsilon_z = \frac{\partial F}{\partial \sigma_z}, \\ \gamma_x &= \frac{\partial F}{\partial \tau_x}, \quad \gamma_y = \frac{\partial F}{\partial \tau_y}, \quad \gamma_z = \frac{\partial F}{\partial \tau_z}, \end{aligned} \tag{4.8.14}$$

wobei F den beiden Forderungen 1 und 2 in Abschnitt 1 genügen muß."

"If we transfer the same reflections to crystals and postulate that the six quantities $\varepsilon_x, \ldots, \gamma_x, \ldots$ (with $\varepsilon_x + \varepsilon_y + \varepsilon_z = 0$) are such linear functions of the five stress quantities $\sigma_x - \sigma_z, \sigma_y - \sigma_z, \tau_x, \tau_y, \tau_z$ which fulfil the invariance conditions of the respective crystal system, then one comes to a very clear and simple result. The six linear functions must be the derivatives of a quadratic form F of the stress components:

$$\begin{aligned} \varepsilon_x &= \frac{\partial F}{\partial \sigma_x}, \quad \varepsilon_y = \frac{\partial F}{\partial \sigma_y}, \quad \varepsilon_z = \frac{\partial F}{\partial \sigma_z}, \\ \gamma_x &= \frac{\partial F}{\partial \tau_x}, \quad \gamma_y = \frac{\partial F}{\partial \tau_y}, \quad \gamma_z = \frac{\partial F}{\partial \tau_z}, \end{aligned} \tag{4.8.14}$$

whereby F must satisfy both requirements 1 and 2 in Section 1."

The essential requirement in Section 1 is:

"Die Funktionswerte (der Fließbedingung, der Autor) bleiben ungeändert, wenn man dem Spannungszustand eine allseits gleiche Normalspannung (einen hydrostatischen Druck) hinzufügt"

"The values of the function (of the yield condition, the author) remain unchanged if one adds an all-round equal normal stress (a hydrostatic pressure)."

It can be stated that von Mises (1928) clearly listed all of the premises which were to become the basis of the "concept of plastic potential", namely that the six strain components had to be linear functions of the five stress quantities introduced and that the yield function had to be independent of the hydrostatic pressure. In particular, the second requirement was not considered by later authors (see Drucker and Prager, 1952).

In a work of unusual depth for its time, Fromm (1933) constructed a general constitutive theory for isotropic continua within the framework of purely mechanical-plastic deformations. After he had described ideal-plastic behavior, Fromm (1933) went on to discuss the behavior of materials under the influence of the loading history. He made a clear distinction between the influence of the loading history on the yield conditions and its influence on the flow rule.

At first, he showed the possibility that the effect of the loading history could be included in the yield condition. He introduced a scalar valued yield function $X(\mathbf{S}, \mathcal{H})$, which depends on the stress tensor $\mathbf{S}$ and on the quantity $\mathcal{H}$ describing the deformation (loading) history. This quantity can be a scalar, a tensor, or something else. The yield condition is only a function of the stresses if the deformation history is prescribed

$$X(\mathbf{S}, \mathcal{H}) = \Psi_{(\mathcal{H})}(\mathbf{S}). \tag{4.8.15}$$

The response function $\Psi_{(\mathcal{H})}$ depends, of course, on the prescribed deformation history $\mathcal{H}$. At the onset of the plastic deformations, at the limit of the elastic range, the flow is independent of the deformation history and therefore $\mathcal{H}$ is equal to zero, so that

$$X(\mathbf{S}, \mathbf{O}) = \Psi_{(\mathbf{O})}(\mathbf{S}) = \Phi(\mathbf{S}) \tag{4.8.16}$$

corresponds with the yield condition Φ at the limit of the elastic range. After these introductory remarks, Fromm (1933) discussed two forms of the yield function

$$X(\mathbf{S},\mathcal{H}) = \Psi_{(\mathcal{H})}(\mathbf{S}) = L\left(\frac{\mathbf{S}}{m}\right) = H(\mathbf{S}-\mathbf{Z}) \tag{4.8.17}$$

with $m = m(\mathbf{S},\mathcal{H})$ as a scalar quantity and $\mathbf{Z} = \mathbf{Z}(\mathbf{S},\mathcal{H})$ as an additional tensor.

Fromm [16] remarked that it is always possible to identify L as well as H with Ψ. This means that the stresses are pulled back to the initial state. This procedure leads to $m(\mathbf{S},\mathbf{O})$ equals unity and $\mathbf{Z}(\mathbf{S},\mathbf{O})$ equals the zero tensor. Thus, one can state that, with (4.8.17), isotropic and kinematic hardening have been introduced. Finally, Fromm (1933) adopted von Mises's yield condition in order to show an application of (4.8.16)

$$X(\mathbf{S},\mathbf{O}) = \Psi_{(\mathbf{O})}(\mathbf{S}) = \frac{1}{2}(\mathbf{S}^D \cdot \mathbf{S}^D) - K^2 \,. \tag{4.8.18}$$

With (4.8.16) and (4.8.17), the scalar-valued yield condition $X(\mathbf{S},\mathcal{H})$ could be reformulated

$$X(\mathbf{S},\mathcal{H}) = \Psi\left(\frac{1}{m}\mathbf{S}\right) = \frac{1}{m^2}\left[\frac{1}{2}\mathbf{S}^D \cdot \mathbf{S}^D - (mK)^2\right] . \tag{4.8.19}$$

Then he chose a specific hardening factor, m_0, and a specific additional tensor, $\mathbf{Z}_A$, and found the relation:

$$X(\mathbf{S},\mathcal{H}) = \frac{1}{m_0^2}\left[\frac{1}{2}(\mathbf{S}^D - \mathbf{Z}_A) \cdot (\mathbf{S}^D - \mathbf{Z}_A) - m_0^2 K^2\right] \tag{4.8.20}$$

which turns at the flow limit, $X = 0$, into

$$(\mathbf{S}^D - \mathbf{Z}_A) \cdot (\mathbf{S}^D - \mathbf{Z}_A) = 2m_0^2 K^2 \,. \tag{4.8.21}$$

This form of the yield condition, describing combined isotropic and kinematic hardening, is sometimes used in the plasticity theory. The scalar hardening factor m_0 was identified by Fromm (1933) as a function of the plastic work A_p:

$$m_0 = m_0(A_p) \,. \tag{4.8.22}$$

For the additional tensor $\mathbf{Z}_A$, he chose the following approach:

$$\mathbf{Z}_A = C_1(\mathbf{S},\mathbf{H})\mathbf{H}^D + C_2(\mathbf{S},\mathbf{H})(\mathbf{H}\cdot\mathbf{I})\mathbf{I} \tag{4.8.23}$$

in which the deformation history $\mathcal{H}$ is described by the second-order tensor $\mathbf{H}$. Fromm (1933) stated that, for the description of the deformation history, only such tensors could be chosen which, according to all experience, depended essentially on the plastic strains. It is true, however, that the plastic strain tensor itself is less suitable for the description of the deformation history because it can happen that it is independent of the deformation history. For example, in the case of rigid-plastic deformations, the plastic strain tensor can be calculated from the displacement field independent of the deformation history. Even if the

[16] Fromm (1933) considered von Mises' yield condition only as an example. A general yield condition was proposed by him in the form

$$\psi(\mathbf{S}) = F(\mathbf{S}^D) - f(\mathbf{S}\cdot\mathbf{I}) \,,$$

which corresponds to the yield (failure) condition of frictional materials

plastic strain tensor is determined by an integration process from the plastic strain rates field, the plastic strain tensor will still be less appropriate for the description of the deformation history. Of course, in this case, the plastic strain tensor depends on the deformation history. However, it may happen that the deformation (loading) history takes the special course that unloading occurs on the same path as loading. Thus, the plastic strain tensor would turn out to be zero and would miss the description of the deformation history.

After discussing various possibilities of including the deformation history, Fromm (1933) proposed two relations for the tensor of deformation history

$$\mathbf{H} = \int_0^t f(\mathbf{S}, \mathbf{E}^{D\prime})\left[\alpha + \beta \frac{\mathbf{H} \cdot \mathbf{E}^{D\prime}}{|\mathbf{H}||\mathbf{E}^{D\prime}|}\right] \frac{\mathbf{E}^{D\prime}}{|\mathbf{E}^{D\prime}|} Dt \tag{4.8.24}$$

and

$$\mathbf{H} = \sqrt{2} \int_0^t (\mathbf{S}^D \mathbf{E}^{D\prime}) Dt\,, \tag{4.8.25}$$

where $(\ldots)'$ marks the material time derivative with respect to the motion of the solid, t the time, and $\mathbf{E}$ the plastic strain tensor. The material dependent constants α and β must be determined by tests; the scalar value function $f(\mathbf{S}, \mathbf{E}^{D\prime})$ is always positive.

At the end of his valuable paper, Fromm (1933) discussed the properties of the flow rule. He stated that the deformation history had to have a significant influence on the flow rule, but that very little was known about this subject in the literature of that time.

It seems that Fromm (1933), in fact, pioneered the field of deformation history and, more specifically, the field of kinematic hardening. It is hard to understand why his paper has been almost completely ignored in the literature.

It seems that Geiringer and Prager (1934) were the first scientists to introduce a non-associated flow rule. Departing from von Mises's procedure (1928), they chose not the yield condition but, without any further foundation, an arbitrary function (plastic potential function) as a potential in the flow rule. Their procedure corresponds exactly to that which is used today by many researchers in the field of geomechanics.

The next step towards a complete theory of kinematic hardening materials was made by Melan (1938). He started with the description of the uni-axial stress-strain relations in the states of loading, unloading, and reloading in the opposite direction, considering the Bauschinger effect. He then extended his results to the three-dimensional stress-strain state and formulated the yield function

$$\Psi(\mathbf{S}) = \Phi(\mathbf{S} - \mathbf{Z}). \tag{4.8.26}$$

Plastic deformations can only occur if

$$\Phi(\mathbf{S} - \mathbf{Z}) = 0 \tag{4.8.27}$$

and a certain loading criterion is fulfilled.

Melan (1938) assumed that $\Phi = 0$ bound a convex region. He stated that the physical character of the problem further stipulated that the yield function be an invariant quantity, remarking that two problems remained open. First, during plastic flow the relation between the strain rates and the stresses and stress rates (flow rule) had to be known. Second, the rate of the tensor **Z** had to be a function of the strain and stress rates. Melan (1938) solved the first question by applying the uniqueness theorem and deriving an associated flow rule for the plastic strain rates **E**′:

$$\mathbf{E}' = \frac{1}{C}\left(\frac{\partial \Phi}{\partial \mathbf{S}} \cdot \mathbf{S}'\right)\frac{\partial \Phi}{\partial \mathbf{S}}, \tag{4.8.28}$$

where C denotes a positive scalar. For the second problem, he discussed the possibilities:

$$\mathbf{Z}' = \mathbf{S}' , \ \mathbf{Z}' = c\mathbf{E}', \tag{4.8.29}$$

where c is a material-dependent number. He pointed out that the first case in (4.8.29) would lead to a displacement of the flow area, but not to plastic strain rates. On the other hand, the second case in (4.8.29) would lead to a displacement of the yield area which is always connected with plastic strain rates.

Melan's paper is distinguished by its great clarity concerning the development of the yield function and the flow rule. Melan (1938) recognized that, during the plastic flow, the yield function shifted in such a way that the stresses always remained on the limit of the yield area. With respect to his investigations of kinematic hardening, the paper has rarely been mentioned in the literature. If the paper is cited at all, then it is mostly in connection with the shake down problem which Melan treated at the end of his paper.

It took 17 years for the problem of kinematic hardening to be again taken into account. In 1955, Prager (1955) published a survey of recent achievements in plasticity. In this paper, Prager first illustrated stress-strain relations with so-called dynamic models of mechanical behavior under uniaxial stress, e.g., springs and "blocks that experience solid friction as they slide along their supports." He criticized these dynamic models "that represent stresses by forces and uses instead kinematic models that represent stresses by displacements." He specifically chose "a kinematic model of mechanical behavior under uniaxial stress." In subsequent considerations, he illustrated the mechanical behavior of sandwich beams in the elastic and plastic range and the two-dimensional Bauschinger effect with this kinematic model. Finally, he extended the illustrations materials of the von Mises' type, including rigid, work-hardening, and elastic and perfectly plastic materials.

It seems that Prager considered the kinematic model as a substitute for a mathematical model which can serve as a base for mathematically and physically founded theories. This can be seen from the fact that, at the end of his general treatment of the stress-strain relations, he stated: "In the older literature on perfectly plastic solids, yield condition and stress-strain law were treated as independent ingredients of a theory of plasticity. Our kinematical models, however, exhibit a definite connexion between the shape of the frame (yield condition) on one hand and the relation between small displacement of pin and frame (stress-strain law) on the other." He remarked further: "All kinematical models considered so far use convex frames. If this convexity is stipulated as a necessary feature of an acceptable yield condition, the models suggest that the energy dissipation in a rigid, perfectly plastic solid is uniquely defined by the history of straining." In a second paper, Prager (1956) repeated his ideas regarding the description of the mechanical behavior of materials in the plastic range.

In Prager's papers (Prager, 1955, 1956), an analytical treatment of the complex field of plastic deformations cannot be found. These articles remain far beyond the works of Fromm (1933) and E. Melan (1938).

Shield and Ziegler (1958) supported the reflections of W. Prager in their paper *On Prager's Hardening Rule.* In their general treatment of "a rigid-work-hardening solid", they introduced "the hardening rule suggested by Prager"

$$F(\sigma_{ik} - \alpha_{ik}) = k^2, \tag{4.8.30}$$

where σ_{ik} denotes the stress components, k^2 is a constant, and "α_{ik} represents the local translation." Furthermore, they introduced the associated flow rule and stated finally: "The definition of a Prager-hardening material is completed by assuming that the surface moves in the direction of $d\varepsilon_{ik}$; more explicitly

$$d\alpha_{ik} = c\, d\varepsilon_{ik} \text{ ."} \tag{4.8.31}$$

Here, $d\varepsilon_{ik}$ is the plastic strain increment and c is a material dependent constant. Eqs. (4.8.30) and (4.8.31) do not exist in Prager's paper (1955, 1956). In a further note, Ziegler (1959) extended the above relation

$$d\alpha_{ik} = (\sigma_{ik} - \alpha_{ik})d\mu\,, \quad d\mu > 0 \tag{4.8.32}$$

without any substantially mechanical foundation; Ziegler explained only: "This rule is a modification of Prager's law, physically acceptable since both sides of (4.8.32) are tensors of the second-order."

In the next decades, several papers on kinematic hardening appeared. Only a few will be mentioned in the following section. Mróz (1967) combined kinematic and isotropic hardening and introduced "the concept of a 'field of work-hardening moduli', " saying that, "This concept is intended to describe to a greater degree the behavior of metals for complex loading histories."

Eisenberg and Phillips (1968) treated non-linear hardening by setting

$$\alpha_{ik} = c(\kappa_1)\varepsilon_{ik}\,, \quad \dot{\kappa}_1 = \sqrt{\dot{\varepsilon}_{ik}\dot{\varepsilon}_{ik}}\,. \tag{4.8.33}$$

The paper by Dafalias and Popov (1975) contains additional historical remarks. Their main aim, however, is "to construct a model describing the material behavior for complex multiaxial loadings."

The next major step towards a plasticity theory for brittle and granular materials was taken by Drucker and Prager (1952). Drucker and Prager tried for the first time to transfer the results and the methods of the classical von Mises plasticity theory to the description of the plastic behavior of soils. They extended von Mises's yield function for ductile materials by adding the hydrostatic pressure to the yield function. Moreover, they took over the associated flow rule, although – as mentioned earlier – von Mises had already remarked in 1928 that the associated flow rule should not be used if the yield function depended on the hydrostatic pressure. Indeed, the associated flow rule yields unrealistic results with respect to the volume changes. However, it is apparent that the paper (Drucker and Prager, 1952) was a source of inspiration for intensive research on the mechanical behavior of soils in the following years, leading to extensive experimental and theoretical investigations in this field. Towards the end of the 1950s, and into the 1960s, the so-called cam clay model was developed by Roscoe *et al.* (see, e.g., Roscoe and Burland, 1968). This model is most suitable for the description of the mechanical behavior of saturated clay, not for sand, since it contains the associated flow rule and does not consider the influence of the third invariant of the stress tensor.

Until this time, the yield and failure conditions had been developed in such a manner that these – interpreting them geometrically – were "open" along the hydrostatic axis. Based on the ideas of Drucker *et al.* (1957), Di Maggio and Sandler (1971) developed the cap model in order to restrict the hardening range of frictional material in the triaxial plane. During the time to follow, the description of the different strength properties in compression and extension states and the investigation of the hardening range became more and more important. The influence of the intermediate principal stress on the plastic behavior was especially taken into account, and yield conditions with smooth functions were derived. In his doctoral thesis, Lade (1972) developed a yield condition for cohesionless soils which contained the first and third invariants of the stresses. On the basis of true triaxial tests, Gudehus (1973) worked out an isotropic hardening model, which contained an initial yield condition and a failure condition. On the basis of extensive tests with soils, Matzuoka and Nakai (1974) derived a failure condition which was similar to Lade's condition from 1972. Another known model for the description of the isotropic hardening behavior of cohesionless sand was formulated by Lade and Duncan (1975, 1976) and extended by Lade (1977) in such a way that it allowed the description of curved meridians of the yield surface in the principal stress space. The so-called hierarchical single surface model developed by Desai and his co-workers since 1980 (see Desai, 1989a, b), describing isotropic and kinematic hardening, contains all three stress invariants. A simplified truncated form of the yield function depends only on the first stress invariant and on the second and third invariants of the stress deviators. The single surface model allows for the description of many strength properties of granular and brittle materials. However, it is rather complex. A rather simple, single surface model was developed by Lade and Kim in two basic papers (see Kim and Lade, 1988, Lade and Kim, 1988).

As will be seen in the following sections, both experimental results and thermodynamic restrictions reveal that the hardening range of plastic materials is influenced by kinematic hardening effects. It seems that Mróz, Norris, and Zienkiewicz (1978), were the first authors to include these effects in the plasticity theory of soils. They used a model consisting of an initial yield surface and the so-called consolidation surface. In a further paper, Mróz, Norris, and

Zienkiewicz (1981) completed their considerations in so far as they introduced an additional parameter which marked the failure state. The hardening models of Mróz *et al.* all have in common the fact that they do not consider the different strength properties in the extension and compression region.

On the basis of numerous test observations on the deformation of granular materials under cyclic loading, Lade (1979) geometrically represented a hardening model which showed the variety of phenomena occurring during cyclic loading. Lade's hardening model corresponds to a multi-surface model, similar to Mróz's *et al.* model.

Hirai (1987) modelled the cyclic behavior of sand with combined hardening/softening. His yield function in the hardening range is rather complex. With his yield model, he is able to precisely describe the characteristics of the plastic behavior of sand under cyclic loading. With increasing hardening, however, the numerical results differ from the test results. In addition, the numerous parameters make it difficult to handle the hardening model.

More recently, de Boer and Brauns (1990) have described the kinematic hardening of granular materials on the basis of a yield function developed by de Boer (1988a, b).

As far as the flow rule is concerned, fewer investigations exist than in the field of yield and failure conditions. As has already been mentioned, the associated flow rule fails to give the volume changes correctly. Therefore, other flow rules, which keep the concept of plastic potential, have been proposed, but the potential is no longer identical with the yield function and thus not associated (see, e.g. Gudehus, 1973, Lade, 1979, and Baker and Desai, 1982). For isotropic frictional materials, de Boer (1988a) avoided the use of a plastic potential and proposed quite a simple flow rule, which is similar to the constitutive equation for a viscous fluid. This flow rule was extended by de Boer and Brauns (1990) to kinematic hardening.

The comments in this sections cannot, of course, consider all aspects of the historical development of the constitutive equations in the elastic and plastic range. Rather, the works discussed represent a subjective choice on the part of the author.

4.9 Modern Continuum Mechanics and Mixture Theory

As was already mentioned at the end of Section 3, mechanics has been developed in this century - in parts exclusively - in the direction of applied (technical) mechanics. Reflections on the fundamentals of mechanics, which were already formulated to a great extent in the eighteenth and nineteenth centuries, have been considered in the last decades, beginning in the 1950s. These results form the basis of modern continuum mechanics, which makes a consistent treatment of gaseous, liquid, and solid bodies possible. The development of continuum mechanics has been decisively influenced by the schools of Truesdell and Rivlin (see, e.g., Truesdell and Toupin, 1960, Truesdell and Noll, 1965, and Green and Zerna, 1954). An essential merit of their works was the removal of errors and inadmissible representations in the continuum theory, which had been brought in by representatives of applied (technical) mechanics. The new efforts aim to construct continuum mechanics on the basis of few mechanical and thermodynamical axioms. On the purely mechanical side, this has been very successful, as can be seen in Section 5, in which all the findings of the last decades will be summarized. The establishment of consistent thermodynamics has been less successful. Whereas the correct form of the first law of thermodynamics (balance of energy) has been proved many times and is in no way doubted, the second law of thermodynamics (entropy principle) is still questioned, in particular, for non-equilibrium processes. In this connection, Balian (1991) is cited:

"Still worse, thermodynamics is based upon many, more or less intuitive, concepts which cannot be readily formulated mathematically ..."

and furthermore:

"Thermostatics deals with *thermal*, and also *osmotic, electric, or chemical, equilibrium* states which remain unchanged with time and which are metastable, since the time for the establishing of absolute equilibrium, where all physical, chemical, or nuclear reactions have come, can be huge... On the other hand, statistical mechanics and thermodynamics of irreversible non-equilibrium processes enable one to explain a large number of phenomena, but they do not constitute a discipline which is as coherent or systematic as the theory of thermostatic equilibria..."

However, if the validity of the entropy inequality is accepted, one has gained an eminently useful instrument for obtaining restrictions for constitutive equations. It seems that the application of the entropy inequality to constitutive approaches is as important as the principles of material objectivity and isotropy. This will be demonstrated in Section 5.

As already mentioned, modern continuum mechanics was essentially formed by Truesdell. In two books (Truesdell and Toupin, 1960, as well as Truesdell and Noll, 1965) and in numerous articles, he and his disciples laid down their ideas and created a closed continuum theory. However, their work is not undisputed. The representative of the other school (Rivlin), who founded the finite elasticity theory, remarked in a sarcastic article (see Garcia-Colin and Uribe, 1991) that the Truesdell school has created a fantasy world with a stream of principles and theorems.

However, one can state that Truesdell brought the ideas of Euler, whose work he had carefully investigated, to life again, avoiding such formalistic theories as the Lagrange and Hamilton formalism.

Moreover, Truesdell was the scientist who reformulated and extended the mixture theory. After the fundamental works of Stefan, Duhem, Gibbs, Reynolds, Jaumann, and Lohr, it was Truesdell (1957a) who introduced local balance equations for mass, momentum, and energy of arbitrarily constituted mixtures. These balance equations are referred to the individual constituents in consideration of all coupling terms. They are extensively represented in Truesdell and Toupin (1960). Truesdell used as a basis for his derivations certain principles, which he later adopted as so-called "metaphysical principles" (Truesdell, 1969):

"(1) All properties of the mixture must be mathematical consequences of properties of the constituents.
(2) So as to describe the motion of a constituent, we may in imagination isolate it from the rest of the mixture, provided we allow properly for the actions of the other constituents upon it.
(3) The motion of the mixture is governed by the same equations as is a single body."

The principles (1) and (3) have given rise to some obscure derivations concerning the kinematics and other mechanical statements of the mixture bodies. For example, the barycentric velocity and its material time derivative, the acceleration, as well as the stress definition, contain material-dependent quantities, namely the densities and density supplies. This contradicts the principles of mechanics where the kinematic and stress quantities are general and independent of any special material (see the extensive discussion in de Boer, 1995). The principle (2) is an extension of Euler's cut principle, one of the fundamental principles in mechanics, applied mechanics, and engineering. The cut principle, in connection with the balance equations, allows in continuum and structural mechanics for closed problems, the calculation of unknown mechanical quantities avoiding the use of such formalism as that of Lagrange and Hamilton.

In Truesdell's and Toupin's (1960) description of mixtures, both a proper statement for the moment of momentum balance equation and a generalization of the entropy principle for mixtures were missing. With respect to Truesdell's mixture theory, Kelly (1964) developed distinct balance laws on the basis of one fundamental balance equation, thus allowing a clear assignment of the effects resulting from the partial balance equations to the mechanical quantities of the mixture. Concerning the moment of momentum balance, Kelly proposed moment of momentum supply terms, thus admitting unsymmetrical partial stress tensors.

In the early 1960s, a thermodynamic approach to the constitutive theory was generally unknown, until Coleman and Noll (1963) as well as Coleman and Mizel (1964) introduced the development of thermodynamic restrictions from the entropy inequality. The application of the entropy principle to heterogeneous materials caused exceptional difficulties (see for e.g., Eringen and Ingram, 1965, and Green and Naghdi, 1965).

It was later pointed out that the entropy inequality postulated by Bowen (1967) was the first correct version of a generalization of the entropy principle for mixtures. Müller (1968) developed the first constitutive theory of mixtures based on thermodynamic principles. His entropy inequality was basically equivalent to that of Bowen (1967). However, one difference was that Bowen postulated a special relation between the entropy flux and the heat supply, whereas Müller determined the entropy flux by a constitutive equation. With regard to the development of the entropy inequality, one must also mention Truesdell

(1968). For the formulation of the entropy inequality, Truesdell postulated that the dissipation principle of a mixture had to be determined as the sum over all constituents, analogous to classical continuum mechanics of one component media. On the basis of this postulate, Truesdell derived an entropy inequality which corresponded to that of Bowen (1967). For this reason, the general form of the entropy inequality of mixtures is nowadays known as the Bowen-Truesdell entropy inequality.

The development of the mixture theory was brought to an end to a certain extent already in the early 1970s, namely in so far as the fundamentals developed up to that time have remained valid up to today.

4.10 Theories of Immiscible Mixtures

It seems that Morland (1972) was the first scientist to use the volume fraction concept in connection with the mixture theory. In 1966, however, Mills (1966) had already used the volume fraction concept for incompressible mixtures of two separated Newtonian fluids. In this paper, Mills also formulated the incompressibility condition in such a way that he assumed the real densities of both constituents to be constant and that he considered this fact in the volume fraction condition, i.e, that the sum of the volume fractions was equal to one. Mills's formulation of the incompressibility condition has been adopted since its inception by several authors, although it only reflects the volume fraction condition as an internal constraint considering the incompressibility of the individual constituents. A similar approach was followed by Craine (1971). For empty porous solids, there was also a continuum mechanical approach using the volume fraction concept. Goodman and Cowin (1972) presented a theory for granular materials with interstitial voids and used formal arguments from continuum mechanics. The motion, deformation gradient, velocity, Jacobian, and the material time derivative were introduced in their familiar forms. In the balance equations, however, the bulk density ρ^S of the material was decomposed into the volume fraction n^S and local real density ρ^{SR}. Moreover, in order to overcome the lacks of their continuum mechanic approach, namely one missing field equation due to the introduction of the volume fraction n^S, they defined a so-called balance of equilibrated forces and of equilibrated inertia. They stated that these balance equations "are analogous to the classical balance equations of linear momentum and mass" and that the "balance of equilibrated force is motivated by a variational analysis." Later, the balance of equilibrated force was interpreted and discussed by Nunziato and Cowin (1979). Such balance equations are not contained among the field equations in the mixture theory and remain in some respects obscure. However, the results of some problems treated with the balance of equilibrated force have been encouraging (see, e.g., Goodman and Cowin, 1972, Nunziato and Walsh, 1980).

Another interesting feature of the paper by Goodman and Cowin (1972) is the correct formulation of the incompressibility condition of the solid material in kinematic terms:

$$\frac{n^S}{n_0^S} J_S = 1 , \tag{4.10.1}$$

where n_0^S denotes the volume fraction of the solid in the reference state and J_S the determinant of the deformation gradient of the porous solid. This constraint was consequently considered in the Clausius-Duhem inequality in order to gain restrictions for the constitutive equations of an incompressible porous granular material. It is amazing that the clear statement of the incompressibility of the solid constituent and the consideration of this constraint in the Clausius-Duhem inequality was obviously overlooked in some further publications. Thus, at the end of the 1970s, and on into the 1980s, the incompressibility conditions of the constituents were introduced and considered in another context (see, e.g., Bowen, 1980, Bedford and Drumheller, 1983).

As has already been mentioned, in an excellent paper from that time, Morland (1972) constructed "a simple constitutive theory for a fluid-saturated porous solid" for the purely mechanical state. At the beginning, he stated that "the gross response of fluid-saturated porous solid is most conveniently described within the framework of a mixture (or interacting continuum) theory".

Before introducing the volume fraction concept, Morland considered the kinematics of the individual constituents, and formulated the state of stress and the balance of momentum for the individual constituents in the usual way. In his section on the constitutive theory, Morland (1972) introduced some interesting and, in parts, crucial features. At the beginning, he expressed the partial density and the partial stress tensor ρ^α and $\mathbf{T}^\alpha$ of the constituent by the volume and surface fractions n^α and m^α, and by the real densities $\rho^{\alpha R}$ and the real stress tensors $\mathbf{T}^{\alpha R}$, which were denoted by "effective":

$$\begin{aligned} \rho^\alpha &= n^\alpha \rho^{\alpha R} , \\ \mathbf{T}^\alpha &= m^\alpha \mathbf{T}^{\alpha R} . \end{aligned} \tag{4.10.2}$$

It should be mentioned that while the decomposition of the partial density can be physically founded, there is, however, no physical foundation for the decomposition of the partial stress tensor. Then, Morland (1972) discussed the deformation state and decomposed the partial deformation gradient $\mathbf{F}_\alpha$ into a spherical part and a partial density preserving part $\mathbf{F}_\alpha^*$

$$\mathbf{F}_\alpha = (J_\alpha)^{1/3} \mathbf{F}_\alpha^* = \Big(\frac{\rho_0^\alpha}{\rho^\alpha}\Big)^{1/3} \mathbf{F}_\alpha^* , \tag{4.10.3}$$

where ρ_0^α is the density in the reference configuration and J_α the determinant of $\mathbf{F}_\alpha$. The effective real deformation was then defined by

$$\mathbf{F}_{\alpha R} = (J_{\alpha R})^{1/3} \mathbf{F}_\alpha^* \tag{4.10.4}$$

where,

$$J_{\alpha R} = \frac{n^\alpha}{n_0^\alpha} J_\alpha \tag{4.10.5}$$

and where n_0^α denotes the volume fractions in the reference configuration. Because the real stress tensor $\mathbf{T}^{\alpha R}$ cannot be determined by a balance equation, Morland (1972) assumed a constitutive equation for $\mathbf{T}^{\alpha R}$. The constitutive equation was assumed to be a functional $\mathcal{F}^\alpha$ depending on the effective deformation gradient

$$\mathbf{T}^{\alpha R} = \mathcal{F}^\alpha(\mathbf{F}_{\alpha R}) \,, \tag{4.10.6}$$

or, by applying the decomposition of $\mathbf{F}_{\alpha R}$ stated above,

$$\mathbf{T}^{\alpha R} = \mathcal{F}^\alpha \left[\left(\frac{n^\alpha}{n_0^\alpha} \right)^{1/3} \mathbf{F}_\alpha \right] . \tag{4.10.7}$$

Then, with $\mathbf{T}^\alpha = m^\alpha \mathbf{T}^{\alpha R}$, the partial stress $\mathbf{T}^\alpha$ can be obtained

$$\mathbf{T}^\alpha = m^\alpha \mathcal{F}^\alpha \left[\left(\frac{n^\alpha}{n_0^\alpha} \right)^{1/3} \mathbf{F}_\alpha \right] . \tag{4.10.8}$$

As a simple illustration, Morland considered an ideal fluid and proposed the following constitutive relation:

$$\mathbf{T}^F = -\rho^F \left(\frac{(n^F)^2}{n_0^\alpha} J_F \right) \mathbf{I}, \tag{4.10.9}$$

where the equivalence of the surface and volume fractions has been assumed.

For an isotropic elastic solid constituent as a one-component continuum, Morland (1972) introduced the constitutive equation

$$\mathbf{T} = \phi_0(\mathrm{I}_k)\mathbf{I} + \phi_1(\mathrm{I}_k)\mathbf{B} + \phi_2(\mathrm{I}_k)\mathbf{B}^2, \quad k = 1, 2, 3, \tag{4.10.10}$$

where $\mathbf{B} = \mathbf{F}\mathbf{F}^T$ is the left Cauchy-Green deformation tensor and ϕ_j are response functions of the invariants I_k of $\mathbf{B}$:

$$\mathrm{I}_1 = \mathbf{B} \cdot \mathbf{I}, \quad \mathrm{I}_2 = \mathrm{I}_3 \mathbf{B}^{-1} \cdot \mathbf{I}, \quad \mathrm{I}_3 = \det \mathbf{B} = J^2. \tag{4.10.11}$$

Based on these relations, Morland (1972) gained the following "effective" expressions for the solid (S) constituent:

$$\begin{aligned} &\mathbf{B}_S^E = r^{2/3} \mathbf{B}_S, \quad \mathrm{I}_{3S}^E = r^{2/3} \mathrm{I}_{1S}, \\ &\mathrm{I}_{2S}^E = r^{4/3} \mathrm{I}_{2S}, \quad \mathrm{I}_{3S}^E = r^2 \mathrm{I}_{3S}, \quad r = \frac{n^S}{n_0^S}. \end{aligned} \tag{4.10.12}$$

These considerations finally led to a partial elastic matrix stress:

$$\mathbf{T}^S = n^S \left[\phi_0(\mathrm{I}_{kS}^E)\mathbf{I} + \phi_1(\mathrm{I}_{kS}^E) r^{2/3} \mathbf{B}_S + \phi_2(\mathrm{I}_{kS}^E) r^{4/3} \mathbf{B}_S^2 \right]. \tag{4.10.13}$$

Moreover, in order to complete the constitutive theory, Morland (1972) assumed that

$$n^S = n^S(J_\alpha, \mathrm{I}_1, \mathrm{I}_2). \tag{4.10.14}$$

Thus, the volume fraction n^S appears as a dependent constitutive variable.

Morland's further treatment of the fluid-saturated porous solid was directed towards the geometrically-linear theory and towards a special problem in the finite theory. Furthermore, the elastic-plastic state of a porous solid was investigated, and a saturated porous tuff model was treated by Morland (1972).

Further contributions to the porous media theory were due to Kenyon (1976a, b), who used the mixture theory and a constitutive equation for the volume fractions; he described the thermostatics of solid-fluid mixtures and constructed a theory of an immiscible solid-fluid mixture. Passman (1977) transferred the results of Goodman and Cowin (1972) to mixtures of granular materials. He wrote the balance laws for each constituent and for the mixture and developed a constitutive theory for a mixture of two dry granular media. Sampaio and Williams (1979) published a rather formal mathematical treatment of the thermodynamic behavior of porous media (see also Bedford and Drumheller, 1983).

In a paper entitled *A Nonlinear Theory of Elastic Materials with Voids*, Nunziato and Cowin (1979) extended the balance of equilibrated force introduced by Goodman and Cowin (1972) by adding an equilibrated inertia which depended on the geometrical features of the voids. Moreover, in order to study the incompressibility condition, they decomposed the deformation gradient in a manner similar to Morland's (1972) approach. In a final section, Nunziato and Cowin tried to give a physical interpretation of the balance of equilibrated force. They stated that "this equation is a special case of an equation which arises in the microstructural theories of elastic materials." Moreover, they pointed out that some terms in their balance equations "can be identified with the singularities in classical linear elasticity known as double force systems without moments. Double force systems without moments are the stress systems equivalent to two oppositely directed forces at the same point; such systems have no net force and no resulting moment. Such a system represents, for example, the stress distribution associated with an elastic sphere forced into a spherical hole of slightly smaller diameter in an infinite elastic medium." Then, Nunziato and Cowin gave detailed explanations for several terms in the balance of equilibrated force. However, they also suggested another possible interpretation of their microstructural variables. This interpretation was based on the idea of removing a sphere from a porous material with spherical voids and replacing it with a hollow sphere (with the same porosity) ".. the response of the remainder of the body to the applied loads should be unaltered by the substitution." Thus, on this basis, one can obtain "explicit formulae for the dependence of elastic moduli on the porosity." It should be mentioned that not all of their arguments are convincing, in particular, the last interpretation.

In an extensive paper entitled *On Ideal Multiphase Mixtures with Chemical Reactions and Diffusion*, Nunziato and Walsh (1980) gave a review of the known theories and extended the multiphase mixture theory to include chemical reacting materials. They adhered to the balance equation of equilibrated force and they justified this by: "The balance equation for equilibrated force does not ordinarily arise in mixture theories; however, it is necessary in the theory of multiphase mixtures in order to account for the dynamical effects associated with changes in the volume fractions of the constituents." The results that they gained with their theory are encouraging.

A new approach to the treatment of saturated porous solids was introduced by Bedford and Drumheller (1978, 1979), namely the creation of a variational theory of porous media. Whether this theory is more useful than the other theories discussed in this section will not be investigated. It may be that it possesses certain advantages in regard to numerical calculations.

It seems that the correct formulation of the amount of constraints in fluid-saturated granular materials with incompressible constituents was due to Nunziato and Passman (1981). They consequently introduced Lagrange multipliers to the entropy inequality, in order to gain restrictions for the constitutive relations. In addition, they considered the volume fraction condition, namely that the sum of the volume fractions had to be equal to one, as a constraint. This constraint provided with a Lagrange multiplier was also added to the entropy inequality. In a paper entitled *A Multiphase Mixture Theory for Fluid-Particle Flow*, Nunziato (1983) discussed several details of the multiphase mixture theory from a physical standpoint. These details also appeared in an article by Passman, Nunziato, and Walsh (1984), and will be reviewed there. In the article from 1983, Nunziato dropped the balance of equilibrated forces and chose, instead, constitutive equations for the extra momentum exchange, without further explanation. He only remarked in a footnote: "Another approach has been to use additional microstructural force balance equations arising in granular material theories."

The theory of multiphase mixtures was reviewed and extended by Passman, Nunziato, and Walsh (1984). As has already been mentioned, this theory is characterized mainly by the additional balance equation introduced, namely the balance of equilibrated force. Again, the authors tried to justify the multiphase mixture theory:

"It is important to emphasize that these equations (the balance equations, the author) differ from those of the classical theory of mixtures in two respects:

a) There is an additional equation of balance... which, in the context of granular materials, GOODMAN & COWIN call the *balance of equilibrated force*. In their terms, $\mathbf{k}_\alpha$ is the equilibrated inertia, $\mathbf{h}_\alpha$ is the equilibrated stress, $\mathbf{f}_\alpha$ is the equilibrated force supply, and $\mathbf{l}_\alpha$ is the equilibrated body force. This equation answers to the fact that the constituent volume fractions can change without affecting the gross motion. Thus, in some sense, this equation models the microstructural force systems operative in multiphase mixtures.
b) There are appropriate additional terms in the energy equation corresponding to work done by the respective terms in the balance of equilibrated force."

Then the authors tried to fulfill Truesdell's "Metaphysical Principles" and they stated "with respect to the equations of balance, the Metaphysical Principles are satisfied." A valuable section in the article by Passman, Nunziato, and Walsh is concerned with internal constraints. They pointed out that "one important constraint" is that of saturation:

$$\sum_{\alpha=1}^{N} n^\alpha = 1, \tag{4.10.15}$$

where n^α are the volume fractions of the α-constituents. The authors made reference to Truesdell and Noll (1965), saying that

"constraints are maintained by reaction forces" and "that the simplest system of reaction forces imaginable are those which do not work. With this in mind, let π be the reaction force for the constraint ... and let the work done (W_α, the author) on constituent α due to all the constituents because of the constraint, be defined as the product of the reaction force π and the rate of change (of the volume fractions, the author) at any spatial position $\mathbf{x}$;"

$$W_\alpha = \pi\left[(n^\alpha)'_\alpha - \mathbf{x}'_\alpha \cdot \operatorname{grad} n^\alpha\right], \tag{4.10.16}$$

where $\mathbf{x}'_\alpha$ is the velocity of a material point of the constituent α in $\mathbf{x}$ and $(n^\alpha)'_\alpha$ the material time derivative of n^α.

"Physically, π may be interpreted as the pressure acting at the interface between phases which is required to maintain contact. Consequently, we refer to π as the *interface pressure*."

It should be mentioned that the formula cited above is not strictly valid. A correct rate formulation of the constraint (4.10.15) was developed by Nunziato (1983).

From investigations of the equations for energy balance, Passman, Nunziato, and Walsh derived an interesting result, namely an extra interaction force:

$$\hat{\mathbf{p}}^{\alpha E} = \hat{\mathbf{p}}^\alpha - \pi \operatorname{grad} n^\alpha, \tag{4.10.17}$$

where $\hat{\mathbf{p}}^\alpha$ is the momentum supply. In the following, they commented on this result:

"In the balance law for linear momentum, the interface pressure gives rise to an additional body force which represents the normal forces acting on constituent α due to all the other constituents. This force, which is often referred to as a 'buoyancy' force, does not arise naturally in the classical theory of mixtures."

In a footnote, the authors remarked, in addition:

"The omission of this force from the classical theory of mixtures provided part of the argument used by Müller (1968) and Bowen (1976) to justify including a dependence on density gradients in their constitutive equations."

Moreover, Passman, Nunziato, and Walsh (1984) pointed out that in the multiphase mixture theory with incompressible constituents, additional constraints had to be considered. They clearly postulated that, in this case, the real material density $\rho^{\alpha R}$ is constant so that the balance of mass reduces to

$$(n^\alpha)'_\alpha + n^\alpha \operatorname{div} \mathbf{x}'_\alpha = \frac{\hat{\rho}^\alpha}{\rho^{\alpha R}}, \tag{4.10.18}$$

where $\hat{\rho}^\alpha$ is the mass supply. The authors assumed the reaction to the incompressibility constraint to be a hydrostatic pressure, and they assumed that the work associated with this constraint to be

$$W^\alpha = p^\alpha \left[(n^\alpha)'_\alpha + n^\alpha \operatorname{div} \mathbf{x}'_\alpha - \frac{\hat{\rho}^\alpha}{\rho^{\alpha R}}\right]. \tag{4.10.19}$$

"Of course, *for each incompressible constituent* α, *the arbitrary hydrostatic pressure* p^α *is distinct from the pressure of any other constituent*."

In a section on constitutive principles, Passman, Nunziato, and Walsh (1984) introduced the following principles: Local Action, Phase Separation, Frame Indifference, and Dissipation. They mentioned that, with the exception of the Principle of Phase Separation, these principles coincided with those in the theory of one-component continua:

"In stating the list of principles governing constitutive equations, we have replaced the Principle of Equipresence in TRUESDELL's original list by the Principle of Phase Separation. In multiphase mixtures, the individual constituents are clearly separated physically, and it is plausible to think of the mixture as being ideal, or phase separated."

Finally, the authors formulated the Principle of Phase Separation as follows:

"*The dependent variables of the* α-th *constituent that are material-specific depend only on the independent variables of the* α*-th constituent. The growth* (supply terms, the author) *dependent variables depend on all of the independent variables*."

Subsequently, the authors applied their theory to a variety of physical phenomena and, finally, they commented on other theories of multiphase mixtures: fractional theories with pressure equilibrium, theories with momentum diffusion, and density rate theories. In this connection, they discussed the variational theory developed by Bedford and Drumheller (1978, 1979) and stated some critical remarks.

In two extensive papers, Bowen (1980, 1982) treated the incompressible and compressible porous media by use of the theory of mixtures restricted by the volume fraction concept. He chose another way of describing the thermodynamic behavior of porous media than that of the exponents of the multiphase mixture theory. He constructed, in both articles, porous media theories without the introduction of an additional balance equation as, e.g., the balance of equilibrated force.

In his first paper, Bowen (1980) outlined his objectives:

"This work concerns the use of the thermodynamics of mixtures to formulate incompressible porous media models. The models which result from the formulation are generally well-known. Thus, the contribution of this work is in showing how the formalism of the theory of mixtures applies to porous media modeling. Because the modern theory of mixtures had its roots in the classical theories of gas mixtures, it is perhaps not obvious that this formalism also applies to porous materials.

This article is concerned with a porous material in which the solid matrix and each pore fluid are incompressible. As is well known, the assumption of incompressibility implies a certain constraint among the variables and causes an indeterminacy to exist among the constitutive variables."

After this, Bowen (1980) summarized all of the findings of the mixture theory which he had - among others - formed substantially in the 1960s and 1970s, and introduced the volume fraction concept which contained the saturation condition that the sum of the volume fractions of all constituents be equal to one as a constraint. Furthermore, without mentioning any reasons, Bowen (1980) replaced the free Helmholtz energy function per unit mass in the Clausius-Duhem inequality with a free energy function per unit volume. This substitution leads to some consequences in the evaluation of the Clausius-Duhem inequality, which will be discussed later.

Following this, he investigated the constitutive equations of an incompressible elastic solid filled with incompressible fluids on the basis of materials of the second-order (the gradient of the deformation gradient is contained in the set of process variables). From the Clausius-Duhem inequality, Bowen (1980) derived many restrictions for the constitutive relations. In addition, he discussed special problems which can be derived from his general results: rigid solid, classical porous media models, and incompressible poroelasticity.

There are some weak points in Bowen's (1980) theory. The most severe point is that Bowen did not consider all constraints. Besides the saturation condition (4.10.15), there are α additional incompressibility conditions. Therefore, in this case, constraints must be expected with $\alpha+1$ reaction forces, i.e., $\alpha+1$ Lagrange multipliers (see, e.g., Passman, Nunziato, and Walsh, 1984). However, Bowen (1980) introduced, obviously influenced by the corresponding treatment of Mills (1966), only one constraint (the volume fraction condition) with one reaction force (one Lagrange multiplier).

Another point concerns the thermodynamic investigation of the incompressible porous media model on the basis of second-order materials. There is no need to introduce such a basis, neither from the mathematical point of view nor from the physical point of view. The necessity of considering second-order materials on the part of Bowen (1980) lay in the fact that Bowen had introduced a free energy function per unit volume in the Clausius-Duhem inequality. It can easily be shown that, in this case, the thermodynamic treatment on the basis of second-order materials, or parts of it, is a must, in order to avoid unreasonable results.

It should be mentioned that, despite these above-mentioned shortcomings, Bowen's theory has led to good results in several applications.

In a second paper, Bowen (1982) extended his porous media theory to compressible porous media. In the introduction to his paper, Bowen (1982) described the aim of his article:

"The objective of this work is to utilize the theory of mixtures to formulate a model of a mixture consisting of a compressible elastic solid and an arbitrary number of compressible immiscible fluids. The volume fraction of each fluid appears as an independent variable in the constitutive equations and is assumed to obey a certain rate type constitutive equation. Simply stated, the volume fractions are treated formally like internal state variables. ... The generality of the assumed rate laws is such that one can specialize the model to one where the volume fractions adjust instantaneously to a value determined by the local state of the mixture. No effort is made in this work to present the most general thermodynamically consistent model possible. One major objective is to show how certain classical porous media models evolve as special cases of a single formulation; a formulation which is based upon no more than the theory of mixtures."

At the beginning of his article, Bowen (1982) again summarized all basic relations of the mixture theory extended by the concept of volume fractions. He mentioned in this context that the mixture is incompressible whenever the true densities of the individual constituents are constant and that in this case the volume fraction concept "implies a constraining relationship between the bulk densities." This statement is, however, not correct. The volume fraction concept is always a constraint independent of the assumption whether the constituents are incompressible or compressible (see Nunziato, 1983). In his reflections on the constitutive theory, Bowen (1982) used an idea exploited by Drumheller (1978), "in adopting a rate law to govern the volume fractions." This rate law (evolution equation) appeared in the dissipation inequality developed by Bowen (1982). It is not easy to understand this concept, namely the introduction of evolution equations in order to overcome the missing field equations in the porous media theory, if one considers, for example, an elastic porous solid made of steel and filled with a gas. It seems that in this case the explanation for the change of the porosity as a dissipative effect is physically not founded.

In addition, Bowen (1982) specified his general theory of compressible porous solids saturated with compressible immiscible fluids. First, he considered a solid-fluid mixture with linear dissipation and center of symmetry. Then, he investigated rigid solids and some special porous media models in order to obtain models of porous materials similar to certain classical ones. The remaining part of Bowen's (1982) treatise was concerned with the development of a linear poroelasticity and, in particular, with the one-dimensional poroelasticity. Bowen summarized his basic ideas again in a review article (see Bowen, 1984).

In 1980 Mow *et al.* (1980) published a paper *Biphasic Creep and Stress Relaxation of Articular Cartilage in Compression: Theory and Experiments* which is mainly concerned with biomechanics problems. However, in this paper an imcompressible binary mixture model was indenpendently developed which is similar to Bowen's (1980) approach.

In the time to follow, in the second half of the 1980s and at the beginning of the 1990s, there seems to have been no substantial attempt to improve the fundamentals of the previously developed porous media theories. The research in porous media theories was mainly focused in three directions: first, the implantation of the developed porous media models into numerical algorithms; second, the incorporation of different material behavior into the developed mathematical models; and third, the investigation of special phenomena which appear in saturated and empty porous solids.

For the numerical treatment of initial- and boundary-value problems, quite different models have been used. These range from improved classical models proposed by Biot (1955, 1956a) to such models which are based on the mixture theory restricted by the volume fraction concept. In this connection, one may see, for example, the extended paper by Zienkiewicz *et al.* (1990) which contains an improved Biot model as well as the treatise by Schrefler *et al.* (1993) in which a model based on the mixture theory is treated.

The incorporation of different material behavior has been performed, for example, by de Boer and Kowalski (1983), de Boer and Ehlers (1986a, b), and de Boer and Lade (1991), namely the consideration of elastic-plastic deformations of the porous solids consisting of metallic and frictional materials and the viscous properties of liquids, although partially achieved on the basis of simplified porous media models.

Great efforts started in the 1980s and the beginning of the 1990s to explain and describe special phenomena which occur in saturated and empty porous solids, which were already partially recognized a long time ago but which had, however, never been fully founded. Within this framework, the paper by Baer and Nunziato (1986), which was concerned with the deflagration-to-detonation transition in reactive saturated granular materials, is mentioned. In this paper, the authors adopted, in addition to the balance equation of equilibrated force, Bowen's (1982) procedure of assuming an evolution equation for the volume fractions.

Another interesting phenomenon in saturated and partially saturated granular media was investigated by different authors on the basis of the mixture theory (see Passman and McTigue, 1984, and McTigue, Wilson and Nunziato, 1983), namely the principle of effective stresses. In the treatise by de Boer and Ehlers (1990b), in which Bowen's (1980) model for incompressible constituents was used to prove this important principle, an extended section about the historical development of the principle of effective stresses can be found.

The effects of uplift, friction, and capillarity in liquid saturated porous solids have been extensively discussed by de Boer and Ehlers (1990a, see also de Boer and Ehlers, 1988) in which the historical development of the discovery of these phenomena has also been traced.

In recent times, there have again been new attempts to improve the fundamentals of the porous media theory in such a way that the basic equations are mathematically and physically better understood. Li and Li (1992) constructed a theory on the thermo-elasticity of multicomponent fluid-saturated reacting porous media. The authors pointed out that Bowen's (1982) theory had to be expanded by additional constitutive equations. The treatise of Li and Li (1992) remains in some parts unclear. In de Boer (1994) and de Boer and Kowalski (1995), the closure of the theory was obtained by assuming α constitutive equations for the interface pressure of a porous medium consisting of α constituents. However, this model considers only the compressibility due to the interface pressure.

Chapter 5: Current State of Porous Media Theory

5.1 Introductory Remarks to Porous Media Theory

In many branches of engineering, for example, in chemical engineering, material science, and soil mechanics, as well as in biomechanics, the different reactions of material systems undergoing external and/or internal loadings must be studied and described precisely in order to be able to predict the responses of these systems. Subsequently, the most important point of the investigation is to determine first the composition of the body, because one must know the physically and chemically differing materials that constitute the system under consideration. The material systems (or bodies) in these fields of engineering can be composed in various ways. On the one hand, solids can consist of different solid components, such as dense concrete, without considerable pores. On the other hand, solids can contain closed and open pores, such as ceramics and soils, as well as concrete. The pores can be filled with fluids and, due to the different material properties and the different motions, there may be interaction between the constituents. As was mentioned in Section 1, this fact makes the description of the mechanical (or the thermodynamic) behavior difficult.

Because the exact description of the location of the pores (empty or filled with fluids) and solid material is nearly impossible, the heterogeneous composition can be investigated by using the volume fraction concept (see Section 5.2). This concept results in the effect that "smeared" substitute continua with reduced densities for the solid and fluid phases arise which can then be treated by the mixture theory.

The combination of the mixture theory with the volume fraction concept touches the microscale. The question arises as to what scale the mechanical or thermodynamic investigations should be performed, on the macro- or microscale? In principle, both strategies are possible. However, the micromechanical approach, with all its averaging processes, leads to a huge formalism (see, e.g., de Boer and Didwania, 1997), revealing some important mechanical relations. Because of the fact that this approach is still in its infancy, only the macroscopic theory, which has only recently come to consistent conclusions, will be discussed in the following pages. Of course, the micromechanical effects which are raised by the volume fraction concept will be considered using macromechanical quantities.

Another major problem arises in macroscopic porous media concerning the closure problem. It can easily be shown that, for porous media consisting of α constituents, $\alpha - 1$ field equations are missing. However, this results only out of the macroscopic point of view. Nevertheless, as was mentioned, the porous media theory touches the microscale due to the volume fraction concept. Thus, concerning the closure problem, the microscale should be taken into consideration. Following this idea, the discussion of the closure problem results in some very reasonable conclusions (see Section 5.6).

The aim of the next sections is to develop a consistent macroscopic porous media theory which would contain the results of the former macroscopic theories as special cases.

5.2 The Volume Fraction Concept

In the volume fraction concept, it is assumed that the porous solid always models a control space and that only the liquids and/or gases contained in the pores can leave the control space. Furthermore, it is assumed that the pores are statistically distributed and that an arbitrary volume element in the reference and the actual placement is composed of the volume elements of the real constituents.

The basis of the description of porous media, using elements of the theory of mixtures restricted by the volume fraction concept, is the model of a macroscopic body, where neither a geometrical interpretation of the pore structure nor the exact location of the individual components of the body (constituents) are considered. We proceed from the assumption that the constituents, which are bound against each other, are "smeared" over the control space (partial bodies) which is shaped by the porous solid, i.e., that each substitute constituent occupies the total volume of space simultaneously with the other constituents.

A porous medium occupying the control space of the porous solid B_S, with the boundary ∂B_S in the *actual placement*, consists of constituents φ^α, with real volumes v^α, where the index α denotes κ individual constituents. The boundary ∂B_S is a material surface for the solid phase and a non-material surface for the liquid and/or gas phases.

In order to formulate the volume fraction concept in *the actual placement*, we proceed from an average theory on the microscale. Average quantities are obtained, for example, by integrating a microscopic quantity over the region of the average volume dv or over the number of particles of an ensemble. The result of the averaging process is a field of macroscopic quantities.

We will consider only the volume averaging; the ensemble averaging leads to similar results (see de Boer and Didwania, 1997).

Let $\mathbf{r}$ describe the position of a constituent with volume elements dv_μ in the control space, $\mathbf{x}$ the position vector at the center of the volume element

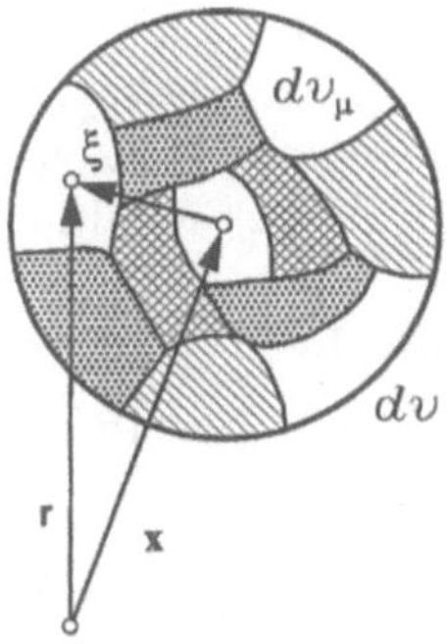

Fig. 5.2.1. Average volume element dv of a porous medium consisting of four constituents

dv (see Fig. 5.2.1), and t the time. In order to determine the average volume element dv^α of each constituent φ^α, an indicator function is defined for all κ constituents in the form

$$\chi^\alpha = \chi^\alpha(\mathbf{r}, t) = \begin{cases} 1 \text{ for } \mathbf{r} \in dv^\alpha \,, \\ 0 \text{ for } \mathbf{r} \in dv^\beta \,, \end{cases} \qquad \beta \neq \alpha \,. \tag{5.2.1}$$

Thus, for the partial volume element dv^α, with (5.2.1), the following relation is obtained:

$$dv^\alpha(\mathbf{x}, t) = \int_{dv} \chi^\alpha(\mathbf{r}, t) dv_\mu = \int_{dv} \chi^\alpha(\mathbf{x} + \boldsymbol{\xi}, t) dv_\mu, \tag{5.2.2}$$

where, in the second equation of (5.2.2), the position vector $\mathbf{r}$ is replaced by $\mathbf{r} = \mathbf{x} + \boldsymbol{\xi}$ (see Fig. 5.2.1). The integration refers to the macroscopic local $\boldsymbol{\xi}$-reference system with the origin $\mathbf{x}$. The volume element dv_μ is represented in Fig. 5.2.1.

With (5.2.2), the concept of volume fractions can be formulated as follows:

$$n^\alpha(\mathbf{x}, t) = \frac{dv^\alpha}{dv} = \frac{1}{dv} \int_{dv} \chi^\alpha(\mathbf{r}, t) dv_\mu . \tag{5.2.3}$$

The volume fractions in (5.2.3) satisfy the volume fraction condition for κ constituents φ^α,

$$\sum_{\alpha=1}^{\kappa} n^\alpha = 1 \,, \tag{5.2.4}$$

as can easily be proven. Moreover, the indicator function allows the replacement of an integral of a function Γ over dv^α by an integral over dv using the indicator function χ^α. Thus,

$$\int_{dv^\alpha} \Gamma(\mathbf{r}, t) dv_\mu = \int_{dv} \Gamma(\mathbf{r}, t) \chi^\alpha(\mathbf{r}, t) dv_\mu \,. \tag{5.2.5}$$

In order to derive average macroscopic quantities from microscopic quantities, averaging operators are necessary. In principle, these are unlimited. However, due to physical reasons, and considering the individual form of the microscopic quantities, the number of such operators is restricted to only a few. The most common of these are the volume averaging operator, the area averaging operator, and the ensemble averaging operator. In the following equations, we will only use the volume averaging operator and apply it to the proper definition of the densities.

Let the real density of the constituent materials be denoted by

$$\rho^{\alpha T} = \rho^{\alpha T}(\mathbf{r}, t) \tag{5.2.6}$$

and let the operators bring this property to the macroscale by

$$\rho^{\alpha R} = \rho^{\alpha R}(\mathbf{x}, t) \, , \; \rho^{\alpha} = \rho^{\alpha}(\mathbf{x}, t) \, . \tag{5.2.7}$$

Then, at the macroscale,

$$\rho^{\alpha R}(\mathbf{x}, t) = \frac{1}{dv^{\alpha}} \int_{dv} \rho^{\alpha T}(\mathbf{r}, t) \chi^{\alpha}(\mathbf{r}, t) dv_{\mu} \tag{5.2.8}$$

and

$$\rho^{\alpha}(\mathbf{x}, t) = \frac{1}{dv} \int_{dv} \rho^{\alpha T}(\mathbf{r}, t) \chi^{\alpha}(\mathbf{r}, t) dv_{\mu} \tag{5.2.9}$$

are the so-called macroscopic real and partial densities which contain the information of the real density referred to the substitute continua.

If a spatial constant microscopic mass density is assumed, Eqn. (5.2.8) can be written, considering (5.2.2), as

$$\rho^{\alpha R}(\mathbf{x}, t) = \rho^{\alpha T}(\mathbf{x}, t) \, . \tag{5.2.10}$$

Between the relations (5.2.8) and (5.2.9), there exists the relation

$$\rho^{\alpha}(\mathbf{x}, t) = n^{\alpha}(\mathbf{x}, t) \rho^{\alpha R}(\mathbf{x}, t), \tag{5.2.11}$$

in which (5.2.3) has been used. Eqn. (5.2.11) is well-known in porous media theory.

We will arrive at the same results if an ensemble average operator, used mainly in suspension theory and in the theory of fluidized beds, is applied.

The volume fraction concept formulated in the *reference state* at $t = t_0$ can be derived from the corresponding expressions in the actual placement.

In the *reference placement*, a porous medium occupying the control space of the porous solid B_{0S} with the boundary ∂B_{0S} denoted by the index 0, consists of κ constituents φ^{α} with real volumes $v_{0\alpha}^{\alpha}$, where the index α denotes the individual constituents. The boundary ∂B_{0S} is a material surface for the solid phase and a non-material surface for the liquid and/or gas phases. According to (5.2.1), the indicator function

$$\mathring{\chi}^{\alpha} = \mathring{\chi}^{\alpha}(\mathring{\mathbf{r}}) = \begin{cases} 1 \text{ for } \mathring{\mathbf{r}} \in dv_{0\alpha}^{\alpha} \, , \\ 0 \text{ for } \mathring{\mathbf{r}} \in dv_{0\alpha}^{\beta} \, , \end{cases} \qquad \beta \neq \alpha \, , \tag{5.2.12}$$

where $\mathring{\mathbf{r}}$ is the position vector of a constituent with volume elements $dv_{0\mu}$ in the control space. Then, for the partial volume elements $dv_{0\alpha}^{\alpha}$, with (5.2.12), the following relation holds:

$$dv_{0\alpha}^{\alpha} = \int_{dv_{0\alpha}} \overset{\circ}{\chi}^{\alpha}(\overset{\circ}{\mathbf{r}})dv_{0\mu} = \int_{dv_{0\alpha}} \overset{\circ}{\chi}^{\alpha}(\mathbf{X}_{\alpha} + \overset{\circ}{\xi})dv_{0\mu}. \tag{5.2.13}$$

The vector $\mathbf{X}_{\alpha}$ is the position vector of the center of the volume element $dv_{0\alpha} = dv_{0\alpha}(\mathbf{X}_{\alpha}, t = t_0)$ and is related to the individual constituents. In the second expression of (5.2.13), the position vector $\overset{\circ}{\mathbf{r}}$ is substituted by $\overset{\circ}{\mathbf{r}} = \mathbf{X}_{\alpha} + \overset{\circ}{\xi}$.

The partial volume element $dv_{0\alpha}^{\alpha}$ in the reference placement depends on $t = t_0$ at the position vector $\mathbf{X}_{\alpha}$, i.e.,

$$dv_{0\alpha}^{\alpha} = dv_{0\alpha}^{\alpha}\left(\mathbf{X}_{\alpha}, t = t_0\right). \tag{5.2.14}$$

With (5.2.13) and (5.2.14), the concept of volume fractions, with respect to the reference state, reads as follows:

$$n_{0\alpha}^{\alpha}(\mathbf{X}_{\alpha}) = \frac{dv_{0\alpha}^{\alpha}}{dv_{0\alpha}} = \frac{1}{dv_{0\alpha}} \int_{dv_{0\alpha}} \overset{\circ}{\chi}^{\alpha}(\overset{\circ}{\mathbf{r}})dv_{0\mu}. \tag{5.2.15}$$

The volume fractions satisfy, at the place $\mathbf{X}_S = \mathbf{X}_{\alpha}$, the local volume fraction condition for κ constituents:

$$\sum_{\alpha=1}^{\kappa} n_{0S}^{\alpha} = 1. \tag{5.2.16}$$

With (5.2.15), the corresponding relations of (5.2.5) through (5.2.9) can easily be obtained. Moreover, a corresponding relation to (5.2.11) is given by

$$\rho_{0\alpha}^{\alpha}(\mathbf{X}_{\alpha}, t = t_0) = n_{0\alpha}^{\alpha}\rho_{0\alpha}^{\alpha R}. \tag{5.2.17}$$

In the special case that in the reference state the pores are homogeneously distributed, the volume fractions in (5.2.15) are independent of $\mathbf{X}_{\alpha}$, i.e., $n_{0\alpha}^{\alpha} = n_{0\alpha}^{\alpha}(t = t_0)$.

With respect to the description of the so-called unsaturated porous media, the pore space is partly material-free and thus

$$v > \sum_{\alpha=1}^{\kappa} v^{\alpha}, \quad \sum_{\alpha=1}^{\kappa} n^{\alpha} < 1 \tag{5.2.18}$$

and

$$v_{0\alpha} > \sum_{\alpha=1}^{\kappa} v_{0\alpha}^{\alpha}, \quad \sum_{\alpha=1}^{\kappa} n_{0\alpha}^{\alpha} < 1 \tag{5.2.19}$$

and where

$$v^{\alpha} = \int_{B_S} n^{\alpha} dv, \quad v_{0\alpha}^{\alpha} = \int_{B_S} n_{0\alpha}^{\alpha} dv_{0\alpha} \tag{5.2.20}$$

have been used. Studies on these porous media were undertaken by Passman, Nunziato, and Walsh (1984).

The concept of volume fractions, introduced in the preceding forms, is an important part of the theory following in the next sections.

Due to the volume fraction concept, all geometric and physical quantities, such as motion, deformation, and stress, are defined in the total control space,

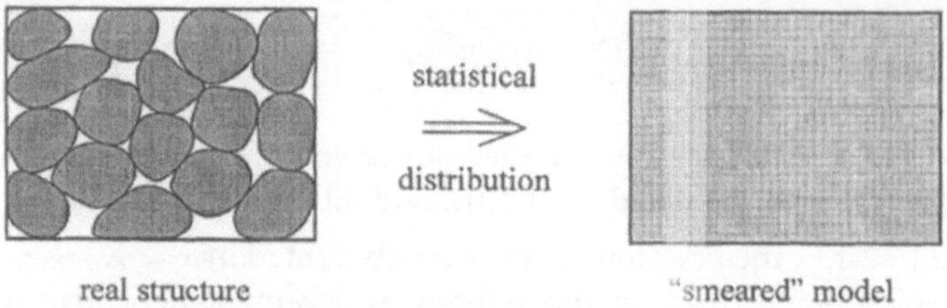

Fig. 5.2.2. Illustration of the statistical distribution of a binary porous medium consisting of a granular solid phase and a liquid phase

and thus, they can be interpreted as the statistical average values of the real quantities. Within the framework of the general porous media theory, a saturated porous medium will be treated as an immiscible mixture of all constituents, with particles X_α. This immiscible mixture, of course, is a substitute model; it can be treated with the methods of continuum mechanics, especially with those of the mixture theory. For the individual constituents, the kinematics and the balance equations will be extensively discussed; the connection between the balance equations and those of the mixture body will also be noted. Moreover, the principle of material objectivity and the entropy inequality, which both yield important restrictions in the constitutive theory, will be developed.

5.3 Kinematics

The kinematics in the porous media theory are based on two fundamental assumptions:

(1) Each spatial point $\mathbf{x}$ of the actual placement is simultaneously occupied by material points X_α of all κ constituents φ^α at the time t. The material points proceed from different reference positions $\mathbf{X}_\alpha$ at time $t = t_0$.
(2) Each constituent is assigned an independent state of motion.

It should, however, be mentioned that the volumetric strains are constrained by the saturation condition (see Section 5.6). If the motion of the constituent is understood as a chronological succession of placements χ_α, then for the spatial position vector $\mathbf{x}$ of the material points X_α, which can be identified with the reference position vector $\mathbf{X}_\alpha$ at time $t = t_0$, the following relation holds at time t:

$$\mathbf{x} = \chi_\alpha(\mathbf{X}_\alpha, t) . \tag{5.3.1}$$

The position vector $\mathbf{x}$ is an element of the control space of the porous solid at time t. In general, it is not necessary to demand that the reference positions $\mathbf{X}_F$ and $\mathbf{X}_G$ of the liquid and gas particles are elements of the reference placement of the solid phase at time $t = t_0$, i.e., $\mathbf{X}_F \not\in B_{0S}$ and $\mathbf{X}_G \not\in B_{0S}$. Only for such deformation processes in which the fluid phases leave the control space of the

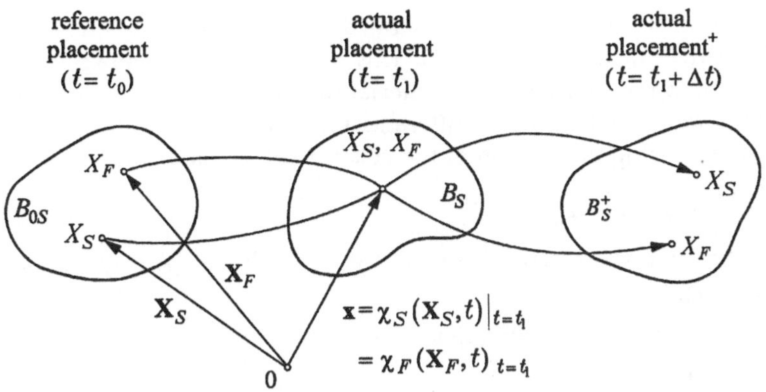

Fig. 5.3.1. Illustration of the motion of a solid and a liquid particle in a liquid saturated porous solid

solid phase are the reference positions $\mathbf{X}_F$ and $\mathbf{X}_G$ elements of B_{0S} (see Bluhm, 1997). A geometrical interpretation of the motion function (5.3.1), concerning the motion of a solid and a fluid particle, is shown in Fig. 5.3.1.

Eqn. (5.3.1) represents the Lagrange description of motion. The function χ_α is postulated to be unique, and uniquely invertible, at any time t. The existence of a function inverse to (5.3.1) leads to the Eulerian description of motion, viz.

$$\mathbf{X}_\alpha = \chi_\alpha^{-1}(\mathbf{x}, t) \,. \tag{5.3.2}$$

A mathematically necessary and sufficient condition for the existence of Eqn. (5.3.2) is given, if the Jacobian

$$J_\alpha = \det \mathbf{F}_\alpha \tag{5.3.3}$$

differs from zero. In (5.3.3), $\mathbf{F}_\alpha$ is the deformation gradient, which is defined as

$$\mathbf{F}_\alpha = \operatorname{Grad}_\alpha \chi_\alpha \,. \tag{5.3.4}$$

The differential operator "$\operatorname{Grad}_\alpha$" denotes a partial differentiation with respect to the reference position $\mathbf{X}_\alpha$ of the constituents φ^α. The inverse of (5.3.4) is given by

$$\mathbf{F}_\alpha^{-1} = \operatorname{grad} \mathbf{X}_\alpha \,, \tag{5.3.5}$$

with the differential operator "grad" referring to the spatial point $\mathbf{x}$. During the deformation process, $\mathbf{F}_\alpha$ is restricted to

$$\det \mathbf{F}_\alpha > 0 \,. \tag{5.3.6}$$

With the Lagrange description of the motion (5.3.1), the velocity and the acceleration of a material point of a constituent φ^α are defined by

$$\mathbf{x}'_\alpha = \frac{\partial \chi_\alpha(\mathbf{X}_\alpha, t)}{\partial t} \,, \quad \mathbf{x}''_\alpha = \frac{\partial^2 \chi_\alpha(\mathbf{X}_\alpha, t)}{\partial t^2} \,. \tag{5.3.7}$$

Using (5.3.2), the Eulerian description is gained for the velocity $\mathbf{v}_\alpha$ and the acceleration $\mathbf{a}_\alpha$:

$$\mathbf{v}_\alpha = \mathbf{x}'_\alpha = \mathbf{x}'_\alpha(\mathbf{x}, t) \,, \quad \mathbf{a}_\alpha = \mathbf{x}''_\alpha = \mathbf{x}''_\alpha(\mathbf{x}, t) \,. \tag{5.3.8}$$

As the individual constituents follow, in general, different motions, different material time derivatives must be formulated. This will be shown for an arbitrary scalar-value function $\Gamma(\mathbf{x}, t)$. Analogous material time derivatives of vector and tensor functions result. If $\Gamma(\mathbf{x}, t)$ is a differentiable function, then its material time derivative, following the motion of the constituent φ^α, is defined by

$$\Gamma'_\alpha = \frac{\partial \Gamma}{\partial t} + \operatorname{grad} \Gamma \cdot \mathbf{x}'_\alpha \, . \tag{5.3.9}$$

With $(5.3.7)_1$, the material velocity gradient of the constituent φ_α is obtained:

$$(\mathbf{F}_\alpha)'_\alpha = \operatorname{Grad}_\alpha \mathbf{x}'_\alpha \, . \tag{5.3.10}$$

The spatial velocity gradient can be calculated from $(5.3.8)_1$ and results in

$$\mathbf{L}_\alpha = \operatorname{grad} \mathbf{x}'_\alpha = \operatorname{grad} \mathbf{v}_\alpha \, , \tag{5.3.11}$$

which is connected to the material velocity gradient and the deformation gradient by

$$\mathbf{L}_\alpha = (\mathbf{F}_\alpha)'_\alpha \mathbf{F}_\alpha^{-1} \, . \tag{5.3.12}$$

Usually, no distinction is made in the literature between $\mathbf{x}'_\alpha$ and $\mathbf{v}_\alpha$, as well as between $\mathbf{x}''_\alpha$ and $\mathbf{a}_\alpha$, because it is, in many cases, obvious in connection with the operator as to whether $\mathbf{x}'_\alpha(\mathbf{X}_\alpha, t)$ or $\mathbf{x}'_\alpha(\mathbf{x}, t)$, as well as $\mathbf{x}''_\alpha(\mathbf{X}_\alpha, t)$ or $\mathbf{x}''_\alpha(\mathbf{x}, t)$, is meant; see, for example, (5.3.10) and (5.3.11).

The additive decomposition of $\mathbf{L}_\alpha$ yields the symmetrical part $\mathbf{D}_\alpha$ of the spatial velocity gradient and the skew-symmetric spin tensor $\mathbf{W}_\alpha$

$$\mathbf{L}_\alpha = \mathbf{D}_\alpha + \mathbf{W}_\alpha \tag{5.3.13}$$

with

$$\mathbf{D}_\alpha = \frac{1}{2}(\mathbf{L}_\alpha + \mathbf{L}_\alpha^T) \, , \quad \mathbf{W}_\alpha = \frac{1}{2}(\mathbf{L}_\alpha - \mathbf{L}_\alpha^T) \, . \tag{5.3.14}$$

Since the local deformations $\mathbf{F}_\alpha$ contain, in general, parts of a rigid body motion, they are less suitable to act as measurements for the deformations in constitutive equations. For this reason, it is convenient to use the line-elements, in the form of the difference of the squares of the line-elements, in the actual and the reference placements for the measurement of the deformation, in order to avoid irrational operations and to take out the rigid body motions. For the evaluation of the squares of the line-elements, the transport mechanism $d\mathbf{x} = \mathbf{F}_\alpha d\mathbf{X}_\alpha$ gained from (5.3.4) will be used. After elementary calculations, the following relations are obtained:

$$d\mathbf{x} \cdot d\mathbf{x} - d\mathbf{X}_\alpha \cdot d\mathbf{X}_\alpha = d\mathbf{X}_\alpha \cdot 2\mathbf{E}_\alpha d\mathbf{X}_\alpha = d\mathbf{x} \cdot 2\mathbf{A}_\alpha d\mathbf{x} \, . \tag{5.3.15}$$

The introduced symmetric strain tensors $\mathbf{E}_\alpha$ and $\mathbf{A}_\alpha$ are known, respectively, as the Green strain tensor and the Almansi strain tensor. They depend on the deformation gradient $\mathbf{F}_\alpha$ in the following way:

$$\mathbf{E}_\alpha = \frac{1}{2}(\mathbf{C}_\alpha - \mathbf{I}) \, , \quad \mathbf{A}_\alpha = \frac{1}{2}(\mathbf{I} - \mathbf{B}_\alpha^{-1}) \, , \tag{5.3.16}$$

where

$$\mathbf{C}_\alpha = \mathbf{F}_\alpha^T \mathbf{F}_\alpha \quad \text{and} \quad \mathbf{B}_\alpha = \mathbf{F}_\alpha \mathbf{F}_\alpha^T \tag{5.3.17}$$

denote the right and left Cauchy-Green deformation tensors, respectively. For further investigations, it is sometimes useful to multiplicatively decompose the deformation gradient $\mathbf{F}_\alpha$ into volume-preserving and spherical parts denoted by the symbols $(\check{\ldots})$ and $(\bar{\ldots})$:

$$\mathbf{F}_\alpha = \bar{\mathbf{F}}_\alpha \check{\mathbf{F}}_\alpha \tag{5.3.18}$$

with

$$\bar{\mathbf{F}}_\alpha = (J_\alpha)^{1/3}\mathbf{I}\,, \quad J_\alpha = \det \mathbf{F}_\alpha\,, \quad \check{J}_\alpha = \det \check{\mathbf{F}}_\alpha = 1\,. \tag{5.3.19}$$

With the decomposition (5.3.18), we obtain for the right Cauchy-Green tensor $(5.3.17)_1$:

$$\mathbf{C}_\alpha = (J_\alpha)^{2/3}\check{\mathbf{C}}_\alpha, \quad \check{\mathbf{C}}_\alpha = \check{\mathbf{F}}_\alpha^T \check{\mathbf{F}}_\alpha\,. \tag{5.3.20}$$

Considering (5.3.3), (5.3.12), and (5.3.13), the material time derivative of the Jacobian, as well as of the right Cauchy-Green deformation tensor, the Green strain tensor, and the volume-preserving part of the right Cauchy-Green tensor, yields

$$\begin{aligned} &(J_\alpha)'_\alpha = J_\alpha(\mathbf{D}_\alpha \cdot \mathbf{I})\,, \quad (\mathbf{C}_\alpha)'_\alpha = 2\mathbf{F}_\alpha^T \mathbf{D}_\alpha \mathbf{F}_\alpha, \quad (\mathbf{E}_\alpha)'_\alpha = \mathbf{F}_\alpha^T \mathbf{D}_\alpha \mathbf{F}_\alpha\,, \\ &(\check{\mathbf{C}}_\alpha)'_\alpha = 2\, J_\alpha^{-2/3}\, \mathbf{F}_\alpha^T \mathbf{D}_\alpha^D \mathbf{F}_\alpha, \end{aligned} \tag{5.3.21}$$

where $\mathbf{D}_\alpha^D$ is the deviatoric part of $\mathbf{D}_\alpha$.

In what follows, some considerations on the microscale are needed in order to describe the compressibility and incompressibility of the real materials. For this purpose, a macroscopic control space filled with a granular solid phase, and a gas without any physical properties, will be considered. The grains in the control space are represented by small balls (see Fig. 5.3.2). It is assumed that the grains are incompressible, i.e., a hydrostatic stress state in the grains produces no volume change of the grains. Although the grains are incompressible, contact forces acting on the grains cause a volume change of the control space; this results from the change of the pore structure and the volume fraction due to the change of the shapes of the individual grains (see Fig. 5.3.2). Therefore, the incompressibility condition cannot be expressed by the deformation gradient $\mathbf{F}_S$ of the partial solid constituent. Rather, the incompressibility condition must be formulated by physical quantities at the microscale. Moreover, statements on the compressibility and other real properties of the constituents must also be expressed by physical quantities at the microscale. In the case of describing compressibility and incompressibility, this means that a motion function at the microscale

$$\mathbf{x}_{SR(\text{micro})} = \chi_{SR(\text{micro})}(\mathbf{X}_S + \xi_{SR}, t) \tag{5.3.22}$$

must be introduced. From (5.3.22), the deformation gradient $\mathbf{F}_{SR(\text{micro})}$ can be determined in a way similar to that in (5.3.4). Then, the incompressibility condition can be reformulated at the microscale:

$$\det \mathbf{F}_{SR(\text{micro})} = J_{SR(\text{micro})} = 1\,. \tag{5.3.23}$$

The crucial point of this procedure is, however, the fact that the motion function $\chi_{SR(\text{micro})}$ in (5.3.22) is completely unknown and cannot be determined by a

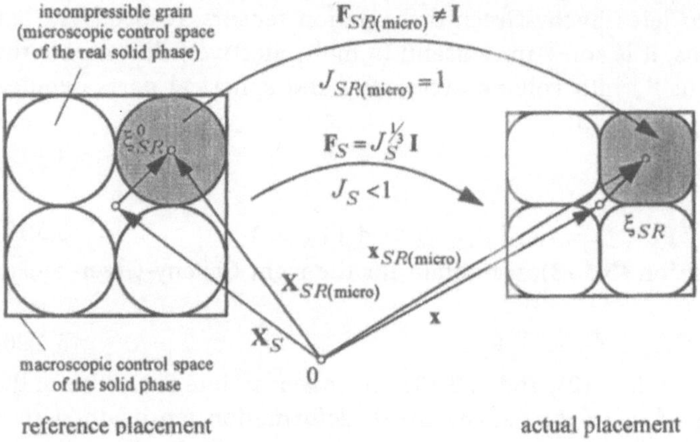

Fig. 5.3.2. Illustration of motivation for a multiplicative decomposition of the deformation gradient of the solid phase

balance equation within the framework of the mixture theory (macroscale). Therefore, it is advisable, in order to describe the phenomena of compressibility and incompressibility, to transfer the microscopic deformation behavior of the real solid phase to the macroscale. For this reason, the deformation tensor $\mathbf{F}_{SR}$ is introduced, which is understood to be a part of the deformation gradient $\mathbf{F}_S$ and which is assumed to reflect the microscopic deformations of the real

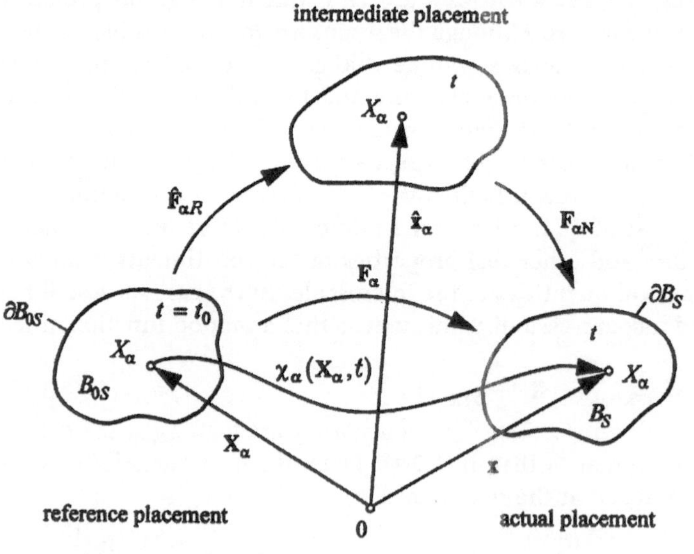

Fig. 5.3.3. Geometrical interpretation of the decomposition $\mathbf{F}_\alpha = \mathbf{F}_{\alpha N}\hat{\mathbf{F}}_{\alpha R}$ for homogeneous deformations of φ^α

solid material at the macroscale. In general, the tensor $\mathbf{F}_{SR}$ is not integrable at the macroscale, i.e., the microscopic deformations $\mathbf{F}_{SR(\text{micro})}$ are represented by incompatible deformations at the macroscale. Since

$$\mathbf{F}_S \neq \mathbf{F}_{SR} , \tag{5.3.24}$$

it is necessary to choose a second tensor $\mathbf{F}_{SN}$, to transfer the relation (5.3.24) into an equation. The part $\mathbf{F}_{SN}$ of the deformation gradient $\mathbf{F}_S$, as well as $\mathbf{F}_{SR}$, is, in general, not integrable. On the contrary, the deformation tensor $\mathbf{F}_S$ is integrable; thus, the deformation gradient at the macroscale must be decomposed multiplicatively into $\mathbf{F}_{SR}$ and $\mathbf{F}_{SN}$.

In the following, some parallels to the theory of the elastic-plastic deformations of metals will be discussed. It is well-known that within the framework of a finite theory, a multiplicative decomposition of the deformation gradient into an elastic and a plastic part is widely used. The plastic part of the deformation at the macroscale results from dislocations at the microscale. These microscopic dislocations are also, in general, represented by incompatible strains at the macroscale. Therefore, the reason for a multiplicative decomposition of the deformation gradient at the macroscale is the same as in the porous media theory, namely to bring physical phenomena from the microscale to the macroscale.

The usefulness of the multiplicative decomposition of the deformation gradient in porous media theory will be revealed in the following paragraphs.

There are two possibilities of decomposing multiplicatively the deformation gradient $\mathbf{F}_\alpha$ of the constituent φ^α, of which only the following one is suitable (see the extensive discussion of this problem and the consequences concerning the kinematics in Bluhm and de Boer, 1997):

$$\mathbf{F}_\alpha = \mathbf{F}_{\alpha N}\hat{\mathbf{F}}_{\alpha R} . \tag{5.3.25}$$

The part $\hat{\mathbf{F}}_{\alpha R}$ is interpreted as that part of $\mathbf{F}_\alpha$ which describes the deformation of the real material, whereas $\mathbf{F}_{\alpha N}$ describes the remaining part of the deformation of the control space, namely the change of the pores in size and shape. The parts $\mathbf{F}_{\alpha N}$ and $\hat{\mathbf{F}}_{\alpha R}$ are to be understood as local mappings of tangent (vector) spaces in each material point of the body. In the case of homogeneous deformations, the multiplicative decomposition (5.3.25) leads to an intermediate state $(\hat{\ldots})$ (see Fig. 5.3.3).

The proof of the multiplicative decomposition (5.3.25) of $\mathbf{F}_\alpha$ into quantities describing properties of the microscale is still awaiting research.

In analogy to (5.3.12) through (5.3.14), material time derivatives of the deformation tensor $\hat{\mathbf{F}}_{\alpha R}$ can be introduced:

$$\begin{aligned} &\hat{\mathbf{L}}_{\alpha R} = (\hat{\mathbf{F}}_{\alpha R})'_\alpha(\hat{\mathbf{F}}_{\alpha R})^{-1} , \quad \hat{\mathbf{L}}_{\alpha R} = \hat{\mathbf{D}}_{\alpha R} + \hat{\mathbf{W}}_{\alpha R} , \\ &\hat{\mathbf{D}}_{\alpha R} = \frac{1}{2}(\hat{\mathbf{L}}_{\alpha R} + \hat{\mathbf{L}}^T_{\alpha R}) , \quad \hat{\mathbf{W}}_{\alpha R} = \frac{1}{2}(\hat{\mathbf{L}}_{\alpha R} - \hat{\mathbf{L}}^T_{\alpha R}) , \\ &\hat{\mathbf{L}}_{\alpha R} = \frac{\partial(\hat{\mathbf{x}}_\alpha)'_\alpha}{\partial \mathbf{X}_\alpha}\frac{\partial \mathbf{X}_\alpha}{\partial \hat{\mathbf{x}}_\alpha} = \frac{\partial(\hat{\mathbf{x}}_\alpha)'_\alpha}{\partial \hat{\mathbf{x}}_\alpha} \quad \text{(homogeneous deformations)} . \end{aligned} \tag{5.3.26}$$

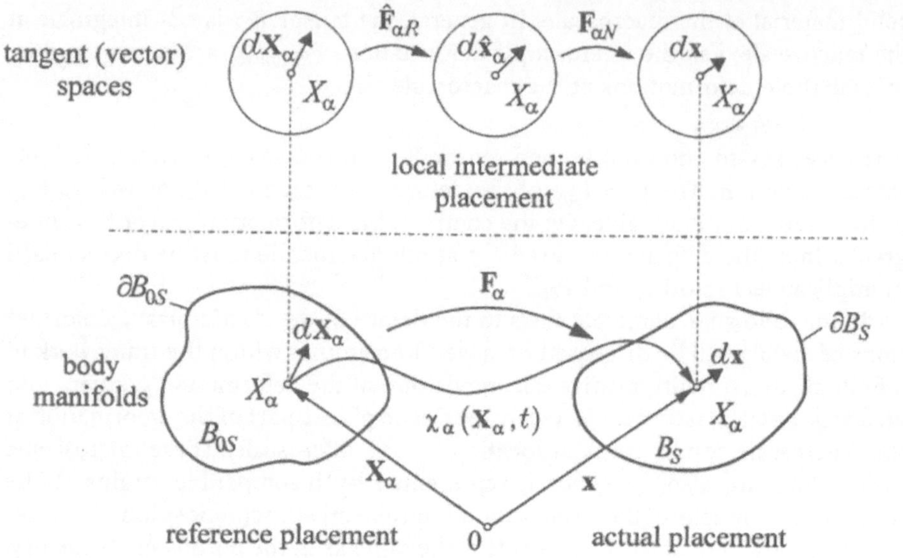

Fig. 5.3.4. Geometrical interpretation of the decomposition of $\mathbf{F}_\alpha = \mathbf{F}_{\alpha N}\hat{\mathbf{F}}_{\alpha R}$ for non-homogeneous deformations with respect to the mapping of line elements of the consistuent φ^α

The introduction of the material time derivative of $\mathbf{F}_{\alpha N}$, namely $\mathbf{L}_{\alpha N}$, is less useful because $\mathbf{L}_{\alpha N}$ is not a spatial velocity gradient (see Bluhm and de Boer, 1997).

Now, the volume fraction concept and the incompressibility condition under consideration of the multiplicative decomposition of the deformation gradient $\mathbf{F}_\alpha$ of the constituent φ^α will be discussed. We proceed from (5.3.25) and split the two parts, $\mathbf{F}_{\alpha N}$ and $\hat{\mathbf{F}}_{\alpha R}$, of the deformation gradient into volume-preserving and spherical parts denoted by the symbols $(\check{\ldots})$ and $(\bar{\ldots})$. Thus,

$$\mathbf{F}_{\alpha N} = \bar{\mathbf{F}}_{\alpha N}\check{\mathbf{F}}_{\alpha N}\,, \quad \hat{\mathbf{F}}_{\alpha R} = \bar{\mathbf{F}}_{\alpha R}\check{\mathbf{F}}_{\alpha R} \tag{5.3.27}$$

with

$$\begin{aligned} \bar{\mathbf{F}}_{\alpha N} &= (J_{\alpha N})^{1/3}\mathbf{I}\,, \quad J_{\alpha N} = \det \mathbf{F}_{\alpha N}\,, \quad \check{J}_{\alpha N} = \det \check{\mathbf{F}}_{\alpha N} = 1\,, \\ \bar{\mathbf{F}}_{\alpha R} &= (\hat{J}_{\alpha R})^{1/3}\mathbf{I}\,, \quad \hat{J}_{\alpha R} = \det \hat{\mathbf{F}}_{\alpha R}\,, \quad \check{J}_{\alpha R} = \det \check{\mathbf{F}}_{\alpha R} = 1\,. \end{aligned} \tag{5.3.28}$$

With these quantities, kinematic expressions corresponding to those of $\bar{\mathbf{F}}_\alpha$ and $\check{\mathbf{F}}_\alpha$ can be formulated. In particular, the following derivatives are valid:

$$(\hat{J}_{\alpha R})'_\alpha = \hat{J}_{\alpha R}(\hat{\mathbf{D}}_{\alpha R} \cdot \mathbf{I})\,, \quad (J_{\alpha N})'_\alpha = J_{\alpha N}(\mathbf{D}_{\alpha N} \cdot \mathbf{I}) \tag{5.3.29}$$

with

$$\mathbf{D}_{\alpha N} \cdot \mathbf{I} = \mathbf{D}_\alpha \cdot \mathbf{I} - \hat{\mathbf{D}}_{\alpha R} \cdot \mathbf{I}\,. \tag{5.3.30}$$

For further investigations, the volume elements in the reference and actual placements must be considered. In continuum mechanics, it is well-known that the following transport theorem concerning the volume elements is valid:

$$dv = J_\alpha dv_{0\alpha} \, , \tag{5.3.31}$$

where

$$dv_{0\alpha} = dv_{0\alpha}(\mathbf{X}_\alpha, t = t_0) \, , \quad dv = dv(\mathbf{x}, t) \tag{5.3.32}$$

are the volume elements in the reference placement at the position $\mathbf{X}_\alpha$, denoted by the subscript index α, and in the actual placement at the position $\mathbf{x}$. In consideration of (5.3.25) through (5.3.28),

$$dv = J_{\alpha N} \hat{J}_{\alpha R} dv_{0\alpha} \tag{5.3.33}$$

is gained. By using (5.3.25), a differential volume $d\hat{v}_\alpha$, at a material point $\mathbf{X}_\alpha$ of a local intermediate placement in the tangent space, is related to the differential volume elements in the reference placement and the actual placements by (see de Boer, 1996, and Bluhm and de Boer, 1997)

$$d\hat{v}_\alpha = \hat{J}_{\alpha R} dv_{0\alpha} \, , \quad dv = J_{\alpha N} d\hat{v}_\alpha \, . \tag{5.3.34}$$

With the relations (5.3.31) through (5.3.34), it is easy to formulate various kinds of volume strains which will be important for the investigations in the following passages. In analogy to the volume strain of the partial material of the constituent φ^α,

$$e_\alpha = \frac{dv - dv_{0\alpha}}{dv_{0\alpha}} = \frac{dv}{dv_{0\alpha}} - 1 = J_\alpha - 1 \, , \tag{5.3.35}$$

where (5.3.31) has been used, the volume strain of the real material of φ^α is defined as

$$\begin{aligned} e_{\alpha R} &= \frac{dv^\alpha - dv^\alpha_{0\alpha}}{dv^\alpha_{0\alpha}} = \frac{n^\alpha dv - n^\alpha_{0\alpha} dv_{0\alpha}}{n^\alpha_{0\alpha} dv_{0\alpha}} = \frac{n^\alpha}{n^\alpha_{0\alpha}} \frac{dv}{dv_{0\alpha}} - 1 \\ &= \frac{n^\alpha}{n^\alpha_{0\alpha}} J_\alpha - 1 = \frac{n^\alpha}{n^\alpha_{0\alpha}} J_{\alpha N} \hat{J}_{\alpha R} - 1 \, , \end{aligned} \tag{5.3.36}$$

where (5.2.3), (5.2.15), (5.3.31), and (5.3.33) have been used. In consideration of the transport theorems (5.3.34), further real volume strains can be formulated:

$$\begin{aligned} \hat{e}_{\alpha R} &= \frac{d\hat{v}^\alpha_\alpha - dv^\alpha_{0\alpha}}{dv^\alpha_{0\alpha}} = \frac{\hat{n}^\alpha_\alpha d\hat{v}_\alpha - n^\alpha_{0\alpha} dv_{0\alpha}}{n^\alpha_{0\alpha} dv_{0\alpha}} = \frac{\hat{n}^\alpha_\alpha}{n^\alpha_{0\alpha}} \frac{d\hat{v}_\alpha}{dv_{0\alpha}} - 1 = \frac{\hat{n}^\alpha_\alpha}{n^\alpha_{0\alpha}} \hat{J}_{\alpha R} - 1 \, , \\ \tilde{e}_{\alpha R} &= \frac{dv^\alpha - d\hat{v}^\alpha_\alpha}{d\hat{v}^\alpha_\alpha} = \frac{n^\alpha dv - \hat{n}^\alpha_\alpha d\hat{v}_\alpha}{\hat{n}^\alpha_\alpha d\hat{v}_\alpha} = \frac{n^\alpha}{\hat{n}^\alpha_\alpha} \frac{dv}{d\hat{v}_\alpha} - 1 = \frac{n^\alpha}{\hat{n}^\alpha_\alpha} J_{\alpha N} - 1 \, , \end{aligned} \tag{5.3.37}$$

where the part $d\hat{v}^\alpha_\alpha$ of the differential volume element $d\hat{v}_\alpha$, in the local intermediate placement in the tangent space, is defined via the volume fraction $\hat{n}^\alpha_\alpha$ as

$$d\hat{v}^\alpha_\alpha = \hat{n}^\alpha_\alpha d\hat{v}_\alpha \, . \tag{5.3.38}$$

The determinants of $\hat{\mathbf{F}}_{\alpha R}$ and $\mathbf{F}_{\alpha N}$ can be expressed depending on the real volume strains $\hat{e}^{\alpha R}$ and $\tilde{e}^{\alpha R}$, namely

$$\hat{J}_{\alpha R} = \frac{n^\alpha_{0\alpha}}{\hat{n}^\alpha_\alpha} (\hat{e}_{\alpha R} + 1) \, , \quad J_{\alpha N} = \frac{\hat{n}^\alpha_\alpha}{n^\alpha} (\tilde{e}_{\alpha R} + 1) \, . \tag{5.3.39}$$

With (5.3.39), the real volume strain (5.3.36) can be reformulated as

$$e_{\alpha R} = \hat{e}_{\alpha R} + \tilde{e}_{\alpha R} + \hat{e}_{\alpha R} \tilde{e}_{\alpha R} \, . \tag{5.3.40}$$

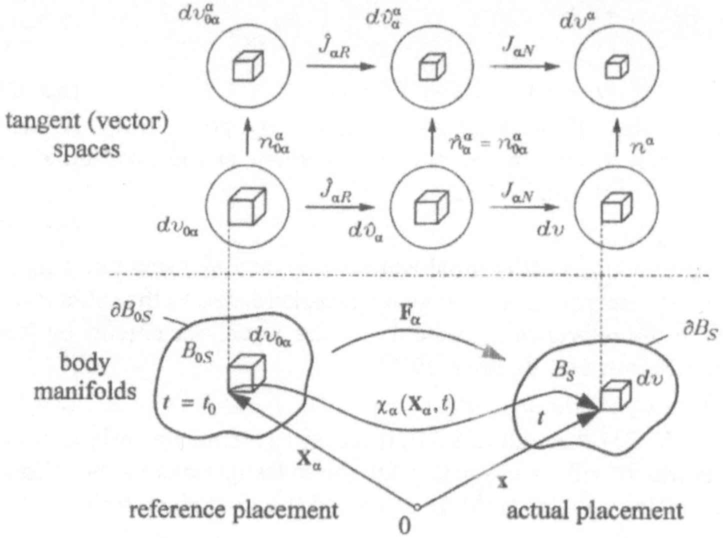

Fig. 5.3.5. Geometrical interpretation of the decomposition of $\mathbf{F}_\alpha = \mathbf{F}_{\alpha N}\hat{\mathbf{F}}_{\alpha R}$ with respect to the mapping of volume elements of the consistuent φ^α

The tensor $\hat{\mathbf{F}}_{\alpha R}$ is interpreted as that part of the deformation gradient which includes the whole deformation of the real material of the constituent φ^α. Thus, the determinant $J_{\alpha R}$ must reflect the volume strain of the real material of φ^α, i.e., the real volume strain is the difference between the part $d\hat{v}_\alpha^\alpha = \hat{n}_\alpha^\alpha \, d\hat{v}_\alpha$ of the differential volume $d\hat{v}_\alpha$ in the local intermediate placement in the vector space, and the part $dv_{0\alpha}^\alpha = n_{0\alpha}^\alpha \, dv_{0S}$ of the volume element dv_{0S} in the reference placement at the position $\mathbf{X}_S$. Therefore, the following relations concerning the real volume strains hold:

$$e_{\alpha R} = \hat{e}_{\alpha R} \, , \quad \tilde{e}_{\alpha R} = 0 \, . \tag{5.3.41}$$

Furthermore, the transport theorem $(5.3.34)_1$ excludes the change of the volume fraction by mapping $dv_{0\alpha}$ from the reference to the local intermediate placement, i.e.,

$$\hat{n}_\alpha^\alpha = n_{0\alpha}^\alpha \, . \tag{5.3.42}$$

With (5.3.41) and (5.3.42), the determinants $\hat{J}_{\alpha R}$ and $J_{\alpha N}$, see (5.3.39), read as follows:

$$\hat{J}_{\alpha R} = \hat{e}_{\alpha R} + 1 = e_{\alpha R} + 1 \, , \quad J_{\alpha N} = \frac{n_{0\alpha}^\alpha}{n^\alpha} \, . \tag{5.3.43}$$

In the case of incompressibility,

$$e_{\alpha R} = 0 \, , \quad \hat{J}_{\alpha R} = 1 \, , \quad \text{and} \quad \hat{\mathbf{D}}_{\alpha R} \cdot \mathbf{I} = 0 \tag{5.3.44}$$

are valid, whereby $(5.3.44)_3$ is the rate formulation of the incompressibility condition (see Bluhm and de Boer, 1997) and where $\hat{\mathbf{D}}_{\alpha R}$ can be interpreted as the Lie derivative of the strain tensor $\mathbf{E}_{\alpha R} = \frac{1}{2}(\mathbf{F}_{\alpha R}^T \mathbf{F}_{\alpha R} - \mathbf{I})$.

With the help of the multiplicative decomposition of the deformation gradient $\mathbf{F}_\alpha$ (5.3.25), it can be shown that the statement of Mills (1967) and Bowen (1980) concerning the incompressibility of the real material, which was described by setting the real densities of the constituents constant, is only identical with the kinematic constraints (5.3.44) in the case of a thermodynamic process without any mass exchange (see Bluhm and de Boer, 1997).

As there is no difference in the volume changes of the real material in the actual and intermediate placements, we will omit the superscript sign $(\hat{\ldots})$ in the following sections.

Now, we will leave the development of kinematics and will turn to the formulation of the balance equations.

5.4 Balance Equations

In mixture theory and porous media theory, balance equations - balance of mass, balance of momentum, and moment of momentum, as well as balance of energy - must be established for each constituent φ^α in consideration of all interaction and external agencies, as had already been called for by von Terzaghi in 1925 (see Chapter 4). This means that all quantities resulting from long- and short-range effects which influence the individual constituents, as well as the interaction effects between the constituents, have to be considered in the balance equations.

In the time following, the balance equations have been discussed at length by Truesdell (1957a, b, 1969, 1984), Truesdell and Toupin (1960), Kelly (1964), Eringen and Ingram (1965), Green and Naghdi (1965), Müller (1968), Bowen and Wiese (1969), and de Boer and Ehlers (1986b).

The interaction effects (supply terms) have to be in the sum equal to zero. These conditions are at least founded by the fact that - in the case of a common velocity $\mathbf{v}$ and acceleration $\mathbf{a}$, a common external acceleration $\mathbf{b}$, as well as a common internal energy ε, and external heat supply r - the sum of the balance equations must formally become the corresponding balance equations of a one-component body.

The following balance equations will be formulated in global and local forms for the individual constituents in consideration of all interaction effects and, finally, some conclusions in view of the comparison with the balance equations of the mixture body will be drawn.

5.4.1 Balance of Mass

As has already been mentioned, two possibilities exist concerning the formulation of the balance equation of mass. On the one hand, this equation can be given for the bulk mixture body; on the other hand, the mass balance equations can be formulated for each individual constituent in such a way that the superposition of the mass balance equations for the individual constituents turn, for special cases, into the balance equation of the mixture body as a one-component body.

The balance of mass for the individual constituents φ^α requires that the rate of the mass M^α equals a mass term $\int_{B_\alpha} \hat{\rho}^\alpha dv$ caused by the other constituents, where $\hat{\rho}^\alpha$ is the mass supply per volume element:

$$(M^\alpha)'_\alpha = \Big(\int_{B_\alpha} \rho^\alpha dv\Big)'_\alpha = \int_{B_\alpha} \hat{\rho}^\alpha dv\,. \tag{5.4.1}$$

The integration in (5.4.1) covers the domain B_α of each individual constituent. With the help of the transport theorem

$$(dv)'_\alpha = \operatorname{div} \mathbf{v}_\alpha dv \tag{5.4.2}$$

from (5.4.1), the local statement

$$(\rho^\alpha)'_\alpha + \rho^\alpha \operatorname{div} \mathbf{v}_\alpha = \hat{\rho}^\alpha \tag{5.4.3}$$

or

$$\frac{\partial \rho^\alpha}{\partial t} + \operatorname{div}(\rho^\alpha \mathbf{v}_\alpha) = \hat{\rho}^\alpha \tag{5.4.4}$$

is derived. Assuming a common velocity $\mathbf{v}$ for all phases φ^α, the summation of (5.4.4) over all κ constituents φ^α leads to

$$\sum_{\alpha=1}^{\kappa} \Big[\frac{\partial \rho^\alpha}{\partial t} + \operatorname{div}(\rho^\alpha \mathbf{v})\Big] = \sum_{\alpha=1}^{\kappa} \hat{\rho}^\alpha\,. \tag{5.4.5}$$

With the statement that the sum of the densities of the individual constituents is equal to the density ρ of the mixture body,

$$\rho = \sum_{\alpha=1}^{\kappa} \rho^\alpha\,, \tag{5.4.6}$$

one obtains

$$\frac{\partial \rho}{\partial t} + \operatorname{div}(\rho \mathbf{v}) = \sum_{\alpha=1}^{\kappa} \hat{\rho}^\alpha\,. \tag{5.4.7}$$

For deriving Eqn. (5.4.7), we have made use of the fact that the summation and the derivative are interchangeable. Eqn. (5.4.7) only results in the form valid for the mixture body if the constraint

$$\sum_{\alpha=1}^{\kappa} \hat{\rho}^\alpha = 0 \tag{5.4.8}$$

is introduced, i.e., if the sum of the local mass supplies of all κ constituents φ^α is equal to zero.

If all mass exchange is excluded, the relation (5.4.3) can be integrated and it follows that:

$$\rho^\alpha = \rho^\alpha_{0\alpha} (\det \mathbf{F}_\alpha)^{-1} \quad \text{with} \quad \rho^\alpha_{0\alpha} = \rho^\alpha_{0\alpha}(\mathbf{X}_\alpha, t = t_0)\,. \tag{5.4.9}$$

The quantity $\rho^\alpha_{0\alpha}$ denotes the partial density of the constituent φ^α (superscript index) in the reference placement at the position $\mathbf{X}_\alpha$ (subscript index).

5.4.2
Balance of Momentum and Moment of Momentum

In this section, the consequences of applying the axioms of the *balance of momentum* to each individual constituent will first be discussed.

The balance equation of momentum states that the material time derivative of the momentum is equal to the sum of external forces. Thus,

$$(\mathbf{l}^\alpha)'_\alpha = \mathbf{k}^\alpha . \tag{5.4.10}$$

Herein, the momentum $\mathbf{l}^\alpha$ for the constituent φ^α is defined by

$$\mathbf{l}^\alpha = \int_{B_\alpha} \rho^\alpha \mathbf{v}_\alpha dv . \tag{5.4.11}$$

The external forces $\mathbf{k}^\alpha$ are given by the sum of the forces $\mathbf{f}^\alpha$, which are caused by long- and short-range effects and are acting on the constituents as volume forces $\rho^\alpha \mathbf{b}^\alpha$ and surface forces $\mathbf{t}^\alpha$, as well as of the interaction forces $\hat{\mathbf{p}}^\alpha$ which belong to the volume forces. The resulting force vector $\mathbf{k}^\alpha$ is thus given by

$$\mathbf{k}^\alpha = \mathbf{f}^\alpha + \int_{B_\alpha} \hat{\mathbf{p}}^\alpha dv , \tag{5.4.12}$$

where $\mathbf{f}^\alpha$ is composed of

$$\mathbf{f}^\alpha = \int_{B_\alpha} \rho^\alpha \mathbf{b}^\alpha dv + \int_{\partial B_\alpha} \mathbf{t}^\alpha da . \tag{5.4.13}$$

Now, all the terms which are contained in the balance equation of momentum have been listed . With Cauchy's theorem,

$$\mathbf{t}^\alpha = \mathbf{T}^\alpha \mathbf{n} , \tag{5.4.14}$$

where $\mathbf{T}^\alpha$ is Cauchy's stress tensor of the constituent φ^α and $\mathbf{n}$ the unit normal at the surface of the individual constituent body. With the divergence theorem, as well as with either mass balance equation, (5.4.3) or (5.4.4), Cauchy's first equation of motion (balance of momentum) for φ^α is obtained from (5.4.10):

$$\operatorname{div} \mathbf{T}^\alpha + \rho^\alpha \mathbf{b}^\alpha + \hat{\mathbf{p}}^\alpha = \rho^\alpha \mathbf{a}_\alpha + \hat{\rho}^\alpha \mathbf{v}_\alpha . \tag{5.4.15}$$

In this equation, the expression $\hat{\rho}^\alpha \mathbf{v}_\alpha$ represents the exchange of linear momentum through the density supply $\hat{\rho}^\alpha$.

The balance of momentum for the mixture body can be gained by superimposition of the momenta of all κ constituents φ^α, assuming a common velocity $\mathbf{v}$ and acceleration $\mathbf{a}$, as well as a common external acceleration $\mathbf{b}$ for all constituents φ^α:

$$\sum_{\alpha=1}^{\kappa} (\operatorname{div} \mathbf{T}^\alpha + \rho^\alpha \mathbf{b} + \hat{\mathbf{p}}^\alpha) = \sum_{\alpha=1}^{\kappa} (\rho^\alpha \mathbf{a} + \hat{\rho}^\alpha \mathbf{v}) . \tag{5.4.16}$$

Introducing the requirements

$$\mathbf{T} = \sum_{\alpha=1}^{\kappa} \mathbf{T}^\alpha, \ \rho\mathbf{b} = \sum_{\alpha=1}^{\kappa} \rho^\alpha \mathbf{b}, \ \dot{\mathbf{l}} = \sum_{\alpha=1}^{\kappa} [\rho^\alpha \mathbf{a} + \hat{\rho}^\alpha \mathbf{v}] = \rho\mathbf{a}, \ \sum_{\alpha=1}^{\kappa} \hat{\mathbf{p}}^\alpha = \mathbf{0}\,, \tag{5.4.17}$$

where $\mathbf{T}, \rho\mathbf{b}$, and $\dot{\mathbf{l}}$ are, respectively, Cauchy's stress tensor, the volume force, and the time rate of the momentum of the mixture body, the balance equation of the mixture body is gained:

$$\operatorname{div} \mathbf{T} + \rho\mathbf{b} = \dot{\mathbf{l}}\,. \tag{5.4.18}$$

The material time derivative $(\dot{\ldots})$ has to be formed with the velocity $\mathbf{v}$. The requirement of $(5.4.17)_4$ is a constraint for the momentum supplies. Note that the second summation in $(5.4.17)_3$ vanishes due to (5.4.8).

The *balance of moment of momentum* for non-polar materials states that the material time derivative of the moment of momentum is equal to the moments of all external forces, where the moments are referred to a fixed point 0:

$$(\mathbf{h}^\alpha_{(0)})'_\alpha = \mathbf{m}^\alpha_{(0)}\,. \tag{5.4.19}$$

We do not consider local moment of momentum supply vectors $\hat{\mathbf{m}}^\alpha$ or the corresponding tensors $\hat{\mathbf{M}}^\alpha$. For the formulation of the moment of momentum balance equation for polar materials, the reader is referred to Ehlers and Volk (1997).

The moment of momentum $\mathbf{h}^\alpha_{(0)}$ for the constituent φ^α is given in consideration of (5.4.11) by:

$$\mathbf{h}^\alpha_{(0)} = \int_{B_\alpha} \mathbf{x} \times \rho^\alpha \mathbf{v}_\alpha dv\,. \tag{5.4.20}$$

The moment of the external forces $\mathbf{m}^\alpha_{(0)}$ can be calculated considering (5.4.12) and (5.4.13) from:

$$\mathbf{m}^\alpha_{(0)} = \int_{B_\alpha} \mathbf{x} \times (\rho^\alpha \mathbf{b}^\alpha + \hat{\mathbf{p}}^\alpha) dv + \int_{\partial B_\alpha} \mathbf{x} \times \mathbf{t}^\alpha da\,. \tag{5.4.21}$$

From (5.4.20), considering the mass balance (5.4.3), the material time derivative of the moment of momentum leads to

$$(\mathbf{h}^\alpha_{(0)})'_\alpha = \int_{B_\alpha} \mathbf{x} \times (\rho^\alpha \mathbf{a}_\alpha + \hat{\rho}^\alpha \mathbf{v}_\alpha) dv\,. \tag{5.4.22}$$

Moreover, the evaluation of the expression for the moment $\mathbf{m}^\alpha_{(0)}$, considering the balance equation of momentum (5.4.15), yields:

$$\mathbf{m}^\alpha_{(0)} = \int_{B_\alpha} \mathbf{x} \times (\rho^\alpha \mathbf{a}_\alpha + \hat{\rho}^\alpha \mathbf{v}_\alpha) dv + \int_{B_\alpha} \mathbf{I} \times \mathbf{T}^\alpha dv \tag{5.4.23}$$

from which, with (5.4.19) and (5.4.21), the local statement

$$\mathbf{I} \times \mathbf{T}^\alpha = \mathbf{0} \tag{5.4.24}$$

is obtained.

The above statement is fulfilled, if

$$\mathbf{T}^\alpha = (\mathbf{T}^\alpha)^T, \tag{5.4.25}$$

i.e., if Cauchy's stress tensor is symmetric.

The requirement that the mixture, as the sum of all κ constituents, should behave as a one-component material contains the condition:

$$\sum_{\alpha=1}^{\kappa} \mathbf{T}^\alpha = \sum_{\alpha=1}^{\kappa} (\mathbf{T}^\alpha)^T \,, \quad \mathbf{T} = \mathbf{T}^T \,, \tag{5.4.26}$$

whereby $(5.4.17)_1$ has been considered. The result of the balance of moment of momentum is the evaluation of the statement that also the stress tensor of the mixture body is symmetric.

5.4.3 Balance of Energy

The first law of thermodynamics (balance of energy), which has been repeatedly proven in the past, is the most fundamental relation in thermodynamics of one-component materials. It states that the sum of the material time derivatives of the internal and kinetic energies equals the rates of the mechanical work and the heat. This balance equation is transferred to the individual constituents. Applying the above statement to the constituents, the following balance equation is obtained:

$$(E^\alpha)'_\alpha + (K^\alpha)'_\alpha = W^\alpha + Q^\alpha + \int_{B_\alpha} \hat{e}^\alpha dv \,, \tag{5.4.27}$$

where $E^\alpha, K^\alpha, W^\alpha, Q^\alpha$, and $\hat{e}^\alpha$ are, respectively, the internal energy, the kinetic energy, the rate of the mechanical energy, the rate of the heat of the constituent φ^α, as well as $\hat{e}^\alpha$, the energy supply to φ^α caused by all other constituents. The internal energy, kinetic energy, and the rate of the mechanical work, as well as the rate of the heat, are given by

$$E^\alpha = \int_{B_\alpha} \rho^\alpha \varepsilon^\alpha dv \,, \quad K^\alpha = \int_{B_\alpha} \frac{1}{2} \rho^\alpha \mathbf{v}_\alpha \cdot \mathbf{v}_\alpha dv \,, \tag{5.4.28}$$

$$W^\alpha = \int_{B_\alpha} \mathbf{v}_\alpha \cdot \rho^\alpha \mathbf{b}^\alpha dv + \int_{\partial B_\alpha} \mathbf{v}_\alpha \cdot \mathbf{t}^\alpha da \,, \tag{5.4.29}$$

and

$$Q^\alpha = \int_{B_\alpha} \rho^\alpha r^\alpha dv - \int_{\partial B_\alpha} \mathbf{q}^\alpha \cdot \mathbf{da} \,. \tag{5.4.30}$$

Here, $\varepsilon^\alpha = \varepsilon^\alpha(\mathbf{x}, t)$ is the specific internal energy, $r^\alpha = r^\alpha(\mathbf{x}, t)$ the partial energy source, and $\mathbf{q}^\alpha = \mathbf{q}^\alpha(\mathbf{x}, t)$ the partial heat flux vector, which is positive when entering the body.

The balance equation of energy (5.4.27) yields, in connection with (5.4.28), (5.4.29), and (5.4.30), the local statement:

$$\rho^\alpha \left(\varepsilon^\alpha\right)'_\alpha + \hat{\rho}^\alpha \varepsilon^\alpha + \rho^\alpha \mathbf{v}_\alpha \cdot \mathbf{a}_\alpha + \frac{1}{2}\hat{\rho}^\alpha \mathbf{v}_\alpha \cdot \mathbf{v}_\alpha = \tag{5.4.31}$$
$$= \rho^\alpha r^\alpha - \operatorname{div} \mathbf{q}^\alpha + \mathbf{x}'_\alpha \cdot \left(\operatorname{div} \mathbf{T}^\alpha + \rho^\alpha \mathbf{b}^\alpha\right) + \mathbf{T}^\alpha \cdot \mathbf{L}_\alpha + \hat{e}^\alpha \,.$$

With Cauchy's first equation of motion (5.4.15), the relation (5.4.31) can be reformulated as:

$$\rho^\alpha \left(\varepsilon^\alpha\right)'_\alpha - \mathbf{T}^\alpha \cdot \mathbf{L}_\alpha - \rho^\alpha r^\alpha + \operatorname{div} \mathbf{q}^\alpha = \hat{e}^\alpha - \hat{\mathbf{p}}^\alpha \cdot \mathbf{v}_\alpha - \hat{\rho}^\alpha \left(\varepsilon^\alpha - \frac{1}{2}\mathbf{v}_\alpha \cdot \mathbf{v}_\alpha\right) . \tag{5.4.32}$$

Excluding all mass and moment of momentum exchanges, the relation (5.4.32) simplifies to

$$\rho^\alpha \left(\varepsilon^\alpha\right)'_\alpha - \mathbf{T}^\alpha \cdot \mathbf{D}_\alpha - \rho^\alpha r^\alpha + \operatorname{div} \mathbf{q}^\alpha = \hat{e}^\alpha - \hat{\mathbf{p}}^\alpha \cdot \mathbf{v}_\alpha \,. \tag{5.4.33}$$

For many practical problems, the balance equation of energy in the form of Eqn. (5.4.33) should be sufficient.

The summation of (5.4.31) over all κ constituents results in

$$\sum_{\alpha=1}^{\kappa} \left[\rho^\alpha (\varepsilon^\alpha)'_\alpha + \hat{\rho}^\alpha \varepsilon^\alpha\right] + \sum_{\alpha=1}^{\kappa} \left(\rho^\alpha \mathbf{v}_\alpha \cdot \mathbf{a}_\alpha + \frac{1}{2}\hat{\rho}^\alpha \mathbf{v}_\alpha \cdot \mathbf{v}_\alpha\right) = \sum_{\alpha=1}^{\kappa} \rho^\alpha r^\alpha - \sum_{\alpha=1}^{\kappa} \operatorname{div} \mathbf{q}^\alpha + \tag{5.4.34}$$
$$+ \sum_{\alpha=1}^{\kappa} \left[\mathbf{v}_\alpha \cdot \left(\operatorname{div} \mathbf{T}^\alpha + \rho^\alpha \mathbf{b}^\alpha\right) + \mathbf{T}^\alpha \cdot \mathbf{L}_\alpha\right] + \sum_{\alpha=1}^{\kappa} \hat{e}^\alpha \,.$$

With the assumptions stated at the beginning of this section, namely $\mathbf{v}_\alpha = \mathbf{v}$, $\mathbf{a}_\alpha = \mathbf{a}$, $\mathbf{b}^\alpha = \mathbf{b}$, $\varepsilon^\alpha = \varepsilon$, and $r^\alpha = r$, and the appropriate statements

$$\rho \dot{\varepsilon} = \sum_{\alpha=1}^{\kappa} \left[\rho^\alpha (\dot{\varepsilon}) + \hat{\rho}^\alpha \varepsilon\right] ,$$

$$\rho \dot{k} = \sum_{\alpha=1}^{\kappa} \left[\rho^\alpha \mathbf{v} \cdot \mathbf{a} + \frac{1}{2}\hat{\rho}^\alpha \mathbf{v} \cdot \mathbf{v}\right] ,$$

$$\rho r = \sum_{\alpha=1}^{\kappa} \rho^\alpha r \,, \quad \mathbf{q} = \sum_{\alpha=1}^{\kappa} \mathbf{q}^\alpha \,, \tag{5.4.35}$$

$$w = \sum_{\alpha=1}^{\kappa} \left[\mathbf{v} \cdot (\operatorname{div} \mathbf{T}^\alpha + \rho^\alpha \mathbf{b}) + \mathbf{T}^\alpha \cdot \mathbf{L}\right] ,$$

$$\sum_{\alpha=1}^{\kappa} \hat{e}^\alpha = 0 \,,$$

we have

$$\rho \dot{\varepsilon} + \rho \dot{k} = w + \rho r - \operatorname{div} \mathbf{q} \tag{5.4.36}$$

for the mixture body, an equation which is formally equivalent to the energy balance equation of a one-component material. Eqn $(5.4.35)_6$ represents a constraint on the energy supplies.

In this section, the introduction of the so-called barycentric velocity - a mass weighted average velocity - and its derivations have been avoided due to the fact that the introduction of this velocity would lead to some obscure statements in the mixture theory (see the extensive discussion of this problem in de Boer, 1995).

It should again be pointed out that, in the global form of the balance equation, the integration process covers the body B_α of each individual constituent φ^α. Only in this case do the local statements of the balance equations given in this section come out. If the global forms of the balance equations are referred to the partial solid body φ^S, an additional term connected with the difference velocity between φ^S and $\varphi^\alpha (\alpha \neq S)$ must be added. This is also valid for the entropy inequality in the next section.

5.5 Entropy Inequality

In order to gain restrictions for constitutive equations, the second law of thermodynamics (entropy principle) has been usefully applied in continuum mechanics, in mixture theory and, in particular, in the theory of porous media. The procedure was created by Coleman and Noll (1963) and modified by Müller and Liu (see Müller, 1985). Coleman and Noll's method has been repeatedly examined in many fields of continuum mechanics and has yielded excellent results. Its contribution to porous media theory has been in the form of restrictions for the constitutive response functions which have helped to formulate a consistent theory for saturated elastic porous solids, where both constituents (solid and fluid) can either be compressible or incompressible (see, e.g., de Boer, 1993, de Boer and Kowalski, 1995, and de Boer, 1996, 1997). The second law of thermodynamics follows from the balance of energy, after some manipulations with the absolute temperature. It can be shown (see, e.g., Planck, 1897) that the absolute temperature *can* serve as an integrating factor for the sum of the rate of internal energy and that of the stress power (see also Müller, 1979). This expression, the above stated sum (or the corresponding heat) divided by the absolute temperature, is denoted as entropy; thermodynamics is based on this notion (introduced by Clausius, 1865). However, this conception seems only to be clear for reversible processes. For irreversible processes, the entropy notion, and the whole procedure of constructing a fundamental inequality, remains in some parts obscure and mysterious (see Balian, 1991/92). For this reason, it is not surprising that it is extremely difficult to impart the entropy notion to students. Baierlein (1992) characterized this situation with the words: "Students find entropy a mysterious concept - and not surprisingly so, for it is a difficult notion". There have been, of course, many attempts to support the entropy notion for irreversible processes by results obtained in other fields of science such as statistical mechanics and information theory (see Balian, 1991/92). However, these efforts are not very convincing and degenerate at times to purely philosophical discussions. There have been, however, attempts to avoid the notion of entropy, which is in general neither accessible nor measurable (see Balian, 1991/92).

This critique of the notion of entropy seems in parts also valid for other concepts in thermodynamics. Balian (1991/92) stated:

" Still worse, thermodynamics is based upon many, more or less intuitive, concepts which cannot be readily formulated mathematically and whose nature is far from clear on a microscopic scale: temperature, pressure, work, heat, entropy, ..."

As it is not the aim of this section to work on the entropy principle and other concepts of thermodynamics, we will use the entropy principle in the classical form in porous media theory for gaining restrictions for constitutive equations (see, e.g., de Boer and Ehlers, 1986b), namely as the sum of all entropy inequalities for the individual constituents. The assumption that the entropy inequality has to be fulfilled for every individual constituent φ^α is indeed a sufficient – though too restrictive – condition. At the same time, the postulate of a common entropy inequality for all constituents is both a necessary and a sufficient condition for the existence of dissipation mechanism within the mixture, and will therefore be preferred:

$$\sum_{\alpha=1}^{\kappa}(H^\alpha)'_\alpha \geq \sum_{\alpha=1}^{\kappa}\int_{B_\alpha}\frac{1}{\Theta^\alpha}\rho^\alpha r^\alpha dv - \sum_{\alpha=1}^{\kappa}\int_{\partial B_\alpha}\frac{1}{\Theta^\alpha}\mathbf{q}^\alpha \cdot d\mathbf{a}\,. \tag{5.5.1}$$

The quantity

$$H^\alpha = \int_{B_\alpha}\rho^\alpha\eta^\alpha dv \tag{5.5.2}$$

denotes the entropy of the constituent φ^α, whereby η^α is the specific entropy. Moreover, Θ^α is the absolute temperature of φ^α. In consideration of the transport theorem (5.3.31) in connection with $(5.3.21)_1$ and the mass balance equation (5.4.3), from (5.5.2), the material time derivative of the entropy is obtained:

$$\sum_{\alpha=1}^{\kappa}(H^\alpha)'_\alpha = \sum_{\alpha=1}^{\kappa}\int_{B_\alpha}[\rho^\alpha(\eta^\alpha)'_\alpha + \hat{\rho}^\alpha\eta^\alpha]dv \geq \sum_{\alpha=1}^{k}\int_{B_\alpha}[\frac{1}{\Theta^\alpha}\rho^\alpha r^\alpha - \operatorname{div}(\frac{1}{\Theta^\alpha}\mathbf{q}^\alpha)]dv\,. \tag{5.5.3}$$

In (5.5.3), the divergence theorem has been used to transform the surface integral into a volume integral. From (5.5.3), the local form of the entropy inequality is gained:

$$\sum_{\alpha=1}^{\kappa}[\rho^\alpha(\eta^\alpha)'_\alpha + \hat{\rho}^\alpha\eta^\alpha - \frac{1}{\Theta^\alpha}\rho^\alpha r^\alpha + \operatorname{div}(\frac{1}{\Theta^\alpha}\mathbf{q}^\alpha)] \geq 0\,. \tag{5.5.4}$$

Considering the balance equation of energy (5.4.32) and the free Helmholtz energy

$$\psi^\alpha = \varepsilon^\alpha - \Theta^\alpha \eta^\alpha \,, \tag{5.5.5}$$

the entropy inequality (5.5.4) can be rewritten as

$$\sum_{\alpha=1}^{\kappa} \frac{1}{\Theta^\alpha}\{-\rho^\alpha[(\psi^\alpha)'_\alpha + (\Theta^\alpha)'_\alpha \eta^\alpha] - \hat{\rho}^\alpha(\psi^\alpha - \frac{1}{2}\mathbf{v}_\alpha \cdot \mathbf{v}_\alpha) + \\ + \mathbf{T}^\alpha \cdot \mathbf{L}_\alpha - \hat{\mathbf{p}}^\alpha \cdot \mathbf{x}'_\alpha - \frac{1}{\Theta^\alpha}\mathbf{q}^\alpha \cdot \operatorname{grad}\Theta^\alpha + \hat{e}^\alpha\} \geq 0 \,. \tag{5.5.6}$$

If all constituents have the same temperature Θ, i.e., $\Theta^\alpha = \Theta$, Inequality (5.5.6) simplifies, by the use of $(5.4.35)_6$, to:

$$\sum_{\alpha=1}^{\kappa} \{-\rho^\alpha[(\psi^\alpha)'_\alpha + (\Theta)'_\alpha \eta^\alpha] - \hat{\rho}^\alpha(\psi^\alpha - \frac{1}{2}\mathbf{v}_\alpha \cdot \mathbf{v}_\alpha) + \\ + \mathbf{T}^\alpha \cdot \mathbf{L}_\alpha - \hat{\mathbf{p}}^\alpha \cdot \mathbf{v}_\alpha - \frac{1}{\Theta}\mathbf{q}^\alpha \cdot \operatorname{grad}\Theta\} \geq 0 \,. \tag{5.5.7}$$

Besides the forms of the entropy inequality of the mixture body represented here, there are further alternative forms possible which, however, do essentially contain only other transformations in energies. In order to gain restrictions for constitutive relations in the constitutive theory, the forms of the entropy inequality (5.5.6) and (5.5.7) are both sufficient and convenient. Thus, we will not mention other forms of the second law of thermodynamics.

5.6 The Closure Problem and the Saturation Constraint

Mixture theory – the basis of porous media theory – is closed, i.e., the number of unknown fields is equal to the sum of the balance equations and the constitutive relations. This can easily be proven. However, by the introduction of the volume fractions n^S and n^F for the real constituents φ^S and φ^F in porous media theory (in order to obtain "smeared" continua which can be treated by continuum mechanical methods), the problem arises that two field equations are missing. This causes a considerable difference between porous media theory and mixture theory, as well as the continuum mechanics of one-component materials. Also, other existing theories in continuum mechanics are closed and every new condition leads to an equation in excess. This condition must be provided with a Lagrange multiplier for the evaluation process of the entropy inequality. If the equation in excess is a constraint of the motion, then the Lagrange multiplier will become an unknown reaction force.

In porous media theory, on the other hand, one has to look for additional equations in order to close the fields. It is, however, difficult to gain additional fields since the volume fractions touch quantities of the microscale for which balance or constitutive equations are not contained in the macroscopic mixture theory. Therefore, much effort has been made to overcome this crucial problem. This effort starts by introducing an additional balance equation to the formulation of an evolution equation for the volume fraction. This procedure solves – from the mathematical point of view – the closure problem. However – from the physical point of view – this method is completely insufficient, because one must be aware of an important constraint, namely the saturation condition (5.2.4). This constraint restricts, in the rate formulation, the rates of the volumetric changes and must, therefore, be considered in the evaluation of the entropy inequality. By differentiating the saturation condition (5.2.4) with respect to the solid phase (the same result can be obtained by differentiating with respect to the fluid phase), we have

$$(n^S)'_S + (n^F)'_S = 0 \tag{5.6.1}$$

or

$$-(n^S)'_S - (n^F)'_F + \operatorname{grad} n^F \cdot (\mathbf{v}_F - \mathbf{v}_S) = 0\,. \tag{5.6.2}$$

Considering (5.2.11), we obtain from (5.6.2):

$$-n^S\frac{(\rho^S)'_S}{\rho^S} + n^S\frac{(\rho^{SR})'_S}{\rho^{SR}} - n^F\frac{(\rho^F)'_F}{\rho^F} + n^F\frac{(\rho^{FR})'_F}{\rho^{FR}} + \operatorname{grad} n^F \cdot (\mathbf{v}_F - \mathbf{v}_S) = 0 \tag{5.6.3}$$

or considering the mass balance equations (5.4.3), in connection with (5.2.11), $(5.3.21)_1$, (5.3.27), (5.3.28), (5.3.29), and (5.3.43), as well as neglecting the mass supplies,

$$n^S(\mathbf{D}_{SN} \cdot \mathbf{I}) + n^F(\mathbf{D}_{FN} \cdot \mathbf{I}) + \operatorname{grad} n^F \cdot (\mathbf{v}_F - \mathbf{v}_S) = 0 \tag{5.6.4}$$

or

$$n^S(\mathbf{D}_S \cdot \mathbf{I}) - n^S(\mathbf{D}_{SR} \cdot \mathbf{I}) + n^F(\mathbf{D}_F \cdot \mathbf{I}) - n^F(\mathbf{D}_{FR} \cdot \mathbf{I}) + \operatorname{grad} n^F \cdot (\mathbf{v}_F - \mathbf{v}_S) = 0 \tag{5.6.5}$$

is obtained. In (5.6.4) and (5.6.5), use is made of the relations

$$\frac{(n^\alpha)'_\alpha}{n^\alpha} + \mathbf{D}_{\alpha N} \cdot \mathbf{I} = 0\,, \quad \mathbf{D}_{\alpha N} \cdot \mathbf{I} = \mathbf{D}_\alpha \cdot \mathbf{I} - (\mathbf{D}_{\alpha R} \cdot \mathbf{I}) \tag{5.6.6}$$

and

$$\frac{(\rho^{\alpha R})'_\alpha}{\rho^{\alpha R}} + \mathbf{D}_{\alpha R} \cdot \mathbf{I} = 0\,, \tag{5.6.7}$$

which result from the balance equations of mass (5.4.3) or (5.4.4), excluding all mass exchanges, considering (5.2.11) and (5.3.29) (see Bluhm, 1997).

The relation (5.6.5) clearly reveals that the rates of the volumetric strains of the partial bodies $\mathbf{D}_\alpha \cdot \mathbf{I}$ and of the real compressible materials $\mathbf{D}_{\alpha R} \cdot \mathbf{I}$ are dependent. Please note in passing that we have, for the sake of simplicity, at this place and in the following sections, omitted the signs on $\mathbf{D}_{SR}$ and $\mathbf{D}_{FR}$ which characterize the intermediate states.

It depends on the problem to be solved as to whether the constraint in the forms (5.6.2), (5.6.3), or (5.6.4), (5.6.5) should be used. If all mass exchanges are neglected, then the constraints in the forms (5.6.4) and (5.6.5) are convenient. If, however, mass exchange occurs, then the forms (5.6.2) and (5.6.3) have to be used in the evaluation of the entropy inequality.

As has already been mentioned, the saturation constraints (5.6.2) and (5.6.3), or (5.6.4) and (5.6.5) have to be considered in the evaluation of the entropy inequality because the rates of either the densities or of the volumetric strains of the solid and fluid phases are dependent. In order to obtain a stress-power-like expression, the constraints (5.6.2) and (5.6.3), or (5.6.4) and (5.6.5), which contain the rates of the volumetric strains, will be multiplied by a hydrostatic interface pressure λ. It is true that the saturation condition (5.2.4) is an equation to further reduce the number of unknown volume fractions, but the grade of indetermination does not change by the introduction of the interface pressure λ. Therefore, it will be postulated that two constitutive equations for λ must be introduced (or two constitutive relations for the hydrostatic pressure in the solid material) which contain properties of both the constituents of the partial solid and of partial fluid phases, in order to achieve closure. This is a reasonable demand from the mechanical point of view, because the interface pressure acts in the solid and the fluid phases. It will be seen that the introduced requirement leads to excellent physical results.

If the materials of the two individual constituents behave as incompressible phases, then additional constraints have to be considered in the evaluation of the entropy inequality. These constraints will be provided with the multipliers κ^{FR} and κ^{SR} for the fluid and solid phases and then added to the entropy inequality. If only one constituent is incompressible, then the corresponding constraint mentioned above must be included in the entropy inequality.

In order to simplify the evaluation of the entropy inequality in the case of a common temperature for both the constituents, the constraints (5.6.2) and (5.6.3), or (5.6.4) and (5.6.5) will, in addition, be multiplied by

$$\bar{\Theta} = \frac{1}{2}\,\frac{\Theta^S + \Theta^F}{\Theta^S \Theta^F}\,. \tag{5.6.8}$$

If $\Theta^S = \Theta^F = \Theta$, Eqn. (5.6.8) leads to $\bar{\Theta} = \dfrac{1}{\Theta}$.

Lastly, the closure problem for empty porous solids will be addressed. In this case, there is no saturation constraint. However, the volume fraction condition (5.2.4) remains valid. Thus, one field equation is missing. However, for this problem, we can also assume two constitutive equations for the hydrostatic pressure in the real compressible solid material. Then, the fields are also closed for this problem. If the real solid material is incompressible, the incompressibility condition is an equation in excess and the corresponding Lagrange multiplier is an undetermined reaction force in the solid material which can, however, be determined by a constitutive equation of the skeleton.

5.7 Principle of Virtual Work

With the equations of motion (5.4.15), we are able to develop the principles of virtual velocities and virtual stresses. These principles are useful for the formulation of general theorems, for example, the development of minimum and maximum principles (see, e.g., de Boer, 1974, de Boer and Ehlers, 1980, and de Boer and Kowalski, 1985), and for proving general theorems, e.g., the uniqueness theorem (see, e.g., de Boer and Kowalski, 1986).

In order to derive the principle of virtual velocities, it is necessary to introduce the virtual velocities $\bar{\mathbf{v}}_S$ (for the partial solid constituent) and $\bar{\mathbf{v}}_F$ (for the partial fluid constituent). These virtual velocities must be *kinematically admissible*, which is defined as follows:

A virtual velocity state is kinematically admissible if it is smooth and satisfies internal and external kinematic constraints.

Please note in passing that, by internal kinematic constraints, such constraints as rigidity, incompressibility, and saturation are denoted. The geometric boundary conditions are denoted as external kinematic constraints.

Multiplying the dynamic equations (5.4.15), where we have neglected the inertia effects and the mass exchange, by the virtual velocities $\bar{\mathbf{v}}_S$ and $\bar{\mathbf{v}}_F$ and integrating over the whole control space B_S, which is shaped by the skeleton, we obtain for the mixture body

$$\int_{B_S} \left[(\operatorname{div} \mathbf{T}^S + \rho^S \mathbf{b} - \hat{\mathbf{p}}^F) \cdot \bar{\mathbf{v}}_S + (\operatorname{grad} p^F + \rho^F \mathbf{b} + \hat{\mathbf{p}}^F) \cdot \bar{\mathbf{v}}_F \right] dv = 0 \,. \tag{5.7.1}$$

In (5.7.1), it is assumed that the external acceleration is the same for all constituents. Bearing in mind that the virtual velocities fulfill the geometrical boundary conditions, we obtain, with the help of Gauß's theorem and some calculation rules,

$$\begin{aligned} &\int_{\partial B_{St}} (\mathbf{t}^S \cdot \bar{\mathbf{v}}_S + \mathbf{t}^F \cdot \bar{\mathbf{v}}_F) da + \int_{B_S} (\rho^F \bar{\mathbf{v}}_F + \rho^S \bar{\mathbf{v}}_S) \cdot \mathbf{b} \, dv = \\ &= \int_{B_S} \left[\mathbf{T}^S \cdot \bar{\mathbf{D}}_S + p^F (\bar{\mathbf{D}}_F \cdot \mathbf{I}) - \hat{\mathbf{p}}^F \cdot \bar{\mathbf{v}}_{FS} \right] dv \,, \quad \bar{\mathbf{v}}_{FS} = \bar{\mathbf{v}}_F - \bar{\mathbf{v}}_S \,. \end{aligned} \tag{5.7.2}$$

The surface integral in $(5.7.2)_1$ covers the area ∂B_{St} with the normal unit vector $\mathbf{n}$, where the surface forces

$$\mathbf{t}^S = \mathbf{T}^S \mathbf{n} \,, \quad \mathbf{t}^F = p^F \mathbf{n} \tag{5.7.3}$$

are prescribed. The tensors $\bar{\mathbf{D}}_\alpha$ stand for the symmetrical parts of the velocity gradients of the virtual velocities $\bar{\mathbf{v}}_S$ and $\bar{\mathbf{v}}_F$, see Eqn. $(5.3.14)_1$.

According to the derivation of the principle of virtual velocities $(5.7.2)_1$, the principle of virtual stresses can be derived. In this case, we are relying on a virtual state of stresses, external accelerations, and interaction forces $\bar{\mathbf{T}}^S, \bar{p}^F, \bar{\mathbf{b}}$, and $\bar{\hat{\mathbf{p}}}^F$, which must be *dynamically admissible:*

A virtual stress state is dynamically admissible if it is smooth and satisfies the internal and external dynamic constraints.

The equations of motion and yield functions belong to the *internal* dynamic constraints, whereas the dynamic boundary conditions form the *external* dynamic constraints.

We commence from the equations of motion (5.4.15), neglecting the inertia effects

$$\operatorname{div} \bar{\mathbf{T}}^\alpha + \rho^\alpha \bar{\mathbf{b}} + \bar{\hat{\mathbf{p}}}^\alpha = \mathbf{0}\,. \tag{5.7.4}$$

We multiply the equations of motion for the virtual stresses (5.7.4) with the real partial velocities $\mathbf{v}_S$ and $\mathbf{v}_F$ and integrate over the whole control space. Manipulating the virtual work, similarly to the aforementioned method, we arrive at

$$\begin{aligned}\int\limits_{\partial B_{Sv}} (\bar{\mathbf{t}}^S \cdot \mathbf{v}_S + \bar{\mathbf{t}}^F \cdot \mathbf{v}_F)\, da + \int\limits_{B_S} (\rho^F \mathbf{v}_F + \rho^S \mathbf{v}_S) \cdot \bar{\mathbf{b}}\, dv = \\ = \int\limits_{B_S} \left[\bar{\mathbf{T}}^S \cdot \mathbf{D}_S + \bar{p}^F (\mathbf{D}_F \cdot \mathbf{I}) - \bar{\hat{\mathbf{p}}}^F \cdot \mathbf{v}_{FS}\right] dv\,,\end{aligned} \tag{5.7.5}$$

whereby ∂B_{Sv} is the part of the boundary where the velocities are prescribed, and $\mathbf{D}_\alpha$ are the symmetric parts of the velocity gradients of $\mathbf{v}_\alpha$. The velocity difference $\mathbf{v}_{FS}$ is determined by $(5.7.2)_2$ with the real velocities $\mathbf{v}_\alpha$.

Next, we reformulate the principles of virtual work, $(5.7.2)_1$ and (5.7.5). From a stress-free placement we come – applying external forces – to the actual placement. By superposition with a virtual velocity state, we arrive at the actual plus virtual state which is marked by an asterix. Now, we replace the virtual velocities $\bar{\mathbf{v}}_S$ and $\bar{\mathbf{v}}_F$ by the differences of the velocities $\overset{*}{\mathbf{v}}_S - \mathbf{v}_S$ and $\overset{*}{\mathbf{v}}_F - \mathbf{v}_F$. With this substitution, we obtain a convenient form of the principle of virtual velocities $(5.7.2)_1$, suitable for the derivation of minimum principles:

$$\begin{aligned}\int\limits_{\partial B_{St}} \left[\mathbf{t}^S \cdot (\overset{*}{\mathbf{v}}_S - \mathbf{v}_S) + \mathbf{t}^F \cdot (\overset{*}{\mathbf{v}}_F - \mathbf{v}_F)\right] da + \int\limits_{B_S} \left[\rho^F (\overset{*}{\mathbf{v}}_F - \mathbf{v}_F) + \rho^S (\overset{*}{\mathbf{v}}_S - \mathbf{v}_S)\right] \cdot \mathbf{b}\, dv = \\ = \int\limits_{B_S} \left[\mathbf{T}^S \cdot (\overset{*}{\mathbf{D}}_S - \mathbf{D}_S) + p^F (\overset{*}{\mathbf{D}}_F - \mathbf{D}_F) \cdot \mathbf{I} - \hat{\mathbf{p}}^F \cdot (\overset{*}{\mathbf{v}}_{FS} - \mathbf{v}_{FS})\right] dv\,.\end{aligned} \tag{5.7.6}$$

In a similar way, Eqn. (5.7.5) will be reformulated. We substitute the virtual boundary forces and the stresses by the differences of the real and virtual plus actual states. By application of a virtual state of forces, we obtain the virtual state. Then, by applying the real force and velocity states, we arrive at the virtual plus actual state which we will mark by a superscript zero. The splitting of the virtual state into a virtual state plus actual state minus actual state, in Eqn. (5.7.5), yields a slightly changed form of the principle of virtual forces:

$$\int_{\partial B_{Sv}} \left[(\overset{\circ}{\mathbf{t}}^S - \mathbf{t}^S) \cdot \mathbf{v}_S + (\overset{\circ}{\mathbf{t}}^F - \mathbf{t}^F) \cdot \mathbf{v}_F\right] da =$$

$$= \int_{B_S} \left[(\overset{\circ}{\mathbf{T}}^S - \mathbf{T}^S) \cdot \mathbf{D}_S + (\overset{\circ}{p}^F - p^F)(\mathbf{D}_F \cdot \mathbf{I}) - (\overset{\circ}{\hat{\mathbf{p}}}{}^F - \hat{\mathbf{p}}^F) \cdot \mathbf{v}_{FS}\right] dv \,. \tag{5.7.7}$$

In obtaining (5.7.7), we assume that the body force densities are not changed in the two stress states.

Both versions of the principle of virtual work, (5.7.6) and (5.7.7), can serve to develop minimum and maximum principles and to prove the uniqueness of the solutions of boundary- and initial-value problems for a certain class of material behavior (see de Boer and Kowalski, 1985, 1986).

5.8 Constitutive Theory

As has already been mentioned, in order to close the system of field equations, it is necessary to introduce constitutive equations. These equations connect certain mechanical or thermodynamic quantities via material-dependent constants which are determined by test observations. Thus, it is ensured that the constitutive relations introduced are able to describe the test results.

In the past, a great number of constitutive equations for empty and saturated porous media have been derived. However, many of them are very complicated due to the use of inadequate mechanical or thermodynamic concepts. Without a doubt, those constitutive equations may closely describe the stress-strain (rate) relations of the special mechanical behavior of materials. However, in many cases this can only be achieved by introducing many parameters and neglecting requirements due to mechanical and thermodynamic "principles". Those constitutive equations are meaningless in view of the calculation of general boundary- and initial-value problems within the framework of geometrically-linear and non-linear theories. The goal should be to formulate relatively simple constitutive equations. This statement becomes even more evident in the field of saturated porous media. In this field, not only the thermodynamic behavior of the skeleton and the content of the pores have to be described, but also various interaction phenomena. Therefore, the idea of formulating relatively simple constitutive equations is of particular relevance. In order to derive consistent constitutive equations that are relatively simple, some strong assumptions have to be introduced at times. For example, for many problems the compressibility of the solid matrix material can be neglected in comparison with the compressibility of the matrix. Thus, the mathematical model reduces to an incompressible model. Another example is the elastic-plastic model, where also some simplifying assumptions have to be introduced in order to predict the essential properties of the porous solids under study by using relatively simple constitutive relations (see Section 5.8.5).

However, it is not sufficient to only fulfill requirements due to test observations; rather more general "principles", which were developed in continuum mechanics in the 1950s and 1960s, should be fulfilled; these are: *Determinism, local action, material objectivity,* and *dissipation.* Some of the above-stated "principles", however, should not be understood as axioms, but should rather be considered as convenient work hypotheses, because the real mechanical or thermodynamic behavior of solids and fluids is, in most cases, too complex to be described by relatively simple "principles".

The most important of the above-stated "principles" are the material objectivity and dissipation principles. The material objectivity principle states that the constitutive equations have to be formulated in such a way that they are not influenced by superimposed rigid body motions. The dissipation principle results from the second law of thermodynamics. Both principles have a big impact on the development of consistent constitutive equations.

Guided by these principles, constitutive equations for compressible and incompressible elastic and plastic porous solids, filled with compressible and incompressible viscous and inviscid fluids, will be derived in the following sections. In this manner, hybrid models will also be treated. The aforementioned individual "principles" will be fulfilled – though not explicitly discussed – here. For example, in the evaluation of the entropy principle, only objective process variables will be used, so that the material objectivity principle will be automatically satisfied. Only in the APPENDIX A will full use be made of the material objectivity principle.

5.8.1 Principle of Material Objectivity

The principles of determinism and local actions assert that the present response functions of a particle are determined by the history of an arbitrary, small neighborhood of the particle (see Truesdell and Noll, 1965). As the history of a small neighborhood of a particle of a rigid body does not change with time, a superimposed rigid body motion cannot influence the response functions. This fact has a strong impact on the development of response functions.

We will consider a motion of the individual constituents which differs only by a superimposed rigid body motion. A particle X_α, which in the actual placement occupies the position $\mathbf{x}$, will be taken by a rigid body motion to the position $\overset{*}{\mathbf{x}}$:

$$\overset{*}{\mathbf{x}} = \overset{*}{\mathbf{c}}\,(\overset{*}{t}) + \mathbf{Q}(t)\left[\mathbf{x} - \mathbf{c}(t)\right] , \tag{5.8.1}$$

with $\overset{*}{\mathbf{c}}(\overset{*}{t}), \mathbf{c}(t)$ as field independent vector-value functions of the times $\overset{*}{t}$ and t, whereby

$$\overset{*}{t} = t + a\,, \qquad a = \text{const.} \tag{5.8.2}$$

(More precisely, $\mathbf{x}$, as well as $\overset{*}{\mathbf{x}}$ in (5.8.1), should be replaced by χ_S and χ_F, as well as by $\overset{*}{\chi}_S$ and $\overset{*}{\chi}_F$).

The vectors $\mathbf{c}(t)$ and $\overset{*}{\mathbf{c}}(\overset{*}{t})$ can be interpreted as position vectors referring to a material point of the body in the actual and in the superimposed actual placement, respectively. Moreover, $\mathbf{Q}(t)$ is a field-independent proper orthogonal tensor.

For further considerations, the following statements are important: The value of a scalar value function $\alpha(\mathbf{x}, t)$ does not change due to a superimposed rigid body motion, this means for an arbitrary vector $\mathbf{u}$:

$$\overset{*}{\mathbf{u}} \cdot (\overset{*}{\mathbf{x}} - \overset{*}{\mathbf{c}}) = \mathbf{u} \cdot (\mathbf{x} - \mathbf{c})\,. \tag{5.8.3}$$

With (5.8.1), we can conclude:

$$\overset{*}{\mathbf{u}} \cdot \mathbf{Q}(\mathbf{x} - \mathbf{c}) = \mathbf{u} \cdot \ (\mathbf{x} - \mathbf{c}), \tag{5.8.4}$$

and

$$\overset{*}{\mathbf{u}} = \mathbf{Q}\mathbf{u}\,. \tag{5.8.5}$$

In a similar way, it follows from (5.8.5) for a second-order tensor $\mathbf{A}$ as a linear mapping of the arbitrary vector $\mathbf{u}$:

$$\overset{*}{\mathbf{A}}\,\overset{*}{\mathbf{u}} = \mathbf{Q}(\mathbf{A}\mathbf{u})\,, \tag{5.8.6}$$

so that

$$\overset{*}{\mathbf{A}} = \mathbf{Q}\mathbf{A}\mathbf{Q}^T\,. \tag{5.8.7}$$

Quantities of the actual placement referred to the base system of the actual placement which, under a superimposed rigid body rotation, show a transformation according to (5.8.5) and (5.8.7), are called *objective quantities.*

With the relations (5.8.5) and (5.8.7), we can define higher-order tensors (see Ehlers, 1989a).

If $\overset{3}{\mathbf{B}}(\mathbf{x}, t)$ is an arbitrary third-order tensor,

$$\overset{*}{\overset{3}{\mathbf{B}}}\overset{*}{\mathbf{A}} = \mathbf{Q}(\overset{3}{\mathbf{B}}\,\mathbf{A}) \tag{5.8.8}$$

is valid, or

$$\overset{*}{\overset{3}{\mathbf{B}}}\,(\mathbf{Q}\mathbf{A}\mathbf{Q}^T) = \mathbf{Q}(\overset{3}{\mathbf{B}}\,\mathbf{A})\,. \tag{5.8.9}$$

After some manipulations, it can be concluded that

$$\overset{*}{\overset{3}{\mathbf{B}}} = \left\{ [(\mathbf{Q}\,\overset{3}{\mathbf{B}})^{\underline{3}}\,\mathbf{Q}^T]^{\overset{23}{\underline{3}^T}}\,\mathbf{Q}^T \right\}^{\overset{23}{\underline{3}^T}}\,. \tag{5.8.10}$$

If $\overset{4}{\mathbf{B}}(\mathbf{x}, t)$ is an arbitrary fourth-order tensor, from (5.8.7) it follows that

$$\overset{*}{\overset{4}{\mathbf{B}}}\overset{*}{\mathbf{A}} = \mathbf{Q}(\overset{4}{\mathbf{B}}\,\mathbf{A})\mathbf{Q}^T \tag{5.8.11}$$

or

$$\overset{*}{\overset{4}{\mathbf{B}}} = (\mathbf{Q}^T \otimes \mathbf{Q}^T)^{\overset{14}{T}} \overset{4}{\mathbf{B}} (\mathbf{Q} \otimes \mathbf{Q})^{\overset{14}{T}} . \tag{5.8.12}$$

Finally, the consequences of a superimposed rigid body motion for kinematic quantities will be revealed. From (5.8.1), we obtain the velocities and accelerations

$$\begin{aligned} \overset{*}{\mathbf{x}}{}'_\alpha &= \overset{*}{\mathbf{c}}{}'_\alpha + \mathbf{Q}(\mathbf{x}'_\alpha - \mathbf{c}'_\alpha) + \mathbf{Q}'_\alpha(\mathbf{x} - \mathbf{c}) , \\ \overset{*}{\mathbf{x}}{}''_\alpha &= \overset{*}{\mathbf{c}}{}''_\alpha + \mathbf{Q}(\mathbf{x}''_\alpha - \mathbf{c}''_\alpha) + 2\mathbf{Q}'_\alpha(\mathbf{x}'_\alpha - \mathbf{c}'_\alpha) + \mathbf{Q}''_\alpha(\mathbf{x} - \mathbf{c}) . \end{aligned} \tag{5.8.13}$$

It should be mentioned that the convective terms in the material time derivatives of $\overset{*}{\mathbf{c}}$, $\mathbf{c}$, and $\mathbf{Q}$ vanish because these quantities are field-independent.

The deformation gradient $\mathbf{F}_\alpha$ is referred to the base system in the actual and the reference placement. Such quantities are objective if

$$\overset{*}{\mathbf{F}}_\alpha = \mathbf{Q}\mathbf{F}_\alpha \tag{5.8.14}$$

is valid. This can be confirmed by applying (5.8.1) to (5.3.4). With (5.8.14), the right Cauchy-Green deformation tensor

$$\overset{*}{\mathbf{C}}_\alpha = \overset{*}{\mathbf{F}}{}^T_\alpha \overset{*}{\mathbf{F}}_\alpha = (\mathbf{Q}\mathbf{F}_\alpha)^T\mathbf{Q}\mathbf{F}_\alpha = \mathbf{F}^T_\alpha\mathbf{F}_\alpha = \mathbf{C}_\alpha \tag{5.8.15}$$

is obtained. Thus, the Green strain tensor $\overset{*}{\mathbf{E}}_\alpha$ is

$$\overset{*}{\mathbf{E}}_\alpha = \mathbf{E}_\alpha . \tag{5.8.16}$$

These quantities, which are referred to the base system of the reference placement, are also objective.

In a similar way we obtain for the left Cauchy-Green deformation tensor $\overset{*}{\mathbf{B}}_\alpha$:

$$\overset{*}{\mathbf{B}}_\alpha = \mathbf{Q}\mathbf{B}_\alpha\mathbf{Q}^T . \tag{5.8.17}$$

Now we turn to the material time derivatives of the introduced kinematic quantities. With (5.8.14), we have

$$(\overset{*}{\mathbf{F}}_\alpha)'_\alpha = \mathbf{Q}'_\alpha\mathbf{F}_\alpha + \mathbf{Q}(\mathbf{F}_\alpha)'_\alpha \tag{5.8.18}$$

and, with (5.8.15) and (5.8.16),

$$(\overset{*}{\mathbf{C}}_\alpha)'_\alpha = (\mathbf{C}_\alpha)'_\alpha , \quad (\overset{*}{\mathbf{E}}_\alpha)'_\alpha = (\mathbf{E}_\alpha)'_\alpha . \tag{5.8.19}$$

The material time derivative of $\overset{*}{\mathbf{B}}_\alpha$ (5.8.17) leads to

$$(\overset{*}{\mathbf{B}}_\alpha)'_\alpha = \mathbf{Q}(\mathbf{B}_\alpha)'_\alpha\mathbf{Q}^T + \mathbf{Q}'_\alpha\mathbf{B}_\alpha\mathbf{Q}^T + \mathbf{Q}\mathbf{B}_\alpha\mathbf{Q}^{T\prime}_\alpha . \tag{5.8.20}$$

The spatial velocity $\overset{*}{\mathbf{L}}_\alpha$ is, with (5.8.18), given by

$$\overset{*}{\mathbf{L}}_\alpha = (\overset{*}{\mathbf{F}})'_\alpha\overset{*}{\mathbf{F}}{}^{-1} = \mathbf{Q}'_\alpha\mathbf{Q}^T + \mathbf{Q}\mathbf{L}_\alpha\mathbf{Q}^T . \tag{5.8.21}$$

It is recognized that $(\mathbf{F}_\alpha)'_\alpha$, $(\mathbf{B}_\alpha)'_\alpha$, and $\mathbf{L}_\alpha$ do not satisfy the condition (5.8.7). Therefore, they are not objective. This is, however, not valid for the symmetric part of the spatial velocity gradient. From $(5.3.14)_1$, it immediately follows, with (5.8.21) and considering that $\mathbf{Q}'_\alpha\mathbf{Q}^T$ and $\mathbf{Q}\mathbf{Q}'^T_\alpha$ are skew-symmetric tensors, that:

$$\overset{*}{\mathbf{D}}_\alpha = \frac{1}{2}(\overset{*}{\mathbf{L}}_\alpha + \overset{*}{\mathbf{L}}{}^T_\alpha) = \frac{1}{2}\mathbf{Q}(\mathbf{L}_\alpha + \mathbf{L}^T_\alpha)\mathbf{Q}^T + \frac{1}{2}\mathbf{Q}'_\alpha\mathbf{Q}^T + \frac{1}{2}\mathbf{Q}\mathbf{Q}'^T_\alpha = \mathbf{Q}\mathbf{D}_\alpha\mathbf{Q}^T. \quad (5.8.22)$$

After $(5.3.14)_2$, and considering (5.8.21), we obtain for the skew-symmetric tensor $\overset{*}{\mathbf{W}}_\alpha$, in a similar way,

$$\overset{*}{\mathbf{W}}_\alpha = \mathbf{Q}\mathbf{W}_\alpha\mathbf{Q}^T + \mathbf{Q}'_\alpha\mathbf{Q}^T . \quad (5.8.23)$$

Thus, it is shown that the decomposition of the spatial velocity gradient $\mathbf{L}_\alpha$ yields the objective symmetric tensor $\mathbf{D}_\alpha$ and the non-objective skew-symmetric tensor $\mathbf{W}_\alpha$.

5.8.2 The Introduction and Evaluation of the Entropy Inequality for a General Binary Porous Medium Model

As has already been pointed out in Section 5.5, the entropy inequality has yielded excellent results in the porous media theory (see the review article of de Boer, 1996). However, in this field, one should be careful when evaluating the entropy inequality purely mathematically, because the mechanical and thermodynamic behavior of saturated porous solids are very complex. If one evaluates the entropy inequality in a purely stereotyped way, without considering the special physical properties of the complex material under study, one can arrive at results which fulfill the entropy inequality but fail to predict physical phenomena arrived at by experiment. The second law of thermodynamics is an inequality and there are many possibilities to satisfy this inequality. Some evaluations may be less restrictive, others more. However, the inequality is not violated.

Therefore, the entropy inequality has to be manipulated in order to include fundamental physical phenomena known from experience, test observations, and theories, which appear independently of the special constitutive behavior of the individual partial constituents, such as elastic, plastic, or viscous behavior. These phenomena mainly concern the concept of effective stresses in different versions (see, e.g., de Boer, 1996, Lade and de Boer, 1997). The concept of effective stresses is caused by the saturation condition and is valid for all kinds of material of the individual partial constituents. However, in order to describe the different versions of the effective stress concept, the closure problem must also be considered, whereby one must distinguish between compressible and incompressible behavior of the real material of the constituents.

As the second law of thermodynamics is an inequality where the entropy is always greater than zero, this inequality can also be considered as a minimum problem which reaches its minimum at zero. All additional constraints, like the saturation or the incompressibility conditions, provided with multipliers, can be taken into consideration by adding to the entropy inequality. It depends on the closure problem as to whether the multipliers are constitutively determined or not.

The entropy inequality will be developed for a binary model consisting of a porous solid and a fluid.

There is one remarkable constraint for this model which is due to the saturation condition (5.6.2) in the versions (5.6.3), (5.6.4), or (5.6.5). This condition will be multiplied with the interface pressure λ. Furthermore, if the temperature is not the same for both the constituents, then it is advisable to multiply the saturation condition by $\bar{\Theta}$ (5.6.8) in order to obtain simple results if, in the limit, the temperatures of all constituents are equal.

The saturation condition in the versions (5.6.2), (5.6.3), (5.6.4), and (5.6.5) provided with the multiplier λ and the temperature $\bar{\Theta}$ (5.6.8), have to be added to the entropy inequality in order to consider the restrictions on the volumetric strains in the evaluation of this basic inequality. Moreover, we will take the constraints for the supply terms $\hat{\rho}^\alpha, \hat{\mathbf{p}}^\alpha, \hat{e}^\alpha$, (5.4.8), (5.4.17)$_4$, and (5.4.35)$_6$ into account. The result is:

$$\begin{aligned}
\sum_{\alpha=S}^{F} \frac{1}{\Theta^\alpha}\{&- \rho^\alpha [(\psi^\alpha)'_\alpha + (\Theta^\alpha)'_\alpha \eta^\alpha] + \\
&+ \mathbf{T}^{\alpha D} \cdot \mathbf{D}^D_\alpha - (p^\alpha + \lambda n^\alpha \Theta^\alpha \bar{\Theta}) \frac{(n^\alpha)'_\alpha}{n^\alpha} - p^\alpha \frac{(\rho^{\alpha R})'_\alpha}{\rho^{\alpha R}} - \\
&- \frac{1}{\Theta^\alpha} \mathbf{q}^\alpha \cdot \operatorname{grad} \Theta^\alpha\} - \frac{1}{\Theta^F}(\hat{\mathbf{p}}^F - \lambda \Theta^F \bar{\Theta} \operatorname{grad} n^F) \cdot \mathbf{v}_{FS} - \\
&- \hat{\rho}^F \left(\frac{\mu^F}{\Theta^F} - \frac{\mu^S}{\Theta^S}\right) - \frac{1}{\Theta^S \Theta^F} (\hat{e}^F - \hat{\mathbf{p}}^F \cdot \mathbf{v}_S) \Theta^{FS} \geq 0 \,,
\end{aligned} \tag{5.8.24}$$

with

$$p^\alpha = \frac{1}{3}(\mathbf{T}^\alpha \cdot \mathbf{I}), \qquad \mathbf{v}_{FS} = \mathbf{v}_F - \mathbf{v}_S, \qquad \Theta^{FS} = \Theta^F - \Theta^S \,. \tag{5.8.25}$$

In (5.8.24), the chemical potentials

$$\mu^\alpha = \psi^\alpha - \frac{p^\alpha}{\rho^\alpha} - \frac{1}{2} \mathbf{v}_\alpha \cdot \mathbf{v}_\alpha \tag{5.8.26}$$

have been introduced.

If mass exchanges are excluded, it is convenient to represent the entropy inequality by kinematic terms instead of the rates of the densities at the microscale. Inequality (5.8.24) simplifies to

$$\begin{aligned}
\sum_{\alpha=S}^{F} \frac{1}{\Theta^\alpha}\{&- \rho^\alpha [(\psi^\alpha)'_\alpha + (\Theta^\alpha)'_\alpha \eta^\alpha] + \\
&+ \mathbf{T}^{\alpha D} \cdot \mathbf{D}^D_\alpha + (p^\alpha + \lambda n^\alpha \Theta^\alpha \bar{\Theta})(\mathbf{D}_{\alpha N} \cdot \mathbf{I}) + p^\alpha (\mathbf{D}_{\alpha R} \cdot \mathbf{I}) - \\
&- \frac{1}{\Theta^\alpha} \mathbf{q}^\alpha \cdot \operatorname{grad} \Theta^\alpha\} - \frac{1}{\Theta^F}(\hat{\mathbf{p}}^F - \lambda \Theta^F \bar{\Theta} \operatorname{grad} n^F) \cdot \mathbf{v}_{FS} - \\
&- \frac{1}{\Theta^S \Theta^F} (\hat{e}^F - \hat{\mathbf{p}}^F \cdot \mathbf{v}_S) \Theta^{FS} \geq 0 \,.
\end{aligned} \tag{5.8.27}$$

For the next step, we must elaborate on the entropy inequalities (5.8.24) and (5.8.27), due to the possible different mechanical behavior of the individual constituents such as compressible and incompressible deformations of the real materials.

The drawback in including the description of incompressibility and compressibility in porous media theory is the fact that these properties touch the microscale, for which the mixture theory does not provide balance equations. Thus, one has to choose other approaches.

Moreover, the incorporation of incompressibility and compressibility effects into the general porous media theories leads, in some parts, to complex considerations. As has already been mentioned, the evaluation of the entropy inequality should not be done in a purely mathematical way; rather, experience and test results should be taken into account. The best known concept in soil mechanics, repeatedly proven by experiment, is the concept of effective stresses in the classical sense (see Section 4). However, this classical concept is only valid for special cases. It does not cover all the different possibilities of incompressibility and compressibility in a binary porous medium model.

The concept of effective stresses is due to the interface pressure; this concept contains the decomposition of the hydrostatic stress state into two parts (in every constituent). The first part reflects the incompressibility or compressibility of the real fluid material; the second part can be interpreted as a hydrostatic stress state due to "intergranular forces". This interpretation goes back to Baer and Nunziato (1986). They called the hydrostatic pressure, caused by "intergranular forces", the *configuration pressure*, and they stated for a binary model (solid skeleton filled with gas):

"... that the pressure in the solid grains equals the pressure in the gas plus the pressure due to contact forces between the grains."

Later, in part **d**) of this section, we will prove this statement. In order to determine the interface pressure λ, constitutive equations are needed, because the hydrostatic pressure is governed by the volumetric strain or the density of the real material in the case of compressibility; the configuration pressure is caused by "intergranular forces" and can be expressed by the volume fractions. However, this point of view is related to elastic behavior. If inelastic behavior is to be taken into account, then the considerations have to be extended (see, e.g., Section 5.8.5). Therefore, we will modify the entropy inequality for the four possible cases, namely where both constituents are incompressible, or compressible, or where one of the two constituents is compressible, while the other is incompressible.

In this place it should, however, be recalled that the above statement of Baer and Nunziato (1986) was already discussed by Rendulic (1936) (see also Section 4.1).

a) The Introduction and Evaluation of the Entropy Inequality for a Binary Porous Medium Model with Incompressible Constituents

We will assume that both constituents are incompressible. This means that the densities of the real materials are constant. The consequences are

$$-\,n^\alpha \bar{\Theta}\frac{(\rho^{\alpha R})'_\alpha}{\rho^{\alpha R}} = 0 \tag{5.8.28}$$

or, if mass exchange processes are excluded so that the volume change of the real material is zero:

$$n^\alpha \bar{\Theta}(\mathbf{D}_{\alpha R}\cdot \mathbf{I}) = 0\,. \tag{5.8.29}$$

The weighting of the incompressibility constraints (5.8.28) and (5.8.29) by the factor $n^\alpha\bar{\Theta}$ is performed for convenience of the evaluation of the entropy inequality. From (5.6.2) and (5.6.3), or (5.6.4) and (5.6.5), it is evident that the motions at the microlevel are restricted. As the motions at the microlevel also influence the motions at the macrolevel, the constraints (5.8.28) or (5.8.29) have also to be taken into account. Therefore, we provide (5.8.28) and (5.8.29) with the Lagrange multiplier $\kappa^{\alpha R}$ and add them to the entropy inequalities (5.8.24) and (5.8.27). In order to describe the configuration pressure, we introduce, for the free Helmholtz energy function ψ^S, the process variables n^S or J_{SN}. One cannot choose the process variable n^S or n^F for the free Helmholtz energy function ψ^F because this ansatz would yield results that cannot be confirmed physically. Thus, only the response function ψ^S depends, among other n process variables s, on n^S or J_{SN}:

$$\psi^S = \hat{\psi}^S(n^S, s) \quad \text{or} \quad \psi^S = \bar{\psi}^S(J_{SN}, s)\,, \quad \psi^F = \hat{\psi}^F(s) = \bar{\psi}^F(s)\,, \tag{5.8.30}$$

where

$$s = \{s_1, \ldots s_i \ldots s_n\} \tag{5.8.31}$$

is the set of common process variables in both phases, which describe additional mechanical and thermical properties (see Appendix A). With (5.8.30) and (5.8.31), the entropy inequalities (5.8.24) and (5.8.27) take a slightly different form:

$$\begin{aligned}
&\sum_{\alpha=S}^{F} \frac{1}{\Theta^\alpha}\{-\rho^\alpha[\sum_{i=1}^{n}\frac{\partial\hat{\psi}^\alpha}{\partial s_i}(s_i)'_\alpha + (\Theta^\alpha)'_\alpha\eta^\alpha] + \\
&\quad +\mathbf{T}^{\alpha D}\cdot\mathbf{D}^D_\alpha - (p^\alpha + \kappa^{\alpha R} n^\alpha\Theta^\alpha\bar{\Theta})\frac{(\rho^{\alpha R})'_\alpha}{\rho^{\alpha R}} - \frac{1}{\Theta^\alpha}\mathbf{q}^\alpha\cdot\operatorname{grad}\Theta^\alpha\} - \\
&\quad -\frac{1}{\Theta^S}(p^S + \lambda n^S\Theta^S\bar{\Theta} + n^S\rho^S\frac{\partial\hat{\psi}^S}{\partial n^S})\frac{(n^S)'_S}{n^S} - \frac{1}{\Theta^F}(p^F + \lambda n^F\Theta^F\bar{\Theta})\frac{(n^F)'_F}{n^F} - \\
&\quad -\frac{1}{\Theta^F}(\hat{\mathbf{p}}^F - \lambda\Theta^F\bar{\Theta}\operatorname{grad} n^F)\cdot\mathbf{v}_{FS} - \hat{\rho}^F(\frac{\mu^F}{\Theta^F} - \frac{\mu^S}{\Theta^S}) - \frac{1}{\Theta^S\Theta^F}(\hat{e}^F - \hat{\mathbf{p}}^F\cdot\mathbf{v}_S)\Theta^{FS} \geq 0
\end{aligned} \tag{5.8.32}$$

or neglecting mass exchanges:

$$
\begin{aligned}
&\sum_{\alpha=S}^{F} \frac{1}{\Theta^\alpha}\{-\rho^\alpha[\sum_{i=1}^{n}\frac{\partial \bar{\psi}^\alpha}{\partial s_i}(s_i)'_\alpha + (\Theta^\alpha)'_\alpha \eta^\alpha] + \\
&+\mathbf{T}^{\alpha D}\cdot \mathbf{D}^D_\alpha + (p^\alpha + \kappa^{\alpha R} n^\alpha \Theta^\alpha \bar{\Theta})(\mathbf{D}_{\alpha R}\cdot \mathbf{I}) - \frac{1}{\Theta^\alpha}\mathbf{q}^\alpha \cdot \operatorname{grad}\Theta^\alpha\} + \\
&+\frac{1}{\Theta^S}(p^S + \lambda n^S \Theta^S \bar{\Theta} - \rho^S J_{SN}\frac{\partial \bar{\psi}^S}{\partial J_{SN}})(\mathbf{D}_{SN}\cdot \mathbf{I}) + \frac{1}{\Theta^F}(p^F + \lambda n^F \Theta^F \bar{\Theta})(\mathbf{D}_{FN}\cdot \mathbf{I}) - \\
&-\frac{1}{\Theta^F}(\hat{\mathbf{p}}^F - \lambda \Theta^F \bar{\Theta}\operatorname{grad} n^F)\cdot \mathbf{v}_{FS} - \frac{1}{\Theta^S\Theta^F}(\hat{e}^F - \hat{\mathbf{p}}^F \cdot \mathbf{v}_S)\Theta^{FS} \geq 0 \,.
\end{aligned}
\tag{5.8.33}
$$

The set (5.8.31) may contain tensor, vector, and scalar value parameters. For the multiplicative connection in the first term of (5.8.32) and (5.8.33), corresponding calculation rules have to be established between scalars, vectors, and tensors.

From the evaluation of the entropy inequalities (5.8.32) and (5.8.33), we obtain important restrictions for the partial hydrostatic stress states p^α,

$$p^\alpha = -\,\kappa^{\alpha R} n^\alpha \Theta^\alpha \bar{\Theta}\,, \tag{5.8.34}$$

and for the interface pressure λ,

$$\lambda = \kappa^{SR} - \frac{\rho^S}{\Theta^S\bar{\Theta}}\frac{\partial \hat{\psi}^S}{\partial n^S} \quad \text{or} \quad \lambda = \kappa^{SR} + \frac{\rho^{SR}}{\Theta^S\bar{\Theta}} J_{SN}\frac{\partial \bar{\psi}^S}{\partial J_{SN}}\,, \tag{5.8.35}$$

if all mass exchange is neglected, and

$$\lambda \Theta^F \bar{\Theta} = \kappa^{FR}\Theta^F \bar{\Theta} = p\,. \tag{5.8.36}$$

In (5.8.36), we have identified the weighted multiplier κ^{FR} with the porefluid pressure p because κ^{FR} is related to the incompressibility of the porefluid.

It is quite natural that we gain two expressions for the interface pressure λ, because λ acts in both the constituents φ^S and φ^F. However, λ remains undetermined, since κ^{SR} and κ^{FR} are undetermined. The reason for this is the fact that, in the incompressibility problem, the saturation condition is an equation in excess. The only way to calculate the Lagrange multiplier κ^{FR}, which is related to the porefluid pressure p, is to solve the equation of motion for the partial fluid, considering the boundary conditions, if the motion of the partial fluid is known. In this case, p or κ^{FR} is known and κ^{SR} can be expressed by p, considering (5.8.35).

Furthermore, the following inequality remains:

$$
\begin{aligned}
&\sum_{\alpha=S}^{F}\frac{1}{\Theta^\alpha}\{-\rho^\alpha[\sum_{i=1}^{n}\frac{\partial \hat{\psi}^\alpha}{\partial s_i}(s_i)'_\alpha + \eta^\alpha(\Theta^\alpha)'_\alpha] + \mathbf{T}^{\alpha D}\cdot \mathbf{D}^D_\alpha - \frac{1}{\Theta^\alpha}\mathbf{q}^\alpha\cdot \operatorname{grad}\Theta^\alpha\} - \\
&-\frac{1}{\Theta^F}(\hat{\mathbf{p}}^F - \lambda\Theta^F\bar{\Theta}\operatorname{grad} n^F)\cdot \mathbf{v}_{FS} - \hat{\rho}^F(\frac{\mu^F}{\Theta^F} - \frac{\mu^S}{\Theta^S}) - \\
&-\frac{1}{\Theta^S\Theta^F}(\hat{e}^F - \hat{\mathbf{p}}^F\cdot \mathbf{v}_S)\Theta^{FS} \geq 0\,.
\end{aligned}
\tag{5.8.37}
$$

In the next section, the consequences of the compressibility of the real materials of the constituents on the entropy inequality will be investigated.

b) The Introduction and Evaluation of the Entropy Inequality for a Binary Porous Model with Compressible Constituents

At first sight, it seems reasonable to choose constitutive equations for $\rho^{\alpha R}$ or $J_{\alpha R}$ (problem without mass exchanges) in order to correctly formulate the problem under study. However, a saturated porous medium is not an ideal mixture and the "smeared" solid matrix and the fluid do not have the same properties regarding their volumetric behavior. Due to experiences in Bluhm (1997), de Boer (1997), and test results in Lade and de Boer (1997), a constitutive equation for ρ^{SR} or J_{SR} can be introduced. In this case, the system of field equations is closed. This procedure has, however, some disadvantages, in particular for numerical problems. In order to take the saturation condition (5.6.2) and (5.6.3), or (5.6.4) and (5.6.5), into consideration as a constraint, concerning the evaluation of the entropy inequality, it is necessary that the multiplier λ be constitutively determined, in order not to violate the closure problem.

With these considerations in mind, we choose $\rho^{\alpha R}$ or $J_{\alpha R}$ and n^α or $J_{\alpha N}$ as process variables and expect two constitutive equations for the interface pressure λ, in order to close the set of field equations. This is indeed the case, as we shall soon come to see.

From the entropy inequality (5.8.24) and from (5.8.27), considering the constitutive assumptions

$$\begin{aligned} &\psi^S = \hat{\psi}^S(\rho^{SR}, n^S, s)\,, \qquad \psi^S = \bar{\bar{\psi}}^S(J_{SR}, J_{SN}, s)\,, \\ &\psi^F = \hat{\psi}^F(\rho^{FR}, s)\,, \quad \psi^F = \bar{\bar{\psi}}^F(J_{FR}, s)\,, \end{aligned} \tag{5.8.38}$$

the following version of the entropy inequality for the compressible binary model is obtained:

$$\begin{aligned} &\sum_{\alpha=S}^{F} \frac{1}{\Theta^\alpha}\{-\rho^\alpha[\sum_{i=1}^{n} \frac{\partial\hat{\psi}^\alpha}{\partial s_i}(s_i)'_\alpha + (\Theta^\alpha)'_\alpha\eta^\alpha] + \\ &+ \mathbf{T}^{\alpha D}\cdot\mathbf{D}^D_\alpha - [p^\alpha + n^\alpha(\rho^{\alpha R})^2\frac{\partial\hat{\psi}^\alpha}{\partial\rho^{\alpha R}}]\frac{(\rho^{\alpha R})'_\alpha}{\rho^{\alpha R}} - \frac{1}{\Theta^\alpha}\mathbf{q}^\alpha\cdot\operatorname{grad}\Theta^\alpha\} - \\ &- \frac{1}{\Theta^S}(p^S + \lambda n^S\Theta^S\bar{\Theta} + n^S\rho^S\frac{\partial\hat{\psi}^S}{\partial n^S})\frac{(n^S)'_S}{n^S} - \frac{1}{\Theta^F}(p^F + \lambda n^F\Theta^F\bar{\Theta})\frac{(n^F)'_F}{n^F} - \\ &- \frac{1}{\Theta^F}(\hat{\mathbf{p}}^F - \lambda\Theta^F\bar{\Theta}\operatorname{grad} n^F)\cdot\mathbf{v}_{FS} - \\ &- \hat{\rho}^F(\frac{\mu^F}{\Theta^F} - \frac{\mu^S}{\Theta^S}) - \frac{1}{\Theta^S\Theta^F}(\hat{e}^F - \hat{\mathbf{p}}^F\cdot\mathbf{v}_S)\Theta^{FS} \geq 0 \end{aligned} \tag{5.8.39}$$

or

$$
\begin{aligned}
&\sum_{\alpha=S}^{F} \frac{1}{\Theta^\alpha}\{-\rho^\alpha[\sum_{i=1}^{n} \frac{\partial \bar{\bar{\psi}}^\alpha}{\partial s_i}(s_i)'_\alpha + (\Theta^\alpha)'_\alpha \eta^\alpha] + \\
&+ \mathbf{T}^{\alpha D} \cdot \mathbf{D}_\alpha^D + (p^\alpha - \rho^\alpha J_{\alpha R} \frac{\partial \bar{\bar{\psi}}^\alpha}{\partial J_{\alpha R}})(\mathbf{D}_{\alpha R} \cdot \mathbf{I}) - \frac{1}{\Theta^\alpha} \mathbf{q}^\alpha \cdot \operatorname{grad} \Theta^\alpha\} + \\
&+ \frac{1}{\Theta^S}(p^S + \lambda n^S \Theta^S \bar{\Theta} - \rho^S J_{SN} \frac{\partial \bar{\bar{\psi}}^S}{\partial J_{SN}})(\mathbf{D}_{SN} \cdot \mathbf{I}) + \frac{1}{\Theta^F}(p^F + \lambda n^F \Theta^F \bar{\Theta})(\mathbf{D}_{FN} \cdot \mathbf{I}) - \\
&- \frac{1}{\Theta^F}(\hat{\mathbf{p}}^F - \lambda \Theta^F \bar{\Theta} \operatorname{grad} n^F) \cdot \mathbf{v}_{FS} - \frac{1}{\Theta^S \Theta^F}(\hat{e}^F - \hat{\mathbf{p}}^F \cdot \mathbf{v}_S)\Theta^{FS} \geq 0 .
\end{aligned}
\tag{5.8.40}
$$

The results of the evaluation of (5.8.39) and (5.8.40) for the hydrostatic stress states p^α and the interface stress λ are:

$$
\begin{aligned}
p^\alpha &= -\, n^\alpha (\rho^{\alpha R})^2 \frac{\partial \hat{\psi}^\alpha}{\partial \rho^{\alpha R}}, \quad p^\alpha = \rho^\alpha J_{\alpha R} \frac{\partial \bar{\bar{\psi}}^\alpha}{\partial J_{\alpha R}}, \\
\Theta^S \bar{\Theta} \lambda &= (\rho^{SR})^2 \frac{\partial \hat{\psi}^S}{\partial \rho^{SR}} - \rho^S \frac{\partial \hat{\psi}^S}{\partial n^S}, \\
\Theta^S \bar{\Theta} \lambda &= -\, \rho^{SR} J_{SR} \frac{\partial \bar{\bar{\psi}}^S}{\partial J_{SR}} + \rho^{SR} J_{SN} \frac{\partial \bar{\bar{\psi}}^S}{\partial J_{SN}}, \\
\Theta^F \bar{\Theta} \lambda &= p = (\rho^{FR})^2 \frac{\partial \hat{\psi}^F}{\partial \rho^{FR}} \quad \text{or} \quad \Theta^F \bar{\Theta} \lambda = p = -\, \rho^{FR} J_{FR} \frac{\partial \bar{\bar{\psi}}^F}{\partial J_{FR}},
\end{aligned}
\tag{5.8.41}
$$

where $(5.8.41)_{2,4,6}$ are only valid for vanishing mass exchanges.

Indeed, the interface pressure is determined by two constitutive equations, as has already been indicated. Furthermore, after the evaluation of the entropy inequalities (5.8.39) and (5.8.40) for special material behavior, the remaining Inequality (5.8.37) has to be treated.

After the investigations of purely incompressible and compressible models, we will treat hybrid models of first and second type, i.e., a model consisting of compressible solid and incompressible fluid phases, and a model with incompressible solid and compressible fluid phases, respectively.

c) The Introduction and Evaluation of the Entropy Inequality for a Binary Porous Medium Model with Compressible Solid and Incompressible Fluid Constituents (Hybrid Model of First Type)

Guided by the same ideas as in Sections **a)** and **b)** we choose ρ^{SR} or J_{SR}, respectively, as the process variable and set ρ^{FR} constant or J_{FR} equal to unity.

The incompressibility of the fluid reads in the rate formulation as

$$- n^F \bar{\Theta} \frac{(\rho^{FR})'_F}{\rho^{FR}} = 0 \quad \text{or} \quad n^F \bar{\Theta} (\mathbf{D}_{FR} \cdot \mathbf{I}) = 0 \,, \tag{5.8.42}$$

where, in $(5.8.42)_2$, the relation with purely kinematic quantitites is only valid if no mass exchange occurs.

Assuming ρ^{FR} as constant (or J_{FR} equals unity), the set of fields is closed. Since the saturation condition must be considered in the evaluation of the entropy inequality, it must be provided with a Lagrange multiplier. However, to ensure closure, this multiplier must be constitutively determined.

The entropy inequality (5.8.24) will be rearranged considering the constraints (5.8.42), which will be provided with a multiplier κ^{FR} and constitutive equations for the free Helmholtz energy functions similar to (5.8.30) and (5.8.38):

$$\psi^S = \hat{\psi}^S(n^S, \rho^{SR}, s) \quad \text{or} \quad \psi^S = \bar{\hat{\psi}}^S(J_{SN}, J_{SR}, s) \,, \quad \psi^F = \hat{\psi}^F(s) \,. \tag{5.8.43}$$

Then,

$$\begin{aligned}
&\sum_{\alpha=S}^{F} \frac{1}{\Theta^\alpha} \{ - \rho^\alpha (\Theta^\alpha)'_\alpha \eta^\alpha + \mathbf{T}^{\alpha D} \cdot \mathbf{D}^D_\alpha - \frac{1}{\Theta^\alpha} \mathbf{q}^\alpha \cdot \operatorname{grad} \Theta^\alpha \} - \\
&\quad - \sum_{i=1}^{n} \frac{\rho^S}{\Theta^S} \frac{\partial \hat{\psi}^S}{\partial s_i} (s_i)'_S - \sum_{i=1}^{n} \frac{\rho^F}{\Theta^F} \frac{\partial \hat{\psi}^F}{\partial s_i} (s_i)'_F - \\
&\quad - \frac{1}{\Theta^S} (p^S + \lambda n^S \Theta^S \bar{\Theta} + n^S \rho^S \frac{\partial \hat{\psi}^S}{\partial n^S}) \frac{(n^S)'_S}{n^S} - \\
&\quad - \frac{1}{\Theta^F} (p^F + \lambda n^F \Theta^F \bar{\Theta}) \frac{(n^F)'_F}{n^F} - \\
&\quad - \frac{1}{\Theta^S} [p^S + n^S (\rho^{SR})^2 \frac{\partial \hat{\psi}^S}{\partial \rho^{SR}}] \frac{(\rho^{SR})'_S}{\rho^{SR}} - \\
&\quad - \frac{1}{\Theta^F} (p^F + \kappa^{FR} n^F \Theta^F \bar{\Theta}) \frac{(\rho^{FR})'_F}{\rho^{FR}} - \\
&\quad - \frac{1}{\Theta^F} (\hat{\mathbf{p}}^F - \lambda \Theta^F \bar{\Theta} \operatorname{grad} n^F) \cdot \mathbf{v}_{FS} - \\
&\quad - \hat{\rho}^F (\frac{\mu^F}{\Theta^F} - \frac{\mu^S}{\Theta^S}) - \frac{1}{\Theta^S \Theta^F} (\hat{e}^F - \hat{\mathbf{p}}^F \cdot \mathbf{v}_S) \Theta^{FS} \geq 0 \,.
\end{aligned} \tag{5.8.44}$$

If mass exchange is neglected,

$$\sum_{\alpha=S}^{F} \frac{1}{\Theta^\alpha}\{-\rho^\alpha(\Theta^\alpha)'_\alpha\eta^\alpha + \mathbf{T}^{\alpha D}\cdot\mathbf{D}_\alpha^D - \frac{1}{\Theta^\alpha}\mathbf{q}^\alpha\cdot\operatorname{grad}\Theta^\alpha\} -$$
$$-\sum_{i=1}^{n}\frac{\rho^S}{\Theta^S}\frac{\partial\bar{\bar{\psi}}^S}{\partial s_i}(s_i)'_S - \sum_{i=1}^{n}\frac{\rho^F}{\Theta^F}\frac{\partial\hat{\psi}^F}{\partial s_i}(s_i)'_F +$$
$$+\frac{1}{\Theta^S}(p^S + \lambda n^S\Theta^S\bar{\Theta} - \rho^S J_{SN}\frac{\partial\bar{\bar{\psi}}^S}{\partial J_{SN}})(\mathbf{D}_{SN}\cdot\mathbf{I}) +$$
$$+\frac{1}{\Theta^F}(p^F + \lambda n^F\Theta^F\bar{\Theta})(\mathbf{D}_{FN}\cdot\mathbf{I}) + \tag{5.8.45}$$
$$+\frac{1}{\Theta^S}(p^S - \rho^S J_{SR}\frac{\partial\bar{\bar{\psi}}^S}{\partial J_{SR}})(\mathbf{D}_{SR}\cdot\mathbf{I}) +$$
$$+\frac{1}{\Theta^F}(p^F + \kappa^{FR}n^F\Theta^F\bar{\Theta})(\mathbf{D}_{FR}\cdot\mathbf{I}) -$$
$$-\frac{1}{\Theta^F}(\hat{\mathbf{p}}^F - \lambda\Theta^F\bar{\Theta}\operatorname{grad}n^F)\cdot\mathbf{v}_{FS} -$$
$$-\frac{1}{\Theta^S\Theta^F}(\hat{e}^F - \hat{\mathbf{p}}^F\cdot\mathbf{v}_S)\Theta^{FS} \geq 0$$

is obtained.

The evaluation of the entropy inequality (5.8.44) or (5.8.45) leads to restrictions for the hydrostatic stress states p^α and the interface pressure λ,

$$p^S = -\,n^S(\rho^{SR})^2\frac{\partial\hat{\psi}^S}{\partial\rho^{SR}} \quad \text{or} \quad p^S = \rho^S J_{SR}\frac{\partial\bar{\bar{\psi}}^S}{\partial J_{SR}}\,, \tag{5.8.46}$$

$$p^F = -\,\kappa^{FR}n^F\Theta^F\bar{\Theta}\,, \tag{5.8.47}$$

$$\lambda = \frac{(\rho^{SR})^2}{\Theta^S\bar{\Theta}}\frac{\partial\hat{\psi}^S}{\partial\rho^{SR}} - \frac{\rho^S}{\Theta^S\bar{\Theta}}\frac{\partial\hat{\psi}^S}{\partial n^S}\,, \quad \lambda = -\frac{\rho^{SR}J_{SR}}{\Theta^S\bar{\Theta}}\frac{\partial\bar{\bar{\psi}}^S}{\partial J_{SR}} + \frac{\rho^{SR}}{\Theta^S\bar{\Theta}}J_{SN}\frac{\partial\bar{\bar{\psi}}^S}{\partial J_{SN}}\,, \tag{5.8.48}$$

$$\Theta^F\bar{\Theta}\lambda = \Theta^F\bar{\Theta}\kappa^{FR} = p\,,$$

where we have again identified κ^{FR} with the porefluid pressure p, see (5.8.36).

Eqns. $(5.8.48)_{1,2}$ clearly show that the interface pressure λ is uniquely determined by quantities of the solid phase. The weighted value of λ in the fluid phase is equal to the unknown porefluid pressure p. The remaining inequality of the entropy inequality is given by (5.8.37).

Note that the widely-used procedure for considering the compressibility of the solid skeleton, namely the reduction of the porefluid pressure p, is approximately consistent with the methods in **b**) and **c**). In the literature (see Biot and Willis, 1957, Suklje, 1969, Nur and Byerlee, 1971, Lade and de Boer, 1997, de Boer, 1996, as well as Bluhm, 1997), the reduction factor is introduced as

$$\eta = 1 - \frac{K^S}{K^{SR}}\,, \tag{5.8.49}$$

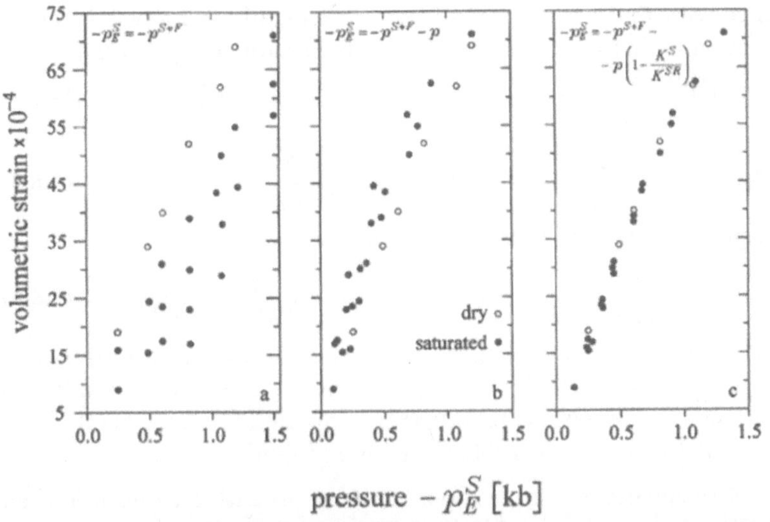

Fig. 5.8.1. Test results by Nur and Byerlee (1971)

where K^S and K^{SR} are the compression moduli of the skeleton and the real solid material, respectively.

The effective stress of the solid phase is then, if $\Theta^S = \Theta^F$, given by (see Lade and de Boer, 1997),

$$\begin{aligned} \mathbf{T}_E^S &= \mathbf{T}^S + n^S p\eta \mathbf{I} \\ &= \mathbf{T}^S + n^S p(1 - \frac{K^S}{K^{SR}})\mathbf{I}\,, \end{aligned} \tag{5.8.50}$$

and for the effective and partial hydrostatic stress states

$$p_E^S = p^S + n^S p(1 - \frac{K^S}{K^{SR}}) \tag{5.8.51}$$

and

$$p^S = p_E^S - n^S p(1 - \frac{K^S}{K^{SR}})\,. \tag{5.8.52}$$

For the total stress state $\mathbf{T}^{S+F}$ we have

$$\mathbf{T}^{S+F} = \mathbf{T}^S + \mathbf{T}^F = -\,p(1 - n^S \frac{K^S}{K^{SR}})\mathbf{I} + \mathbf{T}_E^S \tag{5.8.53}$$

and for the total hydrostatic stress state

$$p^{S+F} = p^S + p^F = -\,p(1 - n^S \frac{K^S}{K^{SR}}) + p_E^S\,. \tag{5.8.54}$$

It will be proven that the second term on the right-hand side of (5.8.50) approximately describes the influence of the compressibility of the real solid material.

We commence with $(5.8.41)_1$ or (5.8.46) and (5.8.48) of the models b) and c) whereby we assume that both constituents have the same temperature.

$$p^S = n^S p^{SR} \,, \quad p^S = -\, n^S p + n^S p^{SC} \,, \quad p^S = -\, n^S p + p_E^S \tag{5.8.55}$$

with

$$p^{SR} = -\,(\rho^{SR})^2 \frac{\partial \hat{\psi}^S}{\partial \rho^{SR}} \quad \text{or} \quad p^{SR} = \rho^{SR} J_{SR} \frac{\partial \bar{\bar{\psi}}^S}{\partial J_{SR}} \tag{5.8.56}$$

as the hydrostatic stress state in the solid material and

$$p^{SC} = -\,\rho^S \frac{\partial \hat{\psi}^S}{\partial n^S} = \frac{1}{n^S} p_E^S \quad \text{or} \quad p^{SC} = \rho^{SR} J_{SN} \frac{\partial \bar{\bar{\psi}}^S}{\partial J_{SN}} = \frac{1}{n^S} p_E^S \tag{5.8.57}$$

as the configuration stress state with p_E^S as the effective hydrostatic stress.

Biot and Willis (1957) chose two experiments in order to determine the coefficient η (5.8.49), namely the jacketed and unjacketed compressibility tests for a binary model with compressible skeleton and fluid.

"In the jacketed compressibility test, a specimen of the material is enclosed in a thin impermeable jacket and then subjected to an external fluid pressure p'."

In this test, only the configure pressure p^{SC} or the effective stress state p_E^S is present; the fluid pressure is equal to zero.

"In the unjacketed compressibility test, a sample of the material is immersed in a fluid to which is applied a pressure p'."

The characteristic of this test is that only the weighted fluid pressure is acting whereas the configuration pressure is equal to zero.

Considering the characteristics of these two tests it can be stated that, according to $(5.8.55)_2$ in the *jacketed test*

$$p^S = n^S p^{SC} = p_E^S \tag{5.8.58}$$

and in the *unjacketed test*

$$p^S = -\, n^S p \tag{5.8.59}$$

are present. Furthermore, we replace, in the *jacketed test*, either the process variables n^S or J_{SN}, within the framework of the geometrically-linear theory, by the volumetric strain e_{SN} and choose a simple constitutive relation for p_E^S, namely

$$p_E^S = n^S p^{SC} = n^S K^S e_{SN} \,, \tag{5.8.60}$$

where K^S is the compression modulus of the skeleton. Moreover, in the unjacketed test, we choose within the framework of the geometrically-linear theory the volumetric strain e_{SR} of the real solid material as a process variable and we gain

$$p^{SR} = -\, p = K^{SR} e_{SR} \tag{5.8.61}$$

with K^{SR} as the compression modulus of the real solid material. Thus, from $(5.8.55)_3$ we gain, considering (5.8.60) and (5.8.57),

$$n^S K^S e_{SN} = p^S - n^S K^{SR} e_{SR} \,. \tag{5.8.62}$$

According to (5.3.30), the volumetric strain e_{SN} can be expressed by the volumetric strain of the partial solid phase e_S and the volume change of the real solid material e_{SR}:

$$e_{SN} = e_S - e_{SR} \, . \tag{5.8.63}$$

Thus, with (5.8.63), Eqn. (5.8.62) can be reformulated to yield

$$\begin{aligned} n^S K^S e_S &= p^S - \ n^S K^{SR} e_{SR} + n^S K^S e_{SR} \\ &= p^S - \ n^S K^{SR} e_{SR} \big(1 - \frac{K^S}{K^{SR}}\big) \, . \end{aligned} \tag{5.8.64}$$

Then, after rearranging (5.8.64) with (5.8.61), we obtain

$$n^S K^S e_S = p^S + n^S p \big(1 - \frac{K^S}{K^{SR}}\big) \, . \tag{5.8.65}$$

The term $K^S e_S$ is identified by Nur and Bylerlee (1971) as the effective hydrostatic stress state $\bar{p}_E^S$. Thus,

$$n^S \bar{p}_E^S = p^S + n^S p \big(1 - \frac{K^S}{K^{SR}}\big) \tag{5.8.66}$$

or

$$p^S = n^S \bar{p}_E^S - n^S p \big(1 - \frac{K^S}{K^{SR}}\big) \tag{5.8.67}$$

is obtained. Then, for the total hydrostatic stress state p^{S+F} we have:

$$p^{S+F} = p^S + p^F = n^S \bar{p}_E^S - p \big(1 - n^S \frac{K^S}{K^{SR}}\big) \, . \tag{5.8.68}$$

Eqns. (5.8.67) and (5.8.68) coincide with (5.8.52) and (5.8.54) with the exception that, in (5.8.67) and (5.8.68), the effective stress state $\bar{p}_E^S$ is connected with the volume fraction n^S. How important this is cannot be answered by the test results of Nur and Byerlee (1971), because they performed their tests on Weber sandstone with the volume fraction $n^S = 0.94$, which equals approximately unity.

Finally, we change the sign of $\bar{p}_E^S$ and p^S and consider only pressures, as Nur and Byerlee (1971) did. Then, we obtain

$$n^S \bar{p}_E^S = p^S - n^S p \big(1 - \frac{K^S}{K^{SR}}\big) \, . \tag{5.8.69}$$

This form corresponds to the results of Nur and Byerlee up to the factor n^S on the left-hand side of (5.8.69).

d) The Introduction and Evaluation of the Entropy Inequality for a Binary Porous Medium Model with Incompressible Solid and Compressible Fluid Constituents (Hybrid Model of Second Type)

This model is of certain importance in the case of gas-filled porous solids. Usually, the compressibility of the solid material can be neglected in comparison with the volume changes of the pores, whereas the compressibility of the real gas can be very dominant. Thus, we introduce (see Section **a)**)

$$- n^S\bar{\Theta}\frac{(\rho^{SR})'_S}{\rho^{SR}} = 0 \quad \text{or} \quad n^S\bar{\Theta}(\mathbf{D}_{SR}\cdot\mathbf{I}) = 0 \tag{5.8.70}$$

and ρ^{FR} or J_{FR}, respectively, as a process variable. In this case, the multiplier λ is also constitutively determined.

Following this step, we elaborate on the entropy inequality, whereby we introduce the incompressibility constraint (5.8.70) into the entropy inequality (5.8.24) and assume the following constitutive equations for the free Helmholtz energy functions:

$$\begin{aligned} \psi^S &= \hat{\psi}^S(n^S, s)\,, \quad \psi^F = \hat{\hat{\psi}}^F(\rho^{FR}, s)\,, \\ \psi^S &= \bar{\psi}^S(J_{SN}, s)\,, \quad \psi^F = \bar{\bar{\psi}}^F(J_{FR}, s)\,. \end{aligned} \tag{5.8.71}$$

Then, we have

$$\begin{aligned} &\sum_{\alpha=S}^{F}\frac{1}{\Theta^\alpha}\,\{-\rho^\alpha(\Theta^\alpha)'_\alpha\eta^\alpha + \mathbf{T}^{\alpha D}\cdot\mathbf{D}^D_\alpha - \frac{1}{\Theta^\alpha}\mathbf{q}^\alpha\cdot\operatorname{grad}\Theta^\alpha\} - \\ &\quad - \sum_{i=1}^{n}\frac{\rho^S}{\Theta^S}\frac{\hat{\psi}^S}{\partial s_i}(s_i)'_S - \sum_{i=1}^{n}\frac{\rho^F}{\Theta}\frac{\partial\hat{\hat{\psi}}^F}{\partial s_i}(s_i)'_F - \\ &\quad - \frac{1}{\Theta^S}(p^S + \lambda n^S\Theta^S\bar{\Theta} + n^S\rho^S\frac{\partial\hat{\psi}^S}{\partial n^S})\frac{(n^S)'_S}{n^S} - \\ &\quad - \frac{1}{\Theta^F}(p^F + \lambda n^F\Theta^F\bar{\Theta})\frac{(n^F)'_F}{n^F} - \\ &\quad - \frac{1}{\Theta^S}(p^S + \kappa^{SR}n^S\Theta^S\bar{\Theta})\frac{(\rho^{SR})'_S}{\rho^{SR}} - \\ &\quad - \frac{1}{\Theta^F}(p^F + n^F(\rho^{FR})^2\frac{\partial\hat{\hat{\psi}}^F}{\partial\rho^{FR}})\frac{(\rho^{FR})'}{\rho^{FR}} - \frac{1}{\Theta^F}(\hat{\mathbf{p}}^F + \lambda\Theta^F\bar{\Theta}\operatorname{grad}n^F)\cdot\mathbf{v}_{FS} - \\ &\quad - \hat{\rho}^F(\frac{\mu^F}{\Theta^F} - \frac{\mu^S}{\Theta^S}) - \frac{1}{\Theta^S\Theta^F}(\hat{e}^F - \hat{\mathbf{p}}^F\cdot\mathbf{v}_S)\Theta^{FS} \geq 0\,, \end{aligned} \tag{5.8.72}$$

or, in the case of vanishing mass exchange,

$$
\begin{aligned}
&\sum_{\alpha=S}^{F} \frac{1}{\Theta^\alpha}\left\{-\rho^\alpha(\Theta^\alpha)'_\alpha\eta^\alpha + \mathbf{T}^{\alpha D}\cdot\mathbf{D}^D_\alpha - \frac{1}{\Theta^\alpha}\mathbf{q}^\alpha\cdot\operatorname{grad}\Theta^\alpha\right\} - \\
&\quad - \sum_{i=1}^{n}\frac{\rho^S}{\Theta^S}\frac{\partial\bar{\psi}^S}{\partial s_i}(s_i)'_S - \sum_{i=1}^{n}\frac{\rho^F}{\Theta^F}\frac{\partial\bar{\bar{\psi}}^F}{\partial s_i}(s_i)'_F + \\
&\quad + \frac{1}{\Theta^S}(p^S + \lambda n^S\Theta^S\bar{\Theta} - \rho^S J_{SN}\frac{\partial\bar{\psi}^S}{\partial J_{SN}})\cdot(\mathbf{D}_{SN}\cdot\mathbf{I}) + \\
&\quad + \frac{1}{\Theta^F}(p^F + \lambda n^F\Theta^F\bar{\Theta})(\mathbf{D}_{FN}\cdot\mathbf{I}) + \\
&\quad + \frac{1}{\Theta^S}(p^S + \kappa^{SR}n^S\Theta^S\bar{\Theta})(\mathbf{D}_{SR}\cdot\mathbf{I}) + \\
&\quad + \frac{1}{\Theta^F}(p^F - \rho^F J_{FR}\frac{\partial\bar{\bar{\psi}}^F}{\partial J_{FR}})(\mathbf{D}_{FR}\cdot\mathbf{I}) - \frac{1}{\Theta^F}(\hat{\mathbf{p}}^F - \lambda\Theta^F\bar{\Theta}\operatorname{grad} n^F)\cdot\mathbf{v}_{FS} - \\
&\quad - \frac{1}{\Theta^S\Theta^F}(\hat{e}^F - \hat{\mathbf{p}}^F\cdot\mathbf{v}_S)\Theta^{FS} \geq 0\,.
\end{aligned}
\tag{5.8.73}
$$

From the evaluation of the entropy inequality (5.8.72) and (5.8.73), we obtain

$$
\begin{aligned}
&p^S = -\,\kappa^{SR}n^S\Theta^S\bar{\Theta}\,,\\
&p^F = -\,n^F(\rho^{FR})^2\frac{\partial\hat{\bar{\psi}}^F}{\partial\rho^{FR}} \quad\text{or}\quad p^F = \rho^F J_{FR}\frac{\partial\bar{\bar{\psi}}^F}{\partial J_{FR}}\,,
\end{aligned}
\tag{5.8.74}
$$

$$
\begin{aligned}
&\lambda = \kappa^{SR} - \frac{\rho^S}{\Theta^S\bar{\Theta}}\frac{\partial\hat{\psi}^S}{\partial n^S} \quad\text{or}\quad \lambda = \kappa^{SR} + \frac{\rho^{SR}}{\Theta^S\bar{\Theta}}J_{SN}\frac{\partial\bar{\psi}^S}{\partial J_{SN}}\,,\\
&\lambda = \frac{(\rho^{FR})^2}{\Theta^F\bar{\Theta}}\frac{\partial\hat{\bar{\psi}}^F}{\partial\rho^{FR}} \quad\text{or}\quad \lambda = -\frac{\rho^{FR}}{\Theta^F\bar{\Theta}}J_{FR}\frac{\partial\bar{\bar{\psi}}^F}{\partial J_{FR}}\,,
\end{aligned}
\tag{5.8.75}
$$

where $(5.8.74)_3$ and $(5.8.75)_{2,4}$ are only valid if no mass exchange occurs. Furthermore, Inequality (5.8.37) remains.

e)
Additional Remarks

With the investigation of the entropy inequality of the hybrid model of second type, the general rearrangements of the entropy inequality, incorporating some properties at the micro-level, have come to an end.

However, the results gained in **a)** through **d)** will be discussed. With

$$
\begin{aligned}
&p^{SR} = -(\rho^{SR})^2\frac{\partial\hat{\bar{\psi}}^S}{\partial\rho^{SR}} && \text{(compressible solid material)}\,,\\
&p^{SR} = -\kappa^{SR} && \text{(incompressible solid material)}\,,\\
&p^{FR} = -(\rho^{FR})^2\frac{\partial\hat{\bar{\psi}}^F}{\partial\rho^{FR}} && \text{(compressible fluid)}\,,\\
&p^{FR} = -\kappa^{FR} = -\,p && \text{(incompressible fluid)}\,,
\end{aligned}
\tag{5.8.76}
$$

the interface pressure λ – independent of compressible or incompressible behavior of the phases – can, in general, be written as:

$$\lambda = -\, p^{SR} - \rho^S \frac{\partial \hat{\psi}^S}{\partial n^S} = -\, p^{SR} - \rho^S \frac{\partial \hat{\hat{\psi}}^S}{\partial n^S}\,, \quad \Theta^S \bar{\Theta} = \Theta^F \bar{\Theta} = 1\,, \qquad (5.8.77)$$
$$\lambda = -\, p^{FR}\,.$$

where, for simplicity, a common temperature $\Theta^S = \Theta^F = \Theta$ for the constituents is chosen.

With both the equations in (5.8.77) for λ, we can express p^{SR} by quantities of the fluid phase:

$$p^{SR} = p^{FR} - \rho^S \frac{\partial \hat{\psi}^S}{\partial n^S} = p^{FR} - \rho^S \frac{\partial \hat{\hat{\psi}}^S}{\partial n^S}\,, \qquad (5.8.78)$$

or considering the saturation condition (5.2.4) and replacing the volume fraction n^S by n^F in $\hat{\psi}^S$ and $\hat{\hat{\psi}}^S$:

$$p^{SR} = p^{FR} + \rho^S \frac{\partial \hat{\psi}^{Sf}}{\partial n^F} = p^{FR} + \rho^S \frac{\partial \hat{\hat{\psi}}^{Sf}}{\partial n^F}\,, \qquad (5.8.79)$$

where we have renamed $\hat{\psi}^S$ and $\hat{\hat{\psi}}^S$ as $\hat{\psi}^{Sf}$ and $\hat{\hat{\psi}}^{Sf}$. As there are no balance equations within the mixture theory for p^{SR} or ρ^{SR}, Eqn. (5.8.79) represents a relation to express p^{SR} or ρ^{SR} by p^{FR} or ρ^{FR} and n^F. This is the missing field equation. Now, ρ^{FR} can be determined from the saturation condition, provided ρ^S and ρ^F are known (from the balance equations of mass and momentum).

A few remarks will be made concerning a porous medium with *empty pores.* The fields of this medium are also closed if two constitutive equations for the hydrostatic stress state exist in the solid material. Thus, the constitutive equations for p^{SR} of a *compressible empty* porous solid can be read from the problems **b)** or **c)**, setting the interface pressure λ equal to zero:

$$p^{SR} = -\,(\rho^{SR})^2 \frac{\partial \hat{\hat{\psi}}^S}{\partial \rho^{SR}} \quad \text{or} \quad p^{SR} = -\,\rho^S \frac{\partial \hat{\hat{\psi}}^S}{\partial n^S}\,. \qquad (5.8.80)$$

Furthermore, the constitutive equations for p^{SR} of an *incompressible empty* porous solid are contained in the corresponding equations of the problems **a)** and **d)**

$$p^{SR} = -\,\kappa^{SR} \quad \text{or} \quad p^{SR} = -\,\rho^S \frac{\partial \hat{\psi}^S}{\partial n^S}\,. \qquad (5.8.81)$$

Moreover, the parameters ρ^{SR} and n^S can be replaced by J_{SR} and J_{SN} (refer to problems **a)** through **d)**).

Preceding investigations have revealed that the method to gain two constitutive relations for the hydrostatic stress state, in the real material of the solid phase, or two constitutive relations for the interface pressure, in the case of saturated porous solids, has led to an approach meeting all the thermodynamical and experimental restrictions and containing all special problems like incompressible, compressible, and hybrid binary models, as well as models with empty pores. It seems that a consistent theory has been created concerning the elastic hydrostatic deformation and stress state, resulting in the fact that additional artificial balance equations or evolution equations are unnecessary.

At this stage of the porous media theory, a clear analysis of the acting stresses (and also the deformations) could only be worked out, however, for the elastic hydrostatic deformation and stress states. A corresponding analysis of the deviatoric deformation and stress states is still awaiting search.

In the following sections, the special thermodynamic behavior of the partial constituents at the macrolevel (like elastic, plastic, and viscous behavior) will be investigated, whereby interaction effects will also be considered.

5.8.3 Thermoelastic Compressible Porous Solid Filled with an Incompressible Viscous Fluid

The forerunner of finite elasticity laws, Hooke's law, which describes linear-elastic behavior, is a creation of the 19th century (see Section 3.2). Later, within the framework of geometrically-linear theory, the generalized Hooke's law was extended in such a way that non-linear deformations could also be included. A well-known non-linear elasticity for small deformations has been introduced by Kauderer (1958).

General non-linear elasticity laws to describe finite distortions have only recently been developed. The treatise of Murnaghan (1937) on finite elastic deformations seems to be incomplete (see Truesdell and Noll, 1965). The efforts of Mooney (1940) and Rivlin (1948), as well as Rivlin and Saunders (1951) were, on the other hand, much more successful. A review of the development of finite elasticity theory can be found in Truesdell and Noll (1965).

In extension of the Neo-Hooke model (Rivlin, 1948), Simo and Pister (1984) developed a new elasticity law which contains two response parameters and permits the description of large volumetric strains. All discussed elasticity laws are related to the elastic behavior of non-porous solids. The conversion of these laws to porous solids was performed by, among others, Morland (1972) and Ehlers (1989a).

The development of linear constitutive equations for viscous fluids is also a creation of the 19th century (see Section 3.4). As we are only interested in linear constitutive relations close to the mixture equilibrium state, we will not discuss the development of non-linear relations for viscous fluids.

We will restrict our investigations to a hybrid binary porous media model of the first type (compressible thermoelastic porous solid φ^S, incompressible viscous fluid φ^F). For the following mathematical model, we introduce the following assumptions:

- The thermodynamic restrictions are valid within the framework of the geometrically non-linear theory;
- all supply terms are unequal to zero;

- in the reference placement of the partial bodies, the volume fractions $n^\alpha_{0\alpha}$ and the real densities $\rho^{\alpha R}_{0\alpha}$ of both constituents are homogeneous, i.e.,

$$\operatorname{Grad}_\alpha n^\alpha_{0\alpha} = \mathbf{0}\,, \quad \operatorname{Grad}_\alpha \rho^{\alpha R}_{0\alpha} = \mathbf{0}\,; \tag{5.8.82}$$

- the solid phase φ^S is compressible, and the fluid phase φ^F incompressible.

Recall that we have already treated the influence of compressibility and incompressibility in Section 5.8.2 c) and that we have gained constitutive equations for

$$p^S = -\, n^S (\rho^{SR})^2 \frac{\partial \hat{\psi}^S}{\partial \rho^{SR}}\,, \quad p^F = -\, n^F \lambda \Theta^F \bar{\Theta} = -\, n^F p\,,$$
$$\lambda = \frac{(\rho^{SR})^2}{\Theta^S \bar{\Theta}} \frac{\partial \hat{\psi}^S}{\partial \rho^{SR}} - \frac{\rho^S}{\Theta^S \bar{\Theta}} \frac{\partial \hat{\psi}^S}{\partial n^S}\,, \quad \Theta^F \bar{\Theta} \lambda = p\,. \tag{5.8.83}$$

The thermodynamic behavior of an elastic porous solid saturated by an incompressible viscous fluid may be characterized by a constitutive assumption, for which the additional response functions

$$\mathcal{R} = \left\{\mathbf{T}^{\alpha D}, \hat{\mathbf{p}}^F - \lambda \Theta^F \bar{\Theta} \operatorname{grad} n^F, \hat{e}^F - \hat{\mathbf{p}}^F \cdot \mathbf{v}_S, \mathbf{q}^\alpha, \eta^\alpha, \psi^\alpha, \hat{\rho}^F\right\} \tag{5.8.84}$$

are assumed to depend on the process variables s:

$$s = \left\{n^S, \rho^{SR}, \check{\mathbf{C}}_S, \mathbf{v}_F - \mathbf{v}_S, \mathbf{D}^D_F, \Theta^S, \Theta^F, \operatorname{grad} \Theta^S, \operatorname{grad} \Theta^F\right\}. \tag{5.8.85}$$

The choice of the process variables are justified as follows: the volume-preserving elastic deformations of the partial solid are described by $\check{\mathbf{C}}_S$. The velocity difference $\mathbf{v}_F - \mathbf{v}_S$ and the deviator $\mathbf{D}^D_F$ govern dissipative effects. Finally, thermal effects are reflected by Θ^α and $\operatorname{grad} \Theta^\alpha$.

The remaining entropy inequality (5.8.37) is listed with $\mathbf{v}_{FS} = \mathbf{v}_F - \mathbf{v}_S$ and $\Theta^{FS} = \Theta^F - \Theta^S$:

$$\begin{aligned}
&- \frac{\rho^S}{\Theta^S} \sum_{i=1}^n \frac{\partial \hat{\psi}^S}{\partial s_i} (s_i)'_S - \frac{\rho^F}{\Theta^F} \sum_{i=1}^n \frac{\partial \hat{\psi}^F}{\partial s_i} (s_i)'_F - \frac{\rho^S}{\Theta^S} \eta^S (\Theta^S)'_S - \frac{\rho^F}{\Theta^F} \eta^F (\Theta^F)'_F + \\
&+ \frac{1}{\Theta^S} \mathbf{T}^{SD} \cdot \mathbf{D}^D_S + \frac{1}{\Theta^F} \mathbf{T}^{FD} \cdot \mathbf{D}^D_F - \frac{1}{\Theta^F} (\hat{\mathbf{p}}^F - \lambda \Theta^F \bar{\Theta} \operatorname{grad} n^F) \cdot \mathbf{v}_{FS} - \\
&- \hat{\rho}^F \Big(\frac{\mu^F}{\Theta^F} - \frac{\mu^S}{\Theta^S}\Big) - \frac{1}{\Theta^S \Theta^F} (\hat{e}^F - \hat{\mathbf{p}}^F \cdot \mathbf{v}_S) \Theta^{FS} - \\
&- \frac{1}{(\Theta^S)^2} \mathbf{q}^S \cdot \operatorname{grad} \Theta^S - \frac{1}{(\Theta^F)^2} \mathbf{q}^F \cdot \operatorname{grad} \Theta^F \geq 0\,.
\end{aligned} \tag{5.8.86}$$

With the constitutive assumptions (5.8.84) and (5.8.85), the entropy inequality can be evaluated. The procedure leads, however, to a lengthy formalism, which appears in the APPENDIX A.

We restrict our investigations for simplicity close to the *mixture equilibrium state*, which is defined in the APPENDIX A (A 41). The main results are summarized for deformation processes in this state:

$$\psi^S = \hat{\psi}^S(\rho^{SR}, n^S, \Theta, \check{\mathbf{C}}_S), \quad \psi^F = \hat{\psi}^F(\Theta)\,, \tag{5.8.87}$$

see (A 61). It should be mentioned that the results (5.8.87) remain valid for a linear viscous fluid even if the time rate of the fluid volumetric strain is considered in the set of process variables.

From (A 24) and $(5.8.55)_2$, in connection with (5.8.57), (A 51), and (A 52), we obtain

$$\mathbf{T}^S = -\, n^S p\mathbf{I} - n^S \rho^S \frac{\partial \hat{\psi}^S}{\partial n^S}\mathbf{I} + 2\rho^S J_S^{-2/3}(\mathbf{F}_S \frac{\partial \hat{\psi}^S}{\partial \check{\mathbf{C}}_S}\mathbf{F}_S^T)^D \tag{5.8.88}$$

and, from (A 91), (A 53), and (5.8.47),

$$\mathbf{T}^F = -\, n^F p\mathbf{I} + 2\mu^F \mathbf{D}_F^D \tag{5.8.89}$$

with

$$p = \kappa^{FR}\,, \tag{5.8.90}$$

see $(5.8.48)_3$.

Moreover, the interaction agencies are determined by the following constitutive relations, see (A 49), (A 91) in connection with (A 56), and (A 57):

$$\hat{\mathbf{p}}^F = p \operatorname{grad} n^F - \beta_S \operatorname{grad} \Theta^S - \beta_F \operatorname{grad} \Theta^F - \beta_{\mathcal{V}}^F \mathbf{v}_{FS}\,, \tag{5.8.91}$$

$$\hat{e}^F = \hat{\mathbf{p}}^F \cdot \mathbf{v}_S - e_\Theta \Theta^{FS}\,. \tag{5.8.92}$$

The response parameters $\mu^F, \beta_S, \beta_F, \beta_{\mathcal{V}}$, and e_Θ are constrained by the restrictions (A 90), and are functions of s_0 (see APPENDIX A).

Moreover, according to (A 91), (A 54), (A 56), and (A 57), we have for the heat flux vectors $\mathbf{q}^\alpha$:

$$\begin{aligned} \frac{1}{(\Theta^S)^2}\mathbf{q}^S &= -\,\alpha_S^S \operatorname{grad} \Theta^S - \alpha_F^S \operatorname{grad} \Theta^F - \alpha_{\mathcal{V}}^S \mathbf{v}_{FS}\,, \\ \frac{1}{(\Theta^F)^2}\mathbf{q}^F &= \alpha_F^S \operatorname{grad} \Theta^S - \alpha_F^F \operatorname{grad} \Theta^F - \alpha_{\mathcal{V}}^F \mathbf{v}_{FS}\,. \end{aligned} \tag{5.8.93}$$

In the remaining part of this section, a non-linear elasticity law for the effective stresses $\mathbf{T}_E^S$ will be developed (see Bluhm, 1999). Recalling (5.8.88):

$$\mathbf{T}_E^S = -\, n^S \rho^S \frac{\partial \hat{\psi}^S}{\partial n^S}\mathbf{I} + 2\rho^S\, J_S^{-2/3}\,(\mathbf{F}_S \frac{\partial \hat{\psi}^S}{\partial \check{\mathbf{C}}_S}\,\mathbf{F}_S^T)^D\,. \tag{5.8.94}$$

The free Helmholtz energy function ψ^S depends on the absolute temperature Θ, on the real density ρ^{SR}, on the volume fraction n^S, and on the volume-preserving part of the right Cauchy-Green deformation tensor $\check{\mathbf{C}}_S$:

$$\psi^S = \hat{\psi}^S(\Theta, \rho^{SR}\; n^S, \check{\mathbf{C}}_S)\,. \tag{5.8.95}$$

The treatment of elastic porous solids will be restricted to *isotropic* material. For this kind of material, the free Helmholtz energy ψ^S is an isotropic scalar value function of the temperature Θ, the real density ρ^{SR}, the volume fraction n^S, and the principal invariants of $\check{\mathbf{C}}_S$. Thus, for isotropic thermoelastic materials, the tensor $\check{\mathbf{C}}_S$ must be replaced by the invariants $\mathrm{I}_{\check{\mathbf{C}}_S}$ and $\mathrm{II}_{\check{\mathbf{C}}_S}$ (see Beatty, 1987) (note that the third invariant $\mathrm{III}_{\check{\mathbf{C}}_S}$ is equal to unity).

$$\psi^S = \bar{\psi}^S(\Theta, \rho^{SR}, n^S, \mathrm{I}_{\check{\mathbf{C}}_S}, \mathrm{II}_{\check{\mathbf{C}}_S}) \,, \tag{5.8.96}$$

where

$$\mathrm{I}_{\check{\mathbf{C}}_S} = \check{\mathbf{C}}_S \cdot \mathbf{I} \,, \quad \mathrm{II}_{\check{\mathbf{C}}_S} = \frac{1}{2}\big[(\check{\mathbf{C}}_S \cdot \mathbf{I})^2 - \check{\mathbf{C}}_S \cdot \check{\mathbf{C}}_S\big] \,. \tag{5.8.97}$$

With (5.8.97), the effective stresses for the solid phase (5.8.94) can be expressed as:

$$\begin{aligned}
\mathbf{T}_E^S &= -\, n^S \rho^S \frac{\partial \bar{\psi}^S}{\partial n^S}\mathbf{I} + 2\rho^S\, J_S^{-2/3}\Big[\mathbf{F}_S\Big(\frac{\partial \bar{\psi}^S}{\partial \mathrm{I}_{\check{\mathbf{C}}_S}}\frac{\partial \mathrm{I}_{\check{\mathbf{C}}_S}}{\partial \check{\mathbf{C}}_S} + \frac{\partial \bar{\psi}^S}{\partial \mathrm{II}_{\check{\mathbf{C}}_S}}\frac{\partial \mathrm{II}_{\check{\mathbf{C}}_S}}{\partial \check{\mathbf{C}}_S}\Big)\mathbf{F}_S^T\Big]^D \\
&= -\, n^S \rho^S \frac{\partial \bar{\psi}^S}{\partial n^S}\mathbf{I} + 2\rho^S J_S^{-2/3}\Big\{\mathbf{F}_S\Big[\frac{\partial \bar{\psi}^S}{\partial \mathrm{I}_{\check{\mathbf{C}}_S}}\mathbf{I} + \frac{\partial \bar{\psi}^S}{\partial \mathrm{II}_{\check{\mathbf{C}}_S}}(\mathrm{I}_{\check{\mathbf{C}}_S}\mathbf{I} - \check{\mathbf{C}}_S)\Big]\mathbf{F}_S^T\Big\}^D \\
&= -\, n^S \rho^S \frac{\partial \bar{\psi}^S}{\partial n^S}\mathbf{I} + 2\rho^S\, J_S^{-2/3}\Big[\frac{\partial \bar{\psi}^S}{\partial \mathrm{I}_{\check{\mathbf{C}}_S}}\mathbf{B}_S + \frac{\partial \bar{\psi}^S}{\partial \mathrm{II}_{\check{\mathbf{C}}_S}}(\mathrm{I}_{\check{\mathbf{C}}_S}\mathbf{B}_S - J_S^{-2/3}\mathbf{B}_S\mathbf{B}_S)\Big]^D \\
&= \alpha_0\, \mathbf{I} + \alpha_1 \mathbf{B}_S^D + \alpha_2 (\mathbf{B}_S\mathbf{B}_S)^D \,,
\end{aligned} \tag{5.8.98}$$

where the response parameters α_0, α_1, and α_2 are given by:

$$\begin{aligned}
\alpha_0 &= -\, n^S \rho^S \frac{\partial \bar{\psi}^S}{\partial n^S} \,, \quad \alpha_1 = 2\rho^S J_S^{-2/3}\Big(\frac{\partial \bar{\psi}^S}{\partial \mathrm{I}_{\check{\mathbf{C}}_S}} + \mathrm{I}_{\check{\mathbf{C}}_S}\frac{\partial \bar{\psi}^S}{\partial \mathrm{II}_{\check{\mathbf{C}}_S}}\Big) \,, \\
\alpha_2 &= -\, 2\rho^S J_S^{-4/3}\frac{\partial \bar{\psi}^S}{\partial \mathrm{II}_{\check{\mathbf{C}}_S}} \,.
\end{aligned} \tag{5.8.99}$$

Using the Cayley-Hamilton theorem, the tensor product $(\mathbf{B}_S\mathbf{B}_S)^D$ found in (5.8.98), can be reformulated:

$$(\mathbf{B}_S\mathbf{B}_S)^D = J_S^{2/3} I_{\check{\mathbf{C}}_S}\mathbf{B}_S^D + J_S^2 (\mathbf{B}_S^{-1})^D \,, \tag{5.8.100}$$

(see de Boer, 1982). Considering (5.8.100), the alternative representation of the constitutive relation (5.8.98) for the effective stresses $\mathbf{T}_E^S$ can be formulated in the following form:

$$\mathbf{T}_E^S = \beta_0 \mathbf{I} + \beta_1 \mathbf{B}_S^D + \beta_{-1}(\mathbf{B}_S^{-1})^D \,, \tag{5.8.101}$$

where

$$\begin{aligned}
\beta_0 &= \alpha_0 = -\, n^S \rho^S \frac{\partial \bar{\psi}^S}{\partial n^S} \,, \\
\beta_1 &= \alpha_1 + \alpha_2 J_S^{2/3}\mathrm{I}_{\check{\mathbf{C}}_S} = 2\rho^S J_S^{-2/3}\frac{\partial \bar{\psi}^S}{\partial \mathrm{I}_{\check{\mathbf{C}}_S}} \,, \\
\beta_{-1} &= \alpha_2 J_S^2 = -\, 2\rho^S J_S^{2/3}\frac{\partial \bar{\psi}^S}{\partial \mathrm{II}_{\check{\mathbf{C}}_S}} \,.
\end{aligned} \tag{5.8.102}$$

Discussion of constitutive relations for $\mathbf{T}_E^S$

The constitutive relation for the effective stresses of the Cauchy stress tensor of a compressible thermoelastic solid phase is given by (5.8.94)

$$\mathbf{T}_E^S = -\, n^S \rho^S \frac{\partial \psi^S}{\partial n^S} \mathbf{I} + 2 \rho^S J_S^{-2/3} (\mathbf{F}_S \frac{\partial \psi^S}{\partial \check{\mathbf{C}}_S} \mathbf{F}_S^T)^D \,. \tag{5.8.103}$$

For simplicity, we omit the superscript notations for the free Helmholtz energy function.

If no mass exchange occurs, the transformation

$$\psi^S = \psi^S(\ldots, n^S, \ldots) \quad \Longrightarrow \quad \psi^S = \psi^S(\ldots, J_{SN} = \frac{n_{0S}^S}{n^S}, \ldots) \tag{5.8.104}$$

is possible. Then, (5.8.94) can be replaced by

$$\mathbf{T}_E^S = \rho^S J_{SN} \frac{\partial \psi^S}{\partial J_{SN}} \mathbf{I} + 2 \rho^S J_S^{-2/3} (\mathbf{F}_S \frac{\partial \psi^S}{\partial \check{\mathbf{C}}_S} \mathbf{F}_S^T)^D \,. \tag{5.8.105}$$

In the following equations, we restrict our investigation to *incompressible* isotropic thermo-hyperelastic porous solids within the framework of finite deformation processes. The assumption of incompressibility has the advantage that the intermediate placement is identical with the reference one. Thus, in the following derivations, an extensive formalism can be avoided. It should, however, be mentioned that it is, in principle, also possible to form the following derivations in the intermediate placement (the interested reader is referred to Bluhm, 1999). Concerning the free Helmholtz energy function ψ^S, the following ansatz is postulated:

$$\begin{aligned} \psi^S = \psi^S(\Theta, \check{\mathrm{C}}, J_{SN}) = \frac{1}{\rho_{0S}^S} \{ & \frac{1}{2} K^{SN} \left[(\log J_{SN})^2 + 2 \log J_{SN} \right] + \\ & + K^{SN} \frac{1 - n_{0S}^S}{n_{0S}^S (n_{0S}^S - 2)} \left[\log \frac{n_{0S}^S - J_{SN}}{J_{SN}(n_{0S}^S - 1) - n_{0S}^S} - \log(1 - n_{0S}^S) \right] + \\ & + \frac{1}{2} \mu^S (\mathrm{I}_{\check{\mathrm{C}}_S} - 3) - 3 \alpha^S K^S (\log J_{SN})(\Theta - \Theta_{0S}) - \\ & - \rho_{0S}^S c^S (\Theta \log \frac{\Theta}{\Theta_{0S}} - \Theta + \Theta_{0S}) \} \,, \end{aligned} \tag{5.8.106}$$

where the relation $J_S = \dfrac{\rho_{0S}^S}{\rho^S}$ has been used. It should be mentioned that, due to the assumed incompressibility, the Jacobian J_S equals J_{SN}.

Moreover,

$$\mu^S = \text{const.}\,, \; K^S = \text{const.}\,, \; K^{SN} = \text{const.}\,, \; \alpha^S = \text{const.}\,, \text{ and } c^S = \text{const.} \tag{5.8.107}$$

are interpreted as macroscopic response parameters of the constituent φ^S. The temperature Θ_{0S}, as well as the density ρ_{0S}^S, belong to the reference placement of the solid.

With the derivatives (and the Karni-Reiner strain tensor $\mathbf{K}_S$)

$$
\begin{aligned}
(\mathbf{F}_S \frac{\partial \psi^S}{\partial \check{\mathbf{C}}_S} \mathbf{F}_S^T)^D &= \frac{1}{2\,\rho_{0S}^S} \mu^S \mathbf{B}_S^D = \frac{1}{\rho_{0S}^S} \mu^S \mathbf{K}_S^D \,, \\
\frac{\partial \psi^S}{\partial J_{SN}} &= \frac{1}{\rho_{0S}^S\, J_{SN}} K^{SN} [\log J_{SN} + 1 - \frac{1}{J_{SN} - \frac{n_{0S}^S}{1 - n_{0S}^S} (\frac{n_{0S}^S}{J_{SN}} - n_{0S}^S)}] - \\
&\quad - \frac{1}{\rho_{0S}^S\, J_{SN}}\, 3\,\alpha^S K^S (\Theta - \Theta_{0S})
\end{aligned}
\tag{5.8.108}
$$

the constitutive relation

$$
\begin{aligned}
\mathbf{T}_E^S = \frac{1}{J_{SN}} \{ & 2\mu^S J_{SN}^{-2/3} \mathbf{K}_S^D + \\
& + K^{SN} [\log\ J_{SN} + 1 - \frac{1}{J_{SN} - \frac{n_{0S}^S}{1 - n_{0S}^S} (\frac{n_{0S}^S}{J_{SN}} - n_{0S}^S)}] \mathbf{I} - \\
& - 3\alpha^S K^S (\Theta - \Theta_{0S}) \mathbf{I} \} \,,
\end{aligned}
\tag{5.8.109}
$$

for the effective Cauchy stress tensor is obtained,

Considering the relations between the symmetric Piola-Kirchhoff, the Kirchhoff, and the Cauchy stress tensor of the effective stresses,

$$
\mathbf{S}_E^S = J_S\, \mathbf{F}_S^{-1}\, \mathbf{T}_E^S\, \mathbf{F}_S^{T-1} \quad , \quad \boldsymbol{\tau}_E^S = J_S\, \mathbf{T}_E^S = \mathbf{F}_S\, \mathbf{S}_E^S\, \mathbf{F}_S^T \,,
\tag{5.8.110}
$$

the effective stresses $\mathbf{S}_E^S$ and $\boldsymbol{\tau}_E^S$, concerning the reference and actual placement, can be derived:

$$
\begin{aligned}
\mathbf{S}_E^S = {} & \mu^S J_S^{-2/3} [\mathbf{I} - \frac{1}{3} (\mathbf{C}_S \cdot \mathbf{I})\, \mathbf{C}_S^{-1}] + \\
& + K^{SN} [\log J_{SN} + 1 - \frac{1}{J_{SN} - \frac{n_{0S}^S}{1 - n_{0S}^S} (\frac{n_{0S}^S}{J_{SN}} - n_{0S}^S)}]\, \mathbf{C}_S^{-1} - \\
& - 3\,\alpha^S K^S (\Theta - \Theta_{0S})\, \mathbf{C}_S^{-1} \,,
\end{aligned}
\tag{5.8.111}
$$

$$
\begin{aligned}
\boldsymbol{\tau}_E^S = {} & 2\,\mu^S J_S^{-2/3}\, \mathbf{K}_S^D + \\
& + K^{SN} [\log J_{SN} + 1 - \frac{1}{J_{SN} - \frac{n_{0S}^S}{1 - n_{0S}^S} (\frac{n_{0S}^S}{J_{SN}} - n_{0S}^S)}]\, \mathbf{I} - \\
& - 3\,\alpha^S K^S (\Theta - \Theta_{0S})\, \mathbf{I} \,.
\end{aligned}
\tag{5.8.112}
$$

Next, constitutive relations for the effective stress of the solid phase concerning the reference and actual placements for small deformation processes will be derived. In order to obtain Hooke type constitutive equations, the linearizations

$$\mathbf{S}^S_{E\text{lin}} = \mathbf{S}^S_E \Big|_{\mathcal{P}_0} + \frac{\partial \mathbf{S}^S_E}{\partial \mathbf{C}_S}\Big|_{\mathcal{P}_0} \Delta\mathbf{C}_S + \frac{\partial \mathbf{S}^S_E}{\partial \Theta}\Big|_{\mathcal{P}_0} \Delta\Theta\,,$$
$$\boldsymbol{\tau}^S_{E\text{lin}} = \boldsymbol{\tau}^S_E \Big|_{\hat{\mathcal{P}}_0} + \frac{\partial \boldsymbol{\tau}^S_E}{\partial \mathbf{B}^{-1}_S}\Big|_{\hat{\mathcal{P}}_0} \Delta\mathbf{B}^{-1}_S + \frac{\partial \boldsymbol{\tau}^S_E}{\partial \Theta}\Big|_{\hat{\mathcal{P}}_0} \Delta\Theta \tag{5.8.113}$$

of the stress tensors $\mathbf{S}^S_E$ and $\boldsymbol{\tau}^S_E$ will be developed. The symbols

$$(\ldots)\Big|_{\mathcal{P}_0} = (\ldots)\Big|_{\mathbf{C}_S = \mathbf{I},\, \Theta = \Theta_{0S}} \quad , \quad (\ldots)\Big|_{\hat{\mathcal{P}}_0} = (\ldots)\Big|_{\mathbf{B}^{-1}_S = \mathbf{I},\, \Theta = \Theta_{0S}} \tag{5.8.114}$$

imply that the quantity $(\ldots)$ must be evaluated for $\mathbf{C}_S = \mathbf{I}$ and $\Theta = \Theta_{0S}$ as well as $\mathbf{B}^{-1}_S = \mathbf{I}$ and $\Theta = \Theta_{0S}$, respectively. With the stresses at the positions $\mathcal{P}_0$ and $\hat{\mathcal{P}}_0$,

$$\mathbf{S}^S_E \Big|_{\mathcal{P}_0} = \mathbf{0} \quad , \quad \boldsymbol{\tau}^S_E \Big|_{\hat{\mathcal{P}}_0} = \mathbf{0}\,, \tag{5.8.115}$$

the elastic tangent operators

$$\frac{\partial \mathbf{S}^S_E}{\partial \mathbf{C}_S}\Big|_{\mathcal{P}_0} = \Big[\frac{\partial \mathbf{S}^S_E(\mathbf{C}_S)}{\partial \mathbf{C}_S} + \frac{\partial \mathbf{S}^S_E(J_{SN})}{\partial J_{SN}} \otimes \frac{\partial J_{SN}}{\partial \mathbf{C}_S} + \Big(\frac{\partial \mathbf{C}^{-1}_S}{\partial \mathbf{C}_S}\Big)^T \frac{\partial \mathbf{S}^S_E(\mathbf{C}^{-1}_S)}{\partial \mathbf{C}^{-1}_S}\Big]\Big|_{\mathcal{P}_0}$$
$$= \mu^S \overset{4}{\mathbf{I}} + \frac{1}{2}\Big(K^{SN}\,\frac{2\,(1 - n^S_{0S}) + (n^S_{0S})^2}{1 - n^S_{0S}} - \frac{2}{3}\,\mu^S\Big)\,\overset{4}{\bar{\mathbf{I}}}\,,$$
$$\frac{\partial \boldsymbol{\tau}^S_E}{\partial \mathbf{B}^{-1}_S}\Big|_{\hat{\mathcal{P}}_0} = \Big[\frac{\partial \boldsymbol{\tau}^S_E(J_{SN})}{\partial J_{SN}} \otimes \frac{\partial J_{SN}}{\partial \mathbf{B}^{-1}_S} + \Big(\frac{\partial \mathbf{B}_S}{\partial \mathbf{B}^{-1}_S}\Big)^T \frac{\partial \boldsymbol{\tau}^S_E(\mathbf{B}_S)}{\partial \mathbf{B}_S}\Big]\Big|_{\hat{\mathcal{P}}_0} \tag{5.8.116}$$
$$= -\,\mu^S \overset{4}{\mathbf{I}} - \frac{1}{2}\Big(K^{SN}\,\frac{2\,(1 - n^S_{0S}) + (n^S_{0S})^2}{1 - n^S_{0S}} - \frac{2}{3}\,\mu^S\Big)\,\overset{4}{\bar{\mathbf{I}}}\,,$$

and the thermal tangent operators

$$\frac{\partial \mathbf{S}^S_E}{\partial \Theta}\Big|_{\mathcal{P}_0} = -\,3\,\alpha^S K^S\,\mathbf{I} \quad , \quad \frac{\partial \boldsymbol{\tau}^S_E}{\partial \Theta}\Big|_{\hat{\mathcal{P}}_0} = -\,3\,\alpha^S K^S\,\mathbf{I}\,, \tag{5.8.117}$$

the linearized effective stress tensors $\mathbf{S}^S_{E\text{lin}}$ and $\boldsymbol{\tau}^S_{E\text{lin}}$ read

$$\mathbf{S}^S_{E\text{lin}} = \Big[\mu^S \overset{4}{\mathbf{I}} + \frac{1}{2}\Big(K^{SN}\,\frac{2\,(1 - n^S_{0S}) + (n^S_{0S})^2}{1 - n^S_{0S}} - \frac{2}{3}\,\mu^S\Big)\,\overset{4}{\bar{\mathbf{I}}}\Big]\,\Delta\mathbf{C}_S - 3\,\alpha^S K^S\,\Delta\Theta\,\mathbf{I}$$
$$= \mu^S\,\Delta\mathbf{C}_S + \frac{1}{2}\Big(K^{SN}\frac{2\,(1 - n^S_{0S}) + (n^S_{0S})^2}{1 - n^S_{0S}} - \frac{2}{3}\mu^S\Big)(\Delta\mathbf{C}_S \cdot \mathbf{I})\,\mathbf{I} - 3\,\alpha^S K^S\,\Delta\Theta\,\mathbf{I}\,,$$
$$\boldsymbol{\tau}^S_{E\text{lin}} = \Big[-\,\mu^S \overset{4}{\mathbf{I}} - \frac{1}{2}\Big(K^{SN}\,\frac{2\,(1 - n^S_{0S}) + (n^S_{0S})^2}{1 - n^S_{0S}} - \frac{2}{3}\,\mu^S\Big)\,\overset{4}{\bar{\mathbf{I}}}\Big]\,\Delta\mathbf{B}^{-1}_S - \tag{5.8.118}$$
$$-\,3\,\alpha^S K^S\,\Delta\Theta\,\mathbf{I}$$
$$= -\,\mu^S\,\Delta\mathbf{B}^{-1}_S - \frac{1}{2}\Big(K^{SN}\,\frac{2\,(1 - n^S_{0S}) + (n^S_{0S})^2}{1 - n^S_{0S}} - \frac{2}{3}\,\mu^S\Big)\,(\Delta\mathbf{B}^{-1}_S \cdot \mathbf{I})\,\mathbf{I} -$$
$$-\,3\,\alpha^S K^S\,\Delta\Theta\,\mathbf{I}\,.$$

Within the framework of the linear theory, the rates of the right and the inverse left Cauchy-Green tensor $\mathbf{C}_S$ and $\mathbf{B}_S^{-1}$ can be expressed with the help of the Green strain tensor $\mathbf{E}_S$ and the Almansian strain tensor $\mathbf{A}_S$,

$$\mathbf{E}_S = \frac{1}{2}(\mathbf{C}_S - \mathbf{I}) \quad , \quad \mathbf{A}_S = \frac{1}{2}(\mathbf{I} - \mathbf{B}_S^{-1}) \, , \tag{5.8.119}$$

viz.:

$$\begin{aligned} \Delta \mathbf{C}_S &= \mathbf{C}_S - \mathbf{C}_S \Big|_{\mathcal{P}_0} = \mathbf{C}_S - \mathbf{I} = 2\,\mathbf{E}_S \, , \\ \Delta \mathbf{B}_S^{-1} &= \mathbf{B}_S^{-1} - \mathbf{B}_S^{-1} \Big|_{\hat{\mathcal{P}}_0} = \mathbf{B}_S^{-1} - \mathbf{I} = -2\,\mathbf{A}_S \, . \end{aligned} \tag{5.8.120}$$

The rate of the temperature, within the framework of the linear theory, is given by

$$\Delta\Theta = \Theta - \Theta\Big|_{\mathcal{P}_0} = \Theta - \Theta\Big|_{\hat{\mathcal{P}}_0} = \Theta - \Theta_{0S} \, . \tag{5.8.121}$$

The insertion of (5.8.120) and (5.8.121) into (5.8.118) yields the following forms of the linearized laws of Hooke-type, with respect to the reference and the actual placements:

$$\begin{aligned} \mathbf{S}^S_{E\mathrm{lin}} &= 2\,\mu^S\,\mathbf{E}_S + \Big[K^{SN}\,\frac{2\,(1-n^S_{0S}) + (n^S_{0S})^2}{1-n^S_{0S}} - \frac{2}{3}\,\mu^S\Big]\,(\mathbf{E}_S \cdot \mathbf{I})\,\mathbf{I} - \\ &\quad - 3\,\alpha^S\,K^S\,(\Theta - \Theta_{0S})\,\mathbf{I} \, , \\ \boldsymbol{\tau}^S_{E\mathrm{lin}} &= 2\,\mu^S\,\mathbf{A}_S + \Big[K^{SN}\,\frac{2\,(1-n^S_{0S}) + (n^S_{0S})^2}{1-n^S_{0S}} - \frac{2}{3}\,\mu^S\Big]\,(\mathbf{A}_S \cdot \mathbf{I})\mathbf{I} - \\ &\quad - 3\,\alpha^S\,K^S\,(\Theta - \Theta_{0S})\,\mathbf{I} \, . \end{aligned} \tag{5.8.122}$$

In accordance with the linear theory of one component elastic material, the response parameters μ^S,

$$\lambda^S = K^{SN}\,\frac{2\,(1-n^S_{0S}) + (n^S_{0S})^2}{1-n^S_{0S}} - \frac{2}{3}\,\mu^S \tag{5.8.123}$$

and

$$K^S = K^{SN}\,\frac{2\,(1-n^S_{0S}) + (n^S_{0S})^2}{1-n^S_{0S}} = \frac{2}{3}\,\mu^S + \lambda^S \tag{5.8.124}$$

will be interpreted as the macroscopic Lamé constants and macroscopic compression modulus of the constituent φ^S. With the aforementioned interpretation of the response parameters μ^S, λ^S, and K^S, the linearized constitutive relations (5.8.122) simplify to

$$\begin{aligned} \mathbf{S}^S_{E\mathrm{lin}} &= 2\,\mu^S\,\mathbf{E}_S + \lambda^S(\mathbf{E}_S \cdot \mathbf{I})\,\mathbf{I} - 3\,\alpha^S\,K^S\,(\Theta - \Theta_{0S})\,\mathbf{I} \\ &= 2\,\mu^S\,\mathbf{E}^D_S + K^S(\mathbf{E}_S \cdot \mathbf{I})\,\mathbf{I} - 3\,\alpha^S\,K^S\,(\Theta - \Theta_{0S})\,\mathbf{I} \, , \\ \boldsymbol{\tau}^S_{E\mathrm{lin}} &= 2\,\mu^S\,\mathbf{A}_S + \lambda^S(\mathbf{A}_S \cdot \mathbf{I})\,\mathbf{I} - 3\,\alpha^S\,K^S\,(\Theta - \Theta_{0S})\,\mathbf{I} \\ &= 2\,\mu^S\,\mathbf{A}^D_S + K^S\,(\mathbf{A}_S \cdot \mathbf{I})\,\mathbf{I} - 3\,\alpha^S\,K^S\,(\Theta - \Theta_{0S})\,\mathbf{I} \, . \end{aligned} \tag{5.8.125}$$

Futhermore, the determination of the strain tensors

$$\mathbf{E}_S = \frac{1}{2\,\mu^S}\,\mathbf{S}^S_{E\mathrm{lin}} - \frac{\lambda^S}{2\,\mu^S\,(2\,\mu^S + 3\,\lambda^S)}\,(\mathbf{S}^S_{E\mathrm{lin}} \cdot \mathbf{I})\,\mathbf{I} + \alpha^S\,(\Theta - \Theta_{0S})\,\mathbf{I},$$

$$\mathbf{A}_S = \frac{1}{2\,\mu^S}\,\boldsymbol{\tau}^S_{E\mathrm{lin}} - \frac{\lambda^S}{2\,\mu^S\,(2\,\mu^S + 3\,\lambda^S)}\,(\boldsymbol{\tau}^S_{E\mathrm{lin}} \cdot \mathbf{I})\,\mathbf{I} + \alpha^S\,(\Theta - \Theta_{0S})\,\mathbf{I} \tag{5.8.126}$$

from (5.8.125) reveals that the response equations for $\mathbf{E}_S$ and $\mathbf{A}_S$ can be compared with the well-known strain relations of the linear theory of one component elastic material. Thus, the response parameter α^S can be interpreted as the thermal expansion coefficient of the constituent φ^S.

It is worth mentioning that the relation $\boldsymbol{\tau}^S_E = \mathbf{F}_S\,\mathbf{S}^S_E\,\mathbf{F}^T_S$ cannot be transferred to the linearized form (5.8.125), i.e.,

$$\boldsymbol{\tau}^S_{E\mathrm{lin}} \neq \mathbf{F}_S\,\mathbf{S}^S_{E\mathrm{lin}}\,\mathbf{F}^T_S. \tag{5.8.127}$$

Concerning the lengthy derivatives in (5.8.116) and, in particular, in (5.8.118) the interested reader is referred to Bluhm (1999).

Finally, for two elastic porous solids with the porosities in the reference placement $n^S_{0S} = 0.8$ and $n^S_{0S} = 0.2$, the free Helmholtz energy function ψ^S and the effective hydrostatic stress state p^S_E in dependence of $J_{SN} = \frac{n^S_{0S}}{n^S}$ for $\check{\mathbf{C}}_S = \mathbf{I}$ and $\Theta = \Theta_{0S}$ are depicted (see Fig. 5.8.2).

5.8.4 Rigid Ideal-Plastic Porous Solid Filled with an Inviscid Compressible Fluid

The following investigations will be restricted to a porous solid filled with gas (ideal-plastically deformed solid φ^S, inviscid (or an ideal) fluid φ^G). For the treatment of this binary model, the following assumptions are introduced:

- The thermodynamic investigations are performed within the framework of geometrically non-linear theory;
- both constituents have the same temperature and the same external acceleration

$$\Theta^S = \Theta^G = \Theta, \quad \mathbf{b}^S = \mathbf{b}^G = \mathbf{b}\,; \tag{5.8.128}$$

- there is no mass exchange, no moment of momentum, and no energy supplies, so that

$$\hat{\rho}^\alpha = 0, \quad \hat{\mathbf{m}}^\alpha = \mathbf{0}, \quad \hat{e}^\alpha = 0\,; \tag{5.8.129}$$

- the fluid (gas) is compressible and the solid is incompressible, i.e.,

$$\rho^{SR} = \rho^{SR}_{0S} = \text{const.}\,, \tag{5.8.130}$$

where ρ^{SR}_{0S} is the density at the reference placement.

The assumption of incompressibility of the solid constituent is reasonable at moderate stress levels, which almost always occur in soil mechanics. At those

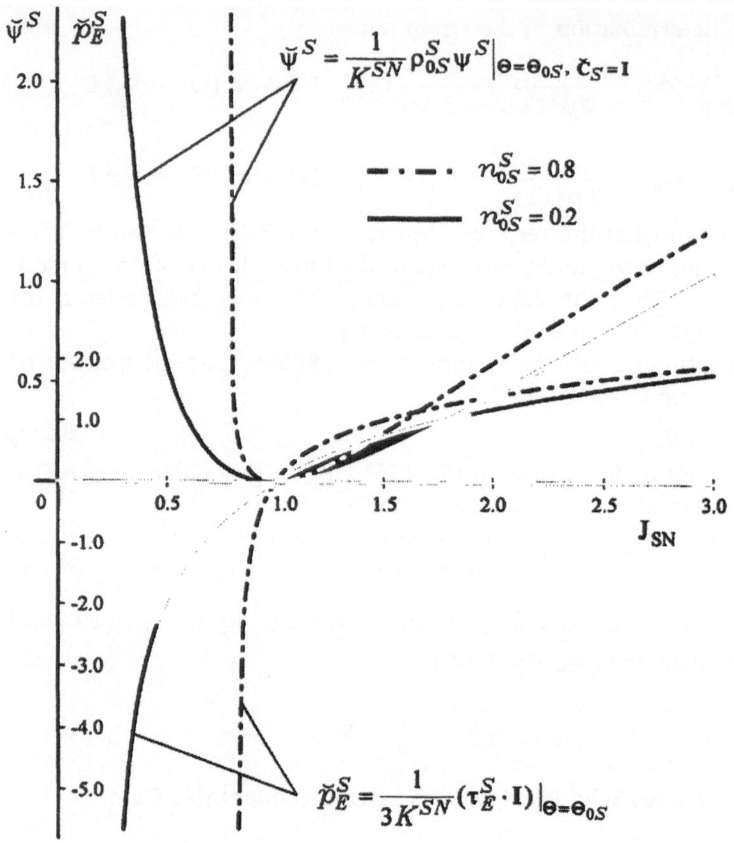

Fig. 5.8.2. Representation of the free energy function and the effective hydrostatic stress state in dependence of J_{SN}

stress levels, the compressibility of the solid material is far less than the compressibility of the whole porous solid.

The present model is defined by the balance equations of mass (5.4.3), the balance of momentum (5.4.15), the balance of energy (5.4.32), as well as the entropy inequality (5.5.6). In addition, the model is defined by the following set $\mathcal{R}$ of constitutive equations which have to be a function of a set s of common variables:

$$\mathcal{R}(s) = \{\psi^\alpha, \eta^\alpha, \mathbf{T}^\alpha + \lambda n^\alpha \mathbf{I}, \mathbf{q}^S + \mathbf{q}^G, \hat{\mathbf{p}}^G - \lambda \operatorname{grad} n^F\} . \qquad (5.8.131)$$

Within the framework of the constitutive theory, several suggestions have been made in the literature for the possible choice of constitutive equations (see e.g., Ehlers, 1989a, b). The principles of determinism and local action state that the response functions in (5.8.131) for a material point X_α are determined in any place $\mathbf{x}$ and time t by the history of an arbitrary small neighborhood of the material point. The history of this small neighborhood is fixed by the process variables $s(\mathbf{x}, t)$, which reflect the history of motion and temperature. For plastically-deformed, gas-filled materials, the following set of independent constitutive variables is introduced:

$$s = \left\{\Theta, \operatorname{grad}\Theta, \rho^{GR}, \mathbf{D}_S, \mathbf{v}_{GS}\right\} . \tag{5.8.132}$$

Due to the rigid ideal-plastic behavior of the skeleton, the symmetric part of the velocity gradient exclusively describes the plastic deformation rates.

Starting with the entropy inequality of the hybrid model of the second type, Section 5.8.2 d), we obtain the following restrictions for the free Helmholtz energy, the entropy, and Cauchy's stress tensor of the fluid

$$\begin{aligned} &\psi^S = \psi^S(\Theta), \quad \psi^G = \psi^G(\Theta, J_{GR}), \quad \eta^\alpha = -\frac{\partial\psi^\alpha}{\partial\Theta}, \\ &\mathbf{T}^G = -\, n^G\lambda\mathbf{I}, \quad \lambda = p = (\rho^{GR})^2\frac{\partial\psi^G}{\partial\rho^{GR}} = \kappa^{SR}, \end{aligned} \tag{5.8.133}$$

where λ is the interface pressure and κ^{SR} is the undetermined factor related to the incompressibility of the solid phase. Moreover, an important inequality remains:

$$\begin{aligned} &\mathbf{D}_S \cdot (\mathbf{T}^S + pn^S\mathbf{I}) - (\hat{\mathbf{p}}^G - \lambda \operatorname{grad} n^G) \cdot \mathbf{v}_{GS} - \\ &-\frac{1}{\Theta}(\mathbf{q}^S + \mathbf{q}^G) \cdot \operatorname{grad}\Theta \geq 0 \,, \quad \mathbf{v}_{GS} = \mathbf{v}_G - \mathbf{v}_S \,. \end{aligned} \tag{5.8.134}$$

Similar to the evaluation of the dissipation inequality in Appendix A, we arrive close to the mixture equilibrium state at

$$\begin{aligned} &\hat{\mathbf{p}}^G = p \operatorname{grad} n^G + \hat{\mathbf{p}}^G_E \,, \\ &\hat{\mathbf{p}}^G_E = -\,\beta_\Theta \operatorname{grad}\Theta - \beta^F_\upsilon \mathbf{v}_{GS} \,, \\ &\mathbf{q}^S + \mathbf{q}^G = -\,\alpha_\Theta \operatorname{grad}\Theta + \Theta\beta_\Theta \mathbf{v}_{GS} \,, \end{aligned} \tag{5.8.135}$$

where $\beta_\Theta = \beta_S + \beta_F = -\,\Theta(\alpha^S_\upsilon + \alpha^F_\upsilon)$, $\alpha_\Theta = \Theta^2(\alpha^S_S + \alpha^F_F)$, and β^F_υ are positive response parameters (see Appendix A (A 90)).

The description of *rigid ideal-plastic behavior* of porous solids is governed by the yield condition, the flow rule, and the loading criteria. The yield condition can be interpreted as a constraint for the effective stresses, which must always be fulfilled during the plastic deformation process. In this section, we would like to describe the mechanical behavior of an artificially created porous solid, which is gained from metallic powder by applying high pressures. We choose, for the yield criterion F, the quadratic function

$$2F = \mathbf{T}^{SD}_E \cdot \mathbf{T}^{SD}_E + \alpha^2 (\mathbf{T}^S_E \cdot \mathbf{I})^2 - \kappa^2 = 0, \tag{5.8.136}$$

where

$$\mathbf{T}^S_E = \mathbf{T}^S + pn^S\mathbf{I} \tag{5.8.137}$$

is the effective stress and α, as well as κ, stand for material response parameters.

As already mentioned, the yield condition (5.8.136) is valid for porous solids filled with fluid whose skeletons consist of ductile materials. It is not applicable to porous solids with skeletons composed of brittle materials such as soil, rock, and concrete (see, e.g., Bluhm *et al.*, 1996).

The flow rule provides the relation between the symmetric part of the velocity gradient $\mathbf{D}_S$ and the effective stresses of the porous solid. We apply the associated plastic potential concept:

$$\mathbf{D}_S = \lambda'_S \frac{\partial F}{\partial \mathbf{T}_E^S}, \qquad \mathbf{D}_S = \lambda'_S \left[\mathbf{T}_E^{SD} + \alpha^2 (\mathbf{T}_E^S \cdot \mathbf{I})\mathbf{I}\right], \tag{5.8.138}$$

together with the loading criteria

$$F = 0 \text{ and } \frac{\partial F}{\partial \mathbf{T}_E^S} \cdot (\mathbf{T}_E^S)'_S \begin{cases} = 0 \text{ neutral state } \mathbf{D}_S \neq \mathbf{0}, \\ < 0 \text{ unloading } \quad \mathbf{D}_S = \mathbf{0}. \end{cases} \tag{5.8.139}$$

From (5.8.136) and (5.8.138), the unknown positive quantity λ'_S can be specified as a scalar value function in the neutral state (loading for ideal-plastic porous solids). Expressing λ'_S by the material response parameters α and κ, as well as the plastic strain rates, we have:

$$\lambda'_S = \frac{1}{\kappa}\sqrt{\mathbf{D}_S^D \cdot \mathbf{D}_S^D + \frac{1}{3}\left(\frac{\mathbf{D}_S \cdot \mathbf{I}}{3\alpha}\right)^2}, \tag{5.8.140}$$

where $\mathbf{D}_S^D$ is the deviator of $\mathbf{D}_S$.

Thus, with the flow rule (5.8.138) and considering (5.8.137), the yield condition (5.8.136), and the constitutive equations (5.8.135), one can show that Inequality $(5.8.134)_1$ is satisfied.

In a similar way, viscoelastic, elastic-plastic (for ductile and frictional materials), and viscoplastic behavior can be described. However, not all problems in this field have been clarified to date.

In particular, the *thermoplastic* behavior of the metallic skeleton (metallic powder) is unknown to date. Also, the interacting agencies between the porous solid in the plastic state and the gas in the pores due to a change of the temperature and to the volumetric strain are still awaiting search. Therefore, the description of the complete thermodynamic behavior of the binary model consisting of the skeleton and of the gas (in the following paragraph) should be understood as a preliminary approach.

Next, we elaborate on the *constitutive equations* for *the inviscid fluid*, see $(5.8.133)_5$, taking an ideal gas into consideration. An *ideal gas* is defined as an inviscid fluid in which the pressure is related to the temperature Θ and the density ρ^{GR} by

$$p = \bar{R}\rho^{GR}\Theta = \bar{R}\frac{\Theta}{\nu^{GR}}, \quad \bar{R} = \frac{R}{M}, \quad \nu^{GR} = \frac{1}{\rho^{GR}}, \tag{5.8.141}$$

where M is the molecular weight of the gas and R, the so-called gas constant (the same for all ideal gases). Often, in the discussion of inviscid fluids, it is convenient to regard the free Helmholtz energy ψ^G and the internal energy ε^G as functions of Θ, ν^{GR} and η^G (entropy), ν^{GR}, respectively,

$$\psi^G = \overset{+}{\psi}{}^G(\Theta, \nu^{GR}), \quad \varepsilon^G = \overset{+}{\varepsilon}{}^G(\eta^G, \nu^{GR}). \tag{5.8.142}$$

From thermodynamic investigations, we obtain, instead of (5.8.141),

$$p = -\frac{\partial \overset{+}{\psi}{}^G}{\partial v^{GR}}, \qquad p = -\frac{\partial \overset{+}{\varepsilon}{}^G}{\partial v^{GR}}, \tag{5.8.143}$$

so that

$$\frac{\partial \overset{+}{\psi}{}^G}{\partial v^{GR}} = -\bar{R}\Theta \frac{1}{v^{GR}}. \tag{5.8.144}$$

It follows, by integrating, that

$$\overset{+}{\psi}{}^G = -\bar{R}\Theta \log v^{GR} + C(\Theta), \tag{5.8.145}$$

where C is a quantity depending on the temperature Θ.

Further, recalling the expressions for the free Helmholtz energy function and the entropy function

$$\psi^G = \varepsilon^G - \eta^G\Theta, \qquad \eta^G = -\frac{\partial \psi^G}{\partial \Theta} = -\left(\frac{\partial \overset{+}{\psi}{}^G}{\partial \Theta}\right)_{v^{GR}}, \tag{5.8.146}$$

we obtain

$$\varepsilon^G = \psi^G - \Theta\frac{\partial \psi^G}{\partial \Theta} = C(\Theta) - \Theta\frac{dC}{d\Theta}, \tag{5.8.147}$$

which shows that the internal energy of the ideal gas ε^G is a function of Θ only. (Note that, in writing (5.8.147), we have omitted the overcross from the various functions). The specific heat at constant volume c_v^G and the specific heat at constant pressure c_p^G for an inviscid fluid are

$$\begin{aligned} c_v^G &= \left(\frac{\partial \varepsilon^G}{\partial \Theta}\right)_{v^{GR}} = \Theta\left(\frac{\partial \eta^G}{\partial \Theta}\right)_{v^{GR}}, \\ c_p^G &= \left(\frac{\partial \varepsilon^G}{\partial \Theta}\right)_p + p\left(\frac{\partial v^{GR}}{\partial \Theta}\right)_p = \Theta\left(\frac{\partial \eta^G}{\partial \Theta}\right)_p, \end{aligned} \tag{5.8.148}$$

respectively, where the indices at the brackets signify that these quantities must be held constant in the differentiation process (see Truesdell and Toupin, 1960). It can also be shown that

$$c_p^G - c_v^G = \left[p + \left(\frac{\partial \varepsilon^G}{\partial v^{GR}}\right)_\Theta\right]\left(\frac{\partial v^{GR}}{\partial \Theta}\right)_p. \tag{5.8.149}$$

Since ε^G for an ideal gas is a function of Θ only, from (5.8.148), (5.8.146)$_2$, and (5.8.145), it is clear that c_v^G is also only a function of Θ, i.e.,

$$c_v^G = -\Theta\frac{d^2C}{d\Theta^2}. \tag{5.8.150}$$

Also, from (5.8.148), (5.8.146)$_2$, and (5.8.141)$_2$, as well as from (5.8.145), we get

$$c_p^G - c_v^G = \bar{R}, \tag{5.8.151}$$

where c_p^G is also only a function of Θ. We also have

$$\varepsilon^G = \int c_V^G d\Theta, \quad \eta^G = -\frac{\partial \psi^G}{\partial \Theta} = \bar{R}\log v^{GR} - \frac{\partial C}{\partial \Theta}$$
$$= \bar{R}\log v^{GR} + \int c_V^G \frac{d\Theta}{\Theta}. \tag{5.8.152}$$

In the special case where c_V^G is a constant, it follows from the equation above that c_p^G is also a constant, and so we write

$$\frac{c_p^G}{c_V^G} = \gamma \,, \quad c_V^G(\gamma - 1) = \bar{R}. \tag{5.8.153}$$

Eqn. (5.8.152) can then be integrated to give

$$\varepsilon^G = c_V^G \Theta + B = \frac{1}{\gamma - 1}\,\frac{p}{\rho^{GR}} + B \,,$$
$$\eta^G = \bar{R}\log v^{GR} + c_V^G \log\Theta + D \,, \tag{5.8.154}$$

where B and D are constants.

5.8.5 Elastic-Plastic Behavior of an Incompressible Porous Solid Filled with an Incompressible Inviscid Fluid

In the following, the mechanical behavior of an elastic-plastically deformed porous solid φ^S, filled with an incompressible inviscid (ideal) fluid φ^F, will be investigated. For the treatment of this binary model, the following assumptions are introduced:

- The thermodynamic investigations will be performed within the framework of the geometrically-linear theory;
- the total Green strain tensor $\mathbf{E}_S$ will be additively decomposed into an elastic and a plastic part $\mathbf{E}_{Se}$ and $\mathbf{E}_{Sp}$:
$$\mathbf{E}_S = \mathbf{E}_{Se} + \mathbf{E}_{Sp} \,; \tag{5.8.155}$$
- only isothermal deformations will be considered, i.e.,
$$\Theta'_\alpha = 0 \,, \quad \text{Grad}_\alpha\, \Theta = 0 \tag{5.8.156}$$
with Θ as the same temperature in both constituents;
- there will be no mass exchange, and no supply of moment of momentum as well as of energy, so that
$$\hat{\rho}^\alpha = 0 \,, \quad \hat{\mathbf{m}}^\alpha = \mathbf{0} \,, \quad \hat{e}^\alpha = 0 \,; \tag{5.8.157}$$
- both constituents are incompressible, i.e.,
$$\rho^{\alpha R} = \rho_{0\alpha}^{\alpha R} = \text{const.} \quad \text{or} \quad J_{\alpha R} = 1 \,. \tag{5.8.158}$$

The assumption of the incompressibility of both the constituents is reasonable at moderate stress levels, which almost always occur in soil mechanics. At those stress levels, liquids are incompressible and the compressibility of the solid material is far less than the compressibility of the whole porous solid.

The binary model presently under study is defined by the balance equations of mass (5.4.3), momentum (5.4.15), moment of momentum (5.4.26), and energy (5.4.33), as well as by the entropy inequality (5.5.7).

From the incompressible model in Section 5.8.2 **a**), we can read the following results for the hydrostatic stress states in the elastic range:

$$p^\alpha = -\, n^\alpha \kappa^{\alpha R}\,, \quad \lambda = \kappa^{SR} - \rho^S \frac{\partial \hat{\psi}^S}{\partial n^S}\,, \quad \lambda = \kappa^{FR}. \tag{5.8.159}$$

With $\kappa^{FR} = p$ (porefluid pressure), Eqn. $(5.8.159)_2$ leads to

$$\kappa^{SR} = p + \rho^S \frac{\partial \hat{\psi}^S}{\partial n^S}\,. \tag{5.8.160}$$

The term $\rho^S \frac{\partial \hat{\psi}^S}{\partial n^S}$ in (5.8.160) must be extended in order to consider the influence of plastic deformations on the stress state. The elastic-plastic model will, therefore, be defined by the following set $\mathcal{R}$ of constitutive equations, which are assumed to be a function of a set s of common variables:

$$\mathcal{R}(s) = \left\{\psi^\alpha, \eta^\alpha, \mathbf{S}^\alpha + n^\alpha p\mathbf{I}, \hat{\mathbf{p}}^F - \lambda \operatorname{grad} n^F\right\}. \tag{5.8.161}$$

The reader may note in passing that we have introduced the second Piola-Kirchhoff stress tensors $\mathbf{S}^\alpha$, which are related to the reference placement. It is true that, in the geometrically-linear theory, the Cauchy stress tensor is approximately equal to the first and second Piola-Kirchhoff stress tensors; but to make it very clear that all response functions in (5.8.161) are related to the reference placement, the introduction of the second Piola-Kirchhoff stress tensor seems to be useful.

For elastic-plastically deformed materials, the following set of additional independent constitutive variables are introduced:

$$s = \left\{\mathbf{E}_{Se}, \mathbf{H}_S, \mathbf{v}_F - \mathbf{v}_S\right\}. \tag{5.8.162}$$

With respect to the plastic range, where all mechanical quantities depend to a great extent on the deformation history (see Bluhm *et al.*, 1996), a constitutive relation (evolution equation) for the tensor of deformation history $\mathbf{H}_S$ has to be added:

$$\begin{aligned} (\mathbf{H}_S)'_S &= (\mathbf{H}_S)'_S\left[s, (\mathbf{E}_{Sp})'_S\right], \\ (\mathbf{E}_{Sp})'_S &= (\mathbf{E}_{Sp})'_S\left[s, \ldots\right]. \end{aligned} \tag{5.8.163}$$

The tensor $(\mathbf{H}_S)'_S$, which is referred to the reference configuration, describes the stress and deformation history (see the following remarks), and is considered as an internal state variable depending on the plastic strain rates $(\mathbf{E}_{Sp})'_S$ and the set of variables s.

Moreover, the elastic strains $\mathbf{E}_{Se}$ are given by the difference of the strains of the partial porous solid and the plastic strains, see (5.8.155):

$$\mathbf{E}_{Se} = \mathbf{E}_S - \mathbf{E}_{Sp}\,. \tag{5.8.164}$$

As mentioned earlier, the plastic behavior in the hardening range depends to a great extent on the deformation history. This dependence is described by the tensor of history $\mathbf{H}_S$. It is well-known that the plastic strain $\mathbf{E}_{Sp}$ has a significant influence on the deformation history. However, the plastic strain tensor $\mathbf{E}_{Sp}$ itself is not suitable for the description of the deformation history because, for example, in rigid-plastic behavior, the (plastic) strain tensor is uniquely determined by the displacement field and in no way reflects the deformation history. Even if the plastic strain is calculated from the strain rates by an integration process, the plastic strain is less appropriate. In this case, it may happen that the unloading path is identical with the loading path, and that the plastic strain turns out to be zero. Therefore, the tensor valued function $\mathbf{H}_S$ is introduced, assuming that this quantity depends on the plastic strain $\mathbf{E}_{Sp}$ and on the set of variables s. The tensor $\mathbf{H}_S$ is zero at the onset of the plastic deformations, and so does not enter the set of constitutive parameters in the case of ideal-plastic behavior of metals (see de Boer and Kowalski, 1983).

For the binary model under study, thermodynamic restrictions result from the entropy inequality (5.8.27) together with the constitutive assumptions (5.8.161), (5.8.162), and (5.8.163). However, for porous media, the procedure for gaining restrictions leads to lengthy derivations and to a large formalism. Therefore, only the relevant results are reported (the interested reader is referred to de Boer and Ehlers, 1986b, and de Boer and Kowalski, 1995, as well as to the APPENDIX A).

Using standard arguments (Bowen, 1976), combined with several symmetry and skew-symmetry conditions, a linear expansion for the mixture equilibrium state

$$s \rightarrow s_0 = \left\{\mathbf{E}_{Se}, \mathbf{H}_S, \mathbf{v}_F - \mathbf{v}_S = \mathbf{0}\right\} \tag{5.8.165}$$

yields, among other things, the following restrictions which have to be fulfilled by the constitutive equations of the liquid and solid constituents.

The thermodynamic investigations give rise to restrictions for the free energy density of the porous solid

$$\psi^S = \tilde{\psi}^S(\mathbf{E}_{Se}, \mathbf{H}_S) \, . \tag{5.8.166}$$

Moreover, the stresses are given by

$$\begin{aligned} \mathbf{S}^S &= -\, n^S p \mathbf{I} + \rho^S_{0S} \frac{\partial \tilde{\psi}^S}{\partial \mathbf{E}_{Se}} \, , \\ \mathbf{S}^F &= -\, n^F p \mathbf{I} \, , \end{aligned} \tag{5.8.167}$$

where (5.8.159) and (5.8.160) have been considered.

Finally, the constitutive relations (5.8.167) can be expressed by

$$\begin{aligned} \mathbf{S}^S &= -\, n^S p \mathbf{I} + \mathbf{S}^S_E \, , \\ \mathbf{S}^F &= -\, n^F p \mathbf{I} \, , \end{aligned} \tag{5.8.168}$$

where the effective stress $\mathbf{S}^S_E$ is determined by

$$\mathbf{S}_E^S = \rho_{0S}^S \frac{\partial \tilde{\psi}^S}{\partial \mathbf{E}_{Se}} \,. \tag{5.8.169}$$

Hence, due to the incompressibility constraint, the partial second Piola-Kirchhoff stress tensor $\mathbf{S}^S$ can be expressed by the weighted true liquid pressure p and the extra or effective stresses which are caused by the motions of the constituents.

Moreover, the entropy inequality yields an important dissipation inequality which restricts the constitutive relation for the rate of the plastic strain tensor:

$$\mathbf{S}_E^S \cdot (\mathbf{E}_{Sp})_S' - \frac{\partial \tilde{\psi}^S}{\partial \mathbf{H}_S} \cdot (\mathbf{H}_S)_S' \geq 0 \,. \tag{5.8.170}$$

For the rate of the tensor of the deformation history $(\mathbf{H}_S)_S'$, a linear dependence on $(\mathbf{E}_{Sp})_S'$ is assumed according to the similar approach of Fromm (1933), where $g(s,\ldots)$ is a scalar value function, namely

$$(\mathbf{H}_S)_S' = \rho_{0S}^S \, g(s,\ldots)(\mathbf{E}_{Sp})_S' \,. \tag{5.8.171}$$

With

$$\mathbf{Z}^S = \rho_{0S}^S \, g(s,\ldots) \frac{\partial \tilde{\psi}^S}{\partial \mathbf{H}_S} \tag{5.8.172}$$

and (5.8.171), Inequality (5.8.170) reads

$$(\mathbf{E}_{Sp})_S' \cdot (\mathbf{S}_E^S - \mathbf{Z}^S) \geq 0 \,. \tag{5.8.173}$$

The results derived contain strong restrictions concerning the constitutive equations, especially for the partial solid constituent, and some fundamental mechanical effects. First, in the constitutive relations for the stresses (5.8.168), the concept of effective stresses, well-known in soil mechanics (see de Boer and Ehlers, 1990b), is recognized. This concept states that constitutive equations are only required for the effective stresses. In soil mechanics, the concept of effective stresses is often formulated for the total stress tensor,

$$\mathbf{S} = \mathbf{S}^S + \mathbf{S}^F \,. \tag{5.8.174}$$

This procedure, however, is only valid for saturated porous solids with vanishing extra stresses for the liquid. Only in this case does $(5.8.168)_1$, together with $(5.8.168)_2$, yield

$$\mathbf{S} = -\,p\mathbf{I} + \mathbf{S}_E^S \,. \tag{5.8.175}$$

In the plastic range, the concept of effective stresses demands that the yield condition, the loading criteria, and the flow rule be formulated with the total stress plus the weighted (with n^S) pressure p. This can result in surprising mechanical consequences, e.g., concerning the failure state. The onset of failure of granular materials depends, to a great extent, on the hydrostatic pressure within the partial solid. If the boundary conditions are such that the entire hydrostatic pressure is carried by the liquid and the hydrostatic effective stress state in the partial solid constituent vanishes, then failure occurs. This takes place at a much lower stress level, at least at a stress state limited by cohesion or by zero (for cohesionless materials).

An important effect is recognized from Inequality (5.8.173). The tensor $\mathbf{Z}^S$ (5.8.172) reflects the total deformation history; it indicates kinematic hardening. This tensor will be discussed in detail, in connection with the yield condition.

With these remarks, the thermodynamic considerations are closed. In the next section, explicit constitutive equations for frictional materials will be discussed. We will not investigate the elastic region. Later, when the constitutive equations are needed, we will refer to Section 5.8.3.

Whereas the description of the elastic behavior of porous media, within the framework of the constitutive theory, is characterized by the consideration of the deformation gradient (or due to invariant requirements by other kinematic quantities), the description of plastic response is governed by the total deformation process. Thus, the total deformation process has to be known, since the response of porous media to different loading processes is different.

It is well-known that the classical plasticity theory is based, first, upon a yield condition that indicates the onset of plastic deformations and describes the hardening range and the failure state, second, upon the consistency condition, third, upon loading criteria that are gained from the consistency condition and the assumption of stable material behavior, and fourth, upon the flow rule that connects strain increments with stresses and stress increments. The constitutive relations must reflect test observations and thermodynamic restrictions. The main feature of the test results and the thermodynamic restrictions is the indication of kinematic hardening, which has to be described by a tensor containing all the information on the deformation history of the skeleton. However, test results also reveal that isotropic hardening is involved. Thus, the main aim of the following investigations is to elaborate on this behavior.

The yield condition marks the onset of plastic deformations when a determined stress state is achieved. This condition is, of course, material-dependent. Therefore, a constitutive relation is necessary and is understood to separate the elastic and plastic ranges at a stage where no plastic deformations have occurred (initial yield condition), as well as in the hardening and failure states. As pointed out, thermodynamic investigations reveal that, for the saturated porous solid, the constitutive equations and therefore also the yield condition have to be formulated with the effective stresses.

The plastic behavior of porous bodies depends strongly on the structure of the solid material. In contrast to metals, granular and brittle materials show a completely different behavior in extension and compression tests. Due to the complex behavior of granular and brittle materials (empty porous solids) in the plastic range, several different yield conditions have been developed (see, e.g., Lade, 1972, Gudehus, 1973, Willam and Warnke, 1975, Ottosen, 1977, Desai, 1980, Lade and Kim, 1988, and de Boer, 1988a). The scope of these yield conditions is mostly restricted to the failure state, or to isotropic hardening. According to Mróz *et al.* (1978), the attempt to develop a kinematic hardening model for soils was pioneered by Prévost and Hoeg (1975). The main feature of the method of Mróz *et al.* (1978) is to apply the concept of a field of hardening moduli, developed by Mróz (1967), to soils. Mróz and Pietruszczak (1983) derived constitutive equations, including kinematic hardening, which are applicable to both dense and loose sands. Further research in the field of soil mechanics concerning kinematic hardening has been done by Hirai (1987).

Yield Condition

The yield function has to be constructed in such a way that it reflects all relevant test observations and invariant requirements. In particular, in the hardening range, it has to depend on the deformation history, which will be denoted by the symbol $\mathcal{H}^S$ (see Fromm, 1933). At the onset of plastic deformations and in the failure state, the yield function does not depend on the deformation history and, therefore, $\mathcal{H}^S$ turns to zero and constant. The yield function, generally assumed to depend on the stress tensor $\mathbf{S}_E^S$, is a scalar function, and is described by

$$Y = Y(\mathbf{S}_E^S, \mathcal{H}^S) \,. \tag{5.8.176}$$

If the deformation history $\mathcal{H}^S$ is known, Y depends only on $\mathbf{S}_E^S$:

$$Y = Y_{(\mathcal{H}^S)}(\mathbf{S}_E^S) \,. \tag{5.8.177}$$

$\mathcal{H}^S$ is equal to zero, especially for the state at the onset of plastic deformation which is free from the deformation history, and

$$Y = Y_{(0)}(\mathbf{S}_E^S) = F(\mathbf{S}_E^S) \,. \tag{5.8.178}$$

The failure state also does not depend on the deformation history. In this case, $\mathcal{H}^S$ is to be considered as a constant C:

$$Y = Y_{(C)}(\mathbf{S}_E^S) = F(\mathbf{S}_E^S) \,. \tag{5.8.179}$$

At the onset of plastic deformations, the yield function $F(\mathbf{S}_E^S)$ (5.8.178) depends on the invariants of $\mathbf{S}_E^S$, if the material under discussion is isotropic. This is not always valid for the other forms of (5.8.176). However, it is possible to express $Y = Y(\mathbf{S}_E^S, \mathcal{H}^S)$ by a scalar valued function H, which is invariant against rotations, that is

$$Y = Y(\mathbf{S}_E^S, \mathcal{H}^S) = H(\mathbf{S}_E^S - \mathbf{Z}^S) \,, \tag{5.8.180}$$

where $\mathbf{Z}^S$ is the so-called backstress or translation tensor depending on the stress $\mathbf{S}_E^S$ and the deformation history $\mathcal{H}^S$, see (5.8.172). It is possible to replace H by F, and in this case $\mathbf{S}_E^S$ is returned by $\mathbf{Z}^S$ to its initial state. Thus, in the following, yield functions of the kind

$$Y = Y(\mathbf{S}_E^S, \mathbf{Z}^S) = F(\mathbf{S}_E^S - \mathbf{Z}^S) \tag{5.8.181}$$

are considered, especially those that are represented by the invariants of the stress difference

$$\bar{\mathbf{S}} = \mathbf{S}_E^S - \mathbf{Z}^S \,. \tag{5.8.182}$$

Then

$$Y = F(\mathrm{I}_{\bar{\mathbf{S}}}, \mathrm{II}_{\bar{\mathbf{S}}}, \mathrm{III}_{\bar{\mathbf{S}}}) \tag{5.8.183}$$

or

$$Y = F(\mathrm{I}_{\bar{\mathbf{S}}}, \mathrm{II}_{\bar{\mathbf{S}}^D}, \mathrm{III}_{\bar{\mathbf{S}}^D}) \tag{5.8.184}$$

are assumed.

The above yield functions are limited if the plastic state is attained, i.e.,

$$Y = F(\mathrm{I}_{\bar{\mathbf{S}}}, \mathrm{II}_{\bar{\mathbf{S}}}, \mathrm{III}_{\bar{\mathbf{S}}}) = 0 \tag{5.8.185}$$

or

$$Y = F(\mathrm{I}_{\bar{\mathbf{S}}}, \mathrm{II}_{\bar{\mathbf{S}}^D}, \mathrm{III}_{\bar{\mathbf{S}}^D}) = 0\,. \tag{5.8.186}$$

It can be seen that, in geometric representation, yield functions that are located inside the Mohr-Coulomb lines and that are adjusted to the plane deviatoric stresses can describe the test results. In this case, the convexity of the yield function is ensured. In geometric representation, the backstress tensor causes a translation of the center of the yield condition (see APPENDIX C), with all the other properties remaining unchanged. Thus, all known yield conditions (some are mentioned at the beginning of this section) for granular and brittle materials can be used for kinematic-hardening, if the stress tensor $\mathbf{S}_E^S$ is replaced by the difference tensor $\bar{\mathbf{S}}$, according to (5.8.182). In this sense, the yield function

$$\begin{aligned} F &= \sqrt{\Phi}\,(1+\gamma\vartheta)^{1/m} + \beta \mathrm{I}_{\mathbf{S}_E^S} + \varepsilon \mathrm{I}^2_{\mathbf{S}_E^S} - \kappa = 0\,, \\ \Phi &= \mathrm{II}_{\mathbf{S}_E^{SD}} + \frac{1}{2}\alpha^2 \mathrm{I}^2_{\mathbf{S}_E^S}\,, \\ \vartheta &= \frac{\mathrm{III}_{\mathbf{S}_E^{SD}}}{\sqrt{\Phi^m}} \sqrt{\mathrm{II}^{m-3}_{\mathbf{S}_E^{SD}}}\,, \end{aligned} \tag{5.8.187}$$

developed by de Boer (1988a), de Boer and Dresenkamp (1989), and extended by Bluhm (1994), can be formulated in the stress subspace $\bar{\mathbf{S}} = \mathbf{S}_E^S - \mathbf{Z}^S$:

$$\begin{aligned} F &= \sqrt{\bar{\Phi}}\,(1+\gamma\bar{\vartheta})^{1/m} + \beta \mathrm{I}_{\bar{\mathbf{S}}} + \varepsilon \mathrm{I}^2_{\bar{\mathbf{S}}} - \kappa = 0\,, \\ \bar{\Phi} &= \mathrm{II}_{\bar{\mathbf{S}}^D} + \frac{1}{2}\alpha^2 \mathrm{I}^2_{\bar{\mathbf{S}}}\,, \\ \bar{\vartheta} &= \frac{\mathrm{III}_{\bar{\mathbf{S}}^D}}{\sqrt{\bar{\Phi}^m}} \sqrt{\mathrm{II}^{m-3}_{\bar{\mathbf{S}}^D}}\,. \end{aligned} \tag{5.8.188}$$

The six quantities α, β, γ, ϵ, κ, and m, in (5.8.187) and (5.8.188), are response parameters which have to be adjusted to test observations. In the special case of $m = 3$, the yield condition results in the approach of de Boer and Dresenkamp (1989).

It is well-known in the classical plasticity theory for metallic materials that the second invariant of the stress state $\mathrm{II}_{\mathbf{S}^D}$ or $\mathrm{II}_{\bar{\mathbf{S}}^D}$ is a circle in the graphic representation in the deviator plane. The term $(1+\gamma\vartheta)^m$, or $(1+\gamma\bar{\vartheta})^m$, governs the typical shape of the yield surface of frictional materials in the deviatoric plane. In particular, the right choice of the parameter m allows the correct approach to the Mohr-Coulomb failure condition (see APPENDIX C). Moreover,

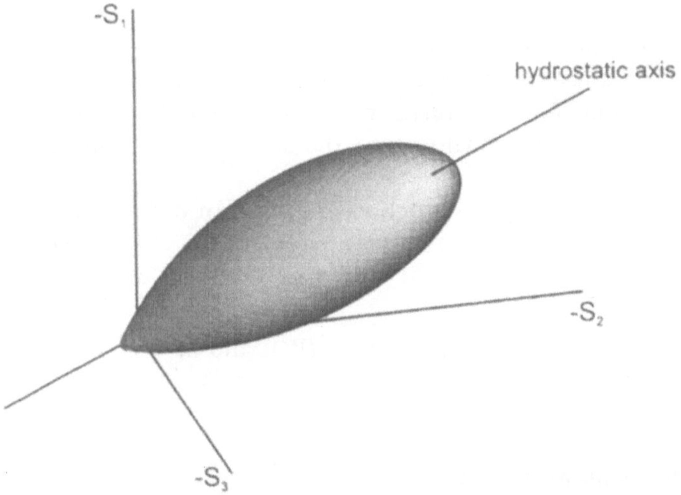

Fig. 5.8.3. Representation of the yield function for $\alpha = 0.0775$, $\beta = 0.274$, $\varepsilon = 0.0003\ \frac{mm^2}{N}$, $\kappa = 10.27\ \frac{N}{mm^2}$, $\gamma = 1.299$, $m = 3$

the incorporation of $\frac{1}{2}\alpha^2 \mathrm{I}_\mathrm{S}^2$ or $\frac{1}{2}\alpha^2 \mathrm{I}_{\bar{\mathrm{S}}}^2$ into Φ or $\bar{\Phi}$ guarantees the influence of the first invariant of the stress tensor on the shape of the yield surface. This is an essential feature in the plasticity theory of frictional materials. It is easily recognized that a hydrostatic pressure has to influence the deviatoric shape of the yield surface. For example, loose sand under small hydrostatic pressure will show another shape in the deviatoric plane than sand under a strong hydrostatic pressure. Furthermore, the parameters β and ϵ, as well as the parameter α^2, govern the form of the yield surface in the hydrostatic plane (see Appendix C).

The response parameters may depend on the plastic deformation history. If this is the case, then the yield condition (5.8.188) represents a combination of kinematic and isotropic hardening. The above-stated yield condition is therefore very general. It is suitable to describe the onset of plastic deformations as well as the plastic hardening range, or the failure state of kinematically and isotropically hardening brittle and granular skeletons, including the effects of compression and extension.

The great advantage of the yield condition (5.8.188) is that this condition represents a "single-surface" condition. Thus, no cap-model is needed to limit the yield curve in the hydrostatic plane. Further discussions of the introduced yield condition are contained in the papers of de Boer (1988a), de Boer and Dresenkamp (1989), de Boer and Lade (1991), as well as Bluhm *et al.* (1996). In these papers, the convexity of the yield condition (5.8.188) is, in particular, investigated.

Finally, it should be mentioned that the general yield condition (5.8.188) contains some special cases of yield conditions. If the response parameters γ

and ϵ are equal to zero, we obtain a yield condition proposed earlier by de Boer (1986). With α, γ, and ϵ being equal to zero, the condition (5.8.188) goes over to the failure condition of Drucker and Prager (1952). Moreover, if γ, β, and ϵ are neglected, the general yield condition turns into the yield condition of Green (1972). Finally, with α, γ, β, and ϵ equal to zero, the famous yield condition of von Mises (1913) is recognized.

Let us now turn again to the yield function (5.8.188). In order to include *isotropic hardening*, we assume that the response parameters $\alpha, \beta, \gamma, \epsilon, \kappa$, and m are functions of the plastic work W. We choose

$$\alpha = \alpha(W),\ \beta = \beta(W), \gamma = \gamma(W), \varepsilon = \varepsilon(W),\ \kappa = \kappa(W),\ \text{and } m = m(W)\,, \tag{5.8.189}$$

where the rate of the plastic work is given by

$$DW = \mathbf{S}_E^S \cdot D\mathbf{E}_{Sp}\,. \tag{5.8.190}$$

In the classical plasticity theory, the constitutive equations are, in part, formulated in the time rates. However, the constitutive relations do not explicitly depend on time. Therefore, in (5.8.190) and in the sequel, we use only the rates, for example, DW and $D\mathbf{E}_{Sp}$.

Kinematic hardening is described by the backstress tensor $\mathbf{Z}^S$. According to Fromm (1933), we propose a general ansatz:

$$\mathbf{Z}^S = C_1\left(\mathbf{S}_E^S, \mathbf{H}_S^D\right)\mathbf{H}_S^D + C_2\left(\mathbf{S}_E^S, \mathbf{H}_S\right)\mathrm{I}_{\mathbf{H}_S}\mathbf{I}\,. \tag{5.8.191}$$

The tensor of deformation history $\mathbf{H}_S$ is strongly influenced by the plastic deformations due to experience. The following relation for the rate of $\mathbf{H}_S$ may be appropriate:

$$D\mathbf{H}_S = f\left(\varphi, D\mathbf{E}_{Sp}\right)D\mathbf{E}_{Sp}\,, \tag{5.8.192}$$

where $f(\varphi, D\mathbf{E}_{Sp})$ has to be a positive scalar value function due to the restriction (5.8.170), depending on the plastic strain rates and on the angle of internal friction φ. Relation (5.8.191) is similar to Fromm's (1933) general approach. However, Fromm's relation depends only on the deviatoric strain rates, while the function f depends on the stresses and the strain rates. It has already been proven by de Boer and Brauns (1990) that the angle of internal friction of the Mohr-Coulomb theory can serve as a parameter for describing the hardening process. Although the Mohr-Coulomb theory deals only with the failure state, the above-procedure is promising.

In the general formulas (5.8.191) and (5.8.192), the approaches of Melan (1938), Eisenberg and Phillips (1968), and de Boer and Brauns (1990) – all of which represent an extended version of the Eisenberg and Phillips formula – are contained as special cases. It has been shown by de Boer and Brauns (1990) that the extended version efficiently describes for the backstress tensor the test results gained by Lade (1979).

Consistency Condition

As has already been mentioned, in order to describe the mechanical behavior of a plastically deformed solid, the consistency condition has to be formulated. If the consistency condition is fulfilled, the plastic flow continues.

The condition of consistency states that, if the yield condition is fulfilled, it must also be satisfied after a step. Thus, the condition of consistency guarantees plastic deformations:

$$\frac{\partial F}{\partial \bar{\mathbf{S}}} \cdot D\bar{\mathbf{S}} + \frac{\partial F}{\partial W} DW = 0\,, \tag{5.8.193}$$

or, using (5.8.182),

$$\frac{\partial F}{\partial \bar{\mathbf{S}}} \cdot D\mathbf{S}_E^S = \frac{\partial F}{\partial \bar{\mathbf{S}}} \cdot D\mathbf{Z}^S - \frac{\partial F}{\partial W} DW\,. \tag{5.8.194}$$

If the stress state satisfies (5.8.188), plastic deformations can occur. If, however,

$$F < 0\,, \tag{5.8.195}$$

then there will be no plastic deformations.

We use (5.8.190), (5.8.191), and (5.8.192) to explicitly determine the consistency condition (5.8.194). After elementary – although lengthy and laborious – calculations, we arrive at

$$\frac{\partial F}{\partial \bar{\mathbf{S}}} \cdot D\mathbf{S}_E^S = (\mathbf{M} - \mathbf{N}) \cdot D\mathbf{E}_{Sp}\,, \tag{5.8.196}$$

where

$$\begin{aligned} \mathbf{N} &= \frac{\partial F}{\partial W} \mathbf{S}_E^S\,, \\ \mathbf{M} &= f(\varphi, D\mathbf{E}_{Sp}) \left(\frac{\partial \mathbf{Z}^S}{\partial \mathbf{H}_S}\right)^T \frac{\partial F}{\partial \bar{\mathbf{S}}}\,, \end{aligned} \tag{5.8.197}$$

$$\begin{aligned} \frac{dF}{\partial \bar{\mathbf{S}}} &= \frac{1}{2} (\bar{\Phi})^{-1/2} (1 + \gamma \bar{\vartheta})^{1/m} (\bar{\mathbf{S}}^D + \alpha^2 \mathrm{I}_{\bar{\mathbf{S}}} \mathbf{I}) + \\ &+ \frac{1}{m} (1 + \gamma \bar{\vartheta})^{\frac{1-m}{m}} \gamma \Big\{ \Big[(\bar{\Phi})^{-\frac{m-1}{2}} (\bar{\mathbf{S}}^D \bar{\mathbf{S}}^D - \frac{2}{3} \mathrm{II}_{\bar{\mathbf{S}}^D} \mathbf{I}) - \\ &- \frac{m}{2} \bar{\Phi}^{-\frac{m+1}{2}} (\bar{\mathbf{S}}^D + \alpha^2 \mathrm{I}_{\bar{\mathbf{S}}}) \, \mathrm{III}_{\bar{\mathbf{S}}^D} \Big] (\mathrm{II}_{\bar{\mathbf{S}}^D})^{\frac{m-3}{2}} + \\ &+ \frac{m-3}{2} \bar{\Phi}^{-\frac{m-1}{2}} \, \mathrm{III}_{\bar{\mathbf{S}}^D} (\mathrm{II}_{\bar{\mathbf{S}}^D})^{\frac{m-5}{2}} \bar{\mathbf{S}}^D \Big\} + (2\varepsilon \mathrm{I}_{\bar{\mathbf{S}}} + \beta) \mathbf{I}\,, \end{aligned} \tag{5.8.198}$$

and

$$\begin{aligned} \frac{\partial F}{\partial W} &= \frac{1}{2} (\bar{\Phi})^{-1/2} \, \mathrm{I}_{\bar{\mathbf{S}}}^2 \, \alpha (1 + \gamma \bar{\vartheta})^{1/m} \alpha_{,W} + \\ &+ \frac{1}{m} (1 + \gamma \bar{\vartheta})^{\frac{1-m}{m}} \Big[\bar{\vartheta} \gamma_{,W} + \bar{\Phi}^{-\frac{m+2}{2}} \sqrt{\mathrm{II}_{\bar{\mathbf{S}}^D}^{m-3}} \, \mathrm{III}_{\bar{\mathbf{S}}^D} \, \mathrm{I}_{\bar{\mathbf{S}}}^2 \, m \alpha \, \alpha_{,W} \Big] + \\ &+ \mathrm{I}_{\bar{\mathbf{S}}} \, \beta_{,W} + I_{\bar{\mathbf{S}}}^2 \, \varepsilon_{,W} + \kappa_{,W}\,, \end{aligned} \tag{5.8.199}$$

with $(\ldots)_{,W}$ marking the partial derivative with respect to W. The tensor $\mathbf{N}$ results from the assumption of isotropic hardening, see (5.8.189) and (5.8.190), and the tensor $\mathbf{M}$ is calculated from the ansätze for kinematic hardening, see (5.8.191) and (5.8.192).

Loading Criteria

Now, the *loading criteria* will be formulated:

$$F = 0 \text{ and } \frac{\partial F}{\partial \bar{\mathbf{S}}} \cdot D\mathbf{S}_E^S \begin{cases} > 0 \text{ loading} & D\mathbf{E}_{Sp} \neq \mathbf{0}\,, \\ = 0 \text{ neutral state} & D\mathbf{E}_{Sp} = \mathbf{0}\,, \\ < 0 \text{ unloading} & D\mathbf{E}_{Sp} = \mathbf{0}\,. \end{cases} \tag{5.8.200}$$

In the case of *ideal-plastic behavior* (critical state), the loading criteria take the forms:

$$F = 0 \text{ and } \frac{\partial F}{\partial \bar{\mathbf{S}}} \cdot D\mathbf{S}_E^S \begin{cases} = 0 \text{ neutral state} & D\mathbf{E}_{Sp} \neq \mathbf{0}\,, \\ < 0 \text{ unloading} & D\mathbf{E}_{Sp} = \mathbf{0}\,. \end{cases} \tag{5.8.201}$$

As has already been mentioned, the yield condition (5.8.188) is very general, with six response parameters, including the description of kinematic and isotropic hardening. It depends on the problem as to which response parameters have to be considered in order to correctly describe the plastic behavior. Moreover, for kinematic hardening, some simplifications can be introduced if the characteristic stress-strain curve is nearly linear and, for isotropic hardening, some response parameters may be constant, or may depend on the problem under study.

The set of constitutive relations will be completed with the development of the flow rule.

Flow Rule

In order to describe the complete motion of an initial- and boundary-value problem, a constitutive equation for the rates $D\mathbf{E}_{Sp}$ of the plastic strains of the partial solid constituent (flow rule) is needed. This constitutive equation must also reflect the deformation history according to test observations. This will be done in such a way that the flow rule will be formulated in the special stress space created with the stress difference $\bar{\mathbf{S}}$, see (5.8.182). In continuum mechanics of isotropic materials, in general, it is stated that the change of the strain deviator and the volume change are completely independent of each other. Here, the same statement is introduced and a relatively simple flow rule is formulated:

$$D\mathbf{E}_{Sp} = D\lambda\left[\bar{\mathbf{S}}^D + \mu(\bar{\mathbf{S}} \cdot \mathbf{I})\mathbf{I}\right]\,, \tag{5.8.202}$$

where $D\lambda$ is a spatial-dependent scalar and

$$\mu = \mu\left(\mathrm{I}_{\bar{\mathbf{S}}}, \mathrm{II}_{\bar{\mathbf{S}}^D}\right) \tag{5.8.203}$$

is a function governing the volume change. There is no reason for the flow rule to have another form. Neither the test results, nor the theoretical investigations, indicate this as necessary. The formulation of the flow rule with a plastic potential, widely used for geomechanical materials, can only be understood from the practical point of view. On the one hand, it is easy to construct such a potential because in the geometric representation the potential lines are always perpendicular to the strain rate directions observed in tests results. On the other hand, a lot of software based on the plastic potential concept already exists. However, such a flow rule is very complex and not useful for complicated boundary-value problems, especially within the framework of the geometrically non-linear theory.

The quantities $D\lambda$ and μ must always be positive due to the restriction (5.8.173) in connection with the relation (5.8.202). The scalar value function $D\lambda$ in (5.8.202) can be determined from the condition of consistency (5.8.196), in connection with (5.8.197), (5.8.198), and (5.8.199). For the function μ, a backstress-dependent equation is required since μ has to reflect the deformation history. The following relation seems to be reasonable:

$$\mu = \varepsilon(\varphi)\mathrm{I}_{\xi} + \xi(\varphi)\sqrt{\mathrm{II}_{\xi^D}}\,, \tag{5.8.204}$$

where the material-dependent parameters $\epsilon(\varphi)$ and $\xi(\varphi)$ are supposed to depend on the angle of internal friction φ. As has been proven by de Boer and Brauns (1990), it seems reasonable to use the angle of internal friction which, although introduced for the failure state as a constant, also exists in the hardening range as a parameter. The following approaches for determining ε and ξ yield excellent results (see de Boer and Brauns, 1990):

$$\varepsilon = -\, m_1\left[\sin(\varphi_B - \varphi)\right]^k\,, \tag{5.8.205}$$

$$\xi = -\, m_2\left[\sin(\varphi_B - \varphi)\right]^k \quad (k = 1, 2, 3, \ldots)\,. \tag{5.8.206}$$

Here, the factors m_1 and m_2 denote parameters which follow from the adjustment of the flow rule to special test results, and φ_B represents the angle of internal friction in the failure state; the exponent k influences the shape of the curve of the volume change in the hardening range. For the special case of exclusive contracted volume changes due to hydrostatic pressure, the parameter ξ has to be dropped. In general, however, the sum of the terms of the right-hand side of (5.8.204) causes a sign change of the function μ with increasing hardening due to $\mathrm{I}_{\xi} < 0$, so that by the appropriate choice of the parameters m_1, m_2, and k the size of the region of the contractant volume changes can be adjusted to the test results (see the extensive discussions in de Boer and Brauns, 1990, as well as in de Boer and Lade, 1991).

Stress-Strain Relations

With the development of the constitutive relations for the elastic and plastic strain rates $D\mathbf{E}_{Se}$ (see Section 5.8.3) and $D\mathbf{E}_{Sp}$ of the solid skeleton, we are able to construct a corresponding relation for $D\mathbf{E}_S$ according to (5.8.155). For the elastic and plastic states, we have the constitutive equations (5.8.126)$_1$ and (5.8.202)

$$\begin{aligned} D\mathbf{E}_{Se} &= (\overset{4}{\mathbf{K}}{}^S)^{-1}\, D\mathbf{S}^S_E \,, \\ D\mathbf{E}_{Sp} &= D\lambda\big[\bar{\mathbf{S}}^D + \mu(\bar{\mathbf{S}}\cdot\mathbf{I})\mathbf{I}\big] \,. \end{aligned} \tag{5.8.207}$$

In (5.8.207), the fourth-order symmetric tensors $\overset{4}{\mathbf{K}}{}^S$ and $(\overset{4}{\mathbf{K}}{}^S)^{-1}$ are given by:

$$\begin{aligned} \overset{4}{\mathbf{K}}{}^S &= 2G^S\big(\overset{4}{\mathbf{I}} + \frac{\nu^S}{1-2\nu^S}\,\overset{4}{\bar{\mathbf{I}}}\big)\,, \quad 2G^S = \frac{E^S}{1+\nu^S}\,, \\ (\overset{4}{\mathbf{K}}{}^S)^{-1} &= \frac{1}{E^S}\big[(1+\nu^S)\,\overset{4}{\mathbf{I}} - \nu^S\,\overset{4}{\bar{\mathbf{I}}}\big]\,, \end{aligned} \tag{5.8.208}$$

where, in (5.8.126)$_1$, the Lamé constants have been replaced by Young's modulus E^S, Poison's ratio ν^S, and the shear modulus G^S, respectively. Moreover, $\overset{4}{\mathbf{I}}$ and $\overset{4}{\bar{\mathbf{I}}}$ denote fourth-order identity tensors (see de Boer, 1982).

The multiplier $D\lambda$ is obtained from the consistency condition. After some calculations, we have

$$D\lambda = \frac{\dfrac{\partial F}{\partial \bar{\mathbf{S}}}\cdot D\mathbf{S}^S_E}{\big[\bar{\mathbf{S}}^D + \mu(\bar{\mathbf{S}}\cdot\mathbf{I})\mathbf{I}\big]\cdot(\mathbf{M}-\mathbf{N})}\,. \tag{5.8.209}$$

Thus, considering (5.8.207) through (5.8.209), we arrive with (5.8.155) at

$$D\mathbf{E}_S = \Big\{(\overset{4}{\mathbf{K}}{}^S)^{-1} + \frac{\big[\bar{\mathbf{S}}^D + \mu(\bar{\mathbf{S}}\cdot\mathbf{I})\mathbf{I}\big]\otimes\dfrac{\partial F}{\partial \bar{\mathbf{S}}}}{\big[\bar{\mathbf{S}}^D + \mu(\bar{\mathbf{S}}\cdot\mathbf{I})\mathbf{I}\big]\cdot(\mathbf{M}-\mathbf{N})}\Big\}D\mathbf{S}^S_E\,. \tag{5.8.210}$$

With the hardening parameter

$$h = \big[\bar{\mathbf{S}}^D + \mu(\bar{\mathbf{S}}\cdot\mathbf{I})\mathbf{I}\big]\cdot(\mathbf{M}-\mathbf{N}) \tag{5.8.211}$$

and with

$$(\overset{4}{\mathbf{P}}{}^S)^{-1} = (\overset{4}{\mathbf{K}}{}^S)^{-1} + \frac{1}{h}\big[\mathbf{S}^D + \mu(\mathbf{S}\cdot\mathbf{I})\mathbf{I}\big]\otimes\frac{\partial F}{\partial \bar{\mathbf{S}}}\,, \tag{5.8.212}$$

we obtain, instead of (5.8.210),

$$D\mathbf{E}_S = (\overset{4}{\mathbf{P}}{}^S)^{-1}\,D\mathbf{S}^S_E\,. \tag{5.8.213}$$

In the next step, we are concerned with the inversion of (5.8.213). We commence with the constitutive relation for the rates of $\mathbf{E}_S$ (5.8.210). With the hardening parameter h (5.8.211), we can write

$$D\mathbf{E}_S = \Big\{(\overset{4}{\mathbf{K}}{}^S)^{-1} + \frac{1}{h}\big[\bar{\mathbf{S}}^D + \mu(\bar{\mathbf{S}}\cdot\mathbf{I})\mathbf{I}\big]\otimes\frac{\partial F}{\partial \bar{\mathbf{S}}}\Big\}D\mathbf{S}^S_E \tag{5.8.214}$$

or

$$D\mathbf{E}_S = (\overset{4}{\mathbf{K}}{}^S)^{-1}\,D\mathbf{S}^S_E + \frac{1}{h}\big[\bar{\mathbf{S}}^D + \mu(\bar{\mathbf{S}}\cdot\mathbf{I})\mathbf{I}\big]\Big(\frac{\partial F}{\partial \bar{\mathbf{S}}}\cdot D\mathbf{S}^S_E\Big)\,. \tag{5.8.215}$$

With $(5.8.207)_2$ and (5.8.211), we can express the loading condition (5.8.201) by the following relation:

$$\frac{1}{h}\frac{\partial F}{\partial \bar{\mathbf{S}}} \cdot D\mathbf{S}_E^{\overset{S}{}} = D\lambda \,. \tag{5.8.216}$$

Now, the stress-strain relation (5.8.215) will be multiplied by $\overset{4}{\mathbf{K}}{}^S \frac{\partial F}{\partial \bar{\mathbf{S}}}$ considering (5.8.216):

$$\begin{aligned} D\mathbf{E}_S \cdot \overset{4}{\mathbf{K}}{}^S \frac{\partial F}{\partial \bar{\mathbf{S}}} &= (\overset{4}{\mathbf{K}}{}^S)^{-1}\, D\mathbf{S}_E^S \cdot \overset{4}{\mathbf{K}}{}^S \frac{\partial F}{\partial \bar{\mathbf{S}}} + \\ &+ D\lambda\big[\bar{\mathbf{S}}^D + \mu(\bar{\mathbf{S}}\cdot\mathbf{I})\mathbf{I}\big]\cdot \overset{4}{\mathbf{K}}{}^S \frac{\partial F}{\partial \bar{\mathbf{S}}} \,. \end{aligned} \tag{5.8.217}$$

The first term on the right-hand side of (5.8.217) reduces, due to the symmetry of $\overset{4}{\mathbf{K}}_S$, to:

$$\begin{aligned} (\overset{4}{\mathbf{K}}{}^S)^{-1}\, D\mathbf{S}_E^S \cdot \overset{4}{\mathbf{K}}{}^S \frac{\partial F}{\partial \bar{\mathbf{S}}} &= \overset{4}{\mathbf{K}}{}^{S^T} (\overset{4}{\mathbf{K}}{}^S)^{-1}\, D\mathbf{S}_E^S \cdot \frac{\partial F}{\partial \bar{\mathbf{S}}} \\ &= \frac{\partial F}{\partial \bar{\mathbf{S}}} \cdot D\mathbf{S}_E^S \,. \end{aligned} \tag{5.8.218}$$

Thus, from (5.8.217), considering (5.8.218) and (5.8.216), we obtain another version of the multiplier $D\lambda$:

$$D\lambda = \frac{D\mathbf{E}_S \cdot \overset{4}{\mathbf{K}}{}^S \dfrac{\partial F}{\partial \bar{\mathbf{S}}}}{h + \overset{4}{\mathbf{K}}{}^S \cdot \big[\bar{\mathbf{S}}^D + \mu(\bar{\mathbf{S}}\cdot\mathbf{I})\mathbf{I}\big] \otimes \dfrac{\partial F}{\partial \bar{\mathbf{S}}}} \,. \tag{5.8.219}$$

Moreover, we again go back to Eqn. (5.8.215), considering, (5.8.216),

$$D\mathbf{E}_S = (\overset{4}{\mathbf{K}}{}^S)^{-1}\, D\mathbf{S}_E^S + D\lambda\big[\bar{\mathbf{S}}^D + \mu(\bar{\mathbf{S}}\cdot\mathbf{I})\mathbf{I}\big] \,, \tag{5.8.220}$$

applying the fourth-order elasticity tensor $\overset{4}{\mathbf{K}}{}^S$, and we get:

$$\overset{4}{\mathbf{K}}{}^S\, D\mathbf{E}_S = \overset{4}{\mathbf{K}}{}^S (\overset{4}{\mathbf{K}}{}^S)^{-1}\, D\mathbf{S}_E^S + \overset{4}{\mathbf{K}}{}^S\big[\bar{\mathbf{S}}^D + \mu(\bar{\mathbf{S}}\cdot\mathbf{I})\mathbf{I}\big]\, D\lambda \,. \tag{5.8.221}$$

Eqn. (5.8.221) will be rearranged and, with (5.8.219), leads to

$$D\mathbf{S}_E^S = \overset{4}{\mathbf{P}}{}^S\, D\mathbf{E}_S \tag{5.8.222}$$

with

$$\overset{4}{\mathbf{P}}{}^S = \overset{4}{\mathbf{K}}{}^S - \frac{\overset{4}{\mathbf{K}}{}^S\big[\bar{\mathbf{S}}^D + \mu(\bar{\mathbf{S}}\cdot\mathbf{I})\mathbf{I}\big] \otimes \overset{4}{\mathbf{K}}{}^S \dfrac{\partial F}{\partial \bar{\mathbf{S}}}}{h + \overset{4}{\mathbf{K}}{}^S \cdot \Big\{\big[\bar{\mathbf{S}}^D + \mu(\bar{\mathbf{S}}\cdot\mathbf{I})\mathbf{I}\big] \otimes \dfrac{\partial F}{\partial \bar{\mathbf{S}}}\Big\}} \,. \tag{5.8.223}$$

The "elastic-plastic tangent" (5.8.223) is important for numerical calculations of initial- and boundary-value problems. The realization will be shown in Section 5.9 as a special example.

Note in passing that for ideal-plastic behavior, e.g., in the critical state of soils, the hardening parameter h disappears. Since in this case, the loading criterion also is equal to zero, see (5.8.201), an undetermined term arises in the constitutive equation for $D\mathbf{E}_S$ (5.8.215). In the constitutive equations for $D\mathbf{S}_E^S$ (5.8.222), along with (5.8.223), however, this indefiniteness disappears.

Finally, it can be stated that the fourth-order tensor $(\overset{4}{\mathbf{P}}{}^S)^{-1}$ in (5.8.212) is indeed the inverse tensor to $\overset{4}{\mathbf{P}}$ in (5.8.223) and that

$$\overset{4}{\mathbf{P}}{}^S\,(\overset{4}{\mathbf{P}}{}^S)^{-1} = \overset{4}{\mathbf{I}} \tag{5.8.224}$$

is valid. This has already been proven by the author. The algebraic rearrangements are, however, lengthy and laborious. The proof will be published elsewhere.

5.8.6 Constitutive Relations and Transport Phenomena in Fluid-Saturated Rigid Porous Solids

In this section, constitutive equations for a binary model consisting of a rigid porous solid saturated with an incompressible viscous fluid will be discussed. Thus, the thermodynamic treatment will be performed with the following assumptions:

- The rigidity of the solid skeleton brings about the fact that the deformation gradient $\mathbf{F}_S$ must always be a proper orthogonal tensor, when the reference placement is suitably selected. Thus,

$$\mathbf{C}_S = \mathbf{F}_S^T \mathbf{F}_S = \mathbf{I}\,, \tag{5.8.225}$$

  and all derivatives of (5.8.225) are equal to zero;
- the fluid is incompressible and viscous;
- the fluid motion will be investigated within the Eulerian description;
- the thermodynamic process is non-isothermal;
- both constituents have the same temperature Θ and acceleration $\mathbf{b}$;
- mass exchange, as well as moment of momentum and energy exchanges, will be neglected.

As a consequence of these assumptions, the entropy inequality (5.8.27) for incompressible constituents reduces to

$$\begin{aligned}
&-\rho^S(\psi^S)'_S - \rho^F(\psi^F)'_F - \rho^S(\Theta)'_S\eta^S - \rho^F(\Theta)'_F\eta^F + \\
&+ (\mathbf{T}^F + \lambda n^F\mathbf{I})\cdot\mathbf{D}_F - (\lambda - \kappa^{FR})n^F(\mathbf{D}_{FR}\cdot\mathbf{I}) - \\
&- \frac{1}{\Theta}\mathbf{q}^S\cdot\operatorname{grad}\Theta - \frac{1}{\Theta}\mathbf{q}^F\cdot\operatorname{grad}\Theta - \\
&- (\hat{\mathbf{p}}^F - \lambda\operatorname{grad} n^F)\cdot\mathbf{v}_{FS} \;\geq\; 0\,.
\end{aligned} \tag{5.8.226}$$

Constitutive relations of the binary model under study have to be formulated for

$$R(s) = \{\psi^\alpha\,,\ \eta^\alpha\,,\ \mathbf{T}^F + \lambda n^F \mathbf{I}\,,\ \hat{\mathbf{p}}^F - \lambda\,\mathrm{grad}\, n^F\,,\ \mathbf{q}^\alpha\}\,. \qquad (5.8.227)$$

The set of process parameters is introduced as

$$s = \{\Theta\,,\ \mathrm{grad}\,\Theta\,,\ \mathbf{D}_F^D\,,\ \mathbf{v}_{FS} = \mathbf{v}_F - \mathbf{v}_S\}\,. \qquad (5.8.228)$$

The parameters Θ and $\mathrm{grad}\,\Theta$ govern the thermal process and the velocity difference $\mathbf{v}_{FS} = \mathbf{v}_F - \mathbf{v}_S$, as well as the deviator $\mathbf{D}_F^D$ dissipative effects, due to friction phenomena.

Only the consideration of the deviator of $\mathbf{D}_F$ needs further explanation. From the saturation constraint (5.2.4), it follows – with the assumption of rigidity of the solid skeleton – that the volume fraction of the porous solid does not change, i.e.:

$$n^F = 1 - n_{0S}^S\,, \qquad (5.8.229)$$

$$\mathrm{grad}\, n^F = -\,\mathrm{Grad}_S\, n_{0S}^S\,. \qquad (5.8.230)$$

The saturation constraint in the rate formulation (5.6.5) yields

$$n^F(\mathbf{D}_F \cdot \mathbf{I}) + \mathrm{grad}\, n^F \cdot \mathbf{v}_{FS} = 0\,. \qquad (5.8.231)$$

Thus, the spherical part of $\mathbf{D}_F$ is uniquely determined by the process variable $\mathbf{v}_{FS}$ and by the known quantities in the reference placement n_{0S}^S and $\mathrm{Grad}_S\, n_{0S}^S$. As a result, only the deviatoric part of $\mathbf{D}_F$ enters the set of process variables s.

For this model, thermodynamic restrictions result from the entropy inequality (5.8.226), together with the constitutive assumptions (5.8.227) and (5.8.228). Using standard arguments combined with several symmetry and skew-symmetry conditions (de Boer and Ehlers, 1986b), it should easily be recognized that a linear expansion for the so-called mixture equilibrium state,

$$s \to s_0 = \{\Theta,\ \mathrm{grad}\,\Theta = \mathbf{0},\ \mathbf{D}_F^D = \mathbf{0},\ \mathbf{v}_{FS} = \mathbf{0}\}\,, \qquad (5.8.232)$$

yields the model to be governed (see APPENDIX A, in particular, (A 91)) by:

$$\begin{aligned}
&\psi^\alpha = \psi^\alpha(\Theta)\,, \quad \eta^\alpha = -\frac{\partial\psi^\alpha}{\partial\Theta}\,, \quad \lambda = p\,,\\
&\hat{\mathbf{p}}^F = p\,\mathrm{grad}\, n^F - \beta_\Theta(\Theta)\,\mathrm{grad}\,\Theta - \beta_v^F(\Theta)\mathbf{v}_{FS}\,,\\
&\mathbf{q}^S + \mathbf{q}^F = -\,\alpha_\Theta(\Theta)\,\mathrm{grad}\,\Theta + \Theta\beta_\Theta(\Theta)\mathbf{v}_{FS}\,,\\
&\mathbf{T}^F = -\,n^F p\mathbf{I} + 2\mu^F(\Theta)\mathbf{D}_F^D\,,
\end{aligned} \qquad (5.8.233)$$

where $\beta_\Theta = \beta_S + \beta_F = -\,\Theta(\alpha_v^S + \alpha_v^F)$, see (A 90) and $\alpha_\Theta = \Theta^2(\alpha_S^S + \alpha_F^F)$ have been introduced for simplicity. In the literature, the coefficient β_v^F in $(5.8.233)_4$ is often denoted with S_v.

Relation $(5.8.233)_6$ denotes the constitutive equation of an incompressible viscous fluid.

In the next section, we shall focus on the heat transport and fluid motion (mass transport) in the above-stated model (rigid porous solid filled with an incompressible viscous fluid).

a)
Heat Conduction

This section, as well as the next section, is based on an already-mentioned report by de Boer and Ehlers (1986b). For the investigation of the heat conduction in a rigid solid skeleton saturated with an incompressible viscous fluid, we shall use the constitutive relations derived in this section. Thus, we must consider all the assumptions introduced in the preceding treatment; i.e., we investigate only such thermodynamic processes for which we can take for granted that, in the presence of an arbitrary temperature Θ, only small temperature gradients grad Θ, small deformation velocities $\mathbf{D}_F^D$, and small relative velocities $\mathbf{v}_{FS} = \mathbf{v}_F - \mathbf{v}_S$ occur.

The equation of heat conduction can be derived from the balance of energy by insertion of the corresponding constitutive relations. Considering (5.8.231) and substituting the internal energy ε by the free Helmholtz energy function (5.5.5) with (5.4.33), the balance of energy for the binary model under study is given by:

$$\begin{aligned}&-\rho^S(\psi^S)'_S-\rho^F(\psi^F)'_F-\rho^S\eta^S\Theta'_S-\rho^F\eta^F\Theta'_F-\\&-\rho^S\Theta(\eta^S)'_S-\rho^F\Theta(\eta^F)'_F+(\mathbf{T}^F+n^F p\mathbf{I})\cdot\mathbf{D}_F+\\&+\rho^S r^S+\rho^F r^F-\operatorname{div}(\mathbf{q}^S+\mathbf{q}^F)-\\&-(\hat{\mathbf{p}}^F-p\operatorname{grad}n^F)\cdot\mathbf{v}_{FS}=0\,.\end{aligned}\tag{5.8.234}$$

With (5.8.233) and (5.8.234), we have

$$\begin{aligned}&-\rho^S\frac{\partial\psi^S}{\partial\Theta}\Theta'_S-\rho^F\frac{\partial\psi^F}{\partial\Theta}\Theta'_F+\rho^S\frac{\partial\psi^S}{\partial\Theta}\Theta'_S+\rho^F\frac{\partial\psi^F}{\partial\Theta}\Theta'_F+\\&+\rho^S\Theta\frac{\partial^2\psi^S}{\partial\Theta^2}\Theta'_S+\rho^F\Theta\frac{\partial^2\psi^F}{\partial\Theta^2}\Theta'_F+2\mu^F\mathbf{D}_F^D\cdot\mathbf{D}_F^D+\\&+\rho^S r^S+\rho^F r^F+\operatorname{div}(\alpha_\Theta\operatorname{grad}\Theta-\Theta\beta_\Theta\mathbf{v}_{FS})+\\&+\beta_\Theta\operatorname{grad}\Theta\cdot\mathbf{v}_{FS}+\beta_{\upsilon}^F\mathbf{v}_{FS}\cdot\mathbf{v}_{FS}=0\,.\end{aligned}\tag{5.8.235}$$

With the assumption that the distribution of the volume fraction in the reference placement is homogeneous ($n_{0S}^S = \text{const.}$, $n_{0F}^F = \text{const.}$), and that the response parameter β_Θ is spatially independent, relation (5.8.235) reduces by use of (2.8.231) to:

$$\begin{aligned}&\rho^S\Theta\frac{\partial^2\psi^S}{\partial\Theta^2}\Theta'_S+\rho^F\Theta\frac{\partial^2\psi^F}{\partial\Theta^2}\Theta'_F+\rho^S r^S+\rho^F r^F+\\&+\operatorname{div}(\alpha_\Theta\operatorname{grad}\Theta)+\phi_D=0\,.\end{aligned}\tag{5.8.236}$$

The quantity

$$\phi_D=2\mu^F\mathbf{D}_F^D\cdot\mathbf{D}_F^D+\beta_{\upsilon}^F\mathbf{v}_{FS}\cdot\mathbf{v}_{FS}\;\geq\;0\tag{5.8.237}$$

denotes the dissipation function caused by the flow of the incompressible, viscous fluid. The condition $\phi_D \geq 0$ results from the positive response parameters μ^F and β_V^F.

In analogy to classical continuum mechanics of one-component continua, the notion specific heat for constant volume c_V^α can be introduced (see Section 5.8.4):

$$c_V^\alpha = \Theta \frac{\partial \eta^\alpha}{\partial \Theta} \,. \tag{5.8.238}$$

With (5.8.233) and (5.8.238), relation (5.8.236) simplifies to

$$-\rho^S c_V^S \Theta_S' - \rho^F c_V^F \Theta_F' + \rho^S r^S + \rho^F r^F + \operatorname{div}(\alpha_\Theta \operatorname{grad} \Theta) + \phi_D = 0 \,. \tag{5.8.239}$$

Equation (5.8.239) can be further reduced if we assume that the velocity of the rigid solid phase $\mathbf{v}_S$ is equal to zero. Then,

$$-\rho^S c_V^S \frac{\partial \Theta}{\partial t} - \rho^F c_V^F \Theta_F' + \rho^S r^S + \rho^F r^F + \operatorname{div}(\alpha_\Theta \operatorname{grad} \Theta) + \phi_D = 0 \,, \tag{5.8.240}$$

where ϕ_D simplifies to

$$\phi_D = 2\mu^F \mathbf{D}_F^D \cdot \mathbf{D}_F^D + \beta_V^F \mathbf{v}_F \cdot \mathbf{v}_F \,. \tag{5.8.241}$$

Furthermore, it may happen that the sum of the external heat supplies $\rho^S r^S + \rho^F r^F$ as well as the sum of the external heat flux $\mathbf{q}^S + \mathbf{q}^F$ vanish. However, the condition $\mathbf{q}^S + \mathbf{q}^F = \mathbf{0}$ is only possible with $\operatorname{grad} \Theta = \mathbf{0}$ (homogeneous distribution of the temperature), compare $(5.8.233)_5$. With these assumptions, (5.8.240) takes a very convenient form:

$$-(\rho^S c_V^S + \rho^F c_V^F) \frac{\partial \Theta}{\partial t} + \phi_D = 0 \,. \tag{5.8.242}$$

This relation shows that an increase in temperature is possible even if the external heat supply and flux are absent. It is recognized that the increase in temperature is caused by the flow of the viscous fluid.

Finally, we will compare the equation of heat conduction (5.8.239) with those of one-component continua – rigid solid and incompressible viscous fluid – in classical continuum mechanics.

In the case of a *rigid* solid ($n^S = 1$), we have

$$\rho^F c_V^F = 0 \,, \quad \rho^F r^F = 0 \,, \quad \alpha_\Theta = \alpha_\Theta^S \,. \tag{5.8.243}$$

Moreover,

$$\phi_D = 0 \tag{5.8.244}$$

and the interaction force $\hat{\mathbf{p}}^F$ is identical to zero. With (5.8.243) and (5.8.244), we obtain the equation of heat conduction in a rigid solid:

$$-\rho^S c_V^S \Theta_S' + \operatorname{div}(\alpha_\Theta^S \operatorname{grad} \Theta) + \rho^S r^S = 0 \,, \tag{5.8.245}$$

see, e.g., Parkus (1976).

In the case of an *incompressible, viscous fluid*,

$$\rho^S c_V^S = 0 \,, \quad \rho^S r^S = 0 \,, \quad \text{and} \quad \alpha_\Theta = \alpha_\Theta^F \tag{5.8.246}$$

are valid and the interaction force $\hat{\mathbf{p}}^F$ is identically zero. The dissipation function simplifies to

$$\phi_D = 2\mu^F \mathbf{D}_F^D \cdot \mathbf{D}_F^D \, . \tag{5.8.247}$$

With (5.8.246) and (5.8.247), we obtain the equation of heat conduction for an incompressible, viscous fluid

$$-\rho^F c_V^F \Theta_F' + \operatorname{div}(\alpha_\Theta^F \operatorname{grad}\Theta) + 2\mu^F \mathbf{D}_F^D \cdot \mathbf{D}_F^D + \rho^F r^F = 0 \, . \tag{5.8.248}$$

This result can be proven by comparison with the corresponding considerations of either Serrin (1959) or Truckenbrodt (1968).

b)
Motion of an Incompressible, Viscous Fluid

The motion of a fluid in a rigid, porous solid can be described with the equation of motion for the fluid (balance of momentum), see (5.4.15) and the constitutive equations developed in this section. The equation of motion (5.4.15) for the partial fluid body reads

$$\operatorname{div}\mathbf{T}^F + \rho^F\left[\mathbf{b} - (\mathbf{v}_F)_F'\right] + \hat{\mathbf{p}}^F = \mathbf{0} \, , \tag{5.8.249}$$

wherein the constitutive equations from $(5.8.233)_6$ and $(5.8.233)_4$,

$$\begin{aligned} \mathbf{T}^F &= -\, n^F p \mathbf{I} + 2\mu^F(\Theta)\mathbf{D}_F^D \, , \\ \hat{\mathbf{p}}^F &= p \operatorname{grad} n^F - \beta_\Theta(\Theta)\operatorname{grad}\Theta - \beta_V^F(\Theta)\mathbf{v}_{FS} \, , \end{aligned} \tag{5.8.250}$$

have to be considered.

With the constitutive equations (5.8.250), the description of the fluid motion is limited - like the description of the heat conduction - to those motions of an incompressible, viscous fluid for which, with arbitrary temperature Θ, small temperature gradients $\operatorname{grad}\Theta$, small velocity gradients $\mathbf{D}_F^D$, and small velocity differences, $\mathbf{v}_{FS} = \mathbf{v}_F - \mathbf{v}_S$ are assumed. Since, in (5.8.249) and (5.8.250), no turbulences are considered, the performance of the fluid motion remains limited to laminar flow with a small Reynolds-number.

For the next step, we determine the unknown porefluid pressure p from the equation of motion for the partial fluid body (5.8.249), neglecting the inertia force and the effective quantities in (5.8.250). With the assumption that the external acceleration $\mathbf{b}$ is equal to the gravity field $\mathbf{g}$, which can be performed by the gradient of a potential U,

$$\begin{aligned} \mathbf{g} &= -\operatorname{grad} U \, , \\ U &= -\,\mathbf{g}\cdot\mathbf{x} \, , \end{aligned} \tag{5.8.251}$$

where $\mathbf{x}$ is the position vector for a place x in the porefluid from an arbitrary stationary reference point beneath the fluid level. The scalar product in (5.8.251) yields

$$U = gx_3 , \quad x_3 = \frac{U}{g} , \tag{5.8.252}$$

where g is the norm of $\mathbf{g}$ and x_3 is the coordinate connected with the unit vector $\mathbf{g}/g$. From the equation of motion for the partial flow (5.8.249) and the constitutive equations (5.8.250), we obtain, neglecting the inertia force and the effective quantities as stated above,

$$\operatorname{grad} p = \rho^{FR}\mathbf{g} \quad \text{or} \quad p = -\rho^{FR} g x_3 + C , \tag{5.8.253}$$

where C is a constant which can be determined from the boundary condition and where ρ^{FR} is assumed to be a spatial constant:

$$x_3 = h : \quad p(x_3 = h) = 0 . \tag{5.8.254}$$

We then have

$$p = \gamma^{FR}(h - x_3) , \quad \gamma^{FR} = \rho^{FR} g , \tag{5.8.255}$$

where γ^{FR} is the real specific weight of the fluid. From $(5.8.255)_1$,

$$h = \frac{p}{\gamma^{FR}} + x_3 , \tag{5.8.256}$$

or considering $(5.8.255)_2$ and $(5.8.252)_2$,

$$h = \frac{p}{\gamma^{FR}} + \frac{U}{g} \tag{5.8.257}$$

is gained. The height h is known as the *pressure head.* With (5.8.257), the pressure gradient can be expressed by the pressure head h:

$$\begin{aligned} \operatorname{grad} p &= \gamma^{FR} \operatorname{grad}\Big(h - \frac{U}{g}\Big), \\ \operatorname{grad} p &= \gamma^{FR} \operatorname{grad} h + \rho^{FR}\mathbf{g}, \\ \operatorname{grad} p &= -\gamma^{FR}\mathbf{i} + \rho^{FR}\mathbf{g} \end{aligned} \tag{5.8.258}$$

with $\mathbf{i}$ equal to $\operatorname{grad} h$ as the gradient of the pressure head.

After these preliminary investigations, the motion of an incompressible, viscous fluid will be described. We proceed from the equation of motion (5.8.249) with the constitutive equations $(5.8.250)_1$. At first, we calculate the divergence of the stress tensor $\mathbf{T}^F$:

$$\begin{aligned} \operatorname{div} \mathbf{T}^F = &- p \operatorname{grad} n^F - n^F \operatorname{grad} p + 2\mu^F \operatorname{div} \mathbf{D}_F^D + \\ &+ 2\frac{\partial \mu^F}{\partial \Theta}\mathbf{D}_F^D \operatorname{grad} \Theta . \end{aligned} \tag{5.8.259}$$

Furthermore,

$$\mathbf{D}_F^D = \frac{1}{2}(\operatorname{grad} \mathbf{v}_F + \operatorname{grad}^T \mathbf{v}_F) - \frac{1}{3}(\operatorname{div} \mathbf{v}_F)\mathbf{I} \tag{5.8.260}$$

is valid, so that

$$\operatorname{div} \mathbf{D}_F^D = \frac{1}{2} \,\triangle\, \mathbf{v}_F + \frac{1}{6} \operatorname{grad} \operatorname{div} \mathbf{v}_F . \tag{5.8.261}$$

The symbol

$$\Delta(\ldots) = \operatorname{div}\operatorname{grad}(\ldots) \tag{5.8.262}$$

is the Laplace-operator. With (5.8.260) and (5.8.261), we obtain, instead of (5.8.259),

$$\begin{aligned} \operatorname{div}\mathbf{T}^F = &- p\operatorname{grad} n^F - n^F \operatorname{grad} p + \mu^F(\Delta \mathbf{v}_F + \frac{1}{3}\operatorname{grad}\operatorname{div}\mathbf{v}_F) + \\ &+ \frac{\partial \mu^F}{\partial \Theta}\left[\operatorname{grad}\mathbf{v}_F + \operatorname{grad}^T\mathbf{v}_F - \frac{2}{3}(\operatorname{div}\mathbf{v}_F)\mathbf{I}\right]\operatorname{grad}\Theta\,. \end{aligned} \tag{5.8.263}$$

Considering (5.8.263) and $(5.8.250)_2$, the equation of motion (5.8.249) yields

$$- n^F \operatorname{grad} p + \rho^F\left[\mathbf{b} - (\mathbf{v}_F)'_F\right] - \beta^F_{\upsilon}(\mathbf{v}_F - \mathbf{v}_S) + \mathbf{z}^F = \mathbf{0}\,. \tag{5.8.264}$$

The quantity

$$\begin{aligned} \mathbf{z}^F = &\,\mu^F(\Delta \mathbf{v}_F + \frac{1}{3}\operatorname{grad}\operatorname{div}\mathbf{v}_F) + \\ &+ \frac{\partial \mu^F}{\partial \Theta}\left[\operatorname{grad}\mathbf{v}_F + \operatorname{grad}^T\mathbf{v}_F - \frac{2}{3}(\operatorname{div}\mathbf{v}_F)\mathbf{I}\right]\operatorname{grad}\Theta \end{aligned} \tag{5.8.265}$$

represents the viscous force. The rearrangement of (5.8.264), together with (5.8.265), in porous media theory represents a pendant to the Navier-Stokes equation of one-component continua:

$$\rho^F(\mathbf{v}_F)'_F = \mathbf{z}^F - n^F \operatorname{grad} p + \rho^F\mathbf{b} - \beta^F_{\upsilon}(\mathbf{v}_F - \mathbf{v}_S)\,. \tag{5.8.266}$$

The general relation (5.8.266) can be simplified in many cases.

For isothermal processes,

$$\rho^F(\mathbf{v}_F)'_F = \mathbf{z}^F - n^F \operatorname{grad} p + \rho^F\mathbf{b} - \beta^F_{\upsilon}(\mathbf{v}_F - \mathbf{v}_S)\,, \tag{5.8.267}$$

with

$$\mathbf{z}^F = \mu^F(\Delta \mathbf{v}_F + \frac{1}{3}\operatorname{grad}\operatorname{div}\mathbf{v}_F)\,, \tag{5.8.268}$$

is valid. Moreover, if the mass distribution of the solid is homogeneous, then $\operatorname{div}\mathbf{v}_F = 0$, see (5.8.231). The viscous force $\mathbf{z}_F$ then reduces to

$$\mathbf{z}^F = \mu^F \Delta \mathbf{v}_F\,. \tag{5.8.269}$$

In the limit for $n^F = 1$, the binary medium under study turns into a one-component model. Since, in this case, $\beta^F_{\upsilon} = 0$, from (5.8.267) we obtain the Navier-Stokes equation for an incompressible fluid:

$$\rho^F(\mathbf{v}_F)'_F = \mu^F \Delta \mathbf{v}_F - \operatorname{grad} p + \rho^F\mathbf{b}\,. \tag{5.8.270}$$

There are other special cases for the viscous force possible if the temperature gradient is neglected. With the calculation rule (see de Boer, 1982),

$$\Delta \mathbf{v}_F = \operatorname{grad}\operatorname{div}\mathbf{v}_F - \operatorname{rot}\operatorname{rot}\mathbf{v}_F\,, \tag{5.8.271}$$

we gain an alternative form of (5.8.268)

$$\mathbf{z}^F = \mu^F(\frac{4}{3}\operatorname{grad}\operatorname{div}\mathbf{v}_F - \operatorname{rot}\operatorname{rot}\mathbf{v}_F)\,. \tag{5.8.272}$$

For an incompressible, viscous fluid within a homogeneous rigid skeleton ($\operatorname{div} \mathbf{v}_F = 0$), the viscous force reduces to

$$\mathbf{z}^F = -\,\mu^F \operatorname{rot} \operatorname{rot} \mathbf{v}_F \,. \tag{5.8.273}$$

If we have an irrotational flow, so that the rotation vector

$$\boldsymbol{\omega}_F = \frac{1}{2} \operatorname{rot} \mathbf{v}_F \tag{5.8.274}$$

disappears; then, from (5.8.273), it is recognized that

$$\mathbf{z}^F = \mathbf{0} \,. \tag{5.8.275}$$

However, that part of the stress tensor $\mathbf{T}^F$, which is caused by the viscosity, remains. From (5.8.250), we can read that the viscous force for the general motion of an incompressible, viscous fluid always occurs. It is only identical to zero if $\mu^F = 0$. Then, also in the constitutive equation for $\mathbf{T}^F$, the part caused by the viscosity vanishes.

Next, we consider only such flows where we can assure

$$\operatorname{div} \mathbf{v}_F = 0, \quad \operatorname{grad} \Theta = \mathbf{0} \,. \tag{5.8.276}$$

Then, from (5.8.267), in connection with the viscous force (5.8.273), we obtain the following equation of motion:

$$\rho^F (\mathbf{v}_F)'_F = -\,\mu^F \operatorname{rot} \operatorname{rot} \mathbf{v}_F - n^F \operatorname{grad} p + \rho^F \mathbf{b} - \beta^F_{\upsilon} (\mathbf{v}_F - \mathbf{v}_S) \,. \tag{5.8.277}$$

If we assume that the external acceleration $\mathbf{b}$ can be expressed by the gradient of a potential U,

$$\mathbf{b} = -\operatorname{grad} \mathrm{U} \,, \tag{5.8.278}$$

and neglecting the translation velocity $\mathbf{v}_S$ of the rigid skeleton; then, instead of (5.8.277), we gain

$$(\mathbf{v}_F)'_F = -\,\nu^F \operatorname{rot} \operatorname{rot} \mathbf{v}_F - \frac{1}{\rho^{FR}} \operatorname{grad} p - \operatorname{grad} U - \frac{\beta^F_{\upsilon}}{\rho^F} \mathbf{v}_F, \tag{5.8.279}$$

where the kinematic viscosity

$$\nu^F = \frac{\mu^F}{\rho^F} \tag{5.8.280}$$

has been introduced.

The acceleration $(\mathbf{v}_F)'_F$ can be represented, after d'Alembert-Euler, in the form

$$(\mathbf{v}_F)'_F = \frac{\partial \mathbf{v}_F}{\partial t} + (\operatorname{grad} \mathbf{v}_F) \mathbf{v}_F, \tag{5.8.281}$$

whereby the second term on the right-hand side can be further rearranged as

$$(\mathbf{v}_F)'_F = \frac{\partial \mathbf{v}_F}{\partial t} + \frac{1}{2} \operatorname{grad} (\mathbf{v}_F \cdot \mathbf{v}_F) - \mathbf{v}_F \times \operatorname{rot} \mathbf{v}_F \,. \tag{5.8.282}$$

With (5.8.282), the equation of motion (5.8.279) results in

$$\frac{\partial \mathbf{v}_F}{\partial t} + \frac{1}{2}\operatorname{grad}(\mathbf{v}_F \cdot \mathbf{v}_F) - \mathbf{v}_F \times \operatorname{rot} \mathbf{v}_F + \\ + \nu^F \operatorname{rot} \operatorname{rot} \mathbf{v}_F + \frac{\beta_\upsilon^F}{\rho^F}\mathbf{v}_F = -\frac{1}{\rho^{FR}} \operatorname{grad} p - \operatorname{grad} U \, . \tag{5.8.283}$$

This relation can be essentially simplified, if one applies the rotation operator to (5.8.283). In addition, a homogeneous fluid ($\operatorname{grad} \rho^{FR} = 0$) is assumed and one is aware that the response parameters ν^F and β_υ^F are only functions of the temperature Θ. Then, we have

$$\frac{\partial \operatorname{rot} \mathbf{v}_F}{\partial t} - \operatorname{rot} (\mathbf{v}_F \times \operatorname{rot} \mathbf{v}_F) + \nu^F \operatorname{rot} \operatorname{rot} \operatorname{rot} \mathbf{v}_F + \\ + \frac{\beta_\upsilon^F}{\rho^F} \operatorname{rot} \mathbf{v}_F = \mathbf{0} \, . \tag{5.8.284}$$

With (5.8.271) and (5.8.274), we can alternatively write

$$\frac{\partial \boldsymbol{\omega}_F}{\partial t} - \operatorname{rot} (\mathbf{v}_F \times \boldsymbol{\omega}_F) - \nu^F \Delta \boldsymbol{\omega}_F + \frac{\beta_\upsilon^F}{\rho^F} \boldsymbol{\omega}_F = \mathbf{0} \, . \tag{5.8.285}$$

We immediately recognize that every irrotational flow ($\boldsymbol{\omega}_F = \mathbf{0}$) is a solution of the equation of motion.

The irrotational flow can be treated with the introduction of a velocity potential W^F. If we assume that

$$\mathbf{v}_F = \operatorname{grad} W^F, \tag{5.8.286}$$

the identity

$$\operatorname{rot} \operatorname{grad} W^F = \mathbf{0} \tag{5.8.287}$$

requires irrotational flow. For an irrotational incompressible flow, Eqn. (5.8.283) reduces to

$$\frac{\partial \mathbf{v}_F}{\partial t} + \frac{1}{2}\operatorname{grad}(\mathbf{v}_F \cdot \mathbf{v}_F) + \frac{\beta_\upsilon^F}{\rho^F}\mathbf{v}_F = \\ = -\frac{1}{\rho^{FR}} \operatorname{grad} p - \operatorname{grad} U. \tag{5.8.288}$$

Alternatively, with (5.8.286),

$$\operatorname{grad} \frac{\partial W^F}{\partial t} + \frac{1}{2}\operatorname{grad}(\mathbf{v}_F \cdot \mathbf{v}_F) + \frac{\beta_\upsilon^F}{\rho^F} \operatorname{grad} W^F + \\ + \frac{1}{\rho^{FR}} \operatorname{grad} p + \operatorname{grad} U = \mathbf{0} \tag{5.8.289}$$

is valid. With (5.2.11) and (5.8.230), we have, for a homogeneous fluid within a homogeneous rigid skeleton,

$$\operatorname{grad} \rho^{FR} = \operatorname{grad} \rho^F = \mathbf{0}, \quad \operatorname{grad} \beta_\upsilon^F = \mathbf{0} \, . \tag{5.8.290}$$

Therefore, the relation (5.8.289) can be transferred to

$$\operatorname{grad}\Big(\frac{\partial W^F}{\partial t}+\frac{1}{2}\mathbf{v}_F\cdot\mathbf{v}_F+\frac{p}{\rho^{FR}}+U+\frac{\beta_{\upsilon}^F}{\rho^F}W^F\Big)=\mathbf{0}\,. \tag{5.8.291}$$

This states that, for irrotational fluid motions, the term

$$\frac{\partial W^F}{\partial t}+\frac{1}{2}\mathbf{v}_F\cdot\mathbf{v}_F+\frac{p}{\rho^{FR}}+U+\frac{\beta_{\upsilon}^F}{\rho^F}W^F=C(t) \tag{5.8.292}$$

does not change provided that the time is fixed. Thus, for a stationary flow, the expression

$$\frac{1}{2}\mathbf{v}_F\cdot\mathbf{v}_F+\frac{p}{\rho^{FR}}+U+\frac{\beta_{\upsilon}^F}{\rho^F}W^F=C \tag{5.8.293}$$

becomes the simplest form of the solution of the equation of motion (5.8.288).

The above-mentioned relation (5.8.293) has been derived at a homogeneous temperature distribution for the irrotational incompressible flow within a rigid solid skeleton with homogeneous porosity, so that the viscous force $\mathbf{z}^F$ equals zero. With n^F equal to unity and β_{υ}^F equal to zero, Eqn. (5.8.293) corresponds to the Bernoulli equation for an incompressible inviscid flow of a one-component fluid.

The homogeneous distribution of the solid mass causes $\operatorname{div}\mathbf{v}_F=0$ so that, for the velocity potential W^F, in stationary and instationary flow processes,

$$\operatorname{div}\mathbf{v}_F=\operatorname{div}\operatorname{grad}W^F \tag{5.8.294}$$

or

$$\Delta W^F=0 \tag{5.8.295}$$

is valid. This is the Laplace differential equation for the velocity potential W^F. In many cases, solutions for (5.8.295) are well-known (see Truckenbrodt, 1968).

In the following paragraph, we will treat those fluid motions where the unknown hydrostatic pressure p can be expressed with the help of the so-called *pressure head*, see (5.8.257). Then, with (5.8.257),

$$\operatorname{grad}p=g\rho^{FR}\operatorname{grad}h+\rho^{FR}\mathbf{g} \tag{5.8.296}$$

is valid, see $(5.8.258)_2$.

Now, we will focus on those motions of incompressible, viscous fluids for which

$$\operatorname{grad}\Theta=\mathbf{0}\,,\quad \operatorname{div}\mathbf{v}_F=0\,,\quad \mathbf{v}_S=\mathbf{0}\,. \tag{5.8.297}$$

Then, the equation of motion (5.8.266), considering (5.2.11) and (5.8.296), reads as:

$$\rho^F(\mathbf{v}_F)'_F=\mathbf{z}^F-g\rho^F\operatorname{grad}h-\beta_{\upsilon}^F\mathbf{v}_F\,. \tag{5.8.298}$$

In (5.8.298), the viscous force is determined by

$$\mathbf{z}^F=-\mu^F\operatorname{rot}\operatorname{rot}\mathbf{v}_F\,. \tag{5.8.299}$$

If one introduces the coefficient of permeability k^F,

$$k^F=:\frac{g\rho^F}{\beta_{\upsilon}^F}\,, \tag{5.8.300}$$

Eqn. (5.8.298) turns into:

$$\rho^F (\mathbf{v}_F)'_F = -\,\mu^F \text{ rot rot } \mathbf{v}_F - k^F \beta^F_\upsilon \operatorname{grad} h - \beta^F_\upsilon \mathbf{v}_F \,. \tag{5.8.301}$$

The coefficient of permeability k^F as well as the response parameter β^F_υ are functions of the absolute temperature Θ:

$$k^F = k^F(\Theta)\,, \quad \beta^F_\upsilon = \beta^F_\upsilon(\Theta)\,. \tag{5.8.302}$$

The coefficient of permeability has to be determined from experiment and depends on the properties of the porous solid (porosity and pore structure), as well as on the properties of the transporting fluid (viscosity).

For the irrotational fluid motion (rot $\mathbf{v}_F = \mathbf{0}$), it follows from (5.8.301) that

$$(\mathbf{v}_F)'_F + \frac{g}{k^F}\mathbf{v}_F = -\,g \operatorname{grad} h \tag{5.8.303}$$

or, alternatively with (5.8.282),

$$\frac{\partial \mathbf{v}_F}{\partial t} + \frac{1}{2}\operatorname{grad}(\mathbf{v}_F \cdot \mathbf{v}_F) + \frac{g}{k^F}\mathbf{v}_F = -\,g \operatorname{grad} h\,. \tag{5.8.304}$$

If the flow is stationary, the above relation reduces to

$$\frac{k^F}{2g}\operatorname{grad}(\mathbf{v}_F \cdot \mathbf{v}_F) + \mathbf{v}_F = -\,k^F \operatorname{grad} h\,, \tag{5.8.305}$$

and if, in addition, the velocity field is homogeneous, i.e.,

$$\operatorname{grad} \mathbf{v}_F = \mathbf{0}\,, \tag{5.8.306}$$

then (5.8.305) simplifies to the well-known Darcy's law of the filter flow:

$$\mathbf{v}_F = -\,k^F \operatorname{grad} h \quad \text{or} \quad \mathbf{v}_F = k^F \mathbf{i}\,. \tag{5.8.307}$$

The assumption of a stationary flow with a homogeneous velocity field is often replaced, in the literature, by the assumption that the inertia force $\rho^F (\mathbf{v}_F)'_F$ can be neglected. It is obvious that both assumptions are identical.

For the irrotational fluid motion, the equations of motion (5.8.303) and the following equations can be represented by the velocity potential W^F (5.8.286) which, with the assumption div $\mathbf{v}_F = 0$, has to fulfill the Laplace differential equation (5.8.295).

If the coefficients of permeability are spatially constant, with (5.8.286), instead of (5.8.305),

$$\operatorname{grad} W^F = -\operatorname{grad}(k^F h) \tag{5.8.308}$$

is valid. In this case, the velocity potential corresponds to the negative pressure head h multiplied by k^F, plus a constant C:

$$W^F = -\,k^F h + C\,. \tag{5.8.309}$$

With (5.8.309), the Laplace differential equation (5.8.295) for the velocity potential W^F can be transferred into a differential equation for the pressure head h, i.e.,

$$\Delta h = 0\,. \tag{5.8.310}$$

Finally, some conclusions will be drawn concerning the flow of water in isotropic porous media. It is well-known that the viscosity of water is much less than that of other viscous fluids. This means the influence of the second term of the partial stress $\mathbf{T}^F$ (5.8.233)$_6$, caused by viscosity, can be neglected for many problems, e.g., for the filter flow of water in soils. With vanishing viscosity, the term $\beta_V^F(\mathbf{v}_F - \mathbf{v}_S)$ in the constitutive equation for the interaction force $\hat{\mathbf{p}}^F$ decreases, i.e., the response parameter becomes smaller according to (5.8.300) with the simultaneously increasing coefficient of permeability k^F. Moreover, neglecting the shear viscosity parameter μ^F, the viscous force $\mathbf{z}^F$ disappears, so that the general equation of the motion (5.8.256) of an incompressible fluid within a rigid skeleton and other derived equations simplify considerably.

c) Further Transport Phenomena: Diffusion, Capillarity, Filtration, and Motion of Moisture

All motion processes of liquids and gases in rigid porous solids are based on the equation of motion (5.8.249) and the appropriate response function for the interaction force $\hat{\mathbf{p}}^F$. For the *diffusion* process, it has been proven by de Boer and Ehlers (1986a) that the starting point for the derivation of Fick's first and second law is a constitutive relation for the interaction $\hat{\mathbf{p}}^F$ which corresponds to the constitutive equation for the effective interaction force in (5.8.233)$_4$ with, however, different response parameters. Fick's first diffusion law for an ideal gas results in the statement that the difference velocity of the gas phase and the solid phase are proportional to the gradient of the gas pressure, whereby the proportionality constant can be expressed by the gas constant, the absolute temperature, the density of the gas, and an additional positive response parameter.

Fick's second diffusion law (see Section 3.3.2) results from his first diffusion law and the mass balance equation for the free gas constituent, assuming the response parameters in Fick's first law to be a constant and the velocity of the skeleton to be identically zero. From the mathematical point of view, the diffusion problem is not substantially different from the flow of liquid in porous solids, discussed in the preceding Section (b), as both motions are based on the same fundamentals. It differentiates from the liquid in so far as the flowing substance which is considered here, the gas, is compressible.

Capillarity is a well-known phenomenon in physics and engineering. It denotes the behavior of liquids in narrow tubes, cracks, and pores caused by the surface tension of the liquid. If a narrow tube contrary to the direction of gravity is plunged into water, for example, water will rise in the tube. This effect does not occur for non-moistening liquids like mercury. Capillarity is caused by the intermolecular forces of cohesion and adhesion. If the forces of adhesion between the liquid and the tube wall are greater than the forces of cohesion between the molecules of the liquid, then it leads to a raise in the liquid.

From this discussion it is evident that the capillary force is an interaction force, which must enter the equation of motion. As has already been pointed out in Section 4.1, Kozeny (1927) seems to have been the first scientist who investigated the dynamic behavior of a water column in a porous solid. However, his attempt is far away from a continuum-mechanical approach. Such an approach obviously has not been performed in the past. Indeed, some difficulties arise in considering the capillary force in the local balance equation of momentum (5.8.249), where interaction effects are extensively described by the volume force $\hat{\mathbf{p}}^F$. One interaction effect, the friction phenomenon, could already be successfully considered in the equation of motion. The incorporation of the friction force, acting between the fluid and solid phases, was performed by Fillunger (1936), who introduced the friction force as a volume force in the direction of the relative velocity between the fluid and the solid constituents. First approaches made to incorporate the capillary forces into the local equations of motions are encouraging. It seems that it is possible to simulate the capillary tension as a volume force acting in the direction of the relative velocity $\mathbf{v}_F - \mathbf{v}_S$.

The problem relating to the description of the capillary phenomena in a rigid porous solid within the macroscopic porous media theory is currently under intensive study at Essen University in Germany.

The effect of *filtration* is closely connected with the effect of capillarity. However, filtration is the capillarity effect in the direction of gravity. Thus, the motion of liquid in the skeleton is governed by the capillary effect and the gravity in the direction of the gravity. Problems arising for filtration are, in parts, other than those for capillarity, which will be clarified for nonviscous and viscous liquids as well as for polluted liquids in the near future. The filtration of liquids is of great importance in environmental engineering.

A problem in building physics is the *motion of moisture* in capillary-porous solids, which can lead to considerable damages in brick and concrete walls. Although the motion of moisture has been under intensive study in building physics for a long time, the results are not always satisfactory because, in many cases, the principles of porous media theory have not been considered. For the description of the transport of moisture, a three-phase model is necessary for the simulation of the solid, liquid, and gas phases. Moreover, the notion of moisture has to be defined, which is not unique in the literature. In order to derive the mathematical description of the problem, one has to start from the balance equations of mass for the partial fluid and gas constituents, the balance equations of momentum for the porous solid, fluid, and gas phases, and appropriate constitutive equations for the interaction forces $\hat{\mathbf{p}}^F$ and $\hat{\mathbf{p}}^G$, which have to include terms for the description of heat, friction, and capillarity effects. Then, by eliminating the velocities of the fluid and the gas phases and neglecting the inertia forces, one arrives at a time-dependent differential equation for the moisture, which is similar to a known equation in building physics; see, e.g., Garrecht (1992). However, the differential equation in Garrecht (1992), and other approaches in building physics, are developed intuitively, not based on the principles of continuum mechanics and thermodynamics. Therefore, the identification of various terms in the relations of Garrecht (1992) is cumbersome. Currently, the whole problem is being studied in order to clarify the basic assumptions, the definitions, and the notions as well as the formulation of the needed principles. For example, when heat is involved, the balance of energy must be incorporated into the discussion of solving the moisture problem. This problem, and the application to practical problems, is also under study at Essen University in Germany. The results of these studies will be published elsewhere.

5.9 Applications

In the past, there have been many attempts to apply existing porous media theories to initial- and boundary-value problems. This has been done by Heinrich (1938b) and Heinrich and Desoyer (1955, 1956, 1961) in order to describe the one- and three-dimensional consolidation processes, the ground-water flow, and the layering of the ground-water level caused by pumping water out of a well, using the theory which they themselves developed.

Frenkel (1944) applied his theory to wave propagation and seismoelectric effects in a moist soil. However, most of the authors have used Biot's theory for describing the mechanical behavior of fluid-filled porous solids as a basis for solving many practical problems, for example, in engineering, geomechanics, and biomechanics. This theory is approximately valid within the framework of the geometrically- and physically-linear theory (see Section 4.7.1). This means that the initial- and boundary-value problems can only undergo small deformations within the range of linear-elastic (constitutive equation of Hookean type) behavior. The attempts on the part of, for example, Biot (1956a through d), Abousleiman *et al.* (1993), and Rice and Cleary (1976), as well as Katsube (1985) or Carroll and Katsube (1983), to reformulate the original version of Biot's theory in order to be able to treat the viscoelastic behavior of the solid-fluid body, and to obtain some stress diffusion solutions, as well as to include micromechanical considerations into Biot's theory, are not very convincing. Therefore, most of the practical problems are investigated with Biot's original poroelasticity theory or with parts of this theory. The solutions of initial- and boundary-value problems are referred, e.g., to uniaxial strain problems, the cylinder problem, the borehole problem, early time evolution of stress near a permeable boundary, and hydraulic fracture (see the extensive review article of Detournay and Cheng, 1993), as well as to dynamic problems (see, e.g., the various articles of Deresiewicz, 1961, 1962, Dziecielak, 1986, Hajra and Mukhopadhyay, 1982 and Bourbié *et al.*, 1987, as well as the discussion by de Boer *et al.*, 1993).

Many papers have already been published in the field of numerical calculations of deformable porous media. An intensive discussion of this topic can be found in the dissertation of Lund (1995), who discussed the numerical treatments of initial- and boundary-value problems during the last 30 years. A more extensive review was given by Lewis and Schreffler (1987). Most of the numerical investigations in the past are based on Biot's theory (see, for example,

Zienkiewicz and Shiomi, 1984). There are, however, contributions which used elements of the mixture theory. In this context, the treatise of Prévost (1981) should be mentioned. It seems that the first numerical applications based on the modern porous media theory were performed by Bluhm and Lund (1993) as well as by Ehlers and Diebels (1995) and Diebels and Ehlers (1996).

Whereas Biot's theory is not developed from the fundamentals of mechanics and thermodynamics and therefore limited in its applicability to various problems which occur, for example, in engineering, geomechanics, and biomechanics, the macroscopic porous media theory (elements of the mixture theory and volume fraction concept) is both thermodynamically and physically well-founded. This theory can consistently describe such different problems as fundamental phenomena in saturated porous solids, stress, and deformation problems in the elastic and elastic-plastic range of the porous solid, dynamic problems, transport phenomena (heat and fluid flow) and phase, as well as energy transitions. However, the number of practical problems which have been investigated within porous media theory seems to be less than the number of those which have been treated with Biot's theory. This is due to the fact that the fundamentals of Biot's theory were already developed in the 1940s and 1950s, whereas the porous media theory has been essentially established in the 1980s and 1990s. Considering the relatively short time, approximately twenty years, there are already a large number of applications that have been worked out within the framework of the porous media theory. For example, the fundamental effects of uplift, friction, and capillarity have been clarified by de Boer and Ehlers (1990a); the consolidation problem within the elastic and plastic ranges has been computed and solved with the finite element method (see Breuer, 1999a, Ehlers and Volk, 1997). In Ehlers and Volk (1997), the kinematics of the partial solid continuum has been extended by the Cosserat-kinematics in order to describe the possible rotation of grains in a granular body.

The porous media theory has also been successfully applied to avalanche problems (by Wu and Hutter, 1999), to drying processes (by Kowalski, 1987, 1990), and to transport phenomena of heat and fluid (by de Boer and Ehlers, 1986b).

The greatest progress has been achieved in the application of the porous media theory to dynamic problems. In a series of papers, dynamic problems have been analytically treated with incompressible models at Essen University in Germany. The one-dimensional transient wave propagation in fluid-saturated porous media has been investigated by de Boer, Ehlers, and Liu (1993). An exact solution is obtained via the Laplace transformation technique. In further publications, plane waves in a semi-infinite fluid-saturated porous medium (de Boer and Liu, 1994), the propagation of acceleration waves (de Boer and Liu, 1995), two types of surface waves, namely Rayleigh- and Love-type waves (Liu and de Boer, 1997), and inhomogeneous plane waves, as well as energy flux and energy dissipation in a binary medium (Liu *et al.*, 1998), have all been treated.

Dynamic problems have also been investigated by Gubaidullin and Kuchugurina (1999) though, in part, however, for simplified porous media models.

Parallel to the analytical investigations of the dynamic behavior of saturated porous solids, the numerical solution of initial- and boundary-value problems in the dynamic range has made great progress (see Breuer, 1997a–d, 1999a, b). However, in this field, there are still many problems left to investigate.

Readers interested in some aspects of modern approaches to dynamic problems of saturated porous media are referred to the special issues of the Journal *Transport in Porous Media* (Kowalski and Kubik, 1992, de Boer, 1999).

In the following section, some selected applications of the porous media theory will be discussed, namely the theoretical foundation of the uplift, friction, and capillarity phenomena, the consolidation problem, the compaction of metallic powder, and the one-dimensional transient wave propagation in fluid-saturated incompressible porous media.

5.9.1 Uplift, Friction and Capillarity: three Fundamental Effects for Liquid-Saturated Porous Solids

The fundamental effects in saturated porous solids – uplift, friction, and capillarity – were the main subjects of investigation into saturated porous media in the first half of this century (see the extensive discussion in Section 4.1). However, the theoretical foundation of these effects was established more or less intuitively and was in no way based on profound mechanical or thermodynamic principles. This was impossible at the time because the porous media theory and the constitutive theory within continuum mechanics, with all the restrictions gained from the objectivity principle and thermodynamic considerations, were all still awaiting further research. With the development of the first porous media theories, on the basis of the mixture theory and the volume fraction concept in the late 1970s and the beginning of the 1980s, a tool was provided to examine the aforementioned effects theoretically. This was done by de Boer and Ehlers (1988, 1990a), who were able to confirm some results which had been intuitively stated in the early stages of the development of the porous media theory, and thereby were able to correct some statements in the older literature. However, the investigations of de Boer and Ehlers were based on a model from Bowen (1980) with incompressible constituents. Even though such a model can describe the basic effects quite accurately, it is advisable to investigate the physical phenomena uplift, friction, and capillarity on the basis of a more general model (developed in the last section). This model contains not only incompressible, but also compressible, constituents.

It is the goal of the following investigations to derive the formulas for the uplift, friction, and capillarity phenomena for a liquid- or gas-saturated porous medium with an incompressible or a compressible solid phase. It will be shown that the structures for the relations for the uplift, friction, and capillarity phenomena are the same, independent of the special behavior of the constituents, such as incompressibility or compressibility.

In the following, the equations of motion (5.4.15) for the individual constituents of the binary model (elastic solid skeleton φ^S filled with an inviscid fluid φ^F) are needed to describe the effects of uplift, friction, and capillarity:

$$\begin{aligned} &\operatorname{div} \mathbf{T}^S + \rho^S(\mathbf{g} - \mathbf{a}_S) + \hat{\mathbf{p}}^S = \mathbf{0}\,, \\ &\operatorname{div} \mathbf{T}^F + \rho^F(\mathbf{g} - \mathbf{a}_F) + \hat{\mathbf{p}}^F = \mathbf{0}\,, \end{aligned} \tag{5.9.1}$$

whereby any mass exchange has been neglected and, as external acceleration $\mathbf{b}^\alpha$, the gravity acceleration $\mathbf{g}$ has been introduced.

Moreover, the main constitutive equations in the so-called mixture equilibrium are summarized from Section 5.8. Assuming a common temperature for both constituents, the following relations are obtained:

$$\begin{aligned} \mathbf{T}^S &= -\, n^S p \mathbf{I} + \mathbf{T}^S_E\,, \\ \mathbf{T}^F &= -\, n^F p \mathbf{I}\,, \\ \hat{\mathbf{p}}^F &= p \operatorname{grad} n^F + \hat{\mathbf{p}}^F_E\,, \end{aligned} \tag{5.9.2}$$

where the effective stresses $\mathbf{T}^S_E$ is determined by the volume fractions and the motion of the partial solid constituent, and where p is the porefluid pressure. The porefluid pressure can be an undetermined reaction force (incompressible fluid), or is determined by a constitutive relation (compressible fluid), see (5.8.76). The effective stresses $\mathbf{T}^S_E$ and forces $\hat{\mathbf{p}}^F_E$ will not be specified, as they do not effect the further considerations.

In order to gain the proper relations for the description of the uplift, friction, and capillarity phenomena, it is convenient to sum up the equations of motion (5.9.1), considering $(5.4.17)_4$:

$$\operatorname{div} \mathbf{T} + (\rho^S + \rho^F)\mathbf{g} = \rho^S \mathbf{a}_S + \rho^F \mathbf{a}_F\,, \tag{5.9.3}$$

where

$$\mathbf{T} = \mathbf{T}^S + \mathbf{T}^F \tag{5.9.4}$$

has been introduced. Using the constitutive Equations (5.9.2), it follows from (5.9.3), in connection with (5.9.4), that

$$-\operatorname{grad} p + (\rho^S + \rho^F)\mathbf{g} + \operatorname{div} \mathbf{T}^S_E = \rho^S \mathbf{a}_S + \rho^F \mathbf{a}_F\,. \tag{5.9.5}$$

According to (5.2.11), the partial densities ρ^α can be replaced by the real densities $\rho^{\alpha R}$. Moreover, the real densities $\rho^{\alpha R}$ are related to the specific weight $\gamma^{\alpha R}$ as follows:

$$\gamma^{\alpha R} = \rho^{\alpha R} |\mathbf{g}|\,. \tag{5.9.6}$$

Thus, considering the saturation condition (5.2.4), as well as (5.2.11) and (5.9.6), the equation of motion for the whole mixture body (5.9.5) can be reformulated as

$$-\operatorname{grad} p + n^S(\gamma^{SR} - \gamma^{FR})\frac{\mathbf{g}}{|\mathbf{g}|} + \gamma^{FR}\frac{\mathbf{g}}{|\mathbf{g}|} + \operatorname{div} \mathbf{T}^S_E = \rho^S \mathbf{a}_S + \rho^F \mathbf{a}_F\,. \tag{5.9.7}$$

Further modifications of this equation are due to the pressure head h_p in the flow zone or to the suction head h_t in the capillary zone:

$$h = h_p = \frac{p}{\gamma^{FR}} + \frac{U}{|\mathbf{g}|}, \quad h = h_t = -\frac{p_t}{\gamma^{FR}} + \frac{U}{|\mathbf{g}|}, \tag{5.9.8}$$

where the gravity field $\mathbf{g}$ can be derived from a potential U,

$$\mathbf{g} = -\operatorname{grad} U, \tag{5.9.9}$$

and where p_t characterizes the effective liquid suction in the capillary zone.

From (5.9.7), with (5.9.8) and (5.9.9), one obtains

$$\operatorname{div} \mathbf{T}_E^S - \gamma^{FR} \operatorname{grad} h + n^S(\gamma^{SR} - \gamma^{FR})\frac{\mathbf{g}}{|\mathbf{g}|} = \rho^S \mathbf{a}_S + \rho^F \mathbf{a}_F, \tag{5.9.10}$$

where the obvious conclusion

$$\operatorname{grad} \gamma^{FR} = \operatorname{grad}(\rho^{FR}|\mathbf{g}|) = \mathbf{0} \tag{5.9.11}$$

has been used.

Visualizing the left-hand side of (5.9.10), it is easily recognized that this relation consists of three parts: The divergence of the effective stress $\mathbf{T}_E^S$ is governed by the motion of the solid partial constituent, the friction force (or capillary force) proportional to the gradient of the pressure head, and the specific weight of the solid reduced by uplift. Following this, the uplift force $\mathbf{k}_u$ in the solid phase is given by

$$\mathbf{k}_u = -\, n^S \gamma^{FR} \frac{\mathbf{g}}{|\mathbf{g}|}. \tag{5.9.12}$$

Eqn. (5.9.12) can be reformulated considering the saturation condition (5.2.4):

$$\mathbf{k}_u = (n^F - 1)\gamma^{FR} \frac{\mathbf{g}}{|\mathbf{g}|}. \tag{5.9.13}$$

The uplift formula (5.9.13) for porous solids was first developed by von Terzaghi (1937) in a paper that was never published and which, in some parts, remains obscure. Von Terzaghi's (1937) paper on the uplift problem has recently been discovered by the author in the Terzaghi Library in Oslo, Norway (see the discussion of this paper in de Boer and Didwania, 1997). The reader may note that, for saturated sand, von Terzaghi (1925a) had already stated the result (5.9.13) in his famous first textbook on soil mechanics.

From (5.9.10), the friction force in the flow zone yields, with $h = h_p$,

$$\mathbf{k}_f = -\, \gamma^{FR} \operatorname{grad} h_p \tag{5.9.14}$$

or, using $(5.9.8)_1$, (5.9.9), and (5.9.11),

$$\mathbf{k}_f = -\operatorname{grad} p + \gamma^{FR} \frac{\mathbf{g}}{|\mathbf{g}|}. \tag{5.9.15}$$

This result corresponds directly to Hoffman's (1929) formula.

Finally, since (5.9.10) is not only valid in the flow zone but also in the capillary zone, one easily concludes that the capillary force yields

$$\mathbf{k}_c = -\, \gamma^{FR} \operatorname{grad} h_t, \tag{5.9.16}$$

or, by using $(5.9.8)_2$ and (5.9.9),

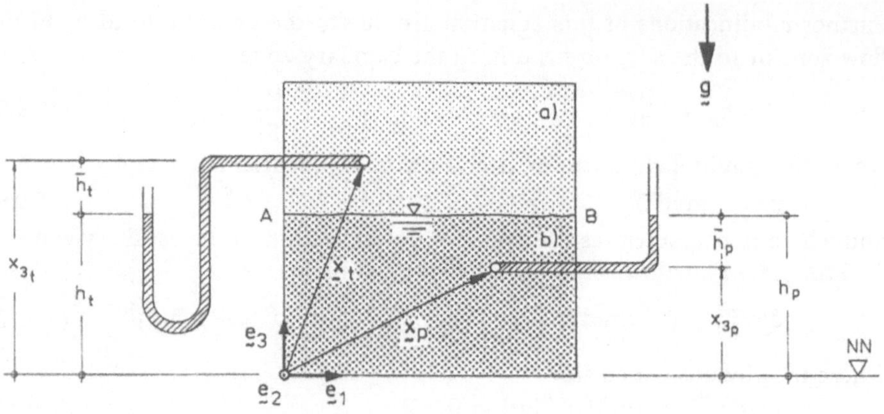

Fig. 5.9.1. Liquid-saturated soil with free ground-water level A–B (**a**) Capillary zone; (**b**) flow zone

$$\mathbf{k}_c = \operatorname{grad} p_t + \gamma^{FR} \frac{\mathbf{g}}{|\mathbf{g}|} \,. \tag{5.9.17}$$

Using the capillary rise $\bar{h}_t$ (de Boer and Ehlers, 1990a), the result of von Terzaghi's approach is gained:

$$p_t = \gamma^{FR} \left(\frac{U}{|\mathbf{g}|} - h_t \right) = \gamma^{FR} \bar{h}_t \,, \tag{5.9.18}$$

where

$$\bar{h}_t = \frac{U}{|\mathbf{g}|} - h_t \,. \tag{5.9.19}$$

Whereas the static case of capillarity is well-founded in porous media theory, the instationary capillary motion is still an open problem. It is, however, also currently under study at Essen University in Germany.

Finally, it should be mentioned that friction and capillary forces only occur if grad $h \neq \mathbf{0}$, in the case of flow processes in the flow zone, or in the case of vaporization processes in the downstream face of the suction zone. The uplift force is, however, always effective.

5.9.2 One-Dimensional Transient Wave Propagation in Fluid-Saturated Incompressible Porous Media

Due to the complexity of the dynamic behavior of saturated porous solids, exact analytical solutions of initial- and boundary-value problems are seldom found. Fortunately, the one-dimensional wave propagation can be solved exactly, see de Boer *et al.* (1993). Thus, one obtains an excellent tool for comparing numerical results for complicated practical problems, gained by the finite or boundary element methods, with this special example of the one-dimensional wave propagation.

The investigation of wave motion phenomena in fluid-saturated porous media is attracting more and more attention because of its significance in a great number of practical engineering problems. Generally, the most widely accepted opinion, that there are two dilatational waves and one rotational wave, has been concluded on the basis of Biot's (1956a) theory. This theory, which is based on the assumption of compressible constituents along with some of his results, has been taken as a standard reference and the basis for much of the subsequent analysis in acoustics, geophysics, geomechanics, and other fields up to the present. The prediction of the existence of two dilatational waves seems to be confirmed by test observations on the part of Plona (1982). However, Plona's test results are not very well worked out. From his results, one can also read that three dilatational waves exist. Indeed, this should be the case from the theoretical point of view. Concerning the dilatancy of the saturated porous solid with two compressible constituents, there are three grades of freedom, namely the compressibility of the real fluid and of the real solid material, and the volume change of the pores. Theoretical investigations of this subject in porous media theory are still awaiting adequate research.

From the discussion above, it is clear that an incompressible model with two incompressible constituents can only have one dilatational wave. This obvious result comes out of theoretical investigations.

In the following paragraphs, an analytical solution for analyzing transient phenomena in fluid-saturated elastic porous media will be presented. The fluid-saturated porous medium is modeled as a two-phase system composed of elastic, incompressible solid, and inviscid, incompressible fluid phases, thus meeting the assumptions of many problems in engineering practice, such as in soil mechanics. On the basis of the fundamental equations developed in the preceding sections, governing equations are given considering linear-elastic deformations of the solid skeleton. An exact solution is obtained via the Laplace transformation technique for the one-dimensional problem, taking into account initial- and boundary-value conditions. Of interest is the result that, as a direct consequence of the incompressibility constraints, only one independent compressible wave in the partial solid and fluid phases exists.

The included material parameters, compare Table 5.1 (see Section 5.9.2 d)), are physically evident and can be taken from simple laboratory tests. It should, furthermore, be noted that the fluid-saturated porous material is endowed with characteristics similar to those occurring in viscoelastic solids.

These results may also be used to make a critical comparison between various numerical and analytical results, as well as to provide an alternative understanding of the mechanism of wave propagation in fluid-saturated porous materials.

All investigations to follow are performed within the framework of the geometrically-linear theory. Because the fields in the actual and reference placements of the solid phase are approximately equal, we will refer the fields (in the representation) to the actual placements. A rigorous linearization of the fields would lead to a lengthy formalism (see de Boer, 1982, and Section 5.8.3).

a) Field Equations

We will commence with field equations from Sections 5.3, 5.4, 5.6, and 5.8.

Excluding all mass and heat exchanges between the solid and the liquid phases, and excluding the supply term of moment of momentum, the saturation condition and the balance equations for the constituents are given as follows:

1. saturation condition
$$n^S + n^F = 1\,, \tag{5.9.20}$$
2. balance of mass
$$(\rho^\alpha)'_\alpha + \rho^\alpha \operatorname{div} \mathbf{v}_\alpha = 0\,, \tag{5.9.21}$$
3. balance of momentum
$$\begin{aligned} \operatorname{div} \mathbf{T}^\alpha + \rho^\alpha(\mathbf{b}^\alpha - \mathbf{a}_\alpha) + \hat{\mathbf{p}}^\alpha &= \mathbf{0}\,, \\ \hat{\mathbf{p}}^S + \hat{\mathbf{p}}^F &= \mathbf{0}\,, \end{aligned} \tag{5.9.22}$$
4. balance of moment of momentum
$$\mathbf{T}^\alpha = (\mathbf{T}^\alpha)^T\,. \tag{5.9.23}$$

Considering incompressible constituents, the combination of
$$\rho^\alpha = n^\alpha \rho^{\alpha R}\,, \quad \rho^{\alpha R} = \text{const.}\,, \tag{5.9.24}$$
and (5.9.21) yields
$$(n^\alpha)'_\alpha + n^\alpha \operatorname{div} \mathbf{v}_\alpha = 0\,, \tag{5.9.25}$$
i.e., the balance of mass equations are reduced to balance equations for the volume fractions. Then, using (5.9.20) and (5.9.25), the relation
$$(n^F)'_S = (n^F)'_F - \operatorname{grad} n^F \cdot (\mathbf{v}_F - \mathbf{v}_S) \tag{5.9.26}$$
holds. Alternatively, one obtains, using (5.9.25),
$$\operatorname{div}(n^S \mathbf{v}_S + n^F \mathbf{v}_F) = 0\,. \tag{5.9.27}$$
In the case of $\mathbf{b} = \mathbf{b}^S = \mathbf{b}^F$, the balance equations of momentum $(5.9.22)_1$ can be rewritten in the forms
$$\operatorname{div} \mathbf{T}^S + \rho^S(\mathbf{b} - \mathbf{a}_S) - \hat{\mathbf{p}}^F = \mathbf{0}\,, \tag{5.9.28}$$
$$\operatorname{div} \mathbf{T}^F + \rho^F(\mathbf{b} - \mathbf{a}_F) + \hat{\mathbf{p}}^F = \mathbf{0}\,, \tag{5.9.29}$$
where, in addition, the constraint $(5.9.22)_2$ has been used.

As a consequence of the saturation condition (5.9.20) (see Section 5.8), the stress tensor of the partial solid phase and the interaction force are additively decomposed into two terms, whereas the stress tensor of the fluid phase is represented by the weighted porefluid pressure, i.e.,

$$\mathbf{T}^S = -\,n^S p\mathbf{I} + \mathbf{T}^S_E\,, \quad \hat{\mathbf{p}}^F = p\,\mathrm{grad}\,n^F + \hat{\mathbf{p}}^F_E\,, \tag{5.9.30}$$

$$\mathbf{T}^F = -\,n^F p\mathbf{I}\,. \tag{5.9.31}$$

In (5.9.30), the index E expresses the effective stress and interaction force for which constitutive equations have to be formulated. Insertion of (5.9.30) and (5.9.31) into (5.9.28) and (5.9.29) produces

$$\mathrm{div}\,\mathbf{T}^S_E - n^S\,\mathrm{grad}\,p + \rho^S(\mathbf{b} - \mathbf{a}_S) - \hat{\mathbf{p}}^F_E = \mathbf{0}\,, \tag{5.9.32}$$

$$-\,n^F\,\mathrm{grad}\,p + \rho^F(\mathbf{b} - \mathbf{a}_F) + \hat{\mathbf{p}}^F_E = \mathbf{0}\,. \tag{5.9.33}$$

For further considerations, it is convenient to replace the velocities $\mathbf{v}_\alpha$ and the accelerations $\mathbf{a}_\alpha$ by the corresponding derivatives of the displacement vectors $\mathbf{u}_\alpha$. From

$$\mathbf{x} = \mathbf{X}_\alpha + \mathbf{u}_\alpha\,, \tag{5.9.34}$$

we obtain

$$\mathbf{v}_\alpha = (\mathbf{u}_\alpha)'_\alpha\,, \quad \mathbf{a}_\alpha = (\mathbf{u}_\alpha)''_\alpha\,. \tag{5.9.35}$$

Within the framework of the physically- and geometrically-linear theory (where the effective Cauchy stress tensor $\mathbf{T}^S_E$ is approximately equal to the symmetric stress tensor $\mathbf{S}^S_E$) we introduce the following constitutive equations for the effective stress and the interaction force:

$$\mathbf{T}^S_E \approx \mathbf{S}^S_E = 2\mu^S\mathbf{E}_S + \lambda^S(\mathbf{E}_S\cdot\mathbf{I})\mathbf{I}\,, \quad \hat{\mathbf{p}}^F_E = -\,\mathbf{S}_V\left[(\mathbf{u}_F)'_F - (\mathbf{u}_S)'_S\right]\,, \tag{5.9.36}$$

where μ^S and λ^S are the macroscopic Lamé constants of the porous solid, and

$$\mathbf{E}_S \approx \mathbf{A}_S = \frac{1}{2}(\mathrm{grad}\,\mathbf{u}_S + \mathrm{grad}^T\,\mathbf{u}_S) \tag{5.9.37}$$

is the linearized Almansi strain tensor which is approximately equal to the linearized Green strain tensor. In the case of isotropic permeability, the tensor $\mathbf{S}_V$, describing the coupled interaction between the solid and the fluid, is given by (Heinrich, 1938a)

$$\mathbf{S}_V = \frac{(n^F)^2\gamma^{FR}}{k^F}\mathbf{I} = S_V\mathbf{I}\,, \tag{5.9.38}$$

where k^F is the Darcy permeability coefficient of the porous medium. It should be remarked that, within the geometrically-linear theory, the differential operators grad (...) and div (...) are approximately equal to the operators Grad_α(...) and Div_α(...), which contain the derivatives with respect to $\mathbf{X}_\alpha$. Now, inserting (5.9.36) through (5.9.38) into (5.9.32) and (5.9.33) as well as into (5.9.27), we may write the field equations as follows:

$$\begin{aligned}(\lambda^S + \mu^S)\,\mathrm{grad}\,\mathrm{div}\,\mathbf{u}_S + \mu^S\,\mathrm{div}\,\mathrm{grad}\,\mathbf{u}^S - n^S\,\mathrm{grad}\,p\, -& \\ + \rho^S\left[\mathbf{b} - (\mathbf{u}_S)''_S\right] + S_V\left[(\mathbf{u}_F)'_F - (\mathbf{u}_S)'_S\right] &= \mathbf{0}\,,\end{aligned} \tag{5.9.39}$$

$$-\,n^F\,\mathrm{grad}\,p + \rho^F\left[\mathbf{b} - (\mathbf{u}_F)''_F\right] - S_V\left[(\mathbf{u}_F)'_F - (\mathbf{u}_S)'_S\right] = \mathbf{0}\,, \tag{5.9.40}$$

$$\mathrm{div}\left[n^S(\mathbf{u}_S)'_S + n^F(\mathbf{u}_F)'_F\right] = 0\,. \tag{5.9.41}$$

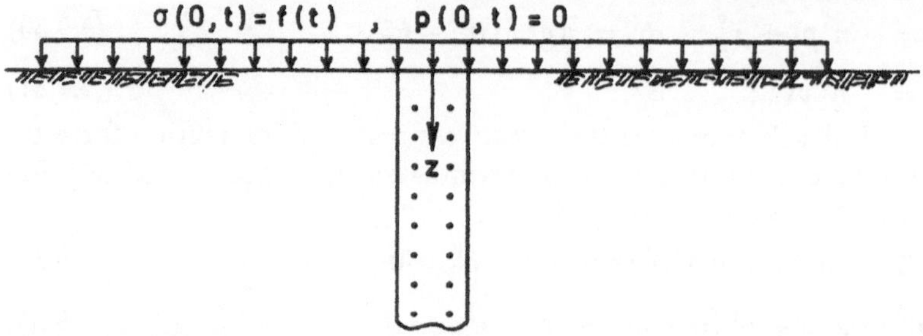

Fig. 5.9.2. Geometry of the investigated problem

Within the framework of the infinitesimal theory, the superposition principle holds, i.e., the loading by body forces and by surface forces can be treated separately. Furthermore, by only considering the loading by surface forces, the equations of motion, (5.9.39) and (5.9.40), can be written as

$$\begin{aligned}(\lambda^S + \mu^S)\,\text{grad div}\,\mathbf{u}_S + \mu^S\,\text{div grad}\,\mathbf{u}_S - n^S\,\text{grad}\,p - \rho^S(\mathbf{u}_S)''_S + \\ + S_V\left[(\mathbf{u}_F)'_F - (\mathbf{u}_S)'_S\right] = \mathbf{0}\,,\end{aligned} \tag{5.9.42}$$

$$- n^F\,\text{grad}\,p - \rho^F(\mathbf{u}_F)''_F - S_V\left[(\mathbf{u}_F)'_F - (\mathbf{u}_S)'_S\right) = \mathbf{0}\,. \tag{5.9.43}$$

Moreover, the volume fractions n^S can be approximated by n^S_{0S}, which is the solid fraction in the reference placement. This fact can be easily proven. From (5.9.25), by integration,

$$n^S = n^S_{0S}(\det \mathbf{F}_S)^{-1} = n^S_{0S}(1 + \mathbf{E}_S \cdot \mathbf{I})^{-1} \tag{5.9.44}$$

is obtained. Within the scope of infinitesimal deformations, all terms of higher-order are neglected. Moreover, because $|\mathbf{E}_S \cdot \mathbf{I}| \ll 1$, the volume fraction n^S may be approximated by n^S_{0S}.

b)
One-Dimensional Transient Wave Propagation Solution

The equations of motion described above are usually solved by numerical methods. However, in the case of one-dimensional small strains and displacements, an analytical solution is possible in which the small variation of the volume fractions is approximately neglected. Now, the focus is put on a one-dimensional infinitely long column (see Fig. 5.9.2) separated from a half-space consisting of a liquid-saturated porous elastic skeleton. The motion of both the solid and the fluid material is constrained to take place in the vertical direction. Loading as a function of time, $\sigma(z = 0, t) = f(t)$ by a permeable punch with ideal permeability, is applied to the half space boundary. The z-axis is taken as the normal line at the boundary and thus, for the one-dimensional problem, the governing equations (5.9.41) through (5.9.43) are directly simplified to:

$$(\lambda^S + 2\mu^S) u_{,zz} - n^S p_{,z} - \rho^S u_{,tt} + S_V (U_{,t} - u_{,t}) = 0 \,, \tag{5.9.45}$$

$$- n^F p_{,z} - \rho^F U_{,tt} - S_V (U_{,t} - u_{,t}) = 0 \,, \tag{5.9.46}$$

$$n^S u_{,tz} + n^F U_{,tz} = 0 \,, \tag{5.9.47}$$

where the vertical components of $\mathbf{u}_S$ and $\mathbf{u}_F$ are replaced by u and U. In (5.9.47), homogeneous pore distribution is assumed. Furthermore, $u_{,z}$ or $u_{,zz}$, respectively, mean the first or the second differentiation of u with respect to the spatial coordinate z; the partial time derivatives are denoted by $u_{,t}$ and $u_{,tt}$, etc.

For the following investigations, the loading function at the free boundary is given by

$$\sigma(0,t) = f(t), \quad p(0,t) = 0 \tag{5.9.48}$$

(Fig. 5.9.2), where $f(t)$ is an arbitrary function of time which describes surface loading onto the skeleton material. The boundary condition $(5.9.48)_2$ implies a free, liquid surface, assuming that the half-space boundary shows an adequate permeability.

The initial conditions are

$$u(z,0) = 0 \,, \quad U(z,0) = 0, \quad u_{,t}(z,0) = 0, \quad U_{,t}(z,0) = 0 \,. \tag{5.9.49}$$

Taking the Laplace transformation of (5.9.45) through (5.9.47), with initial conditions (5.9.49), and using matrix notation, one obtains

$$\mathbf{A}\mathbf{m}_{,zz} + \mathbf{B}\mathbf{m}_{,z} + \mathbf{C}\mathbf{m} = \mathbf{0} \,, \tag{5.9.50}$$

where $\mathbf{m}^T = [L(u), L(U), L(p)]$. The functions $L(u), L(U)$, and $L(p)$ are the Laplace transformations of the solid displacement, the fluid displacement and the porefluid pressure, respectively

$$L(u) = \int_0^\infty e^{-rt} u dt \,, \; L(U) = \int_0^\infty e^{-rt} U dt \,, \; L(p) = \int_0^\infty e^{-rt} p dt \,, \tag{5.9.51}$$

where r is the Laplace transformations parameter. In (5.9.50), the corresponding matrices $\mathbf{A}$, $\mathbf{B}$, and $\mathbf{C}$ are defined by

$$\mathbf{A} = \begin{bmatrix} \lambda^S + 2\mu^S & 0 & 0 \\ 0 & 0 & 0 \\ 0 & 0 & 0 \end{bmatrix} , \tag{5.9.52}$$

$$\mathbf{B} = \begin{bmatrix} 0 & 0 & -n^S \\ 0 & 0 & -n^F \\ -n^S r & -n^F r & 0 \end{bmatrix} , \tag{5.9.53}$$

$$\mathbf{C} = \begin{bmatrix} -\rho^S r^2 - S_V r & S_V r & 0 \\ S_V r & -\rho^F r^2 - S_V r & 0 \\ 0 & 0 & 0 \end{bmatrix} . \tag{5.9.54}$$

Assuming $\mathbf{m}$ to be solved via

$$\mathbf{m} = \mathbf{m}_0 e^{\alpha z} \,, \tag{5.9.55}$$

where $\mathbf{m}_0$ and α are functions of the transformation parameter r, one obtains the eigenvalue problem by insertion of (5.9.55) into (5.9.50), i.e.:

$$(\alpha^2\mathbf{A} + \alpha\mathbf{B} + \mathbf{C})\mathbf{m}_0 e^{\alpha z} = \mathbf{0}\,. \tag{5.9.56}$$

Using (5.9.52) through (5.9.54), the corresponding characteristic equation is

$$(n^F)^2(\lambda^S + 2\mu^S)\alpha^4 - \left[(n^S)^2\rho^F + (n^F)^2\rho^S\right]r^2\alpha^2 - S_v r\alpha^2 = 0\,. \tag{5.9.57}$$

Eqn. (5.9.57) may be solved to yield two pairs of roots,

$$\alpha_{1,2} = \pm\sqrt{ar^2 + br}\,,\ \alpha_{3,4} = 0\,, \tag{5.9.58}$$

where

$$a = \frac{(n^S)^2\rho^F + (n^F)^2\rho^S}{(\lambda^S + 2\mu^S)(n^F)^2}\,,\qquad b = \frac{S_v}{(\lambda^S + 2\mu^S)(n^F)^2}\,. \tag{5.9.59}$$

The eigenvectors associated with $\alpha_{1,2}$ and $\alpha_{3,4}$, respectively, are

$$\mathbf{q}_{1,2}^T = \left[-\frac{n^F}{n^S}\,,\quad 1\,,\quad \mp\frac{n^S\rho^F r^2 + S_v r}{n^S n^F\sqrt{ar^2 + br}}\right], \tag{5.9.60}$$

and

$$\mathbf{q}_{3,4}^T = \left[0\,,\quad 0\,,\quad 1\right]. \tag{5.9.61}$$

Thus, the transformed solution for the transient response of the porous medium is

$$\mathbf{m} = C_1\mathbf{q}_1 e^{\alpha_1 z} + C_2\mathbf{q}_2 e^{\alpha_1 z} + C_3\mathbf{q}_3 + C_4 z\mathbf{q}_4\,. \tag{5.9.62}$$

In order to ensure the limit at infinity, the coefficients C_1 and C_4 are:

$$C_1 = 0\,,\ C_4 = 0\,. \tag{5.9.63}$$

Then, the solution takes the form

$$\mathbf{m} = C_2\mathbf{q}_2 e^{\alpha_2 z} + C_3\mathbf{q}_3\,, \tag{5.9.64}$$

in which C_2 and C_3 are determined from the transformed boundary condition, i.e.,

$$C_2 = \frac{n^S L[f(t)]}{n^F(\lambda^S + 2\mu^S)\sqrt{ar^2 + br}}\,,\qquad C_3 = \frac{-(S_v + n^S\rho^F r)rL[f(t)]}{(n^F)^2(\lambda^S + 2\mu^S)(ar^2 + br)}\,. \tag{5.9.65}$$

Applying the convolution integral and transformation formulation (Abramowitz and Stegun, 1965), the inverse transformation of (5.9.62) produces the exact solution for the transient response problem in the fluid-saturated porous medium:

$$u(z,t) = -\frac{1}{\sqrt{a}(\lambda^S + 2\mu^S)}\int_0^t f(t-\tau)e^{-\frac{b}{2a}\tau}I_0\left(\frac{b\sqrt{\tau^2 - az^2}}{2a}\right)Z(\tau - \sqrt{a}z)\,d\tau\,, \tag{5.9.66}$$

$$U(z,t) = \frac{n^S}{n^F\sqrt{a}(\lambda^S + 2\mu^S)}\int_0^t f(t-\tau)e^{-\frac{b}{2a}\tau}I_0\left(\frac{b\sqrt{\tau^2 - az^2}}{2a}\right)Z(\tau - \sqrt{a}z)\,d\tau\,, \tag{5.9.67}$$

$$p(z,t) = \frac{1}{(n^F)^2(\lambda^S + 2\mu^S)}\left[n^S\rho^F L_{,tt}(z,t) + S_v L_{,t}(z,t)\right], \tag{5.9.68}$$

where

$$L(z,t) = \int_0^t Q(t-\tau)G(z,\tau)d\tau\,, \tag{5.9.69}$$

$$Q(t) = \frac{1}{\sqrt{a}}\int_0^t f(t-\tau)e^{-\frac{b}{2a}\tau} I_0\left(\frac{b}{2a}\tau\right)d\tau\,, \tag{5.9.70}$$

$$G(z,t) = \frac{1}{\sqrt{a}}e^{-\frac{b}{2a}t} I_0\left(\frac{b(\sqrt{t^2-az^2}}{2a}\right) Z(t-\sqrt{a}z) - \frac{1}{\sqrt{a}}e^{-\frac{b}{2a}t} I_0\left(\frac{b}{2a}t\right). \tag{5.9.71}$$

With the results above, it is not difficult to determine the one-dimensional extra stress σ_E^S:

$$\sigma_E^S = \frac{b}{2\sqrt{a}}\int_0^t f(t-\tau)e^{-\frac{b}{2a}\tau} I_1\left(\frac{b(\sqrt{\tau^2-az^2}}{2a}\right)\frac{z}{\sqrt{\tau^2-az^2}} Z(\tau-\sqrt{a}z)d\tau + \tag{5.9.72}$$

$$+f(t-\sqrt{a}z)e^{-\frac{b}{2\sqrt{a}}z}\,.$$

In (5.9.66) through (5.9.71), $I_0(z)$ and $I_1(z)$ are the modified Bessel functions of zero and one order, respectively, and $Z(t)$ is the unit step function (Heaviside function).

c)
General Properties of the Analytical Solution

In the preceding section, a one-dimensional analytical solution via Laplace transformation technique was presented for an incompressible linear-elastic skeleton material saturated by a single, incompressible poreliquid.

The resulting expressions for the solid and the liquid displacements, Eqns. (5.9.66) and (5.9.67), and for the liquid pressure and the solid extra stresses, Eqns. (5.9.68) and (5.9.72), exhibit a strong history-dependence comparable to that in the theory of viscoelasticity. In particular, the different response functions not only depend on time, but also on the previous loading history. This point can easily be understood from the squeezing out of water (when, e.g., the material is subjected to external loads), combined with effects of internal friction included in the momentum supply term $\hat{\mathbf{p}}^F$ (de Boer and Ehlers, 1986a). Thus, a saturated porous skeleton material is provided with certain features similar to those appearing in viscoelastic solids.

The wave motion in the porous medium may be expressed by the solid and the fluid displacements or the solid extra stresses, respectively, but it cannot be expressed by the porefluid pressure which, of course, is nothing other than a multiplier corresponding to the incompressibility constraint of the fluid. The saturation condition, in connection with the incompressibility conditions of the model, see (5.9.27), furthermore produces the ratio of the solid and the liquid displacements, yielding

$$u(z,t)/U(z,t) = -\, n^F/n^S \,, \qquad (5.9.73)$$

i.e., there is only one coupled disturbance propagating in the medium. In order to obtain (5.9.73), it has been assumed for the boundary condition that at a certain depth a non-permeable layer exists. The unit step function (Heaviside function) included in the formulas (5.9.66) through (5.9.72) regulates the relation between the disturbed spatial position and the necessary propagation time. Thus, the propagation velocity c_0 included in the argument of the unit step function Z yields

$$c_0 = \frac{z}{t} = \frac{1}{\sqrt{a}} = \sqrt{\frac{(n^F)^2(\lambda^S + 2\mu^S)}{(n^F)^2\rho^S + (n^S)^2\rho^F}} \,. \qquad (5.9.74)$$

If the pore liquid is absent, or if gas is contained in the pores of the matrix, the term ρ^F will be zero or can be neglected in comparison with ρ^S. Then, since in this case n^F means porosity, one obtains the propagation velocity c' of the dilatational wave in incompressible, empty porous solids as

$$c' = \sqrt{(\lambda^S + 2\mu^S)/\rho^S} \,, \qquad (5.9.75)$$

where the corresponding volume changes are due only to changes in porosity. The expression above can be compared with the well-known result of classical elasticity theories.

Finally, if only the solid constituent is present, and furthermore $n^F \to 0$, which corresponds to a non-porous, incompressible solid material, then the propagation velocity c'' of the dilatational wave tends to infinity

$$c'' = \infty \,. \qquad (5.9.76)$$

This is a direct consequence of the incompressibility constraint.

d)
An Illustrative Example of a One-Dimensional Soil Column Subject to three Different Surface Loadings

In this section, a number of numerical results for the exact solution in a one-dimensional water-saturated soil column are presented. The loading function at the surface is $\sigma(0,t) = f(t)$, where $f(t)$ is chosen to be a sine function, a step function, and an impulse function, respectively. The physical properties of the soil are assumed as shown in Table 5.1.

In contrast to the various discussions on Biot's approach, the following paragraph concerns an illustration of the characteristics of one-dimensional transient wave motion for the incompressible binary model under discussion. In particular, the solid and the liquid displacements, the solid extra stresses, and

Table 5.1. Material properties

$n^S = 0.67$	$n^F = 0.33$
$\rho^S = 1.34\,\mathrm{Mg/m^3}$	$\rho^F = 0.33\,\mathrm{Mg/m^3}$
$E^S = 30.00\,\mathrm{MN/m^2}$	$\nu^S = 0.20$
$\lambda^S = 5.5833\,\mathrm{MN/m^2}$	$\mu^S = 8.3750\,\mathrm{MN/m^2}$
$k^F = 0.01\,\mathrm{m/s}$	$\gamma^{FR} = 10.00\,\mathrm{kN/m^3}$

the pore pressure are given with respect to time and with respect to different spatial positions within the framework of three loading forms, i.e., sinusoidal, step loading and impulsive loading.

Response to Sinusoidal Loading

In the case of sinusoidal loading (Fig. 5.9.3), the responses of the solid and the liquid displacements versus time and versus depth, measured from the free surface, are shown in Figs. 5.9.4–5.9.7. From the comments on (5.9.73), it is clear that there exists only one independent dilatational wave propagating through the medium, i.e., the solid displacement can be given as a function of the fluid displacement and vice versa. It is furthermore concluded from (5.9.73), in the

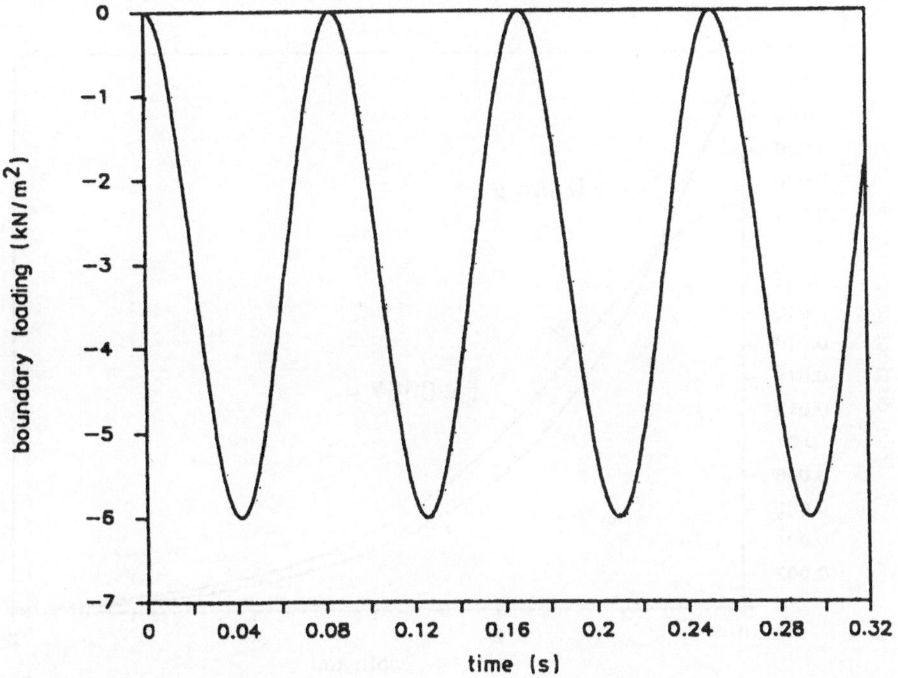

Fig. 5.9.3. Sine loading at the boundary

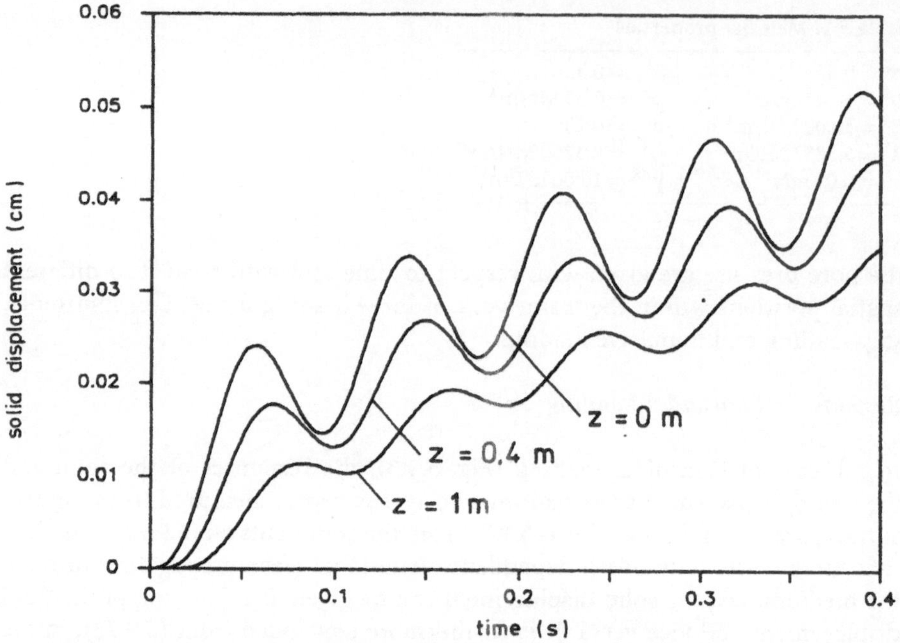

Fig. 5.9.4. Response of solid displacement vs. time to sine loading

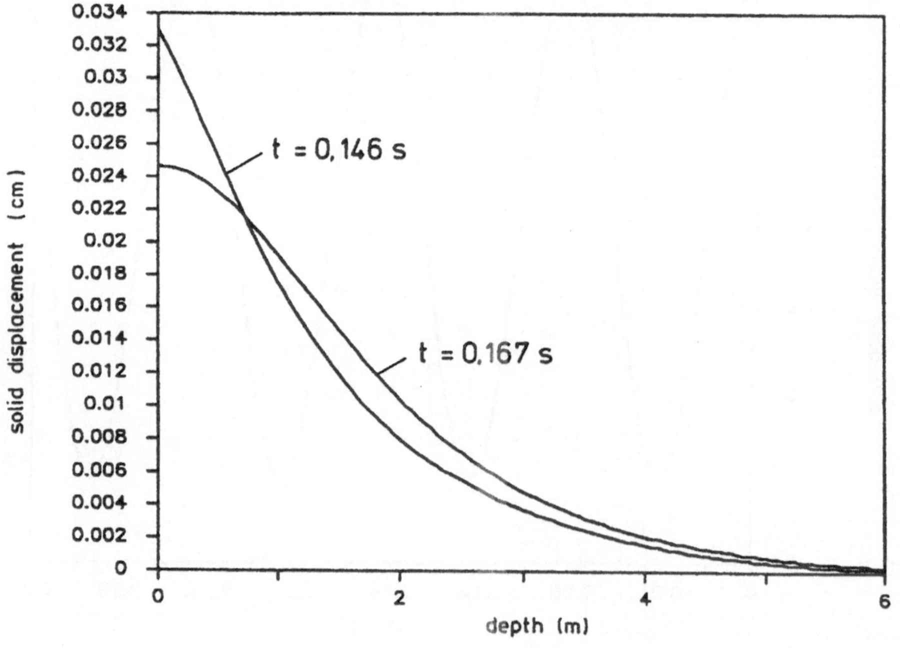

Fig. 5.9.5. Response of solid displacement vs. depth to sine loading

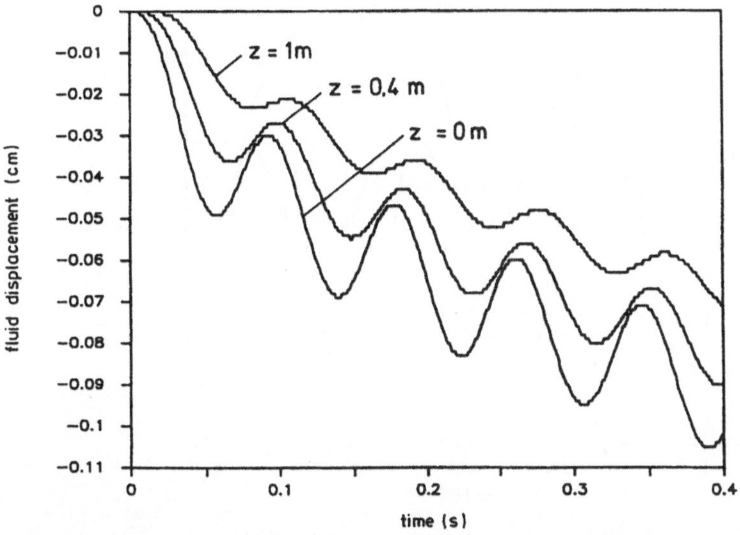

Fig. 5.9.6. Response of fluid displacement vs. time to sine loading

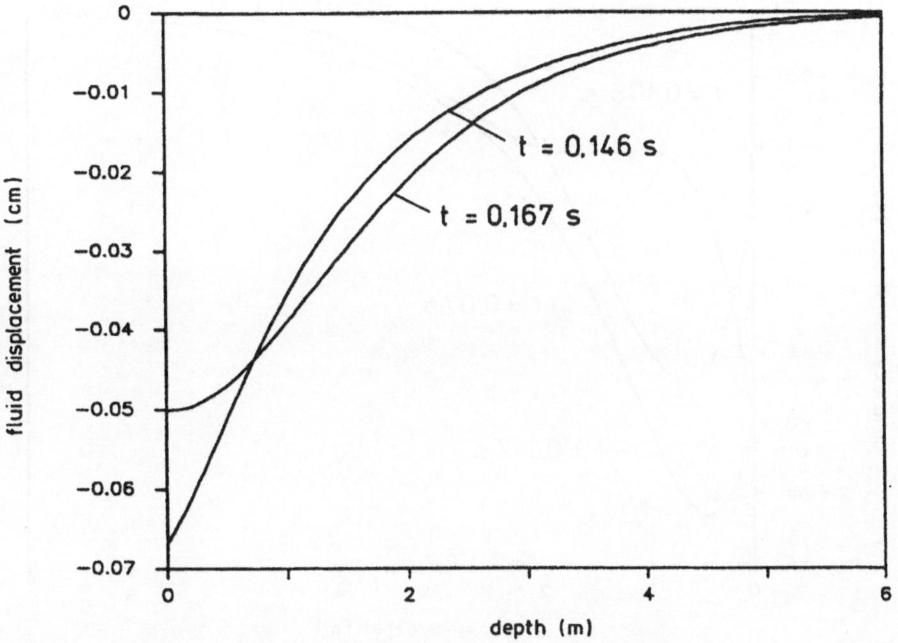

Fig. 5.9.7. Response of fluid displacement vs. depth to sine loading

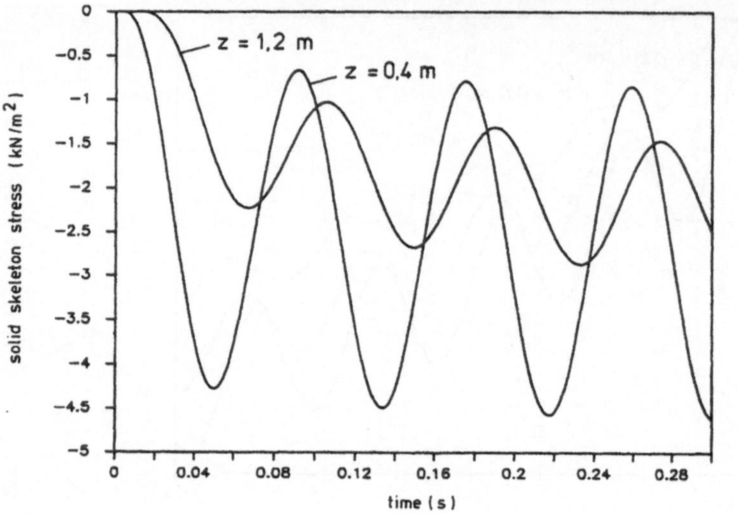

Fig. 5.9.8. Response of solid skeleton effective stress vs. time to sine loading

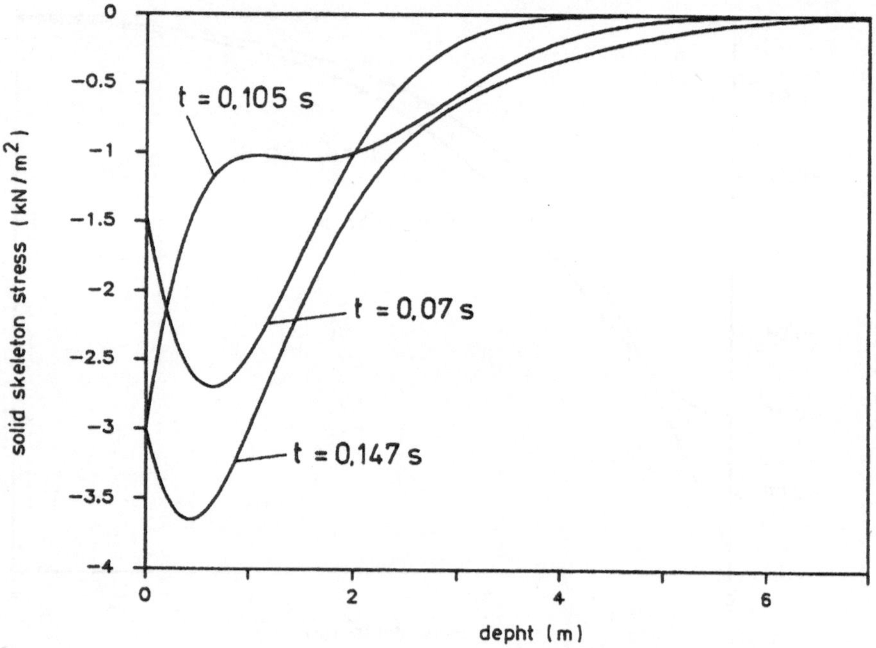

Fig. 5.9.9. Response of solid skeleton effective stress vs. depth to sine loading

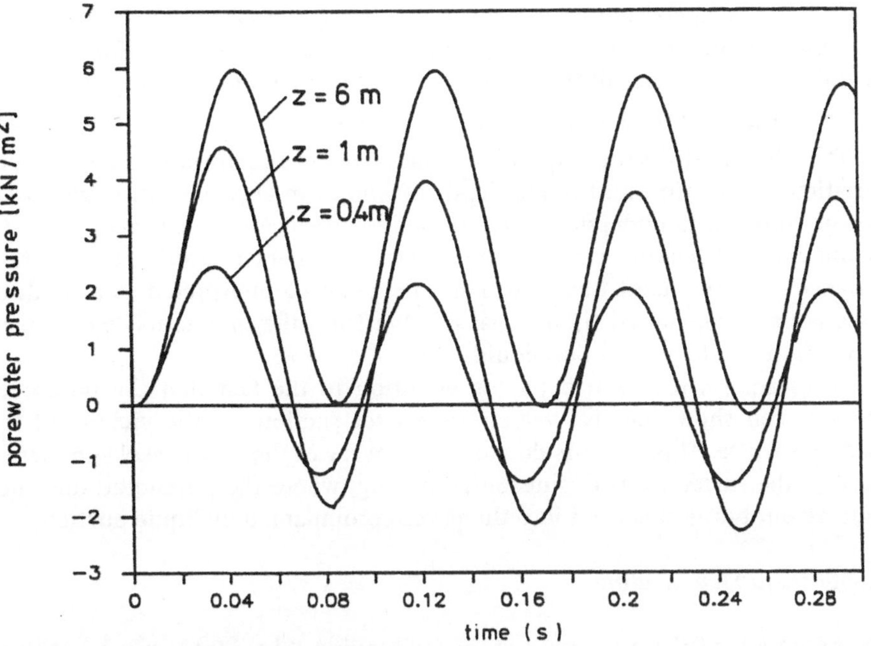

Fig. 5.9.10. Response of porewater pressure vs. time to sine loading

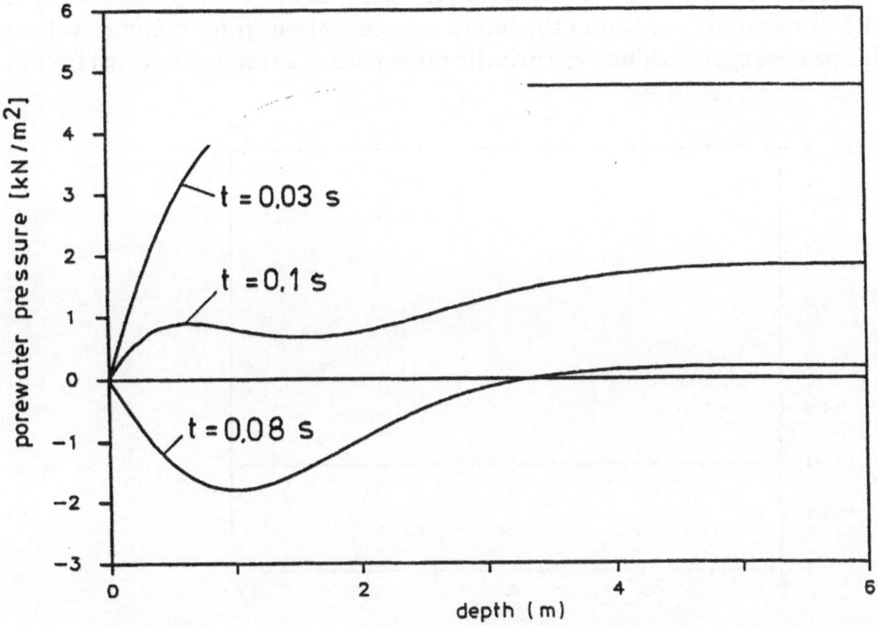

Fig. 5.9.11. Response of porewater pressure vs. depth to sine loading

case of the geometrically-linear theory with approximately constant volume fractions, that the sum of the solid and the liquid volume fluxes vanishes, namely as a direct consequence of the saturation and the incompressibility constraints:

$$n^S u_S' + n^F U_F' = n^S u_S' - n^F (n^S / n^F) u_S' = 0 \, . \qquad (5.9.77)$$

The effective stress functions of the solid skeleton and the porewater pressure variations versus time and versus depth are shown in Figs. 5.9.8 through 5.9.11. It is not difficult to understand that the solid extra stress wave is very sensitive to the external loading close to the surface and tend to vanish at a certain distance from the loaded surface. This effect can be interpreted as a result of viscous damping caused by internal friction from the interaction mechanism between the skeleton and poreliquid.

Furthermore, it is worth paying attention to the fact that the porewater pressure can show negative values (porewater suction) in the vicinity of the loading surface. This result is due to the recovery of the elastic skeleton matrix close to the surface during sinusodial loading, where the poreliquid does not squeeze out but is absorbed into the pores accompanied by liquid suction.

Responses to Step Loading

The responses of the medium due to step loading (Fig. 5.9.12) can be utilized to analyze a consolidation process with a free porewater surface (Figs. 5.9.13 through 5.9.16). In particular, Figs. 5.9.13 – 5.9.20 show the solid and the liquid displacements changing with respect to time and depth. With growing time, the solid moves downwards and the liquid is squeezed out from the pore volume. This process again exhibits viscoelastic properties as a result of internal friction.

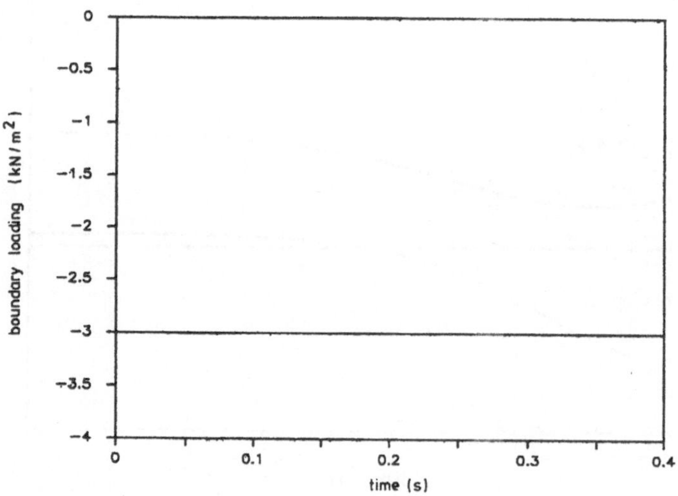

Fig. 5.9.12. Step loading at the boundary

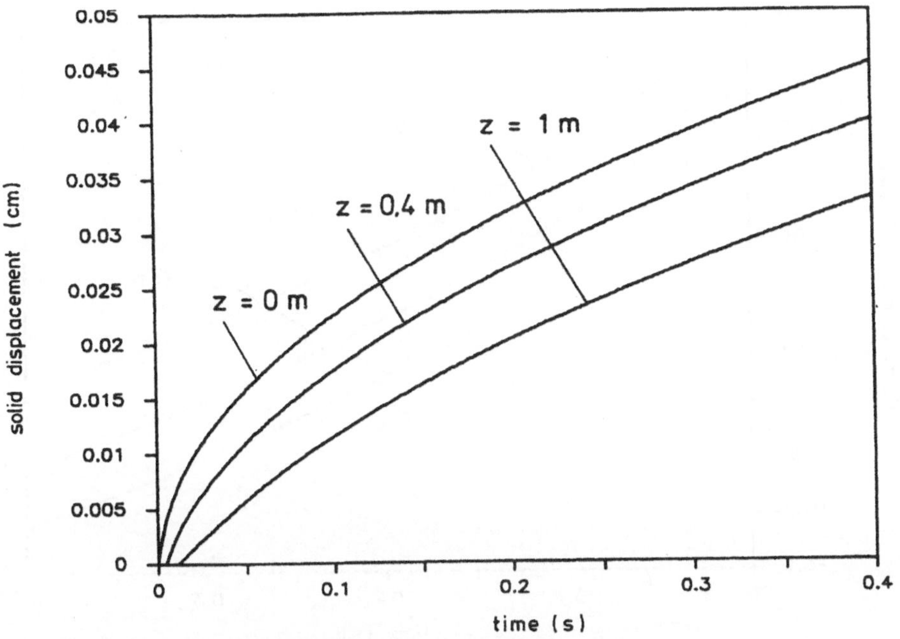

Fig. 5.9.13. Response of solid displacement vs. time to step loading

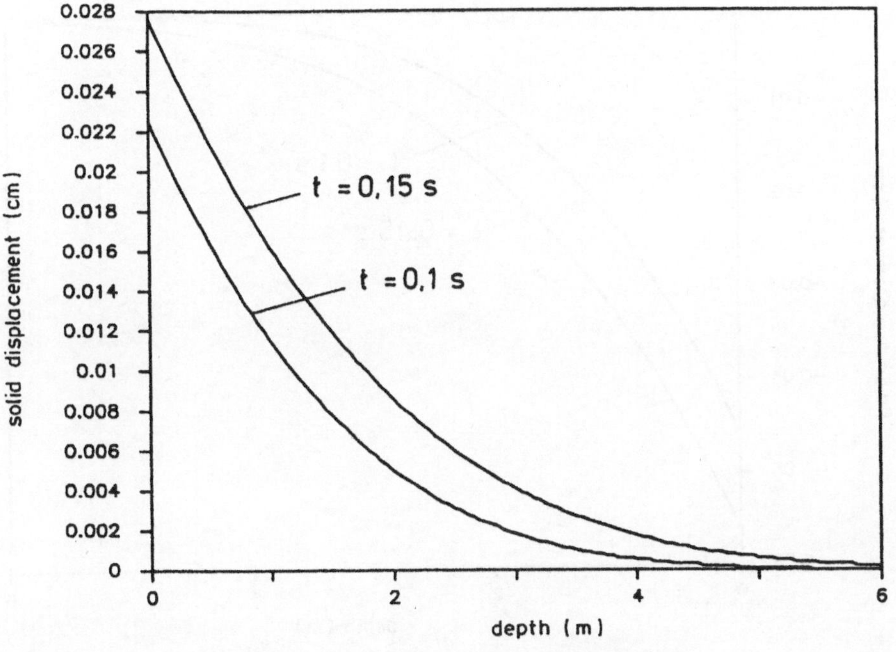

Fig. 5.9.14. Response of solid displacement vs. depth to step loading

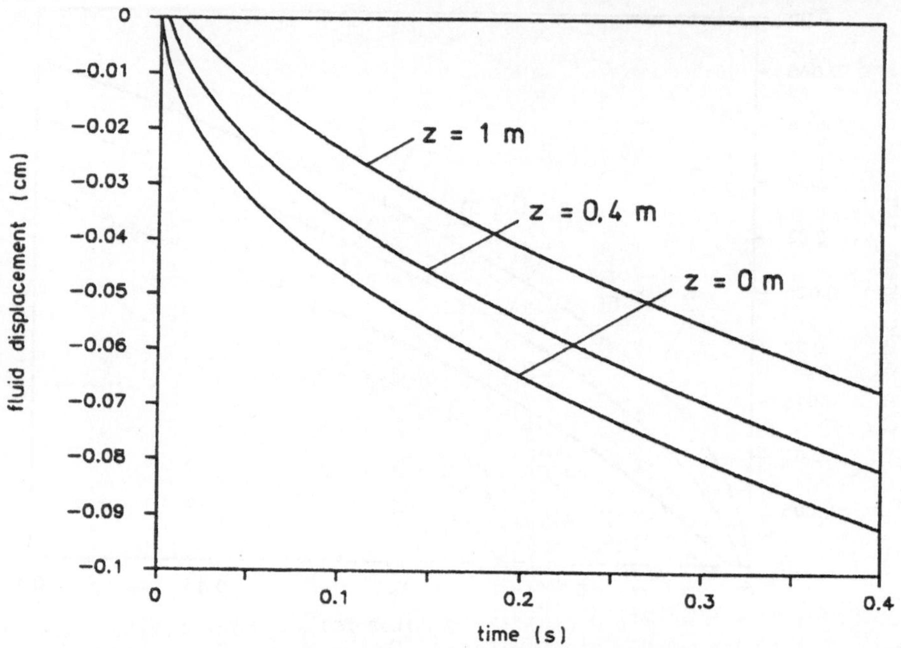

Fig. 5.9.15. Response of fluid displacement vs. time to step loading

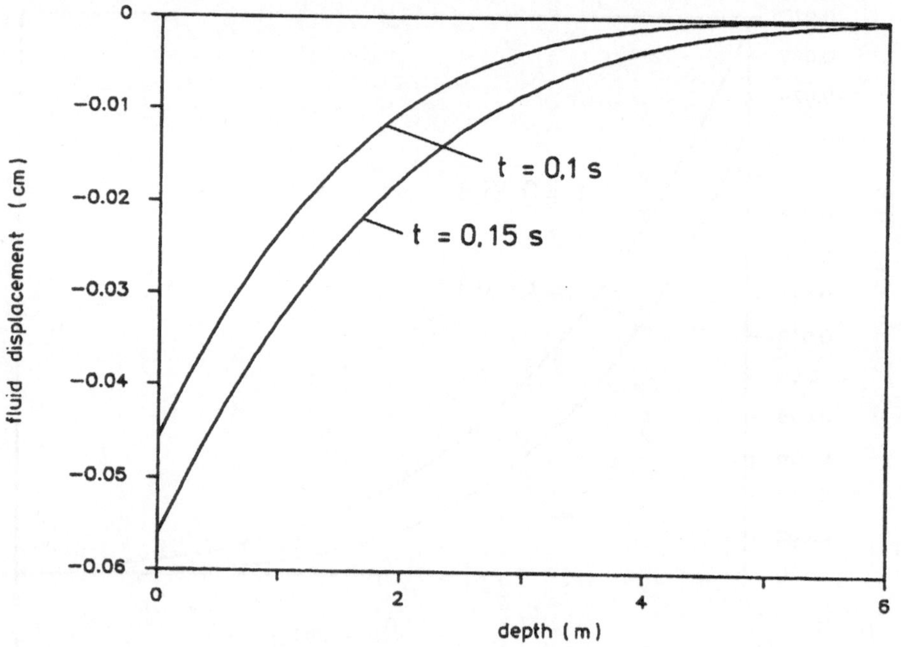

Fig. 5.9.16. Response of fluid displacement vs. depth to step loading

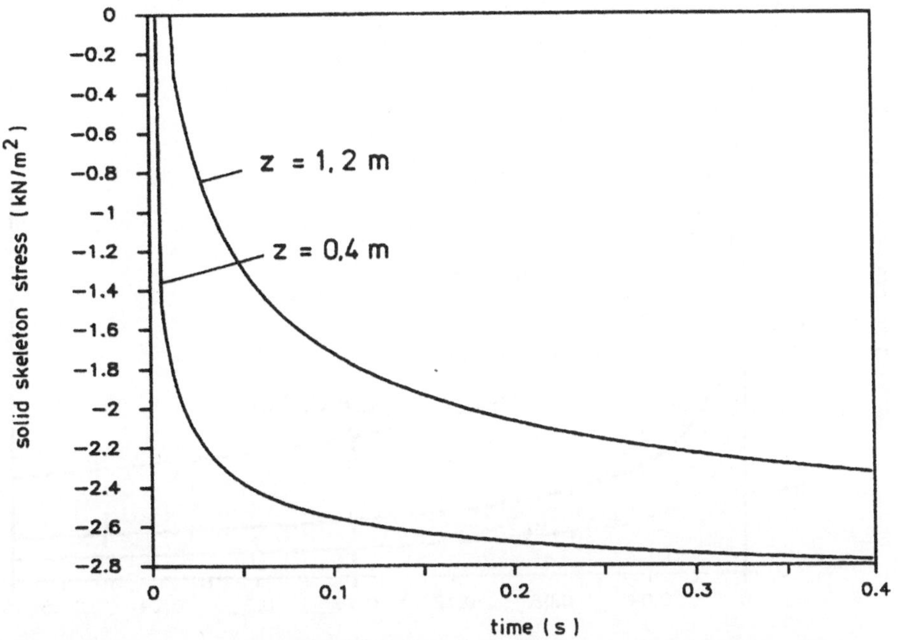

Fig. 5.9.17. Response of solid skeleton effective stress vs. time to step loading

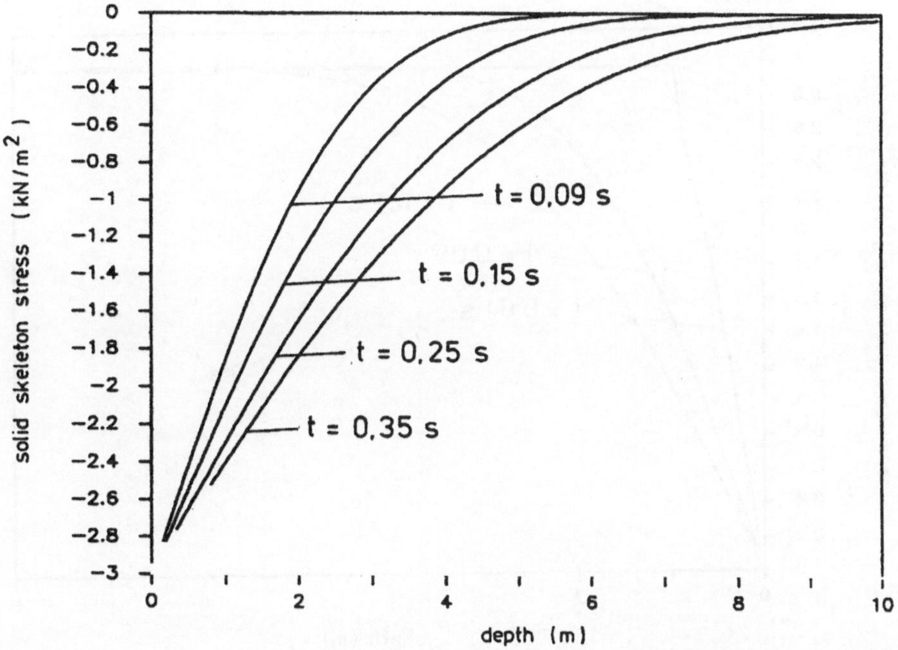

Fig. 5.9.18. Response of solid skeleton effective stress vs. depth to step loading

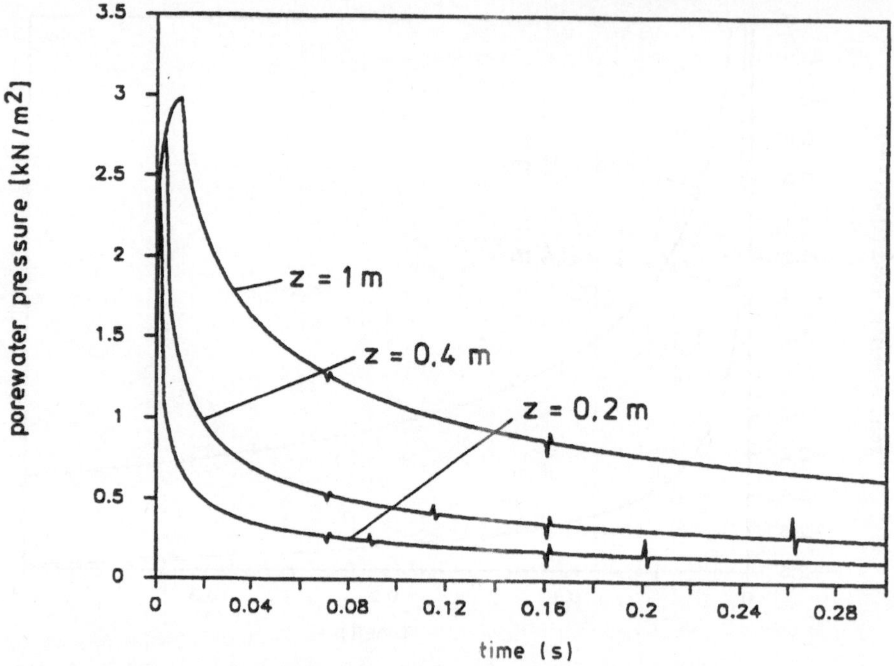

Fig. 5.9.19. Response of solid porewater pressure vs. time to step loading

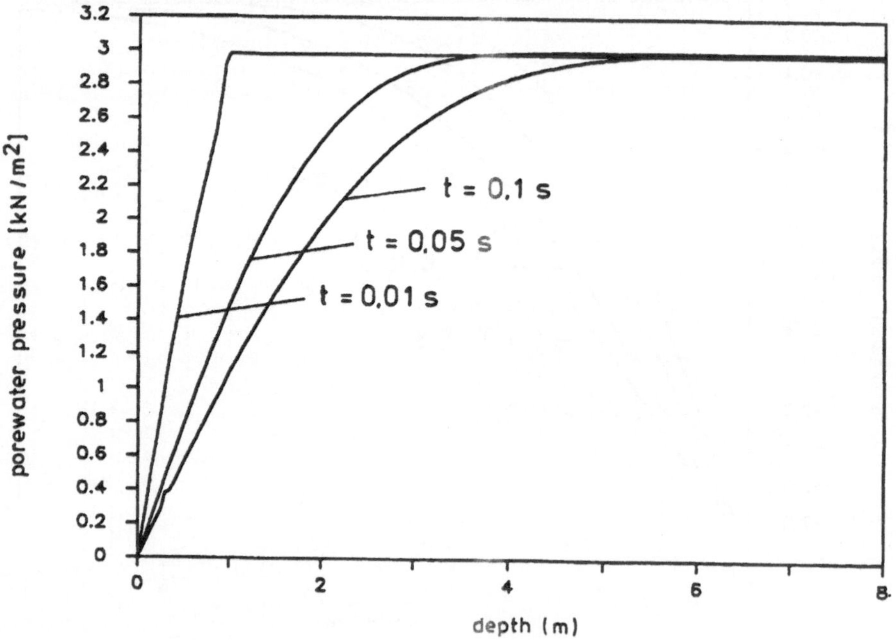

Fig. 5.9.20. Response of solid porewater pressure vs. depth to step loading

During the consolidation process, the solid extra stresses increase with time at a given depth, Fig. 5.9.17, but decrease with the distance from the loading surface at a given time, Fig. 5.9.18. At any depth, the pore pressure decreases to as low as zero, Fig. 5.9.19, when previously, the pressure increasing process has been taken as a function of time and depth, Fig. 5.9.20.

Responses to Impulse Loading

The responses of the medium due to impulse loading (Fig. 5.9.21) can be taken from Figs. 5.9.22 through 5.9.29. As a matter of fact, when an impact is applied to the loading surface, the displacements of the solid and the liquid phases reach their maximum values within a very short time, and then very quickly decrease to smaller values, Figs. 5.9.22 through 5.9.25. The wave propagating process clearly appears in the solid extra stress and the poreliquid pressure curves given with time at different depths and with depth at different times, Figs. 5.9.26 through 5.9.29. Note the sharpness of the pore-liquid pressure functions at different depths at time $t = 0.01$ s, compare with Fig. 5.9.28, which corresponds to the impulsive loading. The variation in poreliquid pressure from positive to negative values close to the surface again exhibits the elastic recovery of the solid skeleton combined with a water absorption process.

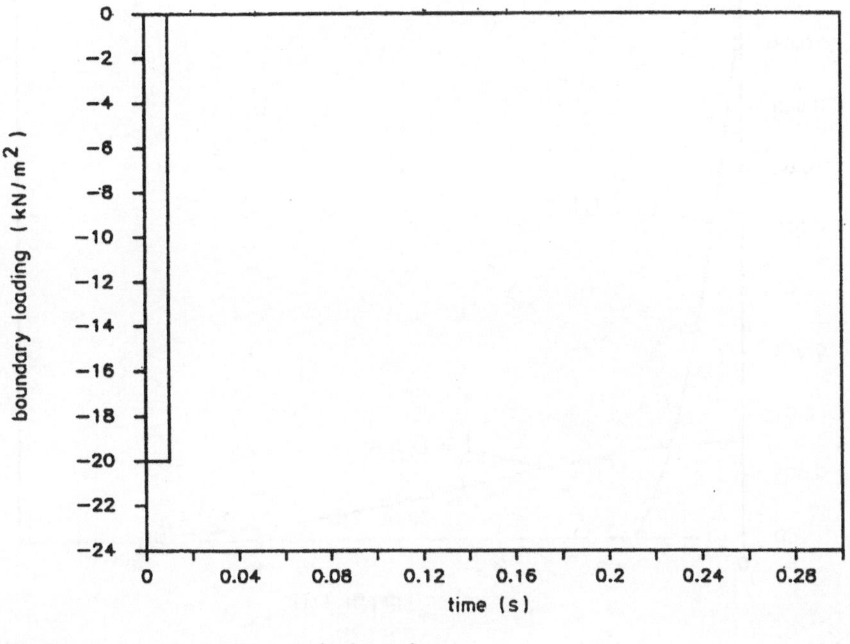

Fig. 5.9.21. Impulse loading at the boundary

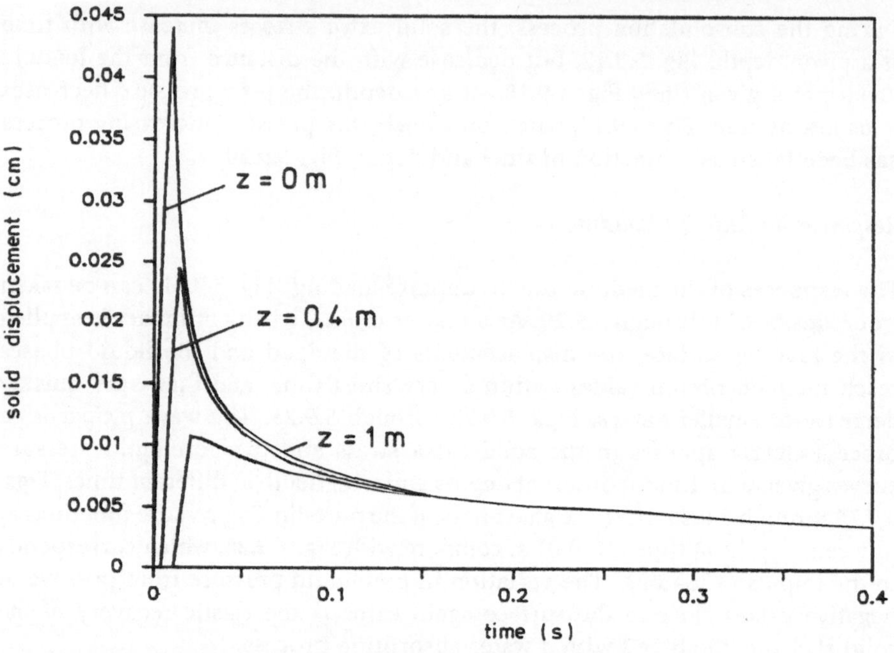

Fig. 5.9.22. Response of solid displacement vs. time to impulse loading

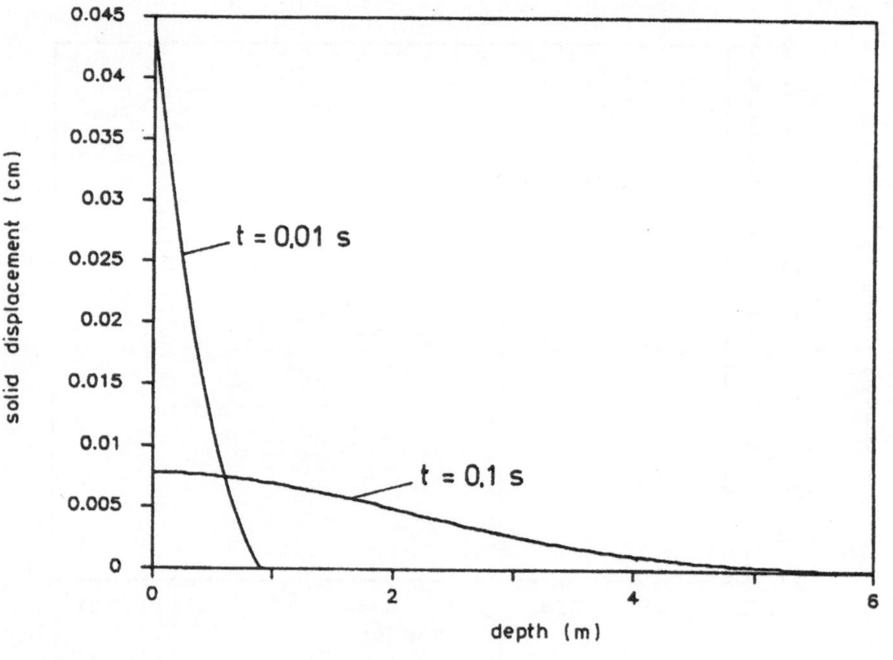

Fig. 5.9.23. Response of solid displacement vs. depth to impulse loading

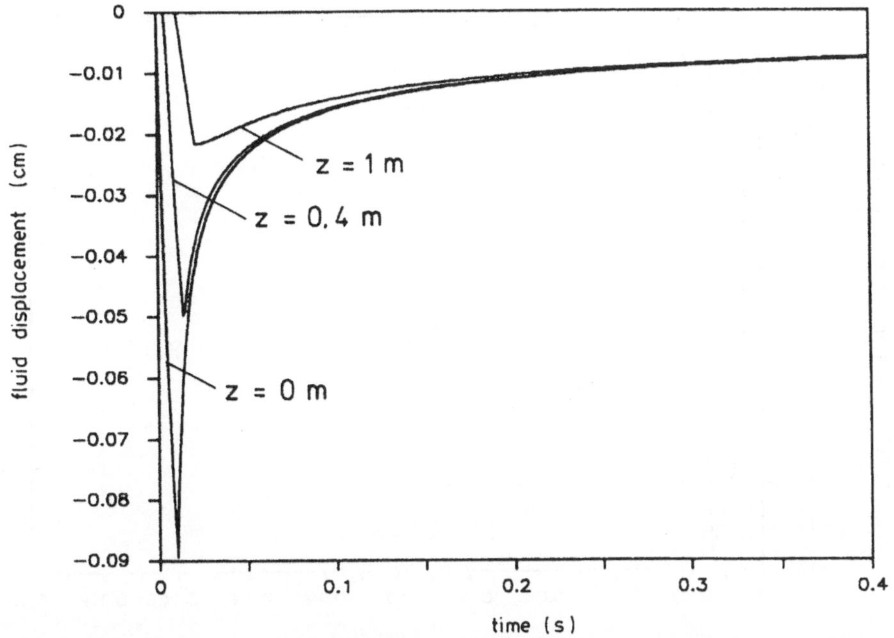

Fig. 5.9.24. Response of fluid displacement vs. time to impulse loading

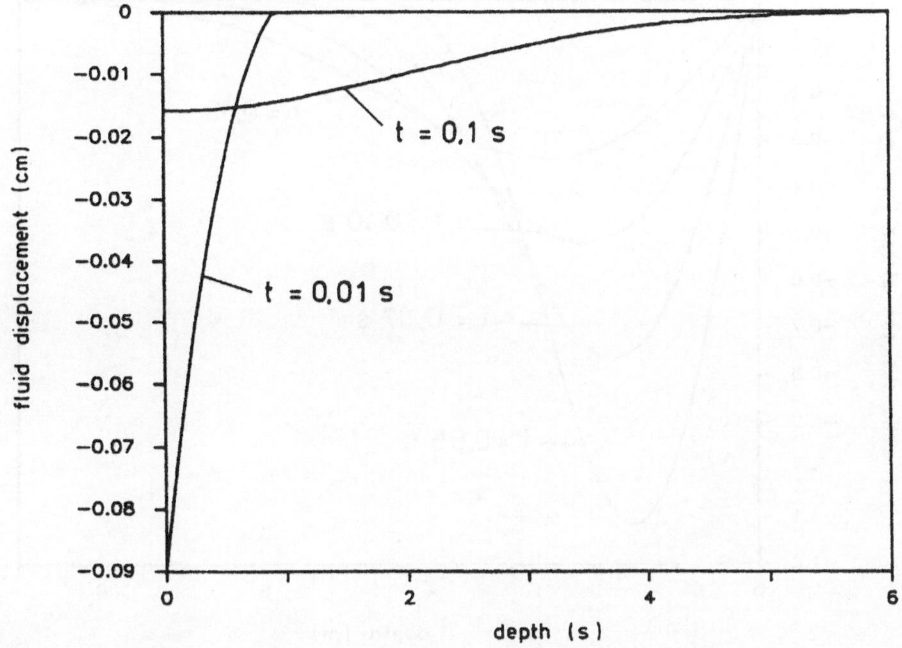

Fig. 5.9.25. Response of fluid displacement vs. depth to impulse loading

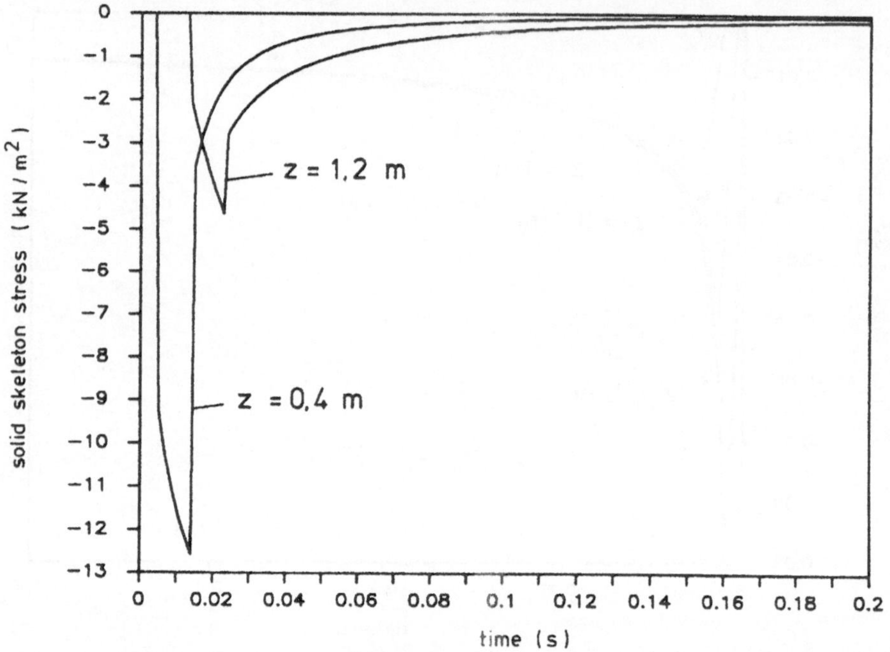

Fig. 5.9.26. Response of solid skeleton effective stress vs. time to impulse loading

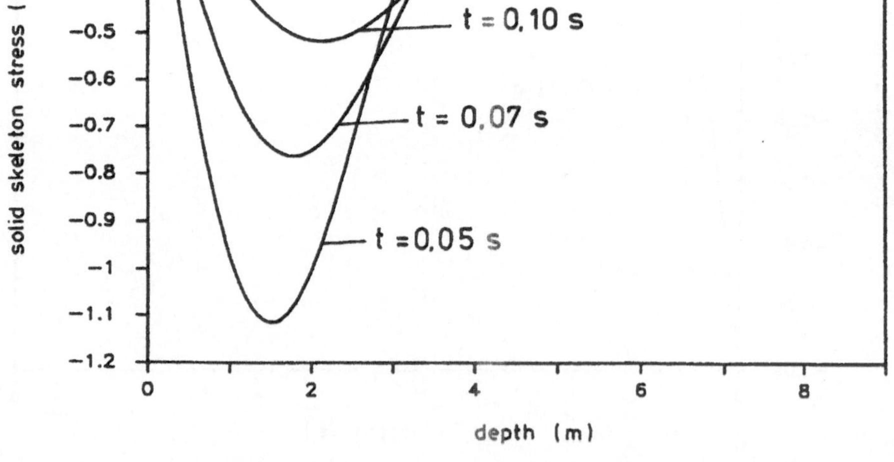

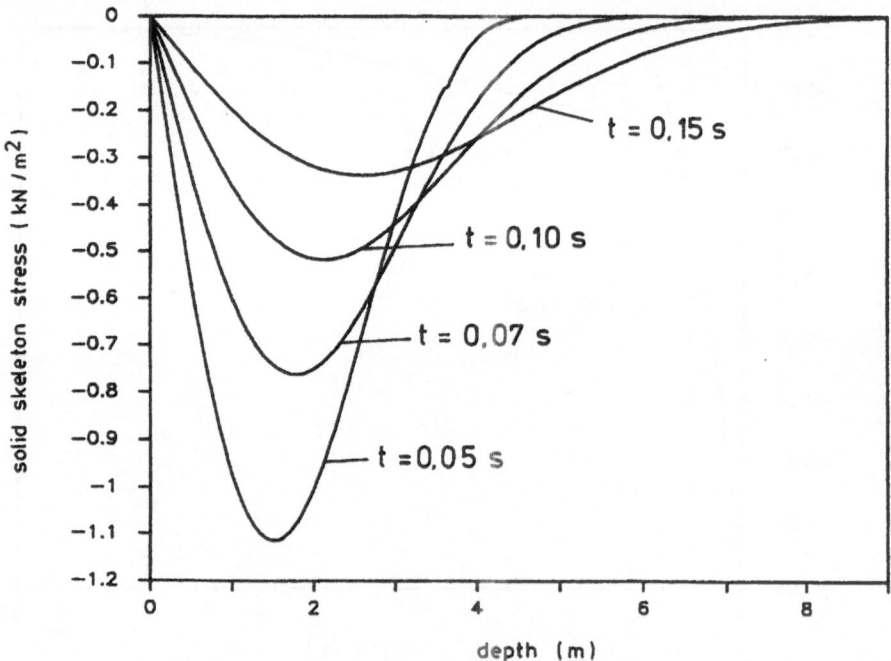

Fig. 5.9.27. Response of solid skeleton effective stress vs. depth to impulse loading

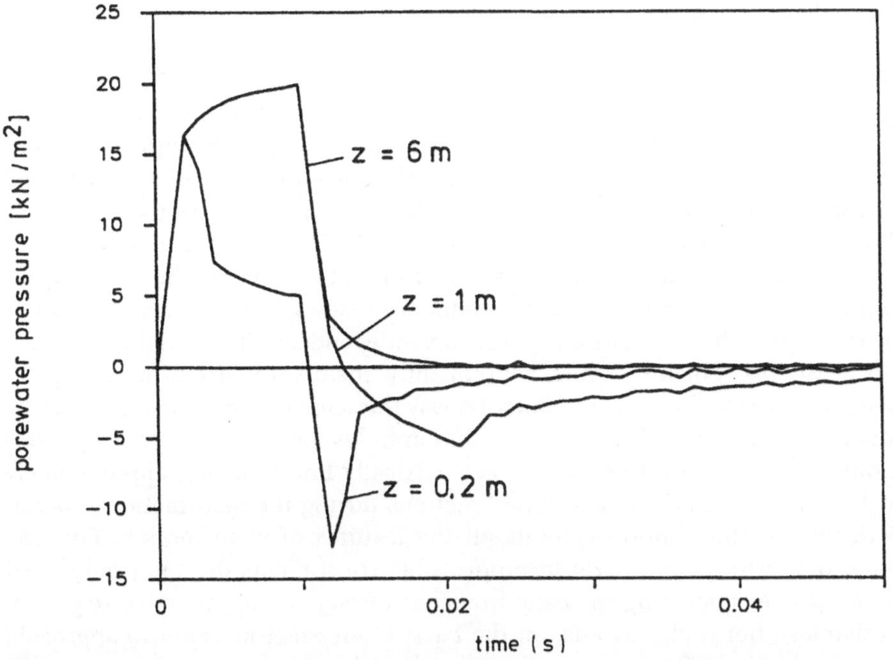

Fig. 5.9.28. Response of porewater pressure vs. time to impulse loading

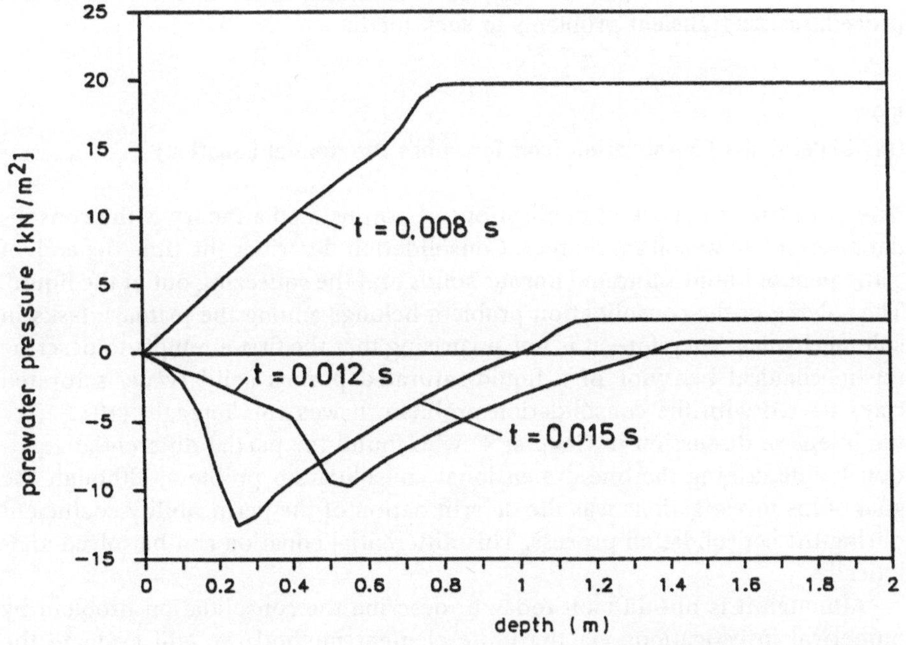

Fig. 5.9.29. Response of porewater pressure vs. depth to impulse loading

Concluding Remarks

An exact solution for a transient analysis of a one-dimensional column of a liquid-saturated elastic porous skeleton has been presented in this section. The saturated porous medium was modeled as a two-phase system with two incompressible constituents, where the general field equations were directly adopted according to the work of de Boer and Ehlers (1990a). The exact solution was obtained by taking the Laplace transformation of the governing equations with the initial- and boundary-conditions. The transient response of the medium was demonstrated with respect to three boundary loading functions. As a result of the incompressibility constraints, only one independent dilatational wave in the two phases was obtained while, consequently, the disturbance propagating velocity included in the unit step function was the same for both the solid and the liquid constituents. Apparently, the solution holds for two incompressible constituents within the framework of geometrically-linear theory, approximately neglecting the variations in volume fractions during the deformation process. Nevertheless, the solution exhibits all the features of wave motion. Furthermore, the assumption of two incompressible constituents does not only meet the properties appearing in many branches of engineering practice (e.g., soil mechanics), but it also avoids, on the basis of an exact mechanical approach, the introduction of many complicated material parameters such as must be considered in the theories of Biot and his subsequent disciples. Moreover, the present solutions allow for the evaluation of the accuracy of some numerical procedures for transient problems in such media.

5.9.3 One-Dimensional Consolidation (von Terzaghi's Differential Equation)

One of the most important applications of porous media theory is the consolidation problem in soil mechanics. Consolidation describes the time-dependent settlement of liquid-saturated porous solids and the squeezing out of the liquid. The solving of the consolidation problem belongs among the primary tasks in soil mechanics. Therefore, it is not surprising that the first attempts to describe the mechanical behavior of a liquid-saturated porous solid (water-saturated clay) started with the consolidation problem. It was von Terzaghi (1923) (see the intensive discussion in Chapter 4) who found the partial differential equation for describing the one-dimensional consolidation problem, although the goal of his investigations was the determination of the permeability coefficient during the consolidation process. This differential equation can be solved analytically.

Although it is not difficult today to describe the consolidation problem by numerical investigations via the finite element method, we will evaluate the basic equations of porous media theory developed previously in order to derive a partial differential equation which corresponds to von Terzaghi's equation. Moreover, we will solve these equations analytically considering the real initial and boundary conditions.

The purpose of the following investigation is to be of interest for the reader in learning how von Terzaghi's famous differential equation, describing the one-dimensional consolidation of clay, comes out from the basic relations of the modern porous media theory and, in particular, which special assumptions have to be introduced.

The derivation of von Terzaghi's (1923) partial differential equation from the basic equations is primarily due to Heinrich (1938a). However, in his paper, there are many ideas from the professors Flamm and, in particular, Lechner, both members of the expert commission in Vienna in 1937 (see Section 4.4). As Heinrich's paper was published in a journal not very well-known in the scientific community, we will refer to his paper primarily word for word. However, we will use the definitions and notations of the modern porous media theory. Concerning the linearization of the fields, we adhere to the statements given in Section 5.9.2.

Before we develop von Terzaghi's (1923) differential equation for the one-dimensional consolidation problem (see Heinrich, 1938a), we will list all the assumptions which are introduced in order to derive the simple differential equation:

- The binary model consists of the incompressible liquid (water) and the incompressible solid (clay) phases;
- in the reference placement the pores are homogeneously distributed and also the real densities are homogeneous;
- as the problem under study will be treated one-dimensionally, only two independent quantities, the time t and the coordinate z, will be introduced (see Fig. 5.9.2);
- the consolidation process runs isothermally;
- the displacements and the spatial derivatives of the displacements are small. Thus, within the scope of these infinitesimal deformations, all terms of higher-order of the displacements and their spatial derivatives will be neglected. This means that we can use the linearized Almansi and Green strain tensors and can approximately replace the volume fractions n^S and n^F by n^S_{0S} and n^F_{0F}, and therefore due to the second assumption, these are approximately constant (see also Section 5.9.2);
- due to all experiences, the changes of the mechanical quantities during the consolidation process occur very slowly. Thus, the velocities $\mathbf{v}_S$ and $\mathbf{v}_F$ and their time derivatives as well as the time derivatives, of stresses and volume fractions, can be considered as small. Therefore, the acceleration $\mathbf{a}_S$ and $\mathbf{a}_F$, e.g., can be replaced by

$$\mathbf{a}_\alpha = \frac{D\,\mathbf{v}_\alpha}{Dt} = \frac{\partial\,\mathbf{v}_\alpha}{\partial t} + (\operatorname{grad}\mathbf{v}_\alpha)\mathbf{v}_\alpha = \frac{\partial\,\mathbf{v}_\alpha}{\partial t}\,; \tag{5.9.78}$$

- viscous effects within the real liquid will not be considered.

We begin our investigations with the equations (5.9.32) and (5.9.33), as well as (5.9.27), which can be written for the assumed one-dimensional case, and with the results of the constitutive theory in Section 5.8.2 (**a**) as:

$$\frac{\partial p}{\partial z} + \frac{1}{n^S}\frac{\partial(n^S p^{SC})}{\partial z} - \gamma^{SR} + \rho^{SR}\frac{\partial v_S}{\partial t} - \frac{1}{n^S} S_v(v_F - v_S) = 0 \tag{5.9.79}$$

and

$$\frac{\partial p}{\partial z} - \gamma^{FR} + \rho^{FR}\frac{\partial v_F}{\partial t} + \frac{1}{n^F} S_v(v_F - v_S) = 0\,, \tag{5.9.80}$$

where the specific weight

$$\gamma^{\alpha R} = \rho^{\alpha R} g \tag{5.9.81}$$

has been introduced. Furthermore, we obtain, from (5.9.27),

$$\frac{\partial(n^S v_S + n^F v_F)}{\partial z} = 0\,. \tag{5.9.82}$$

By integrating (5.9.82), we have

$$n^S v_S + n^F v_F = C(t)\,, \tag{5.9.83}$$

where C(t) is a time-dependent function. However, if we assume that at a certain depth a non-permeable layer exists ($v_S = 0,\ v_F = 0$), it turns out that C(t) is equal to zero and we can write:

$$n^S v_S + n^F v_F = 0\,. \tag{5.9.84}$$

In consideration of the assumptions introduced at the beginning of this section, we see that, from (5.9.84),

$$n^S\frac{\partial v_S}{\partial t} + n^F\frac{\partial v_F}{\partial t} = 0 \tag{5.9.85}$$

is obtained.

The configuration pressure p^{SC} is given in Section 5.8.2, see also $(5.8.57)_1$, i.e.,

$$p^{SC} = -\rho^S\frac{\partial \psi^S}{\partial n^S}\,, \tag{5.9.86}$$

or considering the saturation condition, i.e.,

$$\begin{aligned} p^{SC} &= \rho^S\frac{\partial \psi^S}{\partial n^F} \\ p^S_E &= n^S p^{SC} = F(n^F) \end{aligned} \tag{5.9.87}$$

with

$$F(n^F) = n^S\rho^S\frac{\partial \psi^S}{\partial n^F}\,. \tag{5.9.88}$$

The dependence of the effective partial configuration pressure $n^S p^{SC}$ on the porocity n^F in $(5.9.87)_2$ was already required by Fillunger (1936).

We replace the velocity v_S and the acceleration $\frac{\partial v_S}{\partial t}$ in (5.9.78) and (5.9.79) by v_F and $\frac{\partial v_F}{\partial t}$ as well as p^{SC} by $F(n^F)$ via Eqns. (5.9.84) and (5.9.85), as well as (5.9.88):

$$\frac{\partial p}{\partial z}+\frac{1}{n^S}\frac{\partial F(n^F)}{\partial z}-\gamma^{SR}-\rho^{SR}\frac{1}{n^S}\frac{(\partial n^F v_F)}{\partial t}-S_v\frac{1}{n^F(n^S)^2}(n^F v_F)=0\,, \tag{5.9.89}$$

$$\frac{\partial p}{\partial z}-\gamma^{FR}+\rho^{FR}\frac{1}{n^F}\frac{\partial(n^F v_F)}{\partial t}+S_v\frac{1}{n^S(n^F)^2}(n^F v_F)=0\,. \tag{5.9.90}$$

We differentiate (5.9.89) and (5.9.90) with respect to the coordinate z, neglecting higher-order terms:

$$\frac{\partial^2 p}{\partial z^2}+\frac{1}{n^S}\frac{\partial^2 F(n^F)}{\partial z}-\rho^{SR}\frac{1}{n^S}\frac{\partial}{\partial t}\left[\frac{\partial(n^F v_F)}{\partial z}\right]-S_v\frac{1}{n^F(n^S)^2}\frac{\partial(n^F v_F)}{\partial z}=0\,, \tag{5.9.91}$$

$$\frac{\partial^2 p}{\partial z^2}+\rho^{FR}\frac{1}{n^F}\frac{\partial}{\partial t}\left[\frac{\partial(n^F v_F)}{\partial z}\right]+S_v\frac{1}{n^S(n^F)^2}\frac{\partial(n^F v_F)}{\partial z}=0\,. \tag{5.9.92}$$

From (5.9.25),

$$\frac{\partial n^\alpha}{\partial t}+\operatorname{div}(n^\alpha \mathbf{v}_\alpha)=0 \tag{5.9.93}$$

is obtained, which simplifies, in the one-dimensional case, to

$$\frac{\partial n^\alpha}{\partial t}+\frac{\partial(n^\alpha v_\alpha)}{\partial z}=0\,. \tag{5.9.94}$$

With the relation (5.9.94), we can eliminate the fluid velocity v_F and obtain, instead of (5.9.91) and (5.9.92),

$$\frac{\partial^2 p}{\partial z^2}+\frac{1}{n^S}\frac{\partial^2 F(n^F)}{\partial z^2}+\rho^{SR}\frac{1}{n^S}\frac{\partial^2 n^F}{\partial t^2}+S_v\frac{1}{n^F(n^S)^2}\frac{\partial n^F}{\partial t}=0\,, \tag{5.9.95}$$

$$\frac{\partial^2 p}{\partial z^2}-\rho^{FR}\frac{1}{n^F}\frac{\partial^2 n^F}{\partial t^2}-S_v\frac{1}{n^S(n^F)^2}\frac{\partial n^F}{\partial t}=0\,. \tag{5.9.96}$$

We subtract (5.9.96) from (5.9.95) and gain:

$$(\rho^{SR}\frac{1}{n^S}+\rho^{FR}\frac{1}{n^F})\frac{\partial^2 n^F}{\partial t^2}+S_v\frac{1}{(n^F)^2(n^S)^2}\frac{\partial n^F}{\partial t}+\frac{1}{n^S}\frac{\partial^2 F(n^F)}{\partial z^2}=0\,. \tag{5.9.97}$$

If $F(n^F)$ would be known, Eqn. (5.9.97) could be considered as a differential equation for n^F. However, we would like to derive a differential equation for the partial configuration pressure p^{SC}. Differentiating $(5.9.87)_2$ with respect to the time t, we obtain:

$$\frac{\partial p_E^S}{\partial t}=F'(n^F)\frac{\partial n^F}{\partial t}\,,\quad F'(n^F)=\frac{\partial F}{\partial n^F} \tag{5.9.98}$$

or

$$\frac{\partial n^F}{\partial t}=\frac{1}{F'(n^F)}\frac{\partial p_E^S}{\partial t}\,. \tag{5.9.99}$$

We again differentiate (5.9.99) with respect to the time t and, neglecting the higher-order terms, arrive at:

$$\frac{\partial^2 n^F}{\partial t^2} = \frac{1}{F'(n^F)} \frac{\partial^2 p_E^S}{\partial t^2} \, . \tag{5.9.100}$$

Finally, if we consider (5.9.98), (5.9.99), and (5.9.100) in (5.9.97), we obtain a differential equation for the partial configuration pressure:

$$\left(\frac{\rho^{SR}}{n^S} + \frac{\rho^{FR}}{n^F}\right) \frac{\partial^2 p_E^S}{\partial t^2} + \frac{S_v}{(n^F)^2 (n^S)^2} \frac{\partial p_E^S}{\partial t} + \frac{F'(n^F)}{n^S} \frac{\partial^2 p_E^S}{\partial z^2} = 0 \, . \tag{5.9.101}$$

Note that the coefficients in the differential equation (5.9.100) are variable quantities. From experience, however, it is known that within the practical integration area, the variation of n^F is small (the variation becomes smaller as the applied load and the layer are decreased). Thus, in this case, one can calculate with constant average values of the coefficients.

In order to also derive a differential equation for the porewater pressure p we proceed from (5.9.95) and (5.9.96) and eliminate, at first, p_E^S. This is possible if we differentiate Eqn. $(5.9.87)_2$ two times with respect to z. We obtain, in an analogous way to (5.9.100):

$$\frac{\partial^2 n^F}{\partial z^2} = \frac{1}{F'(n^F)} \frac{\partial^2 p_E^S}{\partial z^2} \, . \tag{5.9.102}$$

From (5.9.95), considering (5.9.102), we have:

$$\frac{\partial^2 p}{\partial z^2} + \frac{F'(n^F)}{n^S} \frac{\partial^2 n^F}{\partial z^2} + \rho^{SR} \frac{1}{n^S} \frac{\partial^2 n^F}{\partial t^2} + S_v \frac{1}{n^F (n^S)^2} \frac{\partial n^F}{\partial t} = 0 \, . \tag{5.9.103}$$

For the further investigations, we reformulate (5.9.96) and (5.9.103):

$$\begin{aligned} \mathcal{D}_F n^F + \frac{\partial^2 p}{\partial z^2} &= 0 \, , \\ \mathcal{D}_S n^F + \frac{\partial^2 p}{\partial z^2} &= 0 \, , \end{aligned} \tag{5.9.104}$$

where the differential operators $\mathcal{D}_F$ and $\mathcal{D}_S$ are defined by

$$\begin{aligned} \mathcal{D}_F &= -\frac{\rho^{FR}}{n^F} \frac{\partial^2}{\partial t^2} - \frac{S_v}{n^S (n^F)^2} \frac{\partial}{\partial t} \, , \\ \mathcal{D}_S &= \frac{\rho^{SR}}{n^S} \frac{\partial^2}{\partial t^2} + \frac{S_v}{n^F (n^S)^2} \frac{\partial}{\partial t} + \frac{F'(n^F)}{n^S} \frac{\partial}{\partial z^2} \, . \end{aligned} \tag{5.9.105}$$

We multiply $(5.9.104)_1$ by $\mathcal{D}_S$, $(5.9.104)_2$ by $\mathcal{D}_F$, and subtract then $(5.9.104)_2$ from $(5.9.104)_1$, thereby obtaining

$$(\mathcal{D}_S - \mathcal{D}_F) \, \triangle \, p = 0 \, , \tag{5.9.106}$$

where we have introduced

$$\triangle p = \frac{\partial^2 p}{\partial z^2} \, . \tag{5.9.107}$$

Finally, we have

$$\left(\frac{\rho^{SR}}{n^S} + \frac{\rho^{FR}}{n^F}\right)\frac{\partial \triangle p}{\partial t^2} + \frac{S_v}{(n^S)^2(n^F)^2}\frac{\partial \triangle p}{\partial t} + \frac{F'(n^F)}{n^S}\frac{\partial \triangle p}{\partial z^2} = 0\,. \quad (5.9.108)$$

This is a fourth-order differential equation for the poreliquid pressure p.

Now, we return to the differential equation (5.9.101) for the effective pressure p_E^S. At first, however, we will determine the original distribution of the effective pressure. It is assumed that, before the load is applied which causes the settlement process, the mixture body is at rest. If $\overset{\circ}{p}{}_E^S$ and $\overset{\circ}{p}$ are the partial configuration and poreliquid pressures in this state, then the equations of motions (5.9.79) and (5.9.80) yield:

$$\frac{1}{n_{0S}^S}\frac{\partial \overset{\circ}{p}{}_E^S}{\partial z} + \frac{\partial \overset{\circ}{p}}{\partial z} - \gamma^{SR} = 0\,, \quad (5.9.109)$$

and

$$\frac{\partial \overset{\circ}{p}}{\partial z} - \gamma^{FR} = 0\,. \quad (5.9.110)$$

From (5.9.109) and (5.9.110), we have

$$\frac{\partial \overset{\circ}{p}{}_E^S}{\partial z} = (\gamma^{SR} - \gamma^{FR})n_{0S}^S\,. \quad (5.9.111)$$

Assuming that the cover at the upper boundary is permeable, where the liquid can flow freely, then, from (5.9.110),

$$\overset{\circ}{p} = \bar{p} + \gamma^{FR} z \quad (5.9.112)$$

is obtained, where $\bar{p}$ is the atmospheric pressure. Moreover, (5.9.111) yields

$$\overset{\circ}{p}{}_E^S = \bar{p}_E^S + (\gamma^{SR} - \gamma^{FR})n_{0S}^S z\,, \quad (5.9.113)$$

with $\bar{p}_E^S$ as the effective pressure at the upper boundary. It is recognized, from (5.9.112) and (5.9.113), that the poreliquid pressure and the partial effective pressure increase linearly with the depth before the settlement starts.

As Eqn. (5.9.113) identically fulfills the differential equation (5.9.100), the effective pressure p_E^S in (5.9.100) can be considered as the additional configuration pressure which is caused by the load. This idea has the advantage that the boundary conditions become easier, because the additional configuration pressure in the control space is zero. Immediately after the abrupt application of the load, the velocities v_S and v_F in the whole region are zero because the velocities v_S and v_F cannot experience a jump. After (5.9.93), $\frac{\partial n^\alpha}{\partial t}$ is also zero, and, from (5.9.97), it is recognized that the effective pressure has not changed. For $t = 0$, the additional effective pressure p_E^S is equal to zero in the total control space and, since $\frac{\partial n^F}{\partial t}$ is equal to zero, the time derivative of p_E^S is also equal to zero, see (5.9.99). Thus, we can state the initial condition,

$$t = 0 : \begin{cases} v_F = 0 \,,\; v_S = 0 \,, \\ p_E^S = 0 \,,\; \dfrac{\partial p_E^S}{\partial t} = 0. \end{cases} \tag{5.9.114}$$

At t equal to zero, the settlement then begins with the velocity equal to zero.

In order to correspond to the usual notions in soil mechanics, the quantities n^F and n^S, S_v and $F'(n^F)$ will be expressed by the void ratio ϵ, the permeability coefficient k^F, and the compression modulus K. The void ratio ϵ is defined by the equation

$$\epsilon = \frac{n^F}{n^S} \,. \tag{5.9.115}$$

Considering the saturation condition, Eqn. (5.9.115) yields

$$n^F = \frac{\epsilon}{1+\epsilon} \,, \quad n^S = \frac{1}{1+\epsilon} \,. \tag{5.9.116}$$

The coefficient k^F can be determined by the permeability test. Water will be stationarily pressed through a clay cylinder at rest with a cross-section of 1 cm^2, length of 1 cm, and a constant pore volume n^F with the pressure gradient $-\dfrac{\partial p}{\partial z}$ in a horizontal direction. The amount of water which flows through the cylinder per second will be denoted by Q. Then, Darcy's law can be written in the following form:

$$Q = -\frac{k^F}{\gamma^{FR}} \frac{\partial p}{\partial z} \,. \tag{5.9.117}$$

In this manner, k^F is defined.

If one applies the equation of motion for the fluid (5.9.80) to the permeability test, then the permeability quantity S_v can be determined. In this case, the acceleration $\dfrac{\partial v_F}{\partial t}$ is equal to zero, due to the vanishing velocity of the solid phase (the clay cylinder is at rest), the stationary flow, and the constant pore volume. Moreover, due to the horizontal flow, the term $-\gamma^{FR} n^F$ can be neglected. Then, we obtain, from (5.9.80),

$$S_v v_F + n^F \frac{\partial p}{\partial z} = 0 \,. \tag{5.9.118}$$

There is a simple relation between v_F and Q. If f_W is that part of the plane cylinder cross-section which is related to the fluid, then the equation $Q = f_W v_F$ is valid, where f_W is the surface porosity. However, for statistically distributed pores, the surface porosity is equal to the volume porosity (see Delesse, 1848). Thus,

$$Q = n^F v_F \tag{5.9.119}$$

is valid. From the comparison of (5.9.117) with (5.9.118), considering (5.9.119) and (5.9.116), S_v is determined:

$$S_{\upsilon} = \frac{\gamma^{FR}}{k^F} \frac{\epsilon^2}{(1+\epsilon)^2} \,. \tag{5.9.120}$$

The compression modulus K can be experimentally determined via the consolidometer test. At first, with this modulus the effective pressure p_E^S is established as a function of the void ratio ϵ. Then, the compression modulus is defined by:

$$K = -\,\frac{d\epsilon}{dp_E^S} \,. \tag{5.9.121}$$

The quantity K can be easily read from the $p_E^S - \epsilon$-curve.

If one differentiates Eqn. $(5.9.87)_2$ with respect to p_E^S, we have

$$1 = F'(n^F)\,\frac{dn^F}{d\epsilon}\,\frac{d\epsilon}{dp_E^S} \,, \tag{5.9.122}$$

or, with (5.9.121) and (5.9.116),

$$F'(n^F) = -\,\frac{(1+\epsilon)^2}{K} \,, \tag{5.9.123}$$

where $F'(n^F)$ denotes the derivative of $F(n^F)$ with respect to n^F. Considering (5.9.116), (5.9.120), and (5.9.123) in Eqn. (5.9.101), we obtain

$$\frac{k^F(\epsilon\rho^{SR} + \rho^{FR})}{\gamma^{FR}\epsilon(1+\epsilon)}\frac{\partial^2 p_E^S}{\partial t^2} + \frac{\partial p_E^S}{\partial t} - \frac{k^F(1+\epsilon)}{\gamma^{FR}K}\frac{\partial^2 p_E^S}{\partial z^2} = 0 \,. \tag{5.9.124}$$

This is a partial differential equation for the effective hydrostatic stress state p_E^S. In the following paragraph, we will investigate the coefficients in (5.9.124) and simplify this differential equation.

With (5.9.113), we have already calculated the configuration pressure in the reference placement:

$$\overset{\circ}{p}{}_E^S = \bar{p}_E^S + (\gamma^{SR} - \gamma^{FR})\frac{1}{1+\epsilon_o} z \,, \tag{5.9.125}$$

where $(5.9.116)_2$ has been used and $\gamma^{FR}\dfrac{1}{1+\epsilon_o}z$ is the uplift force. Now, if Δp_E^S is the increase of the effective pressure and $\Delta\epsilon$, the corresponding change of ϵ, it follows from (5.9.121), replacing the differential quotient by the difference quotient, that

$$\Delta\epsilon = -\,K\Delta p_E^S \,. \tag{5.9.126}$$

From (5.9.126), one can determine for every practical problem the maximum Δp_E^S for the change of ϵ. If one has gained k^F and K from experiments as functions of ϵ, then also the changes of k^F and K are known. Moreover, now one is also able to find out the maximum deviations of the coefficients of Eqn. (5.9.124) and can recognize whether or not they exceed a permissible measurement. Also the change of ϵ with increasing depth can be estimated. The results depend apart from the clay nature, on the thickness of the layer and the specific load.

Table 5.2. Response parameters for different kinds of clay (Heinrich, 1938)

Different kinds of clay	Pore-number ϵ	Compression-modulus K $[cm^2/kg]10^{-2}$	Permeability-coefficient k^F $[cm/min]10^{-7}$	$\frac{k^F(1+\epsilon)}{K\gamma^{FR}}$ $[cm^2/min]10^{-2}$	$\frac{k^F(\epsilon\gamma^{SR}+\gamma^{FR})}{\gamma^{FR}g\epsilon(1+\epsilon)}$ $[min]10^{-13}$
Colorful stiff Viennese clay	0.73	1.9	0.8	0.728	53.3
Blue-grey marl (Vienna)	0.77	2.5	4.5	3.185	2.88
Stiff meager clay (Vienna)	0.71	2.0	17	14.52	0.157
Blue sandy clay (Vienna)	0.68	3.0	65	36.4	0.457
Stiff-plastic clay (Paris)	0.67	2.3	3.8	2.76	2.7
Brown clay (Cairo)	1.07	3.5	2.5	1.48	1.24
Clay from Halikko (Finland)	1.60	17.0	50.0	7.65	0.181

Now, for the estimation of the order of the terms in (5.9.124), for several kinds of clay, the coefficients of Eqn. (5.9.124) will be calculated and listed in Table 5.2. The following common datas will be used for the calculations: $g = 3.53 \cdot 10^6\ cm/(min)^2$, $\gamma^{FR} = 10^{-3}\ kg/cm^2$, and $\gamma^{SR} = 2.7 \cdot 10^{-3}\ kg/cm^2$.

It is recognized that for all kinds of clay the coefficient of $\frac{\partial^2 p_E^S}{\partial t^2}$ is so small that this term can be neglected. This conclusion, and some additional statements in Heinrich's (1938a) paper, were heavily attacked by Flamm (1938), a member of the inquiry commitee in Vienna in 1937 (see Section 4.4). If, in (5.9.124), one neglects the product of the acceleration with the corresponding coefficient, the differential equation

$$\frac{\partial p_E^S}{\partial t} = \frac{k^F(1+\epsilon)}{\gamma^{FR}K}\frac{\partial^2 p_E^S}{\partial z^2} \qquad (5.9.127)$$

remains. In evaluating (5.9.127), one must dispense, however, with the physically-required initial condition that, for t equal to zero, the time derivative of the configuration pressure $\frac{\partial p_E^S}{\partial t}$ is equal to zero. In this case, the formula for the settlement does not start with the velocity equal to zero.

Note that the differential equation (5.9.127) can be obtained directly from the basic equations, if the inertia terms are neglected.

By adding the equations of motion (5.9.79) and (5.9.80), neglecting the inertia terms, we obtain

$$\frac{\partial p}{\partial z} + \frac{\partial p_E^S}{\partial z} - n^S \gamma^{SR} - n^F \gamma^{FR} = 0 . \tag{5.9.128}$$

For a fixed time t, by intergrating (5.9.128) with respect to z, we have

$$p + p_E^S = \int_0^z (n^S \gamma^{SR} + n^F \gamma^{FR}) dz + q(t) . \tag{5.9.129}$$

The integral represents nothing more than the weight of the clay layer of the height z per unit of the surface, and $q(t)$ represents load changing with time on the upper layer. Thus, in the quasi-static range, the sum of the pressures can immediately be determined.

As n^S and n^F change only a little, we can solve (5.9.129):

$$p + p_E^S = (n^S \gamma^{SR} + n^F \gamma^{FR}) z + q(t) . \tag{5.9.130}$$

Finally, in (5.9.113), one can replace n_{0S}^S by n^S because the change of n^S is very small. Subtracting (5.9.112) and (5.9.113) from (5.9.130), we obtain

$$p_E^S - \overset{\circ}{p}{}_E^S = -(p - \overset{\circ}{p}) + q(t) - \bar{p}_E^S - \bar{p} . \tag{5.9.131}$$

We write:

$$w = p - \overset{\circ}{p} , \tag{5.9.132}$$

where w means the additional water pressure caused by the load. As previously shown, the additional effective pressure satisfies the differential equation (5.9.124), and therefore also (5.9.127). Thus, the right-hand side of (5.9.131) must also satisfy the differential equation (5.9.127). Considering (5.9.132), we obtain from (5.9.127) and (5.9.131)

$$\frac{\partial w}{\partial t} - \frac{dq(t)}{dt} = \frac{k^F(1+\epsilon)}{\gamma^{FR} K} \frac{\partial^2 w}{\partial z^2} . \tag{5.9.133}$$

If the load $q(t)$ does not depend on the time, then (5.9.133) can be replaced by

$$\frac{\partial w}{\partial t} = \frac{k^F(1+\epsilon)}{\gamma^{FR} K} \frac{\partial^2 w}{\partial z^2} , \tag{5.9.134}$$

which is nothing more than von Terzaghi's (1923) famous differential equation.

In his conclusions, Heinrich (1938a) wrote that he had tried to treat the settlement problem of clay layers on a scientific basis. This statement was heavily attacked by Flamm (1938), who was of the opinion that Heinrich's statement could create the impression that the derivations of von Terzaghi were not scientifically well-founded.

From the point of view of the modern porous media theory, one has to support Heinrich's view. Indeed, von Terzaghi's derivation of the partial differential equation for describing the consolidation problem is not based on the principles of mechanics but on test observations and was introduced by an analogous conclusion. Such methods, also applied by Fick and Biot (see Chapter 3 and 4), do not support the progress of science; quite the contrary, they can hinder it. It is not known that von Terzaghi's treatment of the behavior of a fluid-saturated porous solid has given rise to some progress in the theories of porous media. There have been no contributions made, e.g., to elastic-plastic deformations, to dynamics, and to phase transitions.

It should be mentioned that with the solution of (5.9.134) for the porewater overpressure, one is able to form the effective stresses. Then, connecting these with Young's modulus, the one-dimensional settlements of porous media within the framework of the geometrically-linear theory can be calculated.

5.9.4
The Elastic-Plastic Compaction of Metallic Powders

For aerospace and other high-tech applications, many powder-metallurgy alloys have been developed in order to achieve better combinations of strength, toughness, fatigue resistance, and resistance to stress-corrosion cracking than those found in alloys produced by ingot metallurgy (see Daraivelu *et al.*, 1984). The success of powder-metallurgy processing depends heavily upon the ability to economically produce a near-net-shape. The forming of this shape is dependent on the success of the die compaction process in delivering defect-free, uniform-density green parts (Lewis *et al.*, 1993). The compaction process for compressing the powder is, without any doubt, the main process in manufacturing engineered products in powder metallurgy. Therefore, this process should be clearly understood from the mechanical and thermodynamic point of view, the more so as there are many difficulties which exist in the compaction process for powders. This concerns, for example, the non-homogeneous density distribution and considerably large residual stresses in the green end product. Hence, a need exists to develop a mathematical model which can predict mechanical phenomena for the compaction process.

This mathematical model must be based on an appropriate plasticity theory, because the compression of the metallic powder during the compaction process is caused by the plastic deformation of the powder. Due to the metallic properties of the powder or the green, a von-Mises-type plasticity theory seems to be appropriate. However, the yield function of von Mises (1913) has to be extended in order to include the considerably large volume changes of the powder during the compaction process due to the porosity. This effect is unknown in the classical plasticity of metals, where incompressibility of the material in the plastic range is assumed (see von Mises, 1913).

In the following section, an elastic-plasticity theory which meets the above statements will be developed. The incorporation of elastic deformation of the powder is justified by the numerical results represented at the end of this section.

The development of the elastoplasticity theory will be performed within the framework of the geometrically-linear theory.

We proceed from Section 5.8.5 and commence with the yield function. The yield function limits the elastic state, not only the deviatoric stress state as in the classical plasticity theory of metals, but also the hydrostatic state. We choose the yield function (5.8.136) represented, however, by the effective second Piola-Kirchhoff stress tensor $\mathbf{S}_E^S$:

$$F = \frac{1}{\sqrt{2}}\sqrt{\mathbf{S}_E^{SD} \cdot \mathbf{S}_E^{SD} + \alpha^2 (\mathbf{S}_E^S \cdot \mathbf{I})^2} - \kappa = 0 \tag{5.9.135}$$

or, formulated with the stress invariants,

$$F = \frac{1}{\sqrt{2}}\sqrt{2\,\mathrm{II}_{\mathbf{S}_E^{SD}} + \alpha^2 \mathrm{I}^2_{\mathbf{S}_E^S}} - \kappa = 0\,. \tag{5.9.136}$$

Equations (5.9.135) and (5.9.136) follow directly from (5.8.188) by setting the response parameter γ, β, and ε equal to zero.

The yield function (5.9.135) or (5.9.136) is, in the geometrical interpretation, a single-surface yield condition which limits the deviatoric plane by the second invariant of the stress deviator, and the hydrostatic plane by the first stress invariant.

It is well-known in the classical plasticity theory of metals that the associated flow rule leads to reasonable predictions of test observations. Thus, we choose

$$D\mathbf{E}_{SP} = D\lambda \frac{\partial F}{\partial \mathbf{S}_E^S}\,. \tag{5.9.137}$$

Considering Eqn. (5.9.135) or (5.9.136), we obtain, for the plastic strain increment,

$$\begin{aligned} D\mathbf{E}_{SP} &= D\lambda \frac{1}{\sqrt{2}} \frac{\mathbf{S}_E^{SD} + \alpha^2 (\mathbf{S}_E^S \cdot \mathbf{I})\mathbf{I}}{\sqrt{\mathbf{S}_E^{SD} \cdot \mathbf{S}_E^{SD} + \alpha^2 (\mathbf{S}_E^S \cdot \mathbf{I})^2}} \\ &= D\lambda \frac{1}{\kappa} \left[\mathbf{S}_E^{SD} + \alpha^2 (\mathbf{S}_E^S \cdot \mathbf{I})\mathbf{I}\right]\,. \end{aligned} \tag{5.9.138}$$

Furthermore, we limit our investigations to isotropic hardening. Thus, the response parameters α and κ are supposed to depend on the plastic work, see (5.8.189) and (5.8.190).

With the introduced simplified yield function (γ, β, and ε equal to zero in (5.8.188)), the associated flow rule (5.9.137) or (5.9.138), instead of (5.8.202), and the assumption of isotropic hardening (the backstress tensor in Section 5.8.5 is vanishing, and also the tensor $\mathbf{M}$ is equal to the zero tensor), we can conclude from (5.8.213), (5.8.211), and (5.8.212).

$$D\mathbf{E}_S = (\overset{4}{\mathbf{P}}{}^S)^{-1}\, D\mathbf{S}_E^S \tag{5.9.139}$$

with

$$(\overset{4}{\mathbf{P}}{}^S)^{-1} = (\overset{4}{\mathbf{K}}{}^S)^{-1} + \frac{1}{h}\frac{\partial F}{\partial \mathbf{S}_E^S} \otimes \frac{\partial F}{\partial \mathbf{S}_E^S} \tag{5.9.140}$$

and

$$h = -\left(\frac{\partial F}{\partial \alpha}\frac{\partial \alpha}{\partial W} + \frac{\partial F}{\partial \kappa}\frac{\partial \kappa}{\partial W}\right)\frac{\partial F}{\partial \mathbf{S}_E^S} \cdot \mathbf{S}_E^S\,, \tag{5.9.141}$$

where, in deriving (5.9.141), Eqn. $(5.8.197)_1$ has been considered.

Furthermore, the inversion of (5.9.139) leads to (see (5.8.222))

$$D\mathbf{S}_E^S = \overset{4}{\mathbf{P}}{}^S\, D\mathbf{E}_S \tag{5.9.142}$$

with

$$\overset{4}{\mathbf{P}}{}^S = \overset{4}{\mathbf{K}}{}^S - \frac{\overset{4}{\mathbf{K}}{}^S \dfrac{\partial F}{\partial \mathbf{S}_E^S} \otimes \overset{4}{\mathbf{K}}{}^S \dfrac{\partial F}{\partial \mathbf{S}_E^S}}{h + \overset{4}{\mathbf{K}}{}^S \cdot \dfrac{\partial F}{\partial \mathbf{S}_E^S} \otimes \dfrac{\partial F}{\partial \mathbf{S}_E^S}} \,. \tag{5.9.143}$$

The reader may note in passing that the fourth-order tensor $\overset{4}{\mathbf{P}}{}^S$ (and, of course, its inversion $(\overset{4}{\mathbf{P}}{}^S)^{-1}$) is symmetric in contrast to the corresponding fourth-order tensor in Section 5.8.5. This fact makes the proof possible that $\overset{4}{\mathbf{P}}_S$ is positive-definite. This proof is based on the requirement that

$$\omega = \overset{4}{\mathbf{P}}{}^S \cdot (\mathbf{A} \otimes \mathbf{A}) > 0 \,, \tag{5.9.144}$$

where $\mathbf{A}$ is an arbitrary second-order tensor.

With the abbreviation

$$L = h + \overset{4}{\mathbf{K}}{}^S \cdot \frac{\partial F}{\partial \mathbf{S}_E^S} \otimes \frac{\partial F}{\partial \mathbf{S}_E^S} \,, \tag{5.9.145}$$

it follows, with (5.9.143), from (5.9.144) that

$$\omega = \overset{4}{\mathbf{K}}{}^S \cdot (\mathbf{A} \otimes \mathbf{A}) - \frac{1}{L}\left(\overset{4}{\mathbf{K}}{}^S \frac{\partial F}{\partial \mathbf{S}_E^S} \otimes \overset{4}{\mathbf{K}}{}^S \frac{\partial F}{\partial \mathbf{S}_E^S}\right) \cdot (\mathbf{A} \otimes \mathbf{A}) \,. \tag{5.9.146}$$

The quantitiy L is always positive if

$$\frac{\partial F}{\partial \mathbf{S}_E^S} \neq \mathbf{0} \tag{5.9.147}$$

and ν^S in $\overset{4}{\mathbf{K}}{}^S$ fulfills the condition

$$\nu^S < \frac{1}{2} \,. \tag{5.9.148}$$

In this case, the elastic response tensor $\overset{4}{\mathbf{K}}{}^S$ is positive definite. The quantity h represents a positive scalar function (see de Boer and Ehlers, 1980). From (5.9.146), we obtain

$$\begin{aligned} L\omega = {} & h\overset{4}{\mathbf{K}}{}^S \cdot (\mathbf{A} \otimes \mathbf{A}) - \left(\overset{4}{\mathbf{K}}{}^S \frac{\partial F}{\partial \mathbf{S}_E^S} \otimes \overset{4}{\mathbf{K}}{}^S \frac{\partial F}{\partial \mathbf{S}_E^S}\right) \cdot (\mathbf{A} \otimes \mathbf{A}) + \\ & + \left(\overset{4}{\mathbf{K}}{}^S \cdot \frac{\partial F}{\partial \mathbf{S}_E^S} \otimes \frac{\partial F}{\partial \mathbf{S}_E^S}\right)\left[\overset{4}{\mathbf{K}}{}^S \cdot (\mathbf{A} \otimes \mathbf{A}\right] \,. \end{aligned} \tag{5.9.149}$$

From (5.9.149),

$$\begin{aligned} L\omega = {} & h\mathbf{A} \cdot \overset{4}{\mathbf{K}}{}^S \mathbf{A} - \left(\overset{4}{\mathbf{K}}{}^S \frac{\partial F}{\partial \mathbf{S}_E^S} \cdot \mathbf{A}\right)\left(\overset{4}{\mathbf{K}}{}^S \frac{\partial F}{\partial \mathbf{S}_E^S} \cdot \mathbf{A}\right) + \\ & + \left(\frac{\partial F}{\partial \mathbf{S}_E^S} \cdot \overset{4}{\mathbf{K}}{}^S \frac{\partial F}{\partial \mathbf{S}_E^S}\right)\left(\overset{4}{\mathbf{K}}{}^S \mathbf{A} \cdot \mathbf{A}\right) \end{aligned} \tag{5.9.150}$$

is gained. Due to the special properties of the parameter h and the elastic response tensor $\overset{4}{\mathbf{K}}{}^S$, the first term in Eqn. (5.9.150) is always positive, if ν_S fulfills (5.9.148). The second and third term in (5.9.150) can be represented in dependence of G_S and ν_S, i.e.,

$$\begin{aligned}
\xi = & - \left(\overset{4}{\mathbf{K}}{}^S \frac{\partial F}{\partial \mathbf{S}_E^S} \cdot \mathbf{A}\right)\left(\overset{4}{\mathbf{K}}{}^S \frac{\partial F}{\partial \mathbf{S}_E^S} \cdot \mathbf{A}\right) + \\
& + \left(\frac{\partial F}{\partial \mathbf{S}_E^S} \cdot \overset{4}{\mathbf{K}}{}^S \frac{\partial F}{\partial \mathbf{S}_E^S}\right)(\overset{4}{\mathbf{K}}{}^S \mathbf{A} \cdot \mathbf{A}) \\
= & -4(G^S)^2 \left(\frac{\partial F}{\partial \mathbf{S}_E^S} \cdot \mathbf{A} + \frac{\nu^S}{1-2\nu^S} \frac{\partial F}{\partial \mathbf{S}_E^S} \cdot \mathbf{A}\right)^2 + \\
& + 4(G^S)^2 \left(\frac{\partial F}{\partial \mathbf{S}_E^S} \cdot \frac{\partial F}{\partial \mathbf{S}_E^S} + \frac{\nu^S}{1-2\nu^S} \frac{\partial F}{\partial \mathbf{S}_E^S} \cdot \frac{\partial F}{\mathbf{S}_E^S}\right)\left(\mathbf{A} \cdot \mathbf{A} + \frac{\nu^S}{1-2\nu^S} \frac{\partial F}{\partial \mathbf{S}_E^S} \mathbf{A} \cdot \mathbf{A}\right)
\end{aligned} \tag{5.9.151}$$

or

$$\frac{1-2\nu^S}{1-\nu^S} \frac{1}{4G^{S2}} \xi = \left(\frac{\partial F}{\partial \mathbf{S}_E^S}\right) \cdot \left(\frac{\partial F}{\partial \mathbf{S}_E^S}\right)(\mathbf{A} \cdot \mathbf{A}) - \left(\frac{\partial F}{\partial \mathbf{S}_E^S} \cdot \mathbf{A}\right)^2 . \tag{5.9.152}$$

We consider Schwarz's inequality for second-order tensors (see de Boer, 1982) and state that

$$\left(\frac{\partial F}{\partial \mathbf{S}_E^S} \cdot \mathbf{A}\right)^2 < \left(\frac{\partial F}{\partial \mathbf{S}_E^S} \cdot \frac{\partial F}{\partial \mathbf{S}_E^S}\right)(\mathbf{A} \cdot \mathbf{A}) , \tag{5.9.153}$$

if $\mathbf{A} \neq \mathbf{O}$. Thus, according to (5.9.152), ξ is always positive and therefore also, according to (5.9.150), the quantity $L\omega$ is positive. Finally, we can state that the elastic-plastic response tensor $\overset{4}{\mathbf{P}}{}^S$ (and, of course, the inversion $(\overset{4}{\mathbf{P}}{}^S)^{-1}$) is always positive definite.

This is an important result in view of the development of general theorems such as, for example, the proof of the uniqueness of the solutions of boundary-value problems or the evidence of minimum/maximum statements. The interested reader is referred to de Boer and Ehlers (1980), where corresponding general theorems are proven in the elastic-plasticity theory of metals, and to de Boer and Kowalski (1985, 1986), where maximum/minimum principles are derived and the uniqueness of solutions of boundary-value problems has been proven on the base of the constitutive relations (5.9.135) and (5.9.137).

At the end of this section, some numerical results for a practical problem, which are gained from a numerical simulation with the finite element method, will be reported. The numerical processing concerning the compaction problem has been worked out in a dissertation by Jägering (1998). In this dissertation, not only the numerical simulation is contained, but also several examples are presented.

In the following, we will study the compaction of copper powder in a T-shaped die (see Fig. 5.9.30). The die is filled with a copper powder, which is

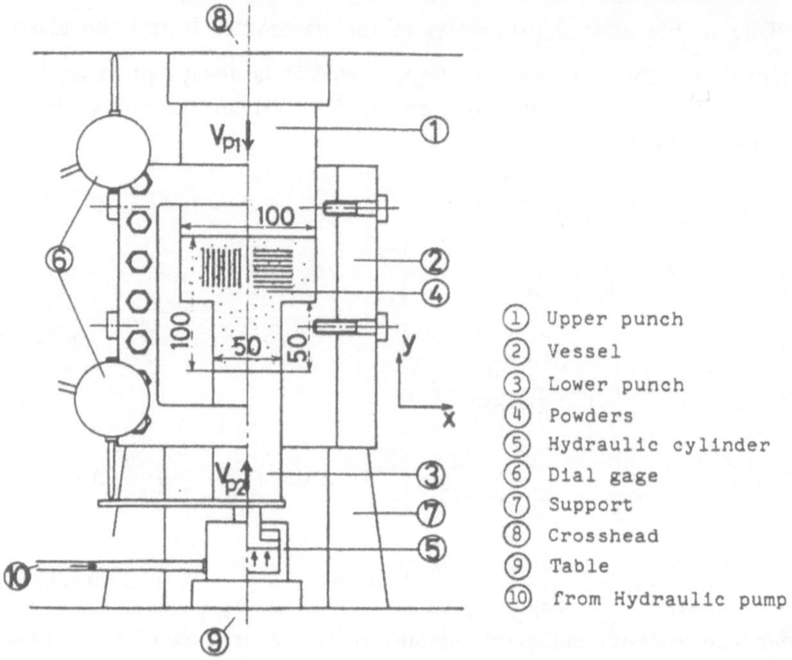

Fig. 5.9.30. Schematic representation of the test layout after Morimoto *et al.* (1982)

compacted up to a relative density of 0.332 and thickness of 20 mm before the real experiment starts. The relative density is defined as the quotient of the momentary (partial) density and the theoretically-possible maximum (real) density. The quotient corresponds to the volume fraction n^S:

$$\rho^{rel} = \frac{\rho^S}{\rho^{SR}} = n^S \, . \tag{5.9.154}$$

The compaction of the copper powder in the T-shaped die will be simulated on the basis of the theory of elastic-plastically deformed gas-filled porous solids, as has been developed earlier in this section. However, we will neglect the physical properties of the gas in the pores. For this reason, the partial density ρ^F of the gas is introduced with $10^{-5}\ kg/m^3$. The calculation shows that the poregas pressure is, in this case, equal to zero. Thus, the simulation can also be interpreted as the description of the mechanical behavior of an empty porous solid.

The Lamé response parameters in the elastic range are given by $\mu^S = 5555.55\,kN/m^2$ and $\lambda^S = 1058.20\,kN/m^2$. The friction coefficient between powder and stamp has the value of $\mu = 0.2$ (see Morimoto et al., 1982). For the description of the friction phenomena, we use Coulomb's friction law, where the friction force is equal to the product of the force, normal to the wall of the die, and the friction coefficient.

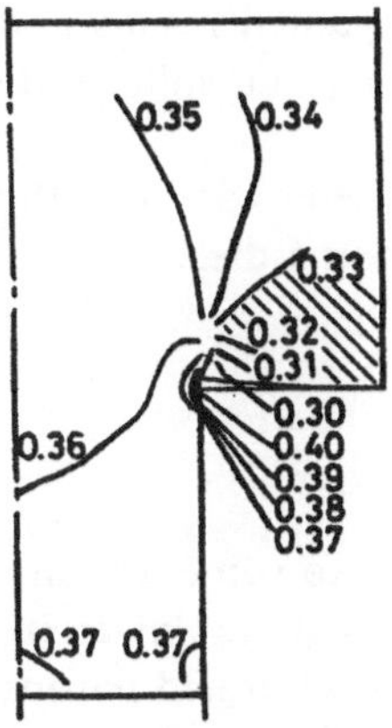

Fig. 5.9.31. Test results for the relative density

The finite element calculation has been performed with 120 elements. In order to control the convergence, a second calculation with 480 elements has been simulated. Moreover, the plastic response parameters α and κ are assumed to be 0.2 and 0.6, respectively. For more information, see Jägering (1998).

In Fig. 5.9.31, the course of the relative density of the experiment is represented first. Fig. 5.9.32 shows the corresponding values obtained by the finite

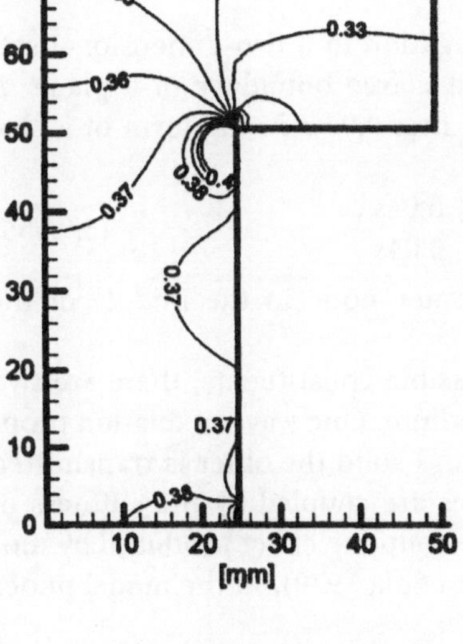

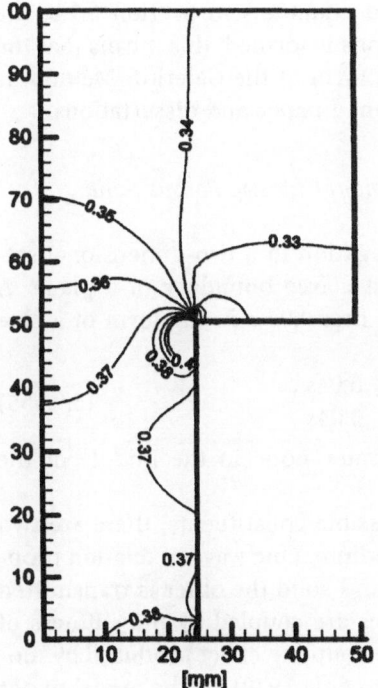

Fig. 5.9.32. Numerical results for the relative density

element method. From the comparison of the numerically gained results and the test results, it becomes apparent that porous media theory can sufficiently predict the test results correctly.

Thus, the porous media theory seems to be excellently suited to describe the mechanical behavior of the compaction of metallic powders in a die.

5.9.5 Further Solutions

The remaining part of this section on applications is devoted to some new numerical results recently gained in a paper and two dissertations at the University of Essen in Germany. These concern the wave propagation in saturated porous media (Breuer, 1999b), the elastic-plastic compaction of metallic powders (see Section 5.9.4) (Jägering, 1998), and elastic-plastic deformations of soils (Skolnik, 2000).

Due to the complexity of the basic field equations and differential equations describing initial- and boundary-value problems, the solution of those problems, within porous media theory, takes place almost exclusively by numerical methods. Analytical solutions are mostly restricted to one-dimensional examples (see Section 5.9.2), where the one-dimensional transient wave propagation problem is solved via a Laplace transformation.

The solution strategy in the aforementioned works is based on the well-known finite element method. With the field equations in Section 5.9.2, the so-called weak formulation of the field equations is formed; this means that the Method of Weighted Residuals, in the special form of the Galerkin-Method, is chosen. For more details, see the aforementioned paper and dissertations.

Wave Propagation in a Saturated Two-Dimensional Elastic Porous Solid

Breuer (1998a, b) investigated the wave propagation in a two-dimensional saturated porous solid and the surface waves at a free boundary of a plane. A time-dependent force, which acts only in the first 0.04 s in the form of a sine load:

$$f(t) = \begin{cases} 100 \, sin \, (78.54) \, [KN] \text{if} & t \leq 0.04s \, , \\ 0 \qquad\qquad\qquad\qquad \text{if} & t \leq 0.04s \, , \end{cases} \tag{5.9.155}$$

is applied in the vertical direction to the center node in the middle of the structure.

In Biot's (1956a) theory with two compressible constituents, there are two longitudinal waves in the saturated porous medium. One wave of dilation propagates due to the compressibility of the fluid and solid the other is transmitted through the elastic structure. These two waves are coupled by the stiffness of the soil and fluid constituents as well as by the coupling effect produced by motions of the solid and fluid phases (see Richart *et al.*, 1970). In the model under

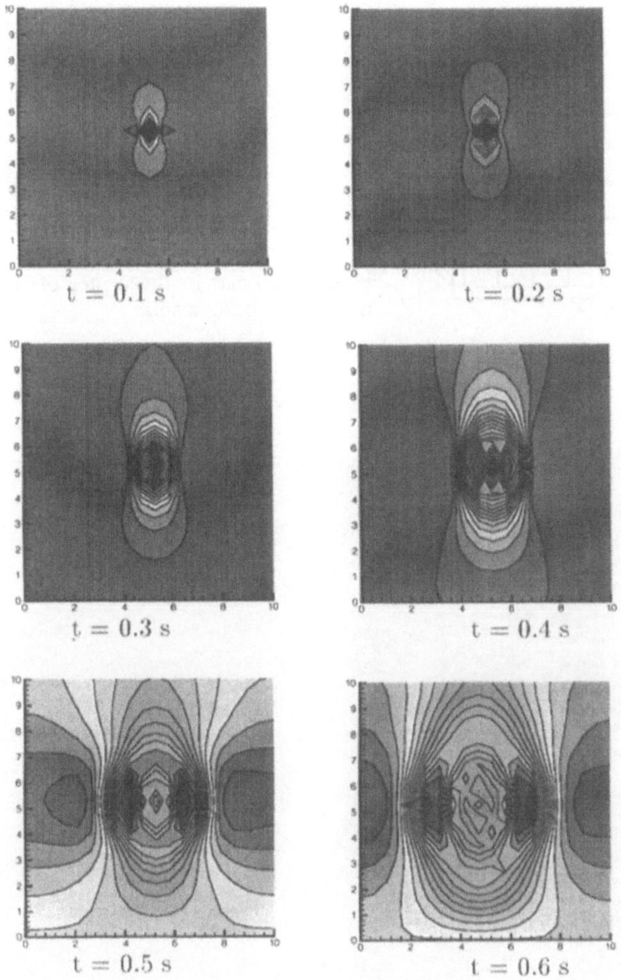

Fig. 5.9.33. Vertical solid displacements

study by Breuer (1998a, b), both constituents are incompressible. Thus, the first dilatational wave propagates with infinite large velocity. The dilatational wave which can be observed is due to the elastic structure of the solid skeleton.

The vertical solid displacements at different intervals of time are represented, where the axes determine the positions of the displacements in [m]. The disturbance, caused by the applied sine-load, propagates in the vertical direction faster than in the horizontal direction. Thus, the disturbance spread in elliptic form. The reason for this phenomenon lies in the fact that in the vertical direction, the disturbance propagates due to the dilatational wave, much faster than in the horizontal direction, due to the slower transversal wave.

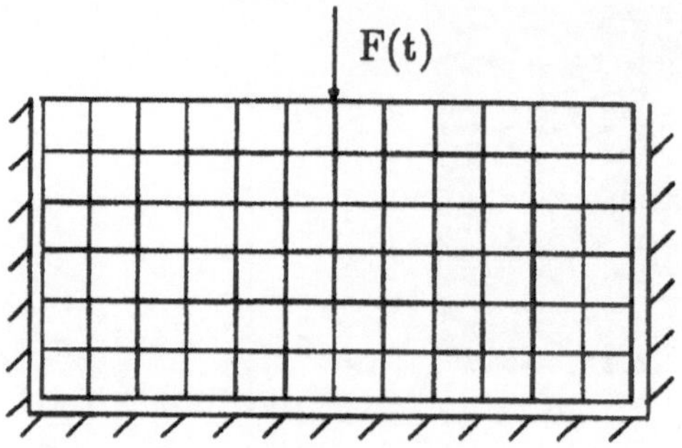

Fig. 5.9.34. Load in the center of the surface of a porous solid

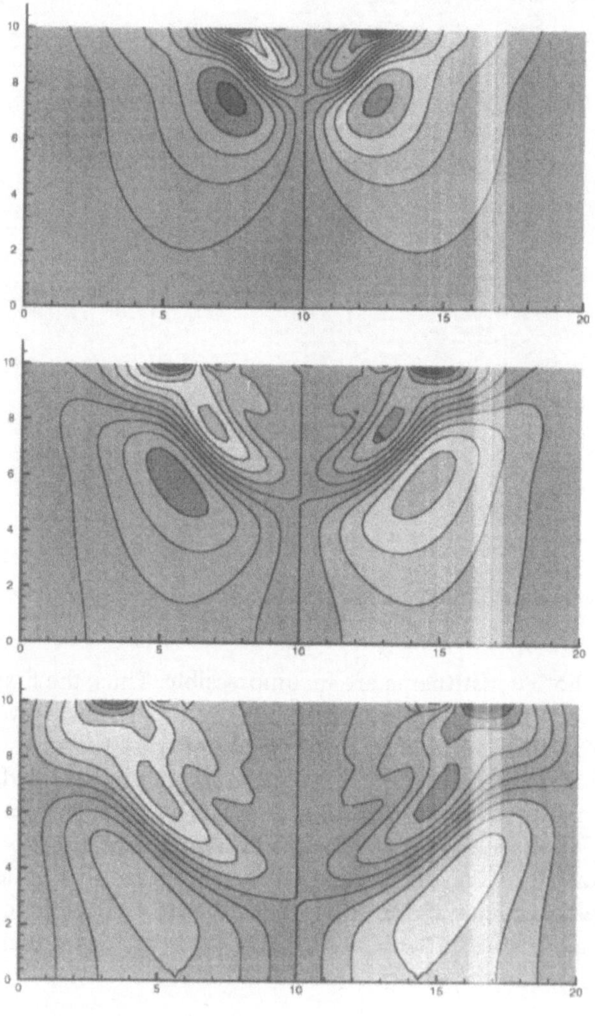

Fig. 5.9.35. Rayleigh-waves at a free boundary

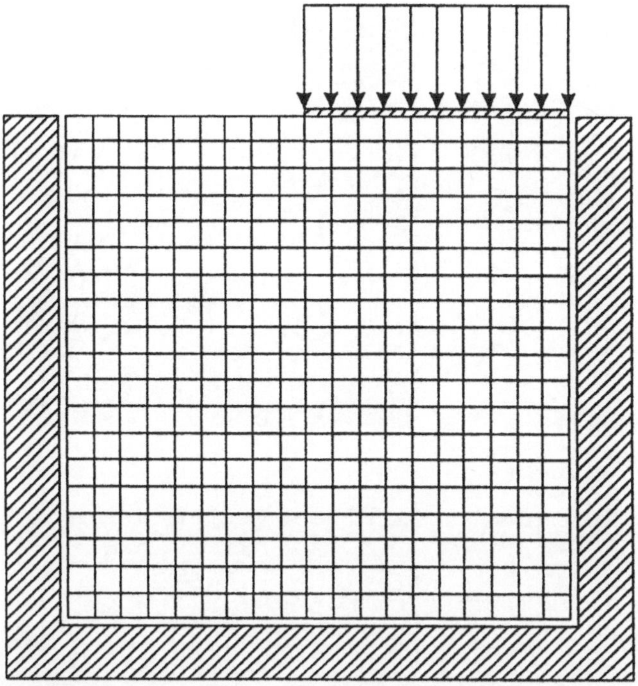

Fig. 5.9.36. Representation of a sample in a die

Next, the surface waves (Rayleigh waves), which occur at a free boundary (Fig. 5.9.34), are the point of interest. Fig. 5.9.35 shows a qualitative plot of the horizontal displacements of the solid at different intervals of time. It is recognized from the plots that there are two extreme peaks travelling at the free boundary. These displacements are caused by the Rayleigh-wave.

Elastic-Plastic Compaction of a Water-Saturated Metallic Porous Medium

A metallic porous solid is simulated in a rigid impermeable die (see Fig. 5.9.36). The fundamental relations of this problem are contained in the Sections 5.8.4 and 5.9.4. The response parameters and the initial porosity can be taken from the dissertation of Jägering (1998).

In the following figures (Fig. 5.9.37 through 5.9.39), the time-dependent development of the porewater pressure and the vertical displacements of the porous solid, as well as the vertical displacements depending on the load and the motion of the fluid, are represented.

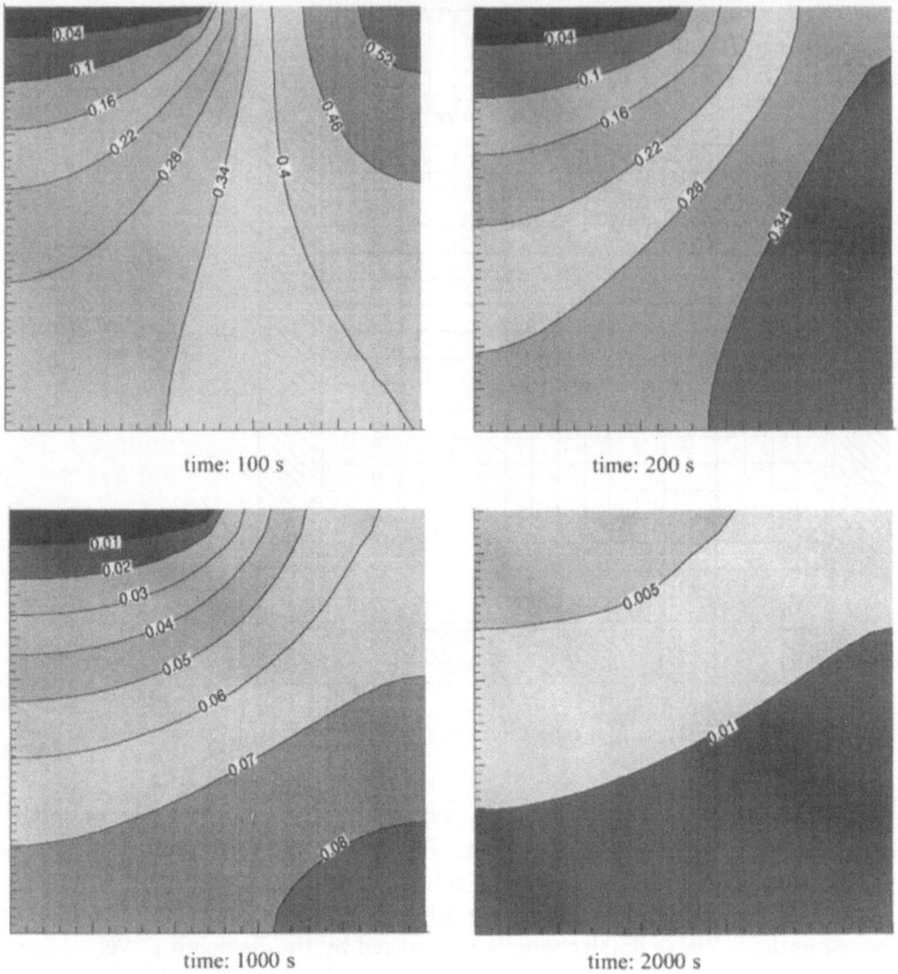

Fig. 5.9.37. Time-dependent development of the porewater pressure in the porespace

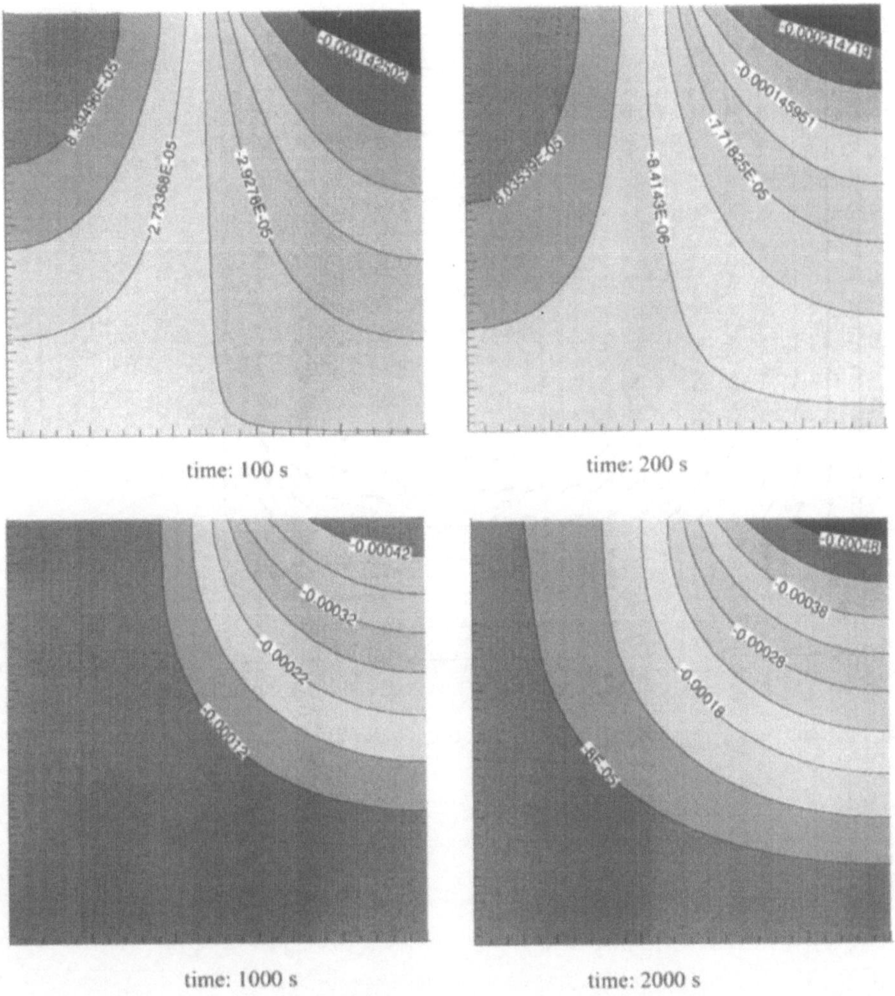

Fig. 5.9.38. Time-dependent development of the vertical displacements

Elastic-Plastic Deformations of Soil

On the basis of the yield condition (5.8.187) and the flow rule (5.8.202) in Section 5.8, the deformation and stress states – as well as the failure of a loaded slope (see Fig. 5.9.40) – has been treated and discussed by Skolnik (1999). All data and further information are given in detail in her dissertation. In the following Fig. 5.9.41, the development of the equivalent stress state depending on the increasing load up to the failure state is depicted.

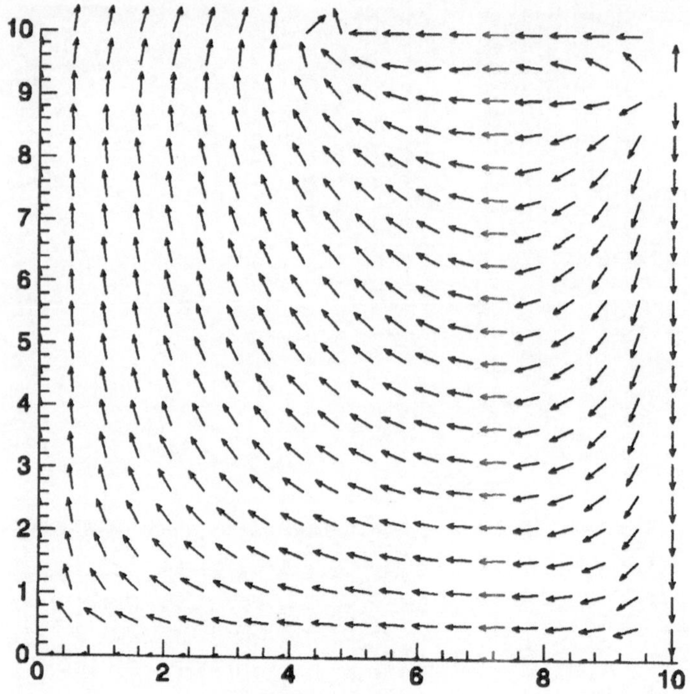

Fig. 5.9.39. Vector-representation of the motion of the fluid

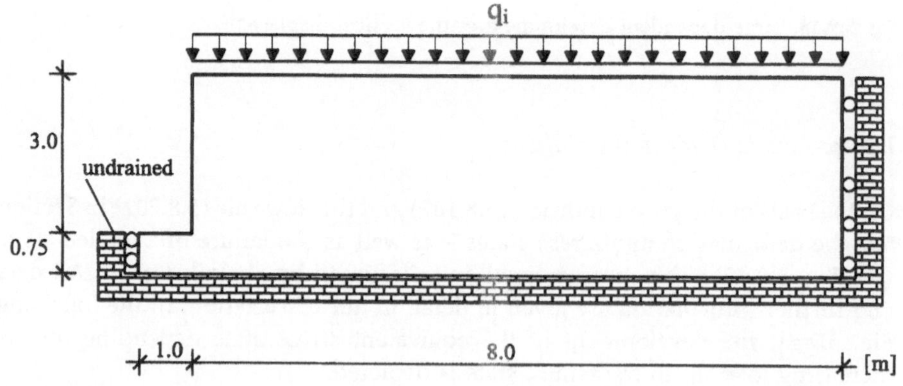

Fig. 5.9.40. Loaded slope

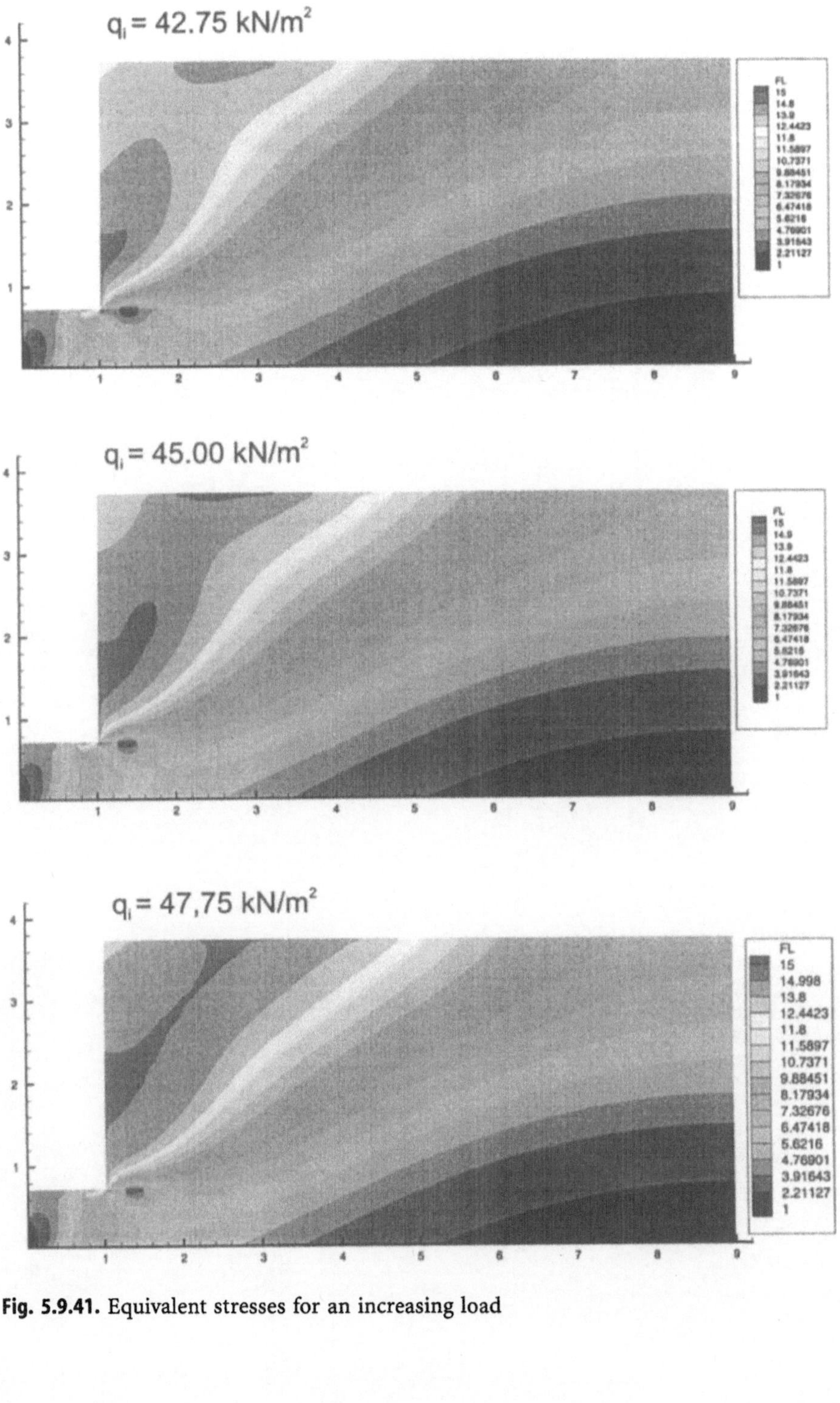

Fig. 5.9.41. Equivalent stresses for an increasing load

Chapter 6: Conclusions and Outlook

In the historical part (Chapter 2 through 4) some highlights of the development of porous media theory have been traced from infancy to their current state. Among the highlights, the bitter dispute between the professors Fillunger and von Terzaghi at the Technische Hochschule in Vienna in the 1930s, with its tragic results, has a particular place. The affair is representative of the unrest and upheaval in middle Europe during this period. For porous media theory, the year 1936 is a landmark, because it is in this year that the precise and general formulation of the one-dimensional mechanical porous media theory by Fillunger came to light. It is clear with hindsight that Fillunger's approach was about 35 years ahead of its time.

The discovery of Fillunger's masterpiece, in the late 1980s, shows how important the investigation into the historical development of science is. Without doubt, the development of porous media theory would have taken another course if Fillunger's, Heinrich's, and Flamm's contributions, as well as the later publications of Heinrich and Desoyer, would have been known to the international science community. The roundabout way that the application of mixture theory to porous media was developed in the 1960s – without the volume fraction concept and other often obscure approaches – could have been avoided. Certainly, the science of porous media would have advanced more swiftly.

As was mentioned in Chapter 5, porous media theory has been developed into a consistent theory in the last decades, meeting the principles of mechanics and thermodynamics, as well as meeting experimental requirements. One can state that this theory has reached the same high standard concerning the basic equations (kinematics, balance equations, and constitutive theory) as, for example, the mixture theory or the elasticity theory. The development of such a consistent theory is a great achievement, as the subject of porous media is very complex. The difficulties arise mainly due to the closure problem (see Section 5.6). It is relatively easy to introduce additional field equations to close the set of known and unknown fields. However, many approaches in the past have failed to correctly describe the mechanics and thermodynamics of empty and saturated porous media. One has to be aware of the consideration of mechanical and thermodynamical phenomena which are known from experience and experiment. Only the incorporation of these effects leads to consistent theories.

Moreover, it has also been shown that the porous media theory is a powerful tool to describe and solve problems in various fields of application. With the results of this theory, fundamental phenomena – namely uplift, friction, and capillarity – have been described and explained. The one-dimensional transient propagation in fluid-saturated, incompressible porous media and the one-dimensional consolidation process are typical examples to gain analytical solutions for certain initial- and boundary-value problems. These analytical solutions can serve to examine numerical results. This has been already done for many problems numerically investigated in the last years (see Section 5.9 and the corresponding reference literature). The treatment of the elastic-plastic compaction of metallic powders is an example of the successful application of porous media theory in the material sciences. The compaction problem was solved for a special problem with the finite element method.

Other problems, for example, boundary-value problems in geomechanics consisting of elastic-plastically deformed frictional material and dynamic problems, are currently under study at places like the University of Stuttgart, the Technical University of Poznan and the University of Essen; also drying processes in porous solids (important in chemical engineering) and moisture and heat transports in capillary-porous solids are being investigated. These transport phenomena are important, for example, in biomechanics, building physics, and environmental engineering. Furthermore, currently the constitutive theory of porous media is being extended to viscoelastic porous solids at the University of Stuttgart, and to the viscoplastic state at the University of Essen. The mathematical models based on these investigations seem to be important and interesting for the field of biomechanics and the material sciences. These problems require a more sophisticated model, namely a ternary, or even a higher-order model, with phase changes. The creation of such models have already been performed in the literature (see, e.g., de Boer and Ehlers, 1986a, b, Kowalski, 1990, Lai *et al.*, 1991, Huyghe and Janssen, 1999, de Boer, 1997). However, these models are partly introduced with simplifying assumptions.

Furthermore, important contributions to various fields of macroscopic porous media theory are currently under study in many other places of the world, such as in France, Italy, Russia, China, and the USA.

Appendix A: Evaluation of the Entropy Inequality

Restrictions for the constitutive equations of a binary model (*compressible elastic porous solid, incompressible viscous fluid*) with phase and energy transitions, as well as with different temperatures for each constituent, will be gained from the entropy inequality (5.8.44) and (5.8.45), respectively:

$$
\begin{aligned}
&-\frac{\rho^S}{\Theta^S}\sum_{i=1}^{n}\frac{\partial\hat{\psi}^S}{\partial s_i}(s^i)'_S-\frac{\rho^F}{\Theta^F}\sum_{i=1}^{n}\frac{\partial\hat{\psi}^F}{\partial s_i}(s_i)'_F-\frac{\rho^S}{\Theta^S}\eta^S(\Theta^S)'_S-\frac{\rho^F}{\Theta^F}\eta^F(\Theta^F)'_F+\\
&+\frac{1}{\Theta^S}\mathbf{T}^{SD}\cdot\mathbf{D}_S^D+\frac{1}{\Theta^F}\mathbf{T}^{FD}\cdot\mathbf{D}_F^D-\\
&-\frac{1}{\Theta^F}(\hat{\mathbf{p}}^F-\lambda\Theta^F\bar{\Theta}\,\operatorname{grad}n^F)\cdot\mathbf{v}_{FS}-\\
&-\hat{\rho}^F\Big(\frac{\mu^F}{\Theta^F}-\frac{\mu^S}{\Theta^S}\Big)-\frac{1}{\Theta^S\Theta^F}(\hat{e}^F-\hat{\mathbf{p}}^F\cdot\mathbf{v}_S)(\Theta^F-\Theta^S)-\\
&-\frac{1}{(\Theta^S)^2}\mathbf{q}^S\cdot\operatorname{grad}\Theta^S-\frac{1}{(\Theta^F)^2}\mathbf{q}^F\cdot\operatorname{grad}\Theta^F\geq 0\,.
\end{aligned}
\tag{A 1}
$$

Bearing in mind that we have already determined the constitutive equations for the hydrostatic pressures p^S and p^F in the partial solid and fluid materials as well as for the interface pressure λ in Section 5.8.2. c), see (5.8.46) through (5.8.48):

$$
\begin{aligned}
&p^S=-\,n^S(\rho^{SR})^2\frac{\partial\hat{\psi}^S}{\partial\rho^{SR}}\quad\text{or}\quad p^S=n^S\rho^{SR}J_{SR}\frac{\partial\bar{\bar{\psi}}^S}{\partial J_{SR}}\,,\\
&p^F=-\,n^F\lambda\,\Theta^F\bar{\Theta}=-\,n^F\kappa^{FR}\Theta^F\bar{\Theta},\\
&\lambda=\frac{(\rho^{SR})^2}{\Theta^S\bar{\Theta}}\frac{\partial\hat{\psi}^S}{\partial\rho^{SR}}-\frac{p^S}{\Theta^S\bar{\Theta}}\frac{\partial\hat{\psi}^S}{\partial n^S}\quad\text{or}\quad\lambda=-\,J_{SR}\frac{\rho^{SR}}{\Theta^S\bar{\Theta}}\frac{\partial\bar{\bar{\psi}}^S}{\partial J_{SR}}+J_{SN}\frac{\rho^{SR}}{\Theta^S\bar{\Theta}}\frac{\partial\bar{\bar{\psi}}^S}{\partial J_{SN}}\,,\\
&\qquad\Theta^F\bar{\Theta}\lambda=\Theta^F\bar{\Theta}\kappa^{FR}=p\,.
\end{aligned}
\tag{A 2}
$$

In (A 1) and (A 2), $\hat{\psi}^S$ and $\bar{\bar{\psi}}^S$ as well as $\hat{\psi}^F$ are the values of the free Helmholtz energy functions containing the response functions n^S, ρ^{SR}, s and J_{SN}, J_{SR}, s as well as s, see (5.8.43).

Constitutive equations are necessary for the remaining set $\mathcal{R}$ of constitutive response functions

$$\mathcal{R} = \left\{ \mathbf{T}^{\alpha D}, \hat{\mathbf{p}}^F - \lambda \Theta^F \bar{\Theta} \operatorname{grad} n^F, \hat{e}^F - \hat{\mathbf{p}}^F \cdot \mathbf{v}_S, \mathbf{q}^\alpha, \eta^\alpha, \psi^\alpha, \hat{\rho}^F \right\}, \tag{A 3}$$

which depends on the additional set s of independent process variables.

$$s = \left\{ \check{\mathbf{C}}_S, \mathbf{v}_{FS}, \mathbf{D}_F^D, \Theta^S, \Theta^F, \operatorname{grad} \Theta^S, \operatorname{grad} \Theta^F \right\}. \tag{A 4}$$

In order to develop thermodynamic restrictions from the entropy inequality (A 1), it is necessary to first calculate the material time derivatives of the free Helmholtz energy functions of both constituents. With

$$\psi^\alpha = \psi^\alpha(\ldots, s), \text{ where } \psi^S = \hat{\psi}^S(\ldots, s), \; \psi^F = \hat{\psi}^F(s), \tag{A 5}$$

we have

$$\begin{aligned} (\psi^\alpha)'_\alpha = {} & \frac{\partial \psi^\alpha}{\partial \check{\mathbf{C}}_S} \cdot \left[(\check{\mathbf{C}}_S)'_S + (\operatorname{grad} \check{\mathbf{C}}_S) \mathbf{v}_{\alpha S} \right] + \\ & + \frac{\partial \psi^\alpha}{\partial \mathbf{v}_{FS}} \cdot \left[(\mathbf{v}_F)'_F - (\mathbf{v}_S)'_S - (\operatorname{grad} \mathbf{v}_S) \mathbf{v}_{\alpha S} + (\operatorname{grad} \mathbf{v}_F) \mathbf{v}_{\alpha F} \right] + \\ & + \frac{\partial \psi^\alpha}{\partial \mathbf{D}_F^D} \cdot \left[(\mathbf{D}_F^D)'_F + (\operatorname{grad} \mathbf{D}_F^D) \mathbf{v}_{\alpha F} \right] + \\ & + \frac{\partial \psi^\alpha}{\partial \Theta^S} \left[(\Theta^S)'_S + \operatorname{grad} \Theta^S \cdot \mathbf{v}_{\alpha S} \right] + \\ & + \frac{\partial \psi^\alpha}{\partial \Theta^F} \left[(\Theta^F)'_F + \operatorname{grad} \Theta^F \cdot \mathbf{v}_{\alpha F} \right] + \\ & + \frac{\partial \psi^\alpha}{\partial \operatorname{grad} \Theta^S} \cdot \left[(\operatorname{grad} \Theta^S)'_S + (\operatorname{grad} \operatorname{grad} \Theta^S) \mathbf{v}_{\alpha S} \right] + \\ & + \frac{\partial \psi^\alpha}{\partial \operatorname{grad} \Theta^F} \cdot \left[(\operatorname{grad} \Theta^F)'_F + (\operatorname{grad} \operatorname{grad} \Theta^F) \mathbf{v}_{\alpha F} \right] \end{aligned} \tag{A 6}$$

with

$$\mathbf{v}_{\alpha S} = \mathbf{v}_\alpha - \mathbf{v}_S, \quad \mathbf{v}_{\alpha F} = \mathbf{v}_\alpha - \mathbf{v}_F. \tag{A 7}$$

The insertion of (A 6) into (A 1) yields the entropy inequality in the following form:

$$
\begin{aligned}
& - \frac{\rho^S}{\Theta^S} \frac{\partial \hat{\psi}^S}{\partial \check{\mathbf{C}}_S} \cdot (\check{\mathbf{C}}_S)'_S - \\
& - \frac{\rho^S}{\Theta^S} \frac{\partial \hat{\psi}^S}{\mathbf{v}_{FS}} \cdot \big[(\mathbf{v}_F)'_F - (\mathbf{v}_S)'_S - (\operatorname{grad} \mathbf{v}_F) \mathbf{v}_{FS} \big] - \\
& - \frac{\rho^S}{\Theta^S} \frac{\partial \hat{\psi}^S}{\partial \mathbf{D}_F^D} \cdot \big[(\mathbf{D}_F^D)'_F + (\operatorname{grad} \mathbf{D}_F^D) \mathbf{v}_{SF} \big] - \\
& - \frac{\rho^S}{\Theta^S} \frac{\partial \hat{\psi}^S}{\partial \Theta^S} (\Theta^S)'_S - \frac{\rho^S}{\Theta^S} \frac{\partial \hat{\psi}^S}{\partial \Theta^F} \big[(\Theta^F)'_F - \operatorname{grad} \Theta^F \cdot \mathbf{v}_{FS} \big] - \\
& - \frac{\rho^S}{\Theta^S} \frac{\partial \hat{\psi}^S}{\partial \operatorname{grad} \Theta^S} \cdot (\operatorname{grad} \Theta^S)'_S - \\
& - \frac{\rho^S}{\Theta^S} \frac{\partial \hat{\psi}^S}{\partial \operatorname{grad} \Theta^F} \cdot \big[(\operatorname{grad} \Theta^F)'_F - (\operatorname{grad} \operatorname{grad} \Theta^F) \mathbf{v}_{FS} \big] - \\
& - \frac{\rho^F}{\Theta^F} \frac{\partial \hat{\psi}^F}{\partial \check{\mathbf{C}}_S} \cdot \big[(\check{\mathbf{C}}_S)'_S + (\operatorname{grad} \check{\mathbf{C}}_S) \mathbf{v}_{FS} \big] - \\
& - \frac{\rho^F}{\Theta^F} \frac{\partial \hat{\psi}^F}{\partial \mathbf{v}_{FS}} \cdot \big[(\mathbf{v}_F)'_F - (\mathbf{v}_S)'_S - (\operatorname{grad} \mathbf{v}_S) \mathbf{v}_{FS} \big] - \frac{\rho^F}{\Theta^F} \frac{\partial \hat{\psi}^F}{\partial \mathbf{D}_F^D} \cdot (\mathbf{D}_F^D)'_F - \qquad \text{(A 8)} \\
& - \frac{\rho^F}{\Theta^F} \frac{\partial \hat{\psi}^F}{\partial \Theta^F} (\Theta^F)'_F - \frac{\rho^F}{\Theta^F} \frac{\partial \hat{\psi}^F}{\partial \operatorname{grad} \Theta^F} \cdot (\operatorname{grad} \Theta^F)'_F - \frac{\rho^F}{\Theta^F} \frac{\partial \hat{\psi}^F}{\partial \Theta^S} \big[(\Theta^S)'_S + \operatorname{grad} \Theta^S \cdot \mathbf{v}_{FS} \big] - \\
& - \frac{\rho^F}{\Theta^F} \frac{\partial \hat{\psi}^F}{\partial \operatorname{grad} \Theta^S} \cdot \big[(\operatorname{grad} \Theta^S)'_S + (\operatorname{grad} \operatorname{grad} \Theta^S) \mathbf{v}_{FS} \big] - \\
& - \frac{\rho^S}{\Theta^S} \eta^S (\Theta^S)'_S - \frac{\rho^F}{\Theta^F} \eta^F (\Theta^F)'_F + \\
& + \frac{1}{\Theta^S} \mathbf{T}^{SD} \cdot \mathbf{D}_S^D + \frac{1}{\Theta^F} \mathbf{T}^{FD} \cdot \mathbf{D}_F^D - \\
& - \frac{1}{\Theta^F} (\hat{\mathbf{p}}^F - \lambda \Theta^F \bar{\Theta} \operatorname{grad} n^F) \cdot \mathbf{v}_{FS} - \\
& - \hat{\rho}^F \Big[\frac{1}{\Theta^F} \psi^F - \frac{1}{\Theta^F} \frac{p^F}{\rho^F} - \frac{1}{\Theta^S} \psi^S + \frac{1}{\Theta^S} \frac{p^S}{\rho^S} - \frac{1}{2\Theta^F} \mathbf{v}_F \cdot \mathbf{v}_F + \frac{1}{2\Theta^S} \mathbf{v}_S \cdot \mathbf{v}_S \Big] - \\
& - \frac{1}{\Theta^S \Theta^F} (\hat{e}^F - \hat{\mathbf{p}}^F \cdot \mathbf{v}_S)(\Theta^F - \Theta^S) - \frac{1}{(\Theta^S)^2} \mathbf{q}^S \cdot \operatorname{grad} \Theta^S - \frac{1}{(\Theta^F)^2} \mathbf{q}^F \cdot \operatorname{grad} \Theta^F \geq 0 \, .
\end{aligned}
$$

Note that we have omitted, for simplicity, the superscript notations on the free Helmholtz energy functions ψ^S and ψ^F in the brackets connected with $\hat{\rho}^F$. With

$$
\begin{aligned}
&\frac{\partial \hat{\psi}^S}{\partial \mathbf{v}_{FS}} \cdot (\operatorname{grad} \mathbf{v}_F)\mathbf{v}_{FS} = \Big(\frac{\partial \hat{\psi}^S}{\partial \mathbf{v}_{FS}} \otimes \mathbf{v}_{FS}\Big) \cdot \operatorname{grad} \mathbf{v}_F \,, \\
&\frac{\partial \hat{\psi}^F}{\partial \mathbf{v}_{FS}} \cdot (\operatorname{grad} \mathbf{v}_S)\mathbf{v}_{FS} = \Big(\frac{\partial \hat{\psi}^F}{\partial \mathbf{v}_{FS}} \otimes \mathbf{v}_{FS}\Big) \cdot \operatorname{grad} \mathbf{v}_S \,, \\
&\frac{\partial \hat{\psi}^S}{\partial (\operatorname{grad} \Theta^F)} \cdot (\operatorname{grad}\operatorname{grad} \Theta^F)\mathbf{v}_{FS} = \Big[\frac{\partial \hat{\psi}^S}{\partial (\operatorname{grad} \Theta^F)} \otimes \mathbf{v}_{FS}\Big] \cdot (\operatorname{grad}\operatorname{grad} \Theta^F) \,, \\
&\frac{\partial \hat{\psi}^F}{\partial (\operatorname{grad} \Theta^S)} \cdot \big[(\operatorname{grad}\operatorname{grad} \Theta^S)\mathbf{v}_{FS}\big] = \Big[\frac{\partial \hat{\psi}^F}{\partial (\operatorname{grad} \Theta^S)} \otimes \mathbf{v}_{FS}\Big] \cdot (\operatorname{grad}\operatorname{grad} \Theta^S) \,,
\end{aligned}
\tag{A 9}
$$

and

$$
\frac{\partial \hat{\psi}^F}{\partial \check{\mathbf{C}}_S} \cdot (\operatorname{grad} \check{\mathbf{C}}_S)\mathbf{v}_{FS} = \Big(\frac{\partial \hat{\psi}^F}{\partial \check{\mathbf{C}}_S} \otimes \mathbf{v}_{FS}\Big) \cdot \operatorname{grad} \check{\mathbf{C}}_S \,,
$$

$$
(\check{\mathbf{C}}_S)'_S = 2\, J_S^{-2/3}\, \mathbf{F}_S^T\, \mathbf{D}_S^D\, \mathbf{F}_S
$$

it follows with $\operatorname{grad} \mathbf{v}_\alpha = \mathbf{L}_\alpha = \mathbf{D}_\alpha + \mathbf{W}_\alpha$, see (5.3.13), from (A 8):

$$
\begin{aligned}
&- \ (\Theta^S)'_S\Big(\frac{\rho^S}{\Theta^S}\eta^S + \frac{\rho^S}{\Theta^S}\frac{\partial \hat{\psi}^S}{\partial \Theta^S} + \frac{\rho^F}{\Theta^F}\frac{\partial \hat{\psi}^F}{\partial \Theta^S}\Big) - \\
&- \ (\Theta^F)'_F\Big(\frac{\rho^F}{\Theta^F}\eta^F + \frac{\rho^F}{\Theta^F}\frac{\partial \hat{\psi}^F}{\partial \Theta^F} + \frac{\rho^S}{\Theta^S}\frac{\partial \hat{\psi}^S}{\partial \Theta^F}\Big) - \\
&- \ (\operatorname{grad} \Theta^S)'_S \cdot \Big[\frac{\rho^S}{\Theta^S}\frac{\partial \hat{\psi}^S}{\partial (\operatorname{grad} \Theta^S)} + \frac{\rho^F}{\Theta^F}\frac{\partial \hat{\psi}^F}{\partial (\operatorname{grad} \Theta^S)}\Big] - \\
&- \ (\operatorname{grad} \Theta^F)'_F \cdot \Big[\frac{\rho^S}{\Theta^S}\frac{\partial \hat{\psi}^S}{\partial (\operatorname{grad} \Theta^F)} + \frac{\rho^F}{\Theta^F}\frac{\partial \hat{\psi}^F}{\partial (\operatorname{grad} \Theta^F)}\Big] + \\
&+ \ \mathbf{v}_{FS} \cdot \frac{\rho^S}{\Theta^S}\frac{\partial \hat{\psi}^S}{\partial \Theta^F} \operatorname{grad} \Theta^F - \mathbf{v}_{FS} \cdot \frac{\rho^F}{\Theta^F}\frac{\partial \hat{\psi}^F}{\partial \Theta^S} \operatorname{grad} \Theta^S + \\
&+ \ \operatorname{grad}(\operatorname{grad} \Theta^F) \cdot \Big\{\frac{\rho^S}{\Theta^S}\Big[\frac{\partial \hat{\psi}^S}{\partial (\operatorname{grad} \Theta^F)} \otimes \mathbf{v}_{FS}\Big]\Big\} - \\
&- \ \operatorname{grad}(\operatorname{grad} \Theta^S) \cdot \Big\{\frac{\rho^F}{\Theta^F}\Big[\frac{\partial \hat{\psi}^F}{\partial (\operatorname{grad} \Theta^S)} \otimes \mathbf{v}_{FS}\Big]\Big\} + \\
&+ \ \mathbf{D}_S^D \cdot \frac{1}{\Theta^S}\Big\{\mathbf{T}^{SD} - 2\rho^S J_S^{-2/3}\Big(\mathbf{F}_S \frac{\partial \hat{\psi}^S}{\partial \check{\mathbf{C}}_S}\mathbf{F}_S^T\Big)^D - \\
&\qquad -2\frac{\Theta^S \rho^F}{\Theta^F} J_S^{-2/3}\Big(\mathbf{F}_S \frac{\partial \hat{\psi}^F}{\partial \check{\mathbf{C}}_S}\mathbf{F}_S^T\Big)^D + \Big(\frac{\Theta^S \rho^F}{\Theta^F}\frac{\partial \hat{\psi}^F}{\partial \mathbf{v}_{FS}} \otimes \mathbf{v}_{FS}\Big)^D\Big\} -
\end{aligned}
\tag{A 10}
$$

$$
\begin{aligned}
&- \operatorname{grad} \check{\mathbf{C}}_S \cdot \Big(\frac{\rho^F}{\Theta^F}\frac{\partial\hat{\psi}^F}{\partial\check{\mathbf{C}}_S} \otimes \mathbf{v}_{FS}\Big) - (\mathbf{D}_F^D)'_F \cdot \Big(\frac{\rho^F}{\Theta^F}\frac{\partial\hat{\psi}^F}{\partial\mathbf{D}_F^D} + \frac{\rho^S}{\Theta^S}\frac{\partial\hat{\psi}^S}{\partial\mathbf{D}_F^D}\Big) + \\
&+ \operatorname{grad} \mathbf{D}_F^D \cdot \Big(\frac{\rho^S}{\Theta^S}\frac{\partial\hat{\psi}^S}{\partial\mathbf{D}_F^D} \otimes \mathbf{v}_{FS}\Big) - \\
&- \big[(\mathbf{v}_F)'_F - (\mathbf{v}_S)'_S\big] \cdot \Big(\frac{\rho^S}{\Theta^S}\frac{\hat{\psi}^S}{\partial\mathbf{v}_{FS}} + \frac{\rho^F}{\Theta^F}\frac{\partial\hat{\psi}^F}{\partial\mathbf{v}_{FS}}\Big) + \\
&+ \mathbf{W}_S \cdot \Big(\frac{\rho^F}{\Theta^F}\frac{\partial\hat{\psi}^F}{\partial\mathbf{v}_{FS}} \otimes \mathbf{v}_{FS}\Big) + \mathbf{D}_F^D \cdot \frac{1}{\Theta^F}\Big[\mathbf{T}^{FD} + \Big(\rho^S\frac{\Theta^F}{\Theta^S}\frac{\partial\hat{\psi}^S}{\partial\mathbf{v}_{FS}} \otimes \mathbf{v}_{FS}\Big)^D\Big] + \\
&+ \mathbf{W}_F \cdot \Big(\frac{\rho^S}{\Theta^S}\frac{\partial\hat{\psi}^S}{\partial\mathbf{v}_{FS}} \otimes \mathbf{v}_{FS}\Big) - \mathbf{v}_{FS} \cdot \frac{1}{\Theta^F}(\hat{\mathbf{p}}^F - \lambda\Theta^F\bar{\Theta}\operatorname{grad} n^F) - \\
&- \hat{\rho}^F\Big(\frac{\mu^F}{\Theta^F} - \frac{\mu^S}{\Theta^S} - \frac{1}{3}\frac{\rho^S}{\rho^F\Theta^S}\frac{\partial\hat{\psi}^S}{\partial\mathbf{v}_{FS}} \cdot \mathbf{v}_{FS} + \frac{1}{3}\frac{\rho^F}{\rho^S\Theta^F}\frac{\partial\hat{\psi}^F}{\partial\mathbf{v}_{FS}} \cdot \mathbf{v}_{FS}\Big) - \\
&- (\Theta^F - \Theta^S)\frac{1}{\Theta^S\Theta^F}(\hat{e}^F - \hat{\mathbf{p}}^F \cdot \mathbf{v}_S) - \operatorname{grad}\Theta^S \cdot \frac{1}{(\Theta^S)^2}\mathbf{q}^S - \operatorname{grad}\Theta^F \cdot \frac{1}{(\Theta^F)^2}\mathbf{q}^F \geq 0\,.
\end{aligned}
$$

The chemical potentials μ^α are defined by

$$\mu^\alpha = \psi^\alpha - \frac{p^\alpha}{\rho^\alpha} - \frac{1}{2}\mathbf{v}_\alpha \cdot \mathbf{v}_\alpha\,. \tag{A 11}$$

The evaluation of the entropy inequality yields:

$$\frac{\rho^S}{\Theta^S}\Big(\eta^S + \frac{\partial\hat{\psi}^S}{\partial\Theta^S}\Big) + \frac{\rho^F}{\Theta^F}\frac{\partial\hat{\psi}^F}{\partial\Theta^S} = 0\,, \tag{A 12}$$

$$\frac{\rho^F}{\Theta^F}\Big(\eta^F + \frac{\partial\hat{\psi}^F}{\partial\Theta^F}\Big) + \frac{\rho^S}{\Theta^S}\frac{\partial\hat{\psi}^S}{\partial\Theta^F} = 0\,, \tag{A 13}$$

$$\frac{\rho^S}{\Theta^S}\frac{\partial\hat{\psi}^S}{\partial(\operatorname{grad}\Theta^S)} + \frac{\rho^F}{\Theta^F}\frac{\partial\hat{\psi}^F}{\partial(\operatorname{grad}\Theta^S)} = \mathbf{0}\,, \tag{A 14}$$

$$\frac{\rho^S}{\Theta^S}\frac{\partial\hat{\psi}^S}{\partial(\operatorname{grad}\Theta^F)} + \frac{\rho^F}{\Theta^F}\frac{\partial\hat{\psi}^F}{\partial(\operatorname{grad}\Theta^F)} = \mathbf{0}\,, \tag{A 15}$$

$$\frac{\partial\hat{\psi}^S}{\partial(\operatorname{grad}\Theta^F)} \otimes \mathbf{v}_{FS} = -\mathbf{v}_{FS} \otimes \frac{\partial\hat{\psi}^S}{(\partial\operatorname{grad}\Theta^F)}\,, \tag{A 16}$$

$$\frac{\partial\hat{\psi}^F}{\partial(\operatorname{grad}\Theta^S)} \otimes \mathbf{v}_{FS} = -\mathbf{v}_{FS} \otimes \frac{\partial\hat{\psi}^F}{\partial(\operatorname{grad}\Theta^S)}\,, \tag{A 17}$$

$$\frac{\rho^S}{\Theta^S}\frac{\partial\hat{\psi}^S}{\partial\mathbf{D}_F^D} + \frac{\rho^F}{\Theta^F}\frac{\partial\hat{\psi}^F}{\partial\mathbf{D}_F^D} = \mathbf{0}\,, \tag{A 18}$$

$$\frac{\partial \hat{\psi}^F}{\partial \check{\mathbf{C}}_S} \otimes \mathbf{v}_{FS} = -\Big(\frac{\partial \hat{\psi}^F}{\partial \check{\mathbf{C}}_S}\Big)^T \otimes \mathbf{v}_{FS} \tag{A 19}$$

or, since $\check{\mathbf{C}}_S$ is a symmetric tensor,

$$\frac{\rho^F}{\Theta^F}\frac{\partial \hat{\psi}^F}{\partial \check{\mathbf{C}}_S} \otimes \mathbf{v}_{FS} = \overset{3}{\mathbf{0}} \tag{A 20}$$

or

$$\frac{\partial \hat{\psi}^F}{\partial \check{\mathbf{C}}_S} = \mathbf{0}\,, \tag{A 21}$$

$$\frac{\partial \hat{\psi}^S}{\partial \mathbf{D}_F^D} \otimes \mathbf{v}_{FS} = -\Big(\frac{\partial \hat{\psi}^S}{\mathbf{D}_F^D}\Big)^T \otimes \mathbf{v}_{FS} \tag{A 22}$$

or, with (A 18),

$$\frac{\partial \hat{\psi}^S}{\partial \mathbf{D}_F^D} = \frac{\partial \hat{\psi}^F}{\partial \mathbf{D}_F^D} = \mathbf{0}\,. \tag{A 23}$$

In a similar way, we obtain

$$\mathbf{T}^{SD} = 2\rho^S J_S^{-2/3}\Big(\mathbf{F}_S \frac{\partial \hat{\psi}^S}{\partial \check{\mathbf{C}}_S}\mathbf{F}_S^T\Big)^D - \Big(\frac{\Theta^S}{\Theta^F}\rho^F \frac{\partial \hat{\psi}^F}{\partial \mathbf{v}_{FS}} \otimes \mathbf{v}_{FS}\Big)^D\,, \tag{A 24}$$

where (A 21) has been used.

Furthermore, the evaluation of (A 10) yields

$$\frac{\partial \hat{\psi}^F}{\partial \mathbf{v}_{FS}} \otimes \mathbf{v}_{FS} = \mathbf{v}_{FS} \otimes \frac{\partial \hat{\psi}^F}{\partial \mathbf{v}_{FS}}\,, \tag{A 25}$$

$$\frac{\partial \hat{\psi}^S}{\partial \mathbf{v}_{FS}} \otimes \mathbf{v}_{FS} = \mathbf{v}_{FS} \otimes \frac{\partial \hat{\psi}^S}{\partial \mathbf{v}_{FS}}\,, \tag{A 26}$$

$$\frac{\rho^S}{\Theta^S}\frac{\partial \hat{\psi}^S}{\partial \mathbf{v}_{FS}} = -\frac{\rho^F}{\Theta^F}\frac{\partial \hat{\psi}^F}{\partial \mathbf{v}_{FS}}\,. \tag{A 27}$$

Moreover, an important inequality remains

$$\begin{aligned}
&-\frac{1}{(\Theta^S)^2}\operatorname{grad}\Theta^S \cdot \Big[\mathbf{q}^S + (\Theta^S)^2\frac{\rho^F}{\Theta^F}\frac{\partial \hat{\psi}^F}{\partial \Theta^S}\mathbf{v}_{FS}\Big] - \\
&-\frac{1}{(\Theta^F)^2}\operatorname{grad}\Theta^F \cdot \Big[\mathbf{q}^F - (\Theta^F)^2\frac{\rho^S}{\Theta^S}\frac{\partial \hat{\psi}^S}{\partial \Theta^F}\mathbf{v}_{FS}\Big] - \\
&-\mathbf{v}_{FS}\cdot\frac{1}{\Theta^F}(\hat{\mathbf{p}}^F - \lambda\Theta^F\bar{\Theta}\operatorname{grad} n^F) - \hat{\rho}^F\Big(\frac{\bar{\mu}^F}{\Theta^F} - \frac{\bar{\mu}^S}{\Theta^S}\Big) - \\
&-(\Theta^F - \Theta^S)\frac{1}{\Theta^S\Theta^F}(\hat{e}^F - \hat{\mathbf{p}}^F\cdot\mathbf{v}_S) + \mathbf{D}_F^D\cdot\frac{1}{\Theta^F}\Big[\mathbf{T}^{FD} + \Big(\rho^S\frac{\Theta^F}{\Theta^S}\frac{\partial \hat{\psi}^S}{\partial \mathbf{v}_{FS}} \otimes \mathbf{v}_{FS}\Big)^D\Big] \geq 0\,,
\end{aligned} \tag{A 28}$$

where

$$\bar{\mu}^F = \mu^F + \frac{1}{3}\frac{\rho^F}{\rho^S}\frac{\partial \hat{\psi}^F}{\partial \mathbf{v}_{FS}}\cdot\mathbf{v}_{FS}\,, \tag{A 29}$$

$$\bar{\mu}^S = \mu^S + \frac{1}{3}\frac{\rho^S}{\rho^F}\frac{\partial\hat{\psi}^S}{\partial \mathbf{v}_{FS}} \cdot \mathbf{v}_{FS} \,. \tag{A 30}$$

With the introduction of the abbreviations

$$\mathbf{m}^S = \frac{1}{(\Theta^S)^2}\mathbf{q}^S + \frac{\rho^F}{\Theta^F}\frac{\partial\hat{\psi}^F}{\partial\Theta^S}\mathbf{v}_{FS} \,, \tag{A 31}$$

$$\mathbf{m}^F = \frac{1}{(\Theta^F)^2}\mathbf{q}^F - \frac{\rho^S}{\Theta^S}\frac{\partial\hat{\psi}^S}{\partial\Theta^F}\mathbf{v}_{FS} \,. \tag{A 32}$$

Inequality (A 28) can be reformulated as

$$\mathcal{D}(s) - \Big(\frac{\bar{\mu}^F}{\Theta^F} - \frac{\bar{\mu}^S}{\Theta^S}\Big)\hat{\rho}^F \geq 0 \,, \tag{A 33}$$

with

$$\begin{aligned}\mathcal{D}(s) = & - \operatorname{grad}\Theta^S \cdot \mathbf{m}^S - \operatorname{grad}\Theta^F \cdot \mathbf{m}^F - \\ & - \mathbf{v}_{FS} \cdot \frac{1}{\Theta^F}(\hat{\mathbf{p}}^F - \lambda\Theta^F\bar{\Theta}\operatorname{grad} n^F) - \\ & - (\Theta^F - \Theta^S)\frac{1}{\Theta^S\Theta^F}(\hat{e}^F - \hat{\mathbf{p}}^F \cdot \mathbf{v}_S) + \\ & + \mathbf{D}_F^D \cdot \frac{1}{\Theta^F}\Big[\mathbf{T}^{FD} + \Big(\rho^S\frac{\Theta^F}{\Theta^S}\frac{\partial\hat{\psi}^S}{\partial\mathbf{v}_{FS}} \otimes \mathbf{v}_{FS}\Big)^D\Big] \geq 0 \,.\end{aligned} \tag{A 34}$$

The relations (A 16), (A 17), as well as (A 25) yield some restrictions in view of the dependencies of the free energies $\hat{\psi}^S$ and $\hat{\psi}^F$ on the various temperature gradients and the velocity differences. Differentiation of (A 17) with respect to $\mathbf{v}_{FS}$ results in

$$\begin{aligned}&\frac{\partial^2\hat{\psi}^F}{\partial(\operatorname{grad}\Theta^S)\otimes\partial\mathbf{v}_{FS}} \otimes \mathbf{v}_{FS} + \frac{\partial\hat{\psi}^F}{\partial(\operatorname{grad}\Theta^S)} \otimes \mathbf{I} = \\ &= - \mathbf{I} \otimes \frac{\partial\hat{\psi}^F}{\partial(\operatorname{grad}\Theta^S)} - \mathbf{v}_{FS} \otimes \frac{\partial^2\hat{\psi}^F}{\partial(\operatorname{grad}\Theta^S)\otimes\partial\mathbf{v}_{FS}} \,.\end{aligned} \tag{A 35}$$

Thus, in the case of the mixture equilibrium state, where the difference velocity $\mathbf{v}_{FS}$ vanishes, it can be concluded from (A 35)

$$\frac{\rho^F}{\Theta^F}\frac{\partial\hat{\psi}^F}{\partial(\operatorname{grad}\Theta^S)}\Big|_{\mathbf{v}_{FS}=\mathbf{0}} = \mathbf{0} \,. \tag{A 36}$$

From (A 16) and (A 36), considering (A 14) and (A 15), we gain, in an analogous way, for the mixture equilibrium state:

$$\frac{\partial \hat{\psi}^S}{\partial(\text{grad}\,\Theta^F)}\,|_{\mathbf{v}_{FS}=\mathbf{0}} = \mathbf{0}\,, \qquad \frac{\partial \hat{\psi}^F}{\partial(\text{grad}\,\Theta^F)}\,|_{\mathbf{v}_{FS}=\mathbf{0}} = \mathbf{0}\,,$$
$$\frac{\partial \hat{\psi}^S}{\partial(\text{grad}\,\Theta^S)}\,|_{\mathbf{v}_{FS}=\mathbf{0}} = \mathbf{0}\,. \tag{A 37}$$

It is recognized that, in the mixture equilibrium state, the dependencies of the free Helmholtz energy functions of the partial constituents on the various temperature gradients vanish. Furthermore, differentiating (A 25), with respect to the velocity difference $\mathbf{v}_{FS}$, gives

$$\frac{\partial^2 \hat{\psi}^F}{\partial \mathbf{v}_{FS} \otimes \partial \mathbf{v}_{FS}} \otimes \mathbf{v}_{FS} + \frac{\partial \hat{\psi}^F}{\partial \mathbf{v}_{FS}} \otimes \mathbf{I} =$$
$$= \mathbf{I} \otimes \frac{\partial \hat{\psi}^F}{\partial \mathbf{v}_{FS}} + \mathbf{v}_{FS} \otimes \frac{\partial^2 \hat{\psi}^F}{\partial \mathbf{v}_{FS} \otimes \partial \mathbf{v}_{FS}}\,. \tag{A 38}$$

Thus, in the case of the mixture equilibrium, it is recognized from (A 38) that

$$\frac{\partial \hat{\psi}^F}{\partial \mathbf{v}_{FS}}\,|_{\mathbf{v}_{FS}=\mathbf{0}} = \mathbf{0}\,. \tag{A 39}$$

Analogously, it follows from (A 26), that:

$$\frac{\partial \hat{\psi}^S}{\partial \mathbf{v}_{FS}}\,|_{\mathbf{v}_{FS}=\mathbf{0}} = \mathbf{0}\,. \tag{A 40}$$

Inequality(A 34) represents the irreversible or dissipative part of the entropy inequality of the respective binary model. Up to (A 35) all derivations and results are valid for arbitrary thermodynamic processes. The constitutive equations must be chosen for the set $\mathcal{R}$ in such a way that all restrictions gained from the entropy inequality must be considered and, in particular, Inequality (A 33) in connection with (A 34) must be fulfilled.

Next, the dissipation inequality (A 34) will be evaluated in the case of the *equilibrium state* for the mixture. This special state is, in some respect, important because many problems in the porous media theory are referred to small temperature gradients, small difference velocities, and small temperature differences as well as small velocity gradients of a fluid. The equilibrium state is reached if inequality (A 34) reaches a minimum. This is the case for

$$s_0 = \left\{\Theta^F - \Theta^S = 0,\ \Theta,\ \text{grad}\,\Theta^S = \mathbf{0},\ \text{grad}\,\Theta^F = \mathbf{0},\ \check{\mathbf{C}}_S,\ \mathbf{D}_F^D = \mathbf{0},\ \mathbf{v}_{FS} = \mathbf{0}\right\}. \tag{A 41}$$

Note that, in the mixture equilibrium state, $\Theta^F = \Theta^S = \Theta$ so that, in connection with (A 34)

$$\mathcal{D}(s_0) = 0 \, . \tag{A 42}$$

If mass exchange occurs ($\hat{\rho}^S = -\hat{\rho}^F \neq 0$), the dissipation inequality (A 28) or (A 33), respectively, for $s = s_0$, yields:

$$-\hat{\rho}^F \left(\frac{\bar{\mu}^F}{\Theta^F} - \frac{\bar{\mu}^S}{\Theta^S} \right) \geq 0 \, . \tag{A 43}$$

Inequality (A 43) marks the difference from the mixture equilibrium due to mass exchanges. That means, if $\hat{\rho}^F \neq 0$, the minimum of (A 33) is determined by the term on the left-hand side of (A 43).

The equilibrium state can be investigated with the relations

$$\begin{aligned} &\Theta^F - \Theta^S = \gamma a \, , \quad \operatorname{grad} \Theta^S = \gamma \mathbf{c} \, , \quad \operatorname{grad} \Theta^F = \gamma \mathbf{d} \, , \\ &\mathbf{v}_{FS} = \mathbf{v}_F - \mathbf{v}_S = \gamma \mathbf{f} \, , \quad \mathbf{D}_F^D = \gamma \mathbf{A}^D \, , \end{aligned} \tag{A 44}$$

where a, γ are arbitrary scalars, $\mathbf{c}$ through $\mathbf{f}$ are arbitrary vectors, and $\mathbf{A}^D$ is the deviatoric part of an arbitrary tensor $\mathbf{A}$. The criteria for the existence of a local minimum are

$$\frac{d}{d\gamma} \mathcal{D}(\xi) \mid_{\gamma=0} = 0 \, , \quad \frac{d^2}{d\gamma^2} \mathcal{D}(\xi) \mid_{\gamma=0} > 0 \, , \tag{A 45}$$

where

$$\xi = \{\gamma a, \gamma \mathbf{c}, \gamma \mathbf{d}, \gamma \mathbf{f}, \check{\mathbf{C}}_S, \gamma \mathbf{A}^D\} \, . \tag{A 46}$$

With (A 44) and (A 46), Inequality (A 34) takes the following form

$$\begin{aligned} \mathcal{D}(\xi) = &- \frac{1}{\Theta^S \Theta^F} \gamma a (\hat{e}^F - \hat{\mathbf{p}}^F \cdot \mathbf{v}_S)(\xi) - \\ &- \gamma \mathbf{c} \cdot \mathbf{m}^S(\xi) - \gamma \mathbf{d} \cdot \mathbf{m}^F(\xi) - \\ &- \frac{1}{\Theta^F} \gamma \mathbf{f} \cdot (\hat{\mathbf{p}}^F - \lambda \Theta^F \bar{\Theta} \operatorname{grad} n^F)(\xi) + \\ &+ \frac{1}{\Theta^F} \gamma \mathbf{A}^D \cdot \mathbf{T}^{FD}(\xi) \geq 0. \end{aligned} \tag{A 47}$$

The necessary condition for a minimum (A 45)$_1$ yields

$$\begin{aligned} \frac{d}{d\gamma} \mathcal{D}(\xi) \mid_{\gamma=0} = \{ &- \frac{1}{\Theta^S \Theta^F} a (\hat{e}^F - \hat{\mathbf{p}}^F \cdot \mathbf{v}_S)(\xi) - \\ &- \mathbf{c} \cdot \mathbf{m}^S(\xi) - \mathbf{d} \cdot \mathbf{m}^F(\xi) - \\ &- \frac{1}{\Theta^F} \mathbf{f} \cdot (\hat{\mathbf{p}} - \lambda \Theta^F \bar{\Theta} \operatorname{grad} n^F)(\xi) + \\ &+ \frac{1}{\Theta^F} \mathbf{A}^D \cdot \mathbf{T}^{FD}(\xi) \} \mid_{\gamma=0} = 0 \, . \end{aligned} \tag{A 48}$$

Hence, from a comparison of the relations (A 48) and (A 41), and since $\mathbf{a}$, $\mathbf{c}$ through $\mathbf{f}$, as well as $\mathbf{A}^D$ are arbitrary quantities, it can be concluded that

$$\begin{aligned}
&\mathbf{m}^S(s_0) = \mathbf{m}_0^S = \mathbf{0}\,, \\
&\mathbf{m}^F(s_0) = \mathbf{m}_0^F = \mathbf{0}\,, \\
&(\hat{\mathbf{p}}^F - \lambda \Theta^F \bar{\Theta}\,\mathrm{grad}\,n^F)(s_0) = (\hat{\mathbf{p}}^F - \lambda \Theta^F \bar{\Theta}\,\mathrm{grad}\,n^F)_0 = \mathbf{0}\,, \\
&(\hat{e}^F - \hat{\mathbf{p}}^F \cdot \mathbf{v}_S)(s_0) = (\hat{e}^F - \hat{\mathbf{p}}^F \cdot \mathbf{v}_S)_0 = 0\,, \\
&\mathbf{T}^{FD}(s_0) = \mathbf{T}_0^{FD} = \mathbf{0}\,.
\end{aligned} \tag{A 49}$$

The relations (A 49) show that $\mathbf{m}^S$, $\mathbf{m}^F$, $(\hat{\mathbf{p}}^F - \lambda \Theta^F \bar{\Theta}\,\mathrm{grad}\,n^F)$, $\mathbf{T}^{FD}$, and $(\hat{e}^F - \hat{\mathbf{p}}^F \cdot \mathbf{v}_S)$ disappear in the equilibrium state for the mixture, independent of the values of the set of variables s_0. The sufficient condition (A 44)$_2$ will not be considered owing to space limitations.

The main results of the investigations concerning the equilibrium state for the mixture are now summarized:

$$\begin{aligned}
&\hat{\psi}^S = \hat{\psi}_0{}^S(\ldots, \Theta\,, \check{\mathbf{C}}_S\,)\,, \\
&\hat{\psi}^F = \hat{\psi}_0^F(\Theta)\,,
\end{aligned} \tag{A 50}$$

where $\Theta = \Theta^F = \Theta^S$,

$$\mathbf{T}_0^{SD} = 2\rho^S J_S^{-2/3} (\mathbf{F}_S \frac{\partial \hat{\psi}_0{}^S}{\partial \check{\mathbf{C}}_S} \mathbf{F}_S^T)^D \tag{A 51}$$

or considering (A 2)$_1$

$$\mathbf{T}_0^S = 2\rho^S J_S^{-2/3} (\mathbf{F}_S \frac{\partial \hat{\psi}_0{}^S}{\partial \check{\mathbf{C}}_S} \mathbf{F}_S^T)^D - n^S (\rho^{SR})^2 \frac{\partial \hat{\psi}_0{}^S}{\partial \rho^{SR}} \mathbf{I}\,, \tag{A 52}$$

$$\mathbf{T}_0^{FD} = \mathbf{0}\,. \tag{A 53}$$

Moreover, from (A 31) and (A 32), and due to $\mathbf{v}_{FS} = \mathbf{0}$, we have

$$\begin{aligned}
&\mathbf{m}_0^S = \frac{1}{(\Theta)^2} \mathbf{q}_0^S = \mathbf{0}\,, \\
&\mathbf{m}_0^F = \frac{1}{(\Theta)^2} \mathbf{q}_0^F = \mathbf{0}\,.
\end{aligned} \tag{A 54}$$

From (A 54),

$$\mathbf{q}_0^S = \mathbf{0}\,, \quad \mathbf{q}_0^F = \mathbf{0} \tag{A 55}$$

are gained.

According to the procedure by de Boer and Ehlers (1986b), all response functions $\mathcal{R}$ and the dissipation inequality $\mathcal{D}$ can be additively decomposed into quantities of the equilibrium state marked by the index 0 and additional terms (non-equilibrium terms) marked by the index n:

$$\begin{aligned}
&\mathcal{R}(s) = \mathcal{R}_0(s_0) + \mathcal{R}_n(s)\,, \\
&\mathcal{D}(s) = \mathcal{D}_0(s_0) + \mathcal{D}_n(s)\,,
\end{aligned} \tag{A 56}$$

where

$$\mathcal{R}_n(s_0) = 0\,,\ \mathcal{D}_n(s_0) = 0\,. \tag{A 57}$$

Thus, the dissipation inequality (A 34) yields, with (A 49), in consideration of (A 53) through (A 55):

$$\begin{aligned}\mathcal{D}_n(s) = &- \operatorname{grad}\Theta^S \cdot \mathbf{m}_n^S - \operatorname{grad}\Theta^F \cdot \mathbf{m}_n^F - \\ &- \mathbf{v}_{FS} \cdot \frac{1}{\Theta^F}(\hat{\mathbf{p}}^F - \lambda\Theta^F\bar{\Theta}\operatorname{grad} n^F)_n - \\ &- \Theta^{FS}\frac{1}{\Theta^S\Theta^F}(\hat{e}^F - \hat{\mathbf{p}}^F \cdot \mathbf{v}_S)_n + \\ &+ \mathbf{D}_F^D \cdot \frac{1}{\Theta^F}\mathbf{T}_n^{FD} \geq 0\,.\end{aligned} \tag{A 58}$$

Next, only thermodynamic processes close to the mixture equilibrium state will be considered, assuming that only small temperature differences, small temperature gradients, and small relative velocities occur. Thus, the non-equilibrium terms of the response functions are expanded in Taylor series, close to $s = s_0$, with respect to the parameter $\Theta^{FS} = \Theta^F - \Theta^S$, $\operatorname{grad}\Theta^S$, $\operatorname{grad}\Theta^F$, $\mathbf{v}_{FS}$, $\mathbf{D}_F^D$, where only the linear terms will be taken into account. For the non-equilibrium terms of the free energies of the porous solid and the fluid, it follows:

$$\begin{aligned}\psi_n^\alpha = \psi_n^\alpha(s_0) &+ \frac{\partial\psi_n^\alpha}{\partial\Theta^{FS}}\bigg|_{s_0}\Theta^{FS} + \frac{\partial\psi_n^\alpha}{\partial(\operatorname{grad}\Theta^S)}\bigg|_{s_0}\cdot\operatorname{grad}\Theta^S + \\ &+ \frac{\partial\psi_n^\alpha}{\partial(\operatorname{grad}\Theta^F)}\bigg|_{s_0}\cdot\operatorname{grad}\Theta^F + \frac{\partial\psi_n^\alpha}{\partial\mathbf{v}_{FS}}\bigg|_{s_0}\cdot\mathbf{v}_{FS} + \\ &+ \frac{\partial\psi_n^\alpha}{\partial(\mathbf{D}_F^D)}\bigg|_{s_0}\cdot\mathbf{D}_F^D\,.\end{aligned} \tag{A 59}$$

Thus, with (A 50) and (A 57)

$$\psi_n^S = 0\,,\quad \psi_n^F = 0 \tag{A 60}$$

is recognized. Hence, with (A 50), (A 56), and (A 57):

$$\begin{aligned}\psi^S &= \hat{\psi}_0{}^S = \hat{\psi}^S(\ldots,\Theta,\check{\mathbf{C}}_S)\,,\\ \psi^F &= \hat{\psi}_0^F = \hat{\psi}^F(\Theta)\,.\end{aligned} \tag{A 61}$$

From (A 12) and (A 13) it follows, considering (A 61),

$$\eta^S = \eta^S(\ldots,\Theta,\check{\mathbf{C}}_S)\,,\quad \eta^F = \eta^F(\ldots,\Theta)\,. \tag{A 62}$$

The Taylor series expansions of the vectors $\mathbf{m}_n^\alpha$, $(\hat{\mathbf{p}}^F - \lambda\Theta^L\bar{\Theta}\operatorname{grad} n^F)_n$ and the scalar $(\hat{e}^F - \hat{\mathbf{p}}^F \cdot \mathbf{v}_S)_n$ close to the mixture equilibrium state, according to the above procedure, results in:

$$\mathbf{m}_n^\alpha(s) = \mathbf{m}_n^\alpha(s_0) - \mathbf{m}_\Theta^\alpha(s_0)\Theta^{FS} - \mathbf{M}_S^\alpha(s_0)\,\text{grad}\,\Theta^S - \\ - \mathbf{M}_F^\alpha(s_0)\,\text{grad}\,\Theta^F - \mathbf{M}_\mathcal{V}^\alpha(s_0)\mathbf{v}_{FS} - \overset{3}{\mathbf{M}}{}_D^\alpha(s_0)\mathbf{D}_F^D\,, \tag{A 63}$$

where

$$\mathbf{m}_\Theta^\alpha(s_0) = -\left.\frac{\partial \mathbf{m}_n^\alpha}{\partial \Theta^{FS}}\right|_{s_0}, \quad \mathbf{M}_S^\alpha(s_0) = -\left.\frac{\partial \mathbf{m}_n^\alpha}{\partial(\text{grad}\,\Theta^S)}\right|_{s_0},$$

$$\mathbf{M}_F^\alpha(s_0) = -\left.\frac{\partial \mathbf{m}_n^\alpha}{\partial(\text{grad}\,\Theta^F)}\right|_{s_0}, \quad \mathbf{M}_\mathcal{V}^\alpha(s_0) = -\left.\frac{\partial \mathbf{m}_n^\alpha}{\partial \mathbf{v}_{FS}}\right|_{s_0}, \tag{A 64}$$

$$\overset{3}{\mathbf{M}}{}_D^\alpha(s_0) = -\left.\frac{\partial \mathbf{m}_n^\alpha}{\partial(\mathbf{D}_F^D)}\right|_{s_0}.$$

Furthermore,

$$(\hat{\mathbf{p}}^F - \lambda\Theta^F\bar{\Theta}\,\text{grad}\,n^F)_n(s) =: \hat{\hat{\mathbf{p}}}{}_n^F(s) = \hat{\hat{\mathbf{p}}}{}_n^F(s_0) - \mathbf{p}_\Theta(s_0)\,\Theta^{FS} - \\ - \mathbf{P}_S(s_0)\,\text{grad}\,\Theta^S - \mathbf{P}_F(s_0)\,\text{grad}\,\Theta^F - \mathbf{P}_\mathcal{V}(s_0)\mathbf{v}_{FS} - \\ - \overset{3}{\mathbf{P}}{}_D\,(s_0)\mathbf{D}_F^D\,, \tag{A 65}$$

with

$$\mathbf{p}_\Theta(s_0) = -\left.\frac{\partial \hat{\hat{\mathbf{p}}}{}_n^F}{\partial \Theta^{FS}}\right|_{s_0}, \qquad \mathbf{P}_S(s_0) = -\left.\frac{\partial \hat{\hat{\mathbf{p}}}{}_n^F}{\partial(\text{grad}\,\Theta^S)}\right|_{s_0},$$

$$\mathbf{P}_F(s_0) = -\left.\frac{\partial \hat{\hat{\mathbf{p}}}{}_n^F}{\partial(\text{grad}\,\Theta^F)}\right|_{s_0}, \qquad \mathbf{P}_\mathcal{V}(s_0) = -\left.\frac{\partial \hat{\hat{\mathbf{p}}}{}_n^F}{\partial \mathbf{v}_{FS}}\right|_{s_0}, \tag{A 66}$$

$$\overset{3}{\mathbf{P}}{}_D\,(s_0) = -\left.\frac{\partial \hat{\hat{\mathbf{p}}}{}_n^F}{\partial \mathbf{D}_F^D}\right|_{s_0}.$$

For the stresses of the fluid phase, the following relations hold:

$$\mathbf{T}_n^{FD}(s) = \mathbf{T}_n^{FD}(s_0) + \mathbf{T}_\Theta(s_0)\Theta^{FS} + \overset{3}{\mathbf{T}}{}_S\,(s_0)\,\text{grad}\,\Theta^S + \overset{3}{\mathbf{T}}{}_F\,(s_0)\,\text{grad}\,\Theta^F + \\ + \overset{3}{\mathbf{T}}{}_\mathcal{V}\,(s_0)\mathbf{v}_{FS} + \overset{4}{\mathbf{T}}{}_D\,(s_0)\mathbf{D}_F^D\,, \tag{A 67}$$

where

$$\mathbf{T}_\Theta(s_0) = \left.\frac{\partial \mathbf{T}_n^{FD}}{\partial \Theta^{FS}}\right|_{s_0}, \qquad \overset{3}{\mathbf{T}}{}_S\,(s_0) = \left.\frac{\partial \mathbf{T}_n^{FD}}{\partial(\text{grad}\,\Theta^S)}\right|_{s_0},$$

$$\overset{3}{\mathbf{T}}{}_F\,(s_0) = \left.\frac{\partial \mathbf{T}_n^{FD}}{\partial(\text{grad}\,\Theta^F)}\right|_{s_0}, \qquad \overset{3}{\mathbf{T}}{}_\mathcal{V}\,(s_0) = \left.\frac{\partial \mathbf{T}_n^{FD}}{\partial \mathbf{v}_{FS}}\right|_{s_0}, \tag{A 68}$$

$$\overset{4}{\mathbf{T}}{}_D\,(s_0) = \left.\frac{\partial \mathbf{T}_n^{FD}}{\partial \mathbf{D}_F^D}\right|_{s_0}.$$

Moreover, the linear expansion of the energy supply leads to

$$(\hat{e}^F - \hat{\mathbf{p}}^F \cdot \mathbf{v}_S)_n(s) =: \bar{\hat{e}}^F_n(s) = \bar{\hat{e}}^F_n(s_0) - e_\Theta(s_0)\Theta^{FS} - \\ - \mathbf{e}_S(s_0) \cdot \operatorname{grad}\Theta^S - \mathbf{e}_F(s_0) \cdot \operatorname{grad}\Theta^F - \\ - \mathbf{e}_{\mathcal{V}}(s_0) \cdot \mathbf{v}_{FS} - \mathbf{E}_D(s_0) \cdot \mathbf{D}^D_F , \tag{A 69}$$

where the following abbreviations have been used

$$e_\Theta(s_0) = -\frac{\partial\,\bar{\hat{e}}^F_n}{\partial\Theta^{FS}}\bigg|_{s_0} , \qquad \mathbf{e}_S(s_0) = -\frac{\partial\,\bar{\hat{e}}^F_n}{\partial(\operatorname{grad}\Theta^S)}\bigg|_{s_0} ,$$

$$\mathbf{e}_F(s_0) = -\frac{\partial\,\bar{\hat{e}}^F_n}{\partial(\operatorname{grad}\Theta^F)}\bigg|_{s_0} , \qquad \mathbf{e}_{\mathcal{V}}(s_0) = -\frac{\partial\,\bar{\hat{e}}^F_n}{\partial\mathbf{v}_{FS}}\bigg|_{s_0} , \tag{A 70}$$

$$\mathbf{E}_D(s_0) = -\frac{\partial\,\bar{\hat{e}}^F_n}{\partial\mathbf{D}^D_F}\bigg|_{s_0} .$$

The linearized relations (A 63), (A 65), (A 67), and (A 69) must obey the principle of material objectivity. This means for (A 63) (see de Boer, 1982 and Section 5.8.1)

$$\mathbf{Q}\mathbf{m}^\alpha_n(s) = \mathbf{m}^\alpha_n(\overset{*}{s}) \tag{A 71}$$

or with (A 64) where, according to (A 57), $\mathbf{m}^\alpha_n(s_0)$ disappears

$$\mathbf{Q}\,\big[\mathbf{m}^\alpha_\Theta(s_0)\Theta^{FS} + \mathbf{M}^\alpha_S(s_0)\operatorname{grad}\Theta^S + \mathbf{M}^\alpha_F(s_0)\operatorname{grad}\Theta^F + \\ \mathbf{M}^\alpha_{\mathcal{V}}(s_0)\mathbf{v}_{FS} + \overset{3}{\mathbf{M}}{}^\alpha_D(s_0)\mathbf{D}^D_F\big] = \\ = \mathbf{m}^\alpha_\Theta(\overset{*}{s}_0)\Theta^{FS} + \mathbf{M}^\alpha_S(\overset{*}{s}_0)\mathbf{Q}\operatorname{grad}\Theta^S + \mathbf{M}^\alpha_F(\overset{*}{s}_0)\mathbf{Q}\operatorname{grad}\Theta^F + \\ + \mathbf{M}^\alpha_{\mathcal{V}}(\overset{*}{s}_0)\mathbf{Q}\mathbf{v}_{FS} + \overset{3}{\mathbf{M}}{}^\alpha_D(\overset{*}{s}_0)\mathbf{Q}\,\mathbf{D}^D_F\,\mathbf{Q}^T . \tag{A 72}$$

The results are

$$\mathbf{m}^\alpha_\Theta(\overset{*}{s}_0) = \mathbf{Q}\mathbf{m}^\alpha_\Theta(s_0) , \quad \mathbf{M}^\alpha_S(\overset{*}{s}_0) = \mathbf{Q}\mathbf{M}^\alpha_S(s_0)\mathbf{Q}^T , \\ \mathbf{M}^\alpha_F(\overset{*}{s}_0) = \mathbf{Q}\mathbf{M}^\alpha_F(s_0)\mathbf{Q}^T , \quad \mathbf{M}^\alpha_{\mathcal{V}}(\overset{*}{s}_0) = \mathbf{Q}\mathbf{M}^\alpha_{\mathcal{V}}(s_0)\mathbf{Q}^T , \\ \overset{3}{\mathbf{M}}{}^\alpha_D(\overset{*}{s}_0) = \Big[\{\mathbf{Q}\,\overset{3}{\mathbf{M}}{}^\alpha_D(s_0)\}^{\overset{3}{-}}\mathbf{Q}^T\}^{\overset{23}{3^T}}\Big]\mathbf{Q}^T\}^{\overset{23}{3^T}} . \tag{A 73}$$

Within the framework of simple isotropic ansätze, Eqn. (A 72) can be fulfilled by:

$$\mathbf{m}^\alpha_\Theta(s_0) = \mathbf{0} , \\ \mathbf{M}^\alpha_S(s_0) = -\,\alpha^\alpha_S(s_0)\mathbf{I} , \quad \mathbf{M}^\alpha_F = -\,\alpha^\alpha_F(s_0)\mathbf{I} , \\ \mathbf{M}^\alpha_{\mathcal{V}} = -\,\alpha^\alpha_{\mathcal{V}}(s_0)\mathbf{I} , \quad \overset{3}{\mathbf{M}}{}^\alpha_D = -\,\alpha_D\,\overset{3}{\mathbf{E}} , \tag{A 74}$$

where $\overset{3}{\mathbf{E}}$ is a fundamental tensor of the third-order (see de Boer, 1982). It can be shown that the axial vector $\overset{A}{\mathbf{d}}_F$ related to $\mathbf{D}_F^D$ vanishes:

$$\overset{3}{\mathbf{E}}\,\overset{D}{\mathbf{D}_F} = \mathbf{I} \times \overset{D}{\mathbf{D}_F} = \ 2\,\overset{A}{\mathbf{d}}_F = \mathbf{0}\ . \tag{A 75}$$

Thus, with (A 73) and (A 74), we obtain from (A 63)

$$\mathbf{m}_n^\alpha(s_0) = -\ \alpha_S^\alpha(s_0)\,\mathrm{grad}\,\Theta^S - \alpha_F^\alpha(s_0)\,\mathrm{grad}\,\Theta^F - \alpha_\upsilon^\alpha(s_0)\mathbf{v}_{FS}\ . \tag{A 76}$$

It should be mentioned that it is not difficult to consider more general ansätze, instead of the introduced simple isotropic ansätze, in order to describe also anisotropic behavior. The corresponding procedure, above-mentioned, would result in positive-definite tensors $\mathbf{M}_S^\alpha, \mathbf{M}_F^\alpha$, and $\mathbf{M}_\upsilon^\alpha$.

In a similar way, we obtain for (see (A 65))

$$\begin{aligned}\hat{\mathbf{p}}_n^F &= (\hat{\mathbf{p}}^F - \lambda\Theta^F\bar{\Theta}\,\mathrm{grad}\,n^F)_n \\ &= -\ \beta_S(s_0)\,\mathrm{grad}\,\Theta^S - \beta_F(s_0)\,\mathrm{grad}\,\Theta^F - \beta_\upsilon^F(s_0)\mathbf{v}_{FS}\ .\end{aligned} \tag{A 77}$$

The quantity β_υ^F is, in the literature, also known as S_υ. For the description of anisotropic permeability, the derivation would lead to a positive-definite tensor $\mathbf{S}_\upsilon$.

The principle of material objectivity, applied to the second-order tensor $\mathbf{T}_n^{FD}(s)$ (A 67), leads to the requirement

$$\mathbf{Q}\mathbf{T}_n^{FD}(s)\mathbf{Q}^T = \mathbf{T}_n^{FD}(\overset{*}{s}) \tag{A 78}$$

or with (A 67), considering (A 57):

$$\begin{aligned}&\mathbf{Q}\left[\ \overset{3}{\mathbf{T}}_S\,(s_0)\,\mathrm{grad}\,\Theta^S + \overset{3}{\mathbf{T}}_F\,(s_0)\,\mathrm{grad}\,\Theta^F + \overset{3}{\mathbf{T}}_\upsilon\,(s_0)\mathbf{v}_{FS} + \right.\\ &\quad \left. + \overset{4}{\mathbf{T}}_D\,(s_0)\mathbf{D}_F^D + \mathbf{T}_\Theta(s_0)\Theta^{FS}\,\right]\mathbf{Q}^T = \\ &= \overset{3}{\mathbf{T}}_S\,(\overset{*}{s}_0)\mathbf{Q}\,\mathrm{grad}\,\Theta^S + \overset{3}{\mathbf{T}}_F\,(\overset{*}{s}_0)\mathbf{Q}\,\mathrm{grad}\,\Theta^F + \overset{3}{\mathbf{T}}_\upsilon\,(\overset{*}{s}_0)\mathbf{Q}\mathbf{v}_{FS} + \\ &\quad + \overset{4}{\mathbf{T}}_D\,(\overset{*}{s}_0)\mathbf{Q}\,\mathbf{D}_F^D\,\mathbf{Q}^T + \mathbf{T}_\Theta(\overset{*}{s}_0)\Theta^{FS}\ .\end{aligned} \tag{A 79}$$

From (A 79), it follows:

$$\begin{aligned}\overset{3}{\mathbf{T}}_S\,(\overset{*}{s}_0) &= \left[\{[\mathbf{Q}\,\overset{3}{\mathbf{T}}_S\,(s_0)]^{\underline{3}}\mathbf{Q}^T\}^{\overset{23}{\underline{3}^T}}\,\mathbf{Q}^T\right]^{\overset{23}{\underline{3}^T}}\ ,\\ \overset{3}{\mathbf{T}}_F\,(\overset{*}{s}_0) &= \left[\{[\mathbf{Q}\,\overset{3}{\mathbf{T}}_F\,(s_0)]^{\underline{3}}\mathbf{Q}^T\}^{\overset{23}{\underline{3}^T}}\,\mathbf{Q}^T\right]^{\overset{23}{\underline{3}^T}}\ ,\\ \overset{3}{\mathbf{T}}_\upsilon\,(\overset{*}{s}_0) &= \left[\{[\mathbf{Q}\,\overset{3}{\mathbf{T}}_\upsilon\,(s_0)]^{\underline{3}}\mathbf{Q}^T\}^{\overset{23}{\underline{3}^T}}\,\mathbf{Q}^T\right]^{\overset{23}{\underline{3}^T}}\ ,\\ \overset{4}{\mathbf{T}}_D\,(\overset{*}{s}_0) &= (\mathbf{Q}^T\otimes\mathbf{Q}^T)^{\overset{14}{T}}\,\overset{4}{\mathbf{T}}_D\,(s_0)(\mathbf{Q}\otimes\mathbf{Q})^{\overset{14}{T}}\ ,\\ \mathbf{T}_\Theta(\overset{*}{s}_0) &= \mathbf{Q}\mathbf{T}_\Theta(s_0)\mathbf{Q}^T\ .\end{aligned} \tag{A 80}$$

Within the framework of simple isotropic ansätze, the relations in (A 80) can be fulfilled with

$$\begin{aligned}
&\overset{3}{\mathbf{T}}_S\,(s_0) = \gamma_S(s_0)\,\overset{3}{\mathbf{E}}\,, \quad \overset{3}{\mathbf{T}}_F\,(s_0) = \gamma_F(s_0)\,\overset{3}{\mathbf{E}}\,, \quad \overset{3}{\mathbf{T}}_{\mathcal{V}}\,(s_0) = \gamma_{\mathcal{V}}(s_0)\,\overset{3}{\mathbf{E}}\,,\\
&\overset{4}{\mathbf{T}}_D\,(s_0) = \gamma_D(s_0)\,\overset{4}{\mathbf{I}} + \gamma_{\overline{\overline{D}}}(s_0)\,\overset{4}{\overline{\overline{\mathbf{I}}}}\,,\\
&\mathbf{T}_\Theta(s_0) = \gamma_\Theta(s_0)\mathbf{I}\,.
\end{aligned} \tag{A 81}$$

It can be shown with the arbitrary vector $\mathbf{a}$ that

$$\overset{3}{\mathbf{E}}\mathbf{a} \;=\; -\;(\overset{3}{\mathbf{E}}\mathbf{a})^T \tag{A 82}$$

is a skew-symmetric tensor, thus showing that $\mathbf{a}$ is an axial vector. As $\mathbf{T}_n^{FD}$ is symmetric and $\operatorname{grad}\Theta^S$, $\operatorname{grad}\Theta^F$, and $\mathbf{v}_{FS}$ in (A 79) are in general not axial vectors, therefore the response parameters $\gamma_S(s_0), \gamma_F(s_0)$, and $\gamma_{\mathcal{V}}(s_0)$, must be chosen as zero:

$$\gamma_S(s_0) = 0\,, \quad \gamma_F(s_0) = 0\,, \quad \gamma_{\mathcal{V}}(s_0) = 0\,. \tag{A 83}$$

Moreover, due to $\mathbf{T}_n^{FD} = (\mathbf{T}_n^{FD})^T$, it follows

$$\overset{4}{\mathbf{T}}_D\,(s_0) = 2\mu^F(s_0)\,\overset{4}{\mathbf{I}}\,, \tag{A 84}$$

with

$$\mu^F(s_0) = \frac{1}{2}\left[\gamma_D(s_0) + \gamma_{\overline{\overline{D}}}(s_0)\right]\,. \tag{A 85}$$

Analogously, since $\mathbf{T}_n^{FD}$ is a deviator, we have

$$\gamma_\Theta = 0\,. \tag{A 86}$$

From (A 67), considering (A 56), (A 81), and (A 83) through (A 86) $\mathbf{T}_n^{FD}(s)$ is given by

$$\mathbf{T}_n^{FD}(s) = 2\mu^F(s_0)\mathbf{D}_F^D\,. \tag{A 87}$$

Finally, analogous to the above derivations, we arrive at:

$$\bar{\hat{e}}_n^F(s) = (\hat{e}^F - \hat{\mathbf{p}}^F\cdot\mathbf{v}_S)_n(s) = -\,e_\Theta(s_0)\Theta^{FS}\,. \tag{A 88}$$

From the dissipation inequality (A 58), restrictions for the response parameters $\alpha_S^\alpha(s_0)$ through $e_\Theta(s_0)$ can be gained. Considering (A 76) through (A 88), the dissipation inequality (A 58) yields:

$$\begin{aligned}
D_n(s) = {} & [\alpha_S^S(s_0)\operatorname{grad}\Theta^S + \alpha_F^S(s_0)\operatorname{grad}\Theta^F + a_{\mathcal{V}}^S(s_0)\mathbf{v}_{FS}]\cdot\operatorname{grad}\Theta^S + \\
& + [\alpha_S^F(s_0)\operatorname{grad}\Theta^S + \alpha_F^F(s_0)\operatorname{grad}\Theta^F + \alpha_{\mathcal{V}}^F(s_0)\mathbf{v}_{FS}]\cdot\operatorname{grad}\Theta^F + \\
& + \frac{1}{\Theta^F}[\beta_S(s_0)\operatorname{grad}\Theta^S + \beta_F(s_0)\operatorname{grad}\Theta^F + \beta_{\mathcal{V}}(s_0)\mathbf{v}_{FS}]\cdot\mathbf{v}_{FS} + \\
& + \frac{1}{\Theta^F}[2\mu^F(s_0)\mathbf{D}_F^D]\cdot\mathbf{D}_F^D + \\
& + \frac{1}{\Theta^S\Theta^F}[e_\Theta(s_0)\Theta^{FS}]\Theta^{FS} \geq 0\,.
\end{aligned} \tag{A 89}$$

For arbitrary $\operatorname{grad}\Theta^S$, $\operatorname{grad}\Theta^F$, $\mathbf{v}_{FS}$, and $\mathbf{D}_F^D$, it follows from (A 89):

$$\alpha_S^S(s_0) \geq 0\,,\ \alpha_F^S(s_0) + \alpha_S^F(s_0) = 0\,,\ \alpha_\mathcal{V}^S + \frac{1}{\Theta^F}\beta_S = 0\,,$$

$$\alpha_F^F(s_0) \geq 0\,,\ \alpha_\mathcal{V}^F(s_0) + \frac{1}{\Theta^F}\beta_F(s_0) = 0\,, \tag{A 90}$$

$$\beta_\mathcal{V}(s_0) \geq 0\,,\ \mu^F(s_0) \geq 0\,,\ e_\Theta(s_0) \geq 0\,.$$

Thus, with (A 90), the linearized response functions can finally be summarized:

$$\mathbf{m}_n^S = -\ \alpha_S^S(s_0)\,\mathrm{grad}\,\Theta^S - \alpha_F^S(s_0)\,\mathrm{grad}\,\Theta^F - \alpha_\mathcal{V}^S(s_0)\mathbf{v}_{FS} = \mathbf{q}^S\frac{1}{(\Theta^S)^2}\,,$$

$$\mathbf{m}_n^F = -\ \alpha_S^F(s_0)\,\mathrm{grad}\,\Theta^S - \alpha_F^F(s_0)\,\mathrm{grad}\,\Theta^F - \alpha_\mathcal{V}^F(s_0)\mathbf{v}_{FS} = \mathbf{q}^F\frac{1}{(\Theta^F)^2}\,,$$

$$\begin{aligned}\hat{\bar{\mathbf{p}}}_n^F &= -\ \beta_S(s_0)\,\mathrm{grad}\,\Theta^S - \beta_F(s_0)\,\mathrm{grad}\,\Theta^F - \beta_\mathcal{V}^F(s_0)\mathbf{v}_{FS} \\ &= (\hat{\mathbf{p}}^F - \lambda\Theta^F\bar{\Theta}\,\mathrm{grad}\,n^F)_n\,,\end{aligned} \tag{A 91}$$

$$\mathbf{T}_n^{FD} = 2\mu^F(s_0)\mathbf{D}_F^D\,,$$

$$\hat{\bar{e}}_n^F = -\ e_\Theta(s_0)\Theta^{FS} = (\hat{e}^F - \hat{\mathbf{p}}^F\cdot\mathbf{v}_S)_n\,.$$

Thus, with (A 49), (A 5), and (A 91), the linearized response functions $\mathbf{m}^\alpha(s)$, $\hat{\bar{\mathbf{p}}}_n^F(s)$, $\mathbf{T}^{FD}(s)$, and $\hat{\bar{e}}^F$ are determined.

In order to ensure the validity of the entropy inequality (A 28) in the case of mass exchange processes, the constitutive equation for $\hat{\rho}^F$ must be chosen in such a way that the fourth term on the left-hand side of (A 28) is always positive. A first approach can be

$$\hat{\rho}^F = -\ \frac{G}{\Theta^S\Theta^F}(\Theta^S\mu^F - \Theta^F\mu^S)\,, \tag{A 92}$$

with G as a positive response parameter.

This approach guarantees the experimental observations that, in the static case, mass exchange does not occur if the temperature of the solid phase equals the temperature of the fluid phase and the chemical potentials are the same for both the constituents (see Elwell and Pointon, 1972).

Appendix B: Introduction to the Vector- and Tensor Calculus for Engineers

B 1. Introduction

Appendix B provides an introduction to the vector and tensor calculus used in this book on the porous media theory. Readers not familiar with the vector and tensor calculus can study this appendix in order to follow the chapters of interest to them. For an in-depth study, the reader is referred to the book of de Boer (1982) which, however, is written in German.

The vector and tensor calculus opens up, in many cases, a direct approach to the formulation of scientific relations in physics and engineering, connecting the vector and tensor quantities, as can be seen in the previous chapters.

The vector notion is introduced at all times by a symbolic notation. Only the explicit calculation of vector operations requires the establishment of the vector in a base (coordinate) system. Concerning the representation of the tensor calculus, there are, in contrast, essentially two opinions. On the one hand, tensors are defined as certain quantities with indices which, due to coordinate transformations, obey determined transformation rules. The representation of this tensor calculus is denoted as an index notation. On the other hand, tensors are introduced by linear mappings; this representation allows a symbolic notation.

The consideration of the tensor as a special linear mapping in the previous chapters seems to be a very useful viewpoint in mechanics and in engineering science. In this field, there are a wide range of relations containing such linear connections – independent of special coordinate systems. With this approach of the tensor calculus, one arrives not only at a coordinate free representation of the tensors but also at a very compact description of the most important calculation rules. However, in the individual sections we will represent the tensors and the tensor operations in base systems so that at each point a comparison with the representation, used in many books, is possible.

The first sections give an introduction to the most important results of the vector and tensor algebra. The vector algebra will be developed graphically; this procedure corresponds to the thinking of engineers. This method cannot be applied to the tensor algebra because a tensor cannot be graphically explained in a simple way. In order to arrive at an exact representation of the tensor algebra, the tensor and tensor operations will be introduced axiomatically by definitions. Within the framework of the vector and tensor algebra, some new considerations on the cross-products of vectors and tensors will be introduced. Also, in the section on the algebra of higher-order tensors, some new definitions will be provided which are necessary for the vector and tensor analysis. In the sections on the vector and tensor analysis, vector and tensor functions - which depend on real scalar parameters - will be considered first; thereafter, the derivatives of vector and tensor functions depending on vector and tensor parameters will be treated. Finally, some transformation rules for volume and surface integrals will be given.

In order to make this step into the tensor calculus easier, first some basic notions as, e.g., indexed quantities, summation rule of Einstein, and the Kronecker symbol will be discussed.

The definitions are denoted by numbers and letters, where the first two numbers refer to the section. Within the same section, the definitions are cited without the numbers referring to the section.

B 2.
Basic Concepts

B 2.1
Symbols

Calculation involving indexed quantities is difficult for many people, especially for engineers still studying mathematics. This is definitely a result of the fact that, in the beginning lectures for engineering students, the explanation of indexed quantities is avoided, e.g., indexed coordinates. Experience shows that students who have a basic knowledge of indexed symbols have a better grasp of tensor calculation. In order to better comprehend the following ideas, and to make the step into the next chapter easier, the following symbols will de used:

a)
Symbols of First-Order

These symbols are real quantities s, having an index. Within the framework of our considerations in three-dimensional Euclidean space, the index described by any arbitrary Latin character, has the number 1, 2 and 3, e.g.,

$$s^i \rightarrow s^1 \,, \quad s^2 \,, \quad s^3 \qquad \text{(B 2.1.1)}$$

or

$$s_i \rightarrow s_1 \,, \quad s_2 \,, \quad s_3 \,. \qquad \text{(B 2.1.2)}$$

In both the cases we get 3^1 elements.

b)
Symbols of Second-Order

The real quantity s has two indices,

$$s^{ik} \rightarrow \begin{matrix} s^{11}, & s^{12}, & s^{13}, \\ s^{21}, & s^{22}, & s^{23}, \\ s^{31}, & s^{32}, & s^{33}. \end{matrix} \qquad \text{(B 2.1.3)}$$

This results in 3^2 elements. The second-order symbols can also be expressed as follows:

$$s_{ik}, \quad s^{i}_{.k}, \quad s_{i}^{.k}, \qquad \text{(B 2.1.4)}$$

whereby, in the last forms, the point before the index indicates an empty space, so that the index k has the second position, e.g.,

$$s^{i}_{.k} \rightarrow s^{i}_{.1}, \quad s^{i}_{.2}, \quad s^{i}_{.3}. \qquad \text{(B 2.1.5)}$$

c)
Symbols of Higher-Order

We can generalize our observations and define the corresponding symbols of higher-order, as:

$$s^{ijk}, \quad s^{ijkm}, \quad s^{ij}_{km}, \quad \text{etc.} \qquad \text{(B 2.1.6)}$$

It should be pointed out that the introduction of the symbols has no further mathematical or physical meaning. The symbols only serve to make one more familiar with certain forms of the tensor calculation.

B 2.2
Einstein's Summation Convention

The representation of a summation in the form

$$s^{1}_{.1} + s^{2}_{.2} + s^{3}_{.3} \qquad \text{(B 2.2.1)}$$

is normally done with the summation symbol $\sum$, in the form

$$\sum_{i=1}^{3} s^{i}_{.i}. \qquad \text{(B 2.2.2)}$$

The use of the summation symbol in derivations and calculations often proves to be troublesome. Therefore, the summation symbol is avoided and instead the following summation convention, which goes back to Einstein, has been agreed upon:

When, in a symbol or in a term having many symbols, an index appears twice, and this index stands above and below (superscript and subscript), then the summation must be done over this index.

We can help to explain the summation convention with two examples:

$$s^i_{.i} = s^1_{.1} + s^2_{.2} + s^3_{.3} \,, \qquad s^i t_i = s^1 t_1 + s^2 t_2 + s^3 t_3 \,. \tag{B 2.2.3}$$

The summation index (in our example, the index i) is denoted as a *silent, or dummy index*, because this does not appear again in its corresponding form after the summation has been carried out. Therefore, it is valid for:

$$s_{ij} t^i = s_{1j} t^1 + s_{2j} t^2 + s_{3j} t^3 = r_j \,, \tag{B 2.2.4}$$

in which the letter r has been chosen for the sum of the products. The index j is named as the *free index*; it does not lead to summation.

In special cases, we will extend the summation convention up to the point that the summation must also be carried out for silent superscript or subscript indices exclusively.

Naturally, the presented summation convention of Einstein in the preceding expressions, are not limited to only three summands. In view of the three-dimensional Euclidean vector space to be taken up in the forthcoming part, it is sufficient to only consider the summation of three dimensions.

B 2.3 Kronecker Symbol

We introduce now the *Kronecker symbol* δ^k_i, the importance of which will be explained later. The Kronecker symbol has the following values:

$$\delta^k_i = \begin{cases} 0 & \text{for} \quad i \neq k \,, \\ 1 & \text{for} \quad i = k \,. \end{cases} \tag{B 2.3.1}$$

It holds that

$$\begin{aligned} &\delta^1_1 = 1 \,, \quad \delta^2_2 = 1 \,, \quad \delta^3_3 = 1 \,, \\ &\delta^1_2 = 0 \,, \quad \delta^2_1 = 0 \,, \quad \delta^1_3 = 0 \,, \\ &\delta^3_1 = 0 \,, \quad \delta^2_3 = 0 \,, \quad \delta^3_2 = 0 \,. \end{aligned} \tag{B 2.3.2}$$

Expressions containing the Kronecker symbol can, for the most part, be simplified. This can be explained with the help of a few examples:

$$s^i \delta^k_i = s^1 \delta^k_1 + s^2 \delta^k_2 + s^3 \delta^k_3 \,. \tag{B 2.3.3}$$

For $k = 1, 2, 3$, we obtain

$$\begin{aligned} k=1: \quad & s^i\delta_i^1 = s^1\delta_1^1 + s^2\delta_2^1 + s^3\delta_3^1 , \\ k=2: \quad & s^i\delta_i^2 = s^1\delta_1^2 + s^2\delta_2^2 + s^3\delta_3^2 , \\ k=3: \quad & s^i\delta_i^3 = s^1\delta_1^3 + s^2\delta_2^3 + s^3\delta_3^3 . \end{aligned} \quad \text{(B 2.3.4)}$$

Taking into consideration the definition for the Kronecker symbols introduced above, we see that the expressions described above can be reduced to

$$s^i\delta_i^1 = s^1 , \quad s^i\delta_i^2 = s^2 , \quad s^i\delta_i^3 = s^3 . \quad \text{(B 2.3.5)}$$

We can therefore write:

$$s^i\delta_i^k = s^k . \quad \text{(B 2.3.6)}$$

With this example, we come to an important statement:

> When the indexed symbols are multiplied with the Kronecker symbol, then the summation index (silent index) by these symbols is exchanged with the free index of the Kronecker symbol, and the Kronecker symbol is set equal to one.

This statement is further explained with the help of the following examples:

$$\begin{aligned} & s^{im}\delta_m^k = s^{ik} , \quad s^{im}\delta_i^s\delta_m^k = s^{sk} , \\ & s^{ik}s_{mn}\delta_i^s\delta_j^m = s^{sk}s_{jn} . \end{aligned} \quad \text{(B 2.3.7)}$$

B 3.
Vector Algebra

In physics and engineering, we come across many quantities which, by stating one single number, are uniquely characterized, as for example, the temperature, density, or power. Such quantities are described as *scalars*. Other quantities, such as velocity and force are first given uniquely by a scalar value and then by a direction in the Euclidean space and are called *vectors*. The investigation of its properties as well as its relation with each other and with scalars is the subject of discussion in the forthcoming chapters. In the following chapters, we will characterize arbitrary scalars with Greek letters and arbitrary vectors with Latin letters (bold face). Important *calculation rules*, that are defined rules will be denoted with capital letters, whereas we shall denote derived rules of calculation, as usual, with the help of numerals.

B 3.1
Vector Notion and Vector Operations

As has already been mentioned, we will limit ourselves to the development of the vector calculus within Euclidean space. We will consider this space as given.

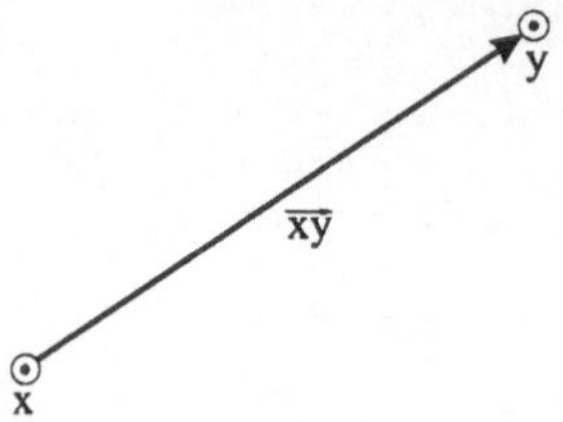

Fig. B 3.1.1. Directional arrow

The geometrical objects (namely points, straight lines, distances, etc., and their properties as length, direction, etc.) as well as their relationship with each other (such as parallelism, perpendicular lines, etc.) can be described with the help of illustrative terms without having the need to go back to formal definitions. Distance is an orderly group of points (x,y) with x being the starting point and y the end point, with the length being the distance between these two points. The direction of the straight line can be explained in such a way that the Euclidean space at point x is taken to be divided into two half-spaces by a plane at right angles to the straight line (x,y), one of the half-spaces containing the point y. One can say that the two distances have the same *distance*, when one of the assigned half-spaces is contained in the other. The direction of the length is shown with the help of an arrow (directional arrow).
The direction arrow has a definite length and a definite direction. These notions are used in order to define the vector.

A vector **u** is defined as a class of all **directional arrows** having the same length and direction. (B 3.1.A1)

We take into consideration two non-zero vectors, **u** and **v**, that are parallel. This means that **v** points in the same direction as **u** or $-$ **u**. With α as the quotient of the *norms* of **v** or **u**, it is reasonable to assume **v** to be a α-time multiple of **u** or $-$ **u**. The multiplication of a vector with a scalar (real numbers) is defined as:

For $\alpha = \dfrac{|\mathbf{v}|}{|\mathbf{u}|} \geq 0$ is $\alpha\mathbf{u} = \mathbf{u}\alpha$

the vector of the length $\alpha|\mathbf{u}|$ with the same direction as **u**; (B 3.1.A2)

for $\alpha = \dfrac{|\mathbf{v}|}{|\mathbf{u}|} < 0$ is $\alpha\mathbf{u} = |\alpha|(-\mathbf{u})$.

We can derive the following relations from the definition (A2):

$$0\mathbf{u} = \mathbf{0}\,, \quad \alpha\mathbf{0} = \mathbf{0}\,, \quad 1\mathbf{u} = \mathbf{u}\,, \quad (-1)\mathbf{u} = 1(-\mathbf{u}) = -\,\mathbf{u}\,, \qquad \text{(B 3.1.1)}$$

and

$$\mathbf{u} = |\mathbf{u}|\,\mathring{\mathbf{u}} \text{ and } \mathring{\mathbf{u}} = \frac{1}{|\mathbf{u}|}\mathbf{u}\,. \qquad \text{(B 3.1.2)}$$

Furthermore, the associative rule is valid,

$$(\alpha\beta)\mathbf{u} = \alpha(\beta\mathbf{u})\,, \qquad \text{(B 3.1.3)}$$

which can be easily proven.

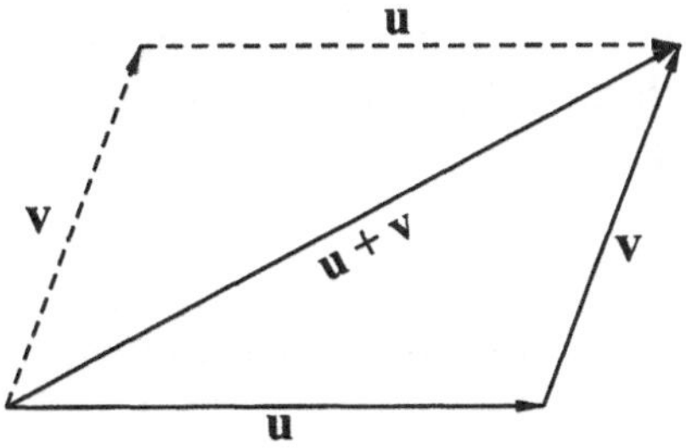

Fig. B 3.1.2. Addition of vectors

We will now define the *vector addition* of the vectors **u** and **v**:

A vector **u** + **v** is the straight line from the starting point of the vector **u** to the end point of the vector **v**; thereby the vector **v** is depicted at the end point of the vector **u**. (B 3.1.B)

Obviously, one can extend the vector triangle in Figure B 3.1.2 to a parallelogram, so that one can interpret the resulting vector $\mathbf{u}+\mathbf{v}$ as a diagonal vector in the parallelogram spanned by **u** and **v**. The immediate result can be seen as:

$$\mathbf{u}+\mathbf{v}=\mathbf{v}+\mathbf{u} \quad \text{(commutative rule)}\,. \tag{B 3.1.4}$$

Further conclusions of the vector addition, which are stated without being proved, are:

$$\begin{aligned} \mathbf{u}+(\mathbf{v}+\mathbf{w})=(\mathbf{u}+\mathbf{v})+\mathbf{w} \quad &\text{(associative rule)}\,,\\ \mathbf{u}+\mathbf{0}&=\mathbf{u}\,,\\ \mathbf{u}+(-\mathbf{u})&=\mathbf{0}\,,\\ \mathbf{u}+(-\mathbf{v})&=\mathbf{u}-\mathbf{v}\,. \end{aligned} \tag{B 3.1.5}$$

For the vector $\alpha\mathbf{u}$ defined in (A2), along with (B) holds:

$$\begin{aligned} (\alpha+\beta)\mathbf{u}=\alpha\mathbf{u}+\beta\mathbf{u} \quad &\text{(distributive rule for the scalar addition)}\,,\\ \alpha(\mathbf{u}+\mathbf{v})=\alpha\mathbf{u}+\alpha\mathbf{v} \quad &\text{(distributive rule for the vector addition)}\,. \end{aligned} \tag{B 3.1.6}$$

These operations, also called affine operations, allow us, with regard to the examining of the geometrical characteristics of the space, to state only whether or not the two vectors are parallel. Besides that, we can only compare the length of two vectors when they are parallel. In order to determine the angle between two nonparallel vectors, and to calculate the ratio of their lengths, it is necessary to introduce another operation, namely the *scalar product* (*dot product*).

Given are two arbitrary vectors **u** and **v**, with φ being the angle between the directional arrows ($0 \le \varphi \le \pi$) that is defined for non-zero vectors. The scalar product of the vectors **u** and **v**, which is written in the form $\mathbf{u}\cdot\mathbf{v}$, is the real number

$$\mathbf{u}\cdot\mathbf{v}=\begin{cases} 0\,, & \text{if} \quad \mathbf{u}=\mathbf{0} \quad \text{or} \quad \mathbf{v}=\mathbf{0}\,,\\ |\mathbf{u}||\mathbf{v}|\cos\varphi\,, & \text{if} \quad \mathbf{u}\neq\mathbf{0} \quad \text{and} \quad \mathbf{v}\neq\mathbf{0}\,. \end{cases} \tag{B 3.1.C}$$

The number $|\mathbf{u}|\cos\varphi$ is the projection of the vector **u** in the direction of the vector **v**. From the definition (C), along with the arbitrarily chosen non-zero vectors **u**, **v**, and **w**, it can be concluded that:

$$\mathbf{u}\cdot\mathbf{v} = \mathbf{v}\cdot\mathbf{u} \text{ (commutative rule)}, \qquad \mathbf{u}\cdot\mathbf{v} = 0 \tag{B 3.1.7}$$

precisely, then, when **u** is perpendicular to **v** (**u** and **v** are *orthogonal*),

$$\begin{aligned} \mathbf{u}\cdot(\mathbf{v}+\mathbf{w}) &= \mathbf{u}\cdot\mathbf{v}+\mathbf{u}\cdot\mathbf{w}\,, \\ (\alpha\mathbf{u})\cdot\mathbf{v} = \mathbf{u}\cdot(\alpha\mathbf{v}) &= \alpha(\mathbf{u}\cdot\mathbf{v})\,, \\ \mathbf{u}\cdot\mathbf{u} &= |\mathbf{u}|^2 > 0. \end{aligned} \tag{B 3.1.8}$$

The positive square root

$$|\mathbf{u}| = \sqrt{\mathbf{u}\cdot\mathbf{u}} \tag{B 3.1.9}$$

from (B 3.1.8)$_3$ gives the amount, or the norm, of the vector **u**. *Schwarz's inequality* is valid for the sum of the scalar product. This states:

The sum of the scalar product of two vectors **u** and **v** is smaller or equal to the product of the norms of both the vectors, i.e.,

$$|\mathbf{u}\cdot\mathbf{v}| \le |\mathbf{u}||\mathbf{v}|. \tag{B 3.1.10}$$

The sign of equality is valid when the vectors **u** and **v** proceed in the same direction. The proof of Schwarz's inequality results directly from (C).

The norms of the vectors $\alpha\mathbf{u}$ and $\mathbf{u}+\mathbf{v}$ possess the following characteristics:

$$\begin{aligned} |\alpha\mathbf{u}| &= |\alpha||\mathbf{u}|\,, \\ |\mathbf{u}+\mathbf{v}| &\le |\mathbf{u}| + |\mathbf{v}|\,. \end{aligned} \tag{B 3.1.11}$$

In order to prove the inequality (B 3.1.11)$_2$, one needs to use Schwarz's inequality (B 3.1.10).

B 3.2 Base System

We consider two vectors, **u** and **v**, that can be represented in Euclidean space E^3 in the following form:

$$\mathbf{u} = u^i\mathbf{g}_i,\ \mathbf{v} = v^k\mathbf{g}_k\,, \tag{B 3.2.1}$$

where $\mathbf{g}_i$ ($\mathbf{g}_1, \mathbf{g}_2, \mathbf{g}_3$) are base vectors. It is said that the vectors **u** and **v** are split into the direction of the base vectors. The scalar product of the vectors **u** and **v** is determined with (B 3.1.8)$_2$ and (B 3.2.1) through

$$\mathbf{u}\cdot\mathbf{v} = u^i v^k \mathbf{g}_i\cdot\mathbf{g}_k\,. \tag{B 3.2.2}$$

We denote the scalar product of the base vectors $\mathbf{g}_i$ and $\mathbf{g}_k$ with

$$\mathbf{g}_i\cdot\mathbf{g}_k = g_{ik} \tag{B 3.2.3}$$

and name (due to the validity of the commutative rule regarding the indices i and k) the symmetrical quantities g_{ik}, *metric coefficients*. The introduction of the term metric coefficient becomes clear with the help of the previous considerations regarding the norms of a vector and the angle between two vectors. For the amount (norm) of the base vector $\mathbf{g}_i$ we obtain with the help of (B 3.1.2)

$$|\mathbf{g}_i| = \sqrt{\mathbf{g}_i \cdot \mathbf{g}_i} = \sqrt{g_{ii}}, \quad \not\Sigma\, i \quad (\not\Sigma\, i \text{ means : no summation}) \qquad \text{(B 3.2.4)}$$

and for the angle $\varphi_{(ik)}$ between both the base vectors $\mathbf{g}_i$ and $\mathbf{g}_k$ according to (B 3.1.C) and (B 3.2.3)

$$\cos\varphi_{(ik)} = \frac{g_{ik}}{\sqrt{g_{ii}}\sqrt{g_{kk}}}, \quad \not\Sigma\, i, \ \not\Sigma\, k\,. \qquad \text{(B 3.2.5)}$$

The norm of the base vector, and the angle between the two base vectors, depend on the coefficient g_{ik}. As in geometry, one can measure the norm of a vector and the angle between two vectors. It seems reasonable to call the coefficient g_{ik} as metric coefficients.

B 3.3
Reciprocal Base System

Given the base vectors $\mathbf{g}_k$ in E^3, we now introduce a second system of base vectors $\mathbf{g}^i$ reciprocal to $\mathbf{g}_k$ in accordance to the following rules:

$$\mathbf{g}^i \cdot \mathbf{g}_k = \delta^i_k, \qquad \text{(B 3.3.1)}$$

keeping in mind that δ^i_k is the Kronecker symbol. The base vectors $\mathbf{g}_k$ and $\mathbf{g}^i$ are termed as *covariant and contravariant* base vectors.

We now form the scalar product of the contravariant base vectors

$$\mathbf{g}^i \cdot \mathbf{g}^k = g^{ik}, \qquad \text{(B 3.3.2)}$$

where g^{ik}, due to the commutative rule, is symmetrical for the indices i and k, i.e.,

$$g^{ik} = g^{ki}. \qquad \text{(B 3.3.3)}$$

Following the above-mentioned description for the base vectors, we name g_{ik} as the *covariant metric coefficients* and g^{ik} as the *contravariant metric coefficients.*

The description of the base vectors in Sections 3.2 and 3.3, namely with an index placed at the bottom and at the top, is arbitrary in nature. One could easily have chosen the opposite way to describe it.

A base, which consists of a system of base vectors $\mathbf{e}_1, \mathbf{e}_2, \mathbf{e}_3$, that are normed (unit vectors) and are reciprocally orthogonal are called *orthonormed* and it is valid that:

$$\mathbf{e}_i \cdot \mathbf{e}_k = 0 \text{ for } i \neq k, \quad \mathbf{e}_i \cdot \mathbf{e}_k = 1 \text{ for } i = k. \qquad \text{(B 3.3.4)}$$

This can also be written in a compact form,

$$\mathbf{e}_i \cdot \mathbf{e}_k = \delta_{ik}, \qquad \text{(B 3.3.5)}$$

where δ_{ik} for $i = k$ takes a value of one and for $i \neq k$ takes the value of zero.

In an orthonormed system, the differences between the covariant and contravariant base vectors vanish. Thus, it is normal to represent coefficients of the vector components as covariant, and to renounce the mixed-variant placement of the indices in the summation convention.

B 3.4
Covariant and Contravariant Coefficients of the Vector Components

A vector **v** is represented in the covariant as well as in the contravariant base system. In the covariant base system, we write it as:

$$\mathbf{v} = v^i \mathbf{g}_i . \qquad \text{(B 3.4.1)}$$

The coefficients v^i can be described (taking in to view the definition in the previous section) as *contravariant coefficients of vector components* of the vectors **v** or shorter (not precise) as contravariant components of the vectors **v**.

Correspondingly, we can express the vector **v** in a contravariant base system:

$$\mathbf{v} = v_i \mathbf{g}^i . \qquad \text{(B 3.4.2)}$$

The coefficients v_i are named as the *covariant coefficients of vector components* of the vector **v** or, more briefly, as the covariant component of the vector **v**.

The coefficients v^i and v_i are also characterized as the associated components of the vector **v**, for the coefficients v^i and v_i are not independent of each other, as we shall soon see. We assume that the components of the vector **v** are represented in the form as depicted in (B 3.4.1). Scalar multiplication with the base vector $\mathbf{g}^k$ gives us

$$v^i = \mathbf{v} \cdot \mathbf{g}^i = v_k \mathbf{g}^k \cdot \mathbf{g}^i = g^{ki} v_k . \qquad \text{(B 3.4.3)}$$

Besides that, the scalar multiplication of (B 3.4.2) with $\mathbf{g}_k$ gives:

$$v_i = \mathbf{v} \cdot \mathbf{g}_i = v^k \mathbf{g}_k \cdot \mathbf{g}_i = g_{ki} v^k ; \qquad \text{(B 3.4.4)}$$

with this, the assumed dependency has been proven.

With the above-mentioned results, in place of (B 3.4.1) and (B 3.4.2) we can write

$$\mathbf{v} = (\mathbf{v} \cdot \mathbf{g}^i)\mathbf{g}_i = (\mathbf{v} \cdot \mathbf{g}_i)\mathbf{g}^i . \qquad \text{(B 3.4.5)}$$

In this equation, when we identify the vector **v** at any time with the covariant and contravariant base vectors, the following relationships can be established:

$$\begin{aligned} \mathbf{g}_k &= g_{ki} \mathbf{g}^i , \\ \mathbf{g}^k &= g^{ki} \mathbf{g}_i . \end{aligned} \qquad \text{(B 3.4.6)}$$

From the Eqns. (B 3.4.3) or (B 3.4.4) we can form further on, with the help of scalar multiplication with the contravariant or covariant base vectors, the following relationship:

$$g_{ki} g^{im} = \delta_k^m . \qquad \text{(B 3.4.7)}$$

As a result of the above considerations, two important rules of calculation are formed:

1. *Raising of an index:*
   A lowered index can be raised with the multiplication of the covariant coefficient of the vector component with the contravariant metric coefficient.
2. *Lowering of an index:*
   A raised index can be lowered with the multiplication of the contravariant coefficient of the vector component with the covariant metric coefficients.

These calculation rules are valid according to (B 3.4.3) and (B 3.4.4) and analogously so for the base vectors.

With the introduction of the covariant and contravariant coefficients of the components, the scalar product can be expressed in the following way:

$$\mathbf{u} \cdot \mathbf{v} = g_{ik} u^i v^k = g^{ik} u_i v_k = u^k v_k = u_k v^k. \tag{B 3.4.8}$$

We consider the position vector $\mathbf{x}$ as a special vector. This vector will be decomposed in the base $\mathbf{g}_i$:

$$\mathbf{x} = x^i \mathbf{g}_i. \tag{B 3.4.9}$$

The metric quantities x^i are related to the base vectors $\mathbf{g}_i$ and to the vector $\mathbf{x}$. They are called the *coordinates* of the base system $\mathbf{g}_i$ at the point $\mathbf{x}$. It must be emphasized that the term "coordinates" has been used solely for the coefficients of the components of the position vectors. The coordinates x^i can be explicitly given by the scalar multiplication with $\mathbf{g}^r$,

$$x^i = \mathbf{x} \cdot \mathbf{g}^i. \tag{B 3.4.10}$$

Consequently, the coordinates x^i depend on the position vector $\mathbf{x}$ and the contravariant base vector $\mathbf{g}^i$.

B 3.5 Physical Coefficients of a Vector

The base vectors $\mathbf{g}_i$ and $\mathbf{g}^i$ are not generally vectors with the amount one. When a vector is represented in such a base system, then the coefficients of the vector components do not give the true quantities of the vector components. These can only be obtained, when the vector is represented in a normed system. The coefficients of the components are then called the *physical coefficients*. The true quantities of the coefficients can be calculated with the help of an arbitrarily chosen base system as follows:

$$\mathbf{u} = \overset{*}{u}{}^1 \frac{\mathbf{g}_1}{|\mathbf{g}_1|} + \overset{*}{u}{}^2 \frac{\mathbf{g}_2}{|\mathbf{g}_2|} + \overset{*}{u}{}^3 \frac{\mathbf{g}_3}{|\mathbf{g}_3|} = \overset{*}{u}_1 \frac{\mathbf{g}^1}{|\mathbf{g}^1|} + \overset{*}{u}_2 \frac{\mathbf{g}^2}{|\mathbf{g}^2|} + \overset{*}{u}_3 \frac{\mathbf{g}^3}{|\mathbf{g}^3|}. \quad \text{(B 3.5.1)}$$

(The physical coefficients are marked with an asterisk). Likewise,

$$\mathbf{u} = u^1\mathbf{g}_1 + u^2\mathbf{g}_2 + u^3\mathbf{g}_3 = u_1\mathbf{g}^1 + u_2\mathbf{g}^2 + u_3\mathbf{g}^3 \quad \text{(B 3.5.2)}$$

is valid. Comparing (B 3.5.1) with (B 3.5.2) we get

$$u^i = \overset{*}{u}{}^i \frac{1}{|\mathbf{g}_i|}, \quad u_i = \overset{*}{u}_i \frac{1}{|\mathbf{g}^i|}, \quad \not\Sigma\, i \quad \text{(B 3.5.3)}$$

or with (B 3.2.4)

$$\overset{*}{u}{}^i = \sqrt{g_{ii}}u^i, \quad \overset{*}{u}_i = \sqrt{g^{ii}}u_i, \quad \not\Sigma\, i. \quad \text{(B 3.5.4)}$$

For further clarification we can consider the following example: We are looking for the physical coefficients of the vector $\mathbf{u} = 3\mathbf{g}_1 + 2\mathbf{g}_2 + \mathbf{g}_3$ in the covariant base $\mathbf{g}_1 = \mathbf{e}_1$, $\mathbf{g}_2 = \mathbf{e}_1 + \mathbf{e}_2$, $\mathbf{g}_3 = \mathbf{e}_1 + \mathbf{e}_2 + \mathbf{e}_3$. With the metric coefficients $g_{11} = 1$, $g_{22} = 2$, $g_{33} = 3$, we get $\overset{*}{u}{}^1 = 3$, $\overset{*}{u}{}^2 = 2\sqrt{2}$, $\overset{*}{u}{}^3 = \sqrt{3}$.

B 4.
Tensor Algebra

We have seen that a vector can be interpreted geometrically in Euclidean space. This is not so easily done for a tensor. For our needs, it is sufficient to interpret the *tensor* as a *linear mapping* in which the vectors **u**, **v** of the vector space are transferred into vectors **w**, **z** of another vector space. In this way, we can explain the tensors independently of definite base systems, with the help of defined calculation rules. Naturally, dealing with definite physical problems, for instance in the creation of a shell theory, as well as carrying out calculations, one must rely upon special base systems. However, as long as physical relationships are to be formulated, one should avoid the introduction of the base systems.

In the following section, the tensors will be denoted with capital (bold faced) letters. With regard to the description of the formula, the remarks in Section B 3 are valid.

B 4.1
Tensor Notion (Linear Mapping)

The quantities **u** and **v** are vectors in E^3 (three-dimensional Euclidean space) as well as α a scalar. A mapping **T** is linear, if

$$\mathbf{T}(\mathbf{u} + \mathbf{v}) = \mathbf{Tu} + \mathbf{Tv}, \quad \text{(B 4.1.D1)}$$

$$\mathbf{T}(\alpha\mathbf{u}) = \alpha(\mathbf{Tu}). \quad \text{(B 4.1.D2)}$$

Such a linear mapping or tensor (of second-order) is a rule that relates every arbitrary vector **u** in E^3 to another vector **Tu** in E^3.

The set of all possible tensors (linear mappings) will be denoted with L (E^3, E^3). We can likewise define linear operations in $\mathrm{L}(E^3, E^3)$ in the following way: For every pair of tensors **T**, **S** is **T** + **S** the only tensor that maps every vector **u** upon the sum of the vectors **Tu** and **Su**. The following should be valid

$$(\mathbf{T}+\mathbf{S})\mathbf{u} = \mathbf{Tu} + \mathbf{Su}. \tag{B 4.1.D3}$$

Likewise the scalar multiple of the tensor T, i.e., αT, is defined by

$$(\alpha\mathbf{T})\mathbf{v} = \alpha(\mathbf{Tv})\,. \tag{B 4.1.D4}$$

In addition to that, we require that

$$\alpha\mathbf{T} = \mathbf{T}\alpha. \tag{B 4.1.D5}$$

The mappings (**T** + **S**) and (α**T**), as they are to be seen in (D3) and (D4), fulfill the definitions (D1) and (D2), i.e., they are tensors. The proofs for which are simple. Let us, for example, consider the tensor sum **T** + **S**. When we observe (D1) and (D3), as well as the rules of the vector algebra, it holds that

$$\begin{aligned}
(\mathbf{T}+\mathbf{S})(\mathbf{u}+\mathbf{v}) &= \mathbf{T}(\mathbf{u}+\mathbf{v}) + \mathbf{S}(\mathbf{u}+\mathbf{v}) = (\mathbf{Tu}+\mathbf{Tv}) + (\mathbf{Su}+\mathbf{Sv}) = \\
&= (\mathbf{Tu}+\mathbf{Su}) + (\mathbf{Tv}+\mathbf{Sv}) = (\mathbf{T}+\mathbf{S})\mathbf{u} + (\mathbf{T}+\mathbf{S})\mathbf{v}, \\
(\mathbf{T}+\mathbf{S})(\alpha\mathbf{u}) &= \mathbf{T}(\alpha\mathbf{u}) + \mathbf{S}(\alpha\mathbf{u}) = \alpha(\mathbf{Tu}) + \alpha(\mathbf{Su}) = \alpha(\mathbf{Tu}+\mathbf{Su}) = \\
&= \alpha[(\mathbf{T}+\mathbf{S})\mathbf{u}].
\end{aligned} \tag{B 4.1.1}$$

Likewise, it can be easily proven that α**T** is a tensor. The zero-element in $\mathrm{L}(E^3, E^3)$ is the mapping, that relates any vector **u** to the zero-vector **0**. We name this mapping as the *zero-tensor* and denote it with **O**, i.e.,

$$\mathbf{Ou} = \mathbf{0}. \tag{B 4.1.D6}$$

The identical mapping, the *identity tensor* **I**, is a tensor that maps any arbitrary vector **u** identically,

$$\mathbf{Iu} = \mathbf{u} \tag{B 4.1.D7}$$

for every vector **u** in E^3.

A space can be specially associated with the Euclidean vector space E^3, which we shall denote as $E^3 \otimes E^3$. This product space is named as the *tensor product space*. When **a** and **b** are elements of E^3, then $\mathbf{a} \otimes \mathbf{b}$ is an element of $E^3 \otimes E^3$, which should possess the following characteristics:

$$(\mathbf{a} \otimes \mathbf{b})\mathbf{u} = (\mathbf{b} \cdot \mathbf{u})\mathbf{a} \tag{B 4.1.D8}$$

for every vector **u** contained in E^3.

The *dyadic product* $\mathbf{a}\otimes\mathbf{b}$ fulfills the definitions (D1) and (D2) and is therefore a tensor, also called the *simple tensor*. For the dyadic product, we can derive the following calculation rules from (D8), where all the listed vectors are included in E^3:

$$\mathbf{a} \otimes (\mathbf{b} + \mathbf{c}) = \mathbf{a} \otimes \mathbf{b} + \mathbf{a} \otimes \mathbf{c} \ \ \text{(distributive rule)}\ ,$$
$$(\alpha\mathbf{a}) \otimes \mathbf{b} = \mathbf{a} \otimes \alpha\mathbf{b} = \alpha(\mathbf{a} \otimes \mathbf{b}) \ \ \text{(associative rule)}\ . \tag{B 4.1.2}$$

B 4.2
Algebra in Base Systems

With the calculation rule (B 4.1.2), we are now in a position to represent the dyadic product by its components. For this purpose, we take both the vectors **a** and **b** from the covariant base system $\mathbf{g}_1, \mathbf{g}_2, \mathbf{g}_3$

$$\mathbf{a} = a^i\mathbf{g}_i, \ \mathbf{b} = b^k\mathbf{g}_k \tag{B 4.2.1}$$

and form the dyadic product

$$\mathbf{a} \otimes \mathbf{b} = a^i\mathbf{g}_i \otimes b^k\mathbf{g}_k. \tag{B 4.2.2}$$

With (B 4.1.2), we get

$$\mathbf{a} \otimes \mathbf{b} = a^i b^k \mathbf{g}_i \otimes \mathbf{g}_k. \tag{B 4.2.3}$$

Thus,

$$\mathbf{T} = \mathbf{a} \otimes \mathbf{b} = a^i b^k \mathbf{g}_i \otimes \mathbf{g}_k. \tag{B 4.2.4}$$

As can be seen from the above-mentioned equation, one can interpret the dyadic product of the base vectors as the base for simple tensors **T** in $E^3 \otimes E^3$. It can be supposed that this is valid for all tensors and that, in general, the dyadic product of the base vectors spans the base in $E^3 \otimes E^3$. In order to confirm this supposition, it is first necessary to prove the linear dependence of the dyadic product $\mathbf{g}_i \otimes \mathbf{g}_k$.

The base $\mathbf{g}_i \otimes \mathbf{g}_k$ is called *linear dependent*, when numbers $\alpha^{11}, \alpha^{22}, ..., \alpha^{33}$ can be given that are not all zero, so that

$$\begin{aligned} \alpha^{11}\mathbf{g}_1 \otimes \mathbf{g}_1 + \alpha^{12}\mathbf{g}_1 \otimes \mathbf{g}_2 + ... + \alpha^{33}\mathbf{g}_3 \otimes \mathbf{g}_3 &= \mathbf{0}\ , \\ \alpha^{ik}\mathbf{g}_i \otimes \mathbf{g}_k &= \mathbf{0} \end{aligned} \tag{B 4.2.5}$$

is valid. We apply this tensor to any arbitrary vector **u** from E^3, and then it follows with (B 4.1.D4)

$$(\alpha^{ik}\mathbf{g}_i \otimes \mathbf{g}_k)\mathbf{u} = \mathbf{0} \tag{B 4.2.6}$$

and with (B 4.1.D8)

$$(\mathbf{u} \cdot \mathbf{g}_k)\alpha^{ik}\mathbf{g}_i = \mathbf{0}. \tag{B 4.2.7}$$

Since **u** is an arbitrary vector, $\mathbf{u} \cdot \mathbf{g}_k \neq 0$ is valid for at least one case, we can conclude therefore that $\alpha^{ik} = 0$, if $\mathbf{g}_1, \mathbf{g}_2$, and $\mathbf{g}_3$ are linear independent vectors.

The above-mentioned requirement of linear dependence can be arrived at only with $\alpha^{ik} = 0$, which contradicts the requirement. With this contradiction it has been proven that the base $\mathbf{g}_i \otimes \mathbf{g}_k$ is linear independent. It follows that the tensor space $E^3 \otimes E^3$ is spread out by the dyadic products $\mathbf{g}_i \otimes \mathbf{g}_k$, and that the dimension of $E^3 \otimes E^3$ is determined through the product of the dimensions of each of the vector spaces E^3. The dyadic product space is nine-dimensional.

Let us take into consideration an arbitrary tensor $\mathbf{T}$ in $E^3 \otimes E^3$. The set of ten tensors $\mathbf{T}, \mathbf{g}_1 \otimes \mathbf{g}_1, \mathbf{g}_1 \otimes \mathbf{g}_2, ..., \mathbf{g}_3 \otimes \mathbf{g}_3$ is necessarily linear dependent, so that ten numbers $\lambda, \alpha^{11}, \alpha^{12}, ..., \alpha^{33}$ exist. Then it holds that,

$$\lambda \mathbf{T} + \alpha^{ik} \mathbf{g}_i \otimes \mathbf{g}_k = \mathbf{O}, \qquad \text{or due to} \qquad \lambda \neq 0,$$
$$\mathbf{T} = -\frac{1}{\lambda} \alpha^{ik} \mathbf{g}_i \otimes \mathbf{g}_k. \tag{B 4.2.8}$$

An arbitrary tensor $\mathbf{T}$ can also be represented with the help of the tensor base. The coefficient $-\frac{1}{\lambda}\alpha^{ik}$ will be denoted with t^{ik} and named as the coefficient of the component of $\mathbf{T}$ in reference to the bases $\mathbf{g}_1 \otimes \mathbf{g}_1, \mathbf{g}_1 \otimes \mathbf{g}_2, ..., \mathbf{g}_3 \otimes \mathbf{g}_3$. This can be written more simply as

$$\mathbf{T} = t^{ik} \mathbf{g}_i \otimes \mathbf{g}_k. \tag{B 4.2.9}$$

The same consideration can be applied to the contravariant base vectors, to form the tensor base

$$\mathbf{g}^i \otimes \mathbf{g}^k \text{ as well as } \mathbf{g}_i \otimes \mathbf{g}^k \text{ and } \mathbf{g}^k \otimes \mathbf{g}_i \tag{B 4.2.10}$$

so that we can represent the arbitrary tensor $\mathbf{T}$ also as

$$\mathbf{T} = t_{ik} \mathbf{g}^i \otimes \mathbf{g}^k,$$
$$\mathbf{T} = t^{i}_{.k} \mathbf{g}_i \otimes \mathbf{g}^k, \tag{B 4.2.11}$$
$$\mathbf{T} = t_{k}^{.i} \mathbf{g}^k \otimes \mathbf{g}_i\ .$$

In (B 4.2.11)$_1$, it should be recognized that the indices of the coefficient t_{ik} immediately indicate the position of the corresponding base vectors. One has to take into consideration the correct placement of the base vectors also in the mixed-variant representation (B 4.2.11)$_2$ and (B 4.2.11)$_3$. Thus, in $t^{i}_{.k}$, it is meant that, in the tensor base, first the covariant base vector $\mathbf{g}_i$, and then the contravariant base vector $\mathbf{g}^k$, appears. Following the characterization in vector algebra we name $\mathbf{g}_i \otimes \mathbf{g}_k$ the *covariant base*, $\mathbf{g}^i \otimes \mathbf{g}^k$ the *contravariant base* and $\mathbf{g}_i \otimes \mathbf{g}^k, \mathbf{g}^k \otimes \mathbf{g}_i$ the *mixed-variant base*, respectively. Likewise, we characterize t^{ik} as the *contravariant coefficient* of the components, t_{ik} as the *covariant coefficient* of the components and $t^{i}_{.k}$ or $t_{k}^{.i}$ as the *mixed-variant coefficients* of the components, respectively.

The covariant and contravariant, as well as the mixed-variant coefficients, are not independent of each other. The tensor $\mathbf{T}$ can be represented in a covariant base as

$$\mathbf{T} = t^{ik} \mathbf{g}_i \otimes \mathbf{g}_k, \tag{B 4.2.12}$$

with (B 3.4.6)$_1$, it can be written as

$$\mathbf{T} = t^{rk} g_{ri} \mathbf{g}^i \otimes \mathbf{g}_k. \tag{B 4.2.13}$$

A comparison with the representation of the tensor **T** in a mixed-variant base,

$$\mathbf{T} = t_i^{\cdot k}\mathbf{g}^i \otimes \mathbf{g}_k \tag{B 4.2.14}$$

leads directly to

$$t_i^{\cdot k} = t^{rk} g_{ri}. \tag{B 4.2.15}$$

Analogous results can be arrived at for the alternative base representations and we can expand the arithmetical rules mentioned in Section B 3.4:

1. *Raising of an index:*

   A lower index can be raised with the multiplication of the covariant or the mixed-variant coefficients of the tensor components with the contravariant metric coefficients.

2. *Lowering of an index:*

   An upper index can be lowered with the multiplication of the contravariant or mixed-variant coefficient of the tensor component with the covariant metric coefficients.

In order to give the true (*physical*) amount of the tensor coefficients, it is necessary to decompose the tensors in a normed base. For example,

$$\mathbf{T} = \overset{*}{t}{}^{11}\frac{\mathbf{g}_1}{|\mathbf{g}_1|} \otimes \frac{\mathbf{g}_1}{|\mathbf{g}_1|} + \overset{*}{t}{}^{12}\frac{\mathbf{g}_1}{|\mathbf{g}_1|} \otimes \frac{\mathbf{g}_2}{|\mathbf{g}_2|} + \ldots + \overset{*}{t}{}^{33}\frac{\mathbf{g}_3}{|\mathbf{g}_3|} \otimes \frac{\mathbf{g}_3}{|\mathbf{g}_3|}, \tag{B 4.2.16}$$

where the physical coefficients have been marked with an asterisk *. For the representation in the covariant base $\mathbf{g}_i \otimes \mathbf{g}_k$ we get

$$\mathbf{T} = t^{ik}\mathbf{g}_i \otimes \mathbf{g}_k \tag{B 4.2.17}$$

so that the physical coefficient $\overset{*}{t}{}^{ik}$ can be calculated from

$$\overset{*}{t}{}^{ik} = |\mathbf{g}_i|\,|\mathbf{g}_k|\,t^{ik} = \sqrt{g_{ii}}\,\sqrt{g_{kk}}\,t^{ik}, \quad \not\Sigma\, i, \quad \not\Sigma\, k\,. \tag{B 4.2.18}$$

By analogy, we can determine that

$$\overset{*}{t}_{ik} = \sqrt{g^{ii}}\,\sqrt{g^{kk}}\;t_{ik}, \; \not\Sigma\, i, \; \not\Sigma\, k, \tag{B 4.2.19}$$

$$\overset{*}{t}{}^{i}_{.k} = \sqrt{g_{ii}}\sqrt{g^{kk}}\;t^{i}_{.k}, \quad \overset{*}{t}{}^{i}_{.k} = \sqrt{g^{kk}}\,\sqrt{g_{ii}}\;t^{i}_{.k}, \quad \not\Sigma\, i, \not\Sigma\, k. \tag{B 4.2.20}$$

Finally, we want to represent the identity tensor **I** in the base system. As the base, we choose the covariant base $\mathbf{g}_i \otimes \mathbf{g}_k$,

$$\mathbf{I} = k^{ik}\mathbf{g}_i \otimes \mathbf{g}_k.$$

Whereby, k^{ik} are the unknown coefficients of the components to be determined next. With the identical mapping (B 4.1.D7) we get, when we have represented all the quantities in the bases, $(k^{ik}\mathbf{g}_i \otimes \mathbf{g}_k)u^r\mathbf{g}_r = u^i\mathbf{g}_i$ or with (B 4.1.D8) $k^{ik}u_k\mathbf{g}_i = u^i\mathbf{g}_i$ and with $u^i = g^{ik}u_k$ (see Section B 3.4) $k^{ik}u_k\mathbf{g}_i = g^{ik}u_k\mathbf{g}_i$. A

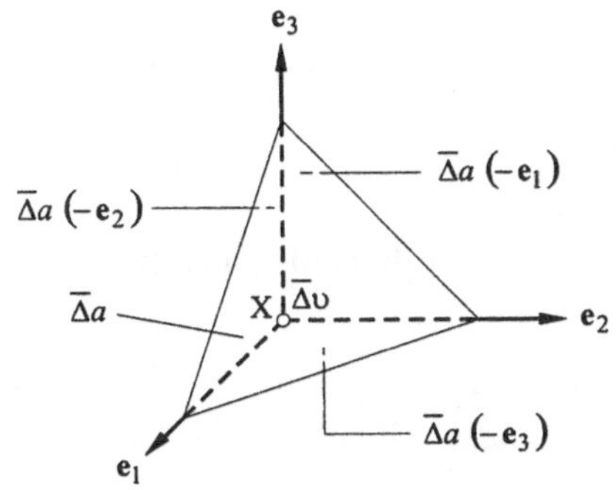

Fig. B 4.2.1. Cauchy's Theorem

comparison of both sides of the last equation gives $k^{ik} = g^{ik}$. Thus, the identity tensor **I** can be represented as

$$\mathbf{I} = g^{ik}\mathbf{g}_i \otimes \mathbf{g}_k \tag{B 4.2.21}$$

or, as can be easily proven,

$$\mathbf{I} = \delta^i_k \mathbf{g}_i \otimes \mathbf{g}^k = \delta^k_i \mathbf{g}^i \otimes \mathbf{g}_k,$$

$$\mathbf{I} = g_{ik}\mathbf{g}^i \otimes \mathbf{g}^k, \tag{B 4.2.22}$$

$$\mathbf{I} = \mathbf{g}_i \otimes \mathbf{g}^i = \mathbf{g}^i \otimes \mathbf{g}_i.$$

The identity tensor **I** is also described as the *fundamental tensor* of the three-dimensional Euclidean space. The previously mentioned arithmetical rules and the description of a tensor in a base system, will be illustrated with the help of an example taken from mechanics. We want to derive *Cauchy's theorem* that states: the stress vector **t** that is assigned to a surface element of a continuum is a linear mapping of the (normal) area vector **n** belonging to this surface:

$$\mathbf{t}(\mathbf{n}) = \mathbf{Tn}. \tag{B 4.2.23}$$

In order to determine the functional dependence of the stress vector on the normal vector, we shall consider a partial continuum body, in the form of a tetrahedron, with the volume $\bar{\Delta}v$ and the marked areas $\bar{\Delta}a$ and $\bar{\Delta}a(-\mathbf{e}_i)$, to which the material point X has been assigned. The derivation is done by taking the orthogonal base system given in the sketch. In principle, it is possible to prove using any base vectors.

We will now apply the equilibrium axiom. This axiom states that the resulting force that influences an arbitrary partial continuum is equal to the zero vector.

$$\begin{aligned}\rho^* \overset{*}{\mathbf{b}}\, \bar{\Delta}v + \mathbf{t}^*(\mathbf{n})\bar{\Delta}a + \mathbf{t}^*(-\,\mathbf{e}_1)\bar{\Delta}a(-\,\mathbf{e}_1) + \mathbf{t}^*(-\,\mathbf{e}_2)\bar{\Delta}a(-\,\mathbf{e}_2) + \\ + \mathbf{t}^*(-\,\mathbf{e}_3)\bar{\Delta}a(-\,\mathbf{e}_3) = \mathbf{0}.\end{aligned} \tag{B 4.2.24}$$

Thereby $\rho^*, \rho^*\mathbf{b}^*$, and $\mathbf{t}^*$ are the averaged quantities of the density ρ, volume force $\rho\mathbf{b}$, and the stress vector **t** as well as **n** is the unit-normal vector on the surface part $\bar{\Delta}a$. The above-mentioned equation can also be written as

$$\begin{aligned} \mathbf{t}^*(\mathbf{n})\bar{\Delta}a = & -\mathbf{t}^*(-\mathbf{e}_1)\bar{\Delta}a(-\mathbf{e}_1) - \mathbf{t}^*(-\mathbf{e}_2)\bar{\Delta}a(-\mathbf{e}_2) - \\ & -\mathbf{t}^*(-\mathbf{e}_3)\bar{\Delta}a(-\mathbf{e}_3) - \rho^* \overset{*}{\mathbf{b}}\,\bar{\Delta}v. \end{aligned} \quad \text{(B 4.2.25)}$$

The volume element $\bar{\Delta}v$ can be expressed with the help of the height h of the tetrahedron and the surface $\bar{\Delta}a$: $\bar{\Delta}v = \frac{1}{3}h\bar{\Delta}a$.

Furthermore, the well-known surface theorem is valid; this shall be developed in Section B 5.4.1,

$$\bar{\Delta}a\mathbf{n} = \bar{\Delta}a(-\mathbf{e}_1)\mathbf{e}_1 + \bar{\Delta}a(-\mathbf{e}_2)\mathbf{e}_2 + \bar{\Delta}a(-\mathbf{e}_3)\mathbf{e}_3. \quad \text{(B 4.2.26)}$$

Scalar multiplication of this equation with $\mathbf{e}_1, \mathbf{e}_2$, and $\mathbf{e}_3$, allows the areas $\bar{\Delta}a(-\mathbf{e}_i)$ to be expressed, with the help of $\bar{\Delta}a$, as

$$\bar{\Delta}a(-\mathbf{e}_1) = \bar{\Delta}a\,\mathbf{n}\cdot\mathbf{e}_1,\quad \bar{\Delta}a(-\mathbf{e}_2) = \bar{\Delta}a\,\mathbf{n}\cdot\mathbf{e}_2,\quad \bar{\Delta}a(-\mathbf{e}_3) = \bar{\Delta}a\,\mathbf{n}\cdot\mathbf{e}_3. \quad \text{(B 4.2.27)}$$

As $\mathbf{n}$ and $\mathbf{e}_i$ are normed, the scalar products $\mathbf{n}\cdot\mathbf{e}_i$ express the cosines between the surface of the normal vector $\mathbf{n}$ and the base vector $\mathbf{e}_i$. For further transformation of (B 4.2.25), it is necessary to mention the Lemma by Cauchy: The stress vectors, that work on the opposite sides of the same material surface, at a given point in the continuum, have the same value but opposite directions, i.e., $\mathbf{t}(+\mathbf{n}) = -\mathbf{t}(-\mathbf{n})$.

The Eqn. (B 4.2.25) can be expressed, with (B 4.2.27) and the above-stated lemma, as:

$$\begin{aligned} \mathbf{t}^*(\mathbf{n})\bar{\Delta}a = & \,[\,\mathbf{t}^*(\mathbf{e}_1)(\mathbf{e}_1\cdot\mathbf{n}) + \mathbf{t}^*(\mathbf{e}_2)(\mathbf{e}_2\cdot\mathbf{n}) + \mathbf{t}^*(\mathbf{e}_3)(\mathbf{e}_3\cdot\mathbf{n})]\bar{\Delta}a - \\ & -\frac{1}{3}h\rho^*\mathbf{b}^*\bar{\Delta}a. \end{aligned} \quad \text{(B 4.2.28)}$$

Division is done with $\bar{\Delta}a$ and we allow the height h of the tetrahedron to go to zero. The stress vector assigned to the material point X is indicated without an asterisk.

$$\mathbf{t}(\mathbf{n}) = \mathbf{t}(\mathbf{e}_1)(\mathbf{e}_1\cdot\mathbf{n}) + \mathbf{t}(\mathbf{e}_2)(\mathbf{e}_2\cdot\mathbf{n}) + \mathbf{t}(\mathbf{e}_3)(\mathbf{e}_3\cdot\mathbf{n}). \quad \text{(B 4.2.29)}$$

When

$$\mathbf{t}(\mathbf{e}_1) = \mathbf{t}_1,\ \mathbf{t}(\mathbf{e}_2) = \mathbf{t}_2,\ \mathbf{t}(\mathbf{e}_3) = \mathbf{t}_3, \quad \text{(B 4.2.30)}$$

then the above-mentioned equation can be expressed simply as

$$\mathbf{t}(\mathbf{n}) = \mathbf{t}_i(\mathbf{e}_i\cdot\mathbf{n}). \quad \text{(B 4.2.31)}$$

This relation can be transformed with the help of (B 4.1.D8):

$$\mathbf{t}(\mathbf{n}) = (\mathbf{t}_i\otimes\mathbf{e}_i)\mathbf{n}. \quad \text{(B 4.2.32)}$$

The tensor $\mathbf{t}_i\otimes\mathbf{e}_i$ contains values that are not dependent on $\mathbf{n}$; we can represent it in a base-free form

$$\mathbf{t}(\mathbf{n}) = \mathbf{T}\mathbf{n}, \quad \text{(B 4.2.33)}$$

where $\mathbf{T}$ is the Cauchy stress tensor. With the relationship (B 4.2.18), which is important for continuum mechanics, we have obtained the Cauchy theorem. Writing (B 4.2.23), using an arbitrary base

$$t^i \mathbf{g}_i = t^i_{.k} n^k \mathbf{g}_i, \qquad \text{(B 4.2.34)}$$

we can see that the first index in the coefficient of the stress component, namely i, shows the direction and the second index k is the corresponding cut surface. Due to the symmetry of the Cauchy stress tensors, the different meanings of the indices are unimportant. The above-mentioned fact concerning the placement of the indices is not often considered in the literature, when the transformation to non-symmetrical stress tensors in non-linear mechanics is considered.

B 4.3
Scalar Product of Tensors

First, we define the scalar product (dot product) of a general tensor $\mathbf{T}$ with a simple tensor $\mathbf{a} \otimes \mathbf{b}$ from $E^3 \otimes E^3$, which can be written as $\mathbf{T} \cdot \mathbf{a} \otimes \mathbf{b}$ and expressed as a real number:

$$\mathbf{T} \cdot (\mathbf{a} \otimes \mathbf{b}) = \mathbf{a} \cdot \mathbf{Tb}. \qquad \text{(B 4.3.E)}$$

With the definitions and relationships mentioned in Section B 4.1 and B 4.2, as well as the numerical calculations for the formulation of scalar products of vectors in Section B 3.1, it follows from (E) for arbitrary tensors $\mathbf{T}$, $\mathbf{S}$, and $\mathbf{R}$:

$$\begin{aligned} &\mathbf{T} \cdot \mathbf{S} = \mathbf{S} \cdot \mathbf{T}, \\ &\mathbf{T} \cdot (\mathbf{S} + \mathbf{R}) = \mathbf{T} \cdot \mathbf{S} + \mathbf{T} \cdot \mathbf{R}, \\ &\alpha\mathbf{T} \cdot \mathbf{S} = \mathbf{T} \cdot (\alpha\mathbf{S}) = \alpha(\mathbf{T} \cdot \mathbf{S}). \end{aligned} \qquad \text{(B 4.3.1)}$$

When, for the arbitrary tensor $\mathbf{T}$, the condition

$$\begin{aligned} &\mathbf{T} \cdot \mathbf{S} = 0 \text{ holds, then } \mathbf{S} = \mathbf{O}, \\ &\mathbf{T} \cdot \mathbf{T} > 0, \text{ for } \mathbf{T} \neq \mathbf{O}. \end{aligned} \qquad \text{(B 4.3.2)}$$

The amount, or the norm, of a tensor $\mathbf{T}$ can be defined as the real value

$$|\mathbf{T}| = \sqrt{\mathbf{T} \cdot \mathbf{T}} \qquad \text{(B 4.3.3)}$$

because of (E).

Furthermore, it is possible to extend Schwarz's inequality (see Section B 3.1) to second-order tensors:

> The value of the scalar product of two tensors T and S is smaller or equal to the product of the norms of both the tensors, i.e.,

$$|\mathbf{T} \cdot \mathbf{S}| \leq |\mathbf{T}||\mathbf{S}| = \sqrt{\mathbf{T} \cdot \mathbf{T}}\sqrt{\mathbf{S} \cdot \mathbf{S}}. \qquad \text{(B 4.3.4)}$$

The equality sign is only valid when the tensor $\mathbf{T}$ is a scalar multiple of the tensor $\mathbf{S}$.

Just as for the vectors, we can also make the following statements for the norms of a tensor:

$$\begin{aligned} |\alpha\mathbf{T}| &= |\alpha||\mathbf{T}|, \\ |\mathbf{T} + \mathbf{S}| &\leq |\mathbf{T}| + |\mathbf{S}|. \end{aligned} \qquad \text{(B 4.3.5)}$$

Finally, taking into consideration Schwarz's inequality and vector algebra, the cosine of the angle φ between the tensors **T** and **S** can be defined as:

$$\cos\varphi = \frac{\mathbf{T}\cdot\mathbf{S}}{|\mathbf{T}||\mathbf{S}|}\,, \qquad \text{(B 4.3.6)}$$

where the angle φ lies within the domain

$$0 \le \varphi \le \pi\,. \qquad \text{(B 4.3.7)}$$

The usefulness of Schwarz's inequality can be seen beside others in the plasticity theory for the development of minimum/maximum statements for solids with rigid ideal-plastic material behavior. The onset of plastic deformations of metallic materials is dependent on the von Mises yield condition $\mathbf{T}^D \cdot \mathbf{T}^D = K^2$, where $\mathbf{T}^D$ is the deviator of the Cauchy stress tensor (see Section B 4.6) and K is a constant, which can be experimentally determined. The constitutive equation $KD\mathbf{E}^D = \sqrt{D\mathbf{E}^D \cdot D\mathbf{E}^D}\mathbf{T}^D$ joins the increase in the strain deviator $D\mathbf{E}^D$ with the stress deviator $\mathbf{T}^D$. Work expressions, such as $\mathbf{T}^D \cdot (D\overset{*}{\mathbf{E}}{}^D - D\mathbf{E}^D)$, come into existence by the development of minimum/maximum statements, where $D\overset{*}{\mathbf{E}}{}^D$ is an estimated value of the increase of the distortion tensor. The aim of developing minimum/maximum statements is to additionally decompose the term $\mathbf{T}^D \cdot (D\overset{*}{\mathbf{E}}{}^D - D\mathbf{E}^D)$ into terms that contain only the estimated and true conditions. This is achieved by Schwarz's inequality; with (B 4.3.7) we get

$$\mathbf{T}^D \cdot (D\overset{*}{\mathbf{E}}{}^D - D\mathbf{E}^D) \le \sqrt{\mathbf{T}^D \cdot \mathbf{T}^D}\sqrt{D\overset{*}{\mathbf{E}}{}^D \cdot D\overset{*}{\mathbf{E}}{}^D} - \mathbf{T}^D \cdot D\mathbf{E}^D \qquad \text{(B 4.3.8)}$$

or, with the yield condition of von Mises and the above-mentioned constitutive equation,

$$\mathbf{T}^D \cdot (D\overset{*}{\mathbf{E}}{}^D - D\mathbf{E}^D) \le K\sqrt{D\overset{*}{\mathbf{E}}{}^D \cdot D\overset{*}{\mathbf{E}}{}^D} - K\sqrt{D\mathbf{E}^D \cdot D\mathbf{E}^D}. \qquad \text{(B 4.3.9)}$$

We now consider the *representation of the scalar product of tensors in base systems.* For simple tensors, (E) along with (B 4.1.D8) yield the calculation rule for the formation of the inner product:

$$(\mathbf{a}\otimes\mathbf{b})\cdot(\mathbf{v}\otimes\mathbf{u}) = (\mathbf{a}\cdot\mathbf{v})(\mathbf{b}\cdot\mathbf{u}). \qquad \text{(B 4.3.10)}$$

In this method, we observe the following calculation rule:

> The scalar product of two simple tensors is to be formed in such a way that the first and the second vector of both the simple tensors are to be scalarly multiplied.

This calculation rule makes it possible to explicitly determine the scalar product of two general tensors **T** and **S**. Therefore, we express **T** and **S** in the co- and contravariant bases:

$$\mathbf{T} = t^{ij}\mathbf{g}_i\otimes\mathbf{g}_j, \quad \mathbf{S} = s_{kl}\mathbf{g}^k\otimes\mathbf{g}^l. \qquad \text{(B 4.3.11)}$$

Considering (B 4.3.1)$_3$, we get, from (B 4.3.10):

$$\mathbf{T} \cdot \mathbf{S} = t^{ij} s_{kl} (\mathbf{g}_i \otimes \mathbf{g}_j) \cdot (\mathbf{g}^k \otimes \mathbf{g}^l) = t^{ij} s_{kl} \delta_i^k \delta_j^l, \qquad \text{(B 4.3.12)}$$
$$\mathbf{T} \cdot \mathbf{S} = t^{ij} s_{ij}.$$

With this, we have gained the calculation rule for the formation of the scalar product of general tensors:

> The scalar product of two common tensors is to be formed in such a way that the first and the second base vectors of both the tensors are to be scalarly multiplied

Further alternatives for the inner product of dual tensors **T** and **S** can be expressed in a simple form:

$$\mathbf{T} \cdot \mathbf{S} = t^{i}_{.j} s_{i}^{.j}, \qquad \text{(B 4.3.13)}$$
$$\mathbf{T} \cdot \mathbf{S} = t_{ij} s^{ij}.$$

In an orthogonal base system, we obtain for the square of a tensor **T**

$$\mathbf{T} \cdot \mathbf{T} = t_{ik} t_{ik} \qquad \text{(B 4.3.14)}$$
$$= t_{11} t_{11} + t_{12} t_{12} + \ldots + t_{33} t_{33},$$

from which one can directly see that (B 4.3.2)$_2$ is fulfilled, because the sum is formed with squared terms.

B 4.4
Tensor Product

Besides the scalar product, there is another possibility to multiplicatively join second-order tensors. As has been done previously, we can get the result of the combination from the term of the product. Thereafter, we define the *tensor* product **TS** of two second-order tensors **T** and **S** by the identity of the mapping:

$$(\mathbf{TS})\mathbf{v} = \mathbf{T}(\mathbf{Sv}) \,. \qquad \text{(B 4.4.F)}$$

This claim is valid for all vectors **v** in E^3. One can immediately discern that **TS** has the characteristics (B 4.1.D1) and (B 4.1.D2), and therefore is a second-order tensor. From the given definition (F), the following calculation rules can be derived for the tensor product:

$$(\mathbf{TS})\mathbf{R} = \mathbf{T}(\mathbf{SR}) \quad \text{(associative rule)},$$
$$\mathbf{T}(\mathbf{R} + \mathbf{S}) = \mathbf{TR} + \mathbf{TS} \quad \text{(distributive rule)},$$
$$(\mathbf{R} + \mathbf{S})\mathbf{T} = \mathbf{RT} + \mathbf{ST},$$
$$\alpha(\mathbf{TS}) = (\alpha\mathbf{T})\mathbf{S} = \mathbf{T}(\alpha\mathbf{S}), \qquad \text{(B 4.4.1)}$$
$$\mathbf{IT} = \mathbf{TI} = \mathbf{T},$$
$$\mathbf{OT} = \mathbf{TO} = \mathbf{O}.$$

It has to be taken into consideration that the commutative rule is, in general, not fulfilled, i.e., **TS** is not the same as **ST**. Exceptions can be seen in the relations (B 4.4.1)$_{5,6}$.

The tensor product, which is of principle importance for further considerations, is now represented in a base system. For this, it is advisable to first construct the tensor product of two simple tensors

$$\mathbf{T} = \mathbf{a} \otimes \mathbf{b}, \quad \mathbf{S} = \mathbf{c} \otimes \mathbf{d}\,. \tag{B 4.4.2}$$

Considering (F), it holds that

$$[(\mathbf{a} \otimes \mathbf{b})(\mathbf{c} \otimes \mathbf{d})]\mathbf{v} = (\mathbf{a} \otimes \mathbf{b})[(\mathbf{c} \otimes \mathbf{d})\mathbf{v}]. \tag{B 4.4.3}$$

From this, taking (B 4.1.D8) and (B 4.1.D2) into account, we get:

$$\begin{aligned}[(\mathbf{a} \otimes \mathbf{b})(\mathbf{c} \otimes \mathbf{d})]\mathbf{v} &= (\mathbf{a} \otimes \mathbf{b})[(\mathbf{d} \cdot \mathbf{v})\mathbf{c}] = (\mathbf{d} \cdot \mathbf{v})[(\mathbf{a} \otimes \mathbf{b})\mathbf{c}] = \\ &= (\mathbf{d} \cdot \mathbf{v})[(\mathbf{b} \cdot \mathbf{c})\mathbf{a}].\end{aligned} \tag{B 4.4.4}$$

Reverting to (B 3.1.3) and (B 4.1.D8), as well as (B 4.1.D4), we obtain the final result:

$$\begin{aligned}[(\mathbf{a} \otimes \mathbf{b})(\mathbf{c} \otimes \mathbf{d})]\mathbf{v} &= (\mathbf{b} \cdot \mathbf{c})[(\mathbf{d} \cdot \mathbf{v})\mathbf{a}] = (\mathbf{b} \cdot \mathbf{c})[(\mathbf{a} \otimes \mathbf{d})\mathbf{v}] = \\ &= [(\mathbf{b} \cdot \mathbf{c})(\mathbf{a} \otimes \mathbf{d})]\mathbf{v},\end{aligned} \tag{B 4.4.5}$$

i.e.,

$$(\mathbf{a} \otimes \mathbf{b})(\mathbf{c} \otimes \mathbf{d}) = (\mathbf{b} \cdot \mathbf{c})(\mathbf{a} \otimes \mathbf{d}). \tag{B 4.4.6}$$

With (B 4.4.6), we have obtained an important calculation rule:

> The tensor product of two simple tensors is to be formed in such a way that the vectors on the inner side are to be scalarly multiplied; the scalar product, with the tensor product of the vectors on the outer side, forms the tensor product of both the simple tensors.

With the help of this calculation rule, it is now also very simple to evaluate the tensor product of general tensors; for example,

$$\mathbf{T} = t^{ij}\mathbf{g}_i \otimes \mathbf{g}_j, \; \mathbf{S} = s_{kl}\mathbf{g}^k \otimes \mathbf{g}^l\,. \tag{B 4.4.7}$$

With the help of (B 4.4.6), considering (B 4.4.1)$_4$, we get

$$\mathbf{TS} = t^{ij}s_{kl}(\mathbf{g}_i \otimes \mathbf{g}_j)(\mathbf{g}^k \otimes \mathbf{g}^l) = t^{ij}s_{kl}\delta^k_j\mathbf{g}_i \otimes \mathbf{g}^l = t^{ij}s_{jl}\mathbf{g}_i \otimes \mathbf{g}^l. \tag{B 4.4.8}$$

The following calculation rules are realized for the general tensors:

> The tensor product of two general tensors must be formed in such a way that the base vectors on the inner side are to be scalarly multiplied; the tensor product of the base vectors located on the outer side forms the base of the tensor product.

As can easily be shown, alternative forms for the tensor product of two tensors **T** and **S** can also be obtained:

$$\mathbf{TS} = t^{i}_{.m} s^{m}_{.k} \mathbf{g}_i \otimes \mathbf{g}^k,$$
$$\mathbf{TS} = t^{i}_{.m} s^{mk} \mathbf{g}_i \otimes \mathbf{g}_k, \tag{B 4.4.9}$$
$$\mathbf{TS} = t_{im} s^{m}_{.k} \mathbf{g}^i \otimes \mathbf{g}^k.$$

Now, the possibility exists to form powers of tensors,

$$\mathbf{T}^0 = \mathbf{I},\ \mathbf{T}^1 = \mathbf{T},\ \mathbf{T}^2 = \mathbf{TT},\ \ldots \tag{B 4.4.10}$$

which again are second-order tensors. These, in turn, obey the following rules for exponential expressions:

$$\mathbf{T}^m \mathbf{T}^n = \mathbf{T}^{m+n} = \mathbf{T}^n \mathbf{T}^m,$$
$$(\alpha \mathbf{T})^m = \alpha^m \mathbf{T}^m, \tag{B 4.4.11}$$
$$(\mathbf{T}^m)^n = \mathbf{T}^{mn}$$

with $m \geq 0$, $n \geq 0$.

B 4.5 Special Tensors and Operations

After discussing the fundamental definitions of the tensor notions and the products of tensors, we come to the next sections, in which various operators and special tensors are described for which, within the framework of physics, there are many useful applications.

B 4.5.1 Inverse Tensor

A tensor **T** can be inverted when, for any arbitrary vector **v** and **w** from E^3 the relationship,

$$\mathbf{w} = \mathbf{Tv} \tag{B 4.5.1}$$

can be solved for the vector **v** from E^3. If the inverse of **T** exists, then **v** is uniquely determined, and can be written as

$$\mathbf{v} = \mathbf{T}^{-1}\mathbf{w}. \tag{B 4.5.2}$$

The mapping $\mathbf{T}^{-1}$ belongs to the definition (B 4.1.D1) and (B 4.1.D2) and is a tensor (second-order), called the *inverse tensor* of **T**. The following relationship is valid for the inverse tensor $\mathbf{T}^{-1}$:

$$\mathbf{TT}^{-1} = \mathbf{T}^{-1}\mathbf{T} = \mathbf{I}\,; \tag{B 4.5.3}$$

namely, if we put the value of **v** from (B 4.5.2) into (B 4.5.1), we have

$$\mathbf{w} = \mathbf{TT}^{-1}\mathbf{w}\,. \tag{B 4.5.4}$$

A comparison with (B 4.1.D7) shows that

$$\mathbf{TT}^{-1} = \mathbf{I}\,. \tag{B 4.5.5}$$

Correspondingly, it follows from the reversed procedure, placing w from (B 4.5.1) in (B 4.5.2) and comparing with (B 4.1.D7) that

$$\mathbf{T}^{-1}\mathbf{T} = \mathbf{I}\,.$$

Besides that,

$$(\mathbf{T}^{-1})^{-1} = \mathbf{T},\ (\alpha\mathbf{T})^{-1} = \alpha^{-1}\mathbf{T}^{-1}\,, \quad (\mathbf{TS})^{-1} = \mathbf{S}^{-1}\mathbf{T}^{-1} \qquad \text{(B 4.5.6)}$$

are valid.

For invertible tensors **T** the rules for exponential expressions ((B 4.4.10) and (B 4.4.11)) can be extended to negative exponents. The explicit description of the inverse tensors will be undertaken in connection with the introduction of the external product (see Section B 4.9.5).

B 4.5.2
Transposed Tensor

The transposed tensor $\mathbf{T}^T$ of a tensor **T** is defined as

$$\mathbf{w}\cdot\mathbf{Tv} = \mathbf{v}\cdot\mathbf{T}^T\mathbf{w} \qquad \text{(B 4.5.G)}$$

for every vector **v**, **w** in E^3.

The transposed tensor $\mathbf{T}^T$ satisfies (B 4.1.D1) and (B 4.1.D2), and is therefore a tensor. The operation joining $\mathbf{T}^T$ with **T** is called *transposition*. Transposition is a linear operation with the following characteristics:

$$(\mathbf{T}+\mathbf{S})^T = \mathbf{T}^T + \mathbf{S}^T,\quad (\alpha\mathbf{T})^T = \alpha\mathbf{T}^T,\quad (\mathbf{TS})^T = \mathbf{S}^T\mathbf{T}^T. \qquad \text{(B 4.5.7)}$$

The transposition of a simple tensor $\mathbf{a}\otimes\mathbf{b}$ can be directly stated through

$$(\mathbf{a}\otimes\mathbf{b})^T = (\mathbf{b}\otimes\mathbf{a}), \qquad \text{(B 4.5.8)}$$

then, according to (G),

$$\mathbf{v}\cdot(\mathbf{a}\otimes\mathbf{b})^T\mathbf{w} = \mathbf{w}\cdot(\mathbf{a}\otimes\mathbf{b})\mathbf{v} = (\mathbf{b}\cdot\mathbf{v})(\mathbf{w}\cdot\mathbf{a}). \qquad \text{(B 4.5.9)}$$

On the other hand, one finds that

$$\mathbf{v}\cdot(\mathbf{b}\otimes\mathbf{a})\mathbf{w} = (\mathbf{a}\cdot\mathbf{w})(\mathbf{b}\cdot\mathbf{v})\,. \qquad \text{(B 4.5.10)}$$

Thus, (B 4.5.8) is confirmed.

In a base system, we can represent the transposed tensor $\mathbf{T}^T$ of a second-order tensor **T** as

$$\mathbf{T}^T = t^{lk}\mathbf{g}_k\otimes\mathbf{g}_l = t_{lk}\mathbf{g}^k\otimes\mathbf{g}^l,\quad \mathbf{T}^T = t_l^{\cdot k}\mathbf{g}_k\otimes\mathbf{g}^l = t^k_{\cdot l}\mathbf{g}^l\otimes\mathbf{g}_k\,, \qquad \text{(B 4.5.11)}$$

which can easily be demonstrated with the help of the base representation of the scalar product contained in (G).

B 4.5.3
Symmetrical and Skew-Symmetrical Tensors

Symmetric and skew-symmetric tensors appear in many areas of physics. With their help, various mathematical and physical relationships can be represented in a more clear form.

The tensor **S** is *symmetric*, when it corresponds with its transposed tensor, i.e.,

$$\mathbf{S} = \mathbf{S}^T . \qquad \text{(B 4.5.H1)}$$

From the considerations in the previous section, it follows for the symmetrical simple tensor $\mathbf{a} \otimes \mathbf{b}$ that

$$\mathbf{a} \otimes \mathbf{b} = \mathbf{b} \otimes \mathbf{a} \qquad \text{(B 4.5.12)}$$

as well for the coefficients of the components of a symmetrical general tensor **S**

$$s^{lk} = s^{kl}, \ s_{lk} = s_{kl}, \ s^{k}_{.l} = s_{l}^{.k} . \qquad \text{(B 4.5.13)}$$

A *skew-symmetrical* tensor **W** is defined as

$$\mathbf{W} = -\,\mathbf{W}^T . \qquad \text{(B 4.5.H2)}$$

The following relation is valid for the skew-symmetrical simple tensor $\mathbf{c} \otimes \mathbf{d}$:

$$\mathbf{c} \otimes \mathbf{d} = -\,\mathbf{d} \otimes \mathbf{c}. \qquad \text{(B 4.5.14)}$$

It can easily be shown that a simple skew-symmetrical tensor can only be formed with **c** or **d** as zero vectors. The coefficients of the components of the skew-symmetrical general tensors have the following characteristics:

$$-\,w^{lk} = w^{kl}, \ w_{lk} = -\,w_{kl}, \ w^{k}_{.l} = -\,w_{l}^{.k} . \qquad \text{(B 4.5.15)}$$

The identity tensor **I** is a simple example for a symmetrical tensor. According to (B 4.1.D7), the following equation holds

$$\mathbf{v} = \mathbf{I}\mathbf{v}, \qquad \text{(B 4.5.16)}$$

scalar multiplication with an arbitrary vector **u**, gives us

$$\mathbf{u} \cdot \mathbf{v} = \mathbf{u} \cdot \mathbf{I}\mathbf{v} \qquad \text{(B 4.5.17)}$$

and, with (B 4.5.G),

$$\mathbf{u} \cdot \mathbf{v} = \mathbf{v} \cdot \mathbf{I}^T \mathbf{u}, \qquad \text{(B 4.5.18)}$$

from which it is immediately apparent that

$$\mathbf{I} = \mathbf{I}^T . \qquad \text{(B 4.5.19)}$$

With the definitions (B 4.5.G), (H1), and (H2), we likewise find that

$$\mathbf{v} \cdot \mathbf{S}\mathbf{w} = \mathbf{w} \cdot \mathbf{S}\mathbf{v}, \qquad \mathbf{v} \cdot \mathbf{W}\mathbf{v} = 0 . \qquad \text{(B 4.5.20)}$$

According to (B 4.5.15)$_2$ with the definitions (B 4.5.G) and (H2), the following relations are valid

$$\mathbf{v} \cdot \mathbf{W}\mathbf{v} = \mathbf{v} \cdot \mathbf{W}^T \mathbf{v} = -\,\mathbf{v} \cdot \mathbf{W}\mathbf{v} \ \text{ or } \ 2\mathbf{v} \cdot \mathbf{W}\mathbf{v} = 0. \qquad \text{(B 4.5.21)}$$

With the assumption that **v** and **Wv** are non-zero vectors, it follows the orthogonality of both vectors.

With regards to the symmetrical tensor **S**, one says that it is *semi-definite*, when

$$\mathbf{v} \cdot \mathbf{S}\mathbf{v} \geq 0 \qquad \text{(B 4.5.22)}$$

is fulfilled for all the vectors and is *positive-definite*, when

$$\mathbf{v} \cdot \mathbf{S}\mathbf{v} > 0\,. \qquad \text{(B 4.5.23)}$$

B 4.5.4
Orthogonal Tensor

An important role is played by the *orthogonal tensor* **Q**, in rigid body mechanics as well as in the mechanics of deformable bodies interrelated with the rotation motion. This distinguishes itself in that its inverse tensor is identical with its transposed tensor, i.e.,

$$\mathbf{Q}^{-1} = \mathbf{Q}^T \quad \text{or} \quad \mathbf{Q}\mathbf{Q}^T = \mathbf{Q}^T\mathbf{Q} = \mathbf{I}. \qquad \text{(B 4.5.I)}$$

For the orthogonal mapping of arbitrary vectors **u** and **v**, with the definition (B 4.5.G) and (I), as well as (B 4.5.3), we get:

$$(\mathbf{Q}\mathbf{u}) \cdot (\mathbf{Q}\mathbf{v}) = \mathbf{v} \cdot \mathbf{Q}^T\mathbf{Q}\mathbf{u} = \mathbf{v} \cdot \mathbf{u}. \qquad \text{(B 4.5.24)}$$

When two vectors **u** and **v** are mapped with the same orthogonal tensor **Q**, then the scalar product of the vectors corresponds with their mappings. In some textbooks, this statement has been taken to be a definition for the orthogonal tensor.

In the following sections, we limit ourselves to those orthogonal tensors having determinants (for the definition of determinants of a tensor, see Section 4.9.5a) with the value $+1$. These orthogonal tensors are called *proper orthogonal tensors*.

The meaning of the orthogonal tensor in mechanics becomes evident immediately, when we identify the vectors **u** and **v** in (B 4.5.24) with two material lines $\bar{\mathbf{x}}$ and $\bar{\mathbf{y}}$ within a body. The mapping with the tensor **Q** causes a change of the position of these lines within the space (see Fig. B 4.5.1). Eqn. (B 4.5.24) states that the angle between the material lines remains constant. Putting $\bar{\mathbf{x}}$ equal to $\bar{\mathbf{y}}$, we can discern that, besides this, the length also remains constant. In the case that mapping is valid for every line element of a body, then **Q** describes a rigid body rotation. This is only the case when we use proper orthogonal tensors, otherwise we would obtain a rotatory reflection.

There are many possibilities to represent the construction of the orthogonal tensor **Q**, out of which one such possibility shall be discussed in this section. We want to construct the orthogonal tensor from the direction cosine of two orthonormed base systems $\bar{\mathbf{e}}_1, \bar{\mathbf{e}}_2, \bar{\mathbf{e}}_3$ and $\overset{*}{\bar{\mathbf{e}}}_1, \overset{*}{\bar{\mathbf{e}}}_2, \overset{*}{\bar{\mathbf{e}}}_3$ that are bound to the rigid body. As a result of the rigid body rotation, the base system $\bar{\mathbf{e}}_1, \bar{\mathbf{e}}_2, \bar{\mathbf{e}}_3$ with the orthogonal tensor **Q** is transferred to the new base $\overset{*}{\bar{\mathbf{e}}}_1, \overset{*}{\bar{\mathbf{e}}}_2, \overset{*}{\bar{\mathbf{e}}}_3$:

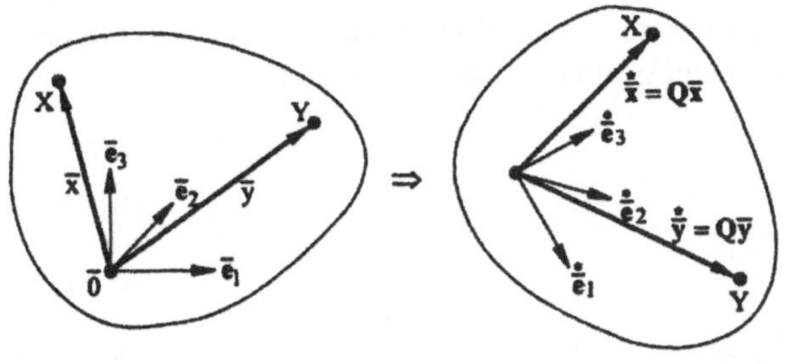

Fig. B 4.5.1. Rigid body rotation

$$\overset{*}{\bar{\mathbf{e}}}_i = \mathbf{Q}\bar{\mathbf{e}}_i. \tag{B 4.5.25}$$

Scalar multiplication with the vector $\bar{\mathbf{e}}_j$ gives us

$$\bar{\mathbf{e}}_j \cdot \overset{*}{\bar{\mathbf{e}}}_i = \bar{\mathbf{e}}_j \cdot \mathbf{Q}\bar{\mathbf{e}}_i \tag{B 4.5.26}$$

or, with (B 4.3.E),

$$\bar{\mathbf{e}}_j \cdot \overset{*}{\bar{\mathbf{e}}}_i = \mathbf{Q} \cdot (\bar{\mathbf{e}}_j \otimes \bar{\mathbf{e}}_i). \tag{B 4.5.27}$$

Now that we can represent the scalar product on the left-hand side with (B 4.3.10) and (B 3.3.5) as

$$\bar{\mathbf{e}}_j \cdot \overset{*}{\bar{\mathbf{e}}}_i = (\bar{\mathbf{e}}_s \cdot \overset{*}{\bar{\mathbf{e}}}_r)(\bar{\mathbf{e}}_s \otimes \bar{\mathbf{e}}_r) \cdot (\bar{\mathbf{e}}_j \otimes \bar{\mathbf{e}}_i)\,, \tag{B 4.5.28}$$

a comparison with (B 4.5.27) gives

$$\mathbf{Q} = (\bar{\mathbf{e}}_i \cdot \overset{*}{\bar{\mathbf{e}}}_k)\bar{\mathbf{e}}_i \otimes \bar{\mathbf{e}}_k. \tag{B 4.5.29}$$

The scalar product $\bar{\mathbf{e}}_i \cdot \overset{*}{\bar{\mathbf{e}}}_k$ contains the directional cosine between the normed unit vectors $\bar{\mathbf{e}}_i$ and $\overset{*}{\bar{\mathbf{e}}}_k$. Denoting this with $\cos\alpha_{(ik)}$, we can also express (B 4.5.29) as

$$\mathbf{Q} = \cos\alpha_{(ik)}\bar{\mathbf{e}}_i \otimes \bar{\mathbf{e}}_k\,. \tag{B 4.5.30}$$

The nine direction angles $\alpha_{(ik)}$ are not independent of each other. Moreover, due to (I), the following interrelations exist

$$\cos\alpha_{(ik)}\cos\alpha_{(jk)} = \delta_{ij}, \tag{B 4.5.31}$$

i.e., there are six relationships (as the relationship (B 4.5.31) is symmetrical for the indices i and j) that have to be fulfilled by the direction angles. The statements of three suitably chosen angles would be sufficient in order to determine the changed location of the base system or of the rigid body. Indeed, the rotation of the rigid body can be described using such angles. This method can be traced back to Euler, and the used angles are named *Euler angles*.

B 4.5.5
Trace of the Tensor

We introduce a special scalar product for tensors, denoted as *trace* (tr):

$$\text{tr}\,\mathbf{T} = \mathbf{I} \cdot \mathbf{T}. \tag{B 4.5.J}$$

Out of this definition, along with the definition of the scalar product from Section B 4.3, we find many calculation rules, which are as follows:

$$\mathrm{tr}\,(\mathbf{a} \otimes \mathbf{b}) = \mathbf{a} \cdot \mathbf{b}, \tag{B 4.5.32}$$

$$\mathrm{tr}\,\mathbf{T}^T = \mathrm{tr}\,\mathbf{T}, \tag{B 4.5.33}$$

$$\mathrm{tr}\,(\mathbf{TS}) = \mathrm{tr}\,(\mathbf{ST}), \tag{B 4.5.34}$$

$$\mathrm{tr}\,(\mathbf{TS}^T) = \mathbf{T} \cdot \mathbf{S} = \mathrm{tr}\,(\mathbf{ST}^T) = \mathbf{T}^T \cdot \mathbf{S}^T. \tag{B 4.5.35}$$

Applying the trace-operation to multiple tensor products, we can draw the conclusion from (B 4.5.34) that the factors can be cyclically interchanged, i.e.,

$$\mathrm{tr}\,(\mathbf{TSR}) = \mathrm{tr}\,(\mathbf{RTS}) = \mathrm{tr}\,(\mathbf{SRT}). \tag{B 4.5.36}$$

This procedure allows the statement of the scalar product in various forms in correlation with tensor products, as for example

$$(\mathbf{TS}) \cdot \mathbf{R}^T = (\mathbf{RT}) \cdot \mathbf{S}^T = (\mathbf{SR}) \cdot \mathbf{T}^T. \tag{B 4.5.37}$$

B 4.6
Decomposition of the Tensor

It is often advantageous to split a tensor into two partial tensors, which show special, characteristic features. This is naturally possible in several ways. We, however, will limit ourselves, in this context, to special additive and multiplicative decompositions.

B 4.6.1
Additive Decomposition

Each arbitrary tensor $\mathbf{T}$ can be additively decomposed into symmetrical tensor $\overset{S}{\mathbf{T}}$ and a skew-symmetrical tensor $\overset{A}{\mathbf{T}}$:

$$\mathbf{T} = \overset{S}{\mathbf{T}} + \overset{A}{\mathbf{T}}, \tag{B 4.6.1}$$

where

$$\begin{aligned} \overset{S}{\mathbf{T}} &= \frac{1}{2}(\mathbf{T} + \mathbf{T}^T) = (\overset{S}{\mathbf{T}})^T, \\ \overset{A}{\mathbf{T}} &= \frac{1}{2}(\mathbf{T} - \mathbf{T}^T) = -\,(\overset{A}{\mathbf{T}})^T. \end{aligned} \tag{B 4.6.2}$$

Following the addition of both the relationships (B 4.6.2), we see that additive decomposition is allowed. Though, it must be proved that $\overset{S}{\mathbf{T}}$ and $\overset{A}{\mathbf{T}}$ are symmetric and skew-symmetric tensors. With the definition (B 4.5.G) for the transposition of a tensor, it immediately follows that

$$\frac{1}{2}(\mathbf{T} + \mathbf{T}^T)^T = \frac{1}{2}(\mathbf{T}^T + \mathbf{T}). \tag{B 4.6.3}$$

As the order of the summands is arbitrary according to (B 4.1.D3), we have

$$\overset{S}{\mathbf{T}} = (\overset{S}{\mathbf{T}})^T. \qquad \text{(B 4.6.4)}$$

Likewise it can be proved that $\overset{A}{\mathbf{T}}$ is a skew-symmetric tensor.

Because of its importance for mechanics, we will further show the possibilities of the decomposition of an arbitrary tensor **T** into a spherical tensor and a deviator. The *spherical tensor* $\overset{K}{\mathbf{T}}$ is defined as

$$\overset{K}{\mathbf{T}} = \frac{1}{3}(\mathbf{T} \cdot \mathbf{I})\mathbf{I}\,. \qquad \text{(B 4.6.5)}$$

We split the tensor **T**, with the help of a spherical tensor, into a sum

$$\mathbf{T} = \mathbf{T} - \overset{K}{\mathbf{T}} + \overset{K}{\mathbf{T}}$$

and denote the difference

$$\overset{D}{\mathbf{T}} = \mathbf{T} - \overset{K}{\mathbf{T}} \qquad \text{(B 4.6.6)}$$

as the *deviator* of **T**. With this, the additive decomposition can be written as

$$\mathbf{T} = \overset{D}{\mathbf{T}} + \overset{K}{\mathbf{T}}. \qquad \text{(B 4.6.7)}$$

It is important to note that the trace of the deviator $\overset{D}{\mathbf{T}}$ is zero:

$$\operatorname{tr} \overset{D}{\mathbf{T}} = \overset{D}{\mathbf{T}} \cdot \mathbf{I} = 0. \qquad \text{(B 4.6.8)}$$

B 4.6.2
Multiplicative Decomposition (Polar Decomposition)

We now take non-singular tensors into consideration. The term *singular* will be explained in Section B 4.9.5. For the moment, we must satisfy ourselves with the comment that a matrix is described as singular, when its determinants vanishes.

Every non-singular, second-order tensor **T** can be explicitly decomposed into positive-definite symmetrical tensors **U**, **V** and into an orthogonal tensor **R**.

$$\mathbf{T} = \mathbf{R}\mathbf{U}, \qquad \text{(B 4.6.9)}$$

$$\mathbf{T} = \mathbf{V}\mathbf{R}. \qquad \text{(B 4.6.10)}$$

We proceed from the identity

$$\mathbf{T} = (\mathbf{T}^T)^{-1}\mathbf{T}^T\mathbf{T} \qquad \text{(B 4.6.11)}$$

and define, using the associative rule of both tensors, the relations:

$$\mathbf{U}\mathbf{U} = \mathbf{T}^T\mathbf{T}, \qquad \mathbf{R} = (\mathbf{T}^T)^{-1}\mathbf{U}. \qquad \text{(B 4.6.12)}$$

The symmetric tensor $\mathbf{T}^T\mathbf{T}$ is positive-definitive and allows to be represented with the square of a likewise symmetric and positive-definitive tensor **U**. The proof for this can be only shown in relation with the eigenvalue problem, which we shall introduce after the external algebra has been treated in Section B 4.9.

We will now show that the tensor **R** is orthogonal, for which we compose the product with (B 4.6.9) as

$$\mathbf{R}\mathbf{R}^T = (\mathbf{T}^T)^{-1}\mathbf{U}\mathbf{U}\mathbf{T}^{-1}. \tag{B 4.6.13}$$

Inserting the definition from (B 4.6.12)$_1$ and taking the relationship (B 4.5.3) into consideration, we find that

$$\mathbf{R}\mathbf{R}^T = \mathbf{I}, \tag{B 4.6.14}$$

which was to be shown (see Section B 4.5.4). With this, the decomposition according to (B 4.6.9) has been proved. To prove it according to (B 4.6.8), we formulate the symmetric and positive definite tensor with (B 4.6.9)

$$\mathbf{T}\mathbf{T}^T = \mathbf{R}\mathbf{U}\mathbf{U}\mathbf{R}^T = (\mathbf{R}\mathbf{U}\mathbf{R}^T)(\mathbf{R}\mathbf{U}\mathbf{R}^T). \tag{B 4.6.15}$$

Analogous to (B 4.6.12)$_1$, we place

$$\mathbf{V}\mathbf{V} = \mathbf{T}\mathbf{T}^T, \tag{B 4.6.16}$$

where

$$\mathbf{V} = \mathbf{R}\mathbf{U}\mathbf{R}^T. \tag{B 4.6.17}$$

From this, we directly find, with (B 4.6.9), that

$$\begin{aligned} \mathbf{V} &= \mathbf{T}\mathbf{R}^T, \\ \mathbf{T} &= \mathbf{V}\mathbf{R}, \end{aligned} \tag{B 4.6.18}$$

which confirms (B 4.6.10).

Finally, we want to show that the decomposition (B 4.6.7) is unique. This shall be proven with a contradictory proof. We assume that there are two different types of decompositions:

$$\mathbf{T} = \mathbf{R}\mathbf{U} = \overset{*}{\mathbf{R}}\overset{*}{\mathbf{U}}. \tag{B 4.6.19}$$

Then the following is valid:

$$\begin{aligned} \mathbf{T}^T\mathbf{T} &= (\mathbf{R}\mathbf{U})^T(\mathbf{R}\mathbf{U}) = \mathbf{U}\mathbf{R}^T\mathbf{R}\mathbf{U} = \mathbf{U}^2 \\ &= (\overset{*}{\mathbf{R}}\overset{*}{\mathbf{U}})^T(\overset{*}{\mathbf{R}}\overset{*}{\mathbf{U}}) = \overset{*}{\mathbf{U}}\,\overset{*}{\mathbf{R}}{}^T\,\overset{*}{\mathbf{R}}\overset{*}{\mathbf{U}} = \overset{*}{\mathbf{U}}{}^2. \end{aligned} \tag{B 4.6.20}$$

As $\overset{*}{\mathbf{U}}$ is positive-definite, it follows that $\mathbf{U} = \overset{*}{\mathbf{R}}\overset{*}{\mathbf{U}}$. Furthermore, we obtain from

$$\mathbf{T} = \mathbf{R}\mathbf{U} = \overset{*}{\mathbf{R}}\overset{*}{\mathbf{U}} \tag{B 4.6.21}$$

with the above result that $\mathbf{U} = \overset{*}{\mathbf{U}}$, the following statement

$$(\mathbf{R} - \overset{*}{\mathbf{R}})\mathbf{U} = \mathbf{O}. \tag{B 4.6.22}$$

As this relationship must be fulfilled for arbitrary **U**, we have

$$\mathbf{R} = \overset{*}{\mathbf{R}}. \tag{B 4.6.23}$$

With this, the uniqueness of the decomposition $\mathbf{T} = \mathbf{R}\mathbf{U}$ has been proven. Analogous to this, one can prove the uniqueness of the decomposition $\mathbf{T} = \mathbf{V}\mathbf{R}$ (B 4.6.10).

B 4.7
Change of the Base

With the arithmetical formulation of problems in engineering sciences it is sometimes advantageous to introduce various base systems. It is in this context that the shell theory is mentioned. In the following paragraphs we want to investigate the effect of *transformation* from one base to another on the relationship between the base vectors and between the coefficients of the components of vectors and tensors.

Given are two arbitrary bases $\mathbf{g}_i$ and $\bar{\mathbf{g}}_i$ in E^3. The corresponding reciprocal bases are $\mathbf{g}^k$ and $\bar{\mathbf{g}}^k$.

Each base vector $\bar{\mathbf{g}}_i$ can be expressed through the base vector $\mathbf{g}_i$ with the help of a non-singular transformation and vice versa; then, according to (B 4.1.D7), with the identity tensor it holds that

$$\begin{aligned}\bar{\mathbf{g}}_i = \mathbf{I}\bar{\mathbf{g}}_i &= (\mathbf{g}_r \otimes \mathbf{g}^r)\bar{\mathbf{g}}_i = (\mathbf{g}^r \cdot \bar{\mathbf{g}}_i)\mathbf{g}_r = (\mathbf{g}^r \cdot \bar{\mathbf{g}}_l \delta^l_i)\mathbf{g}_r \\ &= (\mathbf{g}^r \cdot \bar{\mathbf{g}}_l)(\mathbf{g}_r \otimes \mathbf{g}^l)\mathbf{g}_i,\end{aligned} \tag{B 4.7.1}$$

$$\bar{\mathbf{g}}_i = \mathbf{A}\mathbf{g}_i. \tag{B 4.7.2}$$

The tensor $(\mathbf{g}^r \cdot \bar{\mathbf{g}}_l)\mathbf{g}_r \otimes \mathbf{g}^l$ has been denoted with $\mathbf{A}$. From the above-mentioned transformations, it immediately follows that we can express the transformation explicitly as

$$\bar{\mathbf{g}}_i = (\bar{\mathbf{g}}_i \cdot \mathbf{g}^r)\mathbf{g}_r. \tag{B 4.7.3}$$

As we have assumed both the base systems $\mathbf{g}_i$ and $\bar{\mathbf{g}}_i$, neither (B 4.7.1) nor (B 4.7.2) is a conditional equation for determining $\bar{\mathbf{g}}_i$ from $\mathbf{g}_i$; they serve only as an algebraic link. On the other hand, it is possible, from a given base system $\mathbf{g}_i$, with the introduction of an arbitrary non-singular tensor $\mathbf{A}$, to produce base vectors $\bar{\mathbf{g}}_i$ with linear mappings.

Correspondingly, we get, as alternative possibilities,

$$\begin{aligned}\bar{\mathbf{g}}^i &= (\bar{\mathbf{g}}^i \cdot \mathbf{g}_r)\mathbf{g}^r, \\ \mathbf{g}_i &= (\mathbf{g}_i \cdot \bar{\mathbf{g}}^r)\bar{\mathbf{g}}_r, \\ \mathbf{g}^i &= (\mathbf{g}^i \cdot \bar{\mathbf{g}}_r)\bar{\mathbf{g}}^r.\end{aligned} \tag{B 4.7.4}$$

The relations (B 4.7.3) and (B 4.7.4) can likewise be expressed as linear mappings in the form (B 4.7.2).

Now, the effects of the base transformations on the coefficients of the vector components will be discussed, for which the vector $\mathbf{v}$ will be represented in both the base systems:

$$\mathbf{v} = v^i\mathbf{g}_i = \bar{v}^i\bar{\mathbf{g}}_i. \tag{B 4.7.5}$$

After scalar multiplication with the base vector $\mathbf{g}^k$, we get

$$v^k = (\mathbf{g}^k \cdot \bar{\mathbf{g}}_i)\bar{v}^i. \tag{B 4.7.6}$$

Analogously, we find that

$$\begin{aligned} v_k &= (\mathbf{g}_k \cdot \bar{\mathbf{g}}^i)\bar{v}_i, \\ \bar{v}^k &= (\bar{\mathbf{g}}^k \cdot \mathbf{g}_i)v^i, \\ \bar{v}_k &= (\bar{\mathbf{g}}_k \cdot \mathbf{g}^i)v_i. \end{aligned} \tag{B 4.7.7}$$

One can discern the *rules of transformation* from the given relationships:

> The coefficients of the vector components transform themselves by the change of the base vectors as equal indexed base vectors.

The same consideration can now be applied to the coefficients of the tensor components. For this, we express the component notation of the tensor $\mathbf{T}$ in both the base systems $\mathbf{g}_i \otimes \mathbf{g}^k$ and $\bar{\mathbf{g}}_i \otimes \bar{\mathbf{g}}^k$:

$$\mathbf{T} = t^i_{.k}\mathbf{g}_i \otimes \mathbf{g}^k = \bar{t}^i_{.k}\bar{\mathbf{g}}_i \otimes \bar{\mathbf{g}}^k. \tag{B 4.7.8}$$

Scalar multiplication with $\mathbf{g}^r \otimes \mathbf{g}_s$ gives:

$$t^i_{.k}\delta^r_i\delta^k_s = \bar{t}^i_{.k}(\bar{\mathbf{g}}_i \cdot \mathbf{g}^r)(\bar{\mathbf{g}}^k \cdot \mathbf{g}_s) \tag{B 4.7.9}$$

or

$$t^i_{.k} = (\mathbf{g}^i \cdot \bar{\mathbf{g}}_r)(\mathbf{g}_k \cdot \bar{\mathbf{g}}^s)\bar{t}^r_{.s}. \tag{B 4.7.10}$$

Correspondingly:

$$\begin{aligned} t^{ik} &= (\mathbf{g}^i \cdot \bar{\mathbf{g}}_r)(\mathbf{g}^k \cdot \bar{\mathbf{g}}_s)\bar{t}^{rs}, \\ t_{ik} &= (\mathbf{g}_i \cdot \bar{\mathbf{g}}^r)(\mathbf{g}_k \cdot \bar{\mathbf{g}}^s)\bar{t}_{rs}, \end{aligned} \tag{B 4.7.11}$$

and

$$\begin{aligned} \bar{t}^i_{.k} &= (\bar{\mathbf{g}}^i \cdot \mathbf{g}_r)(\bar{\mathbf{g}}_k \cdot \mathbf{g}^s)t^r_{.s}, \\ \bar{t}^{ik} &= (\bar{\mathbf{g}}^i \cdot \mathbf{g}_r)(\bar{\mathbf{g}}^k \cdot \mathbf{g}_s)t^{rs}, \\ \bar{t}_{ik} &= (\bar{\mathbf{g}}_i \cdot \mathbf{g}^r)(\bar{\mathbf{g}}_k \cdot \mathbf{g}^s)t_{rs}. \end{aligned} \tag{B 4.7.12}$$

From this, we get the following *transformation rule*:

> The coefficients of the tensor components transform themselves as equal indexed tensor bases.

Comparing the transformation rules (B 4.7.6) and (B 4.7.7) with (B 4.7.10) through (B 4.7.12), we can discern a definite regularity. In both groups of formulas, we come across the same transformation values $\bar{\mathbf{g}}^i \cdot \mathbf{g}_r$ and $\bar{\mathbf{g}}_i \cdot \mathbf{g}^r$. We therefore see that the vector and tensor coefficients of the components change from one base to another, according to well-defined rules of transformation. In textbooks, the transformation characteristics are very often used for the definition of tensors, which leads to a considerable formalism. In order to avoid further extending this section, we will consider only a few additional observations.

A special transformation case is to be seen when the scalar product of the base vectors remain unchanged, i.e.,

$$\bar{\mathbf{g}}_{ij} = \bar{\mathbf{g}}_i \cdot \bar{\mathbf{g}}_j = \mathbf{A}\mathbf{g}_i \cdot \mathbf{A}\mathbf{g}_j = \mathbf{g}_i \cdot \mathbf{g}_j = \mathbf{g}_{ij}\,, \tag{B 4.7.13}$$

then $\mathbf{A}$ is an *orthogonal tensor*, which one can read from (B 4.5.24). The substitution of the base vectors in the form

$$\bar{\mathbf{g}}_i = \mathbf{Q}\mathbf{g}_i \tag{B 4.7.14}$$

shows an *orthogonal transformation.*

Transformations between orthogonal base vectors $\mathbf{e}_i$ and $\bar{\mathbf{e}}_i$ are examples for orthogonal transformations. In these cases, the following relations are valid:

$$\bar{\mathbf{e}}_i = \mathbf{Q}\mathbf{e}_i = q_{ki}\mathbf{e}_k, \quad \mathbf{e}_i = \mathbf{Q}^T\bar{\mathbf{e}}_i = q_{ik}\bar{\mathbf{e}}_k, \tag{B 4.7.15}$$

where q_{ki} are the coefficients of the components of the orthogonal tensors in an orthonormed base, containing the direction cosine (B 4.5.30) between the base vectors $\mathbf{e}_i$ and $\bar{\mathbf{e}}_i$. In this context, it should be mentioned that, according to Section B 4.5.4, only proper orthogonal tensors are permitted. Using improper orthogonal tensors for transformation between orthonormed base vectors leads to a *reflection* (change of sign by one of the unit vectors).

B 4.8 Higher-Order Tensors

Until now, we have dealt with vectors as well as linear mappings of vectors, which we have characterized as second-order tensors. As we shall see, in the tensor analysis as well as by its application in continuum mechanics, the introduction of higher-order tensors is very necessary. In our explanations, however, we shall restrict ourselves to a large extent, because the possibilities to define special tensors and operations are very manifold and, therefore, an extensive coverage is practically impossible.

B 4.8.1 Introduction of Higher-Order Tensors

In second-order tensors, we shall extend the validity of the distributive, as well as the associative rules, for the mappings between tensors of arbitrary levels by multiplication with a real scalar quantity. The order of the tensors is marked by a raised natural number $n \geq 1$. With this, we have expressed that we can interpret the vector as a first-order tensor (we dispense the possibility of classifying a scalar value as a zero-order tensor). The first- and second-order tensors will be marked as usual with small and capital boldface letters without superscript numbers, as long as it does not lead to any confusion.

The following relationships are valid ($n, m = 1, 2, ...$) for the nth-order tensors $\overset{n}{\mathbf{L}}$ and $\overset{n}{\mathbf{R}}$, tensors of the mth-order $\overset{m}{\mathbf{T}}$ and $\overset{m}{\mathbf{S}}$, and a real number α:

$$\overset{n}{\mathbf{L}}\,(\overset{m}{\mathbf{T}}+\overset{m}{\mathbf{S}}) = \overset{nm}{\mathbf{LT}}+\overset{nm}{\mathbf{LS}}, \qquad \text{(B 4.8.K1)}$$

$$(\overset{n}{\mathbf{L}}+\overset{n}{\mathbf{R}})\,\overset{m}{\mathbf{T}} = \overset{nm}{\mathbf{LT}}+\overset{nm}{\mathbf{RT}}, \qquad \text{(B 4.8.K2)}$$

$$\overset{n}{\mathbf{L}}\,(\alpha\,\overset{m}{\mathbf{T}}) = (\alpha\,\overset{n}{\mathbf{L}})\,\overset{m}{\mathbf{T}} = \alpha(\overset{nm}{\mathbf{LT}}), \qquad \text{(B 4.8.K3)}$$

$$\alpha\,\overset{n}{\mathbf{L}} = \overset{n}{\mathbf{L}}\,\alpha. \qquad \text{(B 4.8.K4)}$$

In this way, only the definite features of various mappings are defined; to define them adequately, statements have to be made regarding the order of the product, as well as for the tensors that form them. Particularly, in extension of Section B 4.1, we admit that a mapping $\overset{n}{\mathbf{L}}\,\overset{m}{\mathbf{T}}$ with $n \geq m$ exists. This procedure is not necessary, but proves to be advantageous with regard to the associative rule.

For zero tensors the considerations mentioned in Section B 4.1 are valid. In addition to identical mappings additional fundamental tensors also can be defined, which shall be discussed in a separate chapter.

In order to transfer the formalism for the second-order tensors to the higher-order tensors, we introduce the *simple tensors* of the nth- and m^{th}-order:

$$\begin{aligned} \overset{n}{\mathbf{L}} &= \overset{1}{\mathbf{a}}\otimes\overset{2}{\mathbf{a}}\otimes\ldots\otimes\overset{n}{\mathbf{a}}, \\ \overset{m}{\mathbf{T}} &= \overset{1}{\mathbf{b}}\otimes\overset{2}{\mathbf{b}}\otimes\ldots\otimes\overset{m}{\mathbf{b}}, \end{aligned} \qquad \text{(B 4.8.1)}$$

where $\overset{1}{\mathbf{a}}$ through $\overset{n}{\mathbf{a}}$ and $\overset{1}{\mathbf{b}}$ through $\overset{m}{\mathbf{b}}$ are arbitrary vectors in E^3. The tensors $\overset{n}{\mathbf{L}}$ and $\overset{m}{\mathbf{T}}$ satisfy the relations (K1) through (K4). The connection between the simple tensors in (B 4.8.1) emerge as a result of tensor products, as well as tensor products in association with scalar products of the vectors contained in each of the tensors, which are normally referred to as *diminishing products*. Therefore, the possibility exists, depending on the order of the simple tensors, to produce tensors having various orders. We define:

Tensor and diminishing products

$$\begin{aligned}
&(0)\quad (\overset{nm}{\mathbf{LT}})^{n+m} = \overset{1}{\mathbf{a}}\otimes\overset{2}{\mathbf{a}}\otimes\ldots\otimes\overset{n}{\mathbf{a}}\otimes\overset{1}{\mathbf{b}}\otimes\overset{2}{\mathbf{b}}\otimes\ldots\otimes\overset{m}{\mathbf{b}}, \\
&(1)\quad (\overset{nm}{\mathbf{LT}})^{n+m-2} = (\overset{n}{\mathbf{a}}\cdot\overset{1}{\mathbf{b}})\,\overset{1}{\mathbf{a}}\otimes\overset{2}{\mathbf{a}}\otimes\ldots\otimes\overset{n-1}{\mathbf{a}}\otimes\overset{2}{\mathbf{b}}\otimes\ldots\otimes\overset{m}{\mathbf{b}}, \\
&(2)\quad (\overset{nm}{\mathbf{LT}})^{n+m-4} = (\overset{n-1}{\mathbf{a}}\cdot\overset{1}{\mathbf{b}})(\overset{n}{\mathbf{a}}\cdot\overset{2}{\mathbf{b}})\,\overset{1}{\mathbf{a}}\otimes\overset{2}{\mathbf{a}}\otimes\ldots\otimes\overset{n-2}{\mathbf{a}}\otimes\overset{3}{\mathbf{b}}\otimes\ldots\otimes\overset{m}{\mathbf{b}}, \\
&\qquad\vdots \\
&\qquad n>m: \\
&(\mathrm{m})\quad (\overset{nm}{\mathbf{LT}})^{n-m} = (\overset{n-m+1}{\mathbf{a}}\cdot\overset{1}{\mathbf{b}})\ldots(\overset{n}{\mathbf{a}}\cdot\overset{m}{\mathbf{b}})\,\overset{1}{\mathbf{a}}\otimes\overset{2}{\mathbf{a}}\otimes\ldots\otimes\overset{n-m}{\mathbf{a}}, \\
&\qquad n<m: \\
&(\mathrm{n})\quad (\overset{nm}{\mathbf{LT}})^{m-n} = (\overset{1}{\mathbf{a}}\cdot\overset{1}{\mathbf{b}})\ldots(\overset{n}{\mathbf{a}}\cdot\overset{n}{\mathbf{b}})\,\overset{n+1}{\mathbf{b}}\otimes\ldots\otimes\overset{m}{\mathbf{b}}.
\end{aligned} \qquad \text{(B 4.8.K5)}$$

The tensor products, according to (0), and the diminishing products, according to (1) through (m) or (n), are thus determined between arbitrary tensors in association with the associative rule (K3) and the distributive rules. For the special case m = n, we come to the *scalar product of two* n^{th}*-order tensors*:

$$(\overset{1}{\mathbf{a}} \otimes \overset{2}{\mathbf{a}} \otimes ... \otimes \overset{n}{\mathbf{a}}) \cdot (\overset{1}{\mathbf{b}} \otimes \overset{2}{\mathbf{b}} \otimes ... \otimes \overset{n}{\mathbf{b}}) = \\ = (\overset{1}{\mathbf{a}} \cdot \overset{1}{\mathbf{b}})(\overset{2}{\mathbf{a}} \cdot \overset{2}{\mathbf{b}})...(\overset{n}{\mathbf{a}} \cdot \overset{n}{\mathbf{b}}). \qquad \text{(B 4.8.K6)}$$

We can see that herein the scalar product of second-order tensors is contained. Moving on to arbitrary n^{th}-order tensors, we can easily confirm the validity of the commutative rule:

$$\overset{n}{\mathbf{L}} \cdot \overset{n}{\mathbf{T}} = \overset{n}{\mathbf{T}} \cdot \overset{n}{\mathbf{L}} \, . \qquad \text{(B 4.8.2)}$$

Furthermore, the magnitude of an nth-order tensor is given by

$$|\overset{n}{\mathbf{L}}| = \sqrt{\overset{n}{\mathbf{L}} \cdot \overset{n}{\mathbf{L}}} \, . \qquad \text{(B 4.8.3)}$$

B 4.8.2
Special Operations and Tensors

We have defined the products of higher-order tensors in (K5) in such a way that we are able to state the associative rule in a general form:

$$\overset{n}{\mathbf{L}} \, (\overset{m}{\mathbf{T}}\overset{p}{\mathbf{R}}) = (\overset{n}{\mathbf{L}}\overset{m}{\mathbf{T}}) \, \overset{p}{\mathbf{R}} \, . \qquad \text{(B 4.8.4)}$$

In doing this, the order of the product is open, as in (K1) through (K3). This, in association with (K5), has to be determined for the individual case. In order to explain this more clearly, we can consider an example. In (B 4.8.4), we put $n = 2$, $m = 3$ and $p = 2$. The product in the brackets on the left-hand side is a vector:

$$\overset{2}{\mathbf{L}} \, (\overset{3}{\mathbf{T}}\overset{2}{\mathbf{R}})^1 = [(\overset{2}{\mathbf{L}}\overset{3}{\mathbf{T}})^3 \, \overset{2}{\mathbf{R}}]^1 . \qquad \text{(B 4.8.5)}$$

For simple tensors

$$\overset{2}{\mathbf{L}} = \overset{1}{\mathbf{a}} \otimes \overset{2}{\mathbf{a}}, \; \overset{3}{\mathbf{T}} = \overset{1}{\mathbf{b}} \otimes \overset{2}{\mathbf{b}} \otimes \overset{3}{\mathbf{b}}, \; \overset{2}{\mathbf{R}} = \overset{1}{\mathbf{c}} \otimes \overset{2}{\mathbf{c}} \qquad \text{(B 4.8.6)}$$

we apply (K5) on (B 4.8.5), and for the connection of the left-hand side we get:

$$\overset{2}{\mathbf{L}} \, (\overset{3}{\mathbf{T}}\overset{2}{\mathbf{R}})^1 = (\overset{1}{\mathbf{a}} \otimes \overset{2}{\mathbf{a}})(\overset{2}{\mathbf{b}} \cdot \overset{1}{\mathbf{c}})(\overset{3}{\mathbf{b}} \cdot \overset{2}{\mathbf{c}}) \, \overset{1}{\mathbf{b}} = (\overset{2}{\mathbf{b}} \cdot \overset{1}{\mathbf{c}})(\overset{3}{\mathbf{b}} \cdot \overset{2}{\mathbf{c}})(\overset{2}{\mathbf{a}} \cdot \overset{1}{\mathbf{b}}) \, \overset{1}{\mathbf{a}} \, . \qquad \text{(B 4.8.7)}$$

The execution, with the connection chosen on the right-hand side, yields with

$$[(\overset{2}{\mathbf{L}}\overset{3}{\mathbf{T}}) \, \overset{2}{\mathbf{R}}]^1 = (\overset{2}{\mathbf{a}} \cdot \overset{1}{\mathbf{b}})(\overset{1}{\mathbf{a}} \otimes \overset{2}{\mathbf{b}} \otimes \overset{3}{\mathbf{b}})(\overset{1}{\mathbf{c}} \otimes \overset{2}{\mathbf{c}}) = (\overset{2}{\mathbf{a}} \cdot \overset{1}{\mathbf{b}})(\overset{2}{\mathbf{b}} \cdot \overset{1}{\mathbf{c}})(\overset{3}{\mathbf{b}} \cdot \overset{2}{\mathbf{c}}) \, \overset{1}{\mathbf{a}} \qquad \text{(B 4.8.8)}$$

the same result. It is obvious from the definition (K5 (0)), in association with the distributive rule, that the formulation of *higher-order simple tensors* is also possible by using higher-order tensors, as for example

$$\overset{3}{\mathbf{L}} = \mathbf{T} \otimes \mathbf{v}, \quad \overset{3}{\mathbf{L}} = \mathbf{u} \otimes \mathbf{S},$$
$$\overset{4}{\mathbf{L}} = \mathbf{T} \otimes \mathbf{S}, \quad \overset{4}{\mathbf{L}} = \mathbf{v} \otimes \overset{3}{\mathbf{L}}. \tag{B 4.8.9}$$

We now introduce calculation rules, without any proof, for a few definite tensors, which are formed as a consequence of the definitions.

$$(\mathbf{T} \otimes \mathbf{u})\mathbf{v} = (\mathbf{u} \cdot \mathbf{v})\mathbf{T}, \quad (\mathbf{u} \otimes \mathbf{T})\mathbf{R} = (\mathbf{T} \cdot \mathbf{R})\mathbf{u},$$
$$(\mathbf{u} \otimes \mathbf{v} \otimes \mathbf{w})\mathbf{I} = (\mathbf{v} \cdot \mathbf{w})\mathbf{u}, \quad (\mathbf{T} \otimes \mathbf{u})(\mathbf{v} \otimes \mathbf{w}) = (\mathbf{u} \cdot \mathbf{w})\mathbf{T}\mathbf{v},$$
$$(\mathbf{T} \otimes \mathbf{v})\mathbf{R} = \mathbf{T}\mathbf{R}\mathbf{v}, \quad (\mathbf{T} \otimes \mathbf{v})\mathbf{I} = \mathbf{T}\mathbf{v}, \quad (\mathbf{T} \otimes \mathbf{R})\mathbf{S} = (\mathbf{R} \cdot \mathbf{S})\mathbf{T}, \tag{B 4.8.10}$$
$$(\mathbf{T} \otimes \mathbf{R})\mathbf{I} = (tr\mathbf{R})\mathbf{T}, \quad (\mathbf{T} \otimes \mathbf{R})\mathbf{v} = \mathbf{T} \otimes \mathbf{R}\mathbf{v}.$$

Finally, we consider the *transposition* of higher-order tensors. Within the set of second-order tensors, we had defined the transposed tensor in (B 4.5.G) from the special properties by an algebraic transformation. Due to the vast possibilities of transpositions, we will dissociate ourselves from this method in the application on higher-order tensors and visualize that the transposition is determined with the exchange of the vectors of simple products, as in (B 4.5.9), or through the exchange of the base vectors for arbitrary second-order tensors, as in (B 4.5.10). Taking the third-order tensor as an example, we come to the conclusion that the number of possible transpositions are five. In general, in view of the sequence of the order of the base vectors, we can speak of *permutation*. A third-order tensor has, therefore, 6 permutations. Since next we assume a tensor as a definite permutation of the base vectors as given, we shall retain the term transposition. In order to distinguish the transposed higher-order tensor we add the letter T to the tensor as in Section B 4.5.2 and with the help of the letters or numbers at the top of this letter T we can discern which base vectors can be exchanged:

$$\overset{n}{\mathbf{L}}{}^{\overset{ik}{T}} = (\overset{1}{\mathbf{a}} \otimes \ldots \otimes \overset{i}{\mathbf{a}} \otimes \ldots \otimes \overset{k}{\mathbf{a}} \otimes \ldots \otimes \overset{n}{\mathbf{a}})^{\overset{ik}{T}}$$
$$= \overset{1}{\mathbf{a}} \otimes \ldots \otimes \overset{k}{\mathbf{a}} \otimes \ldots \otimes \overset{i}{\mathbf{a}} \otimes \ldots \otimes \overset{n}{\mathbf{a}}. \tag{B 4.8.K7}$$

Further transpositions can be produced with the combination of individual exchanges.

We shall explain this consideration by its application on the third-order tensor. We consider the special transposition for the simple third-order tensor

$$(\overset{1}{\mathbf{a}} \otimes \overset{2}{\mathbf{a}} \otimes \overset{3}{\mathbf{a}})^{\overset{13}{T}} = \overset{3}{\mathbf{a}} \otimes \overset{2}{\mathbf{a}} \otimes \overset{1}{\mathbf{a}}$$

and get, as a consequence of an arbitrary tensor, the following relation:

$$\mathbf{v} \cdot \overset{3}{\mathbf{L}}\mathbf{T} = \mathbf{T}^T \cdot (\overset{3}{\mathbf{L}}{}^{\overset{13}{T}} \mathbf{v}). \tag{B 4.8.11}$$

Likewise for

$$(\overset{1}{\mathbf{a}} \otimes \overset{2}{\mathbf{a}} \otimes \overset{3}{\mathbf{a}})^{\overset{23}{T}} = \overset{1}{\mathbf{a}} \otimes \overset{3}{\mathbf{a}} \otimes \overset{2}{\mathbf{a}} \tag{B 4.8.12}$$

we obtain the consequence

$$(\overset{3}{\mathbf{L}}\,\mathbf{v})\mathbf{u} = (\overset{3}{\mathbf{L}}{}^{\overset{23}{T}}\mathbf{u})\mathbf{v}. \qquad \text{(B 4.8.13)}$$

Lastly, we want to introduce a special *transposition of even-order tensors* $\overset{2n}{\mathbf{L}}$:

$$\overset{n}{\mathbf{T}} \cdot (\overset{2n}{\mathbf{L}}\overset{n}{\mathbf{R}}) = \overset{n}{\mathbf{R}} \cdot (\overset{2n}{\mathbf{L}}{}^{T}\,\overset{n}{\mathbf{T}}). \qquad \text{(B 4.8.K8)}$$

For simple second-order tensors, composed with the tensors $\overset{n}{\mathbf{S}}$ and $\overset{n}{\mathbf{M}}$ according to (K5 (0)), it follows from the definition of the scalar product (K6):

$$(\overset{n}{\mathbf{S}} \otimes \overset{n}{\mathbf{M}})^T = \overset{n}{\mathbf{M}} \otimes \overset{n}{\mathbf{S}}. \qquad \text{(B 4.8.14)}$$

In this, the transposition of the second-order tensor for $n = 1$ is enclosed, according to (B 4.5.G) and (B 4.5.8).

B 4.8.3
Algebra in Base Systems

With reference to Section B 4.2, we presuppose that we can represent the higher-order tensors $\overset{n}{\mathbf{L}}$ in 3^n as linear independent base tensors:

$$\begin{aligned}\overset{n}{\mathbf{L}} &= l^{i_1 i_2 \dots i_n}\mathbf{g}_{i_1} \otimes \mathbf{g}_{i_2} \otimes \dots \otimes \mathbf{g}_{i_n},\\ \overset{n}{\mathbf{L}} &= l_{j_1 j_2 \dots j_n}\mathbf{g}^{j_1} \otimes \mathbf{g}^{j_2} \otimes \dots \otimes \mathbf{g}^{j_n}.\end{aligned} \qquad \text{(B 4.8.15)}$$

We call the scalar values $l^{i_1 i_2 \dots i_n}$ and $l_{j_1 j_2 \dots j_n}$, respectively, the contravariant and convariant coefficients of the components $\overset{n}{\mathbf{L}}$. The tensor base in (B 4.8.15)$_1$ and (B 4.8.15)$_2$ are defined as the covariant and contravariant bases, respectively. Correspondingly, mixed-variant forms are also possible.

Like the vector coefficients, we can project the higher-order tensor coefficients, applying scalar multiplication, in the direction of the base vectors. For this, we scalarly multiply (B 4.8.15) with the contravariant and covariant tensor base and get, taking into consideration the distributive rule, with (K 6):

$$\begin{aligned}l^{i_1 i_2 \dots i_n} &= \overset{n}{\mathbf{L}} \cdot (\mathbf{g}^{i_1} \otimes \mathbf{g}^{i_2} \otimes \dots \otimes \mathbf{g}^{i_n}),\\ l_{j_1 j_2 \dots j_n} &= \overset{n}{\mathbf{L}} \cdot (\mathbf{g}_{j_1} \otimes \mathbf{g}_{j_2} \otimes \dots \otimes \mathbf{g}_{j_n}).\end{aligned} \qquad \text{(B 4.8.16)}$$

Analogously, one can obtain mixed-variant coefficients. By a change of the base for the nth-order tensor, the same rules of transformation are valid as by the second-order tensors.

B 4.9
Cross Product

In mechanics, cross products are very important for problems describing rotational motions. In this context, we would like to mention the balance of moment of momentum and the relationships derived from them. Also, in the general fundamental equations of shell theory and continuum mechanics, we come across

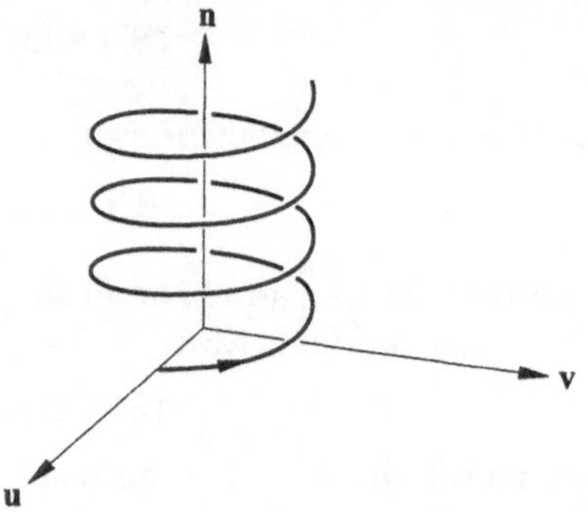

Fig. B 4.9.1. On the cross product of vectors

these products combining vectors with vectors, vectors with tensors as well as tensors with tensors. Due to this fact, it is important to develop and represent the products of vectors and tensors systematically. First, we will explain the cross product of the vectors, then we will be able to introduce this product in a more vivid way. After this, we shall define the cross product of vectors and tensors that can be used in an advantageous way in continuum mechanics. The cross tensor product of tensors at last allows the explicit representation of the adjunct tensors, as well as the invariants of a second-order tensor. The advantage of this products lies in the fact that it is possible to explicitly state the invariants of the tensor sums and the inversion of second-order tensors in a simple way.

B 4.9.1 Cross Product of Vectors

Two non-parallel vectors $\mathbf{u}$ and $\mathbf{v}$ with the enclosed angle φ $(0 \leq \varphi \leq \pi)$ shape a plane when they are plotted at one point. The area enclosed by the parallelogram spanned by the vectors $\mathbf{u}$ and $\mathbf{v}$ is $|\mathbf{u}||\mathbf{v}|\sin\varphi$. Now, we introduce a new vector $\mathbf{u} \times \mathbf{v}$ with the value $|\mathbf{u}||\mathbf{v}|\sin\varphi$ pointing in the direction of the unit normal vector $\mathbf{n}$ of the plane (perpendicular to the plane)

$$\mathbf{u} \times \mathbf{v} = |\mathbf{u}||\mathbf{v}|\sin\varphi\mathbf{n}, \qquad \text{(B 4.9.L)}$$

and call the association $\mathbf{u} \times \mathbf{v}$, as the *cross or vector product of the vectors* $\mathbf{u}$ and $\mathbf{v}$. The unit normal vector $\mathbf{n}$ is explained except for its direction (its sign). We define the vectors $\mathbf{u}, \mathbf{v}$ and $\mathbf{n}$ as a positive mathematical right-handed system. In this way, the association $\mathbf{u} \times \mathbf{v}$ is clearly assigned in the direction $\mathbf{n}$. After this, we obtain from the definition (L)

$$\begin{aligned}(-\mathbf{u}) \times \mathbf{v} &= -\,\mathbf{u} \times \mathbf{v}\,,\\ \mathbf{u} \times \mathbf{v} &= -\,\mathbf{v} \times \mathbf{u}\end{aligned} \qquad \text{(B 4.9.1)}$$

when $\mathbf{u}$ is parallel to $\mathbf{v}$, then

$$\mathbf{u} \times \mathbf{v} = \mathbf{0}. \qquad \text{(B 4.9.2)}$$

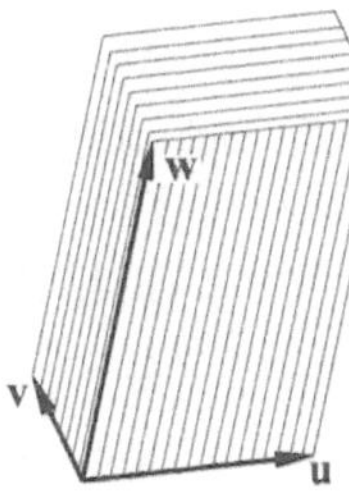

Fig. B 4.9.2. The parallelepiped

Furthermore, it can be easily proven that the following relation is valid:

$$\begin{aligned}\alpha(\mathbf{u}\times\mathbf{v}) &= (\alpha\mathbf{u})\times\mathbf{v} = \mathbf{u}\times(\alpha\mathbf{v}),\\ \mathbf{u}\times(\mathbf{v}+\mathbf{w}) &= \mathbf{u}\times\mathbf{v}+\mathbf{u}\times\mathbf{w}.\end{aligned} \qquad \text{(B 4.9.3)}$$

These two calculation rules explain why the connection $\mathbf{u}\times\mathbf{v}$ can be called a product. Besides this, the calculation rule (B 4.9.2), with (B 4.9.3)$_2$, gives:

$$(\mathbf{u}+\mathbf{v})\times\mathbf{w} = -\,\mathbf{w}\times(\mathbf{u}+\mathbf{v}) = -\,\mathbf{w}\times\mathbf{u}-\mathbf{w}\times\mathbf{v} = \mathbf{u}\times\mathbf{w}+\mathbf{v}\times\mathbf{w}. \qquad \text{(B 4.9.4)}$$

We will now consider the characteristics of *scalar tripel products* (or *box products*) $\mathbf{w}\cdot(\mathbf{u}\times\mathbf{v})$. Geometrically, its value can be interpreted as the volume of a parallelepiped spanned through the vectors $\mathbf{u},\mathbf{v},\mathbf{w}$. The tripel scalar product simply can be written as

$$(\mathbf{u}\times\mathbf{v})\cdot\mathbf{w} = [\mathbf{uvw}]. \qquad \text{(B 4.9.5)}$$

Writing it in a concise form $\mathbf{u}\times\mathbf{v}=\mathbf{z}$, then $\mathbf{w}\cdot\mathbf{z} = |\mathbf{w}||\mathbf{z}|\cos(\mathbf{w},\mathbf{z})$ is the volume of the parallelepiped; then $\mathbf{u}\times\mathbf{v}$ is a vector, whose value is equal to the basic surface area of the parallelepiped, and $|\mathbf{w}|\cos(\mathbf{w},\mathbf{z})$ is its height. The volume can be calculated if we take the parallelogram as the basic surface area, spanned by the vectors $\mathbf{v}$ and $\mathbf{w}$, and multiply scalarly the vector $\mathbf{w}\times\mathbf{u}$ with $\mathbf{v}$. The same volume can also be derived from the scalar product of the product vector $\mathbf{v}\times\mathbf{w}$ and the vector $\mathbf{u}$. For the cyclical exchange of the vectors it holds that:

$$\mathbf{w}\cdot(\mathbf{u}\times\mathbf{v}) = \mathbf{v}\cdot(\mathbf{w}\times\mathbf{u}) = \mathbf{u}\cdot(\mathbf{v}\times\mathbf{w}) \qquad \text{(B 4.9.6)}$$

or

$$[\mathbf{wuv}] = [\mathbf{vwu}] = [\mathbf{uvw}], \qquad \text{(B 4.9.7)}$$

i.e., the parallelepiped product does not change by the cyclical exchange of the vectors. It follows, that the triple scalar product becomes zero, when a linear dependence exists. We make use of this fact, in order to find an explicit form for the vector triple product $\mathbf{u}\times(\mathbf{v}\times\mathbf{w})$. The result is a vector (provided the three vectors $\mathbf{u},\mathbf{v},\mathbf{w}$ represent distances) lying in the area spanned by the vectors $\mathbf{v}$ and $\mathbf{w}$, or parallel to it. From (B 4.9.6) we derive that, with arbitrary real values α and β,

$$\begin{aligned}(\alpha\mathbf{v}+\beta\mathbf{w})\cdot(\mathbf{v}\times\mathbf{w}) \;&=\; 0\,,\\ \mathbf{u}\times(\mathbf{v}\times\mathbf{w})\cdot(\mathbf{v}\times\mathbf{w}) &= 0\,.\end{aligned} \qquad \text{(B 4.9.8)}$$

Comparing both the equations we elucidate that

$$\mathbf{u} \times (\mathbf{v} \times \mathbf{w}) = \mu(\alpha \mathbf{v} + \beta \mathbf{w}), \tag{B 4.9.9}$$

where μ is another arbitrary real number. In addition, we obtain, according to (B 4.9.6),

$$\mathbf{u} \cdot [\mathbf{u} \times (\mathbf{v} \times \mathbf{w})] = \mu\alpha \mathbf{u} \cdot \mathbf{v} + \mu\beta \mathbf{u} \cdot \mathbf{w} = 0. \tag{B 4.9.10}$$

As μ, α, and β are arbitrary real numbers, we put

$$\mu\alpha = \mathbf{u} \cdot \mathbf{w}, \quad \mu\beta = -\,\mathbf{u} \cdot \mathbf{v} \tag{B 4.9.11}$$

fulfilling the above-mentioned equation. The Eqn. (B 4.9.9), along with these values, gives rise to the following relation:

$$\mathbf{u} \times (\mathbf{v} \times \mathbf{w}) = (\mathbf{u} \cdot \mathbf{w})\mathbf{v} - (\mathbf{u} \cdot \mathbf{v})\mathbf{w}. \tag{B 4.9.12}$$

The vector triple product is not associative, as can be seen from (B 4.9.12). However, the *Lagrange identity* is valid

$$\mathbf{u} \times (\mathbf{v} \times \mathbf{w}) + \mathbf{v} \times (\mathbf{w} \times \mathbf{u}) + \mathbf{w} \times (\mathbf{u} \times \mathbf{v}) = \mathbf{0}, \tag{B 4.9.13}$$

which can be easily confirmed by applying Eqn. (B 4.9.12). An important calculation rule is derived from the Lagrange identity (B 4.9.13) by scalar multiplication with the vector $\mathbf{z}$, when (B 4.9.6) and (B 4.9.12) are taken into consideration:

$$\begin{gathered}\mathbf{z} \cdot [\mathbf{u} \times (\mathbf{v} \times \mathbf{w})] = -\,\mathbf{z} \cdot [\mathbf{v} \times (\mathbf{w} \times \mathbf{u})] - \mathbf{z} \cdot [\mathbf{w} \times (\mathbf{u} \times \mathbf{v})],\\ (\mathbf{v} \times \mathbf{w}) \cdot (\mathbf{z} \times \mathbf{u}) = (\mathbf{w} \cdot \mathbf{u})(\mathbf{z} \cdot \mathbf{v}) - (\mathbf{v} \cdot \mathbf{u})(\mathbf{z} \cdot \mathbf{w}).\end{gathered} \tag{B 4.9.14}$$

In addition, we mention the quadruple vector product

$$(\mathbf{u} \times \mathbf{v}) \times (\mathbf{w} \times \mathbf{z}) = [\mathbf{uvz}]\mathbf{w} - [\mathbf{uvw}]\mathbf{z} = [\mathbf{uwz}]\mathbf{v} - [\mathbf{vwz}]\mathbf{u}. \tag{B 4.9.15}$$

Both the forms can be derived from Eqn. (B 4.9.12) when, first, $\mathbf{u} \times \mathbf{v}$ is made equal to $\mathbf{a}$, and second, when $\mathbf{w} \times \mathbf{z}$ is put equal to $\mathbf{b}$.

We have introduced the definition of the cross product and the calculation rules independent of the base. Two vectors $\mathbf{u}$ and $\mathbf{v}$ are now represented in an arbitrary *oriented base system* $\mathbf{g}_1$, $\mathbf{g}_2$, $\mathbf{g}_3$:

$$\mathbf{u} = u^i \mathbf{g}_i, \quad \mathbf{v} = v^k \mathbf{g}_k. \tag{B 4.9.16}$$

The cross product of $\mathbf{u}$ and $\mathbf{v}$, in consideration of (B 4.9.3), is represented as

$$\mathbf{u} \times \mathbf{v} = u^i v^k \mathbf{g}_i \times \mathbf{g}_k\,, \tag{B 4.9.17}$$

and the problem being the evaluation of the cross product of the base vectors. These contain expressions of the form $\mathbf{g}_1 \times \mathbf{g}_1$, $\mathbf{g}_2 \times \mathbf{g}_2$ etc., which according to $(\text{B } 4.9.1)_2$, become zero. Furthermore the value of $\mathbf{g}_1 \times \mathbf{g}_2$ according to (L), is fixed as

$$|\mathbf{g}_1 \times \mathbf{g}_2| \;=\; |\mathbf{g}_1||\mathbf{g}_2|\sin\varphi_{(12)}\,, \tag{B 4.9.18}$$

in which we shall indicate the angle between $\mathbf{g}_1$ and $\mathbf{g}_2$ as $\varphi_{(12)}$. Eqn. (B 4.9.18) can also be written as $|\mathbf{g}_1 \times \mathbf{g}_2| = \sqrt{g_{11}g_{22}(1 - \cos^2\varphi_{(12)})}$ or, taking the definition for the scalar product (B 3.1.C) into consideration, it can be shown that

$$|\mathbf{g}_1 \times \mathbf{g}_2| = \sqrt{g_{11}g_{22} - (g_{12})^2}. \tag{B 4.9.19}$$

The root represents a subdeterminant U_{33} of the coefficient matrix:

$$U_{33} = g_{11}g_{22} - (g_{12})^2, \qquad \text{(B 4.9.20)}$$

so that the following equation is valid:

$$|\mathbf{g}_1 \times \mathbf{g}_2| = \sqrt{U_{33}}. \qquad \text{(B 4.9.21)}$$

According to (L) the cross product of $\mathbf{g}_1$ and $\mathbf{g}_2$ is determined through

$$\mathbf{g}_1 \times \mathbf{g}_2 = \sqrt{U_{33}}\,\mathbf{n}, \qquad \text{(B 4.9.22)}$$

where $\mathbf{n}$ is the unit-normal vector of a surface spanned by the base vectors $\mathbf{g}_1$ and $\mathbf{g}_2$ in a right-handed base system. The unit-normal vector $\mathbf{n}$ expresses itself directly through the contravariant base vector $\mathbf{g}^3$ because this is on $\mathbf{g}_1$ and $\mathbf{g}_2$ and consequently is at right angles to the spanned surface $\mathbf{g}_1$ and $\mathbf{g}_2$:

$$\mathbf{n} = \frac{\mathbf{g}^3}{\sqrt{g^{33}}}, \quad \mathbf{g}_1 \times \mathbf{g}_2 = \sqrt{\frac{U_{33}}{g^{33}}}\mathbf{g}^3. \qquad \text{(B 4.9.23)}$$

With the known relationship from the matrix theory, in consideration of (B 3.4.7),

$$U_{33} = \det \|g_{ik}\| g^{33} \qquad \text{(B 4.9.24)}$$

we get

$$\mathbf{g}_1 \times \mathbf{g}_2 = \sqrt{g}\,\mathbf{g}^3, \qquad \text{(B 4.9.25)}$$

where the following term g has been introduced

$$g = \det \|g_{ik}\|. \qquad \text{(B 4.9.26)}$$

Likewise, the following cross products can be calculated

$$\begin{aligned} &\mathbf{g}_2\times\mathbf{g}_3 = \sqrt{g}\,\mathbf{g}^1, \quad \mathbf{g}_3 \times \mathbf{g}_1 = \sqrt{g}\,\mathbf{g}^2, \\ &\mathbf{g}_2\times\mathbf{g}_1 = -\sqrt{g}\,\mathbf{g}^3, \quad \mathbf{g}_3 \times \mathbf{g}_2 = -\sqrt{g}\,\mathbf{g}^1, \quad \mathbf{g}_1 \times \mathbf{g}_3 = -\sqrt{g}\,\mathbf{g}^2. \end{aligned} \qquad \text{(B 4.9.27)}$$

With the permutation symbol

$$e_{ijk} = e^{ijk} = \begin{cases} 0, & \text{if two indices are equal,} \\ +1 & \text{for } e_{123}, e_{231}, e_{312}, \\ -1 & \text{for } e_{213}, e_{132}, e_{321} \end{cases} \qquad \text{(B 4.9.28)}$$

we can write:

$$\mathbf{g}_i \times \mathbf{g}_j = \sqrt{g}\, e_{ijk}\mathbf{g}^k. \qquad \text{(B 4.9.29)}$$

Analogous observations are valid for the external product of the dual base vectors. This leads to the result

$$\mathbf{g}^i \times \mathbf{g}^j = \frac{1}{\sqrt{g}} e^{ijk}\mathbf{g}_k, \qquad \text{(B 4.9.30)}$$

whereas the contravariant permutation symbol e^{ijk} is explained as the covariant permutation symbol e_{ijk}. With (B 4.9.29) and (B 4.9.30), we are now in a position to explicitly state the scalar triple product (box product) of the base vectors and the dual base vectors:

$$[\mathbf{g}_i\mathbf{g}_j\mathbf{g}_k] = (\mathbf{g}_i \times \mathbf{g}_j)\cdot\mathbf{g}_k = \sqrt{g}\; e_{ijk}\,,$$
$$[\mathbf{g}^i\mathbf{g}^j\mathbf{g}^k] = (\mathbf{g}^i \times \mathbf{g}^j)\cdot\mathbf{g}^k = \frac{1}{\sqrt{g}}e^{ijk}. \qquad \text{(B 4.9.31)}$$

Coming back to our initial problem (B 4.9.17), we can make out that, with (B 4.9.29), the cross product of the two vectors $\mathbf{u}$ and $\mathbf{v}$ is determined in the base $\mathbf{g}_1, \mathbf{g}_2, \mathbf{g}_3$ by

$$\mathbf{u}\times\mathbf{v} = u^i v^j \sqrt{g}\; e_{ijk}\mathbf{g}^k\,. \qquad \text{(B 4.9.32)}$$

Likewise, it follows with (B 4.9.30) for representation in a dual base

$$\mathbf{u}\times\mathbf{v} = u_i v_j \frac{1}{\sqrt{g}}\; e^{ijk}\mathbf{g}_k. \qquad \text{(B 4.9.33)}$$

Together with (B 4.9.31) we obtain for the scalar triple product of the vectors $\mathbf{u}$, $\mathbf{v}$ and $\mathbf{w}$:

$$(\mathbf{u}\times\mathbf{v})\cdot\mathbf{w} = u^i v^j w^k \sqrt{g}\; e_{ijk} = u_i v_j w_k \frac{1}{\sqrt{g}} e^{ijk}. \qquad \text{(B 4.9.34)}$$

The above-mentioned relationships simplify, to a great extent, in an orthonormed base, as in this case g takes the value one and the covariant base vectors coincide with the contravariant base vectors:

$$\mathbf{u}\times\mathbf{v} = u_i v_j e_{ijk}\mathbf{e}_k,$$
$$(\mathbf{u}\times\mathbf{v})\cdot\mathbf{w} = u_i v_j w_k e_{ijk}. \qquad \text{(B 4.9.35)}$$

After evaluation, we get

$$\mathbf{u}\times\mathbf{v} = (u_2v_3 - u_3v_2)\mathbf{e}_1 + (u_3v_1 - u_1v_3)\mathbf{e}_2 + (u_1v_2 - u_2v_1)\mathbf{e}_3,$$
$$(\mathbf{u}\times\mathbf{v})\cdot\mathbf{w} = u_1(v_2w_3 - v_3w_2) + u_2(v_3w_1 - v_1w_3) + u_3(v_1w_2 - v_2w_1), \qquad \text{(B 4.9.36)}$$

and see that these expressions can also be written in a determinant form:

$$\mathbf{u}\times\mathbf{v} = \det\begin{Vmatrix} \mathbf{e}_1 & \mathbf{e}_2 & \mathbf{e}_3 \\ u_1 & u_2 & u_3 \\ v_1 & v_2 & v_3 \end{Vmatrix},$$
$$(\mathbf{u}\times\mathbf{v})\cdot\mathbf{w} = \det\begin{Vmatrix} u_1 & u_2 & u_3 \\ v_1 & v_2 & v_3 \\ w_1 & w_2 & w_3 \end{Vmatrix}. \qquad \text{(B 4.9.37)}$$

We once again turn to the cross product of the base vectors (B 4.9.29). First, we introduce the abbreviation

$$\bar{e}_{ijk} = \sqrt{g}\; e_{ijk}, \quad \bar{e}^{ijk} = \frac{1}{\sqrt{g}}\; e^{ijk}\,. \qquad \text{(B 4.9.38)}$$

With the help of $(\text{B } 4.9.38)_1$, Eqn. (B 4.9.29) can be written as

$$\mathbf{g}_i\times\mathbf{g}_j = \bar{e}_{ijk}\mathbf{g}^k = \bar{e}_{rsk}(\mathbf{g}^r\cdot\mathbf{g}_i)(\mathbf{g}^s\cdot\mathbf{g}_j)\mathbf{g}^k$$
$$= \bar{e}_{rsk}(\mathbf{g}^k\otimes\mathbf{g}^r\otimes\mathbf{g}^s)(\mathbf{g}_i\otimes\mathbf{g}_j), \qquad \text{(B 4.9.39)}$$

where (B 4.8.K5) has been considered by the last transformation. Consequently, the cross product of the base vector can be represented as a linear mapping of the simple tensor $\mathbf{g}_i \otimes \mathbf{g}_j$ with a third-order tensor $\overset{3}{\mathbf{E}}$:

$$\mathbf{g}_i \times \mathbf{g}_j = \overset{3}{\mathbf{E}}\,(\mathbf{g}_i \otimes \mathbf{g}_j)\,. \tag{B 4.9.40}$$

Likewise, it follows from (B 4.9.30) that

$$\mathbf{g}^i \times \mathbf{g}^j = \overset{3}{\mathbf{E}}\,(\mathbf{g}^i \otimes \mathbf{g}^j)\,. \tag{B 4.9.41}$$

In consideration of (B 4.9.25), and the fact that a cyclic interchange of the indices is allowed in $\bar{e}_{rsk}$, the third-order tensor $\overset{3}{\mathbf{E}}$ can be written as

$$\overset{3}{\mathbf{E}} = \bar{e}_{rsk}\mathbf{g}^r \otimes \mathbf{g}^s \otimes \mathbf{g}^k = (\mathbf{g}_r \times \mathbf{g}_s) \cdot \mathbf{g}_k \;\; \mathbf{g}^r \otimes \mathbf{g}^s \otimes \mathbf{g}^k\,. \tag{B 4.9.42}$$

The third-order tensor $\overset{3}{\mathbf{E}}$ is exclusively determined by metric quantities. For this reason, we call the tensor the *fundamental tensor* (see also Section 4.10). Naturally, the fundamental tensor can also be expressed in the covariant base or in the mixed-variant base. For example, we find:

$$\begin{aligned}\overset{3}{\mathbf{E}} &= \bar{e}^{rsk}\mathbf{g}_r \otimes \mathbf{g}_s \otimes \mathbf{g}_k = (\mathbf{g}^r \times \mathbf{g}^s) \cdot \mathbf{g}^k \;\; \mathbf{g}_r \otimes \mathbf{g}_s \otimes \mathbf{g}_k\,,\\ \overset{3}{\mathbf{E}} &= (\mathbf{g}^r \times \mathbf{g}^s) \cdot \mathbf{g}_k \;\; \mathbf{g}_r \otimes \mathbf{g}_s \otimes \mathbf{g}^k.\end{aligned} \tag{B 4.9.43}$$

The calculation rules (B 4.9.40) and (B 4.9.41), valid for the base vectors, can be easily applied to other vectors:

$$\mathbf{u} \times \mathbf{v} = \overset{3}{\mathbf{E}}\,(\mathbf{u} \otimes \mathbf{v}). \tag{B 4.9.44}$$

Lastly, we want to derive a statement valid for a skew-symmetric tensor $\overset{A}{\mathbf{T}}$, stating that, for every skew-symmetric tensor, a vector, called an *axial vector*, is associated with it. In order to prove this, we take a simple tensor $\mathbf{a} \otimes \mathbf{b}$ into consideration, whose skew-symmetric part is formed with $\overset{A}{\mathbf{T}} = \frac{1}{2}(\mathbf{a} \otimes \mathbf{b} - \mathbf{b} \otimes \mathbf{a})$. Applying $\overset{A}{\mathbf{T}}$ on an arbitrary vector $\mathbf{u}$, and taking (B 4.9.12) into consideration, we get

$$\overset{A}{\mathbf{T}}\mathbf{u} = \frac{1}{2}[(\mathbf{u} \cdot \mathbf{b})\mathbf{a} - (\mathbf{u} \cdot \mathbf{a})\mathbf{b}] = \frac{1}{2}\mathbf{u} \times (\mathbf{a} \times \mathbf{b}) = \frac{1}{2}(\mathbf{b} \times \mathbf{a}) \times \mathbf{u}\,. \tag{B 4.9.45}$$

The following relation is valid

$$\overset{A}{\mathbf{T}}\mathbf{u} = \overset{A}{\mathbf{t}} \times \mathbf{u}\,, \tag{B 4.9.46}$$

where the vector

$$\overset{A}{\mathbf{t}} = \frac{1}{2}(\mathbf{b} \times \mathbf{a}) \tag{B 4.9.47}$$

is denoted as the *axial vector*. The considerations carried out for simple tensors can be easily applied to any arbitrary tensor $\mathbf{T}$. From (B 4.9.46) and (B 4.9.43), we can discern that the axial vector is determined, in general, through

$$\overset{A}{\mathbf{t}} = \frac{1}{2} \overset{3}{\mathbf{E}} \mathbf{T}^T \, . \tag{B 4.9.48}$$

The relation (B 4.9.48), in association with (B 4.9.43), directly shows that the axial vector is a zero vector of the symmetrical tensor.

B 4.9.2
Cross Tensor Product of Vector and Tensor

In the cross product of vectors, there exists a vector as a linear mapping with a second-order tensor, e.g., $\mathbf{u} \times (\mathbf{T}\mathbf{v})$. It is then useful, for applications, to introduce a *cross tensor product of the vector* $\mathbf{u}$ *with the tensor* $\mathbf{T}$, and to represent the cross product $\mathbf{u} \times (\mathbf{T}\mathbf{v})$ as a linear mapping. The definition of the cross tensor product of a vector with a tensor can be obtained from the demand that one may associate the vector and tensor in different ways. Taking the arbitrary vectors $\mathbf{u}$ and $\mathbf{v}$ as well as the arbitrary tensors $\mathbf{T}$ and $\mathbf{S}$, we demand:

$$(\mathbf{u} \times \mathbf{T})\mathbf{v} = \mathbf{u} \times (\mathbf{T}\mathbf{v}) \, , \tag{B 4.9.M1}$$

$$(\mathbf{T} \times \mathbf{u})\mathbf{v} = (\mathbf{T}\mathbf{v}) \times \mathbf{u} \, . \tag{B 4.9.M2}$$

Furthermore, the following relation should be valid:

$$\mathbf{T} \cdot (\mathbf{S} \times \mathbf{v}) = - \, \mathbf{S} \cdot (\mathbf{T} \times \mathbf{v}) \, , \tag{B 4.9.M3}$$

where $\mathbf{u} \times \mathbf{T}$ and $\mathbf{T} \times \mathbf{u}$ are second-order tensors, having the following properties:

$$\begin{aligned} \mathbf{u} \times \mathbf{T} &= - \, \mathbf{T} \times \mathbf{u} \, , \\ \mathbf{u} \times (\mathbf{T} + \mathbf{S}) &= \mathbf{u} \times \mathbf{T} + \mathbf{u} \times \mathbf{S} \, , \\ (\mathbf{u} + \mathbf{v}) \times \mathbf{T} &= \mathbf{u} \times \mathbf{T} + \mathbf{v} \times \mathbf{T} \, , \\ \alpha(\mathbf{u} \times \mathbf{T}) &= (\alpha\mathbf{u}) \times \mathbf{T} = \mathbf{u} \times (\alpha\mathbf{T}) \, , \\ (\mathbf{v} \times \mathbf{S}) \cdot \mathbf{T} &= - \, (\mathbf{v} \times \mathbf{T}) \cdot \mathbf{S} \, . \end{aligned} \tag{B 4.9.49}$$

It holds that the tensor $\mathbf{u} \times \mathbf{I}$ is a skew-symmetric tensor, i.e.,

$$(\mathbf{u} \times \mathbf{I}) = - \, (\mathbf{u} \times \mathbf{I})^T \, , \tag{B 4.9.50}$$

because, with the definition of the transposed tensor (B 4.5.G), we can write

$$\mathbf{w} \cdot (\mathbf{u} \times \mathbf{I})\mathbf{v} = \mathbf{v} \cdot (\mathbf{u} \times \mathbf{I})^T\mathbf{w} \, . \tag{B 4.9.51}$$

For the left-hand side, taking into consideration (M1), we get

$$\mathbf{w} \cdot (\mathbf{u} \times \mathbf{I})\mathbf{v} = \mathbf{w} \cdot (\mathbf{u} \times \mathbf{I}\mathbf{v}) = \mathbf{w} \cdot (\mathbf{u} \times \mathbf{v}) \, . \tag{B 4.9.52}$$

Considering (B 4.9.6), as well as (M1), it follows

$$\mathbf{w} \cdot (\mathbf{u} \times \mathbf{I})\mathbf{v} = - \, \mathbf{v} \cdot (\mathbf{u} \times \mathbf{w}) = - \, \mathbf{v} \cdot (\mathbf{u} \times \mathbf{I}\mathbf{w}) = - \, \mathbf{v} \cdot (\mathbf{u} \times \mathbf{I})\mathbf{w} \, . \tag{B 4.9.53}$$

Compared with (B 4.9.51), we come to the result seen in (B 4.9.50), because the vectors $\mathbf{v}$ and $\mathbf{w}$ are arbitrary and the tensor $\mathbf{u} \times \mathbf{I}$ is independent of these vectors. In the skew-symmetric tensor $\mathbf{u} \times \mathbf{I}$, $\mathbf{u}$ is the axial vector. According to (B 4.9.46), the following relation is valid:

$$(\mathbf{u} \times \mathbf{I})\mathbf{v} = \overset{A}{\mathbf{t}} \times \mathbf{v} \,. \qquad \text{(B 4.9.54)}$$

Taking (M1) into consideration, we get

$$\mathbf{u} \times \mathbf{v} = \overset{A}{\mathbf{t}} \times \mathbf{v} \text{ and } \overset{A}{\mathbf{t}} = \mathbf{u} \,. \qquad \text{(B 4.9.55)}$$

The definitions (M1) and (M2) likewise contain

$$\mathbf{u} \times \mathbf{T} = [\overset{3}{\mathbf{E}} \, (\mathbf{u} \otimes \mathbf{T})]^2 \,, \qquad \text{(B 4.9.56)}$$

$$-\,\mathbf{T} \times \mathbf{u} = [\overset{3}{\mathbf{E}} \, (\mathbf{u} \otimes \mathbf{T})]^2 \,, \qquad \text{(B 4.9.57)}$$

$$\mathbf{u} \times (\mathbf{a} \otimes \mathbf{b}) = (\mathbf{u} \times \mathbf{a}) \otimes \mathbf{b} \,, \qquad \text{(B 4.9.58)}$$

$$(\mathbf{a} \otimes \mathbf{b}) \times \mathbf{u} = (\mathbf{a} \times \mathbf{u}) \otimes \mathbf{b} \,. \qquad \text{(B 4.9.59)}$$

The validity of the above-mentioned calculation rules can be confirmed easily with the definitions (M1) through (M3), as well as with some calculation rules of the cross products of vectors.

In conclusion of this section, the cross tensor products of vectors and tensors shall be represented in a base system $\mathbf{g}_1, \mathbf{g}_2, \mathbf{g}_3$. The calculation rule (B 4.9.56), in association with (B 4.9.42), along with the calculation rules of higher-order tensors (B 4.8.K5), gives

$$\begin{aligned} \mathbf{u} \times \mathbf{T} &= \bar{e}_{ikj} u^r t^{sl} (\mathbf{g}^i \otimes \mathbf{g}^k \otimes \mathbf{g}^j)(\mathbf{g}_r \otimes \mathbf{g}_s \otimes \mathbf{g}_l) \\ &= \bar{e}_{ikj} u^k t^{jl} g^{ir} \mathbf{g}_r \otimes \mathbf{g}_l \,. \end{aligned} \qquad \text{(B 4.9.60)}$$

Corresponding representations in the dual or mixed-variant bases can be obtained without difficulty.

B 4.9.3
Cross Tensor Product of Tensors

Once again we come back to the train of thought described in the previous section that in the cross product of vectors, a vector exists as a linear mapping. We also have seen that in this case one attains a new cross product. Continuing with our thought process, we can inquire as to what results we would get when both vectors in the cross product are represented by linear mappings. This, in fact, leads to the definition of a new cross product, namely the *cross tensor product of two tensors*. The cross tensor product of two arbitrary second-order tensors $\mathbf{T}$ and $\mathbf{S}$ is defined with the following mapping rule

$$(\mathbf{T} \# \mathbf{S})(\mathbf{u}_1 \times \mathbf{u}_2) = \mathbf{T}\mathbf{u}_1 \times \mathbf{S}\mathbf{u}_2 - \mathbf{T}\mathbf{u}_2 \times \mathbf{S}\mathbf{u}_1 \,. \qquad \text{(B 4.9.N1)}$$

The product $\mathbf{T} \# \mathbf{S}$ is explicitly a second-order tensor, with $\mathbf{u}_1$ and $\mathbf{u}_2$ being arbitrary vectors. The cross tensor product can be productively used for the formulation of the invariants and the inversion of tensors, which we shall come back to in Section B 4.9.5. From the definition (N1), along with the results of the vector calculus, we get calculation rules, a few of which shall be shown without being proven:

$$\begin{aligned}
\mathbf{T} \# \mathbf{S} &= \mathbf{S} \# \mathbf{T} , \\
\mathbf{R} \# (\mathbf{T} + \mathbf{S}) &= \mathbf{R} \# \mathbf{T} + \mathbf{R} \# \mathbf{S} , \\
\alpha(\mathbf{R} \# \mathbf{T}) &= (\alpha\mathbf{R}) \# \mathbf{T} = \mathbf{R} \# (\alpha\mathbf{T}) , \\
(\mathbf{T} \# \mathbf{S})^T &= \mathbf{T}^T \# \mathbf{S}^T , \\
\mathbf{I} \# \mathbf{I} &= 2\mathbf{I} , \\
(\mathbf{T} \# \mathbf{S})(\mathbf{R} \# \mathbf{U}) &= (\mathbf{TR} \# \mathbf{SU}) + (\mathbf{TU} \# \mathbf{SR}) , \\
(\mathbf{a} \otimes \mathbf{b}) \# (\mathbf{c} \otimes \mathbf{d}) &= (\mathbf{a} \times \mathbf{c}) \otimes (\mathbf{b} \times \mathbf{d}) .
\end{aligned} \tag{B 4.9.61}$$

Furthermore, the scalar triple product of arbitrary tensors **T**, **S**, and **R** can be formed with the definition (N1), and the scalar product of tensors , which satisfies the following equation:

$$(\mathbf{T} \# \mathbf{S}) \cdot \mathbf{R} = \frac{1}{[\mathbf{u}_1\mathbf{u}_2\mathbf{u}_3]} e^{ijk} (\mathbf{T}\mathbf{u}_i \times \mathbf{S}\mathbf{u}_j) \cdot \mathbf{R}\mathbf{u}_k . \tag{B 4.9.62}$$

The scalar triple product of the tensors **T**, **S**, and **R** posesses the following properties:

$$\mathbf{R} \cdot (\mathbf{T} \# \mathbf{S}) = \mathbf{T} \cdot (\mathbf{R} \# \mathbf{S}) = \mathbf{S} \cdot (\mathbf{T} \# \mathbf{R}) . \tag{B 4.9.63}$$

The tensor cross product in (N1), as well as the scalar triple tensor product in (B 4.9.62), can be represented with the use of scalar tensor products. They are individually represented as:

$$\begin{aligned}
\mathbf{T} \# \mathbf{I} &= (\mathbf{T} \cdot \mathbf{I})\mathbf{I} - \mathbf{T}^T , \\
\mathbf{T} \# \mathbf{S} &= (\mathbf{T} \cdot \mathbf{I})(\mathbf{S} \cdot \mathbf{I})\mathbf{I} - (\mathbf{T}^T \cdot \mathbf{S})\mathbf{I} - (\mathbf{T} \cdot \mathbf{I})\mathbf{S}^T - \\
&\quad -(\mathbf{S} \cdot \mathbf{I})\mathbf{T}^T + \mathbf{T}^T\mathbf{S}^T + \mathbf{S}^T\mathbf{T}^T , \\
\mathbf{T} \# \mathbf{T} &= [(\mathbf{T} \cdot \mathbf{I})^2 - \mathbf{T}^T \cdot \mathbf{T}]\mathbf{I} - 2(\mathbf{T} \cdot \mathbf{I})\mathbf{T}^T + 2\mathbf{T}^T\mathbf{T}^T ,
\end{aligned} \tag{B 4.9.64}$$

$$\begin{aligned}
(\mathbf{T} \# \mathbf{S}) \cdot \mathbf{R} &= (\mathbf{T} \cdot \mathbf{I})(\mathbf{S} \cdot \mathbf{I})(\mathbf{R} \cdot \mathbf{I}) - (\mathbf{T} \cdot \mathbf{I})(\mathbf{S}^T \cdot \mathbf{R}) - \\
&\quad - (\mathbf{S} \cdot \mathbf{I})(\mathbf{T}^T \cdot \mathbf{R}) - (\mathbf{R} \cdot \mathbf{I})(\mathbf{T}^T \cdot \mathbf{S}) + \\
&\quad + (\mathbf{T}^T\mathbf{S}^T) \cdot \mathbf{R} + (\mathbf{S}^T\mathbf{T}^T) \cdot \mathbf{R} , \\
(\mathbf{T} \# \mathbf{S}) \cdot (\mathbf{R} \# \mathbf{U}) &= (\mathbf{T} \cdot \mathbf{R})(\mathbf{S} \cdot \mathbf{U}) + (\mathbf{T} \cdot \mathbf{U})(\mathbf{S} \cdot \mathbf{R}) - \\
&\quad - (\mathbf{T}\mathbf{R}^T\mathbf{S}\mathbf{U}^T) \cdot \mathbf{I} - (\mathbf{T}\mathbf{U}^T\mathbf{S}\mathbf{R}^T) \cdot \mathbf{I} .
\end{aligned} \tag{B 4.9.65}$$

It is advantageous to use the calculation rule (B 4.9.61)$_7$, in order to represent the tensor cross product $\mathbf{T} \# \mathbf{S}$ of the tensors **T** and **S** in base systems. After long algebraic transformations (as done in Section B 4.9.1) one comes to the expression

$$\mathbf{T} \# \mathbf{S} = \overset{6}{\mathbf{E}} (\mathbf{T} \otimes \mathbf{S}) . \tag{B 4.9.66}$$

Here, $\overset{6}{\mathbf{E}}$ is a sixth-order tensor, which is solely defined by the metric and represents a fundamental tensor:

$$\overset{6}{\mathbf{E}} = \bar{e}_{ikm}\bar{e}_{jln}\mathbf{g}^i \otimes \mathbf{g}^j \otimes \mathbf{g}^k \otimes \mathbf{g}^l \otimes \mathbf{g}^m \otimes \mathbf{g}^n \,. \tag{B 4.9.67}$$

Similar representations are possible in a covariant base or mixed-variant base. We shall return to this point in Section B 4.10. When the tensors **T** and **S**, for example, are given in a covariant base system, then from (B 4.9.66) we get:

$$\mathbf{T} \# \mathbf{S} = \bar{e}_{irl}\bar{e}_{jsm}t^{ij}s^{rs}\mathbf{g}^l \otimes \mathbf{g}^m \,. \tag{B 4.9.68}$$

The utility of the introduction of the external tensor product of tensors can be particularly seen by the formation of invariants and the calculation of determinants (see Section B 4.9.5).

B 4.9.4
Cross Vector Product of Tensors

The vector product of two tensors is of less importance, however, occasionally one comes about expressions containing such vector products in analysis. We introduce the vector product of two tensors **T** and **S** as follows:

$$\mathbf{v} \cdot (\mathbf{T} \times \mathbf{S}) = -\,\mathbf{T} \cdot (\mathbf{v} \times \mathbf{S}) \,, \tag{B 4.9.N2}$$

where **T** × **S** is an unique vector with the following characteristics:

$$\mathbf{T} \times \mathbf{S} = -\ \mathbf{S} \times \mathbf{T} \,, \tag{B 4.9.69}$$

and

$$\mathbf{T} \times (\mathbf{S} + \mathbf{R}) = \mathbf{T} \times \mathbf{S} + \mathbf{T} \times \mathbf{R} \,, \tag{B 4.9.70}$$

as well as

$$\alpha(\mathbf{T} \times \mathbf{S}) = (\alpha\mathbf{T}) \times \mathbf{S} = \mathbf{T} \times (\alpha\mathbf{S}). \tag{B 4.9.71}$$

The validity of the above-mentioned calculations can easily be confirmed.

In order to calculate the vector product of tensors in a base system, the definition (N2) gives the following mapping rules with the fundamental tensor $\overset{3}{\mathbf{E}}$:

$$\mathbf{T} \times \mathbf{S} = \overset{3}{\mathbf{E}}(\mathbf{T}\mathbf{S}^T) \,. \tag{B 4.9.72}$$

Based on the definition (N2), along with (B 4.9.56), and the calculation rules for higher-order tensors, we get:

$$\begin{aligned} \mathbf{v} \cdot (\mathbf{T} \times \mathbf{S}) &= -\,\mathbf{T} \cdot [\overset{3}{\mathbf{E}}(\mathbf{v} \otimes \mathbf{S})]^2 = \mathbf{T} \cdot [(\overset{3}{\mathbf{E}}\mathbf{v})\mathbf{S}] = (\overset{3}{\mathbf{E}}\mathbf{v}) \cdot \mathbf{T}\mathbf{S}^T \\ &= \overset{3}{\mathbf{E}} \cdot (\mathbf{T}\mathbf{S}^T \otimes \mathbf{v}) = \mathbf{v} \cdot \overset{3}{\mathbf{E}}(\mathbf{T}\mathbf{S}^T) \,. \end{aligned} \tag{B 4.9.73}$$

Here, the special characteristics of the fundamental tensor $\overset{3}{\mathbf{E}}$, from Section B 4.9.1, has been taken into consideration. The validity of (B 4.9.72) for an arbitrary **v** is apparent from the last expression. Besides this, we can see, from (B 4.9.72), that the axial vector, related to a tensor **T**, can be expressed as given in (B 4.9.51), through the vector product of the identity tensor **I** with the tensor **T** or $\overset{A}{\mathbf{T}}$, respectively.

$$\overset{A}{\mathbf{t}} = \frac{1}{2}(\mathbf{I} \times \mathbf{T}) = \frac{1}{2}(\mathbf{I} \times \overset{A}{\mathbf{T}}) \,. \tag{B 4.9.74}$$

In the last step, we have considered the axial vector of a symmetrical tensor as a zero vector.

B 4.9.5
Special Tensors and Operations

Certain problems of the tensor calculation, as for example, the inversion of a tensor, the introduction of some special tensors and operations are necessary, which can be formulated with the help of cross tensor products. This algebraic operation also gives good results for treating the eigenvalue problem and the formulation of determinants. In particular, we shall see that the cross tensor product allows the explicit description of the tensor value determinants. Not only do the calculation rules simplify the cross algebra, but also make the description of the complex mechanical relationships by the rotation of rigid bodies easy.

a)
The Adjunct Tensor and the Determinants

Assuming the definition (N1) and putting S equal to T, we get the relationship

$$\frac{1}{2}(\mathbf{T} \# \mathbf{T})(\mathbf{u}_1 \times \mathbf{u}_2) = \mathbf{T}\mathbf{u}_1 \times \mathbf{T}\mathbf{u}_2 \,. \tag{B 4.9.75}$$

We denote the second-order tensor

$$\overset{+}{\mathbf{T}} = \frac{1}{2}(\mathbf{T} \# \mathbf{T}) \tag{B 4.9.76}$$

as an *adjunct tensor.* The calculation rule (B 4.9.62), with $\mathbf{S} = \mathbf{R} = \mathbf{T}$, leads to

$$\frac{1}{6}(\mathbf{T} \# \mathbf{T}) \cdot \mathbf{T} = \frac{(\mathbf{T}\mathbf{u}_1 \times \mathbf{T}\mathbf{u}_2) \cdot \mathbf{T}\mathbf{u}_3}{[\mathbf{u}_1\mathbf{u}_2\mathbf{u}_3]} \,. \tag{B 4.9.77}$$

The scalar value on the left-hand side of (B 4.9.77) is named as the *determinant* of the tensor T, for which the symbol det is introduced:

$$\det \mathbf{T} = \frac{1}{6}(\mathbf{T} \# \mathbf{T}) \cdot \mathbf{T} \,. \tag{B 4.9.78}$$

With the relationship (B 4.9.78), we immediately find an important calculation rule for finding the determinant of the tensor sum of two tensors T and S; namely, with (B 4.9.78), we get

$$\det(\mathbf{T} + \mathbf{S}) = \frac{1}{6}[(\mathbf{T} + \mathbf{S}) \# (\mathbf{T} + \mathbf{S})] \cdot (\mathbf{T} + \mathbf{S}) \tag{B 4.9.79}$$

or with $(B\ 4.9.61)_2$, and the corresponding calculation rules for the scalar product of tensors, we obtain:

$$\det(\mathbf{T}+\mathbf{S}) = \frac{1}{6}(\mathbf{T}\,\#\,\mathbf{T})\cdot\mathbf{T} + \frac{1}{6}(\mathbf{T}\,\#\,\mathbf{T})\cdot\mathbf{S} + \frac{1}{3}(\mathbf{T}\,\#\,\mathbf{S})\cdot\mathbf{T} + \frac{1}{3}(\mathbf{T}\,\#\,\mathbf{S})\cdot\mathbf{S} + \frac{1}{6}(\mathbf{S}\,\#\,\mathbf{S})\cdot\mathbf{T} + \frac{1}{6}(\mathbf{S}\,\#\,\mathbf{S})\cdot\mathbf{S}\,. \tag{B 4.9.80}$$

Considering (B 4.9.63), (B 4.9.76), and (B 4.9.78), the previous term simplifies to

$$\det(\mathbf{T}+\mathbf{S}) = \det\mathbf{T} + \overset{+}{\mathbf{T}}\cdot\mathbf{S} + \mathbf{T}\cdot\overset{+}{\mathbf{S}} + \det\mathbf{S}\,. \tag{B 4.9.81}$$

Furthermore, an important relationship for the *inversion* of tensors has been developed. Scalar multiplication of (B 4.9.76) with the tensor **T** gives

$$3\det\mathbf{T} = \overset{+}{\mathbf{T}}\cdot\mathbf{T} = (\mathbf{T}\overset{+}{\mathbf{T}}{}^{T})\cdot\mathbf{I} = (\overset{+}{\mathbf{T}}{}^{T}\mathbf{T})\cdot\mathbf{I}\,. \tag{B 4.9.82}$$

With the statement $\mathbf{I}\cdot\mathbf{I} = 3$, valid in E^3, we attain

$$(\det\mathbf{T})\mathbf{I}\cdot\mathbf{I} = (\mathbf{T}\overset{+}{\mathbf{T}}{}^{T})\cdot\mathbf{I} = (\overset{+}{\mathbf{T}}{}^{T}\mathbf{T})\cdot\mathbf{I} \tag{B 4.9.83}$$

or

$$\mathbf{T}\overset{+}{\mathbf{T}}{}^{T} = \overset{+}{\mathbf{T}}{}^{T}\mathbf{T} = (\det\mathbf{T})\mathbf{I}\,. \tag{B 4.9.84}$$

From this, we gain an explicit form for the inversion of the tensor **T**:

$$\mathbf{T}^{-1} = \frac{\overset{+}{\mathbf{T}}{}^{T}}{\det\mathbf{T}}\,. \tag{B 4.9.85}$$

In addition, we mention a few more calculation rules without proof:

$$\begin{aligned}
\overset{+}{(\mathbf{TS})} &= \overset{+}{\mathbf{T}}\overset{+}{\mathbf{S}}\,,\\
\overset{+}{\mathbf{T}}{}^{T} &= (\overset{+}{\mathbf{T}{}^{T}})\,,\\
\det(\alpha\mathbf{T}) &= \alpha^3\det\mathbf{T}\,,\\
\det\mathbf{I} &= 1\,,\\
\det(\mathbf{TS}) &= \det\mathbf{T}\det\mathbf{S}\,,\\
\det\mathbf{T}^{T} &= \det\mathbf{T}\,,\\
(\det\mathbf{Q})^2 &= 1\,,
\end{aligned} \tag{B 4.9.86}$$

when **Q** is an orthogonal tensor,

$$\begin{aligned}
\det\mathbf{T}^{-1} &= (\det\mathbf{T})^{-1}\,,\\
\det\overset{+}{\mathbf{T}} &= (\det\mathbf{T})^2\,.
\end{aligned} \tag{B 4.9.87}$$

b)
The Eigenvalue Problem and the Invariants

Given an arbitrary vector **v** and an arbitrary second-order tensor **T**, we formulate an *eigenvalue problem of a tensor* **T**

$$\mathbf{T}\mathbf{v} = \gamma\mathbf{v}\,, \tag{B 4.9.88}$$

where γ is a real scalar quantity. Instead of (B 4.9.88), we can also write

$$(\mathbf{T} - \gamma \mathbf{I})\mathbf{v} = \mathbf{0} \,. \tag{B 4.9.89}$$

This equation can be interpreted as a conditional equation for **v**; it is homogeneous and possesses non-trivial solutions only when $\mathbf{T} - \gamma\mathbf{I}$ is singular, i.e., when

$$\det(\mathbf{T} - \gamma \mathbf{I}) = 0. \tag{B 4.9.90}$$

With (B 4.9.81), we find that

$$\det(\mathbf{T} - \gamma \mathbf{I}) = \det \mathbf{T} + \overset{+}{\mathbf{T}} \cdot (-\gamma \mathbf{I}) + \mathbf{T} \cdot (-\gamma \mathbf{I})^{+} + \det(-\gamma \mathbf{I}) = 0 \tag{B 4.9.91}$$

or, in consideration of previously stated calculation rules,

$$\begin{aligned} \det(\mathbf{T} - \gamma \mathbf{I}) = \det\mathbf{T} - \gamma \frac{1}{2} (\mathbf{T} \# \mathbf{T}) \cdot \mathbf{I} + \\ + \frac{1}{2} \gamma^2 \mathbf{T} \cdot (\mathbf{I} \# \mathbf{I}) - \gamma^3 \det \mathbf{I} = 0 \,. \end{aligned} \tag{B 4.9.92}$$

This can be written in a more compact form as

$$\det(\mathbf{T} - \gamma \mathbf{I}) = \phi(\gamma) = -\gamma^3 + \mathrm{I}_{\mathrm{T}} \gamma^2 - \mathrm{II}_{\mathrm{T}} \gamma + \mathrm{III}_{\mathrm{T}} = 0 \tag{B 4.9.93}$$

with

$$\begin{aligned} \mathrm{I}_{\mathrm{T}} &= \frac{1}{2} (\mathbf{T} \# \mathbf{I}) \cdot \mathbf{I} \,, \\ \mathrm{II}_{\mathrm{T}} &= \frac{1}{2} (\mathbf{T} \# \mathbf{T}) \cdot \mathbf{I} \,, \\ \mathrm{III}_{\mathrm{T}} &= \frac{1}{6} (\mathbf{T} \# \mathbf{T}) \cdot \mathbf{T} = \det \mathbf{T} \,. \end{aligned} \tag{B 4.9.94}$$

The coefficients $\mathrm{I}_{\mathrm{T}}, \mathrm{II}_{\mathrm{T}},$ and $\mathrm{III}_{\mathrm{T}}$ are the three *principal scalar invariants* of a tensor **T**; these play an important role in mechanics. They can likewise be represented (using the calculation rules described in Section B 4.9.3) through scalar products:

$$\begin{aligned} \mathrm{I}_{\mathrm{T}} &= \mathbf{T} \cdot \mathbf{I} \,, \\ \mathrm{II}_{\mathrm{T}} &= \frac{1}{2} [(\mathbf{T} \cdot \mathbf{I})^2 - \mathbf{T}^T \cdot \mathbf{T}] \,, \\ \mathrm{III}_{\mathrm{T}} &= \frac{1}{6} (\mathbf{T} \cdot \mathbf{I})^3 - \frac{1}{2} (\mathbf{T} \cdot \mathbf{I})(\mathbf{T}^T \cdot \mathbf{T}) + \frac{1}{3} \mathbf{T}^T \mathbf{T}^T \cdot \mathbf{T} \,. \end{aligned} \tag{B 4.9.95}$$

The Eqn. (B 4.9.93) is described as the *characteristic equation* of **T**. The solution of the cubic equation $\phi(\gamma)$ yields the *principal values* or *eigenvalues* of **T**, namely $\gamma_1, \gamma_2,$ and γ_3. The vectors related to the eigenvalues, according to (B 4.9.89), are called the *eigenvectors* of **T**. We assume that the three roots $\gamma_1, \gamma_2,$ and γ_3 are known, and consider the following cases:

α) The roots $\gamma_1, \gamma_2,$ and γ_3 are all distinct from each other. The accompanying eigenvectors are $\mathbf{v}_1, \mathbf{v}_2,$ and $\mathbf{v}_3$, so that the following relation holds:

$$\mathbf{T}\mathbf{v}_i = \gamma_i \mathbf{v}_i \,, \not\Sigma\, i \,. \tag{B 4.9.96}$$

When **T** is symmetric, the following equation can be derived from (B 4.9.96):

$$(\gamma_i - \gamma_j)\mathbf{v}_i \cdot \mathbf{v}_j = 0\,, \; \Sigma i\,, \; \Sigma j\,. \tag{B 4.9.97}$$

For $i \neq j$, the first factor in (B 4.9.97) is unequal to zero so that $\mathbf{v}_i \cdot \mathbf{v}_j$ must vanish in order to fulfill this equation. In the case of i equals j, however, the first factor vanishes and $\mathbf{v}_i \cdot \mathbf{v}_j$ is unequal to zero. Furthermore, presuming that the vector **v** is an unit vector, in the case for i equals j, the scalar product $\mathbf{v}_i \cdot \mathbf{v}_j$ takes the value of one and it holds that:

$$\mathbf{v}_i \cdot \mathbf{v}_j = \delta_{ij}\,,$$

i.e., $\mathbf{v}_i$ forms an orthonormed base. The coefficients of **T** have the values:

$$t_{ik} : \begin{Bmatrix} \gamma_1 & 0 & 0 \\ 0 & \gamma_2 & 0 \\ 0 & 0 & \gamma_3 \end{Bmatrix}\,. \tag{B 4.9.98}$$

β) When $\gamma_1 = \gamma_2 = \gamma \neq \gamma_3$, a single eigenvector $\mathbf{v}_3$ is assigned to γ_3 and each vector orthogonal to $\mathbf{v}_3$ is associated with γ.

γ) When $\gamma_1 = \gamma_2 = \gamma_3 = \gamma$, it holds that $\mathbf{T} = \gamma\mathbf{I}$ and all vectors **v** are eigenvectors.

Next, we want to demonstrate that the three roots are real for symmetric tensors. This proof is carried out with the help of a contradictory proof. Thus we assume that a pair of roots from (B 4.9.93) is conjugate complex: $\gamma = \alpha + i\beta$. Then, also the corresponding eigenvectors **v** are complex: $\mathbf{v} = \mathbf{p} + i\mathbf{q}$. Therefore, according to (B 4.9.89), it holds that:

$$\begin{aligned} &[\mathbf{T} - (\alpha + i\beta)\mathbf{I}](\mathbf{p} + i\mathbf{q}) = 0\,, \\ &\mathbf{T}\mathbf{p} - \alpha\mathbf{p} + \beta\mathbf{q} = 0\,, \quad \mathbf{T}\mathbf{q} - \beta\mathbf{p} - \alpha\mathbf{q} = 0\,. \end{aligned} \tag{B 4.9.99}$$

By scalar multiplication with the vectors **q** and **p** of $(\text{B } 4.9.99)_{2,3}$ and subtracting both equations

$$\mathbf{q} \cdot \mathbf{T}\mathbf{p} - \mathbf{p} \cdot \mathbf{T}\mathbf{q} + \beta(\mathbf{q} \cdot \mathbf{q} + \mathbf{p} \cdot \mathbf{p}) = 0 \tag{B 4.9.100}$$

is obtained. When the tensor **T** is symmetric then the following equation remains:

$$\beta(\mathbf{q} \cdot \mathbf{q} + \mathbf{p} \cdot \mathbf{p}) = 0\,. \tag{B 4.9.101}$$

The terms in the brackets contain squared terms and are therefore unequal to zero. Therefore, β has to vanish, when (B 4.9.101) is to be fulfilled. It follows that for a symmetrical tensor, all three roots are real.

Now we shall represent the three invariants $\mathrm{I}_\mathbf{T}, \mathrm{II}_\mathbf{T}$, and $\mathrm{III}_\mathbf{T}$ through the three solutions γ_1, γ_2, and γ_3 of (B 4.9.93). The first invariant is calculated with (B 4.9.62):

$$\begin{aligned} (\mathbf{T} \# \mathbf{I}) \cdot \mathbf{I}[\mathbf{v}_1\mathbf{v}_2\mathbf{v}_3] = {} & 2\,(\mathbf{T}\mathbf{v}_1 \times \mathbf{v}_2) \cdot \mathbf{v}_3 + 2(\mathbf{v}_1 \times \mathbf{T}\mathbf{v}_2) \cdot \mathbf{v}_3 + \\ & + 2(\mathbf{v}_1 \times \mathbf{v}_2) \cdot \mathbf{T}\mathbf{v}_3\,. \end{aligned} \tag{B 4.9.102}$$

Now, with (B 4.9.96), we have

$$\frac{1}{2}(\mathbf{I} \# \mathbf{I}) \cdot \mathbf{T}[\mathbf{v}_1\mathbf{v}_2\mathbf{v}_3] = (\gamma_1 + \gamma_2 + \gamma_3)[\mathbf{v}_1\mathbf{v}_2\mathbf{v}_3] \,, \tag{B 4.9.103}$$

so that, with (B 4.9.94)$_1$, we come to

$$\mathrm{I}_\mathbf{T} = \frac{1}{2}(\mathbf{T} \# \mathbf{I}) \cdot \mathbf{I} = \gamma_1 + \gamma_2 + \gamma_3, \tag{B 4.9.104}$$

when the scalar triple product is distinct from unity. Likewise follows for the second and third invariants of the tensor **T**:

$$\begin{aligned} \mathrm{II}_\mathbf{T} &= \frac{1}{2}(\mathbf{T} \# \mathbf{T}) \cdot \mathbf{I} = \gamma_1\gamma_2 + \gamma_2\gamma_3 + \gamma_3\gamma_1 \,, \\ \mathrm{III}_\mathbf{T} &= \frac{1}{6}(\mathbf{T} \# \mathbf{T}) \cdot \mathbf{T} = \gamma_1\gamma_2\gamma_3 \,. \end{aligned} \tag{B 4.9.105}$$

The *theorem of Cayley-Hamilton* can be easily developed with the help of the adjunct tensor, according to the premise that a tensor **T** is invertible. With this, we proceed from (B 4.9.76) and form

$$\overset{+}{\mathbf{T}}{}^T = \frac{1}{2}(\mathbf{T} \# \mathbf{T})^T \,. \tag{B 4.9.106}$$

An alternative form for the adjunct tensor can be found with the help of (B 4.9.64)$_3$ as well as (B 4.9.95)

$$\overset{+}{\mathbf{T}}{}^T = \mathrm{II}_\mathbf{T}\mathbf{I} - \mathrm{I}_\mathbf{T}\mathbf{T} + \mathbf{T}^2 \,. \tag{B 4.9.107}$$

Placing this statement into (B 4.9.84), we directly yield the theorem of Cayley-Hamilton

$$\mathbf{T}^3 - \mathrm{I}_\mathbf{T}\mathbf{T}^2 + \mathrm{II}_\mathbf{T}\mathbf{T} - \mathrm{III}_\mathbf{T}\mathbf{I} = \mathbf{0} \,, \tag{B 4.9.108}$$

and comparing with (B 4.9.93), we can discern that each tensor **T** fulfills its characteristic equation.

B 4.10
Fundamental Tensors

In the previous section, we came to know various fundamental tensors, such as the identity tensor **I**, as well as the third- and sixth-order fundamental tensors $\overset{3}{\mathbf{E}}$ and $\overset{6}{\mathbf{E}}$. They are characterized by the fact that they are independent of every scalar, vector, and tensor variable; moreover, they are solely defined by the metric. For the representation of diagrams, scalar, and cross products in base systems, the fundamental tensors can be used in a very advantageous way, as we have already seen in a few cases. Due to their importance, the fundamental tensors will be systematically put together and a few fundamental features will be shown. With the vectors **u** and **v**, as well as the second-order tensors **T** and **S**, we define the fundamental tensors, from the second-order through the sixth-order, by the following relationships:

$$\mathbf{v} = \mathbf{I}\mathbf{v} ,$$
$$\mathbf{u} \times \mathbf{v} = \overset{3}{\mathbf{E}} (\mathbf{u} \otimes \mathbf{v}) ,$$
$$\mathbf{T} = \overset{4}{\mathbf{I}}\, \mathbf{T} , \quad (\mathbf{T} \cdot \mathbf{I})\mathbf{I} = \overset{4}{\bar{\mathbf{I}}}\, \mathbf{T} = (\mathbf{I} \otimes \mathbf{I})\mathbf{T} , \tag{B 4.10.1}$$
$$\mathbf{T}^T = \overset{4}{\bar{\bar{\mathbf{I}}}}\, \mathbf{T} ,$$
$$\mathbf{T} \# \mathbf{S} = \overset{6}{\mathbf{E}} (\mathbf{T} \otimes \mathbf{S}) .$$

By the implementation of tensor calculus, a definite base is introduced. Depending on the chosen base, the metric is defined with the help of fundamental tensors. The coefficients of the fundamental tensors form the metric via the scalar product of the base vectors. In the case of a pure covariant base system, the base-representation of the fundamental tensors is given on the basis of the relations contained in (B 4.10.1) by:

$$\mathbf{I} = (\mathbf{g}^i \cdot \mathbf{g}^j) \quad \mathbf{g}_i \otimes \mathbf{g}_j ,$$
$$\overset{3}{\mathbf{E}} = \mathbf{g}^i \cdot (\mathbf{g}^j \times \mathbf{g}^k)\, \mathbf{g}_i \otimes \mathbf{g}_j \otimes \mathbf{g}_k ,$$
$$\overset{4}{\mathbf{I}} = (\mathbf{g}^i \otimes \mathbf{g}^j) \cdot (\mathbf{g}^k \otimes \mathbf{g}^l)\, \mathbf{g}_i \otimes \mathbf{g}_j \otimes \mathbf{g}_k \otimes \mathbf{g}_l , \tag{B 4.10.2}$$
$$\overset{4}{\bar{\mathbf{I}}} = (\mathbf{g}^i \otimes \mathbf{g}^k) \cdot (\mathbf{g}^j \otimes \mathbf{g}^l)\, \mathbf{g}_i \otimes \mathbf{g}_j \otimes \mathbf{g}_k \otimes \mathbf{g}_l ,$$
$$\overset{4}{\bar{\bar{\mathbf{I}}}} = (\mathbf{g}^i \otimes \mathbf{g}^j) \cdot (\mathbf{g}^l \otimes \mathbf{g}^k)\, \mathbf{g}_i \otimes \mathbf{g}_j \otimes \mathbf{g}_k \otimes \mathbf{g}_l ,$$

and

$$\overset{6}{\mathbf{E}} = \left[(\mathbf{g}^i \times \mathbf{g}^j) \cdot \mathbf{g}^k\right]\left[(\mathbf{g}^l \times \mathbf{g}^m) \cdot \mathbf{g}^n\right] \mathbf{g}_i \otimes \mathbf{g}_l \otimes \mathbf{g}_j \otimes \mathbf{g}_m \otimes \mathbf{g}_k \otimes \mathbf{g}_n . \tag{B 4.10.3}$$

Alternative forms can easily be formed through the exchange of contravariant base vectors against covariant in the coefficients, with the change of the corresponding base vectors in the tensor base taking place at the same time. The fundamental tensors have certain symmetrical properties which can be seen in (B 4.10.2) if one considers the calculation rules for the scalar products for the coefficients. Then, the coefficients can be written in the following way:

$$(\mathbf{I})^{ij} = g^{ij} , \; (\mathbf{I})_{ij} = g_{ij} , \; (\mathbf{I})^i_j = \delta^i_j ,$$
$$(\overset{3}{\mathbf{E}})^{ijk} = \bar{e}^{ijk} , \; (\overset{3}{\mathbf{E}})_{ijk} = \bar{e}_{ijk} , \; (\overset{3}{\mathbf{E}})^{i.k}_{.j} = \bar{e}^{i.k}_{.j} = g_{jr}\bar{e}^{irk} ,$$
$$(\overset{4}{\mathbf{I}})^{ijkl} = g^{ik} g^{jl} , \; (\overset{4}{\mathbf{I}})_{ijkl} = g_{ik} g_{jl} , \; (\overset{4}{\mathbf{I}})^{ij}_{..kl} = \delta^i_k \delta^j_l , \tag{B 4.10.4}$$
$$(\overset{4}{\bar{\mathbf{I}}})^{ijkl} = g^{ij} g^{kl} , \; (\overset{4}{\bar{\mathbf{I}}})_{ijkl} = g_{ij} g_{kl} , \; (\overset{4}{\bar{\mathbf{I}}})^{i.k}_{.j.l} = \delta^i_j \delta^k_l ,$$
$$(\overset{4}{\bar{\bar{\mathbf{I}}}})^{ijkl} = g^{il} g^{jk} , \; (\overset{4}{\bar{\bar{\mathbf{I}}}})_{ijkl} = g_{il} g_{jk} , \; (\overset{4}{\bar{\bar{\mathbf{I}}}})^{ij}_{..kl} = \delta^i_l \delta^j_k ,$$

and

$$\begin{aligned}
&(\overset{6}{\mathbf{E}})^{ijklmn} = \bar{e}^{ijklmn} = \bar{e}^{ikm}\bar{e}^{jln} \,, \\
&(\overset{6}{\mathbf{E}})_{ijklmn} = \bar{e}_{ijklmn} = \bar{e}_{ikm}\bar{e}_{jln} \,, \\
&(\overset{6}{\mathbf{E}})^{i.k.m.}_{.j.l.n} = \bar{e}^{i.k.m}_{.j.l.n} = \bar{e}^{ikm}\bar{e}_{jln} \,.
\end{aligned} \tag{B 4.10.5}$$

From this arrangement (as well as the explanation in Section B 4.9), it is clear that characteristic results for the coefficients of the fundamental tensors can be gained only by a special choice of the base. We ascertain that:

$$\begin{aligned}
&\bar{e}_{ijk} = \sqrt{g}e_{ijk} \,, \; \bar{e}^{ijk} = \frac{1}{\sqrt{g}}e^{ijk} \,, \\
&\bar{e}_{ijklmn} = g\, e_{ikm}e_{jln} \,, \; \bar{e}^{ijklmn} = \frac{1}{g}e^{ikm}e^{jln} \,, \\
&\bar{e}^{i.k.m}_{.j.l.n} = e^{ikm}e_{jln} \,.
\end{aligned} \tag{B 4.10.6}$$

The scalar products of vectors and tensors can be expressed in a simple way with the help of the fundamental tensors:

$$\begin{aligned}
&\mathbf{u}\cdot\mathbf{v} = \mathbf{I}\cdot(\mathbf{u}\otimes\mathbf{v}) \,, \\
&(\mathbf{u}\times\mathbf{v})\cdot\mathbf{w} = \overset{3}{\mathbf{E}}\cdot(\mathbf{u}\otimes\mathbf{v}\otimes\mathbf{w}) \,, \\
&\mathbf{T}\cdot\mathbf{S} = \mathbf{I}\cdot(\mathbf{T}\mathbf{S}^T) = \overset{4}{\mathbf{I}}\cdot(\mathbf{T}\otimes\mathbf{S}) \,, \\
&(\mathbf{T}\,\#\,\mathbf{S})\cdot\mathbf{R} = \overset{6}{\mathbf{E}}\cdot(\mathbf{T}\otimes\mathbf{S}\otimes\mathbf{R}) \,.
\end{aligned} \tag{B 4.10.7}$$

We shall further emphasize a few more characteristics of the fundamental tensors. They distinguish themselves in that, independent of the choice of the base, they possess definite norms.

$$\begin{aligned}
&|\mathbf{I}| = \sqrt{\mathbf{I}\cdot\mathbf{I}} = \sqrt{3} \,, \; |\overset{3}{\mathbf{E}}| = \sqrt{\overset{3}{\mathbf{E}}\cdot\overset{3}{\mathbf{E}}} = \sqrt{3!} = \sqrt{6} \,, \\
&|\overset{4}{\mathbf{I}}| = \sqrt{\overset{4}{\mathbf{I}}\cdot\overset{4}{\mathbf{I}}} = \sqrt{3\cdot 3} = 3 \,, \; |\overset{4}{\bar{\mathbf{I}}}| = |\overset{4}{\bar{\bar{\mathbf{I}}}}| = |\overset{4}{\mathbf{I}}| \,, \\
&|\overset{6}{\mathbf{E}}| = \sqrt{\overset{6}{\mathbf{E}}\cdot\overset{6}{\mathbf{E}}} = \sqrt{6\cdot 6} = 6 \,.
\end{aligned} \tag{B 4.10.8}$$

Another feature is that, by a proper orthogonal exchange of the base vectors, the coefficients remain unchanged. This characteristic will be proven in the following equations: the fundamental tensor **I**, according to (B 4.10.2)$_1$, can be expressed in a covariant base system by

$$\mathbf{I} = \mathbf{g}^i\cdot\mathbf{g}^j\,\mathbf{g}_i\otimes\mathbf{g}_j \,. \tag{B 4.10.9}$$

On the other hand, when we carry out an orthogonal substitution of the base, we get

$$\mathbf{I} = (\mathbf{Q}\mathbf{g}^i)\cdot(\mathbf{Q}\mathbf{g}^j)\mathbf{Q}\mathbf{g}_i\otimes\mathbf{Q}\mathbf{g}_j \,. \tag{B 4.10.10}$$

Considering the definition for the orthogonal tensor in this, we can also write

$$\mathbf{I} = \mathbf{g}^i \cdot \mathbf{g}^j \, \mathbf{Q}\mathbf{g}_i \otimes \mathbf{Q}\mathbf{g}_j \, . \qquad \text{(B 4.10.11)}$$

Likewise, we proceed, with the fundamental tensor $\overset{3}{\mathbf{E}}$:

$$\begin{aligned} \overset{3}{\mathbf{E}} &= \mathbf{g}^i \cdot (\mathbf{g}^j \times \mathbf{g}^k)\mathbf{g}_i \otimes \mathbf{g}_j \otimes \mathbf{g}_k \\ &= \mathbf{Q}\mathbf{g}^i \cdot (\mathbf{Q}\mathbf{g}^j \times \mathbf{Q}\mathbf{g}^k)\mathbf{Q}\mathbf{g}_i \otimes \mathbf{Q}\mathbf{g}_j \otimes \mathbf{Q}\mathbf{g}_k \, . \end{aligned} \qquad \text{(B 4.10.12)}$$

With (B 4.9.77), considering (B 4.9.78) and (B 4.9.86)$_7$, we find:

$$\begin{aligned} \mathbf{Q}\mathbf{g}^i \cdot (\mathbf{Q}\mathbf{g}^j \times \mathbf{Q}\mathbf{g}^k) &= \frac{1}{6}[(\mathbf{Q} \# \mathbf{Q}) \cdot \mathbf{Q}][\mathbf{g}^i \cdot (\mathbf{g}^j \times \mathbf{g}^k)] \\ &= \det \mathbf{Q}[\mathbf{g}^i \cdot (\mathbf{g}^j \times \mathbf{g}^k)] \\ &= \mathbf{g}^i \cdot (\mathbf{g}^j \times \mathbf{g}^k) \, . \end{aligned} \qquad \text{(B 4.10.13)}$$

Consequently,

$$\overset{3}{\mathbf{E}} = \mathbf{g}^i \cdot (\mathbf{g}^j \times \mathbf{g}^k)\mathbf{Q}\mathbf{g}_i \otimes \mathbf{Q}\mathbf{g}_j \otimes \mathbf{Q}\mathbf{g}_k \, . \qquad \text{(B 4.10.14)}$$

The same considerations of the fundamental tensor $\overset{4}{\mathbf{I}}$ lead to

$$\begin{aligned} \overset{4}{\mathbf{I}} &= (\mathbf{g}^i \otimes \mathbf{g}^j) \cdot (\mathbf{g}^k \otimes \mathbf{g}^l)\mathbf{g}_i \otimes \mathbf{g}_j \otimes \mathbf{g}_k \otimes \mathbf{g}_l \\ &= g^{ik} g^{jl} \mathbf{g}_i \otimes \mathbf{g}_j \otimes \mathbf{g}_k \otimes \mathbf{g}_l \, , \end{aligned} \qquad \text{(B 4.10.15)}$$

$$\begin{aligned} \overset{4}{\mathbf{I}} &= (\mathbf{Q}\mathbf{g}^i \otimes \mathbf{Q}\mathbf{g}^j) \cdot (\mathbf{Q}\mathbf{g}^k \otimes \mathbf{Q}\mathbf{g}^l)\mathbf{Q}\mathbf{g}_i \otimes \mathbf{Q}\mathbf{g}_j \otimes \mathbf{Q}\mathbf{g}_k \otimes \mathbf{Q}\mathbf{g}_l \\ &= g^{ik} g^{jl} \mathbf{Q}\mathbf{g}_i \otimes \mathbf{Q}\mathbf{g}_j \otimes \mathbf{Q}\mathbf{g}_k \otimes \mathbf{Q}\mathbf{g}_l \, . \end{aligned} \qquad \text{(B 4.10.16)}$$

The validity of the statement for the remaining fundamental tensors is directly apparent.

In view of some questions formulated in mechanics, the following identities, in correlation with the fundamental tensors, are of importance. With the arbitrary second-order tensor **S**, it holds that:

$$\begin{aligned} &\mathbf{I} = \mathbf{Q}\mathbf{I}\mathbf{Q}^T \, , \\ &\overset{3}{\mathbf{E}} \, (\mathbf{Q}\mathbf{S}\mathbf{Q}^T) = \mathbf{Q}(\overset{3}{\mathbf{E}} \, \mathbf{S}) \, , \\ &\overset{4}{\mathbf{I}} \, (\mathbf{Q}\mathbf{S}\mathbf{Q}^T) = \mathbf{Q}(\overset{4}{\mathbf{I}} \, \mathbf{S})\mathbf{Q}^T \, , \\ &\overset{4}{\bar{\mathbf{I}}} \, (\mathbf{Q}\mathbf{S}\mathbf{Q}^T) = \mathbf{Q}(\overset{4}{\bar{\mathbf{I}}} \, \mathbf{S})\mathbf{Q}^T \, , \end{aligned} \qquad \text{(B 4.10.17)}$$

and

$$\overset{4}{\bar{\bar{\mathbf{I}}}} \, (\mathbf{Q}\mathbf{S}\mathbf{Q}^T) = \mathbf{Q}(\overset{4}{\bar{\bar{\mathbf{I}}}} \, \mathbf{S})\mathbf{Q}^T . \qquad \text{(B 4.10.18)}$$

We denote scalar multiples of the fundamental tensors as *isotropic tensors*. These play an excellent role in the formulation of constitutive equations for isotropic materials. As an example, we can consider the constitutive equation of an isotropic linear-elastic material. This gives the symmetrical stress tensor, depending on Green's linear symmetrical strain measure $\overset{L}{\mathbf{E}}$:

$$\mathbf{T} = \overset{4}{\mathbf{K}}\overset{L}{\mathbf{E}} \; . \qquad \text{(B 4.10.19)}$$

The constitutive response $\overset{4}{\mathbf{K}}$ will be discussed in the following equations. The isotropic condition regarding the material behavior comprises the demand:

$$\overset{4}{\mathbf{K}}\,(\mathbf{Q}\,\overset{L}{\mathbf{E}}\,\mathbf{Q}^T) = \mathbf{Q}(\overset{4}{\mathbf{K}}\overset{L}{\mathbf{E}})\mathbf{Q}^T \; . \qquad \text{(B 4.10.20)}$$

From the comparison of (B 4.10.20) with (B 4.10.17)$_3$ through (B 4.10.18) we find, with the scalar quantities α, β, and γ, that

$$\overset{4}{\mathbf{K}} = \alpha\,\overset{4}{\mathbf{I}} + \beta\,\overset{4}{\bar{\mathbf{I}}} + \gamma\,\overset{4}{\bar{\bar{\mathbf{I}}}} \; . \qquad \text{(B 4.10.21)}$$

As Green's strain tensor $\overset{L}{\mathbf{E}}$ is symmetrical and, due to (B 4.10.1)$_4$ and (B 4.10.1)$_5$, it holds that

$$(\alpha\,\overset{4}{\mathbf{I}} + \gamma\,\overset{4}{\bar{\bar{\mathbf{I}}}})\,\overset{L}{\mathbf{E}} = \; (\alpha + \gamma)\,\overset{L}{\mathbf{E}} = \; \mu\,\overset{L}{\mathbf{E}}\,, \quad \mu \; = \; \alpha + \gamma \qquad \text{(B 4.10.22)}$$

so that the constitutive Equation (B 4.10.19) with (B 4.10.21) and (B 4.10.1)$_4$, reduces to

$$\mathbf{T} = \mu\,\overset{L}{\mathbf{E}} + \beta(\overset{L}{\mathbf{E}} \cdot \mathbf{I})\mathbf{I} \qquad \text{(B 4.10.23)}$$

or with (B 4.10.1)$_4$, to

$$\mathbf{T} = (\mu\,\overset{4}{\mathbf{I}} + \beta\,\overset{4}{\bar{\mathbf{I}}})\,\overset{L}{\mathbf{E}} \; . \qquad \text{(B 4.10.24)}$$

In the section on analysis we shall come back to the fundamental tensors in connection with the development of the derivatives and the introduction of differential operations. We shall see that they allow an explicit statement of the differential operators and consequently, to a large extent, simplify the calculation of the differential operation.

B 5.
Vector and Tensor Analysis

The scalars, vectors, and tensors occurring in physics can be functions of real scalar, vector, and tensor parameters. The determination of the change of these functions, due to an infinitesimal increase of the parameters, is an integral part of the development of the laws of physics. In this context, one is led, beside others, to the differentiation of functions, differential quotients, as well as the partial and the total differentiation. The aim of this chapter is to discuss and then clarify these notions. At the same time we will leave aside the scalar functions, which depend on real scalar parameters, presuming that the analysis of these functions as known. The considerations used will be applied to the vector and tensor analysis, whereas we shall have to introduce modified differentiation definitions for the scalar, vector, and tensor functions, depending on the vector and the tensor parameters.

Due to their great importance in mechanics and engineering, we shall first of all treat vector and tensor functions that depend on real scalar parameters.

In Section B 5.2, the field theory will be treated as has been used in Section 5 of the book. Thereafter, functions will be considered that depend on the arbitrary vector and tensor variables. In conclusion, we shall discuss those integrals which allow the conversion of surface integrals into volume integrals and line integrals into surface integrals.

B 5.1
Functions of Scalar Parameters

It is assumed that the vector **u** is a function of a real scalar variable α. Moreover, only unique functions are considered; besides that, the same symbol for the function, as well as for the value of the function will be used, as long as no confusion occurs.

The *derivative* of this function **u**, when it exists in any open field in the Euclidean space, is the value of the function **v**. It is defined with the following limes consideration:

$$\lim_{\tau\to\alpha}\frac{\mathbf{u}(\tau)-\mathbf{u}(\alpha)-\mathbf{v}(\alpha)(\tau-\alpha)}{|\tau-\alpha|}=\mathbf{0}\,, \qquad \text{(B 5.1.O1)}$$

where τ is likewise a real scalar variable. The derivative $\mathbf{v}(\alpha)$ is normally also written as:

$$\mathbf{v}(\alpha)=\frac{d\mathbf{u}(\alpha)}{d\alpha}\ ,\ \ \mathbf{v}(\alpha)=\mathbf{u}'(\alpha)\,. \qquad \text{(B 5.1.1)}$$

The *differential* of the function **u** for a given value α and a given value of the differential $d\alpha$ is equal to the product $\mathbf{u}'(\alpha)$ and $d\alpha$:

$$d\mathbf{u}(\alpha)=\mathbf{u}'(\alpha)d\alpha\,. \qquad \text{(B 5.1.O2)}$$

Furthermore, we can define the *derivatives* and *higher-order differentials*. The *second derivative* of **u** is written as:

$$\frac{d^2\mathbf{u}}{d\alpha^2}=\mathbf{u}''(\alpha)=\frac{d}{d\alpha}\left(\frac{d\mathbf{u}}{d\alpha}\right). \qquad \text{(B 5.1.O3)}$$

Therefore, $\mathbf{u}''$ is again a vector depending on α. Similarly, the higher derivatives are determined. The *second differential* of a vector function **u** depending on a variable α is the differential of the first differential:

$$d^2\mathbf{u}=d(d\mathbf{u})=\mathbf{u}''(\alpha)d\alpha^2\,. \qquad \text{(B 5.1.O4)}$$

Analogous to this, the *higher-order differentials* will be introduced:

$$d^3\mathbf{u}=d(d^2\mathbf{u})=\mathbf{u}'''(\alpha)d\alpha^3\,. \qquad \text{(B 5.1.O5)}$$

The *partial derivative* of **w** which in turn, is a function of the real scalar variables $\alpha, \beta, \gamma, \ldots$, with respect to one of these variables, for example to α, will be defined as:

$$\lim_{\tau \to \alpha} \frac{\mathbf{u}(\tau, \beta, \gamma, ...) - \mathbf{u}(\alpha, \beta, \gamma, ...) - \mathbf{w}(\alpha, \beta, \gamma, ...)(\tau - \alpha)}{|\tau - \alpha|} = \mathbf{0} . \quad \text{(B 5.1.06)}$$

The partial derivative $\mathbf{w}$ which again is a function of the real scalar variables $\alpha, \beta, \gamma, ...,$ can be written as:

$$\mathbf{w} = \frac{\partial \mathbf{u}}{\partial \alpha} . \quad \text{(B 5.1.2)}$$

The *total differential* of a vector function $\mathbf{u}$, which depends on several real scalar variables, is expressed as:

$$d\mathbf{u} = \frac{\partial \mathbf{u}}{\partial \alpha} d\alpha + \frac{\partial \mathbf{u}}{\partial \beta} d\beta + \frac{\partial \mathbf{u}}{\partial \gamma} d\gamma + ... \quad \text{(B 5.1.07)}$$

The *second-order partial derivative* allows various possibilities for differentiation. Either this can be formed according to the same variable as the first derivative,

$$\frac{\partial^2 \mathbf{u}}{\partial \alpha^2} , \quad \frac{\partial^2 \mathbf{u}}{\partial \beta^2} , \quad ... , \quad \text{(B 5.1.08)}$$

or according to another variable, such as

$$\frac{\partial^2 \mathbf{u}}{\partial \alpha \partial \beta} , \quad \frac{\partial^2 \mathbf{u}}{\partial \beta \partial \gamma} , \quad ... . \quad \text{(B 5.1.09)}$$

The expressions mentioned above are also denoted as mixed derivatives. The total n^{th}-order differential for vector functions which depend on several variables can be represented as

$$d^n\mathbf{u} = (\frac{\partial}{\partial \alpha} d\alpha + \frac{\partial}{\partial \beta} d\beta + ...)^n \mathbf{u} . \quad \text{(B 5.1.O10)}$$

In a similar way, we can define the derivatives and differentials for tensor functions of one or more real scalar variables. The derivative of a tensor function $\mathbf{T}$, with respect to a real scalar variable, is again a tensor function. Therefore, we do not need to develop any special calculus for tensor functions; in the definitions for the vector functions, we need only to replace the vectors with tensors.

The conventional rules of differential calculus for scalar functions can easily be extended to cover vector and tensor functions. For example, λ being a scalar, $\mathbf{a}$, $\mathbf{b}$, and $\mathbf{u}$ vectors, as well as $\mathbf{T}$, $\mathbf{S}$, and $\overset{3}{\mathbf{L}}$ tensors, all depend on a real scalar variable. The definitions, and the corresponding analysis for tensors, furnish rules of differentiation, from which we are obliged to mention a few special cases:

$$(\lambda \mathbf{a})' = \lambda' \mathbf{a} + \lambda \mathbf{a}' , \quad \text{(B 5.1.3)}$$

$$\begin{aligned}
(\mathbf{a} \otimes \mathbf{b})' &= \mathbf{a}' \otimes \mathbf{b} + \mathbf{a} \otimes \mathbf{b}' , \\
(\mathbf{a} \cdot \mathbf{b})' &= \mathbf{a}' \cdot \mathbf{b} + \mathbf{a} \cdot \mathbf{b}' , \\
(\mathbf{a} \times \mathbf{b})' &= \mathbf{a}' \times \mathbf{b} + \mathbf{a} \times \mathbf{b}' , \\
(\mathbf{u} \times \mathbf{T})' &= \mathbf{u}' \times \mathbf{T} + n \times \mathbf{T}' ,
\end{aligned} \quad \text{(B 5.1.4)}$$

$$(\mathbf{T}\mathbf{u})' = \mathbf{T}'\mathbf{u} + \mathbf{T}\mathbf{u}' \,, \tag{B 5.1.5}$$

$$(\mathbf{T}^{-1})' = -\mathbf{T}^{-1}\mathbf{T}'\mathbf{T}^{-1} \,, \quad (\mathbf{T}^T)' = \mathbf{T}'^T \,, \tag{B 5.1.6}$$

$$\begin{gathered}(\mathbf{T}\mathbf{S})' = \mathbf{T}'\mathbf{S} + \mathbf{T}\mathbf{S}' \,, \quad (\mathbf{T}\cdot\mathbf{S})' = \mathbf{T}'\cdot\mathbf{S} + \mathbf{T}\cdot\mathbf{S}' \,, \\ (\mathbf{T} \# \mathbf{S})' = \mathbf{T}' \# \mathbf{S} + \mathbf{T} \# \mathbf{S}' \,,\end{gathered} \tag{B 5.1.7}$$

$$(\overset{3}{\mathbf{L}}\mathbf{u})' = \overset{3}{\mathbf{L}}{}'\mathbf{u} + \overset{3}{\mathbf{L}}\mathbf{u}' \,, \tag{B 5.1.8}$$

$$\begin{gathered}I'_{\mathbf{T}} = \mathbf{I}\cdot\mathbf{T}', \qquad II'_{\mathbf{T}} = (\mathbf{T} \# \mathbf{I})\cdot\mathbf{T}', \qquad III'_{\mathbf{T}} = \overset{+}{\mathbf{T}}\cdot\mathbf{T}' \,, \\ \overset{+}{\mathbf{T}}{}' = (\mathbf{T} \# \mathbf{T}') \,.\end{gathered} \tag{B 5.1.9}$$

A prerequisite for the validity of the relationship $(\text{B 5.1.6})_1$ is the invertibility of the tensor **T**.

We can discern that the product rules for scalar functions, depending on real scalar parameters known from the literature, is likewise also valid for vector and tensor functions. This, for example, is also valid for the chain rule. With the help of the definition (O1), the calculation rules (B 5.1.1) through (B 5.1.9) can be easily proven.

B 5.2 Field Theory

We shall discuss vectors and tensors assigned to a point x in Euclidean point space which will be described with the help of the position vector **x**. Such functions are also defined as *fields*. In general, one differentiates between *scalar*, *vector*, and *tensor fields*, which are all functions of the position vector **x**. These field functions play an important role in physics. In this context, we mention temperature (which represents a scalar field), the force and displacement fields (which are vector fields) and the stress field (which serve as an example of a tensor field); these fields are in general also time-dependent. We can free ourselves from the introduction of special coordinates for the analysis of such functions and assume the functional dependence on the position vector, which shall be an advantage in a more concise and clear representation of the derivatives. For the field functions, we shall assume that they are unique and smooth.

B 5.2.1 Gradient

Given ϕ, a scalar function of the position vector **x** that is defined in an open area; then ϕ is differentiable in this area, when a vector field **w**(**x**) exists, so that with **x** and **y** as arbitrary position vectors, we get

$$\lim_{\mathbf{y}\to\mathbf{x}} \frac{\phi(\mathbf{y}) - \phi(\mathbf{x}) - \mathbf{w}(\mathbf{x}) \cdot (\mathbf{y} - \mathbf{x})}{|\mathbf{y} - \mathbf{x}|} = 0. \tag{B 5.2.P1}$$

When the relationship $(P1)$ is valid, then $\mathbf{w}$ is uniquely determined. We call this vector the *gradient* (grad) *of the scalar field* $\phi(\mathbf{x})$, and write it as

$$\mathbf{w} = \operatorname{grad} \phi(\mathbf{x}) \,. \tag{B 5.2.1}$$

For the gradient, we also permit the following denotion:

$$\mathbf{w} = \frac{d\phi(\mathbf{x})}{d\mathbf{x}} = \nabla\phi(\mathbf{x}) \,. \tag{B 5.2.2}$$

The *differential of the scalar field* $\phi(\mathbf{x})$, for a given value $\mathbf{x}$ and a given value of the differential $d\mathbf{x}$, is equal to the scalar product of $\operatorname{grad} \phi$ and $d\mathbf{x}$:

$$d\phi = \operatorname{grad} \phi \cdot d\mathbf{x} \,. \tag{B 5.2.P2}$$

The *partial derivative of the scalar function* ϕ, depending on the position vector $\mathbf{x}$ and a real scalar variable α, with respect to the position vector $\mathbf{x}$, is defined with the relationship

$$\lim_{\mathbf{y}\to\mathbf{x}} \frac{\phi(\mathbf{y},\alpha) - \phi(\mathbf{x},\alpha) - \mathbf{u}(\mathbf{x},\alpha) \cdot (\mathbf{y} - \mathbf{x})}{|\mathbf{y} - \mathbf{x}|} = 0 \tag{B 5.2.P3}$$

The partial derivative $\mathbf{u}(\mathbf{x},\alpha)$ is written as

$$\mathbf{u} = \frac{\partial\phi(\mathbf{x},\alpha)}{\partial\mathbf{x}} \,. \tag{B 5.2.3}$$

For the partial derivative of the scalar function $\phi = \phi(\mathbf{x},\alpha)$ with respect to the position vector $\mathbf{x}$, the following expressions $\operatorname{grad} \phi$ and$\nabla\phi$, respectively, have established themselves. The same is true for the partial derivatives, introduced in the following equations, of the vector and tensor functions, which depend on the position vector $\mathbf{x}$ and the real scalar variables α.

The *total differential* of $\phi(\mathbf{x},\alpha)$, in consideration of the observations in Section 5.1, can be written as

$$d\phi(\mathbf{x},\alpha) = \frac{\partial\phi}{\partial\mathbf{x}} \cdot d\mathbf{x} + \frac{\partial\phi}{\partial\alpha} d\alpha \,. \tag{B 5.2.P4}$$

We now consider the vector function $\mathbf{v}$ of the position vector $\mathbf{x}$, which should also exist in an open area. Similarly, the *gradient of the vector field* $\mathbf{v}(\mathbf{x})$ can be defined as the tensor field $\mathbf{R}(\mathbf{x})$ through

$$\lim_{\mathbf{y}\to\mathbf{x}} \frac{\mathbf{v}(\mathbf{y}) - \mathbf{v}(\mathbf{x}) - \mathbf{R}(\mathbf{x})(\mathbf{y} - \mathbf{x})}{|\mathbf{y} - \mathbf{x}|} = 0 \tag{B 5.2.P5}$$

for every $\mathbf{x}$ in the open area

$$\mathbf{R}(\mathbf{x}) = \operatorname{grad} \mathbf{v}(\mathbf{x}) = \frac{d\mathbf{v}(\mathbf{x})}{d\mathbf{x}} = \nabla\mathbf{v}(\mathbf{x}) \,. \tag{B 5.2.4}$$

The *differential of the vector field* $\mathbf{v}(\mathbf{x})$ for a given value of the differential $d\mathbf{x}$, is determined with the help of a linear transformation

$$d\mathbf{v} = (\operatorname{grad} \mathbf{v})d\mathbf{x} = \nabla\mathbf{v}d\mathbf{x} \,. \tag{B 5.2.P6}$$

Furthermore, we can determine the *partial derivative of the vector field* **v** (which depends on the position vector **x** and a real scalar quantity α) with respect to the position vector **x** through

$$\lim_{\mathbf{y}\to\mathbf{x}} \frac{\mathbf{v}(\mathbf{y},\alpha) - \mathbf{v}(\mathbf{x},\alpha) - \mathbf{S}(\mathbf{x},\alpha)(\mathbf{y}-\mathbf{x})}{|\mathbf{y}-\mathbf{x}|} = \mathbf{0} \qquad \text{(B 5.2.P7)}$$

with

$$\mathbf{S}(\mathbf{x},\alpha) = \frac{\partial \mathbf{v}(\mathbf{x},\alpha)}{\partial \mathbf{x}} = \operatorname{grad} \mathbf{v} = \nabla \mathbf{v}\,. \qquad \text{(B 5.2.5)}$$

With this we get the *total differential* of $\mathbf{v}(\mathbf{x},\alpha)$

$$d\mathbf{v}(\mathbf{x},\alpha) = \frac{\partial \mathbf{v}}{\partial \mathbf{x}} d\mathbf{x} + \frac{\partial \mathbf{v}}{\partial \alpha} d\alpha\,. \qquad \text{(B 5.2.P8)}$$

Finally, we examine the tensor function **T** depending on the position vector **x**, which also should be present in an open area. The *gradient of the tensor field* $\mathbf{T}(\mathbf{x})$ is then defined as the tensor field $\overset{3}{\mathbf{S}}(\mathbf{x})$ through

$$\lim_{\mathbf{y}\to\mathbf{x}} \frac{\mathbf{T}(\mathbf{y}) - \mathbf{T}(\mathbf{x}) - \overset{3}{\mathbf{S}}(\mathbf{x})(\mathbf{y}-\mathbf{x})}{|\mathbf{y}-\mathbf{x}|} = \mathbf{O} \qquad \text{(B 5.2.P9)}$$

for every **x** in the open area, where we write

$$\overset{3}{\mathbf{S}} = \operatorname{grad} \mathbf{T}(\mathbf{x}) = \frac{d\mathbf{T}(\mathbf{x})}{d\mathbf{x}} = \nabla \mathbf{T}(\mathbf{x})\,. \qquad \text{(B 5.2.6)}$$

The *differential of the tensor field* $\mathbf{T}(x)$ is defined, with the help of the linear mapping, as

$$d\mathbf{T}(\mathbf{x}) = (\operatorname{grad} \mathbf{T})d\mathbf{x} = \nabla \mathbf{T} d\mathbf{x}\,. \qquad \text{(B 5.2.P10)}$$

Similar to the procedure used for partial derivatives of scalar and vector fields, we introduce the *partial derivative of the tensor field* **T**, which is a function of the position vector **x** and the real quantity α, with respect to the position vector **x** and the real quantity α:

$$\lim_{\mathbf{y}\to\mathbf{x}} \frac{\mathbf{T}(\mathbf{y},\alpha) - \mathbf{T}(\mathbf{x},\alpha) - \overset{3}{\mathbf{R}}(\mathbf{x},\alpha)(\mathbf{y}-\mathbf{x})}{|\mathbf{y}-\mathbf{x}|} = \mathbf{O}\,. \qquad \text{(B 5.2.P11)}$$

Thereby is

$$\overset{3}{\mathbf{R}}(\mathbf{x},\alpha) = \frac{\partial \mathbf{T}(\mathbf{x},\alpha)}{\partial \mathbf{x}} = \operatorname{grad} \mathbf{T} = \nabla \mathbf{T}\,. \qquad \text{(B 5.2.7)}$$

For the *total differential* of $\mathbf{T}(\mathbf{x},\alpha)$, it follows from the previous explanation, that

$$d\mathbf{T}(\mathbf{x},\alpha) = \frac{\partial \mathbf{T}}{\partial \mathbf{x}} d\mathbf{x} + \frac{\partial \mathbf{T}}{\partial \alpha} d\alpha\,. \qquad \text{(B 5.2.P12)}$$

The transfer of the considerations, with regard to the formation of gradients and total differentials, to such scalar, vector, and tensor functions, depending on the position vector **x** and various real scalar variables, is possible without problem using the results of Section 5.1.

With the help of the given definitions, the following rules for the formation of gradients from product terms can be derived:

$$\nabla(\phi\psi) = \phi\nabla\psi + \psi\nabla\phi \,, \tag{B 5.2.8}$$

$$\nabla(\phi\mathbf{v}) = \mathbf{v} \otimes \nabla\phi + \phi\nabla\mathbf{v} \,, \tag{B 5.2.9}$$

$$\nabla(\phi\mathbf{T}) = \mathbf{T} \otimes \nabla\phi + \phi\nabla\mathbf{T} \,, \tag{B 5.2.10}$$

$$\nabla(\mathbf{u} \cdot \mathbf{v}) = (\nabla\mathbf{u})^T\mathbf{v} + (\nabla\mathbf{v})^T\mathbf{u} \,, \tag{B 5.2.11}$$

$$\nabla(\mathbf{u} \times \mathbf{v}) = \mathbf{u} \times \nabla\mathbf{v} + \nabla\mathbf{u} \times \mathbf{v} \,, \tag{B 5.2.12}$$

$$\nabla(\mathbf{a} \otimes \mathbf{b}) = [\nabla\mathbf{a} \otimes \mathbf{b} + \mathbf{a} \otimes (\nabla\mathbf{b})^T]^{\overset{23}{T}} \,, \tag{B 5.2.13}$$

$$\nabla(\mathbf{T}\mathbf{v}) = (\nabla\mathbf{T})^{\overset{23}{T}}\mathbf{v} + \mathbf{T}\nabla\mathbf{v} \,, \tag{B 5.2.14}$$

$$\nabla(\mathbf{T}\mathbf{S}) = [(\nabla\mathbf{T})^{\overset{23}{T}}\mathbf{S}]^{3^{\overset{23}{T}}} + (\mathbf{T}\nabla\mathbf{S})^3 \,, \tag{B 5.2.15}$$

$$\nabla(\mathbf{T} \cdot \mathbf{S}) = (\nabla\mathbf{T})^{\overset{13}{T}}\mathbf{S}^T + (\nabla\mathbf{S})^{\overset{13}{T}}\mathbf{T}^T \,. \tag{B 5.2.16}$$

Furthermore, it holds for the gradient of the position vector:

$$\nabla\mathbf{x} = \mathbf{I} \,. \tag{B 5.2.17}$$

B 5.2.2
Derivatives of Higher-Order

Firstly, we define the *second derivative of a scalar field* $\phi(\mathbf{x})$:

$$\frac{d^2\phi}{d\mathbf{x} \otimes d\mathbf{x}} = \frac{d}{d\mathbf{x}}\left(\frac{d\phi}{d\mathbf{x}}\right) = \nabla\nabla\phi \,, \tag{B 5.2.P13}$$

where $\nabla\nabla\phi$ is a second-order tensor, likewise depending on $\mathbf{x}$. Correspondingly, the higher derivatives can be defined.

The *second differential of a scalar field function* $\phi(\mathbf{x})$ is the differential of the first differential:

$$d^2\phi = d(d\phi) = \nabla\nabla\phi \cdot (d\mathbf{x} \otimes d\mathbf{x}) \,. \tag{B 5.2.P14}$$

In a similar way, the *second derivative of a vector field* $\mathbf{v}(\mathbf{x})$ can be determined:

$$\frac{d^2\mathbf{v}}{d\mathbf{x} \otimes d\mathbf{x}} = \frac{d}{d\mathbf{x}}\left(\frac{d\mathbf{v}}{d\mathbf{x}}\right) = \nabla\nabla\mathbf{v} \,, \tag{B 5.2.P15}$$

in which $\nabla\nabla\mathbf{v}$ is a third-order tensor. Likewise, we get the higher derivatives. The *second differential of a vector field function* $\mathbf{v}(\mathbf{x})$ reads as

$$d^2\mathbf{v} = d(d\mathbf{v}) = (\nabla\nabla\mathbf{v})(d\mathbf{x} \otimes d\mathbf{x}) \,. \tag{B 5.2.P16}$$

Finally, we also mention the *second derivative of a tensor field* $\mathbf{T}(\mathbf{x})$:

$$\frac{d^2\mathbf{T}}{d\mathbf{x}\otimes d\mathbf{x}} = \frac{d}{d\mathbf{x}}\left(\frac{d\mathbf{T}}{d\mathbf{x}}\right) = \nabla\nabla\mathbf{T}\,, \qquad \text{(B 5.2.P17)}$$

where $\nabla\nabla\mathbf{T}$ is a fourth-order tensor. For the *second differential of a tensor field function* $\mathbf{T}(\mathbf{x})$, we write:

$$d^2\mathbf{T} = d(d\mathbf{T}) = (\nabla\nabla\mathbf{T})(d\mathbf{x}\otimes d\mathbf{x})\,. \qquad \text{(B 5.2.18)}$$

In a corresponding manner, the higher differentiation of tensor fields are defined. We will avoid the extension of the definition, with regard to the formation of higher derivatives of scalar, vector, and tensor functions, that variably depend on the position vector **x** and the real scalars, as they are of little interest. Furthermore, the definitions can easily be formulated corresponding to the above-mentioned method, and using the results of the Section B 5.1.

B 5.2.3
Special Operations (Divergence, Rotation, and the Laplace-Operator)

In physics, it has proved to be helpful to introduce certain operators that allow the use of abbreviations and partly make physical interpretation possible. We firstly devote ourselves to the operator *divergence* (div). This is defined for a vector field $\mathbf{v}(\mathbf{x})$ in the following manner:

$$\operatorname{div}\mathbf{v}(\mathbf{x}) = \nabla\mathbf{v}(\mathbf{x})\cdot\mathbf{I}\,. \qquad \text{(B 5.2.Q1)}$$

Consequently, the *divergence of a vector field* $\mathbf{v}(\mathbf{x})$ is a scalar field. The divergence of a tensor field $\mathbf{T}(\mathbf{x})$ is given by the definition

$$\operatorname{div}\mathbf{T}(\mathbf{x}) = \nabla\mathbf{T}(\mathbf{x})\mathbf{I}\,. \qquad \text{(B 5.2.Q2)}$$

The d*ivergence of a tensor field* is therefore a vector field.

Taking into consideration the above-mentioned definitions and the calculations rules of Section 5.2.1, the following calculation rules can be derived directly:

$$\operatorname{div}(\phi\mathbf{v}) = \mathbf{v}\cdot\nabla\phi + \phi\operatorname{div}\mathbf{v}\,, \qquad \text{(B 5.2.19)}$$

$$\operatorname{div}(\mathbf{T}\mathbf{v}) = (\operatorname{div}\mathbf{T}^T)\cdot\mathbf{v} + \mathbf{T}^T\cdot\nabla\mathbf{v}\,, \qquad \text{(B 5.2.20)}$$

$$\operatorname{div}(\nabla\mathbf{v})^T = \nabla\operatorname{div}\mathbf{v}\,, \qquad \text{(B 5.2.21)}$$

$$\operatorname{div}(\mathbf{a}\otimes\mathbf{b}) = (\nabla\mathbf{a})\mathbf{b} + (\operatorname{div}\mathbf{b})\mathbf{a}\,, \qquad \text{(B 5.2.22)}$$

$$\operatorname{div}(\mathbf{u}\times\mathbf{v}) = (\nabla\mathbf{u}\times\mathbf{v})\cdot\mathbf{I} - (\nabla\mathbf{v}\times\mathbf{u})\cdot\mathbf{I}\,. \qquad \text{(B 5.2.23)}$$

Using the operation rot (*rotation*), which will be introduced in (Q3) and (Q4), we can instead of (B 5.2.23) write:

$$\operatorname{div}(\mathbf{u}\times\mathbf{v}) = \mathbf{v}\cdot\ \operatorname{rot}\mathbf{u} - \mathbf{u}\cdot\ \operatorname{rot}\mathbf{v}\,, \qquad \text{(B 5.2.24)}$$

$$\operatorname{div}(\phi\mathbf{T}) = \mathbf{T}\nabla\phi + \phi\operatorname{div}\mathbf{T}\,, \qquad \text{(B 5.2.25)}$$

$$\operatorname{div}(\mathbf{TS}) = (\nabla\mathbf{T})\mathbf{S} + \mathbf{T}\operatorname{div}\mathbf{S}\,, \tag{B 5.2.26}$$

$$\operatorname{div}(\mathbf{v}\times\mathbf{T}) = \mathbf{v}\times\operatorname{div}\mathbf{T} + \nabla\mathbf{v}\times\mathbf{T}\,, \tag{B 5.2.27}$$

$$\operatorname{div}(\nabla\mathbf{v})^{+} = \mathbf{0}\,,\ \operatorname{div}(\nabla\mathbf{v}\ \#\ \mathbf{I}) = \mathbf{0}\,. \tag{B 5.2.28}$$

We determine the explicit form of the operation *rotation* (rot) of a *vector* and *tensor field* with the help of the fundamental tensor $\overset{3}{\mathbf{E}}$:

$$\operatorname{rot}\mathbf{v}(\mathbf{x}) = \overset{3}{\mathbf{E}}\,(\nabla\mathbf{v})^{T}\,, \tag{B 5.2.Q3}$$

$$\operatorname{rot}\mathbf{T}(\mathbf{x}) = [\overset{3}{\mathbf{E}}\,(\nabla\mathbf{T})^{\overset{13}{T}}]^{2}\,. \tag{B 5.2.Q4}$$

where rot **v** is a unique vector field, and rot **T** a unique tensor field. The following identities are useful for applications. Given a scalar field ϕ, a vector field **v**, and a tensor field **T**, where all the fields are smooth and can be differentiated, then the following relationships hold:

$$\operatorname{rot}\nabla\phi = \mathbf{0}\,, \tag{B 5.2.29}$$

$$\operatorname{div}\ \operatorname{rot}\mathbf{v} = 0\,, \tag{B 5.2.30}$$

$$\operatorname{rot}\nabla\mathbf{v} = \mathbf{O}\,, \tag{B 5.2.31}$$

$$\operatorname{rot}(\nabla\mathbf{v})^{T} = \nabla\operatorname{rot}\mathbf{v}\,,\ \operatorname{rot}(\phi\mathbf{v}) = \phi\operatorname{rot}\mathbf{v} + \nabla\phi\times\mathbf{v}\,, \tag{B 5.2.32}$$

$$\operatorname{rot}(\mathbf{u}\times\mathbf{v}) = \mathbf{u}\operatorname{div}\mathbf{v} - \nabla\mathbf{v}\mathbf{u} - \mathbf{v}\operatorname{div}\mathbf{u} + \nabla\mathbf{u}\mathbf{v} = \operatorname{div}(\mathbf{u}\otimes\mathbf{v} - \mathbf{v}\otimes\mathbf{u})\,, \tag{B 5.2.33}$$

$$\operatorname{div}\ \operatorname{rot}\mathbf{T} = \operatorname{rot}\ \operatorname{div}\mathbf{T}^{T}\,, \tag{B 5.2.34}$$

$$\operatorname{div}(\operatorname{rot}\mathbf{T})^{T} = \mathbf{0}\,, \tag{B 5.2.35}$$

$$(\operatorname{rot}\operatorname{rot}\mathbf{T})^{T} = \operatorname{rot}\operatorname{rot}\mathbf{T}^{T}\,, \tag{B 5.2.36}$$

$$\operatorname{rot}(\phi\mathbf{I}) = -[\operatorname{rot}(\phi\mathbf{I})]^{T}\,, \tag{B 5.2.37}$$

$$\operatorname{rot}(\mathbf{T}\mathbf{v}) = (\operatorname{rot}\mathbf{T}^{T})\mathbf{v} + (\nabla\mathbf{v})^{T}\times\mathbf{T}\,. \tag{B 5.2.38}$$

When **T** is symmetric, then,

$$\operatorname{rot}\mathbf{T}\cdot\mathbf{I} = 0\,. \tag{B 5.2.39}$$

Given that **T** is a skew-symmetric tensor, and **u** the affiliated axial vector, then it can be proven that:

$$\operatorname{rot}\mathbf{T} = (\operatorname{div}\mathbf{u})\mathbf{I} - \nabla\mathbf{u}\,. \tag{B 5.2.40}$$

The alternative forms of the rotation of the vector field $\mathbf{v}(\mathbf{x})$ are:

$$\operatorname{rot}\mathbf{v}(\mathbf{x}) = \operatorname{div}(\mathbf{I}\times\mathbf{v}) = \mathbf{I}\times\nabla\mathbf{v}\,. \tag{B 5.2.41}$$

Furthermore, we shall introduce the *Laplace-operator* Δ. Given the scalar field $\phi(\mathbf{x})$, then the Laplace-operator applied to $\phi(\mathbf{x})$ is defined as

$$\Delta\phi = \nabla\nabla\phi \cdot \mathbf{I} = \operatorname{div}\operatorname{grad}\phi\,, \tag{B 5.2.Q5}$$

where $\Delta\phi$ is a scalar quantity.

Accordingly, we define the *Laplace-operator* for the *vector field* $\mathbf{v}(\mathbf{x})$ and the *tensor field* $\mathbf{T}(\mathbf{x})$ as:

$$\Delta\mathbf{v} = (\nabla\nabla\mathbf{v})\mathbf{I}\,, \tag{B 5.2.Q6}$$

$$\Delta\mathbf{T} = (\nabla\nabla\mathbf{T})\mathbf{I}\,. \tag{B 5.2.Q7}$$

Here, $\Delta\mathbf{v}$ is a vector field and $\Delta\mathbf{T}$ a tensor field. In connection with the Laplace-operator, the following identities can be stated:

$$\operatorname{div}[\nabla\mathbf{v} \pm (\nabla\mathbf{v})^T] = \Delta\mathbf{v} \pm \nabla\operatorname{div}\mathbf{v}\,, \tag{B 5.2.42}$$

$$\operatorname{rot}\operatorname{rot}\mathbf{v} = \nabla\operatorname{div}\mathbf{v} - \Delta\mathbf{v}\,, \tag{B 5.2.43}$$

$$\Delta\operatorname{tr}\mathbf{T} = \operatorname{tr}\,\Delta\mathbf{T}\,, \tag{B 5.2.44}$$

$$\begin{aligned}\operatorname{rot}\operatorname{rot}\mathbf{T} = {}& -\Delta\mathbf{T} + \nabla\operatorname{div}\mathbf{T} + (\nabla\operatorname{div}\mathbf{T})^T - \nabla\nabla\operatorname{tr}\mathbf{T} + \\ & + \mathbf{I}[\Delta(\operatorname{tr}\mathbf{T}) - \operatorname{div}\operatorname{div}\mathbf{T}]\,,\end{aligned} \tag{B 5.2.45}$$

when $\mathbf{T}$ is symmetrical and $\mathbf{T} = \mathbf{S} - \mathbf{I}\operatorname{tr}\mathbf{S}$, then the following relation is valid

$$\operatorname{rot}\operatorname{rot}\mathbf{T} = -\Delta\mathbf{S} + \nabla\operatorname{div}\mathbf{S} + (\nabla\operatorname{div}\mathbf{S})^T - \mathbf{I}\operatorname{div}\operatorname{div}\mathbf{S}\,. \tag{B 5.2.46}$$

B 5.3 Functions of Vector and Tensor Variables

In Section B 5.2.1, we had used the distance between two points y and x, having the given norm $|\mathbf{x}-\mathbf{y}|$, for the definition of the derivatives of scalar, vector, and tensor fields. The idea behind this procedure being that for scalar, vector, and tensor functions, which supposedly depend on the arbitrary vector and tensor variables, the norm of a variable should likewise be used for the definition of the derivative. In what follows, we shall make use of this idea.

Given a scalar function $\phi(\mathbf{q})$ of a vector $\mathbf{q}$, defined in an open area, then ϕ can be differentiated in this area when a vector function $\mathbf{w}(\mathbf{q})$ exists; so that, with $\mathbf{q}$ and $\mathbf{p}$ as arbitrary vectors, we get

$$\lim_{\mathbf{p}\to\mathbf{q}} \frac{\phi(\mathbf{p}) - \phi(\mathbf{q}) - \mathbf{w}(\mathbf{q})\cdot(\mathbf{p}-\mathbf{q})}{|\mathbf{p}-\mathbf{q}|} = 0 \tag{B 5.3.R1}$$

and we write

$$\mathbf{w}(\mathbf{q}) = \frac{d\phi(\mathbf{q})}{d\mathbf{q}}\,. \tag{B 5.3.1}$$

When the vector $\mathbf{w}(\mathbf{q})$ exists, then it is uniquely determined through (R1). The *differential of the scalar function* $\phi(\mathbf{q})$, namely $d\phi$ for given values $\mathbf{q}$ and $d\mathbf{q}$, is equal to the scalar product

$$d\phi = \frac{d\phi}{d\mathbf{q}} \cdot d\mathbf{q} \,. \tag{B 5.3.R2}$$

We now devote our attention to the vector function $\mathbf{v}(\mathbf{q})$, which is likewise defined in an open area; we can then introduce the derivative of $\mathbf{v}$ with respect to $\mathbf{q}$ through the definition

$$\lim_{\mathbf{p}\to\mathbf{q}} \frac{\mathbf{v}(\mathbf{p}) - \mathbf{v}(\mathbf{q}) - \mathbf{R}(\mathbf{q})(\mathbf{p}-\mathbf{q})}{|\mathbf{p}-\mathbf{q}|} = \mathbf{0} \,. \tag{B 5.3.R3}$$

Thereby we write, for the tensor function $\mathbf{R}(\mathbf{q})$,

$$\mathbf{R} = \frac{d\mathbf{v}}{d\mathbf{q}} \,. \tag{B 5.3.2}$$

The *differential of the vector function* $\mathbf{v}(\mathbf{q})$ can be stated by the linear mapping as

$$d\mathbf{v} = (\frac{d\mathbf{v}}{d\mathbf{q}})d\mathbf{q} \,. \tag{B 5.3.R4}$$

We shall now consider the tensor function $\mathbf{T}(\mathbf{q})$ in an open area. The derivative, with respect to $\mathbf{q}$, as a third-order tensor function $\overset{3}{\mathbf{S}}(\mathbf{q})$ is defined through

$$\lim_{\mathbf{p}\to\mathbf{q}} \frac{\mathbf{T}(\mathbf{p}) - \mathbf{T}(\mathbf{q}) - \overset{3}{\mathbf{S}}(\mathbf{q})(\mathbf{p}-\mathbf{q})}{|\mathbf{p}-\mathbf{q}|} = \mathbf{O} \,. \tag{B 5.3.R5}$$

We state

$$\overset{3}{\mathbf{S}} = \frac{d\mathbf{T}}{d\mathbf{q}} \,. \tag{B 5.3.3}$$

The *differential* $d\mathbf{T}(\mathbf{q})$ is, with (B 5.3.3) determined through linear mapping

$$d\mathbf{T}(\mathbf{q}) = \overset{3}{\mathbf{S}}(\mathbf{q})d\mathbf{q} \,. \tag{B 5.3.R6}$$

We now introduce scalar, vector, and tensor functions which depend on a tensor variable $\mathbf{M}$, namely $\phi(\mathbf{M})$, $\mathbf{v}(\mathbf{M})$ and $\mathbf{T}(\mathbf{M})$. We define the derivatives of these functions, with respect to the variable $\mathbf{M}$ ($\mathbf{M}$ and $\mathbf{N}$ are arbitrary tensors) by

$$\lim_{\mathbf{N}\to\mathbf{M}} \frac{\phi(\mathbf{N}) - \phi(\mathbf{M}) - \mathbf{U}(\mathbf{M}) \cdot (\mathbf{N}-\mathbf{M})}{|\mathbf{N}-\mathbf{M}|} = 0 \,, \tag{B 5.3.R7}$$

$$\lim_{\mathbf{N}\to\mathbf{M}} \frac{\mathbf{v}(\mathbf{N}) - \mathbf{v}(\mathbf{M}) - \overset{3}{\mathbf{V}}(\mathbf{M})(\mathbf{N}-\mathbf{M})}{|\mathbf{N}-\mathbf{M}|} = \mathbf{0} \,, \tag{B 5.3.R8}$$

$$\lim_{\mathbf{N}\to\mathbf{M}} \frac{\mathbf{T}(\mathbf{N}) - \mathbf{T}(\mathbf{M}) - \overset{4}{\mathbf{W}}(\mathbf{M})(\mathbf{N}-\mathbf{M})}{|\mathbf{N}-\mathbf{M}|} = \mathbf{O} \tag{B 5.3.R9}$$

and write for the derivatives

$$\mathbf{U}(\mathbf{M}) = \frac{d\phi(\mathbf{M})}{d\mathbf{M}} \,, \quad \overset{3}{\mathbf{V}}(\mathbf{M}) = \frac{d\mathbf{v}(\mathbf{M})}{d\mathbf{M}} \,, \quad \overset{4}{\mathbf{W}}(\mathbf{M}) = \frac{d\mathbf{T}(\mathbf{M})}{d\mathbf{M}} \,. \tag{B 5.3.4}$$

The *differentials of the scalar, vector, and tensor functions* $\phi(\mathbf{M})$, $\mathbf{v}(\mathbf{M})$, and $\mathbf{T}(\mathbf{M})$ for the given values are defined by

$$d\phi = \frac{d\phi}{d\mathbf{M}} \cdot d\mathbf{M} \,, \tag{B 5.3.R10}$$

$$d\mathbf{v} = \frac{d\mathbf{v}}{d\mathbf{M}} d\mathbf{M} \,, \tag{B 5.3.R11}$$

$$d\mathbf{T} = \frac{d\mathbf{T}}{d\mathbf{M}} d\mathbf{M} \,. \tag{B 5.3.R12}$$

Finally, we consider those scalar, vector, and tensor functions which depend on real scalar variables and on vector and tensor variables

$$\phi = \phi(\alpha, \beta, ..., \mathbf{q}, \mathbf{r}, ..., \mathbf{M}, \mathbf{P}, ...) \,, \tag{B 5.3.5}$$

$$\mathbf{v} = \mathbf{v}(\alpha, \beta, ..., \mathbf{q}, \mathbf{r}, ..., \mathbf{M}, \mathbf{P}, ...) \,, \tag{B 5.3.6}$$

$$\mathbf{T} = \mathbf{T}(\alpha, \beta, ..., \mathbf{q}, \mathbf{r}, ..., \mathbf{M}, \mathbf{P}, ...) \,. \tag{B 5.3.7}$$

The *partial differentiation* of these functions, with respect to the vector and tensor variables (the partial differentiation with respect to the scalar variables are determined in Section B 5.1) are the following definitions, where **p** and **q** are arbitrary vectors and **N** and **M** arbitrary tensors. Here, we shall restrict ourselves to the statement of the definitions for the partial derivatives of the scalar functions ϕ; corresponding definitions are valid for the partial differentiation of the vector and tensor functions **v** (B 5.3.6) and **T**(B 5.3.7).

$$\lim_{\mathbf{p}\to\mathbf{q}} \frac{\phi(\mathbf{p}, ...) - \phi(\mathbf{q}, ...) - \mathbf{u}(\mathbf{q}, ...) \cdot (\mathbf{p} - \mathbf{q})}{|\mathbf{p} - \mathbf{q}|} = 0 \,, \tag{B 5.3.R13}$$

$$\lim_{\mathbf{N}\to\mathbf{M}} \frac{\phi(\mathbf{N}, ...) - \phi(\mathbf{M}, ...) - \mathbf{U}(\mathbf{M}, ...) \cdot (\mathbf{N} - \mathbf{M})}{|\mathbf{N} - \mathbf{M}|} = 0 \,. \tag{B 5.3.R14}$$

We shall denote the partial derivations (the vector and tensor functions **u** and **U**) through

$$\mathbf{u} = \frac{\partial\phi}{\partial\mathbf{q}} \text{ and } \mathbf{U} = \frac{\partial\phi}{\partial\mathbf{M}} \,. \tag{B 5.3.8}$$

Furthermore, the partial differentiation of the vector and tensor functions **v** and **T**, with respect to the vector and tensor variables **q** and **M**, can be written as:

$$\frac{\partial\mathbf{v}}{\partial\mathbf{q}} \text{ and } \frac{\partial\mathbf{v}}{\partial\mathbf{M}} \,, \quad \frac{\partial\mathbf{T}}{\partial\mathbf{q}} \text{ and } \frac{\partial\mathbf{T}}{\partial\mathbf{M}} \,. \tag{B 5.3.9}$$

These are second- and third- as well as fourth-order tensor functions. For the *total differentials*, we find

$$d\phi = \frac{\partial\phi}{\partial\alpha} d\alpha + ... + \frac{\partial\phi}{\partial\mathbf{q}} \cdot d\mathbf{q} + ... + \frac{\partial\phi}{\partial\mathbf{M}} \cdot d\mathbf{M} + ... \,, \tag{B 5.3.R15}$$

$$d\mathbf{v} = \frac{\partial\mathbf{v}}{\partial\alpha} d\alpha + ... + \frac{\partial\mathbf{v}}{\partial\mathbf{q}} d\mathbf{q} + ... + \frac{\partial\mathbf{v}}{\partial\mathbf{M}} d\mathbf{M} + ... \,, \tag{B 5.3.R16}$$

$$d\mathbf{T} = \frac{\partial \mathbf{T}}{\partial \alpha} d\alpha + \ldots + \frac{\partial \mathbf{T}}{\partial \mathbf{q}} d\mathbf{q} + \ldots + \frac{\partial \mathbf{T}}{\partial \mathbf{M}} d\mathbf{M} + \ldots . \tag{B 5.3.R17}$$

We shall go no further into the discussion of higher derivatives of scalar, vector, and tensor functions which depend on scalar, vector, and tensor variables. These can be formed immediately according to the guidelines shown in Section B 5.2.2. The given definitions allow the development of calculation rules. Thereby we can dispense with those rules which contain the derivative with respect to a vector because the rules (B 5.2.8) through (B 5.2.17) can be analogously applied to arbitrary scalar, vector, and tensor functions. Consequently, we restrict ourselves to a few important rules for the differentiation of product terms of the scalar, vector, and tensor functions ϕ, ψ and $\mathbf{v}$, $\mathbf{u}$ as well as $\mathbf{T}$, $\mathbf{S}$ which, beside others, depend on the tensor variable $\mathbf{M}$, with respect to the variable $\mathbf{M}$:

$$\frac{\partial(\phi\psi)}{\partial \mathbf{M}} = \phi \frac{\partial \psi}{\partial \mathbf{M}} + \psi \frac{\partial \phi}{\partial \mathbf{M}} , \tag{B 5.3.10}$$

$$\frac{\partial(\phi \mathbf{v})}{\partial \mathbf{M}} = \mathbf{v} \otimes \frac{\partial \phi}{\partial \mathbf{M}} + \phi \frac{\partial \mathbf{v}}{\partial \mathbf{M}} , \tag{B 5.3.11}$$

$$\frac{\partial(\phi \mathbf{T})}{\partial \mathbf{M}} = \mathbf{T} \otimes \frac{\partial \phi}{\partial \mathbf{M}} + \phi \frac{\partial \mathbf{T}}{\partial \mathbf{M}} , \tag{B 5.3.12}$$

$$\frac{\partial(\mathbf{u} \cdot \mathbf{v})}{\partial \mathbf{M}} = \mathbf{v} \frac{\partial \mathbf{u}}{\partial \mathbf{M}} + \mathbf{u} \frac{\partial \mathbf{v}}{\partial \mathbf{M}} , \tag{B 5.3.13}$$

$$\frac{\partial(\mathbf{T} \cdot \mathbf{S})}{\partial \mathbf{M}} = (\frac{\partial \mathbf{T}}{\partial \mathbf{M}})^T \mathbf{S} + (\frac{\partial \mathbf{S}}{\partial \mathbf{M}})^T \mathbf{T} . \tag{B 5.3.14}$$

Furthermore, the following derivative rules are valid:

$$\frac{\partial \mathbf{M}}{\partial \mathbf{M}} = \overset{4}{\mathbf{I}} , \quad \frac{\partial (\mathbf{M} \cdot \mathbf{I})\mathbf{I}}{\partial \mathbf{M}} = \overset{4}{\bar{\mathbf{I}}} , \quad \frac{\partial \mathbf{M}^T}{\partial \mathbf{M}} = \overset{4}{\bar{\bar{\mathbf{I}}}} , \tag{B 5.3.15}$$

$$\frac{\partial \mathrm{I}_{\mathbf{M}}}{\partial \mathbf{M}} = \mathbf{I} , \quad \frac{\partial \mathrm{II}_{\mathbf{M}}}{\partial \mathbf{M}} = \mathbf{M} \# \mathbf{I} , \quad \frac{\partial \mathrm{III}_{\mathbf{M}}}{\partial \mathbf{M}} = \overset{+}{\mathbf{M}} , \quad \frac{\partial \overset{A}{\mathbf{t}} (\mathbf{M})}{\partial \mathbf{M}} = - \frac{1}{2} \overset{3}{\mathbf{E}} . \tag{B 5.3.16}$$

The definitions given in this section make it possible for us (the definitions given in Section B 5.1) to state the stepwise differentiation (chain rule) for composed functions. So, for example, it holds for the scalar function $\phi(\mathbf{q}, \mathbf{M})$, where $\mathbf{q}$ and $\mathbf{M}$ depend on the real scalar variable α, that

$$\frac{d\phi}{d\alpha} = \frac{\partial \phi}{\partial \mathbf{q}} \cdot \frac{d\mathbf{q}}{d\alpha} + \frac{\partial \phi}{\partial \mathbf{M}} \cdot \frac{d\mathbf{M}}{d\alpha} . \tag{B 5.3.17}$$

Analogously, we can give the derivative (with respect to α for the vector and tensor functions $\mathbf{v}(\mathbf{q}, \mathbf{M})$, when $\mathbf{q}$ and $\mathbf{m}$ are functions of α) as follows:

$$\frac{d\mathbf{v}}{d\alpha} = \frac{\partial \mathbf{v}}{\partial \mathbf{q}} \frac{d\mathbf{q}}{d\alpha} + \frac{\partial \mathbf{v}}{\partial \mathbf{M}} \frac{d\mathbf{M}}{d\alpha} , \tag{B 5.3.18}$$

$$\frac{d\mathbf{T}}{d\alpha} = \frac{\partial \mathbf{T}}{\partial \mathbf{q}} \frac{d\mathbf{q}}{d\alpha} + \frac{\partial \mathbf{T}}{\partial \mathbf{M}} \frac{d\mathbf{M}}{d\alpha} . \tag{B 5.3.19}$$

This procedure can also be applied to spatial variable functions. Consider the scalar, vector, and tensor functions $\phi(\alpha, \mathbf{q}, \mathbf{M})$, $\mathbf{v}(\alpha, \mathbf{q}, \mathbf{M})$, and $\mathbf{T}(\alpha, \mathbf{q}, \mathbf{M})$, where α, $\mathbf{q}$ and $\mathbf{M}$ should be spatial variable functions. Then we get, for the gradients of these functions,

$$\nabla\phi = \frac{\partial\phi}{\partial\alpha}\nabla\alpha + (\nabla\mathbf{q})^T\frac{\partial\phi}{\partial\mathbf{q}} + (\nabla\mathbf{M})^{\overset{13}{T}}(\frac{\partial\phi}{\partial\mathbf{M}})^T , \tag{B 5.3.20}$$

$$\nabla\mathbf{v} = \frac{\partial\mathbf{v}}{\partial\alpha} \otimes \nabla\alpha + \frac{\partial\mathbf{v}}{\partial\mathbf{q}}\nabla\mathbf{q} + (\frac{\partial\mathbf{v}}{\partial\mathbf{M}}\nabla\mathbf{M})^2 , \tag{B 5.3.21}$$

$$\nabla\mathbf{T} = \frac{\partial\mathbf{T}}{\partial\alpha} \otimes \nabla\alpha + (\frac{\partial\mathbf{T}}{\partial\mathbf{q}}\nabla\mathbf{q})^3 + (\frac{\partial\mathbf{T}}{\partial\mathbf{M}}\nabla\mathbf{M})^3 . \tag{B 5.3.22}$$

B 5.4
Integral Theorems

By the evaluation of the conservation laws in mechanics, it is necessary to transform the surface integrals into volume integrals, and also to transform line integrals into surface integrals, when one takes the global view by the formulation of the conservation theorems. These transformations cover scalar, vector, and tensor fields. The integral theorems also play a role in other areas of geometry and mechanics, for example, by the numerical calculation of geometrical moments of inertia and mass moments of inertia. In the following section, we want to develop the most important integral theorems.

B 5.4.1
Transformation of Surface Integrals into Volume Integrals

Firstly, we transform the surface integral of the tensor product of the vector field $\mathbf{u}(\mathbf{x})$, having the area vector $\mathbf{d}a$ equal to $da\mathbf{n}$ (where $\mathbf{n}$ is the normal unit vector at the surface and da the surface element), into a volume integral.

$$\int\limits_{\partial U} \mathbf{u} \otimes \mathbf{d}a = \int\limits_{U} \frac{\partial\mathbf{u}}{\partial\mathbf{x}}\, dv \quad \text{or} \quad \int\limits_{\partial U} \mathbf{u} \otimes \mathbf{n}\, da = \int\limits_{U} \frac{\partial\mathbf{u}}{\partial\mathbf{x}}\, dv , \tag{B 5.4.1}$$

where U is a closed volume, and ∂U its surface.

More integrals can easily be developed from this integral theorem, with respect to the transformation of surface integrals into volume integrals which contain a scalar field $\phi(\mathbf{x})$, a vector field $\mathbf{u}(\mathbf{x})$, and a tensor field $\mathbf{T}(\mathbf{x})$:

$$\int\limits_{\partial U} \phi\mathbf{n}\, da = \int\limits_{U} \frac{\partial\phi}{\partial\mathbf{x}}\, dv , \tag{B 5.4.2}$$

$$\int\limits_{\partial U} \mathbf{u} \cdot \mathbf{n}\, da = \int\limits_{U} \operatorname{div}\mathbf{u}\, dv , \tag{B 5.4.3}$$

$$\int_{\partial U} \mathbf{n} \times \mathbf{u} \, da = \int_{U} \operatorname{rot} \mathbf{u} \, dv \,, \tag{B 5.4.4}$$

$$\int_{\partial U} \mathbf{T}\mathbf{n} \, da = \int_{U} \operatorname{div} \mathbf{T} \, dv \,, \tag{B 5.4.5}$$

$$\int_{\partial U} \mathbf{u} \times \mathbf{T}\mathbf{n} \, da = \int_{U} (\mathbf{u} \times \operatorname{div} \mathbf{T} + \nabla\mathbf{u} \times \mathbf{T}) \, dv \,. \tag{B 5.4.6}$$

In the literature, the transformation (B 5.4.3) is also called as the *Gauß integral theorem*. From (B 5.4.2), with ϕ as a constant field, we directly get the so-called surface theorem:

$$\int_{\partial U} \mathbf{n} \, da = \mathbf{0} \,, \quad \int_{\partial U} \mathbf{da} = \mathbf{0} \,. \tag{B 5.4.7}$$

B 5.4.2
Transformation of Line Integrals into Surface Integrals

The premise of Section B 5.4.1 is valid. The aim of this chapter is to transform the closed line integrals $\mathbf{dx} \otimes \oint_L \mathbf{u}$ into a surface integral. The line L will be described by the position vector $\mathbf{x}$. The derivation yields (see de Boer, 1982).

$$\oint_L \mathbf{dx} \otimes \mathbf{u} = \int_A \mathbf{da} \times (\frac{\partial \mathbf{u}}{\partial \mathbf{x}})^T = \int_A \mathbf{n} \times (\frac{\partial \mathbf{u}}{\partial \mathbf{x}})^T \, da \,. \tag{B 5.4.8}$$

In addition, we give the following integral theorems derived from (B 5.4.8):

$$\oint_L \phi \, \mathbf{dx} = \int_A \mathbf{n} \times \frac{\partial \phi}{\partial \mathbf{x}} \, da \,, \tag{B 5.4.9}$$

$$\oint_L \mathbf{u} \cdot \mathbf{dx} = \int_A \operatorname{rot} \mathbf{u} \cdot \mathbf{n} \, da \,, \tag{B 5.4.10}$$

$$\oint_L \mathbf{u} \times \mathbf{dx} = \int_A \operatorname{div} \mathbf{u} \, \mathbf{n} \, da - \int_A (\frac{\partial \mathbf{u}}{\partial \mathbf{x}})^T \mathbf{n} \, da \,, \tag{B 5.4.11}$$

$$\oint_L \mathbf{T}\mathbf{dx} = \int_A (\operatorname{rot} \mathbf{T})^T \, \mathbf{n} \, da \,, \tag{B 5.4.12}$$

$$\oint_L \mathbf{u} \times \mathbf{T}\mathbf{dx} = \int_A \{[\mathbf{u} \times (\operatorname{rot} \mathbf{T})^T]\mathbf{n} + (\mathbf{n} \times \mathbf{T}^T)^T \times \nabla\mathbf{u}\} \, da \,. \tag{B 5.4.13}$$

The relationship (B 5.4.10) is known as the *Stokes integral theorem*. For every closed surface, the surface integrals disappear in (B 5.4.8) through (B 5.4.12), which directly follow from the integral theorems in Section 5.4.1 in association with the calculation rules in Section B 5.2.3. From (B 5.4.10), we can read that, for every closed curve, $\oint_L \mathbf{u} \cdot d\mathbf{x}$ is equal to zero, when rot **u** vanishes within a specified area. Then, according to (B 5.2.31), it is necessary that **u** is equal to grad ϕ, where ϕ represents a scalar field. Given the vector field **u** as equal to grad ϕ, then rot **u** is equal to zero (necessary and sufficient). As in hydrodynamics, one calls $\oint_L \mathbf{u} \cdot d\mathbf{x}$, in general, the *circulation*.

Appendix C: Geometric Representation of the Principal Stresses, the Stress Invariants and the Mohr-Coulomb Theory

C 1. Preliminaries

In order to study the directions and the norms of the stress and strain rate vectors and to compare them with those observed in experiments, it is useful to represent these vectors geometrically. In the treatment to follow, this will be worked out for the stress and for the shifted stress (by the backstress tensor). The results can be easily transferred to the principal strain rates.

Assuming the principal stresses s_1, s_2, and s_3 of the stress tensor S to be coordinates of the three-dimensional stress space with the orthogonal base vectors $\mathbf{e}_1, \mathbf{e}_2$, and $\mathbf{e}_3$, an arbitrary stress state can be represented by a stress mapping point P. This point P is fixed by the position vector

$$\mathbf{s} = s_i \; \mathbf{e}_i \qquad \text{(C 1.1)}$$

(see Fig. C 1.1). Generally, the triaxial and the octahedral plane, both derived from the principal stress space, are used to visualize the directions and norms of the stress and strain rate vectors. In order to obtain these planes, it is necessary

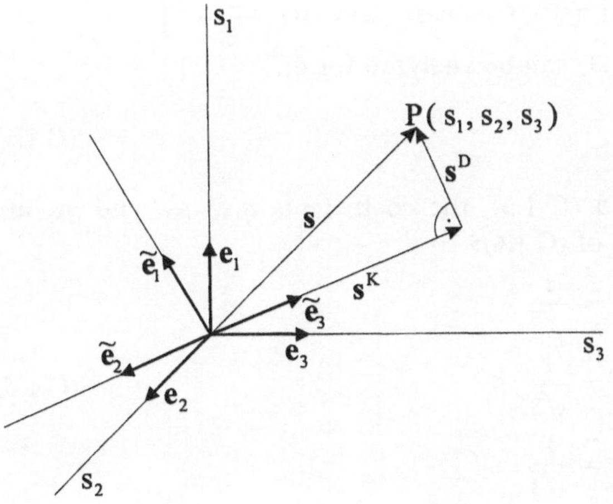

Fig. C 1.1. Stress point in the principal stress space

to introduce the so-called hydrostatic axis, which is determined by the unit vector

$$\tilde{\mathbf{e}}_3 = \frac{1}{\sqrt{3}} (\mathbf{e}_1 + \mathbf{e}_2 + \mathbf{e}_3) . \tag{C 1.2}$$

For the further considerations, it is convenient to rotate the given base system $\mathbf{e}_1$, $\mathbf{e}_2$, and $\mathbf{e}_3$ in such a way that $\mathbf{e}_3$ coincides with $\tilde{\mathbf{e}}_3$. The new base system can be gained by an orthogonal transformation with the orthogonal tensor $\mathbf{Q}$:

$$\tilde{\mathbf{e}}_i = \mathbf{Q}\, \mathbf{e}_i . \tag{C 1.3}$$

The coefficients of $\mathbf{Q}$ can be generally determined by the angles α_1, α_2, and α_3. Thereby α_1 denotes the rotationangle of the base system around the $\mathbf{e}_1$-axis, α_2 the rotationangle around the new $\mathbf{e}_2$- axis and, finally, α_3 the rotation around the new $\mathbf{e}_3$-axis, obtained from the α_2-rotation. Hence, in our case, the base system $\tilde{\mathbf{e}}_i$ is fixed by the angles α_1 and α_2; the angle α_3 is equal to zero.

The coefficients of the orthogonal tensor are given, in general, by (see Magnus, 1971):

$$q_{mn} = \begin{bmatrix} \begin{array}{c} q_{11} \\ \cos\alpha_3 \cos\alpha_2 - \end{array} & \begin{array}{c} q_{12} \\ -\sin\alpha_3 \cos\alpha_2 \end{array} & \begin{array}{c} q_{13} \\ \sin\alpha_2 \end{array} \\[2ex] \begin{array}{c} q_{21} \\ \cos\alpha_3 \sin\alpha_2 \sin\alpha_1 + \sin\alpha_3 \cos\alpha_1 - \end{array} & \begin{array}{c} q_{22} \\ -\sin\alpha_3 \sin\alpha_2 \sin\alpha_1 + \cos\alpha_3 \cos\alpha_1 \end{array} & \begin{array}{c} q_{23} \\ -\cos\alpha_2 \sin\alpha_1 \end{array} \\[2ex] \begin{array}{c} q_{31} \\ \cos\alpha_3 \sin\alpha_2 \cos\alpha_1 + \sin\alpha_3 \sin\alpha_1 + \end{array} & \begin{array}{c} q_{32} \\ +\sin\alpha_3 \sin\alpha_2 \cos\alpha_1 + \cos\alpha_3 \sin\alpha_1 \end{array} & \begin{array}{c} q_{33} \\ \cos\alpha_2 \cos\alpha_1 \end{array} \end{bmatrix} . \tag{C 1.4}$$

With (C 1.4), the relation (C 1.3) can be analyzed for $\tilde{\mathbf{e}}_3$:

$$\begin{aligned} \tilde{\mathbf{e}}_3 &= \mathbf{Q}\, \mathbf{e}_3 , \\ \tilde{\mathbf{e}}_3 &= q_{m3}\, \mathbf{e}_m . \end{aligned} \tag{C 1.5}$$

By comparison of (C 1.2) and (C 1.5), the coefficients q_{13}, q_{23}, and q_{33} are obtained, under consideration of (C 1.4):

$$\begin{aligned} q_{13} &= \sin\alpha_2 &&= \frac{1}{\sqrt{3}} , \\ q_{23} &= -\cos\alpha_2 \sin\alpha_1 &&= \frac{1}{\sqrt{3}} , \\ q_{33} &= \cos\alpha_2 \cos\alpha_1 &&= \frac{1}{\sqrt{3}} . \end{aligned} \tag{C 1.6}$$

The relations (C 1.6) yield the values of the angles α_1 and α_2:

$$\alpha_1 = -45^\circ, \qquad \alpha_2 = 35.26^\circ . \tag{C 1.7}$$

Thus, the coefficients of the orthogonal tensor (C 1.4) take the following values:

$$q_{mn} = \begin{bmatrix} \frac{2}{\sqrt{6}} & 0 & \frac{1}{\sqrt{3}} \\ -\frac{1}{\sqrt{6}} & \frac{1}{\sqrt{2}} & \frac{1}{\sqrt{3}} \\ -\frac{1}{\sqrt{6}} & -\frac{1}{\sqrt{2}} & \frac{1}{\sqrt{2}} \end{bmatrix} . \tag{C 1.8}$$

With (C 1.8), it follows from (C 1.3):

$$\begin{aligned} \tilde{\mathbf{e}}_1 &= \frac{1}{\sqrt{6}} (2\,\mathbf{e}_1 - \mathbf{e}_2 - \mathbf{e}_3) , \\ \tilde{\mathbf{e}}_2 &= \frac{1}{\sqrt{2}} (\mathbf{e}_2 - \mathbf{e}_3) , \\ \tilde{\mathbf{e}}_3 &= \frac{1}{\sqrt{3}} (\mathbf{e}_1 + \mathbf{e}_2 + \mathbf{e}_3) . \end{aligned} \tag{C 1.9}$$

With the above results, it is easy to introduce two distinct plane surfaces, namely the triaxial and the octahedral planes.

C 2. Triaxial Plane

The triaxial plane is defined as a plane containing one principal stress-axis (e.g., the $\mathbf{e}_1$-axis) and the hydrostatical axis, see Fig. C 2.1. In Fig. C 2.1 new base vectors are introduced: $\hat{\mathbf{e}}_1$ (equal to $\mathbf{e}_1$), $\hat{\mathbf{e}}_2$, and $\hat{\mathbf{e}}_3$ which lie in the $\mathbf{e}_2$-$\mathbf{e}_3$-plane. Moreover, the base vector $\mathbf{e}_3$ follows the trace of the hydrostatic axis in the $\mathbf{e}_2$-$\mathbf{e}_3$-plane. Thus,

$$\begin{aligned} \hat{\mathbf{e}}_3 &= -\sin\alpha_1\,\mathbf{e}_2 + \cos\alpha_1\,\mathbf{e}_3 , \\ \hat{\mathbf{e}}_2 &= \cos\alpha_1\,\mathbf{e}_2 + \sin\alpha_1\,\mathbf{e}_3 . \end{aligned} \tag{C 2.1}$$

With

$$\sin\alpha_1 = -\frac{1}{\sqrt{2}} , \qquad \cos\alpha_1 = \frac{1}{\sqrt{2}} , \tag{C 2.2}$$

see $(\text{C }1.7)_1$

$$\begin{aligned} \hat{\mathbf{e}}_3 &= \frac{1}{\sqrt{2}}\,\mathbf{e}_2 + \frac{1}{\sqrt{2}}\,\mathbf{e}_3 , \\ \hat{\mathbf{e}}_2 &= \frac{1}{\sqrt{2}}\,\mathbf{e}_2 - \frac{1}{\sqrt{2}}\,\mathbf{e}_3 , \end{aligned} \tag{C 2.3}$$

are obtained.

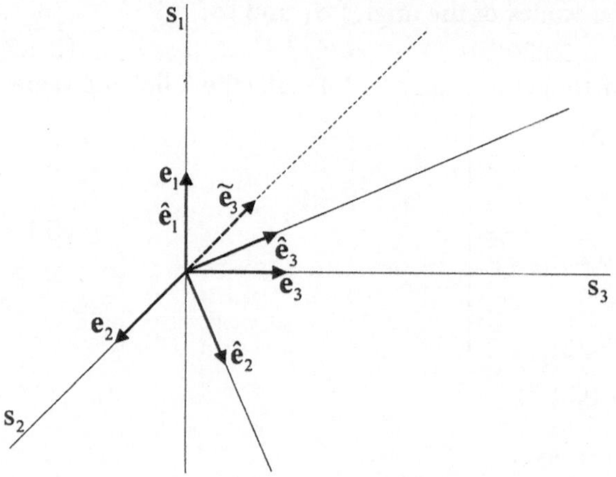

Fig. C 2.1. New base system

The stress vector $\mathbf{s}$ can be represented in the $\hat{\mathbf{e}}_1$, $\hat{\mathbf{e}}_2$, $\hat{\mathbf{e}}_3$-base system:

$$\mathbf{s} = \hat{s}_1\hat{\mathbf{e}}_1 + \hat{s}_2\hat{\mathbf{e}}_2 + \hat{s}_3\hat{\mathbf{e}}_3 \,. \tag{C 2.4}$$

The unknowns coefficients are calculated with the help of following scalar products:

$$\hat{s}_2 = \mathbf{s} \cdot \hat{\mathbf{e}}_2 \,, \qquad \hat{s}_3 = \mathbf{s} \cdot \hat{\mathbf{e}}_3 \,. \tag{C 2.5}$$

With (C 2.3), Eqn. (C 2.5) yields

$$\begin{aligned} \hat{s}_2 &= \frac{1}{\sqrt{2}}\,(s_2 - s_3)\,, \\ \hat{s}_3 &= \frac{1}{\sqrt{2}}\,(s_2 + s_3)\,. \end{aligned} \tag{C 2.6}$$

Hence, from (C 2.4), and under consideration of $\hat{s}_1 = s_1$

$$\mathbf{s} = s_1\,\hat{\mathbf{e}}_1 + \frac{s_2 - s_1}{\sqrt{2}}\,\hat{\mathbf{e}}_2 + \frac{s_2 - s_1}{\sqrt{2}}\,\hat{\mathbf{e}}_3 \tag{C 2.7}$$

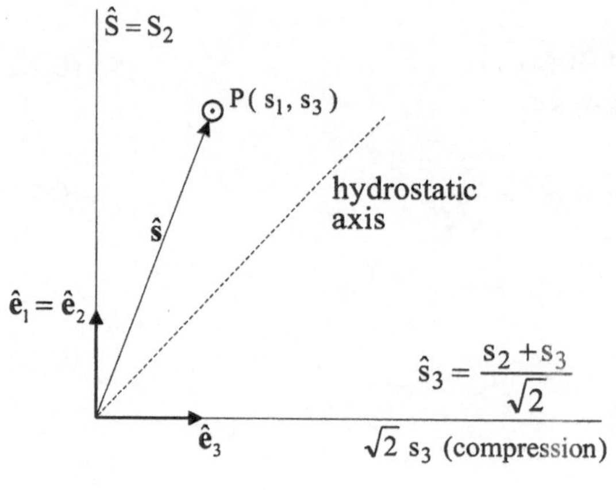

Fig. C 2.2. Traxial plane

is obtained (see Fig. C 2.2). In the case of a compression stress state, (C 2.7) simplifies to

$$\mathbf{s} = s_1\, \hat{\mathbf{e}}_1 + \sqrt{2}\, s_3\, \hat{\mathbf{e}}_3\,. \qquad \text{(C 2.8)}$$

C 3.
Octahedral Plane

The octahedral plane is spread by the $\tilde{\mathbf{e}}_1$ and $\tilde{\mathbf{e}}_2$-base vectors (see Fig. C 1.1). The stress vector $\mathbf{s}$ can be represented by the vectors $\mathbf{s}^K$ and $\mathbf{s}^D$ (see Fig. C 1.1), whereby the vector $\mathbf{s}^K$ lies in the hydrostatic axis and $\mathbf{s}^D$ perpendicular to it:

$$\mathbf{s} = \mathbf{s}^K + \mathbf{s}^D, \qquad \text{or} \qquad \mathbf{s} = s^K\, \tilde{\mathbf{e}}_3 + \tilde{s}_1^D\, \tilde{\mathbf{e}}_1 + \tilde{s}_2^D\, \tilde{\mathbf{e}}_2\,. \qquad \text{(C 3.1)}$$

The unknown coefficients s^K, $\tilde{s}_1^D$, and $\tilde{s}_2^D$ can be determinated by the following scalar products:

$$s^K = \mathbf{s} \cdot \tilde{\mathbf{e}}_3\,, \qquad \tilde{s}_1^D = \mathbf{s} \cdot \tilde{\mathbf{e}}_1\,, \qquad \tilde{s}_2^D = \mathbf{s} \cdot \tilde{\mathbf{e}}_2\,. \qquad \text{(C 3.2)}$$

Under consideration of (C 1.1) and (C 1.9), the relations (C 3.2) yield:

$$\begin{aligned} s^K &= \frac{1}{\sqrt{3}}\,(s_1 + s_2 + s_3)\,,\\ \tilde{s}_1^D &= \frac{1}{\sqrt{6}}\,(2\, s_1^D - s_2^D - s_3^D)\\ &= \frac{1}{\sqrt{6}}\,(2\, s_1 - s_2 - s_3)\,, \qquad \text{(C 3.3)}\\ \tilde{s}_2^D &= \frac{1}{\sqrt{2}}\,(s_2^D - s_3^D)\\ &= \frac{1}{\sqrt{2}}\,(s_2 - s_3)\,. \end{aligned}$$

The vector $\mathbf{s}^D$ represents the deviatoric stress state. This can easily be proven by considering (C 3.1), (C 1.9), and (C 3.3)

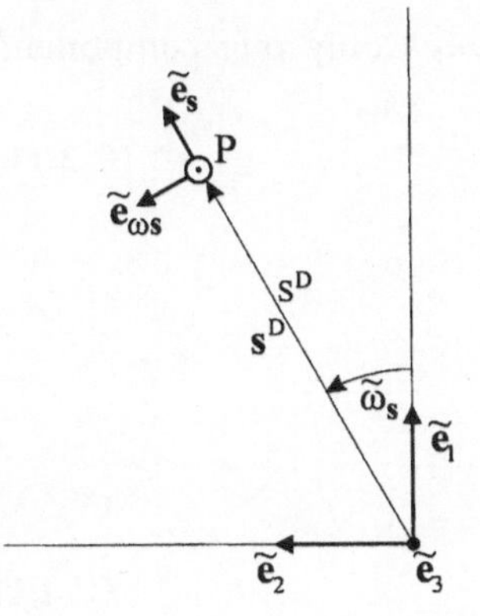

Fig. C 3.1. The octahedral plane

$$\mathbf{s}^D = \mathbf{s} - \frac{1}{\sqrt{3}} (s_1 + s_2 + s_3)\, \tilde{\mathbf{e}}_3 \,,$$
$$\mathbf{s}^D = s_i^D \, \mathbf{e}_i \,, \qquad s_i^D = s_i - \frac{1}{3} (s_1 + s_2 + s_3) \,. \tag{C 3.4}$$

The norm of the vector $\mathbf{s}^D$ is given by

$$|\mathbf{s}^D| = \mathbf{s}^D = \sqrt{s_i^D \, s_i^D} \,, \tag{C 3.5}$$

and the angle $\tilde{\omega}_s$ between the stress vector $\mathbf{s}^D$ and the base vector $\tilde{\mathbf{e}}_1$ (see Fig. C 3.1) can be calculated from:

$$\cos \tilde{\omega}_s = \frac{\mathbf{s}^D \cdot \tilde{\mathbf{e}}_1}{|\mathbf{s}^D||\tilde{\mathbf{e}}_1|} \,,$$
$$\cos \tilde{\omega}_s = \frac{1}{\sqrt{6}} \frac{(2\, s_1 - s_2 - s_3)}{\sqrt{s_i^D \, s_i^D}}$$
$$= \sqrt{\frac{3}{2}} \frac{s_1^D}{\sqrt{s_i^D \, s_i^D}} \,, \tag{C 3.6}$$

where

$$0 \le \tilde{\omega}_s \le 60° \,. \tag{C 3.7}$$

In two papers on experimental results concerning the influence of the intermediate principal stress on the yield limit of metals, Lode (1925, 1926) introduced, without any further foundations, the parameter

$$\mu = \frac{(2\, s_2 - s_1 - s_3)}{s_1 - s_3} \,. \tag{C 3.8}$$

It can easily be shown by simple trigonometry that

$$\mu = \sqrt{3} \operatorname{cotg} (120° - \tilde{\omega}_s) \,. \tag{C 3.9}$$

It is sometimes convenient to use a polar coordinate system. The base vectors $\tilde{\mathbf{e}}_s$ and $\tilde{\mathbf{e}}_{\omega s}$ (see Fig. C 3.1), after some calculations, are as follows

$$\tilde{\mathbf{e}}_s = \cos \tilde{\omega}_s \, \tilde{\mathbf{e}}_1 + \sin \tilde{\omega}_s \, \tilde{\mathbf{e}}_2 \,,$$
$$\tilde{\mathbf{e}}_{\omega s} = -\, s^D \sin \tilde{\omega}_s \, \tilde{\mathbf{e}}_1 + s^D \cos \tilde{\omega}_s \, \tilde{\mathbf{e}}_2 \,. \tag{C 3.10}$$

The stress vector $\mathbf{s}^D$ in the octahedral plane contains only one component, namely the component in the $\tilde{\mathbf{e}}_s$ direction:

$$\mathbf{s}^D = s^D \, \tilde{\mathbf{e}}_s \,, \tag{C 3.11}$$

with $\tilde{\mathbf{e}}_s$, given in (C 3.10).

C 4. Geometric Representation of Yield Functions

In order to adjust yield functions of the form

$$F = F(\mathrm{I_S}, \mathrm{II_{S^D}}, \mathrm{III_{S^D}}) \quad \text{or} \quad F = F(\mathrm{I_S}, \mathrm{II_S}, \mathrm{III_S}) \tag{C 4.1}$$

(where I_S, II_{S^D}, and III_{S^D} are the first invariant of the stress tensor S and the second and third invariants of the deviator of S, as well as II_S and III_S being the second and third invariants of S) to test results, the graphic representation in the principal stress space is very important. Moreover, it has become usual to study the properties of such yield functions at two distinct planes, namely the triaxial and the octahedral planes, already considered in the last two sections. In order to evaluate the graphic representation of (C 4.1), it is necessary to express the invariants I_S, II_{S^D}, and III_{S^D} by geometric quantities, namely straight lines and angles.

The first invariant of the stress tensor

$$I_S = \mathbf{S} \cdot \mathbf{I} = s_1 + s_2 + s_3 \tag{C 4.2}$$

is represented by the component of the stress vector at the hydrostatic axis, see (C 3.3)$_1$:

$$s^K = \frac{1}{\sqrt{3}} I_S . \tag{C 4.3}$$

The second invariant of the stress deviator

$$II_{S^D} = \frac{1}{2} \mathbf{S}^D \cdot \mathbf{S}^D = \frac{1}{2} s_i^D s_i^D \tag{C 4.4}$$

can be expressed by the norm of the stress component $|\mathbf{s}^D|$, see (C 3.5). A comparison of (C 4.4) with (C 3.5) leads to

$$|\mathbf{s}^D| = s^D = \sqrt{2\, II_{S^D}} . \tag{C 4.5}$$

Finally, the third invariant of the stress deviator

$$III_{S^D} = \frac{1}{3} \mathbf{S}^D \mathbf{S}^D \cdot \mathbf{S}^D = \frac{1}{3} \left[\left(s_1^D\right)^3 + \left(s_2^D\right)^3 + \left(s_3^D\right)^3 \right] = s_1^D s_2^D s_3^D \tag{C 4.6}$$

can be represented by the angle $\tilde{\omega}_s$. Eqn. (C 4.6) can be reformulated to an invariant form by means of the trigonometric identity:

$$\cos\,(3\tilde{\omega}_s) = 4\cos^3 \tilde{\omega}_s - 3\cos\tilde{\omega}_s . \tag{C 4.7}$$

Thus, due to (C 3.6), (C 3.5), and (C 4.5)

$$\cos\,(3\tilde{\omega}_s) = \frac{3\sqrt{3}}{2} \frac{s_1^D}{(II_{S^D})^{3/2}} \left[\left(s_1^D\right)^2 - II_{S^D} \right] \tag{C 4.8}$$

is obtained. Under consideration of (C 4.4), and the fact that the first invariant of the stress deviator vanishes, Eqn. (C 4.8) yields:

$$\cos\,(3\tilde{\omega}_s) = \frac{3\sqrt{3}}{4} \frac{1}{(II_{S^D})^{3/2}} \left\{ \left(s_1^D\right)^3 + \left(s_2^D\right)^3 + \left(s_3^D\right)^3 + s_2^D \left(s_3^D\right)^2 + \left(s_2^D\right)^2 s_3^D \right\} . \tag{C 4.9}$$

After some rearrangements and under consideration of (C 4.6)

$$\cos\,(3\tilde{\omega}_s) = \frac{3\sqrt{3}}{2} \frac{III_{S^D}}{(II_{S^D})^{3/2}} \tag{C 4.10}$$

is obtained. This relation expresses the representation of the third invariant of the stress deviator by the angle $\tilde{\omega}_s$.

C 5. Subspace of the Stress State

As mentioned, test results and thermodynamic investigations lead directly to the suggestion that, in the hardening range, the yield condition and the flow rule must be formulated in a subspace of the stress state, namely in the shifted stress space, $\bar{\mathbf{S}} = \mathbf{S} - \mathbf{Z}$, where $\mathbf{S}$ is the stress tensor and $\mathbf{Z}$ the translation or the backstress tensor. Following the previous considerations, the shifted stress state will be represented geometrically.

Introducing the orthogonal base system $\mathbf{e}_1$, $\mathbf{e}_2$, and $\mathbf{e}_3$, the base vectors of the principal stress space – the principal stress vector $\mathbf{s}$ – which describes the place of the stress point P (see Fig. C 5.1), can be additively decomposed into the shifted principal stress vector $\bar{\mathbf{s}}$ and the principal backstress vector $\mathbf{z}$. The stress vector $\bar{\mathbf{s}}$, as well as the backstress vector $\mathbf{z}$, are represented (in Fig. C 5.1) in the base system $\mathbf{e}_1$, $\mathbf{e}_2$, and $\mathbf{e}_3$ of the principal stress space, which is always possible.

The point Q is determined by the backstress vector $\mathbf{z}$ which can be represented in the $\mathbf{e}_1$, $\mathbf{e}_2$, $\mathbf{e}_3$ base system,

$$\mathbf{z} = z_i\, \mathbf{e}_i \,. \tag{C 5.1}$$

The representation of $\mathbf{z}$ in the *triaxial plane* is given by

$$\mathbf{z} = \hat{z}_1 \hat{\mathbf{e}}_1 + \hat{z}_2 \hat{\mathbf{e}}_2 + \hat{z}_3 \hat{\mathbf{e}}_3 \,, \tag{C 5.2}$$

with

$$\begin{aligned} \hat{z}_1 &= z_1 \,, \\ \hat{z}_2 &= \frac{1}{\sqrt{2}}\,(z_2 - z_3) \,, \\ \hat{z}_3 &= \frac{1}{\sqrt{2}}\,(z_2 + z_3) \,. \end{aligned} \tag{C 5.3}$$

For the special compression state, if z_2 is equal to z_3, Eqn. (C 5.2) simplifies in connection with (C 5.3) to

$$\mathbf{z} = \hat{z}_1 \hat{\mathbf{e}}_1 + \sqrt{2}\; z_3 \hat{\mathbf{e}}_3 \,. \tag{C 5.4}$$

In the $\tilde{\mathbf{e}}_1$, $\tilde{\mathbf{e}}_2$, $\tilde{\mathbf{e}}_3$ base system, the following representation

$$\mathbf{z} = \mathbf{z}^D + \mathbf{z}^K = \tilde{z}_1^D \tilde{\mathbf{e}}_1 + \tilde{z}_2^D \tilde{\mathbf{e}}_2 + z^K \tilde{\mathbf{e}}_3 \tag{C 5.5}$$

is obtained with

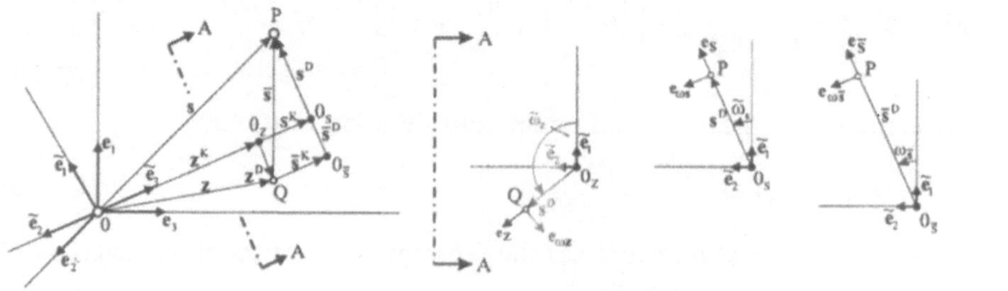

Fig. C 5.1. Geometric representation of the stress subspace

$$
\begin{aligned}
\bar{z}_1^D &= \frac{1}{\sqrt{6}}\left(2z_1^D - z_2^D - z_3^D\right) \\
&= \frac{1}{\sqrt{6}}\left(2z_1 - z_2 - z_3\right), \\
\bar{z}_2^D &= \frac{1}{\sqrt{2}}\left(2z_1^D - z_2^D - z_3^D\right) \\
&= \frac{1}{\sqrt{2}}\left(2z_1 - z_2 - z_3\right), \\
\bar{z}^K &= \frac{1}{\sqrt{3}}\left(z_1 + z_2 + z_3\right),
\end{aligned}
\tag{C 5.6}
$$

where $\mathbf{z}^D$ lies in the *octahedral plane*.

Although the backstress point Q in the octahedral plane is already determined, see (C 5.6), it is sometimes convenient to use a polar coordinate system. With (see (C 3.10))

$$
\begin{aligned}
\tilde{\mathbf{e}}_z &= \cos\tilde{\omega}_z \mathbf{e}_1 + \sin\tilde{\omega}_z \mathbf{e}_2, \\
\tilde{\mathbf{e}}_{\omega z} &= -z^D \sin\tilde{\omega}_z \mathbf{e}_1 + z^D \cos\tilde{\omega}_z \mathbf{e}_2,
\end{aligned}
\tag{C 5.7}
$$

$$
\mathbf{z}^D = z^D \tilde{\mathbf{e}}_z
\tag{C 5.8}
$$

is found where z^D is the norm of $\mathbf{z}^D$.

With the determination of the reference point 0_z, it is easy to represent the shifted stress vector $\bar{\mathbf{s}}$ in the $\mathbf{e}_1$, $\mathbf{e}_2$, $\mathbf{e}_3$-system in Q. According to the above derivations for the translation tensor, the stress vector $\bar{\mathbf{s}}$ can be represented in the *triaxial plane* by

$$
\bar{\mathbf{s}} = \bar{s}_1 \hat{\mathbf{e}}_1 + \frac{\bar{s}_2 - \bar{s}_3}{\sqrt{2}} \hat{\mathbf{e}}_2 + \frac{\bar{s}_2 - \bar{s}_3}{\sqrt{2}} \hat{\mathbf{e}}_3 .
\tag{C 5.9}
$$

For the compression state

$$
\bar{\mathbf{s}} = \bar{s}_1 \hat{\mathbf{e}}_1 + \sqrt{2}\, \bar{s}_3 \hat{\mathbf{e}}_3
\tag{C 5.10}
$$

is obtained.

Moreover, in the *octahedral plane* the shifted stress vector $\bar{\mathbf{s}}$ is described by

$$
\bar{\mathbf{s}} = \bar{s}^K \tilde{\mathbf{e}}_3 + \bar{\bar{s}}_1^D \tilde{\mathbf{e}}_1 + \bar{\bar{s}}_2^D \tilde{\mathbf{e}}_2 ,
\tag{C 5.11}
$$

with

$$
\begin{aligned}
\bar{s}^K &= \frac{1}{\sqrt{3}}\left(\bar{s}_1 + \bar{s}_2 + \bar{s}_3\right), \\
\bar{\bar{s}}_1^D &= \frac{1}{\sqrt{6}}\left(2\bar{s}_1^D - \bar{s}_2^D - \bar{s}_3^D\right) \\
&= \frac{1}{\sqrt{6}}\left(2\bar{s}_1 - \bar{s}_2 - \bar{s}_3\right), \\
\bar{\bar{s}}_2^D &= \frac{1}{\sqrt{2}}\left(\bar{s}_2^D - \bar{s}_3^D\right) \\
&= \frac{1}{\sqrt{2}}\left(\bar{s}_2 - \bar{s}_3\right).
\end{aligned}
\tag{C 5.12}
$$

According to (C 3.10), the shifted stress vector $\bar{\mathbf{s}}^D$ can be described in a polar coordinate system

$$\begin{aligned} \tilde{\mathbf{e}}_{\bar{s}} &= \cos \tilde{\omega}_{\bar{s}} \tilde{\mathbf{e}}_1 + \sin \tilde{\omega}_{\bar{s}} \tilde{\mathbf{e}}_2 \\ \tilde{\mathbf{e}}_{\omega \bar{s}} &= -\bar{s}^D \sin \tilde{\omega}_{\bar{s}} \tilde{\mathbf{e}}_1 + \bar{s}^D \cos \tilde{\omega}_{\bar{s}} \tilde{\mathbf{e}}_2 \end{aligned} \qquad \text{(C 5.13)}$$

($\bar{s}^D$ is the norm of $\bar{\mathbf{s}}^D$) by

$$\bar{\mathbf{s}}^D = \bar{s}^D \tilde{\mathbf{e}}_{\bar{s}} \,. \qquad \text{(C 5.14)}$$

The geometric representation of the yield function

$$F = F\left(\mathrm{I}_{\bar{\mathbf{s}}}\,, \mathrm{II}_{\bar{\mathbf{s}}^D}\,, \mathrm{III}_{\bar{\mathbf{s}}^D}\right) \qquad \text{(C 5.15)}$$

can be achieved according to the treatment in Section (C 4.). Thus,

$$\begin{aligned} \bar{s}^K &= \frac{1}{\sqrt{3}}\, \mathrm{I}_{\bar{\mathbf{s}}}\,, \\ \bar{s}^D &= \sqrt{2\, \mathrm{II}_{\bar{\mathbf{s}}^D}}\,, \\ \cos(3\tilde{\omega}_{\bar{s}}) &= \frac{3\sqrt{3}}{2}\, \frac{\mathrm{III}_{\bar{\mathbf{s}}^D}}{\left(\mathrm{II}_{\bar{\mathbf{s}}^D}\right)^{3/2}}\,. \end{aligned} \qquad \text{(C 5.16)}$$

It is only for special stress states or special states of the backstress tensor, respectively, that an additive decomposition of $\bar{s}^K$ and $\bar{s}^D$ is possible in some yield functions (see de Boer and Brauns, 1990).

C 6. Mohr-Coulomb Theory

After Gollub (1989), it will be shown that, with the help of a maximum/minimum calculation from Coulomb's condition[17],

$$t \leq -\, p \tan \varphi + c\,, \qquad \text{(C 6.1)}$$

which is based on Amontons' (1699) investigations (φ is the angle of internal friction and c the cohesion) and which is assumed to be valid for each cut surface in a solid, an invariant failure condition can be derived. The analytical derivation by Gollub (1989) represents an alternative method to the graphical solution by Mohr (1900).

C 6.1 Coulomb's Strength Hypothesis for Arbitrarily Oriented Surfaces

In order to formulate Coulomb's condition (C 6.1) for arbitrarily oriented surfaces, it is necessary to determine the normal vector $\mathbf{p}$ as a function of the surface orientation. The surface orientation will be described by the surface unit vector $\mathbf{n}$ referring to the base vectors $\mathbf{e}_i$ of the principal stress space (see Fig. C 6.1.1):

[17] In his original paper, Coulomb (1773) used a friction coefficient. The introduction of the angle of internal friction is due to Woltman (1794/99).

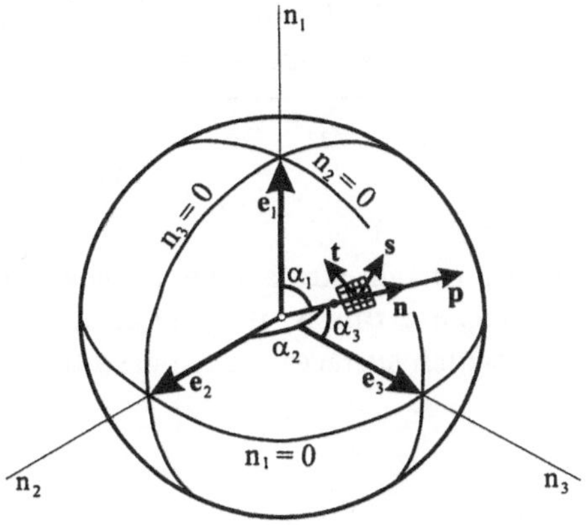

Fig. C 6.1.1. Surface orientation in the principal stress space

$$\begin{aligned} &\mathbf{n} = n_1\mathbf{e}_1 + n_2\mathbf{e}_2 + n_3\mathbf{e}_3\,, \\ &n_1 = \cos\alpha_1\,, \qquad n_2 = \cos\alpha_2\,, \qquad n_3 = \cos\alpha_3\,, \\ &n_1^2 + n_2^2 + n_3^2 = 1\,. \end{aligned} \tag{C 6.1.1}$$

With the stress tensor S referring to the base system $\mathbf{e}_i$

$$\mathbf{S} = \begin{bmatrix} s_1 & 0 & 0 \\ 0 & s_2 & 0 \\ 0 & 0 & s_3 \end{bmatrix} \mathbf{e}_i \otimes \mathbf{e}_k\,, \tag{C 6.1.2}$$

the stress vector s can easily be determined by the linear mapping

$$\begin{aligned} &\mathbf{s} = \mathbf{S}\mathbf{n}\,, \\ &\mathbf{s} = s_1 n_1 \mathbf{e}_1 + s_2 n_2 \mathbf{e}_2 + s_3 n_3 \mathbf{e}_3\,, \end{aligned} \tag{C 6.1.3}$$

where (C 6.1.1)$_1$ has been used. The magnitude of the normal component p of the stress vector s is obtained from the scalar product

$$\mathbf{s}\cdot\mathbf{n} = p = s_1 n_1^2 + s_2 n_2^2 + s_3 n_3^2\,, \tag{C 6.1.4}$$

and the normal stress vector is then given by

$$\mathbf{p} = \left(s_1 n_1^2 + s_2 n_2^2 + s_3 n_3^2\right)\mathbf{n}\,. \tag{C 6.1.5}$$

The tangential stress vector **t** is equal to the vector difference

$$\begin{aligned} \mathbf{s} - \mathbf{p} = \mathbf{t} = &\left[\left(1 - n_1^2\right) s_1 - n_2^2 s_2 - n_3^2 s_3\right] n_1 \mathbf{e}_1 + \\ &+ \left[- n_1^2 s_1 + \left(1 - n_2^2\right) s_2 - n_3^2 s_3\right] n_2 \mathbf{e}_2 + \\ &+ \left[- n_1^2 s_1 - n_2^2 s_2 + \left(1 - n_3^2\right) s_3\right] n_3 \mathbf{e}_3\,. \end{aligned} \tag{C 6.1.6}$$

The norm of the tangential stress vector yields

$$\begin{aligned} t &= \sqrt{\mathbf{t}\cdot\mathbf{t}} \\ &= \sqrt{(s_1 - s_2)^2\; n_1^2 n_2^2 + (s_3 - s_1)^2\; n_3^2 n_1^2 + (s_2 - s_3)^2\; n_2^2 n_3^2}\,. \end{aligned} \tag{C 6.1.7}$$

With (C 6.1.4) and (C 6.1.7), Coulomb's condition (C 6.1) can be expressed by an explicit function of the principal stress s_i and the surface orientations n_i:

$$f\,(s_i,\, n_i) = \sqrt{(s_1 - s_2)^2\ n_1^2 n_2^2 + (s_3 - s_1)^2\ n_3^2 n_1^2 + (s_2 - s_3)^2\ n_2^2 n_3^2}\ + \\ + \left(s_1 n_1^2 + s_2 n_2^2 + s_3 n_3^2\right) \tan\varphi - c \le 0\,. \tag{C 6.1.8}$$

In order to prove whether a solid will fail at a certain stress state, all possible surface orientations must be investigated. Because this is, in general, a very laborious task, it is reasonable to investigate which surfaces are exposed to the maximum failure danger.

C 6.2 Failure Surfaces

The problem of finding the surfaces at which the material is in danger of failing can mathematically be formulated as follows: such surfaces will be calculated in which the function (C 6.1.8) takes its maximum. The condition (C 6.1.1)$_3$ can thereby be considered by adding it to the maximum demand as an auxiliary condition (Lagrange method). The necessary conditions for the surfaces of maximum failure danger are:

$$\begin{aligned}
&\frac{\partial}{\partial n_1}\left[f\,(s_i,\, n_i) + \varepsilon\psi\,(n_i)\right] = 0\,,\\
&\frac{\partial}{\partial n_2}\left[f\,(s_i,\, n_i) + \varepsilon\psi\,(n_i)\right] = 0\,,\\
&\frac{\partial}{\partial n_3}\left[f\,(s_i,\, n_i) + \varepsilon\psi\,(n_i)\right] = 0\,,\\
&\psi\,(n_i) = n_1^2 + n_2^2 + n_3^2 - 1 = 0\,,
\end{aligned} \tag{C 6.2.1}$$

where ε is the Lagrange multiplier. Applying (C 6.2.1) to (C 6.1.8) yields, after some calculations:

$$\begin{aligned}
&2n_1\left[0\ n_1^2 + \frac{(s_1 - s_2)^2}{2t}\ n_2^2 + \frac{(s_3 - s_1)^2}{2t}\ n_3^2 + \varepsilon + s_1 \tan\varphi\right] = 0\,,\\
&2n_2\left[\frac{(s_1 - s_2)}{2t}\ n_1^2 + 0\ n_2^2 + \frac{(s_2 - s_3)^2}{2t}\ n_3^2 + \varepsilon + s_2 \tan\varphi\right] = 0\,,\\
&2n_3\left[\frac{(s_3 - s_1)^2}{2t}\ n_1^2 + \frac{(s_2 - s_3)^2}{2t}\ n_2^2 + 0\ n_3^2 + \varepsilon + s_3 \tan\varphi\right] = 0\,,\\
&n_1^2 + n_2^2 + n_3^2 - 1 = 0\,.
\end{aligned} \tag{C 6.2.2}$$

In order to determine the maximum values, the interior area ($n_1 \neq 0$, $n_2 \neq 0$, $n_3 \neq 0$) and the boundaries ($n_1 = 0$, $n_2 = 0$, $n_3 = 0$) are subsequently considered (see Fig. C 6.1.1), thereby having control over the dependence between the principal stress differences

$$s_1 - s_3 = (s_1 - s_2) + (s_2 - s_3)\,, \qquad s_1 \le s_2 \le s_3 \tag{C 6.2.3}$$

the maximum values in the interior area are investigated.

a)
Maximum Values in the Interior Area ($n_1 \neq 0,\ n_2 \neq 0,\ n_3 \neq 0$)

The condition (C 6.2.2), to calculate the maximum values, demands that the terms in the brackets vanish. The Lagrange multiplier ε is eliminated by subtraction of the brackets of (C 6.2.2). The subtractions $(\mathrm{C}\ 6.2.2)_{1,2}$ and $(\mathrm{C}\ 6.2.2)_{1,3}$ yield:

$$\begin{aligned}
&\frac{1}{2t}\left[-(s_1-s_2)^2\, n_1^2 + (s_1-s_2)^2\, n_2^2 + \left\{(s_1-s_3)^2 - (s_2-s_3)^2\right\} n_3^2\right] + \\
&\qquad + (s_1 - s_2)\tan\varphi = 0\,,\\
&\frac{1}{2t}\left[-(s_3-s_1)^2\, n_1^2 + \left\{(s_1-s_2)^2 - (s_2-s_3)^2\right\} n_2^2 + (s_3-s_1)^2\, n_3^2\right] + \\
&\qquad + (s_1 - s_3)\tan\varphi = 0\,,\\
&\qquad n_1^2 + n_2^2 + n_3^2 - 1 = 0\,.
\end{aligned} \tag{C 6.2.4}$$

The substitution of n_3 from the auxiliary condition $(\mathrm{C}\ 6.2.4)_3$ and the consideration of the dependence between the principal stress differences (C 6.2.3) simplify the equation system (C 6.2.4) to:

$$\begin{aligned}
&(s_1-s_2)\left[\frac{s_1-s_2}{2t}\left(1-2n_1^2\right) + 2\frac{s_2-s_3}{2t}\left(1-n_1^2-n_2^2\right) + \tan\varphi\right] = 0\,,\\
&(s_1-s_3)\left[\frac{s_1-s_2}{2t}\left(1-2n_1^2\right) + 2\frac{s_2-s_3}{2t}\left(1-n_1^2-n_2^2\right) + \tan\varphi\right] = 0\,.
\end{aligned} \tag{C 6.2.5}$$

For all stress states which contain shear stresses, the difference of the principal stresses $(s_1 - s_3)$ is not equal to zero. In this case the term in the right bracket of $(\mathrm{C}\ 6.2.5)_2$ must vanish. By introduction of this relation into $(\mathrm{C}\ 6.2.5)_1$, a necessary condition for a maximum in the interior of the considered area is obtained

$$(s_1 - s_2)(s_2 - s_3) = 0\,, \tag{C 6.2.6}$$

which cannot be satisfied by general stress states ($s_1 \neq s_2 \neq s_3$). Thus, in general, no maximum exists in the interior area.

In the case when the intermediate principal stress state coincides with the smallest or the largest principal pressure stress-compression ($s_2 = s_3$) and extension ($s_2 = s_1$) states, all directions which are perpendicular to the distinct principal stress, are principal stress directions. Thus, a unique determination of a boundary and interior area is no longer possible. The maximum locations for these special stress states are obtained from (C 6.2.5). For the compression state ($s_2 = s_3$) both equations (C 6.2.5) represent identical requirements. They lead via the relation[18]

$$\frac{1 - 2n_1^2}{\pm\sqrt{1 - n_1^2}\, 2n_1} + \tan\varphi = 0 \qquad \text{(C 6.2.7)}$$

to the surfaces

$$n_1 = \pm\cos\left(45^\circ + \varphi/2\right), \quad \sqrt{n_2^2 + n_3^2} = \pm\cos\left(45^\circ - \varphi/2\right). \qquad \text{(C 6.2.8)}$$

For the extension state ($s_2 = s_1$) Eqn. (C 6.2.5)$_1$, is always satisfied. Eqn.(C 6.2.5)$_2$ yields, under consideration of the auxiliary condition (C 6.2.4)$_3$, to the relation

$$\frac{1 - 2n_3^2}{\pm\sqrt{1 - n_3}\, 2n_3} + \tan\varphi = 0 \qquad \text{(C 6.2.9)}$$

and to the surfaces

$$\sqrt{n_1^2 + n_2^2} = \pm\cos\left(45^\circ + \varphi/2\right), \quad n_3 = \pm\cos\left(45^\circ - \varphi/2\right). \qquad \text{(C 6.2.10)}$$

b)
Maximum Values at the Boundaries

In the investigations to follow, the boundaries $n_1 = 0$, $n_2 = 0$, and $n_3 = 0$ are considered. For $n_1 = 0$, the system (C 6.2.2) is reduced to the requirements:

$$-\frac{s_2 - s_3}{2}\frac{n_3}{n_2} + \varepsilon + s_2\tan\varphi = 0,$$

$$-\frac{s_2 - s_3}{2}\frac{n_2}{n_3} + \varepsilon + s_3\tan\varphi = 0, \qquad \text{(C 6.2.11)}$$

$$n_2^2 + n_3^2 - 1 = 0.$$

The subtraction of the second equation from the first in (C 6.2.11) under consideration of (C 6.2.11)$_3$, with the assumption that shear stress free states ($s_2 = s_3$) should be excluded, leads to the relations

$$\frac{1 - 2n_2^2}{\pm\sqrt{1 - n_2^2}\, 2n_2} + \tan\varphi = 0,$$

$$\mathbf{n} = 0\,\mathbf{e}_1 \pm \cos\left(45^\circ + \varphi/2\right)\mathbf{e}_2 \pm \cos\left(45^\circ - \varphi/2\right)\mathbf{e}_3. \qquad \text{(C 6.2.12)}$$

At the boundary $n_2 = 0$, the maximum requirement (C 6.2.2) can be rearranged in the same manner as for the boundary $n_1 = 0$. The requirement is satisfied by the two surfaces:

$$\mathbf{n} = \pm\cos\left(45^\circ - \varphi/2\right)\mathbf{e}_1 + 0\mathbf{e}_2 \pm \cos\left(45^\circ - \varphi/2\right)\mathbf{e}_3. \qquad \text{(C 6.2.13)}$$

Similarly, the maximum requirement for $n_3 = 0$ leads to the surfaces:

$$\mathbf{n} = \pm\cos\left(45^\circ + \varphi/2\right)\mathbf{e}_1 \pm \cos\left(45^\circ - \varphi/2\right)\mathbf{e}_2 + 0\mathbf{e}_3. \qquad \text{(C 6.2.14)}$$

[18] Only solutions leading to a maximum of $f\ (s_i, n_i)$ are listed.

c)
Failure Surfaces of the Absolute Maximum

In order to determine the surfaces for which the function (C 6.1.8) takes its absolute maximum, the function values of the relative maximum surfaces must be compared. For the maximum values in the interior area (see (C 6.2.8) and (C 6.2.10)), the value

$$f = -\frac{s_1}{2}\tan\left(45^\circ - \varphi/2\right) + \frac{s_3}{2}\tan\left(45^\circ + \varphi/2\right) - c \qquad \text{(C 6.2.15)}$$

is obtained. The function values at the boundaries, see (C 6.2.12) through (C 6.2.14), are calculated as follows:

$$\begin{aligned} &f\left[\,0, \pm\cos\left(45^\circ + \varphi/2\right), \pm\cos\left(45^\circ - \varphi/2\right)\,\right] = \\ &= -\frac{s_2}{2}\tan\left(45^\circ - \varphi/2\right) + \frac{s_3}{2}\tan\left(45^\circ + \varphi/2\right) - c\,, \\ &f\left[\,\pm\cos\left(45^\circ + \varphi/2\right), 0, \pm\cos\left(45^\circ - \varphi/2\right)\,\right] = \\ &= -\frac{s_1}{2}\tan\left(45^\circ - \varphi/2\right) + \frac{s_3}{2}\tan\left(45^\circ + \varphi/2\right) - c\,, \\ &f\left[\,\pm\cos\left(45^\circ + \varphi/2\right), \pm\cos\left(45^\circ - \varphi/2\right), 0\,\right] = \\ &= -\frac{s_1}{2}\tan\left(45^\circ - \varphi/2\right) + \frac{s_2}{2}\tan\left(45^\circ + \varphi/2\right) - c\,. \end{aligned} \qquad \text{(C 6.2.16)}$$

The maximum value (C 6.2.15) is equal to (C 6.2.16)$_2$. From the difference of (C 6.2.16)$_2$ or (C 6.2.15), and (C 6.2.16)$_1$ and (C 6.2.16)$_3$, it becomes evident that (C 6.2.16)$_2$ or (C 6.2.15) represents the absolute maximum of the function (C 6.1.8). The possible failure surfaces are, therefore, in general determined by (C 6.2.13) and for the special symmetric stress states of rotation by (C 6.2.8) and (C 6.2.10). Thus, in the cases of compression and extension, the failure surfaces can "rotate" around the s_1- and the s_2-axis.

C 7.
Failure Condition in Invariant Formulation

In the limit state, if failure occurs in relation (C 6.1.8), only the equal sign appears. With (C 6.2.15),

$$-\frac{s_1}{2}\tan\left(45^\circ - \varphi/2\right) + \frac{s_3}{2}\tan\left(45^\circ + \varphi/2\right) - c = 0 \qquad \text{(C 7.1)}$$

is valid. This limit condition, well-known as the Mohr-Coulomb failure condition, can easily be transformed to the general stress state. By use of some trigonometric formulas, the following results are obtained:

$$-s_1 + s_3\tan^2\left(45^\circ - \varphi/2\right) - 2\,c\tan\left(45^\circ + \varphi/2\right) = 0\,,$$

$$-\frac{1}{2}(s_1 - s_3) + \frac{1}{2}(s_1 + s_3)\sin\varphi - c\cos\varphi = 0\,, \tag{C 7.2}$$

$$\sqrt{\left(\frac{s_{11} - s_{33}}{2}\right)^2 + s_{13}^2} + \frac{1}{2}(s_{11} + s_{33})\sin\varphi - c\cos\varphi = 0\,.$$

The failure condition (C 7.1) can also be expressed by the stress invariants $\mathrm{I_S}$, $\mathrm{II_{S^D}}$, and $\mathrm{III_{S^D}}$. In order to obtain this new formulation of the failure condition, it is convenient to decompose the principal stress additively in their deviatoric and hydrostatic parts.

$$s_i = s_i^D + \frac{1}{3}\mathrm{I_S}\,. \tag{C 7.3}$$

Thus, the dependence of s_i on $\mathrm{I_S}$ is already ensured by (C 7.3). In order to gain the relationship between the deviatoric principal stresses and the second and third invariants of the stress deviator the eigenvalue problem of the deviatoric stress $\mathbf{S}^D$ must be investigated:

$$\mathbf{S}^D\,\mathbf{n} = s^D\,\mathbf{n}\,, \tag{C 7.4}$$

where s^D is a real scalar quantity and $\mathbf{n}$ an arbitrary surface unit vector. The non-trivial solution of the eigenvalue problem (C 7.4) leads to an equation of third grade:

$$\left(s^D\right)^3 - \mathrm{II_{S^D}} - \mathrm{III_{S^D}} = 0\,. \tag{C 7.5}$$

From (C 7.5), three different principal stress values are obtained

$$s_1^D = \frac{2}{\sqrt{3}}\sqrt{\mathrm{II_{S^D}}}\,\sin\,(\Theta - 60°)\,,$$

$$s_2^D = \frac{2}{\sqrt{3}}\sqrt{\mathrm{II_{S^D}}}\,\sin\,(-\,\Theta)\,, \tag{C 7.6}$$

$$s_3^D = \frac{2}{\sqrt{3}}\sqrt{\mathrm{II_{S^D}}}\,\sin\,(\Theta + 60°)$$

with

$$\Theta = \frac{1}{3}\arcsin\left(\frac{3\sqrt{3}}{2}\,\frac{\mathrm{III_{S^D}}}{(\mathrm{II_{S^D}})^{3/2}}\right), \tag{C 7.7}$$

$$-30° \le \Theta \le +30°.$$

With (C 7.3) and (C 7.6) the principal stresses in (C 7.1) can be substituted by the stress invariants and the representation of the Mohr-Coulomb failure condition by the stress invariants is obtained:

$$\sqrt{\mathrm{II_{S^D}}}\left(\cos\Theta + \frac{\sin\varphi}{\sqrt{3}}\sin\Theta\right) + \mathrm{I_S}\frac{\sin\varphi}{3} - c\cos\varphi = 0\,. \tag{C 7.8}$$

In the geometric representation with

$$s^K = -\frac{1}{\sqrt{3}}\mathrm{I_S}\,,\quad s^D = \sqrt{2\,\mathrm{II_{S^D}}} \tag{C 7.9}$$

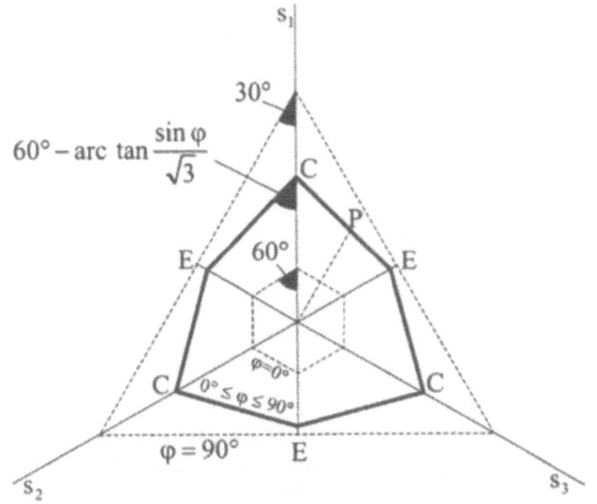

Fig. C 7.1. Mohr-Coulomb's failure condition in the octahedral plane

and Θ, according to (C 7.7), the Mohr-Coulomb failure condition (C 7.8) forms a non-regular hexagon with symmetric properties referring to the three principal stress axis in the octahedral plane. Thus, it is sufficient to consider only an area spread over 60°. This is in accordance with the previously chosen restriction $s_1 \leq s_2 \leq s_3$.

From (C 7.8),

$$s^D = \frac{\sqrt{\frac{2}{3}} \sin\varphi\, s^K + \sqrt{2}\, c \cos\varphi}{\cos\Theta + \frac{\sin\varphi}{\sqrt{3}} \sin\Theta} \tag{C 7.10}$$

is obtained. It is recognized that in the octahedral plane ($s^K = \text{const}$) the numerator in (C 7.10) is a constant and, therefore, (C 7.10) describes a straight line in polar coordinates (see Fig. C 7.1).

Due to the sixfold symmetry in the octahedral plane, relation (C 7.10) yields six straight lines which form, as has already been mentioned, a non-regular hexagon which can be constructed in an easy way if, from (C 7.10), the radius of the compression, or extension, states are known. The ratio of the radius of the compression ($\Theta = -30°$), the plane deviatoric stress, and the extension state in the octahedral plane are calculated as:

$$\begin{aligned} &s^D(\Theta = -30°) \;:\; s^D(\Theta = 0°) \;:\; s^D(\Theta = 30°) = \\ &= \frac{\sqrt{12}}{3 - \sin\varphi} \;:\; 1 \;:\; \frac{\sqrt{12}}{3 + \sin\varphi}\,. \end{aligned} \tag{C 7.11}$$

Thus, the shape of the Mohr-Coulomb failure condition in the octahedral plane depends exclusively on the angle of internal friction. The shape varies between a regular hexagon for $\varphi = 0$ and a regular triangle for $\varphi = 90°$ (see Fig. C 7.1).

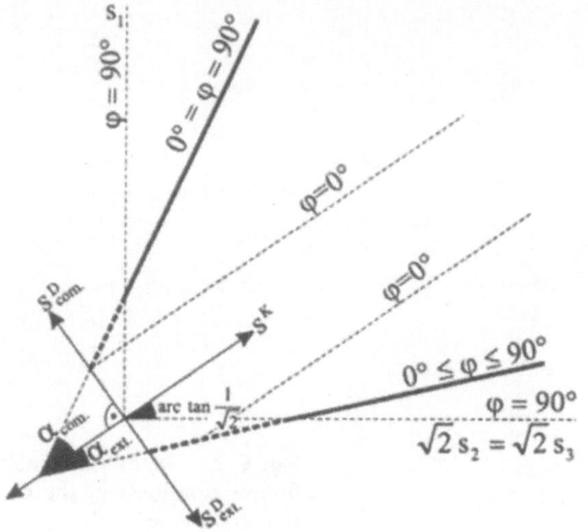

Fig. C 7.2. Mohr-Coulomb's failure condition in the triaxial plane

In the triaxial plane, the relation between the norm of the deviator stress, measured by s^D, and the hydrostatic pressure, measured by s^K, becomes evident. From (C 7.10), it can be read that this relation is linear (see Fig. C 7.2).

In order to depict the Mohr-Coulomb failure condition in the biaxial plane ($s_2 = \text{const.}$), the relation (C 7.2)$_2$ is reconsidered, which takes with s_{min} as the smallest and s_{max} as the largest principal stresses the following form

$$-\, s_{min} + s_{max} \tan^2\left(45^\circ + \varphi/2\right) - 2\,c \tan\left(45^\circ + \varphi/2\right) = 0\,. \qquad \text{(C 7.12)}$$

It is recognized from (C 7.12) that the Mohr-Coulomb failure condition is described by six straight lines in the biaxial plane (see Fig. C 7.3).

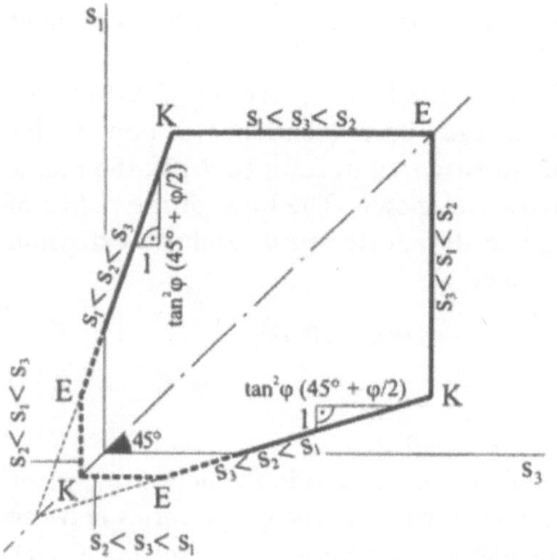

Fig. C 7.3. Mohr-Coulomb's failure condition in the biaxial plane ($s_2 =$ const.)

The Mohr-Coulomb theory is not only valuable for the determination of the failure of special boundary-value problems. It also provides a general idea of the flow mechanism of frictional materials (see Gollub, 1989). In the Mohr-Coulomb theory, all deformations occur in a plane which contains the axis of the intermediate stress and which is perpendicular to the plane of maximum and minimum principal stresses.

References

1638 Galilei, G.: Discorsi e demostrationi mathematiche intorno a due nuove scienze attenenti alla meccanica e ai movimenti locali, Leiden, translated into German by A. von Oettingen, Oswalds Klassiker Nr. 11, 24, 25, Verlag von Wilhelm Engelmann, Leipzig (1913, 1917, 1921).

1668 Hobbes, T.: Problemata physica, Liviathan, cap. IV, Amstelodami.

1677 Descartes, R.: Principia philosophiae, Amstelodami, apud D. Elsevirium, Pars equarta, in: OEUVRES de Descartes publiées par Charles Adam & Paul Tannery, Principia philosophiae 8,1, Paris 1982, Pars Quarta, 203–329.

1678 Hooke, R.: Lectures de potentia restitutiva, or of spring, explaining the power of springing bodies, London. (also: R. T. Guntler: Early Science in Oxford 8, 331–356, 1931).

1687 Newton, I.: Philosophiae naturalis principia mathematica, Jussu Soc. Reg. ac Typis J. Streater, London.

1691 Bernoulli, Jakob: Demonstratio centri oscillationis ex natura vectis, *Acta Eruditorum* (Acta Erudit), Lipsiae, 317–321, Opera, Birkhäuser, Basel • Boston • Berlin (1991), 460–465.

1694 Bernoulli, Jakob: Curvatura laminae elasticae ..., *Acta Eruditorum* (Acta Erudit), Lipsiae, 262–276, Opera, Birkhäuser, Basel • Boston • Berlin (1991), 576–600.

1699 Amontons: De la résistance causée dans les machines ... *Mémoires de l'Académie Royal des Sciences*, 206–266.

1713 Parent, A.: De la véritable méchanique des résistances des solides et réfléxions sur le système de M. Bernoulli de Bâle, *Essais et Recherches de Mathematique et de Physique* 3, Paris, 187–201.

1716 Hermann, J.: Phoronomia sive de viribus et motibus corporum solidorum et fluidorum, libri duo, Amstelodami.

1722 Locke, J.: Elements of natural philosophy of collection of several pieces of J. Locke, London.

1726 Couplet: De la poussée des terres contre leurs revestements et la force qu'on leur doit opposer, Historie de l' academie royale des sciences, Années MDCC XX VI–MDCC X X V III, Paris 1726, 1727, 1728.

1732 Bernoulli, J.: Hydraulica nunc primum detecta ac demonstrata directe ex fundamentis pure mechanics, Sumptibus Marci - Michaelis Bousquet & Sociorum Lausannae & Genevae, Opera omnia, 1742,
Reprint: Georg Olms Verlagsbuchhandlung, Hildesheim (1968).

1733 Zedler, J. H.: Großes vollständiges Universal-Lexikon aller Wissenschaften und Künste. Halle und Leipzig, ed.: Johann Heinrich Zedler. Neuauflage: Akademische Druck- und Verlagsanstalt, Graz - Austria (1961).

1734 Bernoulli, D.: Hydrodynamica sive de viribus et motibus fluidorum comentarii, Sumptibus Johannis Reinhold Dulseckeri, Straßburg (1738).

1736 Euler, L.: Mechanica sive motus scientia analytice, T. I, II Petropoli, ex typogr. Acad. Sci. (also: Opera omnia II 1).

1737–1753 de Bélidor, B.F.: Architecture Hydraulique, Paris;
German edition: Verlag des Eberhard Kletts, Augsburg, 1764.

1745 Euler, L.: Neue Grundzüge der Artillerie, aus dem Englischen des Herrn Benjamin Robins übersetzt und mit vielen Anmerkungen versehen, A. Haude, Berlin (also: Opera omnia II 14).

1749 Euler, L.: Scientia navalis seu tractatus de construendis..., Typis Academiae Scientiarum, Petersburg (also: Opera omnia II 18; II 19).

1752a Euler, L.: Découverte d'un nouveau principe de méchanique, *Mémoires de l'Academie des Sciences de Berlin* (Mém Acad Sci de Berlin) **6** (for the year 1750, printed 1752), 185–217 (also: Opera omnia II 5).

b Euler, L.: Sur le mouvement de l' eau par des anyaux de conduite, *Mémoires de l'Academie des Sciences de Berlin* (Mém Acad Sci de Berlin) **8**, 111–148.

1757a Euler, L.: Principes généraux de l'etat d'équilibre des fluides, *Mémoires de l'Academie des Sciences de Berlin* (Mém Acad Sci de Berlin) **11**, 217–273 (also: Opera omnia II 12).

b Euler, L.: Principes généraux du mouvement des fluides, *Mémoires de l'Academie des Sciences de Berlin* (Mém Acad Sci de Berlin) **11**, 274–315 (also: Opera omnia II 12).

1758 Euler, L.: Du mouvement de rotation des corps solides autor d'un axe variable, *Mémoires de l'Academie des Sciences de Berlin* (Mém Acad Sci de Berlin) **14** (for the year 1758, printed 1765), 154–193 (also: Opera omnia II 8).

1762 Euler, L.: Anleitung zur Naturlehre, Opera postuma mathematica et physica, Bd. 1–2, New York 1969 (Repr. of the edit. St. Petersbourg, 1862).

1765 Euler, L.: Theoria motus corporum solidorum seu rigidorum, litteris et impensis A.F. Röse, Rostochi et Gryhiswaldiae (also: Opera omnia II 3).

1768 Euler, L.: Lettres á une princesse d'Allemagne sur divers sujets de physique & de philosophie, T. premier. L'imprimerie de l'Academie Impériale des Sciences, Saint Petersbourg (also: Opera omnia III 11).

1769a Euler, L.: Briefe an eine deutsche Prinzessin über verschiedene Gegenstände aus der Physik und der Philosophie, erster Teil, Bey Johann Friedrich Junius, Leipzig (also: Friedrich Vieweg & Sohn, Braunschweig/Wiesbaden, 1986).

b Euler, L.: Sectio prima de statu aequilibri fluidorum, *Novi commentarii academie scientarium Petropolitanae* **13**, 345-416 (also: Opera omnia II 13).

1770a Euler, L.: Sectio secundo de principii motus fluidorum, *Novi commentarii academie scientarium Petropolitanae* **14**, 270–386 (also: Opera omnia II 13).

b Euler, L.: Vollständige Anleitung zur Algebra, St. Petersburg gedruckt bey der Kays. Akad. der Wissenschaften, 1770. Appeared in a Russian translation 1768/69 (also: Opera omnia I 1).

1771 Euler, L.: Sectio tertia de motu fluidorium lineari potissimum aquae. *Novi commentarii academie scientarium Petropolitanae* **15**, 219–360 (also: Opera ommia II 13).

1773 Coloumb, C.A.: Versuch einer Anwendung der Methode des Größten und Kleinsten auf einige Aufgaben der Statik, die in die Baukunst einschlagen. Erster Teil: Von der Stärke der Futtermauern, from the Mémoires de Mathematique & de Physique presentés à l'Academie royale des Sciences par divers Savans & au les Assemblées, Anées 1773, Paris 1776, *Böhms Magazin für Ingenieure und Artilleristen*, Frankfurt, 1779.

1776a Euler, L.: Formulae generales pro translatione quacumque corporum rigidorum, *Novi commentarii academiae scientiarium Petropolitanae* **20** (1775, printed 1776), 189–207 (also: Opera omnia II 9).

b Euler, L.: Nova methodus motum corporum rigidorum determinandi, *Novi commentarii academiae scientarium Petropolitanae* **20** (1775, printed 1776), 208–238 (also: Opera omnia II 9).

1785 Coulomb, C. A.: Theorie des machines simples, en ayant égard au frottement de leurs parties et la voideur des cordages, *Mémoires de mathematique et de physique, présentés à l'Académie Royale de Sciences* **10**, Paris, 163–332.

1794/99 Woltman, R.: Beyträge zur Hydraulischen Architectur, Dritter Band, Vierter Band, Johann C. Dietrich, Göttingen.

1799 Davy, H.: An essay on heat, light and the combinations of light, in: The collected works of Davy, Bd. II (1839), London.

1802 de Prony, G.: Recherches sur la poussée des terres, et sur la forme et les dimensions à donner aux murs de revêtement, Imprimerie de la République, Paris.

1805 Von Rumford, B. Graf: Kleine Schriften, vierter Band, erste Abtheilung, Verlag des Landes-Industrie-Comptoirs, Weimar.

1806 Laplace: On capillary attraction, in: Supplement to the tenth book of the Méchanic Céleste. Translated with a Commentary by Nathaniel Bowditch, LL. D. **4**, Chelsea Publishing Company. Inc. Bronx • New York.

1807 Young, Th.: A course of lectures on natural philosophy and the mechanical arts, Johnson, London.

1808 Mayniel, J.-H.: Traité expérimental analytique et pratique de la poussée des terres et des murs de revêtement, Colas, Paris.
Poisson, S.D.: *Journal d'École Polytechnique* **14**.

1820 Francais, J.F.: Recherches sur la poussée des terres, sur la forme les dimensions des murs de revêtements et sur les talus d'excavation, *Mémorial de l' officier du genie* **4**, 157–206.

1823 Cauchy, A.L.: Recherches sur l' équilibre et le mouvement intérieur des corps solides ou fluides, élastiques ou non élastiques, *Bulletin des Sciences par la Société Philomatique*, Paris, 9–13.
Navier, C. L. M. H.: Mémoir sur les lois du mouvement des fluides, printed in the *Mémoires de l'Académie Royal des Sciences* **6**, 389–440.

1824 Carnot, S.: Réflexions sur la puissance motrice du feu et sur les machines propres a développer cette puissance, Chez Bachelier, Libraire Paris (also: Mendoza, 1960).

1827 Navier, C. L. M. H.: Mémoire sur les lois de l'équilibre et du mouvement des corps solides élastiques, *Mémoires de l'Académie Royal des Sciences* **7**, Paris, 375–393.

1828a Cauchy, A.L.: Sur les équations qui experiment les conditions d'équilibre, ou les lois du mouvement intérieur d'un corps solide, élastique, ou non élastique, *Exercises de mathématiques* **3**, 160–187.

b Cauchy, A.L.: Sur l' équilibre et le mouvement d' un système de points materiéls sollicités par des forces d'attraction ou de répulsion mutuelle, *Exercises de mathématiques* **3**, (Contribution from October 1. 1827), 188–212.

1829 Cauchy, A.L.: Sur, l'équilibre et le mouvement intérieur des corps considérés comme des masses continues, *Exercises de mathématiques* **4**, 293–319.
Poisson, S.D.: Mémoires sur l'équilibre et le mouvement des corps élastiques, *Mémoires de l'Académie Royal des Sciences* **8**, Paris, 357–570 (Addition to the mémoir, 623–627).

1830 Poisson, S.D.: Mémoire sur la propagation du mouvement dans les milieux élastique, *Mémoires de l'Académie Royal des Sciences* **10**, 549–605.

1831 Poisson, S.D.: Mémoire sur les équations generales de l'équilibre et du mouvement des corps solides élastiques et des fluides, *Journal de l'École Polytechnique* **13**, Cahier 20, Paris, 1–174.

1833 Lamé, G.; Clapeyron, B.P.E.: Mémoire sur l'équilibre intérieur des corps solides homogènes, *Mémoires présentés par divers Savants* **4**, Paris, 465–562.

1834 Clapeyron, É.: Mémoire sur la puissance motrice de la chaleur, *Journal de l'École Polytechnique* **14**, Cahier 23, 153–190 (also: Mendoza, 1960).

1839 Green, G.: On the laws of reflexion and refraction of light at the common surface of two non-crystallized media, *Transactions of the Cambridge Philosophical Society* (Trans Camb Phil Soc) **7**, 1–24.

1840 Poncelet, J.V.: Mémoire sur la stabilité des revêtements et de leurs foundation..., Bachelier, Paris, Memorial de l'officier du génre XIII, 7–261, 262–270. German translation by Lohmeyer: Über die Stabilität der Erdbekleidungen und deren Fundamente, Braunschweig 1844.

1842 Green, G.: On the propagation of light in crystallized media, *Transactions of the Cambridge Philosophical Society* (Trans Camb Phil Soc) **7**, 113–120.

Mayer, J. R.: Bemerkungen über die Kräfte der unbelebten Natur, *Annalen der Chemie und Pharmazie* (Ann Chem Pharm) **42**, 233–240 (also: Julius Robert Mayer, Die Mechanik der Wärme, ed. by H.P. Münzenmayer, 1978, Stadtarchiv Heilbronn).

1843 De Saint-Venant, B.: Mémoire sur la dynamique des fluides, note in *Comptes Rendus* **17**, Paris, 1240–1243.

1845 Holtzmann, C.: Wärme und Elastizität der Gase und Dämpfe, Verlag von Tobias Loeffer, Mannheim.

Stokes, G.G.: On the theories of the internal friction of fluids in motion, and of equilibrium and motion of elastic solids, *Transactions of the Cambridge Philosophical Society* (Trans Camb Phil Soc) **8**, 287–319.

1846 Holtzmann, C.: On the heat and elasticity of gases and vapours, *Philosophical Magazine and Journal of Science* (Phil Mag J Sci) **4**, London, 189–217.

Stokes, G.G.: Report on recent researches in hydrodynamics, Report of the British Association for 1846, Part I, 1–31.

1847 Helmholtz, H.: Über die Erhaltung der Kraft, Druck und Verlag von G. Reimer, Berlin (also: Oswald's Klassiker der exakten Wissenschaft Nr. 1, Verlag von Wilhelm Engelmann, Leipzig 1915).

1848 Delesse, A.: Pour déterminer la composition des roches, *Annales des Mines Paris* 4. Ser. **13**, 379–388.

Wolfers, J. Ph.: Leonard Euler's Mechanik oder analytische Darstellung der Wissenschaft von der Bewegung, Erster Theil, Zweiter Theil (1850), C. A. Koch's Verlagshandlung, Greifswald.

1849 Stokes, G.G.: On the dynamical theory of diffraction, *Transaction of the Cambridge Philosophical Society* (Trans Camb Phil Soc) **9**, 1–62.

Thomson, W.: An account of Carnot's theory of the motive power of heat, with numerical results deduced from Regnault's experiments on steam, *Transaction of the Royal Society of Edingburgh* (Trans Soc Edinb) **16**, 541–574.

1850 Clausius, R.: Über die bewegende Kraft der Wärme und die Gesetze, welche sich daraus für die Wärmelehre selbst ableiten lassen, *Poggendorffs Annalen der Physik (und Chemie)* (Pogg Ann) **79**, 368–397, 500–524 (also: Ostwalds Klassiker der exakten Wissenschaften, ed. by M. Planck, 1921).

Joule, J.P.: On the mechanical equivalent of heat, *Philosophical Transaction of the Royal Society of London* (Philos Trans R Soc), 61–82 (also:

Das mechanische Aequivalent der Wärme, German by Spengel, Vieweg und Sohn 1872, Braunschweig).

1851 Macquorn Rankine, W. J.: On the mechanical action of heat, especially in gases and vapours, *Transaction of the Royal Society of Edingburgh* (Trans Soc Edingb) **20**, 147.

Mayer, J. R.: Bemerkungen über das mechanische Äquivalent der Wärme, Heilbronn. (also: Julius Robert Mayer, Die Mechanik der Wärme, (ed. by H.P. Münzenmayer, 1978, Stadtarchiv Heilbronn).

Scheffler, H.: Über den Druck im Innern einer Erdmasse, *Crelle's Journal der Baukunst* **30**, 185–222.

Thomson, W.: On the dynamical theory of heat, with numerical results deduced from Joule's equivalent and Regnault's experiments on steam, *Transaction of the Royal Society of Edingburgh* (Trans Soc Edingb) **20**, 261 and *Philosophical Magazine and Journal of Science* (Phil Mag J Sci) **4** (1852), 8–21, 105–117, 168–176.

1852 Lamé, G.: Leçons sur la théorie mathématique de l'élasticité des corps solides, Paris.

Thomson, W.: On an universal tendency in nature to the dissipation of mechanical energy, *Philosophical Magazine and Journal of Science* (Phil Mag J Sci), London **4**, 304–306.

1853 Wolfers, J. Ph.: Leonhard Euler's Theorie der Bewegung fester und starrer Körper, C. A. Koch's Verlagshandlung, Greifswald.

1854 Clausius, R.: Über eine veränderte Form des zweiten Hauptsatzes der mechanischen Wärmetheorie, *Annalen der Physik und Chemie* (Ann Ph Ch) **18**, 481–506.

Lamé, G.: Mémoire sur l'équilibre d'élasticité des enveloppes shériques, published in *Lionville's Journal de mathématiques* **19**.

Macquorn Rankine, W.J.: On the mechanical action of heat, *Philosophical Magazine and Journal of Science* (Phil Mag J Sci), London **7**, 1–21, 111–122, 172–185.

1855 Fick, A.: Ueber Diffusion, *Annalen der Physik und Chemie* (Ann Ph Ch) **94**, 59–86.

1856 Darcy, H.: Les fontaines publiques de la ville Dijon, Dalmont, Paris.

Holtzmann, C.: Die Vertheilung des Drucks im Innern eines Körpers, in: Einladungs-Schrift der K. Polytechnischen Schule in Stuttgart zu der Feier des Geburtsfestes Seiner Majestät des Königs Wilhelm von Württenberg auf den 27. September 1856.

1856/57 Macquorn Rankine, W.J.: On the mathematical theory of the stability of earthwork and masonry, in a letter to Prof. Stokes, Received February 19, 1856, *Proceedings of the Royal Society of London* (Proc Lond) **8**, 60–61.

1857 Macquorn Rankine, W.J.: On the stability of loose earth, *Philosophical Transactions of the Royal Society of London* (Philos Trans R Soc) **147**, London, 9–27.

1859 Kirchhoff, G.: Über das Gleichgewicht und die Bewegung eines unendlich dünnen elastischen Stabes, *Journal für die reine und angewandte Mathematik* (J R A M) **56**, 285–313.

Thomson, W.: On the thermo-elastic and thermo-magnetic properties of matter, *Quarterly Journal of Mathematics* (Q J Math) **1**, 57–77.

1862 Clausius, R.: Über die Anwendungen des Satzes von der Aequivalenz der Verwandlungen auf die innere Arbeit, *Poggendorffs Annalen der Physik (und Chemie)* (Pogg Ann) **116**, 73–112.

Clebsch, A.: Theorie der Elastizität fester Körper, Druck und Verlag von B. G. Teubner, Leipzig.

1863 Airy, G.B.: On the strains in the interior of beams, *Philosophical Transaction of the Royal Society of London* (Philos Trans R Soc) **153**, 49–80.

Dupuit, J.: Etudes théoriques et practiques sur le mouvement des eaux dans les canaux découverts et a travers les terrains perméables, 2 ed. Dunod, Paris.

1864 Tresca, H.: Mémoire sur l'ecoulement des corps solides soumis á de fortes pressions, *Comptes Rendus Acad. Sci.* Paris **59**, 754–758.

1865 Clausius, R.: Über verschiedene für die Anwendung bequeme Formen der Hauptgleichungen der mechanischen Wärmetheorie, *Annalen der Physik und Chemie* (Ann Ph Ch) **125**, 353–400.

1868 De Saint Venant, B.: Two chapters on the history of the equations of elasticity, in: Moigno: Leçons de Mècanique Analytique, Statique, Paris.

1870 Considère, A.: Note sur la poussée des terres, *Annales des ponts e chaussées* **1**, 547–594.

De Saint-Venant, B.: Report in: *Comptes rendus*, Tome LXX, 217–228.

1871a De Saint-Venant, B.: Mémoire sur l'établissement des équations différentielles des mouvements intérieurs opéres dans les corps solides ductiles au delà des limites où l' élasticité pourrait les ramener à leur premier état, *Journal de Mathématiques*, Tome XVI, Paris, 308–316.

b De Saint-Venant, B.: Complément aux mémoires du 7 mars 1870 de M. de Saint-Venant et du 19 juin 1870 de M. Lévy sur les équations différentielles 'indéfinies' du mouvement intérieur des solides ductiles etc,... Equations défines ou relatives aux limites de ces corps; – Applications, *Journal de Mathématiques*, Tome XVI, Paris, 373–382.

Lévy, M.: Extrait du mémoire sur les équations générales des mouvements intérieurs des corps solides ductiles au delà des limites où l'élasticité pourrait les ramener à leur premier état, Par M. Maurice Lévy, *Journal de Mathématiques*, Tome XVI, Paris, 369–372.

Mohr, O.: Beitrag zur Theorie des Erddrucks, *Zeitschrift des Architekten- und Ingenieurvereines für das Königreich Hannover* **17**, 344–372.

Stefan, J.: Über das Gleichgewicht und die Bewegung, insbesondere die Diffusion von Gasmengen, *Kaiserliche Akademie der Wissenschaften in Wien, mathematisch-naturwissenschaftliche Klasse, Abteilung IIa* **63**, Wien, 63–124.

1872 Betti, E.: Nuovo Cimento, Bologna (2) **7**, 89.
Mohr, O.: Zur Theorie des Erddrucks, *Zeitschrift des Architekten- und Ingenieurvereines für das Königreich Hannover* **18**, 67–74 and 246–248.
a Stefan, J.: Untersuchung über die Wärmeleitung in Gasen, *Kaiserliche Akademie der Wissenschaften in Wien, mathematisch-naturwissenschaftliche Klasse, Abteilung IIa* **64**, Wien 45–69.
b Stefan, J.: Über die dynamische Theorie der Diffusion der Gase, *Kaiserliche Akademie der Wissenschaften in Wien, mathematisch-naturwissenschaftliche Klasse, Abteilung IIa* **64**, Wien 323–363.
Winkler, E.: Neue Theorie des Erddruckes, R. v. Waldheim, Wien.
1873a Gibbs, J.W.: Graphical methods in the thermodynamics of fluids, *Transactions of the Connecticut Academy of Arts and Sciences* (Trans Conn Acad Arts Sci) **2**, 309–342.
b Gibbs, J.W.: A method of geometrical representation of the thermodynamic properties of substances by means of surfaces, *Transactions of the Connecticut Academy of Arts and Sciences* (Trans Conn Acad Arts Sci) **2**, 382–404.
Lévy, M.: Essai sur une theorie rationelle de l'équilibre des terres fraîchement remués, et ses applications au calcul de la stabilité des murs de soutènement, *Journal de Liouville* **18**, 241–300.
1875 Castigliano, A.: Nuova teoria intoro all equilibro dei sistemi elastici, *Memorie Accademia Torino.*
Gibbs, J.W.: On the equilibrium of heterogeneous substances, *Transactions of the Connecticut Academy of Arts and Sciences* (Trans Conn Acad Arts Sci) **3**, 108–248, 343–524.
1876 Clausius, R.: Die mechanische Wärmetheorie, Friedrich Vieweg und Sohn, Braunschweig.
Kirchhoff, G.: Vorlesungen über Mathematische Physik. Mechanik, Verlag von B.G. Teubner, Leipzig.
Rühlmann, R.: Handbuch der mechanischen Wärmetheorie, Friedrich Vieweg und Sohn, Braunschweig.
1877/78 Lord Rayleigh: Theory of Sound, II Vol., London.
1879 Castigliano, A.: Théorie de l'equilibre des systèmes élastiques et ses applications, ed.: August Frédéric Negro, Torino.
1880 Rühlmann, M.: Hydromechanik oder die technische Mechanik flüssiger Körper, 2. ed., Hahn'sche Buchhandlung, Hannover.
Seelheim, F.: Methoden zur Bestimmung der Durchlässigkeit des Bodens, *Zeitschrift für analytische Chemie* (Z Anal Chem) **19**, 387–418.
1882 Mohr, O.: Über die Darstellung des Spannungszustandes und des Deformationszustandes eines Körperelementes und über die Anwendung derselben in der Festigkeitslehre, *Civilingenieur* **28**, 113–155.
1883 Reynolds, O.: An experimental investigation of the circumstances, which determine whether the motion of water shall be direct or sinuous, and

of the law of resistance in parallel channels, *Philosophical Transaction of the Royal Society of London* (Philos Trans R Soc) **174**, 935–982.

1884 Kröber, C.: Versuche über die Bewegung des Wassers durch Sandschichten, *Zeitschrift des Vereins deutscher Ingenieure* (Z Ver Dt Ing) **28**, 593–595, 617–619.

1885 Boussinesq, J.V.: Sur l'integration par approximations successive d'une equation ... dont dependent les pressions intérieurs d'un massif des sable á l'état ébouleux. In: Application des potentials á l'étude de l'equilibre et du mouvement des solides élastiques, Gauthier–Villas (No.27), 705–712.

Rühlmann, M.: Vorträge über Geschichte der Technischen Mechanik und Theoretischen Maschinenlehre und der damit im Zusammenhang stehenden mathematischen Wissenschaften (Erster Theil: Technische Mechanik), Baumgärtner's Buchhandlung, Leipzig, Nachdruck Georg Olms Verlag Hildesheim • New York (1979).

Beltrami, H.: *Rendiconti*, 704 – (1885); *Mathematische Annalen* (Math Ann), 94–(1903).

1886/1893 Todhunter, I.; Pearson, K.: A history of the theory of elasticity and of the strength of materials, Vol. I and II, Cambridge University Press. New edition: Dover Publications, Inc. New York (1960).

1887 Voigt, W.: Bestimmung der Elasticitätskonstanten von Beryll und Bergkrystall, *Annalen der Physik und Chemie* (Ann Ph Ch) **31**, 701–720.

1888 Cerruti, V.: Sulla deformazione di un corpo elastico isotropo per alcune speciali condizioni ai limiti, *Roma Accademia dei Lincei* Memorie della Classe di Scienze fisiche, matematiche e naturali **13**, 81.

a Voigt, W.: Bestimmung der Elasticitätskonstanten von Topas und Baryt, *Annalen der Physik und Chemie* (Ann Ph Ch) **34**, 981–1028.

b Voigt, W.: Bestimmung der Elasticitätskonstanten von Flussspath, Pyrit, Steinsalz, Syloin, *Annalen der Physik und Chemie* (Ann Ph Ch) **35**, 642–661.

1892 Beltrami, E.: Osservazioni sulla nota precendente, *Rendiconti della Reale Accademia nazionale dei Lincei Rom* (R R Ac Naz) (5), 141–142.

1893 Kötter, F.: Die Entwicklung der Lehre vom Erddruck, Jahresbericht der deutschen Mathematiker-Vereinigung, Berlin.

1894 Stäckel R. (ed.): Abhandlungen über Variations-Rechnung (Erster Theil: Abhandlungen von Joh. Bernoulli 1696, Jac. Bernoulli 1697 und Leonhard Euler 1744), Oswald' s Klassiker der exakten Wissenschaften, Nr. 46, Verlag von Wilhelm Engelmann in Leipzig.

Voigt, W.: Beobachtungen über die Festigkeit bei homogener Deformation, *Annalen der Physik und Chemie* (Ann Ph Ch), 43–46.

1895 Hazen, A.: The filtration of public water-supplies, 24th Annual Report of the State Board of Health of Massachusetts, New York.

Lamb, H.: Hydrodynamics, Cambridge, Leibniz-Verlag, München.

1897 Planck, M.: Vorlesung über Thermodynamik, Walter de Gruyter & Co, Berlin (11th edition: 1964).

1898 Föppl, A.: Vorlesungen über technische Mechanik, B.G. Teubner, Leipzig.

1899 Voigt, W.: Beobachtungen über die Festigkeit bei homogener Deformation, angestellt von L. Januskiewicz, *Annalen der Physik und Chemie* (Ann Ph Ch) **67**, 452–458.

1900 Mitchell, J.H.: On the direct termination of stress in an elastic solid, with applications to the theory of plates, *Proceedings of the London Mathematical Society - London* (Proc Lond Math Soc) **31**, 100–124.

Mohr, O.: Welche Umstände bedingen die Elastizitätsgrenze und den Bruch eines Materials?, *Zeitschrift des Vereines deutscher Ingenieure* (Z Ver Dt Ing) **44**, 1524–1530, 1572–1577.

1901 Forchheimer, P.: Wasserbewegung durch Boden, *Zeitschrift des Vereines deutscher Ingenieure* (Z Ver Dt Ing) **45**, 1736–1788.

1903 Lord Rayleigh: Obituary notice – Sir George Gabriel Stokes, Bart. 1819–1903, *Proceedings of the Royal Society of London* (Pr Lond).

1904 Huber, M.T.: Czasopismo tech. **15**, Lwóv.

1905a Volterra, V.: Un teorema sulla teoria della elasticà, *Rendiconti delle sedute della Reale Accademia dei Lincei, Classe di Scienze fisiche, matematiche e naturali* (Rend Accad Sci fis Mat) **14**, 127–146.

b Volterra, V.: Sulle distorsioni dei solidi elastici piu volte connessi, *Rendiconti delle sedute della Reale Accademia dei Lincei, Classe di Scienze fisiche, matematiche e naturali* (Rend Accad Sci fis Mat) **14**, 351-371.

1907 Love, A. E. H.: Timpe, A.: Lehrbuch der Elastizität, Verlag von B. G. Teubner, Leipzig and Berlin.

1907/14 Love, A.: Hydrodynamik: Physikalische Grundlegung, in: Enzyklopädie der Mathematischen Wissenschaften, Band IV, 15. Teilband, Verlag B.G. Teubner, Leipzig.

1909 Caratheodory, C.: Untersuchungen über die Grundlagen der Thermodynamik, *Mathematische Annalen* (Math Ann) **67**, 355–386.

Ritz, W.: Theorie der Transversalschwingungen einer quadratischen Platte, *Annalen der Physik* (Ann Phys) **28**, 89.

1911 Jaumann, G.: Geschlossenes System physikalischer und chemischer Differenzialgesetze, *Kaiserliche Akademie der Wissenschaften in Wien, mathematisch-naturwissenschaftliche Klasse, Abteilung IIa* **120**, 385–530.

von Kármán, T., Festigkeitsversuche unter allseitigem Druck, *Zeitschrift des Vereines deutscher Ingenieure* (Z Ver Dt Ing) **55**, Nr. 42, 1749–1757.

1912 Rudeloff; Panzerbieter: Versuche über den Porendruck des Wassers im Mauerwerk, Mitt. a. d. kgl. Materialprüfungsamt zu Gr. Lichterfeld, Erg. H. 1.

1913 Fillunger, P.: Der Auftrieb in Talsperren, *Österreichische Wochenschrift für den öffentlichen Baudienst* (Ö W Ö B) **19**, 532–556, 567–570.

Lorenz, H.: Lehrbuch der Technischen Physik, Verlag R. Oldenbourg, München und Berlin.

von Mises, R.: Mechanik der festen Körper im plastisch-deformablen Zustand, *Nachrichten von der königlichen Gesellschaft der Wissenschaft zu Göttingen* (Nachr kgl Ges WG), Göttingen, 582–592.

1914 Fillunger, P.: Neuere Grundlagen für die statische Berechnung von Talsperren, *Zeitschrift des Österreichischen Ingenieur- und Architekten-Vereines* (Z ö I A V), **23**, 441–447.

Mohr, O.: Abhandlungen aus dem Gebiet der Technischen Mechanik, Wilhelm Ernst und Sohn, Berlin.

1915 Böker, R.: Die Mechanik der bleibenden Formänderung in kristallisch aufgebauten Körpern, Forschungsarbeiten Heft 175/176, Berlin.

Fillunger, P.: Versuche über die Zugfestigkeit bei allseitigem Wasserdruck, *Österreichische Wochenschrift für den öffentlichen Baudienst* (Ö W Ö B) **29**, 443–448.

1917 Lohr, E.: Entropieprinzip und geschlossenes Gleichungssystem, *Kaiserliche Akademie der Wissenschaften in Wien, mathematisch-naturwissenschaftliche Klasse, Abteilung IIa* **93**, 339–421.

1921 Washburn, E. W.: The dynamics of capillary flow, *American Physical Society: the physical review*, **17**, 273–283.

1923 von Terzaghi, K.: Die Berechnung der Durchlässigkeitsziffer des Tones aus dem Verlauf der hydrodynamischen Spannungerscheinungen. *Sitzungsberichte der Akademie der Wissenschaften in Wien, mathematisch-naturwissenschaftliche Klasse, Abteilung IIa* **132**, (No 3/4), 125–138.

1924 von Terzaghi, K.: Die Theorie der hydrodynamischen Spannungserscheinungen und ihr erdbautechnisches Anwendungsgebiet, Proceedings of the first International Congress for Applied Mechanics, 1. Delft, 288–294.

1925 Lode, W.: Versuche über den Einfluß der mittleren Hauptspannung auf die Fließgrenze, *Zeitschrift für angewandte Mathematik und Mechanik* (ZAMM) **5**, 142–144.

a von Terzaghi, K.: Erdbaumechanik auf bodenphysikalischer Grundlage, Franz Deuticke, Leipzig/Wien.

b von Terzaghi, K.: Principles of soil mechanics, *Engineering News-Record* **19**, 742–746, **20**, 796–800, **21**, 832–936, **22**, 974–978, **23**, 912–915, **25**, 987–999, **26**, 1026–1029, **27**, 1064–1068.

1926 Lode, W.: Versuche über den Einfluß der mittleren Hauptspannung auf das Fließen der Metalle Eisen, Kupfer und Nickel, *Zeitschrift für Physik* (Z Phys) **36**, 913–939.

Schleicher, F.: Der Spannungszustand an der Fließgrenze (Plastizitätsbedingung), *Zeitschrift für angewandte Mathematik und Mechanik* (ZAMM) **6**, 199–216.

1927 Kozeny, J.: Über kapillare Leitung des Wassers im Boden (Aufstieg, Versickerung und Anwendung auf die Bewässerung), *Sitzungsberichte der Akademie der Wissenschaften in Wien, mathematisch-naturwissenschaftliche Klasse, Abteilung IIa* **136**, 271–309.

1928 Fillunger, P.: Zur Frage der Betonpfahlgründungen, *Zeitschrift des Österreichischen Ingenieur- und Architekten-Vereines* (Z ö I A V), Heft 41/42, 395–396, Heft 45/46, 430–431.
von Mises, R.: Mechanik der plastischen Formänderungen von Kristallen, *Zeitschrift für angewandte Mathematik und Mechanik* (ZAMM) **8**, 161–185.
Schleicher, F.: Über die Sicherheit gegen Überschreiten der Fließgrenze bei statischer Beanspruchung, *Der Bauingenieur* (Bauing) **9**, 253–261.
Stern, O.: Zur Frage der Betonpfahlgründungen, *Zeitschrift des Österreichischen Ingenieur- und Architekten-Vereines* (Z ö I A V), Heft 43/44, 416, Heft 47/48, 452.
1929 Fillunger, P.: Auftrieb und Unterdruck in Talsperren, *Wasserwirtschaft* **22**, 334–336, 371–377, 380–390.
Hoffman, O.: Zur Frage des Auftriebs in Talsperren, *Wasserwirtschaft* **22**, 562–566.
1930a Fillunger, P.: Zur Frage des Auftriebs in Talsperren, *Wasserwirtschaft* **23**, 63–66.
b Fillunger, P.: Auftrieb und Unterdruck in Staumauern, Transactions second World Power Conference **9**, VDI-Verlag, Berlin, 323–329.
Forchheimer, P.: Hydraulik, 3. ed., B.G. Teubner, Leipzig/Berlin.
Ortenblad, A.: Mathematical theory of the process of consolidation of mud deposits, *Journal of Mathematics and Physics* **9**, 73–149.
1931 von Terzaghi, K.: Festigkeitseigenschaften der Schüttungen, Sedimente und Gele, in: Auerbach & Hort (eds.): Handbuch der physikalischen und technischen Mechanik, Band IV, 513–578.
1932 von Terzaghi, K.: Mein Lebensweg und meine Ziele, Selbstverlag des Verfassers, Wien.
1933 Fromm, H.: Stoffgesetze des isotropen Kontinuums, insbesondere bei zähplastischem Verhalten, *Ingenieur-Archiv* (Ing-Arch) **4**, 432–466.
von Terzaghi, K.: Auftrieb und Kapillardruck an betonierten Talsperren, *Wasserwirtsch.* **26**, 397–399.
1934a Fillunger, P.: Nochmals der Auftrieb in Talsperren, *Zeitschrift des Österreichischen Ingenieur- und Architekten-Vereines* (Z ö I A V), Heft 5/6.
b Fillunger, P.: Die wirksame Flächenporosität Prof. Terzaghis, *Zeitschrift des Österreichischen Ingenieur- und Architekten-Vereines* (Z ö I A V), Heft 7/8.
c Fillunger, P.: Der Kapillardruck in Talsperren. *Wasserwirtschaft* **27**, Heft 13/14.
Geiringer, H.; Prager, W.: Mechanik isotroper Körper im plastischen Zustand, Ergebnisse der exakten Naturwissenschaften **13**, Berlin, Julius Springer, 310–362.
Lehr, E.: Spannungsverteilung in Konstruktionselementen, V.D.I.-Verlag, Berlin.

a von Terzaghi, K.: Zuschrift zum Aufsatz Fillunger: "Zur Bestimmung der maximalen Flächenporosität des Betons" and "Erwiderung" by Fillunger, *Der Bauingenieur* (Bauing) **15**, 413.

b von Terzaghi, K.: Zuschrift zum Aufsatz P. Fillunger, *Zeitschrift des Österreichischen Ingenieur- und Architekten-Vereines* (Z ö I A V), Heft 5/6, 30–32.

von Terzaghi, K.; Rendulic, L.: Die wirksame Flächenporosität des Betons, *Zeitschrift des Österreichischen Ingenieur- und Architekten-Vereines* (Z ö I A V), Heft 1/2, 1–9.

1935 Biot, M. A.: Le problème de la consolidation des matières argileuses sous une charge, *Annales de la Societé scientifique de Bruxelles* (Ann Soc sci Brux) **B 55**, 110–113.

Fillunger, P.: Das Delessesche Gesetz, *Monatshefte für Mathematik und Physik* (Mh M Ph) **42**, 87–96.

1936 Fillunger, P.: Erdbaumechanik?, Selbstverlag des Verfassers, Wien.

Rendulic, L.: Porenziffer und Porenwasserdruck in Tonen, *Der Bauingenieur* (Bauing) **17**, 559–564.

von Terzaghi, K.: The shearing resistance of saturated soils and the angle between the planes of shear. *Internationale Conference on Soil Mechanics and Foundation Engineering, 1.* Cambridge, Mass. Proceedings, Vol. I, 54–56.

von Terzaghi, K.; Fröhlich, O. K.: Theorie der Setzung von Tonschichten, Franz Deuticke, Leipzig /Wien.

1937 Fillunger, P.: Erdbaumechanik und Wissenschaft: Eine Erwiderung (ed. Erwin Fillunger), Selbstverlag, Wien.

Murnaghan, F.D.: Finite deformation of an elastic solid, *American Journal of Mathematics* (Amer J Math) **59**, 235–260.

von Terzaghi, K.: Die mechanische Wirkung des Auftriebs in porösen Körpern. Unpublished manuscript. Terzaghi Library, Norwegian Geotechnical Institute, Oslo.

von Terzaghi, K.; Fröhlich, O. K.: Erdbaumechanik und Baupraxis : Eine Klarstellung, Franz Deuticke, Leipzig • Wien.

1938 Flamm, L.: Beitrag zur Theorie der Setzung von Tonschichten, *Wasserkraft und Wasserwirtschaft* (Wass Kr) **33**, 97–98.

a Heinrich, G.: Wissenschaftliche Grundlagen der Theorie der Setzung von Tonschichten, *Wasserkraft und Wasserwirtschaft* (Wass Kr) **33**, 5–10.

b Heinrich, G.: Theorie eines mechanischen Modells zur Veranschaulichung der Setzung von Tonschichten, *Deutsche Wasserwirtschaft* (D Wass W) **33**, 317–321.

Melan, E.: Zur Plastizität des räumlichen Kontinuums, *Ingenieur-Archiv* (Ing-Arch) **9**, 116–126.

1940 Mooney, M.: A theory of large elastic deformation, *Journal of Applied Physics* (J Appl Phys) **11**, 582–592.

1941a Biot, M.A.: General theory of three-dimensional consolidation, *Journal of Applied Physics* (J Appl Phys) **12**, 155–164.

b Biot, M.A.: Consolidation settlement under a rectangular load distribution, *Journal of Applied Physics* (J Appl Phys) **12**, 426–430.

Biot, M.A.; Clingan, F.M.: Consolidation settlement of a soil with an impervious top surface, *Journal of Applied Physics* (J Appl Phys) **12**, 578–581.

1942 Biot, M.A.; Clingan, F.M.: Bending settlement of a slab resting on a consolidating foundation, *Journal of Applied Physics* (J Appl Phys) **13**, 35–40.

1943 von Terzaghi, K.: Theoretical soil mechanics, Wiley, New York.

1944 Frenkel, I.: On the theory of seismic and seismoelectric phenomena in a moist soil, *Journal of physics* (J Phys) **8**, 230–241.

1948 Rivlin, R. S.: Large elastic deformations of isotropic materials, I. Fundamental concepts, *Philosophical Transactions of the Royal Society of London* (Philos Trans R Soc) A **240**, 459–490.

1951 Rivlin, R. S.; Saunders, D. W.: Large elastic deformations of isotropic materials VII. Experiments on the deformation of rubber, *Philosophical Transactions of the Royal Society of London* (Philos Trans R Soc) A **243**, 251–288.

1952 Drucker, D. C.; Prager, W.: Soil mechanics and plastic analysis of limit design, *Quarterly of Applied Mathematics* (Quart Appl Math) **10**, 157–165.

1953 Timoshenko, S. P.: History of Strength of Materials, Mc Graw-Hill Company Inc, New York • Toronto • London.

1954 Green, A. E.; Zerna, W.: Theoretical Elasticity, Oxford: Clarendon Press.

Truesdell, C.: Rational fluid mechanics, 1687–1765 (Editor' s introduction to Euleri Opera omnia II 12, IX - CXXV).

1955 Biot, M. A. : Theory of elasticity and consolidation for a porous anisotropic solid, *Journal of Applied Physics* (J Appl Phys) **26**, 182–185.

Heinrich, G.; Desoyer, K.: Hydromechanische Grundlagen für die Behandlung von stationären und instationären Grundwasserströmungen, *Ingenieur-Archiv* (Ing-Arch) **23**, 73–84.

Prager, W.: The theory of plasticity: A survey of recent achievements, *Proceedings of the Institution of Mechanical Engineers* (Proc Inst Mech Eng) **169**, 41–57.

1956a Biot, M. A.: Theory of propagation of elastic waves in a fluid-saturated porous solid. I. Low-frequency range, *Journal of Acoustical Society of America* (J Acoust Soc Am) **28**, 168–178.

b Biot, M. A.: Theory of propagation of elastic waves in a fluid-saturated porous solid. II. Higher frequency range, *Journal of Acoustical Society of America* (J Acoust Soc Am) **28**, 179–191.

c Biot, M. A.: General solutions of the equations of elasticity and consolidation for a porous material, *Journal of Applied Mechanics* (J Appl Mech) **23**, 91–96.

d Biot, M. A.: Theory of deformations of a porous viscoelastic anisotropic solid, *Journal of Applied Physics* (J Appl Phys) **27**, 459–467.

Doyle, T. C.; Erickson, J. L.: Nonlinear Elasticity, Advances in Appl. Mech. IV, Academic Press, New York.
Heinrich, G.; Desoyer, K.: Hydromechanische Grundlagen für die Behandlung von stationären und instationären Grundwasserströmungen, II. Mitteilung, *Ingenieur-Archiv* (Ing-Arch) **24**, 81–84.
Prager, W.: A new method of analyzing stresses and strains in work-hardening plastic solids, *Journal of Applied Mechanics* (J App Mech) **23**, 493–496.
a Truesdell, C.: Zur Geschichte des Begriffs innerer Druck, *Physikalische Blätter* (Phys Bl) **12**, 315–326.
b Truesdell, C.: Das ungelöste Hauptproblem der endlichen Elastizitätstheorie, *Zeitschrift für angewandte Mathematik und Mechanik* (ZAMM) **36**, 97–103.
1957 Biot, M. A.; Willis, D. G.: The elastic coefficients of the theory of consolidation, *Journal of Applied Mechanics* (J App Mech) **24**, 594–601.
Drucker, D. C.; Gibson, R. E.; Henkel, D. J.: Soil mechanics and work-hardening theories of plasticity, *American Society of Civil Engineers Transaction* Transaction (ASCE) **122**, 338–346.
a Truesdell, C.: Sulle basi della termomeccanicia *Rendiconti della Reale Accademia nazionale dei Lincei, Roma Classe di Scienze fisiche, matematiche e naturali* (R R Ac Naz) (8) **22**, 33–38, 158–166.
b Truesdell, C.: Euler's Leistungen in der Mechanik, *L'Enseignement Mathematique* **3**, 251–262.
1958 Kauderer, H.: Nichtlineare Mechanik, Springer, Berlin • Göttingen • Heidelberg.
Shield, R. T.; Ziegler, H.: On Prager's hardening rule, *Zeitschrift für angewandte Mathematik und Physik* (Z A M P) **9a**, 260–276.
1959 Fatt, I.: The Biot-Willis elastic coefficients for a sandstone, *Journal of Applied Mechanics* (J Appl Mech) **26**, 296–297.
Serrin, J.: Mathematical principles of classical fluid mechanics, in: Flügge, S. (ed.), Handbuch der Physik, Band VIII/1, Springer-Verlag, Berlin • Göttingen • Heidelberg.
Ziegler, H.: A modification of Prager's hardening rule, *Quarterly of Applied Mathematics* (Quart Appl Math) **17**, 55–65.
1960 Casagrande, A.: Karl von Terzaghi – His Life and Achievements, in: From theory to practice in soil mechanics (eds. Bjerrum, L., Casagrande, A., Peck, R. B., Skempton, A. W.), Wiley, New York • London.
Mendoza, E.: Reflections on the motive power of fire by Sadi Carnot and other papers on the second law of thermodynamics by E. Clapeyron and R. Clausius (ed. by E. Mendoza). Dover Publications Inc., New York.
Skempton, A. W.: Significance of Terzaghi's concept of effective stress (Terzaghi's discovery of effective stress) in: From theory to practice in soil mechanics (eds. Bjerrum, L., Casagrande, A., Peck, R. B., Skempton, A. W.), Wiley, New York • London.

Truesdell, C.; Toupin, R.: The Classical Field Theories, in: Handbuch der Physik (ed. S. Flügge) Vol. III/1, Springer-Verlag Berlin • Göttingen • Heidelberg.

1961 Deresiewicz, H.: The effect of boundaries on wave propagation in a liquid filled porous solid, II. Love waves in a porous layer, *Bulletin of the Seismologital Society of America* (Bull Seism Soc Am) **51**, 51–59.

Heinrich, G. und Desoyer, K.: Theorie dreidimensionaler Setzungsvorgänge in Tonschichten, *Ingenieur-Archiv* (Ing-Arch) **30**, 225–253.

1962a Biot, M. A.: Mechanics of deformation and acoustic propagation in porous media, *Journal of Applied Physics* (J Appl Phys) **33**, 1482–1498.

b Biot, M. A.: Generalized theory of acoustic propagation in porous dissipative media, *Journal of Acoustical Society of America* (J Acoust Am) **34**, 1254–1264.

Deresiewicz, H.: The effect of boundaries on wave propagation in a liquid filled porous solid, IV. surface waves in a half-space, *Bulletin of the Seismological Society of America* (Bull Seism Soc Am) **52**, 627-638.

Heinrich, G.: Der Zusammenhang zwischen Anfangs- und Endgrößen bei dreidimensionalen Setzungsvorgängen in Tonschichten, *Sitzungsberichte der Akademie der Wissenschaften in Wien, mathematisch-naturwissenschaftliche Klasse, Abteilung IIa* **171**, 245–252.

1963 Cammerer, W. F.: Die kapillare Flüssigkeitsbewegung in porösen Körpern, VDI Forschungsheft 500: Beilage zu "Forschung auf dem Gebiet des Ingenieurwesens" Ausgabe B **29**.

Coleman, B. D.; Noll W.: The thermodynamics of elastic materials with heat conduction and viscosity, *Archives for Rational Mechanics and Analysis* (Arch Rat Mech An) **13**, 167–178.

Scheidegger, A. E.: Hydrodynamics in porous media, in: Handbuch der Physik (ed. S. Flügge), Bd. VIII/2, Springer-Verlag, Berlin • Göttingen • Heidelberg.

1964 Coleman, B. D.; Mizel, V. J.: Existence of caloric equations of state in thermodynamics, *Journal of Chemical Physics* (J Chem Phys) **40**, 1116–1125.

Kelly, P. D.: A reacting continuum, *International Journal of Engineering Science* (Int J Eng Sci) **2**, 129–153.

Truesdell, C.: Die Entwicklung des Drallsatzes, *Zeitschrift für angewandte Mathematik und Mechanik* (ZAMM) **44**, 149–158.

1965 Abramowitz, M.; Stegun, J. A.: Handbook of Mathematical Functions, National Bureau of Standards, Washington D. C.

Eringen, A. C.; Ingram, J. D.: A continuum theory of chemically reacting media I, *International Journal of Engineering Science* (Int J Eng Sci) **3**, 231-241.

Green, A. E.; Naghdi, P. M.: A dynamical theory of interacting continua, *International Journal of Engineering Science* (Int J Eng Sci) **3**, 231–241.

Truesdell, C.; Noll, W.: The Non-Linear Field Theories of Mechanics. In: Handbuch der Physik Band III/3 (ed. Flügge, S.), Springer-Verlag, Berlin • Heidelberg • New York.

1966 Mills, N.: Incompressible mixture of Newtonian fluids, *International Journal of Engineering Science* (Int J Eng Sci) **4**, 97–112.

1967 Bowen, R. M.: Toward a thermodynamics and mechanics of mixtures, *Archive for Rational Mechanics and Analysis* (Arch Rat Mech An) **24**, 370–403.

Mills, N.: On a theory of multi-component mixtures, *Quarterly Journal of Mechanics and Applied Mathematics* (Quart J Mech appl Math) **20**, 449–508.

Mróz, Z.: On the description of anisotropic workhardening, *Journal of the Mechanics and Physics of Solids* (J Mech Phys Sol) **15**, 163–175.

1968 Blanc, C.: Préface des volumes II 8 et II 9, Opera Omnia (Leonardi Euleri), Basel.

Eisenberg, M. A.; Phillips, A.: On non-linear kinematic hardening, *Acta Mechanica* (Acta mech) **5**, 1–13.

Macvean, D. B.: Die Elementararbeit in einem Kontinuum und die Zuordnung von Spannungs- und Verzerrungstensoren, *Zeitschrift für angewandte Mathematik und Physik* (Z A M P) **19**, 157–185.

Müller, I.: A thermodynamic theory of mixtures of fluids, *Archive for Rational Mechanics and Analysis* (Arch Rat Mech An) **28**, 1–39.

Roscoe, K. H.; Burland, J. B.: On the generalized stress-strain behaviour of "wet" clay, in Engineering Plasticity, (eds.: Heyman and Leckie), Cambridge University Press, 535–609.

Truckenbrodt, E.: Strömungsmechanik, Springer-Verlag, Berlin • Heidelberg • New York.

Truesdell, C.: Essays in the history of mechanics, Springer-Verlag, Berlin • Heidelberg • New York.

1969 Bjerrum, L.: The young Terzaghi and his way to soil mechanics: Talk presented at the occasion of the unveiling of a memorial tablet on Terzaghi's birthplace, 22nd October, 1969 (unpublished).

Bowen, R. M.; Wiese, J. C.: Diffusion in mixtures of elastic materials, *International Journal of Engineering Science* (Int J Eng Sci) **7**, 689–722.

Suklje, L.:Rheological aspects of soil mechanics, Wiley Interscience, New York.

Truesdell, C.: Rückwirkungen der Geschichte der Mechanik auf die moderne Forschung, *Humanismus und Technik* **13**, 12–13.

1969/84 Truesdell, C.: Historical Introit. The origins of rational thermodynamics. Appendix to the Historical Introit. Failure of Carathéodory's attempt to set the house in order. In Rational Thermodynamics (ed.: C. Truesdell), Springer-Verlag, New York • Berlin • Heidelberg • Tokyo (2. ed. 1984).

1970 Richart, F.E., Hall, I.R. and Woods, R.D.: Vibrations of soils and foundations, Prentice-Hall. New Jersey.

1971 Craine, R. E.: Oscillations of a plate in a binary mixture of incompressible Newtonian fluids, *International Journal of Engineering Sciences* (Int J Eng Sci) **9**, 1177–1192.

Di Maggio, F. L.; Sandler, I. S.: Material model for granular soils, *Proceedings of the American Society of Civil Engineers* Division **97**, 935–950.

Magnus, K.; Kreisel: Theorie mit Anwendungen, Springer-Verlag, Berlin • Heidelberg • New York.

Nur, A.; Byerlee, J. D.: An exact effective stress law for elastic deformation of rock with fluid, *Journal of Geophysical Research* (J Geoph Res) **76**, 6414–6419.

Truesdell, C.: The tragicomedy of classical thermodynamics, Springer-Verlag, Wien • New York.

1972 Biot, M. A.: Theory of finite deformations of porous solids, *Indiana University Mathematics Journal* **21**, 597–620.

Elwell, D.; Pointon, A. J.: Classical thermodynamics, Penguin Books Ltd., Harmondsworth, Middlesex, England • Baltimore • Ringwood.

Goodman, M. A.; Cowin, S. C.: A continuum theory for granular materials, *Archive for Rational Mechanics and Analysis* (Arch Rat Mech An) **44**, 249–266.

Green, R. J.: A plasticity theory for porous solids, *International Journal of Mechanics Science* (Int J Mech Sci) **14**, 215–224.

Lade, P. V.: The stress-strain and strength characteritics of cohesionless soils, dissertation submitted in partial satisfaction of the requirement for the degree of Doctor Philosophy in Engineering in the Graduate Division of the University of California, Berkeley.

Morland, L. W.: A simple constitutive theory for a fluid-saturated porous solid, *Journal of Geophysical Research* (J Geoph Res) **77**, 890–900.

1973 Gudehus, G.: Elastoplastische Stoffgleichungen für trockenen Sand, *Ingenieur-Archiv* (Ing-Arch) **42**, 151–169.

Wang, C. C.; Truesdell, C.: Introduction to rational elasticity, Noordhoff International Publishing, Leyden.

1974 de Boer R.: Zur Theorie der viskoplastischen Stoffe, *Zeitschrift für angewandte Mathematik und Physik* (Z A M P) **25**, 195–208.

Matzuoka, H.; Nakai, T.: Stress-deformation and strength characteristics of soil under three different principal stresses, *Soils and Foundations* (Soils Found) **14**, 59–70.

1975 Dafalias, Y. F.; Popov E. P.: A model of nonlinearly hardening materials for complex loading, *Acta Mechanica* (Acta mech) **21**, 173–192.

Lade, P. V.; Duncan, J. M.: Elastoplastic stress-strain theory for cohesionless soil, *Journal of Geotechnical Engineering Division. Proceedings of the American Society of Civil Engineers* (J Geotech Eng) **101**, No. GT10, 1037–1053.

Päsler, M.: Phänomenologische Thermodynamik – mit einer Einführung in die Thermodynamik irreversibler Prozesse von Jürgen U. Keller, Walter de Gruyter – Berlin • New York.

Prévost, J.; Hoeg, K.: Effective stress-strain strength model for soils, *Journal of Geotechnical Engineering Division. Proceedings of the American Society of Civil Engineers,* (J Geotech Eng) **101**, 237–278.

Willam, K. J.; Warnke, E. P.: Constitutive models for the triaxial behaviour of concrete, *International Association Bridge Structure Engineering Proceedings* **19**, 1–30.

1975/76 Truesdell, C.: Early kinetic theories of gases, *Archive for History of Exact Sciences* (Arch Hist ex Sci) **15**, 1–65.

1976 Bowen, R. M.: Theory of mixtures, in: Continuum Physics (ed.: Eringen, A. C.), Vol. III, Academic Press, New York • San Francisco • London.

a Kenyon, D. E.: Thermostatics of solid-fluid mixtures, *Archive for Rational Mechanics and Analysis* (Arch Rat Mech An) **62**, 117–129.

b Kenyon, D. E.: The theory of an incompressible solid-fluid mixture, *Archive for Rational Mechanics and Analysis* (Arch Rat Mech An) **62**, 131–147.

Lade, P. V.; Duncan, J. M.: Stress-path dependent behaviour of cohesionless soil, *Journal of Geotechnical Engineering Division. Proceedings of the American Society of Civil Engineers.* (J Geotech Eng) **102**, No. GT1, 51–68.

Parkus, H.: Thermoelasticity, 2. ed., Springer-Verlag, Wien • New York.

Rice, J. R.; Cleary, M. P.: Some basic stress diffusion solutions for fluid-saturated elastic porous media with compressible constituents, *Reviews of Geophysics and Space Physics* **14**, 227–241.

1977 Lade, P. V.: Elasto-plastic stress-strain theory for cohesionless soil with curved yield surfaces, *International Journal of Solids and Structures* (Int J Solids Structures) **13**, 1019–1035.

Ottosen, N. S.: A failure criterion for concrete, *Journal of Engineering Mechanics,* (ASCE) **103**, 527–535.

Passmann, S. L.: Mixtures of granular materials, *International Journal of Engineering Science* (Int J Eng Sci) **15**, 117–129.

1978 Bedford, A.; Drumheller, D. S.: A variational theory of immiscible mixtures, *Archive for Rational Mechanics and Analysis* (Arch Rat Mech An) **68**, 37–51.

Derski, W.: Equations of motion for a fluid-saturated porous solid, *Bulletin de l'Academie polonaise des Sciences – Warschau* (Bull Acad pol Sci), Ser. Sci. Techn. **26**, 11–16.

Drumheller, D. S.: The theoretical treatment of a porous solid using a mixture theory, *International Journal of Solids and Structures* (Int J Solids Structures) **14**, 441–456.

Mróz, Z.; Norris, V. A.; Zienkiewicz, O. C.: An anisotropic hardening model for soils and its application to cyclic loading, *International Journal for Numerical and Analytical Methods in Geomechanics* (Int J Numer Anal Methods Geomech) **2**, 203–221.

Truesdell, C.: Absolute temperatures as a consequence of Carnot's general axiom, *Archive for History of exact Sciences* (Arch Hist ex Sci) **19**, 357–380.

1979 Bedford, A.; Drumheller, D. S.: A variational theory of porous media, *International Journal of Solids and Structures* (Int J Solids Structures) **15**, 967–980.

Kowalski, S. J.: Comparison of Biot equations of motion for a fluid-saturated porous solid with those of Derski, *Bulletin de l'Academie polonaise des sciences – Warschau* (Bull Acad pol Sci), Ser. Sci. Techn. **27**, 455–461.

Lade, P. V.: Three-dimensional stress-strain behaviour and modeling of soils, Schriftenreihe des Instituts für Grundbau, Wasserwesen und Verkehrswesen, Ruhr-Universität Bochum, Bochum.

Müller, I.: Entropy in non-equilibrium – a challenge to mathematicians, in: Trends in Applications of Pure Mathematics in Mechanics (H. Zorski, ed.) (Kozubuik, 1977), Pitman, 281–295.

Nunziato, J. W.; Cowin, S. C.: A nonlinear theory of elastic materials with voids, *Archive for Rational Mechanics and Analysis* (Arch Rat Mech An) **72**, 175 –201.

Sampaio, R.; Williams, W. O.: Thermodynamics of diffusing mixtures, *Journal de méchanique* **18**, 19–45.

1980 de Boer, R.; Ehlers, W.: Grundlagen der isothermen Plastizitäts- und Viskoplastizitätstheorie, Forschungsberichte aus dem Fachbereich Bauwesen **14**, Universität-GH Essen.

Bowen, R. M.: Incompressible porous media models by use of the theory of mixtures, *International Journal of Engineering Science* (Int J Eng Sci) **18**, 1129–1148.

Desai, C. S.: A general basis for yield, failure and potential functions in plasticity, *International Journal for Numerical and Analytical Methods in Geomechanics* (Int J Numer Anal Methods Geomech) **4**, 361–375.

Mow, V. C.; Knei, S. C.; Lai, W. M.; Amstrong, C. G.: Biphasic creep and stress relaxation of articular cartilage in compression: theory and experiments, *Journal of Biomechanical Engineering* (J Biomech Eng) (ASME) **102**, 73–84.

Nunziato, J. W.; Walsh, E. K.: On ideal multiphase mixtures with chemical reactions and diffusion, *Archive for Rational Mechanics and Analysis* (Arch Rat Mech An) **73**, 285–311.

a Truesdell, C.: An essay review of Geschichte der mechanischen Prinzipien und ihrer wichtigsten Anwendungen, *Centaurus* **23**; no. 2, 163–175.

b Truesdell, C.: The tragicomical history of thermodynamics 1822-1854, Springer-Verlag New York • Heidelberg • Berlin.

1981 Mróz, Z.; Norris, V. A.; Zienkiewicz, O. C.: An anisotropic critical state model for soils subject to cyclic loading, *Géotechnique* **31**, 451–469.

Nunziato, J. W.; Passman, S. L.: A multiphase mixture theory for fluid-saturated granular materials, in: Mechanics of Structured Media A (ed.: A. P. S. Selvadurai), Elsevier, Amsterdam, 243–254.

Prévost, J. H.: Consolidation of anelastic porous media, *Journal of the Engineering Mechanics Division* (J Eng Mech) Div., (ASCE) 107, EM1, 169–186.

Zammattio, C., Marinomi, A.; Brizio, A. M.: Leonardo der Forscher, Belzer Verlag, Stuttgart und Zürich.

1982 Baker, R.; Desai, C. S.: Consequences of deviatoric normality in plasticity with isotropic strain hardening, *International Journal for Numerical and Analytical Methods in Geomechanics* (Int J Numer Anal Methods Geomech) **6**, 383–390.

de Boer, R.: Vektor- und Tensorrechnung für Ingenieure, Springer, Berlin • Heidelberg • New York.

Bowen, R. M.: Compressible porous media models by use of the theory of mixtures, *International Journal of Engineering Science* (Int J Eng Sci) **20**, 697–735.

Hajra, S.; Mukhopadhyay, A.: Reflection and refraction of seismic waves indicident obliquely at the boundary of a liquid-saturated porous solid, *Bulletin of the Seismological Society of America* (Bull Seism Soc Am) **72**, 1509–1533.

Morimoto, Y.; Hayaski, T.; Takei, T.: Mechanical behavior of powders during compaction in a mold with variable cross sections, *International Journal of Powder Metallurgy and Powder Technology* **18**, 129–145.

Plona, T. J.: Acoustic of fluid-saturated porous media, Ultrasomic Symposium, IEEE, New York, 1044–1048.

1983 Bedford, A.; Drumheller, D. S.: Recent advances: Theories of immiscible and structured mixtures, *International Journal of Engineering Science* (Int J Eng Sci) **21**, 863–960.

de Boer, R.; Kowalski, S.: A plasticity theory for fluid-saturated porous solids, *International Journal of Engineering Science* (Int J Eng Sci) **21**, 1143–1357.

Carroll, M. M.; Katsube, N.: The role of Terzaghi effective stress in linearly elastic deformation, *Journal of Energy Resources Technology* **105**, 509–511.

Marsden, J. E.; Hughes, T. J. R.: Mathematical foundations of elasticity, Prentice-Hall, Englewood Cliffs, New Jersey.

McTigue, D. F.; Wilson, R. K.; Nunziato, J. M.: An effective stress principle for partially saturated granular media, in: Mechanics of Granular Materials: New Models and Constitutive Relations (ed.: J. T. Jenkins and M. Satake), Elsevier Science Publishers B.V., Amsterdam.

Mróz, Z.; Pietruszcak, St.: A constitutive model for sand with anisotropic hardening rule, *International Journal for Numerical and Analytical Methods in Geomechanics* (Int J Numer Anal Methods Geomech) 7, 305–320.
Nunziato, J. W.: A multiphase mixture theory for fluid-particle flows, in: Theory of Dispersed Multiphase Flow (ed. R. E. Meyer), Academic Press, New York, 191–226.
1984 Bowen, R. M.: Diffusion models implied by the theory of mixtures, in: C. Truesdell (ed.): Rational Thermodynamic, sec. ed., Springer-Verlag, New York • Berlin • Heidelberg • Tokio, 237–263.
Doraivelu, S. M.; Gegel, H. L.; Grunasekara, J.S.; Malas, J. C.; Morgan, J. T.; Thomas, Y. F.: A new yield function for compressible P/M materials, *International Journal of Mechanics Science* (Int J Mech Sci) **26**, 527–535.
Passmann, S. L.; McTigue, D. F.: A new approach to the effective stress principle, in: Compressibility phenomena in subsidence (ed. S. K. Saxena), Engineering Foundation 345 East 47th Street, New York. NY 10017, 79–91.
Passman, S. L.; Nunziato, J. W.; Walsh, E. K.: A theory of multiphase mixtures, in: Rational Thermodynamics (ed. C. Truesdell), Mc Graw-Hill, New York, 286–325.
Simo, J. C.; Pister, K. S.: Remarks on rate constitutive equations for finite deformation problems: Computational implications, *Computer Methods in Applied Mechanics and Engineering* (Comput Methods Appl Mech Eng) **46**, 201–215.
Truesdell, C.: Rational Thermodynamics, second ed., Springer, New York • Berlin • Tokyo.
Zienkiewitz, O. C.; Shiomi, T.: Dynamic behavior of saturated porous media; the generalised Biot formulation and its numerical solution, *International Journal for Numerical and Analytical Methods in Geomechanics* (Int J Numer Anal Methods Geomech) **8**, 71–96.
1985 de Boer, R.; Kowalski, S. J.: Extremum principles in the theory of plasticity for fluid-saturated porous media, *Ingenieur-Archiv* (Ing-Arch) **55**, 134–146.
Katsube, N.: The constitutive theory for fluid-filled porous materials, *Journal of Applied Mechanics* (J Appl Mech) **52**, 185–189.
Müller, I.: Thermodynamics, Pitman, Boston, etc.
Steiding, I.: Prof. Dr.-Ing. h. c. Christian Otto Mohr zum 150. Geburtstag, *Bauplanung und Bautechnik* (Baupl) **39**, 395–397.
Truesdell, C.: Classical thermodynamics cleansed and cured. Revised text (1987) of the lecture delivered at the meeting on Finite Thermoelasticity at the Academia dei Lincei, Rome, May 30–June 1, 1985, published on pages 265–291 of Contributi del centro Linceo interdisciplinaire di Scienze Matematiche e loro applicazione No 76, Rome, Accademia dei Lincei, 1986.

1986 Baer, M. R.; Nunziato, J. W.: A two-phase mixture theory for the deflagration-to-detonation transition (DDT) in reactive granular materials, *International Journal of Multiphase Flow* (Int J Multiphase Flow) **12**, 861–889.

de Boer, R.: Failure conditions for brittle and granular materials, *Quarterly of Applied Mathematics* (Quart appl Math) **44**, 71–79.

a de Boer, R.; Ehlers, W.: On the problem of fluid- and gas-filled elasto-plastic solids, *International Journal of Solids and Structures* (Int J Solids Structures) **22**, 1231–1242.

b de Boer, R.; Ehlers, W.: Theorie der Mehrkomponentenkontinua mit Anwendung auf bodenmechanische Probleme, Teil I, Forschungsberichte aus dem Fachbereich Bauwesen der Universität-GH Essen, Heft 40, Essen.

de Boer, R.; Kowalski, S. J.: The uniqueness theorem for the solution of boundary-value problems in the plasticity theory for fluid-saturated porous solids, *Zeitschrift für angewandte Mathematik und Mechanik* (ZAMM) **66**, 119–121.

Dziecielak, R.: Propagation and decay of acceleration waves in thermo-consolidating media, *Acta Mechanica* (Acta Mech) **59**, 213–231.

Truesdell, C.: What did Gibbs and Carathéodory leave us about thermodynamics?, in: New Perspectives in Thermodynamics (ed. James Serrin), Springer-Verlag.

1987 Beatty, M. F.: Topics in finite elasticity: hyperelasticity of rubber, elastomers and biological tissues – with examples, *Applied Mechanics Review* (Appl Mech Rev) **40**, 1699–1734.

Bourbié, T.; Coussy, O.; Zinszner, B.: Acoustic of porous media, Editions Technip, Paris.

Hirai, H.: Modelling of cyclic behaviour of sand with combined hardening, *Soils and Foundation* (Soils Found) **27**, 1–11.

Kowalski, S.J.: Thermomechanics of constant drying rate period, *Archiwum Mechaniki stosowanej* (Arch Mech stosow) **39**, 157–176.

Lewis, R. W.; Schrefler, B. A.: The finite element method in the deformation and consolidation of porous media, John Wiley & Sons.

Szabó, I.: Geschichte der mechanischen Prinzipien – und ihrer wichtigsten Anwendungen, Dritte Auflage, Birkhäuser Verlag, Basel • Bosten • Stuttgart.

1988a de Boer, R.: On plastic deformation of soils, *International Journal of Plasticity* (Int J Plasticity) **4**, 371–391.

b de Boer, R.: Constitutive equations for granular and brittle materials in the plastic range – a kinematic hardening model, Report Mech. 88/6, FB 10/Mechanik, University-GH Essen, Essen.

de Boer, R.; Ehlers, W.: Auftrieb und Reibung in flüssigkeitsgefüllten porösen Körpern – eine Klarstellung, *Zeitschrift für Angewandte Mathematik und Mechanik* (ZAMM) **68**, 567–572.

Kim, M. K.; Lade, P. V.: Single hardening constitutive model for frictional materials: I. Plastic potential functions, *Computers and Geotechnics* (Comput Geotech) **5**, 307–324.

Lade, P. V.; Kim, M. K.: Single hardening constitutive model for frictional materials: II. Yield criterion and plastic work contours, *Computers and Geotechnics* (Comput Geotech) **6**, 13–29; III. Comparisons with experimental data, *Computers and Geotechnics* (Comput Geotech) **6**, 31–47.

1989 de Boer, R.; Dresenkamp, H. T.: Constitutive equations for concrete in failure state, *Journal of Engineering Mechanics (Division)* (J Eng Mech Div) **115**, ASCE, 1591–1608.

a Desai, C. S.: Single surface yield and potential function plasticity models: A review, Letter to Editor, *Computers and Geotechnics* (Comput Geotech) **7**, 319–333.

b Desai, C. S.: Notes for advanced school numerical methods in geomechanics including constitututive modelling July 10–14, 1989, Intern. Centre for Mechanical Science Udine, Italy.

a Ehlers, W.: PORÖSE MEDIEN – ein kontinuumsmechanisches Modell auf der Basis der Mischungstheorie, Forschungsberichte aus dem Fachbereich Bauwesen **47**, Universität-GH Essen.

b Ehlers, W.: On the thermodynamics of elasto-plastic porous media, *Archiwum mechaniki stosowanej* (Arch Mech stosow) **41**, 73–93.

Gollub, W.: Grenzen und Möglichkeiten der Mohr-Coulombschen Bruchbedingung. Dissertation, Universität Essen. Printed in: Fortschrittberichte VDI Reihe **18**: Mechanik/Bruchmechanik, Nr. 68, Verlag des Vereins Deutscher Ingenieure, Düsseldorf, 1989.

1990 de Boer, R.; Brauns, W.: Kinematic hardening of granular materials, *Ingenieur-Archiv* (Ing-Arch) **60**, 463–480.

a de Boer, R.; Ehlers, W.: Uplift, friction and capillarity: Three fundamental effects for liquid-saturated porous solids, *International Journal of Solids and Structures* (Int J Solids Structures) **26**, 43–51.

b de Boer, R.; Ehlers, W.: The development of the concept of effective stresses, *Acta Mechanica* (Acta Mech) **83**, 77–92.

Kowalski, S.J.: Thermomechanics of dried materials, *Archiwum Mechaniki stosowanej* (Arch Mech stosow) **42**, 123–149.

Zienkiewicz, O. C.; Chan, A. H. C.; Pastor, M.; Paul, D. H.; Shiomi, T.: Static and dynamic behaviour of soils: a rational approach to quantitive solutions. I. Fully saturated problems, (Proc R Soc Lond), A **429**, 285–309.

1991/92 Balian, R.: From Microphysics to Macrophysics, Vol. I, II (1992), Springer-Verlag Berlin • Heidelberg • New York.

1991 de Boer, R.; Ehlers, W.; Kowalski, S.; Plischka, J.: Porous Media – a survey of different approaches, Forschungsberichte aus dem Fachbereich Bauwesen **54**, Universität-GH Essen.

de Boer, R.; Lade, P. V.: Towards a general plasticity theory for empty and saturated porous solids, Forschungsberichte aus dem Fachbereich Bauwesen **55**, Universität-GH Essen.

Garcia-Colin, L. S.; Uribe, F. J.: Extended irreversible thermodynamics beyond the linear regime: a critical overview, *Journal of Non-Equilibrium Thermodynamics* (J Non Equilib Thermodyn) **16**, No.2., 89–128.

Lai, W. M.; Hou, Y. S.; Mow, V. C.: A triphasic theory for the swelling and deformation behavior of articular cartilage, *Journal of Biomechanical Engineering* (J Biomech Eng) (ASME) **113**, 245–258.

1992 Baierlein, R.: How entropy got its name, *American Journal of Physics* (Am J Phys) **60**, 1151.

Garrecht, H.: Porenstrukturmodelle für den Feuchtehaushalt von Baustoffen mit und ohne Salzbefrachtung und rechnerische Anwendung auf Mauerwerk, Schriftenreihe des Instituts für Massivbau und Baustofftechnologie, eds.: Prof. Dr.-Ing. J. Eibel, Prof. Dr.-Ing. H. K. Hilsdorf, Universität Fridericiana zu Karlsruhe (TH).

Kowalski, S. J.; Kubik, J. (eds.): Waves in saturated porous media, *Transport in Porous Media* (T I P M) **9** (special issue).

Li, Xiangyue; Li, Xiangwei: On the thermoelasticity of multicomponent fluid-saturated reacting porous media, *International Journal of Engineering Science* (Int J Eng Sci) **30**, 891–912.

1993 Abousleiman, Y.; Cheng, A. H.-D.; Yiang, C.; Roegiers, J.-C.: A micromecally consistent poroviscoelasticity theory for rock mechanics applications, *Int. J. Rock Mech. Min. Sci.* & *Geomech Abstr.* **30**, 1177–1180.

Bluhm. J.; Lund, T.: Ein Modell zur Berechnung des Konsolidationsproblems in Bereich grosser elastischer Deformationen, *Zeitschrift für angewandte Mathematik und Mechanik* (ZAMM) **73**, T446–T449.

de Boer, R.: Thermodynamics of phase transitions in porous media, Report Mech. 93/1, FB 10 / Mechanik, Universität-GH Essen.

de Boer, R.; Ehlers W.; Liu Z.: One-dimensional transient wave propagation in fluid-saturated incompressible porous media, *Archive of Applied Mechanics* **63**, 59–72.

Detournay, E.; Cheng, A. H.-D.: Fundamentals of poroelasticity, chapter 5 in: Comprehensive Rock Engineering: Principles, Practice & Projects, Vol. II, Analysis and Design Method (ed. D. Fairlurst), Pergamon Press.

Lewis, R. W.; Jinka, A. G. K.; Gethin, D. T.: Computer-aided simulation of metal powder die compaction processes, PMI Vol. 25, No. 6, 287–293.

Schrefler, B. A.; Simoni, L.; Turska, E.; Zhan, X. Y.: Zur Berechnung von ungesättigten Konsolidationsproblemen, *Bauingenieur* (Bauing) **68**, 375–384.

1994 Bluhm, J.: Plastic strain directions and back-stress paths for frictional materials, in: Bingye, X.; Wei, Y. (eds.), Advances in Engineering Plasticity and its Application. A E P A 94, International Academic Publishers, 323–328.

Bluhm, J.; de Boer, R.: Endliche Elastizitätsgesetze für poröse Medien, *Zeitschrift für angewandte Mathematik und Mechanik* (ZAMM) **74**, 224–225.

de Boer, R.: Phasenübergänge in porösen Medien, *Zeitschrift für angewandte Mathematik und Mechanik* (ZAMM) **74**, 4, T 186–T 188.

de Boer, R.; Liu, Z.: Plane waves in a semi-infinite fluid saturated porous medium, *Transport in Porous Media* **16**, 147–173.

1995 de Boer, R.: Some issues in the macroscopic porous media theory, in: Festschrift für Professor Siekmann, Universität-GH Essen.

de Boer, R.; Kowalski, S. J.: Thermodynamics of fluid-saturated porous media with a phase change, *Acta Mechanica* (Acta Mech) **109**, 167–189.

de Boer, R.; Liu, Z.: Propagation of acceleration waves in incompressible saturated porous solids, *Transport in Porous Media* **21**, 163–173.

Ehlers, W.; Diebels, S.: Dynamic deformations in the theory of fluid-saturated porous solid materials. *Proceedings of the IUTAM Symposium on Anisotropy, Inhomogeneity and Nonlinearity in Solid Mechanics* (ed.: Parker, D.F. and A. H. England) Nottingham (U. K.), Kluwer Akad. Publ., 241–246.

Lund, T.: Ein Beitrag zur numerischen Behandlung saturierter poröser Festkörper, Dissertation, Fachbereich 10 – Bauwesen, Universität-GH Essen.

1996 Bluhm, J.; de Boer, R.; Skolnik, J.: Allgemeine Plastizitätstheorie für poröse Medien, Forschungsberichte aus dem Fachbereich Bauwesen **73**, Universität-GH Essen.

de Boer, R.: Highlights in the historical development of the porous media theory - toward a consistent macroscopic theory, *Applied Mechanics Review* (Appl Mech Rev) **49**, 201–262.

Diebels, S.; Ehlers, W.: Dynamic analysis of a fully saturated porous medium accounting for geometrical and material non-linearities. *International Journal for Numerical Methods in Engineering* **39**, 81–97.

1997 Bluhm, J.: A consistent model for saturated and empty porous media, Forschungsberichte aus dem Fachbereich Bauwesen **74**, Universität-GH Essen.

Bluhm, J.; de Boer, R.: The volume fraction concept in the porous media theory, *Zeitschrift für angewandte Mathematik und Mechanik* (ZAMM) **77**, 563–577.

de Boer, R.: Compressible porous media: toward a general theory, Proceedings of the IUTAM Symposium on Mechanics of Granular and Porous Materials (Cambridge ed. by N.A. Fleck and A.C.F. Cocks), Kluwer Academic Publishers Dordrecht • Boston • London.

de Boer, R.; Didwania, A. K.: The effect of uplift in liquid-saturated porous solids – Karl Terzaghi's contributions and recent findings, *Géotechnique* **47**, 289–298.

a Breuer, S.: Reflection and refraction of longitudinal waves in a fluid-saturated porous solid, in: "Problems of Enviromental and Damage Mechanics". Proceedings of SolMec '96 Conference (eds.: W. Kosiński, R. de Boer, D. Gross), Mierkie, September 9–14, 1996, Poland, 27–37.

b Breuer, S.: Dynamic response of a fluid-saturated elastic porous solid, *Archiwum mechaniki stosowanej* (Arch Mech stosow) **49**, 791–804.

c Breuer, S.: Zur Berechnung des mechanischen Verhaltens von porösen Materialien mit kompressiblen und inkompressiblen Phasen, *Zeitschrift für angewandte Mathematik und Mechanik* (ZAMM) **77**, 59–60.

d Breuer, S.: Numerical calculation of wave propagation in a saturated porous medium. Proceeding of the *International Conference of Composites Engineering* (ICCE/4) (ed. D. Hui), Big Island of Hawaii, Juli 6–12, 1997, USA, 183–184.

Ehlers, W.; Volk, W.: On shear band localization phenomena of liquid-saturated granular elastoplastic porous solid materials accounting for fluid viscosity and micropolar solid rotations, *Mechanics of cohesive-frictional materials* **2**, 301–320.

Höckel, V.: Die Kinetik starrer Körper – Leonhard Eulers fundamentale Beiträge, Diplomarbeit (unpublished), Institute of Mechanics, University of Essen.

Lade, P. V.; de Boer, R.: The concept of effective stress for soil, concrete and rock, *Géotechnique* **47**, 61–78.

Liu, Z.; de Boer, R.: Dispersion and attenuation of surface waves in a fluid-saturated porous medium, *Transport in Porous Media* **29**, 207–223.

1998 Jägering, S.: Ein Beitrag zur numerischen Behandlung elastoplastischer Prozesse in porösen Medien, Dissertation, Fachbereich Bauwesen, Universität-GH Essen, Shaker Verlag, Aachen.

Liu, Z.; Bluhm, J.; de Boer, R.: Inhomogenious plane waves, mechanical energy flux, and energy dissipation in a two-phase porous medium, *Zeitschrift für Angewandte Mathematik und Mechanik* (ZAMM) **70**, 617–625.

1999 Bluhm, J.: Constitutive relations for thermoelastic porous solids within the framework of finite deformations, Report Mech. 99/1, FB 10/ Mechanik, Universität-GH Essen.

a Breuer, S.: Quasi-static and dynamic behavior of saturated porous media with incompressible constitutents, *Transport in Porous Media* (T I P M).

b Breuer, S.: Numerische Berechnung der Wellenausbreitung, to appear in: *Zeitschrift für angewandte Mathematik und Mechanik* (ZAMM).

de Boer, R.: Special Edition, Porous Media: theory and experiments, *Transport in Porous Media* (T I P M).

Gubaidullin, A. A.; Kuchugurina, O. Yu.: The peculiarities of linear wave propagation in double porous media, *Transport in Porous Media* (T I P M).

Huyghe, J.M.; Jansen, J.D.: Thermo-chemo-electro-mechanical formulation of saturated charged porous solids, *Transport in Porous Media* (T I P M).

Skolnik, J.: Numerische Simulation elastisch-plastischer Deformationen fluid-saturierter poröser Medien, Dissertation, Fachbereich Bauwesen, Universität-GH Essen, Shaker Verlag, Aachen.

Wu, T.; Hutter, K.: On the role of the interface mechanical interaction in a gravity-driven shear flow of an ice-till mixture, *Transport in Porous Media* (T I P M)

Additional literature, which has been used in this treatise, however, has not been cited:

1882 Ersch J. S.; Gruber, J. G.: Allgemeine Enzyklopädie der Wissenschaften und Künste, Johann Friedrich Gleyditsch, Leipzig.
1845 Meyer, J.: Das große Conversations-Lexicon für die gebildeten Stände, Druck und Verlag des Bibliographischen Instituts, Hildburghause, Amsterdam, Paris und Philadelphia.
1874 Larousse, P.: Grand dictionaire universal du XIXE siècle, Administration du grand dictionaire universal, Larousse, Paris.
1875 Historische Kommission bei der bayrischen Akademie der Wissenschaften: Allgemeine Deutsche Biographie, Duncker und Humblot, Berlin.
1883 Clebsch, A.: Théorie de l'élasticité des corps solides de Clebsch. Traduite par MM. Barré de Saint-Venant et Flamant, avec des Notes étendnes de M. de Saint-Venant, Paris.
1887 Hinrichsen, A.: Das literarische Deutschland, Verlag der Album-Stiftung, - XXII, Berlin.
1966 Heimpel, H.; Heuss, Th.; Reifenberg, B.: Die großen Deutschen, Propyläen Verlag, Berlin.
1968 Fachredaktion des Bibliographischen Instituts: Meyers großes Personenlexikon, Bibliographisches Institut, Mannheim/Zürich.
1971 Historische Kommission bei der bayrischen Akademie der Wissenschaften: Neue deutsche Biographie, Duncker und Humblot, Berlin.
1977 Fassmann, K. *et al.:* Die Großen, Kinder, Zürich.
Dictionary of National Biography **16**, 733–734.

Author Index

Subject Index